Twenty-Second Symposium
on Biotechnology for Fuels and Chemicals

Presented as Volumes 91–93
of *Applied Biochemistry and Biotechnology*

Proceedings of the Twenty-Second Symposium
on Biotechnology for Fuels and Chemicals
Held May 7–11, 2000, in Gatlinburg, Tennessee

Sponsored by

US Department of Energy's Office of Fuels Development and the Office
of Industrial Technologies (Agriculture and Chemical Industries)
Oak Ridge National Laboratory
National Renewable Energy Laboratory
Idaho National Engineering and Environmental Laboratory
Lockheed Martin Energy Research
American Chemical Society's Division of Biochemical Technology
Royal Nedalco B. V.
Raphael Katzen Associates
Tate and Lyle (A. E. Staley Manufacturing Co.)
National Resources Canada
E. I. DuPont de Nemours & Co. Inc.
Iogen Corporation
Corn Refiners Association, Inc.
Dow Chemical Company
Argonne National Laboratory
National Institute of Standards and Technology
Idaho National Engineering and Environmental Laboratory
Tembec, Inc.
Pure Energy Corporation
Tennessee Valley Authority Public Power Institute
Cargill, Inc.

Editors

Brian H. Davison
Oak Ridge National Laboratory

James McMillan and Mark Finkelstein
National Renewable Energy Laboratory

 **Humana Press • Totowa, New Jersey**

Applied Biochemistry and Biotechnology
Volumes 91–93, Complete, Spring 2001
Copyright © 2001 Humana Press Inc.
All Rights Reserved.

Applied Biochemistry and Biotechnology is abstracted or indexed regularly in *Chemical Abstracts, Biological Abstracts, Current Contents, Science Citation Index, Excerpta Medica, Index Medicus*, and appropriate related compendia.

Introduction to the Proceedings of the Twenty-Second Symposium on Biotechnology for Fuels and Chemicals

Brian H. Davison

Oak Ridge National Laboratory

Mark Finkelstein

National Renewable Energy Laboratory

The Twenty-Second Symposium on Biotechnology for Fuels and Chemicals was held May 7–11, 2000 in Gatlinburg TN. This field is clearly in an expansion mode, with a window of opportunity for implementing many of the approaches presented at the meeting. The focus of the meeting is to improve the technology and the economics of producing the fuels and chemicals vital to many industrial sectors. This meeting allowed representatives of the industrial, academic, and government sectors to exchange information and ideas in formal and informal settings. Attendance at the conference is split almost equally among industrial, US academic, foreign, and national lab participants, totaling over 200 attendees. The conference, affectionately known as the Gatlinburg conference, has been the longest-standing and best meeting to attend to find out current trends in bioprocessing to produce fuels and chemicals and, equally important, to meet active participants and companies in the field. The conference has an inter-disciplinary focus on the bioprocessing, but has multidisciplinary interests from plant production to utilization. The conference also attempts to cover basic scientific research, engineering development, and the bridging to deployment efforts in bioprocessing for both fuels and chemicals.

"The Technology Roadmap Plant/Crop-based Renewables Resources 2020" identifies biotechnology as a critical approach to moving into new technology. We broadened the meeting scope in Session 1, "Feedstocks Production, Modification, and Characterization," to encompass plant genetics and metabolism for altered composition as well as the production, collection, consistency, and availability of renewable feedstocks (agricultural and energy crops) for fuels and chemicals. Sessions 2 and 3 are the core of the meeting—"Applied Biological Research" examined new bio/catalysts using enzyme, microbial, and plant biochemistries and genetic engineering and "Processing Research" described the conversion of plant components via integration of microbiology, biochemistry, and chemistry with engineering, separations, and hybrid systems. The "Enzymatic Processes and Enzyme Production" session focused on the manufacture and use of enzymes. The "Industrial Chemicals" session emphasized recent developments in the integrated production and scale-up of chemicals from biological rather than petrochemical routes. Special interest was on separation methods and their integration into new fermentation or hybrid processes. The technical program consisted of 35 oral presentations, a roundtable forum, two special topic discussions, and a poster session of 135 posters.

We continued a successful informal roundtable series with "Bioenergy and Bioproducts: Forum on Recent Government Initiatives," which discussed the President's Executive Order, the Bioenergy Initiative, the Technology Roadmap for Renewables Vision 2020, and other thrusts. These events continue the strong industrial focus and active industrial participation in the organizing committee. This has become very popular because it allows industrial and government participants to speak more openly.

A special Topics Discussion Group was held on "CO_2 Sequestration," led by James W. Lee. Another on was held on "Commercialization of Biomass-to-Ethanol" where chairs Jack N. Saddler and David J. Gregg made the goal of this workshop to show participants that we are close to demonstrating the technical viability of an integrated biomass-to-ethanol process and that progressive technical advances and policy decisions will likely greatly enhance the economic attractiveness of the process.

To stretch our perspectives and new biotechnology and to better appreciate our surroundings, Dr. Frank Harris of "Discover Life in America" spoke at the banquet on the "All-Taxa Survey of the Great Smokies National Park: What it is and why."

The 2000 Charles D. Scott Award for Distinguished Contributions in the field of Biotechnology for Fuels and Chemicals was presented to Dr. Karl Grohmann, Lead Scientist for US Department of Agriculture at the US Citrus and Subtropical Products Research Laboratory in Winter Haven, FL. His research focus is the production of value-added commodities from peel and other by-products of citrus processing. Prior to 1991, Dr. Grohmann worked at the National Renewable Energy Laboratory (NREL), then the Solar Energy Research Institute (SERI), in Golden, CO. During his 12-year stay at NREL, Dr. Grohmann was responsible for leading research efforts in biological conversion of cellulosic biomass to ethanol and biogas. Dr. Grohmann earned a chemical engineering diploma from the Institute of Chemical Technology in Prague, Czechoslovakia, and a Ph.D. in chemistry from the University of Houston, TX. He has attended the annual Symposium on Biotechnology for Fuels and Chemicals since 1984 and has shared in organizing and chairing a number of oral and poster sessions. Dr. Grohmann has co-authored over 110 peer-reviewed publications and 12 patents dealing primarily with various aspects of biotechnology for biomass conversion. He has also co-authored over 40 technical reports and has made numerous presentations at national and international scientific meetings.

In addition, he served as the US technical representative for the International Energy Agency Network on Biotechnology for the Conversion of Lignocellulosics. This award is named in honor of Dr. Charles D. Scott, the founder of this Symposium and its chair for the first ten years.

Organization of the Symposium was as follows:

Organizing Committee

Brian Davison, *Conference Chair, Oak Ridge National Laboratory, Oak Ridge, TN*

Mark Finkelstein, *Conference Co-Chair, National Renewable Energy Laboratory, Golden, CO*

William Apel, *Idaho National Engineering and Environmental Laboratory, Idaho Falls, ID*

Marion Bradford, *A. E. Staley, Decatur, IL*

Doug Cameron, *Cargill, Minneapolis, MN*

Bruce Dale, *Michigan State University, East Lansing, MI*

Mark Donnelly, *Argonne National Laboratory, Argonne, IL*

Renae Humphrey, *Oak Ridge National Laboratory, Oak Ridge, TN*

Thomas Jeffries, *USDA Forest Service, Madison, WI*

Raphael Katzen, *Consultant, Bonita Springs, FL*

Hugh Lawford, *University of Toronto, Mississauga, Ontario, Canada*

James Lee, *Oak Ridge National Laboratory, Oak Ridge, TN*

Lee Lynd, *Dartmouth College, Hanover, NH*

James D. McMillan, *National Renewable Energy Laboratory, Golden, CO*

Jonathan Mielenz, *Eastman Chemical Company, Kingsport, TN*

Jack Saddler, *University of British Columbia, Vancouver, British Columbia, Canada*

Valerie Sarisky-Reed, *US Department of Energy, Washington, DC*

Sharon Shoemaker, *University of California, Davis, CA*

Liz Willson, *National Renewable Energy Laboratory, Golden, CO*

Charles Wyman, *Dartmouth College, Hanover, NH*

Guido Zacchi, *Lund University, Lund Sweden*

Gisella Zanin, *State University of Maringá, Maringá, PR, Brazil*

Session Chairpersons and Co-Chairpersons

Session 1: Feedstocks, Production, Modification, and Characterization
Vincent Chiang, *Michigan Technological University, Hancock MI*
J. S. McLaren, *Ph.D., Inverizon International Inc., Chesterfield, MO*

Session 2: Applied Biological Research
Peter Rogers, *University of New South Wales, Sydney, Australia*
Barbara R. Evans, *Oak Ridge National Laboratory, Oak Ridge, TN*

Session 3: Bioprocessing Research
Thomas Hanley, *Ph.D., University of Louisville, Louisville, KY*

Session 3: Bioprocessing Research (continued)
David N. Thompson, *Idaho National Engineering
   and Environmental Laboratory, Idaho Falls, ID*

*Session 4: Bioenergy and Bioproducts:
   Forum on Recent Government Initiatives*
Robert A. Harris, *US Department of Energy, Washington, DC*
Bruce E. Dale, *Michigan State University, East Lansing, MI*

Session 5: Industrial Chemicals
Michael Cockrem, *Ph.D., KiwiChem International, Madison, WI*
Manoj Kumar, *Genencor International Palo Alto, CA*

Session 6: Enzymatic Processes and Enzyme Production
Jeff Tolan, *Iogen Corporation, Ottawa, Ontario*
David Short, *DuPont, Inc., Newark, DE*

Poster Session
Nhuan P. Nghiem, *Oak Ridge National Laboratory, Oak Ridge, TN*

Acknowledgments

The able assistance of Renae Humphrey, Symposium Secretary, Liz Willson, Assistant Symposium Secretary, Norma Cardwell, Conference Coordinator, Linda Puckett, Symposium Treasurer, Marsha Savage, Proceedings Editor, and John Barton, web page assistance.

Oak Ridge National Laboratory is managed by UT-Battelle for the US Department of Energy under Contract DEAC05-00OR22725.

National Renewable Energy Laboratory is operated by Midwest Research Institute for the US Department of Energy under Contract DEAC36-99GO10337.

The submitted manuscript has been authored by a contractor of the US Government under contract DE-AC05-00OR22725. Accordingly, the US Government retains a nonexclusive, royalty-free license to publish or reproduce the published form of this contribution, or allow others to do so, for US Government purposes.

Other Proceedings in This Series

1. "Proceedings of the First Symposium on Biotechnology in Energy Production and Conservation" (1978), *Biotechnol. Bioeng. Symp.* **8**.
2. "Proceedings of the Second Symposium on Biotechnology in Energy Production and Conservation" (1980), *Biotechnol. Bioeng. Symp.* **10**.
3. "Proceedings of the Third Symposium on Biotechnology in Energy Production and Conservation" (1981), *Biotechnol. Bioeng. Symp.* **11**.
4. "Proceedings of the Fourth Symposium on Biotechnology in Energy Production and Conservation" (1982), *Biotechnol. Bioeng. Symp.* **12**.
5. "Proceedings of the Fifth Symposium on Biotechnology for Fuels and Chemicals" (1983), *Biotechnol. Bioeng. Symp.* **13**.
6. "Proceedings of the Sixth Symposium on Biotechnology for Fuels and Chemicals" (1984), *Biotechnol. Bioeng. Symp.* **14**.
7. "Proceedings of the Seventh Symposium on Biotechnology for Fuels and Chemicals" (1985), *Biotechnol. Bioeng. Symp.* **15**.

8. "Proceedings of the Eighth Symposium on Biotechnology for Fuels and Chemicals" (1986), *Biotechnol. Bioeng. Symp.* **17**.

9. "Proceedings of the Ninth Symposium on Biotechnology for Fuels and Chemicals" (1988), *Appl. Biochem. Biotechnol.* **17,18**.

10. "Proceedings of the Tenth Symposium on Biotechnology for Fuels and Chemicals" (1989), *Appl. Biochem. Biotechnol.* **20,21**.

11. "Proceedings of the Eleventh Symposium on Biotechnology for Fuels and Chemicals" (1990), *Appl. Biochem. Biotechnol.* **24,25**.

12. "Proceedings of the Twelfth Symposium on Biotechnology for Fuels and Chemicals" (1991), *Appl. Biochem. Biotechnol.* **28,29**.

13. "Proceedings of the Thirteenth Symposium on Biotechnology for Fuels and Chemicals" (1992), *Appl. Biochem. Biotechnol.* **34,35**.

14. "Proceedings of the Fourteenth Symposium on Biotechnology for Fuels and Chemicals" (1993), *Appl. Biochem. Biotechnol.* **39,40**.

15. "Proceedings of the Fifteenth Symposium on Biotechnology for Fuels and Chemicals" (1994), *Appl. Biochem. Biotechnol.* **45,46**.

16. "Proceedings of the Sixteenth Symposium on Biotechnology for Fuels and Chemicals" (1995), *Appl. Biochem. Biotechnol.* **51/52**.

17. "Proceedings of the Seventeenth Symposium on Biotechnology for Fuels and Chemicals" (1996), *Appl. Biochem. Biotechnol.* **57/58**.

18. "Proceedings of the Eighteenth Symposium on Biotechnology for Fuels and Chemicals" (1997), *Appl. Biochem. Biotechnol.* **63–65**.

19. "Proceedings of the Nineteenth Symposium on Biotechnology for Fuels and Chemicals" (1998), *Appl. Biochem. Biotechnol.* **70–72**.

20. "Proceedings of the Twentieth Symposium on Biotechnology for Fuels and Chemicals" (1999), *Appl. Biochem. Biotechnol.* **77–79**.

21. "Proceedings of the Twenty-First Symposium on Biotechnology for Fuels and Chemicals" (2000), *Appl. Biochem. Biotechnol.* **84–86**.

This symposium has been held annually since 1978. We are pleased to have the proceedings of the Twenty-Second Symposium currently published in this special issue to continue the tradition of providing a record of the contributions made.

The Twenty-Third Symposium is planned for May 6–10, 2001, in Breckenridge, CO. More information on the 22nd and the 23rd Symposia are available at their websites—[www.ct.ornl.gov/symposium] and [www.nrel.gov/biotech_symposium]. We encourage comments or discussions relevant to the format or content of the meetings.

CONTENTS

---

*For papers with multiple authorship, the asterisk identifies the author to whom correspondence and reprint requests should be addressed.

SESSION 4—BIOENERGY AND BIOPRODUCTS
FORUM ON RECENT GOVERNMENT INITIATIVES

SESSION 6—ENZYMATIC PROCESSES AND ENZYME PRODUCTION

SESSION 1

Feedstocks Production, Modification, and Characterization

Feedstocks Production, Modification, and Characterization

James S. McLaren[1] and Vincent Chiang[2]

[1]Inverizon International Inc, Chesterfield, MO 63017
and [2]Michigan Technological University, Hancock, MI 49931

With world demand for consumer goods, materials, and energy continuing to spiral upwards, there is a growing recognition that new or modified bio-based feedstocks can play a major role in supplementing the finite and diminishing reserves of fossil fuels. In addition, the use of bio-based materials will help in mitigating the anthropogenic impact on the atmosphere via lower carbon dioxide emissions. The focal theme of this session was on the production of primary input feedstocks and on the range of modifications that could be beneficial for enhanced use.

Today, bio-based inputs are utilized in only a small portion (<3%) of the manufacturing and energy production processes in the United States. Cellulosic components from forestry and lignocellulosic stems are extracted and used in paper, a variety of packaging materials, and other fiber-rich products. Proteins and carbohydrates, such as starch, from processed crops have constituted a large portion of the bio-based polymers and building blocks for various materials.

Goals for future contributions from bio-based inputs call for an increase in the proportional contribution by at least fivefold. Many projects are underway to meet these goals. The potential for research to impact this area is immense due to the multiple possible sources and the broad range of scientific disciplines that must interact in order to address the practical aspects of the issue.

In some cases, traditional biomass (typical lignocellulosic material) may be used with appropriate processing and improvements can be achieved in production, harvesting, and delivery to the processing facility (for example, achieving higher useful yield per unit land area or the development of alternate crop types for specific situations). In other cases, different or modified feedstocks may be appropriate, such as oils or natural polymeric substances. More recently, developments in biotechnology have broken down previous barriers to the manipulation of natural molecular structures. It is now possible to further improve endogenous substances, or to enhance the production of desirable feedstocks in high-yielding plant species. All of these approaches provide opportunities to create sustainable inputs from renewable resources.

In addition to the scientific challenge in expanding the use of bio-based feedstocks, we must ensure a solid economic base for the new materials and/or new technologies. These dual challenges will require considerable research and much future success in moving the research toward the marketplace. The papers in this session cover several of these aspects and are good examples of the range of technical enquiry necessary to meet future demands. Bioethanol has been in the works for many years, yet production advances continue. Genetic modification of woody feed-stocks and improvement of grasses for biomass have relied on conventional breeding for many years, but now sit on the brink of biotechnology tools that will allow advanced breeding. Separations of component streams are important to further processing, and the paper in this area covered the use of ion-exchange resins to remove toxic substances. New sources of materials are worth exploring, and two papers reported on rapeseed cake and the production of essential oils.

In addition to the breadth of research covered, the papers and posters in this session clearly demonstrated the intensity of the R&D effort required if we are to be successful in developing a more sustainable future.

Twenty Years of Trials, Tribulations, and Research Progress in Bioethanol Technology

Selected Key Events Along the Way

Charles E. Wyman

*Thayer School of Engineering, Dartmouth College,
8000 Cummings Hall, Hanover, NH 03755,
E-mail: charles.e.wyman@dartmouth.edu*

Abstract

The projected cost of ethanol production from cellulosic biomass has been reduced by almost a factor of four over the last 20 yr. Thus, it is now competitive for blending with gasoline, and several companies are working to build the first plants. However, technology development faced challenges at all levels. Because the benefits of bioethanol were not well understood, it was imperative to clarify and differentiate its attributes. Process engineering was invaluable in focusing on promising opportunities for improvements, particularly in light of budget reductions, and in tracking progress toward a competitive goal. Now it is vital for one or more commercial projects to be successful, and improving our understanding of process fundamentals will reduce the time and costs for commercialization. Additionally, the cost of bioethanol must be cut further to be competitive as a pure fuel in the open market, and aggressive technology advances are required to meet this target.

Index Entries: Biomass; biotechnology; ethanol; fuel; hydrolysis.

Introduction

Through sustained research, mostly funded by the Biofuels Program of the US Department of Energy (DOE), the cost of production of ethanol from low-cost cellulosic biomass has been made competitive for blending with gasoline, and several companies are working to commercialize this technology. However, the journey to reach this point has been challenging at all levels, and continuation was severely threatened many times. The intent of this article is to retrace some of this perilous path to present a perspective on key events, advancements, and remaining challenges for this powerful but historically underappreciated route to making a sustainable transportation fuel. More in-depth information on bioethanol technology, feedstock features, benefits, and other details can be found through a number of sources (e.g., *1–4*) as well as historic and more recent work funded by the US Department of Agriculture (e.g., *5,6*).

Benefits of Bioethanol Technology

Although more widely recognized now, the dramatic environmental, economic, strategic, and infrastructure advantages offered by the production of ethanol from abundant sources of lignocellulosic biomass were not appreciated in the past. Perhaps the most unique of these important attributes is the very low greenhouse gas emissions for the production and use of bioethanol, particularly when compared with other liquid transportation fuel options *(3,7,8)*. Because nonfermentable and unconverted materials left after making bioethanol can be burned or gasified to provide all the heat and power to run the process, and lignocellulosic crops require low levels of fertilizer and cultivation, fossil energy inputs are minimized if not eliminated for mature technology *(9–11)*, and net release of carbon dioxide is very low when evaluated in a cradle-to-grave (often called a full fuel cycle) analysis *(3,7,8)*. Bioethanol can also be important in helping meet the growing demand for energy in the developing world as these countries improve the living standards of more and more people *(12)*. An added benefit is that bioethanol could be made in many countries, including the United States, that have limited petroleum resources and rely heavily on imported oil, helping them to reduce their trade deficit and grow their economies. Furthermore, the substitution of bioethanol for fossil fuels will help reduce dependence on the imported oil that makes the United States and other countries susceptible to disruptions and price instabilities and could virtually cripple a transportation sector that almost totally relies on oil *(13)*. Bioethanol production can provide an attractive route to dispose of problematic wastes such as rice straw and wood wastes as mounting regulations limit their historic disposal method—burning *(14)*. In addition to augmenting the fuel supply, adding ethanol to gasoline increases octane and provides oxygen to promote more complete combustion, particularly in older vehicles *(1,3,15,16)*, but neat ethanol provides the greatest benefits with respect to both air and water pollution *(1,16)*. Most studies estimate that enough biomass could be available from wastes and dedicated energy crops to significantly decrease the huge amount of gasoline consumed in the United States *(3,15)*, and the cost of biomass itself is competitive with fossil resources. Because only biomass of the sustainable resources can be readily converted into liquid fuels such as ethanol and a wide range of organic chemicals in addition to food and animal feed (Lynd, L. R., personal communication), it is of paramount importance to develop this truly unique and powerful route to meeting the needs of society on an ongoing basis.

Changing Climate for Energy Technologies

The development of alternative sources of energy became a national priority in the early 1970s in response to the Libyan and later Arab oil embargoes *(17)*. Although this was really a petroleum crisis resulting from controlled production of oil by the Organization of Petroleum Exporters, it was labeled an "energy crisis," with efforts directed at developing any new source of energy. Included were government programs directed at converting abundant domestic resources such as coal into petroleum replacements, and big projects such as the Great Plains gasifier were funded through government grants and loan guarantees to accelerate technology applications.

The energy crisis of the 1970s also fed the development of technologies for utilizing renewable energy resources such as wind, solar, and biomass with the goal of immediate use, and the Office of Alcohol Fuels was created in the US DOE to accelerate scale up of ethanol and methanol production. Technical and economic evaluations of processes for making alcohols were completed, and loan guarantees and other forms of government assistance were awarded for construction of processes that were judged to be promising. Projects were funded to rapidly develop dilute acid, concentrated acid, cellulase enzyme, direct microbial conversion, and other techniques for converting cellulosic biomass into ethanol, and projects were also supported to produce methanol from biomass syngas and biodiesel from plant oils. However, of the biomass-related options, only ethanol production from corn was found to offer the near-term potential viewed critical at that time, and several plants were constructed to produce ethanol from corn starch through state and federal financial assistance, price subsidies, and other incentives. Consequently, ethanol and corn became synonymous.

Because some of these quick fix large projects for corn as well as other energy sources such as coal were poorly executed and not well conceived, costly failures resulted, leaving a bad taste for big government-funded projects. In addition, even though the protein in corn was concentrated in a valuable animal feed coproduct, many viewed conversion to ethanol as competing with food supplies, sparking controversy over food vs fuel. In the haste to build plants and because of large government subsidies, many of these plants used inefficient ethanol recovery equipment, feeding the perception that ethanol recovery is inefficient, even though modern distillation systems perform quite well. In addition, loan guarantees and subsidies were controversial. Some fuel problems were experienced when ethanol blends were used in older vehicles, which sparked more controversy; for example, the different solvent properties of ethanol would release built-up deposits, plugging fuel filters when first used.

In 1980, the climate for energy projects changed dramatically. While loan guarantees, subsidies, and other government incentives were common during the 1970s, the shift was to a free market approach in 1980, and the US federal government pushed to support only long-term, high-risk research that would be far too risky for industry to pursue. Funding for research on other than defense and a few other areas was reduced significantly, and attempts were made to dismantle the DOE. In this new climate, the development of bioethanol technology was threatened with elimination because it was confused with corn ethanol technology and judged to not have the long-term, high-risk profile favored for the new government philosophy. In addition, its promotion during the prior period as being ready for commercial use further jeopardized its continuation. Consequently, budgets were cut significantly almost yearly.

In the late 1980s, the philosophy shifted to a middle ground between immediate commercial use favored during the 1970s and long-term, high-risk research promoted in the early 1980s. Now technology development was motivated by market potential and the fit to economic and commercial needs. In this context, bioethanol was viewed as offering technology that could achieve very low costs. Just as important, the use of ethanol for transportation was recognized to have the potential to reduce the use of imported oil for transportation, the sector that consumed about two-thirds of all oil in the United States and that was almost totally dependent on this single energy source *(13)*. Bioethanol research budgets now increased significantly.

Unfortunately, the favorable position for bioethanol did not last long, and in the early 1990s the budgets dropped again. The shift was now to very immediate projects, particularly in the energy conservation area, and the time frames for bioethanol apparently were judged to be too long. However, this gradually changed over a 4-yr period, and funding began to swing more favorably in the latter part of the 1990s as interest mounted in commercial applications to address mounting waste problems in the agricultural and forestry sectors and heightened interest in oxygenates triggered by the Clean Air Act.

Bioethanol Process Identification

Against this background, process engineering evaluations of bioethanol technology were initially undertaken with the goal of identifying approaches for immediate applications in the 1970s. These culminated in several process designs by selected engineering and consulting firms using several enzymatic and dilute acid–based pathways *(18–21)*. However, none was judged to be competitive for immediate application, and the focus shifted to commercializing corn ethanol technologies to meet immediate energy needs.

Faced with the perception that bioethanol was not a high-risk, high-payback technology, bioethanol research was threatened with elimination in the early 1980s. At this point, John Wright at the then Solar Energy Research Institute, now National Renewable Energy Laboratory (NREL), extended the process-engineering evaluations to other systems such as the use of concentrated acids to hydrolyze biomass to sugars, seeking to identify options that have high potential for substantial cost reductions *(22,23)*. These studies were built on the engineering analyses conducted earlier *(18–21)* integrated with Icarus costing algorithms, vendor quotes, and other tools to upgrade the material and energy balances and costing. The result was the first consistent basis for costing of bioethanol technologies, allowing meaningful comparisons among the various options. This tool was particularly useful for benchmarking the current status of each option and defining opportunities to improve the technologies and their lower costs. These results were integrated into a comprehensive Fuel Alcohol Technology Evaluation (FATE) study in the mid-1980s, and based on these cost projections, tightening federal research budgets, and the emphasis on long-term, high-risk research in the 1980s, a decision was made to focus on enzymatically based bioethanol production technology *(24,25)*. Such technology was judged to be too risky for industry to pursue at that time and offered the promise for significant advances through application of the emerging field of biotechnology that could dramatically reduce costs and make bioethanol competitive.

The general process configuration for enzymatic hydrolysis begins with a material-handling operation that brings feedstock into the plant, where it is stored and prepared for processing. Next, biomass is milled and pretreated to open up its structure and overcome its natural resistance to biologic degradation. The resulting pretreated biomass liquid hydrolysate is neutralized and conditioned to remove or inactivate any compounds naturally released from the material (e.g., acetic acid, lignin) or formed by degradation of biomass (e.g., furfural) that are inhibitory to fermentation. Once technology was developed to convert the five-carbon sugars derived by hemicellulose hydrolysis, the liquid hydrolysate was sent to a fermentation step; otherwise, it had to be treated in waste disposal prior to discharge from the plant. A portion of the pretreated solids and possibly some of the liquid hydrolysate is sent to a separate enzyme production step in which a small portion of the total sugars is consumed by an organism such as the fungus *Trichoderma reesei* to make cellulase. The cellulase is then added back to the bulk of the pretreated solids to catalyze the breakdown of cellulose to release glucose, which many organisms, including common yeast, ferment to ethanol. Next, the fermentation broth is transferred to a series of distillation columns to recover ethanol. The lignin, water, enzymes, organisms, and other components leave with the column bottoms, and the solids are concentrated to feed the boiler that provides the heat and electricity for the entire process with any excess electricity sold. The liquid not retained with the solids is processed through a combined anaerobic and aerobic waste treatment process, with the clean water discharged from the plant or recycled to the process, the sludge disposed of, and the methane fed to the boiler. The ash from the boiler is landfilled *(9–11,24,25)*.

Technology Progress

In addition to identifying promising processes, the technoeconomic models allowed identification of research opportunities and tracking of research progress, keys to the reemergence of bioethanol development *(24,25)*. Initially, a sequential hydrolysis and fermentation route was employed for breakdown of cellulose to glucose and subsequent fermentation to ethanol, with a projected selling price of $3.60/gallon for 1979 technology based on the use of a fungal strain known as QM9414 for cellulase production. Three years later, a strain known as Rut C30 could be used with a cost of about $2.66/gallon, owing to a better balance in enzyme activity components and lower end product inhibition. A different cellulase known as 150L, developed by Genencor, improved hydrolysis results further, lowering the projected cost to $2.25/gal for the year 1985. When this same cellulase enzyme was used in a simultaneous saccharification and fermentation (SSF) configuration, the estimated cost of bioethanol manufacture dropped to $1.78/gal with the year taken as 1986. If the biomass feed rate is kept constant with more efficient cellulase rather than reducing the plant size to maintain a fixed ethanol capacity, the cost drops to about $1.65/gal *(24,25)*.

Additional process advancements and simplifications were incorporated into the bioethanol process later, and the technology was reassessed through parallel studies by NREL and Chem Systems to determine the status and identify opportunities for further improvements *(9,10)*. The most significant change was the incorporation of a newly invented genetically engineered organism into the process that allowed fermentation of all sugars to ethanol for the first time. The projected cost of production including cash costs and capital recovery dropped to only $1.22/gal. Note that there were also several additional differences in the basis for this more recent projection compared with the historic cost projections reported above with the use of a capital recovery factor of 0.20 instead of 0.13 to annualize capital costs being the most significant. Therefore, the projected costs from the historic studies would increase when adjusted to the same capital recovery factor and year dollars as for the NREL and Chem Systems studies. Recently, Wooley et al. (11) updated the cost projections based on further refinements in the cost methodology and more detailed engineering designs.

It is important to recognize the many caveats that apply to these cost estimates for bioethanol technology and that such cost projections should only be used to gage relative progress and identify opportunities to reduce costs further. Bioethanol costs are site specific and will change with many local factors. In addition, costs depend strongly on what type of organization (e.g., small company vs major operating company) does the project, how it is financed (e.g., debt vs equity), the technology used (e.g., risk and costs), and other factors. Also, these projections assume that the technology is for an *n*th plant that benefits from a substantial learning curve and has well-defined risk. Furthermore, such assessments are not likely to have access to important advanced technologies and the know-how that are proprietary or involve trade secrets. Thus, no one should expect to build a plant, particularly a first plant, based on such projections.

Although not obvious in the above economic summary, a key element underlying bioethanol cost reductions was improvements in pretreatment technology. Without pretreatment, sugar yields are low because cellulose is not readily accessible to the large cellulase enzyme protein structures. Over the years, various biologic, chemical, and physical pretreatment approaches have been studied in an attempt to increase the susceptibility of cellulose to attack by enzymes *(26,27)*, and several appear promising. However, building off early work on plug-flow systems by Grethlein and Converse *(28,29)*, researchers chose dilute sulfuric acid because of its relatively low cost and high hemicellulose sugar yields *(9,10,30)*. Steady progress has been made over the years in refining the technology further to remove hemicellulose with high yields and achieve good digestibility of cellulose, and the process has been demonstrated to be effective on a variety of biomass feedstocks *(31,32)*. High yields of about 85–90% or more of the sugars can be recovered from the hemicellulose fraction with temperatures around 160°C, reaction times of about 10 min, and acid levels of about 0.7%, and about 85 to >90% of the remaining solid cellulose can be enzymatically digested to produce glucose. However, dilute acid pretreatment is still a major cost element that introduces technically significant challenges to the process *(33)*.

Without a profitable use of the five-carbon sugars xylose and arabinose, bioethanol is too expensive, at \$1.65/gal, to compete in commercial markets *(23,24)*. Natural organisms do not achieve high enough ethanol yields to be economically viable and typically require careful control of dissolved oxygen levels, which is difficult to accomplish in gigantic commercial fermentors. In addition, alternative products could not be identified that had a sufficient market to be compatible with large-scale ethanol production from cellulose *(34)*. The critical achievement in reducing the costs of ethanol production to the lower value projected by NREL was the genetic engineering of several bacteria to allow these organisms to ferment all five sugars (arabinose, galactose, glucose, mannose, and xylose) found in biomass to ethanol *(6,35,36)*. These organisms achieved excellent ethanol yields from all five of these sugars, a requirement critical to commercial success.

Because cellulose is the largest single fraction of biomass, one of the major challenges in the development of bioethanol technology is to improve the technology for hydrolysis of recalcitrant cellulose. In fact, most of the historic cost reductions reported from 1979 to 1986 resulted from improvements in dilute acid pretreatment and enzymatic hydrolysis of cellulose based on the cellulase-producing organism *T. reesei*, discovered during World War II *(24,25,37)*. In particular, the fungus evolved through classic mutations and strain selection from the earlier strains such as QM9414 to improved varieties such as Rut C30 developed at Rutgers University *(38)*. Later a cellulase known as 150L, produced by Genencor, was quite effective at cellulose hydrolysis because of enhanced levels of β-glu-cosidase that converted cellobiose into glucose *(39–41)*. Furthermore, even though the fermentation temperature must be reduced below that considered optimum for cellulase action to accommodate temperature limitations of known fermentative organisms, accumulation of glucose and cellobiose was minimized when 150L was employed in an SSF configuration, further reducing end product inhibition of the enzyme and improving the rates, yields, and concentrations of ethanol while also reducing the possibility of invasion by unwanted microorganisms *(42,43)*. Nonetheless, cellulase action is still slow, with SSF reaction times of about 5–7 d reported to achieve modest ethanol concentrations *(42,43)*, although others claim shorter times of 2 to 3 d *(44,45)*.

Following the identification of the SSF configuration for cellulose conversion by Takagi et al. *(46)* and Gauss et al. *(47)* in the mid-1970s, it became important to identify fermentative organisms that could tolerate the greater stress associated with the combined effects of high temperatures desired to increase rates of enzymatic hydrolysis, low glucose levels owing to rapid sugar metabolism by the fermenting organism, and high ethanol concentrations. A number of investigations followed to find the best organism-enzyme combinations with particular emphasis on thermotolerant yeast. Several organisms were identified that improved the rates, yields, and concentrations of ethanol formation *(39–43)*. However, it was found that rapid conversion of cellobiose to glucose was more important than the fermentation temperature. Thus, the best results were with a cellulase, such as Genencor 150L, that is higher than many in β-glucosidase *(39,43)*. An organism, such as *Brettanomyces custerii*, that can ferment cellobiose into ethanol either directly or in coculture with a more ethanol-tolerant yeast also enhances performance *(39,41)*. Some of the bacteria genetically engineered to ferment xylose to ethanol also have the ability to ferment cellobiose to ethanol, and genes have been inserted in others to impart this trait *(36)*.

Cellulase is produced commercially, but existing preparations are directed at low-volume, high-value specialty markets such as stone-washed jeans with the primary interest in providing carefully balanced properties that command high prices. Furthermore, research on cellulase production has been very limited for applications to production of low-cost sugars from cellulose for conversion to fuels and commodity chemicals *(48)*. Recent investigations project higher costs of about $0.30–0.50/gal of ethanol produced if cellulase is manufactured on site or $3.00/gal if it is purchased *(49,50)*. Such costs are higher than those estimated in the studies reported, pointing out the significant uncertainty in cellulase production technology and costs.

Features that differentiate cellulase production for bioethanol applications from current markets include the substrate used and the direct addition of whole broth to the SSF process. Production of cellulase on mixed liquid/solid hydrolysate from pretreatment instead of lactose and other more costly and limited carbon sources typically used commercially shows promise to reduce the cost of cellulase production and simplify the integrated production system *(51,52)*. In contrast to enzyme production for specialty markets in which cellulase is typically removed from the fungal source and then concentrated prior to shipment to the user, adding the entire cellulase production broth to SSF vessels improves performance because fungal bodies retain some cellulase and particularly β-glucosidase activity *(46,53)*. This approach also saves on capital investment by eliminating costly equipment and reduces the opportunity for microbial invasion by simplifying the process. Furthermore, any substrate not used for enzyme production passes to the SSF process and is converted to ethanol, increasing yields. The team who originally developed the SSF process termed whole-broth cellulase addition as a koji technique *(46)*.

Product recovery in all these studies is based on conventional distillation technology *(9–11,24,25)*. As pointed out earlier, there has been controversy about high-energy use for ethanol purification, but such concerns were based on inefficient, outdated technology employed by some companies during the emergence of the corn ethanol industry. The cost of and energy use by new distillation equipment is not significant in the production of bioethanol, and given the tremendous experience curve for distillation, the prospect for advances that will have a significant impact on bioethanol production costs is not high *(15)*.

These advancements can be viewed as falling into two major categories. The first can be summarized as progress in overcoming the recalcitrance of biomass and includes advances in pretreatment, cellulase properties, and integrated fermentations (SSF). The second can be described as overcoming the diversity of biomass sugars and centers on achieving fermentation of all five biomass sugars to ethanol with high yields. Gradual progress has been realized in the former while genetic engineering led to a major step forward for the latter.

Competitive Cost Goal

The other key to the reemergence of bioethanol research was the definition of a cost target that would make bioethanol technology competitive as a pure fuel in an open market. In fact, there is little hope that research on bioethanol would be of interest if the technology cannot offer a competitive position and would require continued subsidies. A goal of $0.60/gal of ethanol was set by the DOE in the mid-1980s and increased to $0.67/gal in about 1990, to be consistent with the National Energy Strategy being drafted at that time. At such a price, biomass ethanol could compete with gasoline derived from petroleum costing $25/barrel.

Although it is critical to have a competitive cost goal to justify research on bioethanol technology, it is just as important to verify that competitive costs can be achieved. Four approaches were applied to evaluate the ability to improve the technology to meet the cost goals. In one, sensitivity studies were used to determine the impact of key performance parameters on process economics. Combining all the possibilities identified with a slightly lower feedstock cost of $34/dry ton resulted in a 40% cost reduction to about $0.74/gal of ethanol, a value competitive with gasoline selling for about $0.92/gal at the plant gate assuming bioethanol is used in a properly optimized spark ignition, internal combustion engine *(9,10)*.

Detailed process designs and economic evaluations such as those described provide useful estimates of the cost of production of bioethanol and identify targets for continued cost reductions, but they are confined by the process configuration selected initially. In addition, such studies are complex, making them time-consuming to apply and understand, and different studies can show quite different results, with poor economics possibly reflecting design rather than technology limitations. An alternative measure of the economic viability of bioethanol technology can be gained by a macroscopic evaluation, with one approach estimating an allowable capital cost based on estimates of revenues and all process costs and benchmarking the result against capital costs typical for corn ethanol plants *(54)*. The result was a capital cost allowance similar to that expected for a modern corn ethanol plant, supporting the notion that bioethanol technology could achieve a low enough cost to compete with gasoline through continued research.

Process studies were taken further to define specific technical opportunities to lower bioethanol production costs and estimate the resulting cost of production *(33)*. For this analysis, an advanced process configuration was chosen that focused on improved pretreatment technology fashioned after many features of hot-water pretreatment in conjunction with consolidated bioprocessing that combined the cellulase production, cellulose hydrolysis, cellulose sugar fermentation, and hemicellulose sugar fermentation steps in a single fermentor *(33,55)*. No other improvements relative to the NREL base case were included. Two levels of performance parameters were integrated into the system: one for the best performance conceivable and the other representative of advanced technology that is believed to be the most likely achievable by analogy with similar systems. Higher yields of hemicellulose sugars were also forecast for this approach, and lower-cost materials of construction and other cost reductions were applied. The consolidated biologic processing operations were projected to increase cellulose hydrolysis yields to 92% with subsequent fermentation to ethanol at a 90% yield. The ethanol concentration was set at 5% by weight, and the fermentation time was taken as 36 h. Continuous fermentation was employed, and as a result, costly seed fermentors were eliminated. Material and energy balances were calculated just as for the other studies and used in the estimation of capital and operating costs. Combining these advances resulted in a projected total bioethanol cost including return on investment of about $0.50/gal in the advanced technology scenario for a plant using about 2.74 million dry tons/yr of feedstock costing $38.60/delivered dry ton. More aggressive performance taken for the best possible technology reduced the projected total cost to about $0.34/gal.

The latter study and the sensitivity studies for the base case process clearly indicate that significant advances in biologic processing and pretreatment are vital to low-cost bioethanol and that these areas even outweigh substantial scale-up in plant capacity. Enhancement of technical performance also reduces the cost but would not be sufficient to realize low-cost bioethanol without developing advanced process configurations. These results also reveal that even though advances in pretreatment can have one of the most significant effects on bioethanol economics of all the technology options considered, pretreatment remains by far the most costly step of the advanced process, suggesting that even lower cost options should be pursued.

Differentiation of Bioethanol

The quantification of research progress, definition of a competitive goal, and justification of that goal were quite similar to the approach followed by the Wind and highly successful Photovoltaics programs within the US DOE and were all vital to establishing that bioethanol offered economic promise. However, many still confused bioethanol with ethanol for corn. For example, corn ethanol was the subject of considerable controversy for many years because the amount of fossil fuels used, particularly for early corn fuel ethanol technology, resulted in few energy or greenhouse gas benefits. In addition, because production of bioethanol releases carbon dioxide during manufacture and use, many assumed it would have few greenhouse gas benefits. Several early studies showed this to be irrelevant because very little fossil fuel would be needed in the overall production cycle, with the result that the manufacture of bioethanol is one of the lowest greenhouse gas impact options available for the transportation sector *(7,8)*. Note that modern corn ethanol production based on state-of-the-art technology does reduce greenhouse gas emissions significantly relative to gasoline although not to the degree bioethanol production would.

Another point of confusion was the amount of feedstock available for bioethanol production and the cost of the raw material. Although the ultimate supply can always be debated in much the same fashion that the availability of petroleum has been debated for years, several studies show that cellulosic biomass is sufficiently abundant to make a sizeable impact in the transportation fuel market. Furthermore, if the efficiency of vehicles is improved to levels such as targeted by programs such as the Partnership for New Generation Vehicles, enough ethanol could be made from biomass to meet the total light-duty vehicle market demand in the United States *(15)*.

The cost of biomass is also competitive on a weight or energy content basis. As shown in Fig. 1, biomass costing $42/t would compete with petroleum at about $6/barrel on a weight basis or at about $12.70/barrel on an energy content basis *(56)*. Thus, the primary challenge for bioethanol competitiveness is to reduce the cost of biomass processing to convert this low-cost raw material into a competitive product.

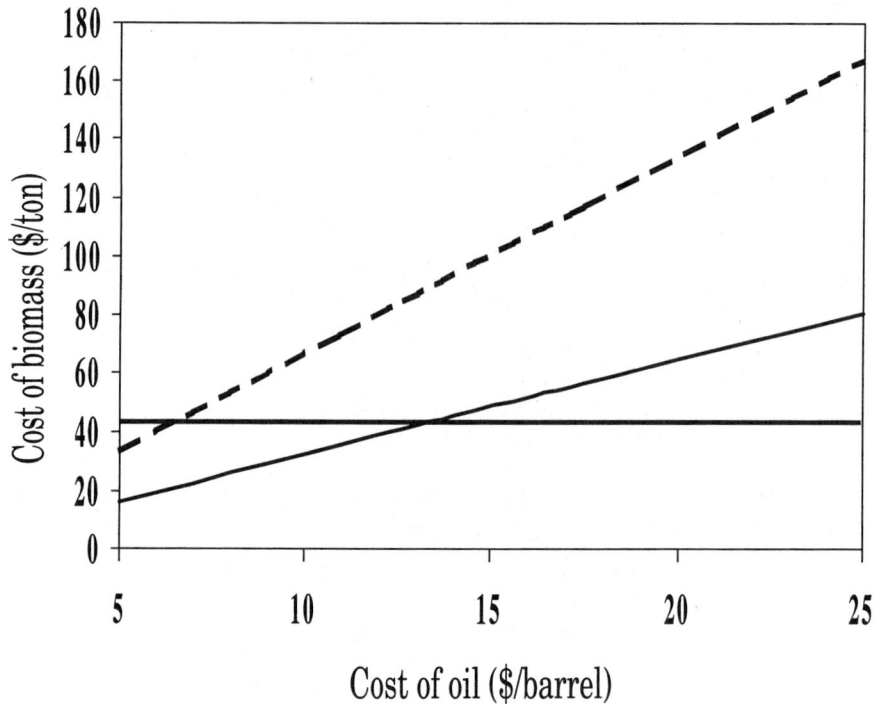

Fig. 1. Cost of biomass compared with the price of petroleum on an equivalent weight (dashed line) and energy content (diagonal line) basis (based on ref. *56*). The horizontal line represents the cost of biomass at $42/t.

The term *bioethanol* was adapted to ensure that the unique attributes owing to the use of cellulosic biomass could be appreciated *(1)*; some also apply the term *cellulosic ethanol* to differentiate the product. However, it is important to recognize the importance of corn ethanol to the development of bioethanol for use as a renewable transportation fuel. Corn growers and processors have done an outstanding job of developing a market for ethanol starting from virtually no ethanol use in the late 1970s, and there would be no established market for bioethanol without the corn ethanol industry, making it almost impossible to enter the market. Furthermore, all major automakers now warranty their vehicles to be compatible with ethanol, and flexible-fueled vehicles that can use any mixture of ethanol and gasoline below 85% ethanol are marketed because of the efforts of corn ethanol producers. In addition, corn ethanol producers have made major improvements in the energy efficiency of ethanol production and established the viability of large-scale fermentations. Both of these improvements facilitate the introduction of technology and improvements for bioethanol that would otherwise face a major hurdle. Thus, corn ethanol and bioethanol are complementary in both product and process evolution and together lead to a sustainable energy future.

Broadening the Product Slate

Historically, the US DOE Biofuels Program has focused almost exclusively on fuels for high-volume product markets and was not chartered to integrate technology for the production of chemicals from biomass in a biorefinery concept that could take advantage of synergies between the production of both bioethanol and chemicals. The program could only include the production of heat and electricity from residuals because they were needed to run the process. This was somewhat a manifestation of the fact that different congressional committees fund the fuels and chemicals programs. However, this has recently changed with the initiation of a new Bioenergy Initiative within the DOE that is hoped will lead to a comprehensive biorefinery process such as that depicted in Fig. 2 (p. 18) *(57)*. The recent introduction of the Sustainable Fuels and Chemicals Act championed by Senator Richard Lugar of Indiana and the Executive Order issued by President Clinton could accelerate progress toward this end and address the critical applied fundamental research needed to attain competitive technologies *(4)*.

Conclusion

Several companies are now striving to commercialize bioethanol technology *(58)*, and it is vital that one or more be successful if we are to enjoy the benefits at a scale that will make a significant difference. In addition, there is substantial promise that the technology can be improved to a point that the cost of bioethanol production will be competitive with fossil sources *(33)*. The great potential of bioethanol is finally being recognized at levels influential to technology funding as evidenced by the recent Foreign Affairs article on bioethanol published by Senator Richard Lugar from Indiana and former CIA director James Woolsey *(59)*. The Sustainable Fuels and Chemicals Act and the Executive Order are further indications of the recent recognition of the potential of biomass processing to a wide range of products in addition to bioethanol. In recent years, two National Research Council studies also illustrate the new importance finally being placed on biomass conversion *(60,61)*. The result is growing budgets for biomass conversion research and development that can catalyze a transition to a new platform for meeting our needs for organic fuels and chemicals on a sustainable basis.

This change also presents new challenges. Enhanced funding for biomass conversion means enhanced expectations for progress and realization of real benefits. Thus, the key now is to focus on critical needs: successfully commercializing technologies now and developing next-generation technologies that can substantially reduce the cost of biomass processing. In the end, what we really need most is the benefits biofuels offer, and such benefits can only come through large-scale commercial use. However, it is also important to realize that commercialization of new technology presents difficulties even greater than overcome in the past for the

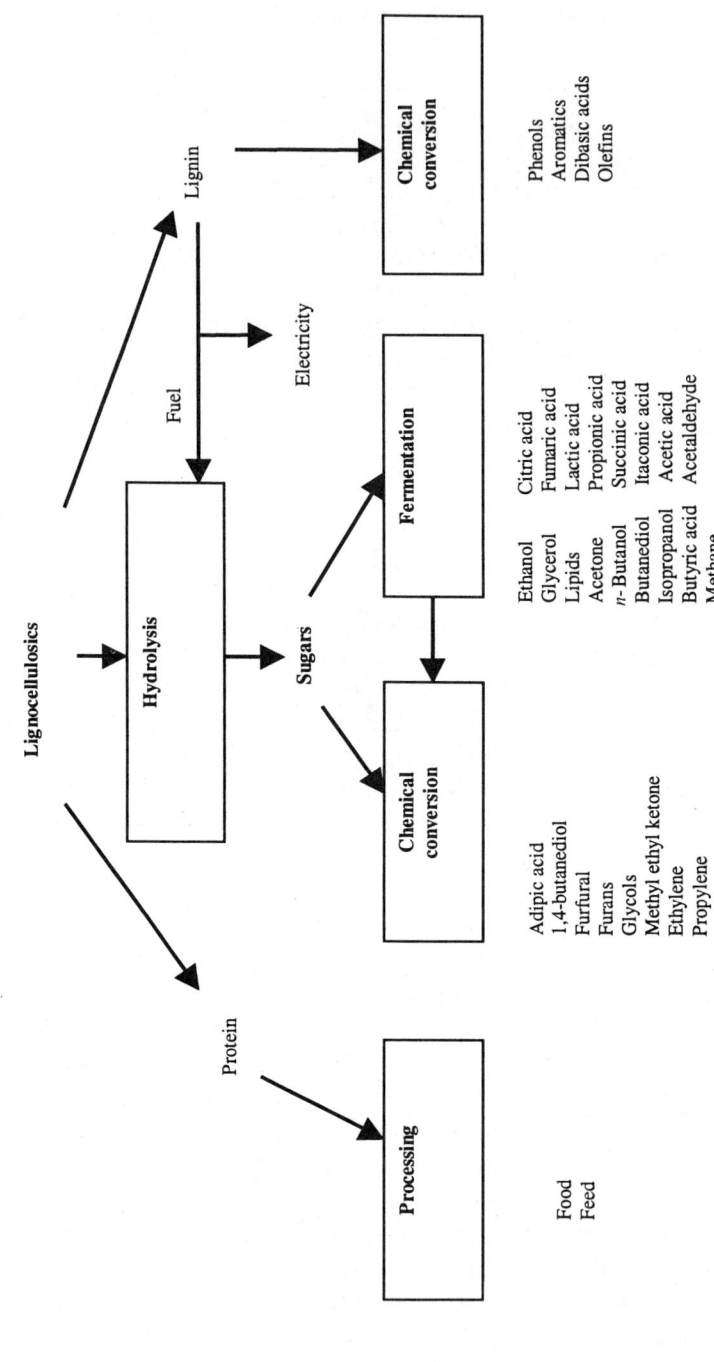

Fig. 2. Biorefinery for utilization of all the major fractions of biomass to make a wide range of products by fermentation and chemical conversion technologies (57).

development of bioethanol technology, and tremendous dedication, persistence, and financial strength are required to clear this last remaining hurdle *(4)*.

In closing, this brief overview can only touch on some of the important challenges faced and progress made in developing bioethanol technology. It shows that although many envy the US position on bioethanol, considerable persistence was required to overcome countless obstacles, and the opportunity now afforded certainly did not emerge overnight or without considerable dedication and effort.

Acknowledgments

The work reported herein is primarily based on extensive analyses and research funded by the Biochemical Conversion Element of the Biofuels Program of the US DOE and reported in the open literature. I wish to thank the Thayer School of Engineering at Dartmouth College for making it possible to participate in the meeting and develop this article.

References

1. Wyman, C. E., ed. (1996), *Handbook on Bioethanol: Production and Utilization*, Applied Energy Technology Series, Taylor & Francis, Washington, DC.
2. Wyman, C. E. (1995), *Bioresour. Technol.* **50,** 3–16.
3. Lynd, L. R., Cushman, J. H., Nichols, R. J., and Wyman, C. E. (1991), *Science* **251,** 1318–1323.
4. Wyman, C. E. (1999), *Annu. Rev. Energy Environ.* **24,** 189–226.
5. Harris, E. E. and Beglinger, E. (1946), *Ind. Eng. Chem.* **38(9),** 890–895.
6. Ingram, L. O., Conway, T., Clark, D. P., Sewell, G. W., and Preston, J. F. (1987), *Appl. Environ. Microbiol.* **53,** 2420–2425.
7. Wyman, C. E. (1994), *Appl. Biochem. Biotechnol.* **45/46,** 897–915.
8. Tyson, K. S. (1993), *Fuel Cycle Evaluations of Biomass-Ethanol and Reformulated Gasoline,* vol. 1, NREL/TP-463-4950, DE94000227, National Renewable Energy Laboratory, Golden, CO.
9. US Department of Energy (1993), Technical Report 11, DOE/EP-0004, US Department of Energy, Washington, DC.
10. Hinman, N. D., Schell, D. J., Riley, C. J., Bergeron, P. W., and Walter, P. J. (1992), *Appl. Biochem. Biotechnol.* **34/35,** 639–649.
11. Wooley, R., Ruth, M., Glassner, D., and Sheehan, J. (1999), *Biotech. Prog.* **15,** 794–803.
12. Beck, R. J. (1997), *Worldwide Petroleum Industry Outlook: 1998–2002 Projections to 2007,* 14th ed., Pennwell, Tulsa, OK.
13. US Department of Energy (1998), Report DOE/EIA-0384(97), Energy Information Administration, US Department of Energy, Washington, DC.
14. Potts, L. W. (1998), *Ethanol: Collins Pine Steps up to New Development Idea*, Chester Progressive, Chester, CA.
15. Lynd, L. R. (1996), *Annu. Rev. Energy Environ.* **21,** 403–465.
16. Bailey, B. K. (1996), Performance of ethanol as a transportation fuel. Ref. 1, pp. 37–58.
17. Yergin, D. (1991), *The Prize.* Simon & Schuster, New York.
18. Chem Systems Inc. (1985), Report ZX-3-03098-1, Solar Energy Research Institute, Golden, CO.
19. Nystrom, J. M., Greenwald, C. G., Hagler, R. W., and Stahr, J. J. (1985), NYSERDA Report no. 85–9, New York State Research and Development Authority, Albany.

20. Stone and Webster Engineering Corporation (1985), Report ZX-3-03096-1, Solar Energy Research Institute, Golden, CO.
21. Stone and Webster Engineering Corporation (1985), Report ZX-3-03097-1, Solar Energy Research Institute, Golden, CO.
22. Wright, J. D. and Power, A. J. (1985), *Biotechnol. Bioeng. Symp.* **15,** 511–532.
23. Wright, J. D. and D'Agincourt, C. G. (1984), *Biotechnol. Bioeng. Symp.* **14,** 105–123.
24. Wright, J. D. (1988), *Chem. Eng. Prog.* 62-74.
25. Wright, J. D. (1988), *Energy Prog.* **8(2),** 71–78.
26. Hsu, T.-A. (1996), Ref. 1, pp. 179–195.
27. McMillan, J. D. (1994), in *Enzymatic Conversion of Biomass for Fuels Production*, ACS Symposium Series 566, Himmel, M. E., Baker, J. O., and Overend, R. P., eds., American Chemical Society, Washington, DC, pp. 292–324.
28. Converse, A. O., Kwarteng, I. K., Grethlein, H. E., and Ooshima, H. (1989), *Appl. Biochem. Biotechnol.* **20/21,** 63–78.
29. Knappert, H., Grethlein, H., and Converse, A. (1980), *Biotechnol. Bioeng. Symp.* **11,** 67–77.
30. Schell, D., Torget, R., Power, A., Walter, P. J., Grohmann, K., and Hinman, N. D. (1991), *Appl. Biochem. Biotechnol.* **28/29,** 87–97.
31. Torget, R., Walter, P., Himmel, M., and Grohmann, K. (1991), *Appl. Biochem. Biotechnol.* **28/29,** 75–86.
32. Grohmann, K., Himmel, M., Rivard, C., Tucker, M., and Baker, J. (1984), *Biotechnol. Bioeng. Symp.* **14,** 137–157.
33. Lynd, L. R., Elander, R. T., and Wyman, C. E. (1996), *Appl. Biochem. Biotechnol.* **57/58,** 741–761.
34. Parker, S., Calnon, M., Feinberg, D., Power, A., and Weiss, L. (1983), Report SERI/TR-231-2000, DE84000007, Solar Energy Research Institute, Golden, CO.
35. Ingram, L. O., Conway, T., and Alterthum, F. (1991), US Patent 5,000,000.
36. Wood, B. E. and Ingram, L. O. (1992), *Appl. Environ. Microbiol.* **58(7),** 2103–2110.
37. Reese, E. T. (1976), *Biotechnol. Bioeng. Symp.* **6,** 9–20.
38. Montencourt, B. S., Kelleher, T. J., and Eveleigh, D. E. (1980), *Biotechnol. Bioeng. Symp.* **10,** 15–26.
39. Wyman, C. E., Spindler, D. D., Grohmann, K., and Lastick, S. M. (1986), *Biotechnol. Bioeng.* **17,** 221–238.
40. Spindler, D. D., Wyman, C. E., Grohmann, K., and Mohagheghi, A. (1988), *Appl. Biochem. Biotechnol.* **21,** 529–540.
41. Spindler, D. D., Wyman, C. E., Grohmann, K., and Philippidis, G. P. (1992), *Biotechnol. Lett.* **14(5),** 403–407.
42. Spindler, D. D., Wyman, C. E., Mohagheghi, A., and Grohmann, K. (1987), *Appl. Biochem. Biotechnol.* **17,** 279–293.
43. Spindler, D. D., Wyman, C. E., and Grohmann, K. (1989), *Biotechnol. Bioeng.* **34(2),** 189–195.
44. Emert, G. H., Katzen, R., Fredrickson, R. E., and Kaupisch, K. F. (1980), *Chem. Eng. Prog.* **76(9),** 47–52.
45. Emert, G. H. and Katzen, R. (1980), *Chemtech* 610–614.
46. Takagi, M., Abe, S., Suzuki, S., Emert, G. H., and Yata, N. (1977), in *Proceedings of the Bioconversion Symposium*, Indian Institute of Technology, Delhi, pp. 551–571.
47. Gauss, W. F., Suzuki, S., and Takagi, M. (1976), US Patent no. 3,990,944.
48. Kadam, K. L. (1996), Cellulase production. Ref. 1, pp. 213–252.
49. Hettenhaus, J. R. and Glassner, D. (1997), *Enzyme Hydrolysis of Cellulose: Short-Term Commercialization Prospects for Conversion of Lignocellulosics to Ethanol*, National Renewable Energy Laboratory, Golden, CO.
50. Himmel, M. E., Ruth, M. F., and Wyman, C. E. (1999), *Curr. Opin. Biotechnol.* **10(4),** 358–364.
51. Mohagheghi, A., Grohmann, K., and Wyman, C. E. (1990), *Biotechnol. Bioeng.* **35,** 211–216.
52. Mohagheghi, A., Grohmann, K., and Wyman, C. E. (1987), *Appl. Biochem. Biotechnol.* **17,** 263–277.

53. Schell, D. J., Hinman, N. D., and Wyman, C. E. (1990), *Appl. Biochem. Biotechnol.* **24/25,** 287–297.
54. Wyman, C. E. (1995), in *Enzymatic Degradation of Insoluble Carbohydrates*, ACS Symposium Series 618, Saddler, J. N. and Penner, M. H., eds., American Chemical Society, Washington, DC, pp. 272–290.
55. van Walsum, G. P., Allen, S. G., Spencer, M. J., Laser, M. S., Antal, M. J., and Lynd, L. R. (1996), *Appl. Biochem. Biotechnol.* **57/58,** 157–170.
56. Lynd, L. R., Wyman, C. E., and Gerngross, T. U. (1999), *Biotechnol. Prog.* **15,** 777–793.
57. Wyman, C. E. (1990), *Biological Production of Chemicals from Renewable Feedstocks*, National Meeting, American Chemical Society, Washington, DC.
58. McCoy, M. (1998), *Chem. Eng. News* **76(49),** December 7, 29–32.
59. Lugar, R. G. and Woolsey, R. J. (1999), *Foreign Affairs* **78(1),** 88–102.
60. National Research Council (1999), *Review of the Research Strategy for Biomass-Derived Transportation Fuels*, National Academy Press, Washington, DC.
61. National Research Council (1999), *Biobased Industrial Products: Research and Commercialization Priorities*, National Academy Press, Washington, DC.

Genetic Improvement of Poplar Feedstock Quality for Ethanol Production

Ronald J. Dinus

*Department of Wood Science, University of British Columbia,
2490 Goshen Road, Bellingham, WA 98226-9556,
E-mail: dinus@telcomplus*

Abstract

Opportunities for matching chemical and physical properties of woody feedstocks to ethanol production process requirements via genetic improvement have long been recognized. Exploitation is now feasible owing to advances in trait measurement, breeding, and gene transfer technologies. Poplar genetic parameters are favorable largely for reducing lignin and increasing cellulose contents and specific gravity. Transgenic poplars with decreased lignin and increased cellulose contents, but otherwise normal growth and development, have been produced via genetic transformation. The long-standing debate on feasibility has thus become one of when, not if, designer varieties will become available.

Index Entries: Biomass feedstock quality; lignin; cellulose; wood specific gravity; selection and breeding; genetic transformation.

Introduction

The Bioenergy Feedstock Development Program, US Department of Energy, Oak Ridge National Laboratory (BFDP) is developing short-rotation poplars (*Populus* spp. and hybrids) as feedstock for ethanol production. Substantial gains in adaptability, growth, and pest/stress resistance have been achieved via classic breeding and intensified cultural practices. Given these and anticipated accomplishments, consideration is being given to fostering research and development on genetically modifying feedstock chemical and physical properties.

This article reviews opportunities for and feasibilities of improving short-rotation poplar feedstock quality via classic breeding and genetic transformation and is derived, in part, from a comprehensive feasibility study performed for BFDP *(1)*. Information was collected via analysis of poplar literature, with emphasis on findings published since 1995; personal communications with prominent scientists, breeders, and growers; and a BFDP-sponsored workshop held in December 1999. Results from research on other short-rotation hardwoods are used in the few situations in which poplar data could not be found.

Ethanol Conversion Processes and Feedstock Quality

Producing ethanol from wood involves using heat, acids, and enzymes to separate constituents and split cellulose and hemicellulose into individual sugar molecules. Simultaneously or subsequently sugars are fermented to ethanol. To improve process economics, some investigators have advocated using five-carbon hemicellulosic sugars to manufacture coproducts (e.g., biodegradable polymers), a reflection of past concerns that such sugars could not be fermented efficiently to ethanol. Recently, however, genetically engineered organisms have been developed to perform this function, and conversion is no longer considered an economic obstacle. Lignin and other residuals are used as fuel to generate process heat, steam, and electricity. Other markets for lignin have been and are likely to remain limited.

All processes likely used for conversion of woody feedstocks to ethanol are sensitive to feedstock quality, some more than others, with concentrated acid hydrolysis the least sensitive *(2)*. This process is well suited to handling difficult feedstocks (i.e., those having variable composition and containing large quantities of contaminants, e.g., extractives and ash). By contrast, enzymatic hydrolysis is sensitive not only to contaminants but also to lignin content. Lignin binds and inactivates enzymes, thereby necessitating large initial enzyme quantities and limiting enzyme recycling *(3,4)*. Lignin content, rather than composition, seems to be the major factor *(5)*. Lowering lignin content should reduce operating costs substantially, given the high costs of enzymes. All processes are sensitive to carbohydrate content. Increased carbohydrate content, especially cellulose, should therefore also rank high as a target for genetic modification. The specific gravity of wood warrants attention as well, given its generally positive correlation with cellulose content. Improving other properties seems likely to yield only incremental benefits. For example, reducing extractives, ash, and bark contents, naturally low in poplars, could have favorable effects, but far less than those of lower lignin and greater cellulose contents. Moreover, information is limited on the means for modifying such constituents. Lignin content, nonetheless, seems to be the prime target, since lignin is the least useful of all major feedstock components, and reduction equates to more cellulose per unit mass. Moreover, lowering lignin content should raise the profitability of enzymatic hydrolysis.

Some research has been done to quantify the effects of feedstock qualities on process efficiency, but analyses generally have been done via simulation models and have addressed only growing, harvesting, handling, transport, and storage costs. Few have addressed directly the effects of changes in quality, but sensitivity of enzymatic processes to enzyme costs (6), and therefore lignin content, and proportionality of ethanol yield to cellulose content (7) have been projected. Because changes in feedstock composition can affect processes at a variety of points and in numerous ways, definitive trials with feedstocks of differing qualities are needed to determine the impacts of changes on each of the steps in individual processes. Individuals engaged in improving and growing feedstocks must know which traits to modify, how to modify them for what process, and how modification will affect the cost/benefit ratio.

Opportunities for Improving Poplar Feedstock Quality

Viewed against the aforementioned improvements in poplar productivity, the time seems ripe for undertaking the research, development, and technology transfer necessary to exploit opportunities for improving feedstock quality. Indeed, the shortened rotations made possible by past improvements mean that such opportunities can be more easily and rapidly exploited. Also, many poplars are propagated clonally, thereby permitting rapid, inexpensive multiplication of valuable variants. Poplars seem an ideal venue for testing and applying techniques to improve feedstock quality.

Feedstock Quality Assessment

Efficient genetic improvement requires that traits be measured accurately and inexpensively. Such measurements constitute one of the largest costs of improvement, and high measurement costs have impeded efforts to improve wood properties. As an illustration, volume, a typical measure of productivity, can be measured in minutes for pennies per tree. By contrast, traditional wet chemistry assays of lignin content take several days and cost $500 or more per sample.

Recognizing these difficulties, various research organizations have developed improved methods for assaying both chemical and physical properties. For chemical assays, techniques such as reflectance near infrared (8), Fourier transform infrared (9), and pyrolysis molecular beam mass spectrometry (10) offer reduced costs as well as speed and convenience. Results correlate strongly with baseline values obtained by traditional methods. Assays require only a few grams or less of wood derived from nondestructively collected increment cores and can be completed in minutes at costs many times less than those of traditional analyses.

Concerning specific gravity, various densitometric techniques have long been used in research, but recent improvements have increased utility and reduced costs (10). Automated systems are now available for not only determining specific gravity but also for elucidating the impact of contributing factors (e.g., cell number and cell wall thickness) (11). Such techniques rely on increment cores and can analyze thousands of samples per year. Sampling costs have been reduced by the use of hydraulically driven increment corers mounted on all-terrain vehicles.

Accurate, rapid, and inexpensive measurement techniques that rely on small, nondestructively collected samples are opening the way to improving wood properties, specifying processing conditions for different feedstocks, and compensating growers for improvements.

Classic Breeding

Feedstock quality improvement via classic breeding requires that valuable traits possess usable levels of genetic variation, meaningful degrees of genetic control, and favorable correlations with other important traits. Stability across ages and environments is also important. Quantity and quality of genetic information on wood chemical and physical properties has increased recently as a result of intensified interest in facilitating the manufacture of traditional wood and paper products, as well as the availability of convenient measurement techniques. The following discussion focuses on lignin and cellulose contents, given their importance to ethanol production. Specific gravity, given its association with cellulose content, is also discussed.

Past reviews indicated the existence of genetic variation in lignin content among and within poplar and many other hardwood species *(12)*. Most recent data generally derives from research on short-rotation *Eucalyptus* species, but similar patterns are likely to be found in most short-rotation poplars. Genetic variation within species is small but statistically significant *(13)*, often only a few percentage points but sometimes as large as nine. Heritabilities are modest to high and typically exceed those for growth. Clone-by-site interactions are sometimes significant. Information on correlations between lignin content and other traits is sparse, but some reports *(14)* suggest unfavorable, though weak, relationships with specific gravity. The high heritabilities already noted infer that lignin content can be reduced by classic breeding, but the extent and pace of reduction would be limited by the narrow range of genetic variability as well as probable unfavorable relationships with growth and interactions across environments. Variability in poplars might be expanded via hybridization of adapted species. In continuing BFDP-sponsored research, 16 interspecific poplar hybrids, derived from several species, exhibited a 9.3 percentage point range in lignin content, as measured by traditional wet chemistry methods (unpublished data). Given the variety of interfertile poplar species, interspecific hybridization is an attractive tack *(15)*.

Variation in cellulose content is roughly 7 to 15 percentage points, a range larger than those for lignin or hemicellulose. Heritabilities typically are low. Olson et al. *(16)* found significant genetic variation in cellulose contents of 75 3-yr-old *Populus deltoides* grown at a single location. Broad-sense heritability (i.e., the proportion of genetic to phenotypic variation) was 0.34, a value greater than that for growth but smaller than those for most wood properties. Both variation and heritability were considered sufficient for improvement. Correlations with growth, however, were negative, and simultaneous improvement seemed impractical at the time. Olson et al. *(16)* nevertheless recommended preserving clones having high cellulose content for possible use in later years. This recommendation recognized that correlations are imperfect and may change with age, and that trait valuations may likewise change with time. Low heritabilities imply that cellulose content may be increased, but only by mating outstanding trees and clonally propagating and planting selected offspring— practices that are well known to and frequently used by poplar breeders. The generally negative, though weak correlation with growth, however, implies that breeding for cellulose content could impede breeding for growth. Decisions to breed or not breed for increased cellulose content would be facilitated by more information on strength of trait and age-age correlations as well as stability across environments.

Data on genetic variation in and control of hemicellulose content are scarce. Early literature disclosed statistically significant variation, but too limited for genetic improvement *(12)*. More recently, assay of 17 poplar clones representing two species plus F_1 and F_2 hybrids disclosed phenotypic variation similar to that in lignin content (unpublished data). Significant variation and strong genetic control has also been found in short-rotation *Eucalyptus* species *(13)*. Further research on poplars seems warranted. Lower five-carbon and higher six-carbon sugar contents as well as fewer hemicellulose-lignin linkages would benefit the efficiency of ethanol conversion.

Although often viewed as a single trait, specific gravity is a composite of several properties, including proportions of juvenile to mature wood and early to late wood; cell types, numbers, and sizes; and cell wall thicknesses. The situation in hardwoods is especially complicated, given the wide variety of cell types. Since characteristics influencing specific gravity can vary together or independently, individual trees can have similar or different specific gravities for different reasons. Improving specific gravity therefore depends on understanding what to change and how changes, in turn, affect process and product.

The considerable literature on specific gravity indicates abundant genetic variation and moderate to strong heritabilities for most hardwood species *(17)*. Olson et al. *(16)* found a broad-sense heritability of 0.62 for specific gravity in 3-yr-old *P. deltoides* and noted a negative genetic correlation (–0.65) with growth. The correlation with cellulose content, however, was positive (0.54). They deemed simultaneous selection for the several traits unprofitable in their situation but advocated preserving trees outstanding for each trait in breeding populations. In short-rotation *Eucalyptus grandis*, specific gravity was positively related to fiber wall thickness and volume of cell wall substance, but negatively correlated with lignin content *(14)*. Concerning age-age correlations, Yanchuk et al. *(18)* found that specific gravity in *Populus tremuloides* declined gradually through the tenth annual ring, and then rose across subsequent rings. Herpka *(19)* documented a similar pattern in other poplar species and hybrids and considered values at 2 or 4 yr of age predictive of those at older ages.

As indicated, genetic variation in and heritability of specific gravity is more than adequate for improvement via classic breeding. The trait seems stable across ages and environments. Favorable correlations with cellulose content and cell wall substance are advantages. Often negative correlations with growth are inconvenient but not insurmountable problems. From tests of several poplar species and hybrids, Herpka *(19)* concluded that variabilities and heritabilities were so large that trees having increased dry wood substance production per unit area and time (i.e., combining rapid growth and high specific gravity) could be selected at early ages with ease. In the end, finding trees excelling in two or more traits is a numbers game and is decided by how much additional time and expense can be devoted to testing and selecting the extra trees needed to identify outliers.

Having achieved substantial gains in survival and growth, poplar breeding programs seem sufficiently mature to pursue broader objectives. Growth, after all, is but one element governing the costs of ethanol production; chemical and physical properties of feedstock are also important. Viewed in this more inclusive context, future breeding efforts should be designed to reduce costs and raise efficiencies at all significant leverage points in production. Accordingly, breeders must generate more and better information on genetic parameters, particularly on correlations among traits. Process specialists must document how and to what extent changes in quality traits affect process efficiency and determine economic weights associated with change. These data can then be combined in selection indices for use by breeders and growers in optimizing genetic gain per unit time. The outcome should be poplar varieties adapted to process requirements, and more likely to reduce production costs than those with only improved growth.

Genetic Transformation

Improvement via genetic transformation has several advantages, and coupling it with classic selection and breeding makes for a potent, complementary combination. Should genes for a trait not exist in the species of interest, they can be procured from other, even unrelated, species and inserted into trees selected for breeding and planting. Perhaps more important, transformation can be used to alter expression of existing genes. For example, so-called antisense gene constructs can be used to suppress expression of genes affecting lignin biosynthesis or sense constructs can be used to increase cellulose biosynthesis. By virtue of cosuppression, sense constructs can sometimes be used to reduce gene activity. Not to be overlooked is the probability that gene transfer can save time via bypassing the sexual cycle and often long generation intervals.

Ideally, traits considered for modification via transformation should have significant economic value, be modified such that change is large relative to that attainable by other means, be the product of a reasonably well-understood biochemical pathway, and be controlled by one or a small family of related genes *(20)*. Moreover, modification should not adversely affect survival and growth. Lignin content meets these criteria rather well and cellulose content comes close.

Basic features of and many enzymes in the lignin biosynthetic pathway have been known for some years *(21)*. Knowledge has increased greatly in recent years, often owing to the use of genetic transformation as a research tool *(22)*. Roughly a decade ago information on biochemistry and genetics of enzymes involved in cellulose biosynthesis was considered inadequate to support research on modification *(12)*. Advances since then have been dramatic *(23)*, and several laboratories have achieved notable transformation successes. Most modifications have been made near the end of the biosynthetic pathways, where changes in enzymatic activity are less likely to have adverse effects on other metabolic processes.

Several methods for poplar transformation are available, and numerous genes have been inserted and expressed in various species and hybrids. Indeed, poplars are often used as models in transformation research, given the ease with which many can be manipulated in and regenerated from cell and tissue cultures. Improvements in transformation efficiency and extension to wider arrays of genotypes, however, remain important research needs *(24)*.

Considerable information is available on genetic modification of enzymes involved in poplar lignin biosynthesis. Most research concerns manipulation of enzymes catalyzing synthesis and interconversion of lignin precursors. Some of the first research was conducted on phenylalanine-ammonia lyase (PAL), the enzyme at the gateway for carbon entry to lignin biosynthesis. Emphasis has since switched to enzymes functioning later in biosynthesis owing to concerns over adverse effects on other pathways. Intermediates through at least the thioester stage are known to be involved in the biosynthesis of metabolites in addition to those slated for lignification. In addition, determining which of the many forms of PAL are active when and where is difficult *(21)*.

O-methyltransferases and caffeate *O*-methyltransferases (COMTs) are thought to catalyze the ortho-methylation of caffeate to ferulate and 5-hydroxyferulate to sinapate in some plants and have also been implicated in methylation of other intermediates. van Doorsselaere et al. *(25)* suppressed COMT activity by transferring an antisense construct into hybrid poplars. Lignin content was not reduced but syringyl/guaiacyl ratios were reduced; a novel lignin monomer, 5-hydroxyguaiacyl, was produced; and xylem was rose colored. Similar results were obtained in *P. tremuloides* by Tsai et al. *(26)* via cosuppression with a COMT sense construct.

In angiosperms, 4-coumarate 3-hydroxylases (4CLs) convert 4-coumarate, caffeate, ferulate, 5-hydroxyferulate, and sinapate to their respective thioesters. Hu et al. *(27)* discovered two different 4CL genes (*Pt4CL* and *Pt4CL2*) in *P. tremuloides*. The activity of *Pt4CL* was specific to lignifying xylem tissues, implying involvement mainly in lignin biosynthesis. Transformation with antisense constructs yielded trees with a 45% decrease in lignin and 15% increase in cellulose contents *(22)*, without changes in lignin composition or structure. Hemicellulose composition was also altered; arabinose, galactose, and rhamnose contents were increased. Transgenic trees are growing faster than nontransgenics, but are morphologically and anatomically normal. This is the first clear-cut reduction in lignin content apparent in the literature. Douglas et al. *(28)* identified similar genes in *Populus trichocarpa* and its hybrid with *P. deltoides* and are using antisense constructs to explore the effects of suppression.

Caffeoyl CoA *O*-methyltransferase (CCoAOMT) is considered responsible for converting caffeoyl-CoA to feruloyl-CoA and, perhaps, 5-hydroxyferuloyl-CoA to sinapoyl-CoA. Several European laboratories are contemplating *(29)* or initiating *(30)* manipulation of CCoAOMT activity.

The enzyme cinnamoyl-CoA reductase (CCR) converts thioester forms of lignin precursors to the corresponding aldehydes, e.g., feruloyl CoA to coniferyl aldehyde. Transgenic poplars with modified CCR activities are being evaluated *(31)*.

Cinnamyl alcohol dehydrogenase (CAD) catalyzes the last step in biosynthesis of lignin precursors, conversion of aldehydes to alcohols. Transformation of poplars with antisense CAD constructs had little or no effect on lignin content but caused changes in lignin composition (i.e., incorporation of aldehydes into lignin) and red-brown xylem *(29)*. Baucher et al. *(30,32)* confirmed these findings and also observed vanillin and syringaldehyde accumulations in cell walls. In related research, Lapierre et al. *(33)* reported that lignin contents of 2-yr-old transgenic poplars were only slightly lower than that of nontransformed controls, but that frequency of free phenolic groups was greater and that coloration was caused by syringaldehyde accumulation. In a somewhat more positive vein, Pilate et al. *(34)* inserted a CAD antisense construct into a poplar different than that used by Lapierre et al. *(33)*. CAD activity was suppressed to a greater extent; outcomes were a 10–15% reduction in lignin content, lower syringyl/guaiacyl ratios, and atypical compounds in lignin.

Information on biochemistry and genetics of enzymes involved in transport, storage, and polymerization of lignin precursors is not nearly as abundant as that on their biosynthesis. Data nevertheless are accumulating rapidly, and various laboratories are attempting improvement via gene transfer. Dinus *(1)* summarized recent research on glucosidases, glucosyltransferases, peroxidases, laccases, and an alcohol oxidase. Modifying functions this far into lignin biosynthesis seems attractive, since intervention should have minimal effects on other metabolic processes.

Concerning improvement in cellulose biosynthesis, Loopstra et al. *(35)* have isolated and sequenced a family of cellulose synthase (celA) genes from poplar. Transgenic *P. tremuloides* containing a celA sense construct have been produced at Michigan Technological University (Chiang, V. L., personal communication). The trees are morphologically normal but are growing faster than nontransformed controls. Transformation of hybrid poplar with a sense construct for a UDP-pyrophosphorylase gene is being attempted (Ellis, D. D., personal communication). This undertaking follows earlier research in which transformation of tobacco (*Nicotiana tabacum*) with a similar construct increased enzyme activity, cellulose synthesis, and biomass (36). Prospects for increasing cellulose content via genetic transformation look almost as good as those for reducing lignin content.

Much transformation research has been done with constitutive promoters, e.g., the CV35S promoter or a strengthened version thereof. This is useful in early stages of research, since the approach enhances probabilities of observable expression. Practical utility, however, demands that research progress beyond this proof-of-principle phase to ensure that transgenes are expressed in appropriate tissues and developmental stages. A few investigators are using sense/antisense constructs fused to xylem-specific promoters. Even greater specificity, nevertheless, seems necessary. As an example, reducing lignin content of secondary cell walls in fibrous cells seems quite useful. Reduction in vessel elements, on the other hand, could disrupt water transport and support functions. Accelerated research on tissue and developmental stage specific promoters (21) remains imperative.

An obstacle to commercialization of transgenic trees is public concern over the possibility that widespread planting will spread so-called foreign genes to natural populations (37). Rendering transgenic trees sexually sterile is one means of minimizing, if not avoiding, this risk. Genetic constructs for sterility in poplar should be available in 5–10 yr. Useful side effects may also accrue; eliminating reproductive structures could channel more energy, water, and nutrients into wood production. Concern also might be eased by the availability of genetic markers to monitor potential for gene flow. In addition, confidence and acceptance can be built by responsible, transparent testing along with clear, consistent communication of risks and benefits. Sterile, transgenic trees, after all, can be expected to lessen the environmental impacts of growing and harvesting.

Conclusion

Poplar breeders and growers have achieved substantial improvements in adaptability, growth, and pest/stress resistance. Over the long run, however, productivity is just one element in the matrix needed to enhance the efficiency of ethanol production. Chemical and physical properties of wood also govern cost and yield. Viewed in this more exclusive context, bioenergy projects would be well served by enlarging research, development, and technology transfer efforts to include genetic improvement of important feedstock qualities. Improvements can be made by both classic breeding and genetic transformation. Pursuit of these approaches will be eased by newly available, convenient, and inexpensive trait measurement methods. Progress will be slow, however, until the costs and benefits of modification are clarified, and economic weights are translated into guidelines for breeding, transformation, and compensating feedstock producers.

Most feedstock quality traits can be improved via classic breeding. Lignin, the least valued of feedstock chemical components, can be reduced, but limited variability will limit progress. Even small changes, however, could improve efficiency of conversion processes, especially that of enzymatic hydrolysis. Raising cellulose content could prove more difficult, given its rather modest heritability and negative association with growth. Even so, useful changes in these traits seem obtainable in perhaps two or three generations of breeding and testing. As an alternative, breeding might be better focused on traits likely to decrease lignin and increase cellulose contents indirectly. For example, simultaneously improving specific gravity and growth (i.e., increased wood substance production per unit area and time) should positively affect productivity, harvesting, and transportation, as well as processing. This tack seems particularly desirable in that the wood would be attractive to a variety of customers. Regardless of strategy, continued research is needed to acquire better genetic information and more economic data for construction of selection indices.

The considerable research done in recent years on genetic transformation has produced transgenic trees with modified lignin content and composition as well as increased cellulose contents. Antisense suppression of most enzymes involved in lignin precursor biosynthesis has changed only composition. Suppression of a few (e.g., CAD) has not only altered composition but also somewhat reduced quantity. Compositional changes noted to date involve incorporation of atypical precursors in lignin, deposition of free compounds in cell walls, and coloration of xylem. Most such changes, unfortunately, are likely to have detrimental effects under the acidic conditions of ethanol conversion; that is, they could raise contaminant levels in process sugar streams, thereby complicating capture of residuals and increasing both effluent quantity and color. That compositional changes have been induced, however, indicates that lignin biosynthesis is quite plastic and suggests that lignin could be bioengineered to favor ethanol conversion. Suppression of only one enzyme, 4CL, gave a 45% reduction in lignin content, without any adverse effects on composition or growth and development. Several laboratories have also increased cellulose content with sense contructs. Modifying lignin and cellulose contents simultaneously could yield commercial varieties with significantly enhanced conversion efficiency. Transformation works, and commercialization is only a matter of time, perhaps 5–10 yr. Ensuring that dividends from this exciting technology are realized in a reasonable time frame nevertheless requires continued, balanced research on transformation methods; biochemistry and genetics of lignin, cellulose, and hemicellulose biosynthesis; gene regulation and control; sexual sterility; and detection and resolution of any adverse side effects.

Poplar feedstock bred for rapid growth and high wood substance production and genetically transformed to have decreased lignin and increased cellulose contents stands to have tremendous beneficial effects on ethanol production.

Acknowledgments

I wish to thank Vincent Chiang, Linda Belles Dinus, Peggy Payne, Vincent Sewalt, Mitchell Sewell, Jerry Tuskan, Lynn Wright, and two anonymous reviewers for their thoughtful contributions. I also acknowledge the US Department of Energy, Bioenergy Feedstock Development Program, Oak Ridge National Laboratory, for funding to analyze literature, collect information via personal communications, conduct a review workshop, and prepare and present this manuscript.

References

1. Dinus, R. J. (2000) US Department of Energy, Biomass Feedstock Development Program Report, Oak Ridge National Laboratory, Environmental Sciences Division, Oak Ridge, TN.
2. Goldstein, I. S. and Easter, J. M. (1992), *TAPPI J.* **75(8),** 135–140.
3. Gregg, D. and Saddler, J. N. (1996), *Appl. Biochem. Biotechnol.* **57/58,** 711–726.
4. Sewalt, V. J. H., Glasser, W. G., and Beauchemin, K. A. (1997), *J. Agric. Food Chem.* **45(5),** 1823–1828.
5. Sewalt, V. J. H., Ni, W., Jung, H. G., and Dixon, R. A. (1997), *J. Agric. Food Chem.* **45(5),** 1977–1983.
6. Gregg, D., Boussaid, A., and Saddler, J. N. (1998), *Bioresour. Technol.* **63,** 7–12.
7. Wooley, R., Ruth, M., Sheehan, J., and Ibsen, K. (1999), National Renewable Energy Laboratory Technical Report NREL/TP-580-26157, US Department of Energy, Golden, CO.
8. Clarke, C. R. E. and Wessels, A. E. (1995), in *Eucalypt Plantations: Improving Fibre Yield and Quality*, Proceedings of the CRC-IUFRO Conference, Potts, B. M., Borralho, N. M. G., Reid, J. B., Cromer, R. N., Tibbits, W. N., and Raymond, C. A., eds., CRC for Temperate Hardwood Forestry, Hobart, Australia, pp. 93–100.
9. Costa, E., Silva, J., Wellendorf, H., and Pereira, H. (1998), *Silvae Genetica* **47(1),** 20–33.
10. Tuskan, G., West, D., Bradshaw, H. D., Neale, D., Sewell, M., Wheeler, N., Megraw, B., Jech, K., Wiselogel, A., Evans, R., Elam, C., Davis, M., and Dinus, R. (1999), *Appl. Biochem. Biotechnol.* **77–79,** 55–65.
11. Evans, R. and Downes, G. (1995), in *Eucalypt Plantations: Improving Fibre Yield and Quality*, Proceedings of the CRC-IUFRO Conference, Potts, B. M., Borralho, N. M. G., Reid, J. B., Cromer, R. N., Tibbits, W. N., and Raymond, C. A., eds., CRC for Temperate Hardwood Forestry, Hobart, Australia, pp. 101–105.
12. Dinus, R. J., Dimmel, D. R., Feirer, R. P., Johnson, M. A., and Malcolm, E. W. (1990), Biomass Production Program Report No. ORNL/Sub/88-SC006/1, US Department of Energy, Oak Ridge National Laboratory, Environmental Sciences Division, Oak Ridge, TN.
13. Bertolucci, F. L. G., Demuner, B. J., Garcia, S. L. R., and Ikemori, Y. K. (1995), in *Eucalypt Plantations: Improving Fibre Yield and Quality*, Proceedings of the CRC-IUFRO Conference, Potts, B. M., Borralho, N. M. G., Reid, J. B., Cromer, R. N., Tibbits, W. N., and Raymond, C. A., eds., CRC for Temperate Hardwood Forestry, Hobart, Australia, pp. 33, 34.
14. Malan, F. S. and Arbuthnot, A. L. (1995), in *Eucalypt Plantations: Improving Fibre Yield and Quality*, Proceedings of the CRC-IUFRO Conference, Potts, B. M., Borralho, N. M. G., Reid, J. B., Cromer, R. N., Tibbits, W. N., and Raymond, C. A., eds., CRC for Temperate Hardwood Forestry, Hobart, Australia, pp. 19–24.

15. Stettler, R. F., Zsuffa, L., and Wu, R. (1996), in *Biology of Populus and Its Implications for Management and Conservation*, Stettler, R. F., Bradshaw, H. D., Jr., Heilman P. E., and Hinckley, T. M., eds., NRC-CNRC Research Press, Ottawa, Canada, pp. 87–112.

16. Olson, J. R., Jourdain, C. R., and Rousseau, R. J. (1985), *Can. J. Forest. Res.* **15**, 393–396.

17. Zobel, B. J. and Jett, J. B. (1995), *Genetics of Wood Production*, Springer-Verlag, Berlin.

18. Yanchuck, A. D., Dancik, B. P., and Micko, M. M. (1983), *Wood Fiber Sci.* **15(4)**, 387–394.

19. Herpka, I. (1990), in *Improvement of Feedstock Quality*, Proceedings of the Second IEA/BA Task V Workshop, University of Toronto, Toronto, Canada, pp. 49–85.

20. Timmis, R. and Trotter, P. C. (1989), in Proceedings of the International Workshop on Applications of Biotechnology in Forestry and Horticulture, Plenum, New York, pp. 349–367.

21. Campbell, M. M. and Sederoff, R. R. (1996), *Plant Physiol.* **110**, 3–13.

22. Hu, W.-J., Harding, S. A., Lung, J., Popko, J. L., Ralph, J., Stokke, D. D., Tsai, C. J., and Chiang, V. L. (1999), *Nat. Biotechnol.* **17**, 808–812.

23. Delmer, D. P. (1999), *Annu. Rev. Plant Physiol. Plant Mol. Biol.* **50**, 245–276.

24. Han, K.-H., Gordon, M. P., and Strauss, S. H. (1996), in *Biology of Populus and Its Implications for Management and Conservation*, Stettler, R. F., Bradshaw, H. D., Jr., Heilman, P. E., and Hinckley, T. M., eds., NRC-CNRC Research Press, Ottawa, Canada, pp. 201–222.

25. van Doorsselaere, J., Baucher, M., Chognot, E., Chabbert, B., Tollier, M. T., Petit-Conil, M., Leple, J. C., Pilate, G., Cornu, D., and Monties, B. (1995), *Plant J.* **8(6)**, 855–864.

26. Tsai, C.-J., Popko, J. L., Mielke, M. R., Hu, W.-J., Podila, G. K., and Chiang, V. L. (1998), *Plant Physiol.* **117**, 101–112.

27. Hu, W.-J., Akiyoshi, K., Tsai, C.-J., Lung, J., Osakabe, K., Ebinuma, H., and Chiang, V. L. (1998), *Proc. Natl. Acad. Sci. USA* **9**, 5407–5412.

28. Douglas, C. J., Cukovic, D., Pereira, J., and Allina, S. (1999), in Program International Poplar Symposium II, I.N.R.A., Orleans, FR, p. 28.

29. Boudet, A. M. and Grima-Pettenati, J. (1996), *Mol. Breeding* **2(1)**, 25–39.

30. Baucher, M., Christensen, J. H., van Doorsselaere, J., Meyermans, H., Chen, C., Burggraeve, B., Leple, J.-C., Pilate, G., Petit-Conil, M., Jouanin, L., Chabbert, B., Monties, B., van Montagu, M., and Boerjan, W. (1997), *Med. Facul. Landbouwk. Toegepaste Biol. Wetenschappen Univ. Gent.* **62(4AB)**, 1403–1411.

31. Boerjan, W., Meyermans, H., Chen, C., Christensen, J., Leple, J.-C., Pilate, G., Lapierre, C., Pollet, B., Jouanin, L., Baucher, M., van Doorsselaere, J., Petit-Conil, M., and van Montagu, M. (1999), in *Program International Poplar Symposium II*, I.N.R.A., Orleans, FR, p. 18.

32. Baucher, M., Chabbert, B., Pilate, G., van Doorsselaere, J., Tollier M.-T., Petit-Conil, M., Cornu, D., Monties, B., van Montagu, M., Inze, D., Jouanin, L., and Boerjan, W. (1996), *Plant Physiol.* **112**, 1479–1490.

33. Lapierre, C., Pollet, B., Petit-Conil, M., Toval, G., Romero, J., Pilate, G., Leple J.-C., Boergan, W., Ferret, V., DeNadai, V., and Jouanin, L. (1999), *Plant Physiol.* **119(1)**, 153–163.

34. Pilate, G., Leple, J. C., Noel, N., deNadai, V., Jouanin, L., Pollet, B., Mila, I., Vallet, C., and Lapierre, C. (1999), in *Program International Poplar Symposium II*, I.N.R.A., Orleans, FR, p. 71.

35. Loopstra, C., Eun-Gyu, N., Zho, Y., Puryear, J., Hongyan, W., and Pawlak, D. (1998), in *1998 Information Exchange Group-40 Workshop: Wood and Wood Fibers: Properties and Genetic Improvement*, Institute of Paper Science and Technology, Atlanta.

36. Xue, B., Ellis, D., Newton, D., Gawley, B., Gibbert, M., and Sutton, B. (1997), *Plant Physiol.* **114(Suppl. 3)**, 300.

37. Meilan, R. and Strauss, S. H. (1997), in *Micropropagation, Genetic Engineering and Molecular Biology of Populus*, Klopfenstein, N. B., Chun, Y. W., Kim, M.-S., and Ahuja, M. R., eds., USDA Forest Service General Technical Report RM-GTR-297, USDA, Forest Service, Rocky Mountain Forest and Range Experiment Station. Fort Collins, CO, pp. 212–219.

Detoxification of Lignocellulose Hydrolysates with Ion-Exchange Resins

Nils-Olof Nilvebrant,[1] Anders Reimann,[1] Simona Larsson,[2] and Leif J. Jönsson*,[2]

[1]STFI, Swedish Pulp and Paper Research Institute, PO Box 5604, SE-114 86 Stockholm, Sweden; and [2]Applied Microbiology, Lund University/Lund Institute of Technology, PO Box 124, SE-221 00 Lund, Sweden, E-mail: leif.jonsson@tmb.lth.se

Abstract

Lignocellulose hydrolysates contain fermentation inhibitors causing decreased ethanol production. The inhibitors include phenolic compounds, furan aldehydes, and aliphatic acids. One of the most efficient methods for removing inhibiting compounds prior to fermentation is treatment of the hydrolysate with ion-exchange resins. The performance and detoxification mechanism of three different resins were examined: an anion exchanger, a cation exchanger, and a resin without charged groups (XAD-8). A dilute acid hydrolysate of spruce was treated with the resins at pH 5.5 and 10.0 prior to ethanolic fermentation with *Saccharomyces cerevisiae*. In addition to the experiments with hydrolysate, the effect of the resins on selected model compounds, three phenolics (vanillin, guaiacol, and coniferyl aldehyde) and two furan aldehydes (furfural and hydroxymethyl furfural), was determined. The cation exchanger increased ethanol production, but to a lesser extent than XAD-8, which in turn was less effective than the anion exchanger. Treatment at pH 10.0 was more effective than at pH 5.5. At pH 10.0, the anion exchanger efficiently removed both anionic and uncharged inhibitors, the latter by hydrophobic interactions. The importance of hydrophobic interactions was further indicated by a substantial decrease in the concentration of model compounds, such as guaiacol and furfural, after treatment with XAD-8.

Index Entries: Detoxification; inhibition; ethanol production; *Saccharomyces cerevisiae*; softwood; ion exchange.

Introduction

Acid hydrolysates of lignocellulosic materials contain inhibitors that cause decreased productivity in ethanolic fermentations by microbes such

*Author to whom all correspondence and reprint requests should be addressed.

as baker's yeast. In the production of fuel ethanol from lignocellulose, the problem with fermentation inhibition can be overcome by detoxification of the hydrolysate prior to fermentation. In a comparison of different methods for detoxification of a dilute-acid hydrolysate of spruce *(1)*, treatment with anion-exchange resin was one of the most efficient. The chemical effect of the different detoxification methods on the hydrolysate was analyzed, and it was concluded that the anion exchanger affected all the different types of inhibitors measured (phenolics, furan aldehydes, and aliphatic acids), as well as the concentration of fermentable sugars in the hydrolysate. The poor specificity observed for the anion-exchange treatment was tentatively explained by the occurrence of hydrophobic interactions with the matrix of the used anion-exchange resin.

Ion-exchange resins have been employed previously in the detoxification of lignocellulose hydrolysates (reviewed in ref. 2). A birch wood hemicellulose hydrolysate was detoxified using a cation exchanger prior to fermentation with *Gluconobacter oxydans (3)*. A mixed-bed ion resin was used in combination with overliming for improving the fermentability of a hemicellulose hydrolysate of red oak with the yeast *Pichia stipitis (4)*. A bagasse hydrolysate was treated by a combination of anion and cation exchangers prior to fermentation with *Pachysolen tannophilus (5)*. Treatment with ion-exchange resin was also found to decrease the toxicity of a waste paper hydrolysate *(6)*, although the mechanism of the detoxification was not further elucidated. Anion exchange was used in combination with active carbon and evaporation to improve the fermentability of a dilute acid hemicellulose hydrolysate from corncobs *(7)*. The alternating use of anion, cation, and mixed resins to treat the lignocellulose hydrolysates in these studies indicates a lack of knowledge of the nature of the target compounds for the detoxification treatment. This stresses the importance of a systematic approach to determine the type of ion-exchange resin that most efficiently provides the desired effect.

The aims of the present study was to investigate whether anion- or cation-exchange resins provide the best detoxification effect, to elucidate whether interactions between the fermentation inhibitors and the matrix material have to be considered in addition to the interactions with the charged groups, and to examine the effects of pH and the amounts of ion-exchange resin. We used three different resins (an anion exchanger, a cation exchanger, and a resin without charged groups) under different conditions for the detoxification of a spruce dilute-acid hydrolysate. In addition, we explored the interactions between the three resins and selected model compounds representing fermentation inhibitors and fermentable sugars. Three phenolic model compounds with different pK_a values were selected (guaiacol, vanillin, and coniferyl aldehyde) together with two furan aldehydes (furfural and hydroxymethyl furfural [HMF]) generally present in lignocellulose hydrolysates in substantial amounts (Fig. 1).

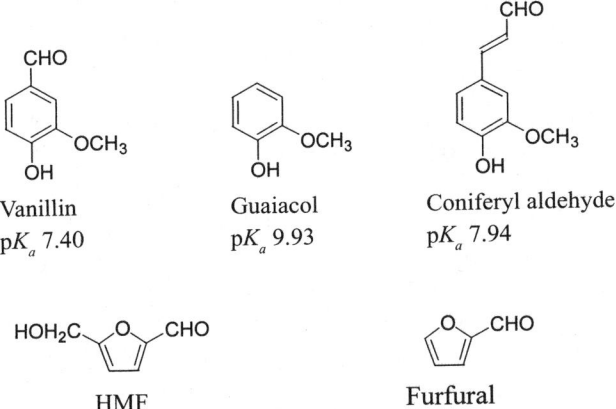

Fig. 1. Model compounds used and the dissociation constants of the phenols *(12)*.

Materials and Methods

Hydrolysate

The raw material, its composition, and the conditions for hydrolysis were as previously reported *(1)*. Chipped Norway spruce, *Picea abies*, was impregnated with 0.5% (w/w) sulfuric acid prior to treatment with saturated steam at 222°C (23 bar) for 7 min. The solid fraction was removed by filtration and the liquid fraction, referred to as the hydrolysate, had a pH of 1.9.

Model Compounds

Vanillin, guaiacol, and coniferyl aldehyde were chosen as representatives for phenols with inhibiting properties. The furan aldehydes were HMF and furfural (see Fig. 1). All model compounds were obtained from Sigma-Aldrich (St. Louis, MO). A solution of the five compounds (0.2 mM each) was prepared in double-deionized water (Millipore, Bedford, MA). The model experiments were performed under a slow stream of helium to prevent oxidation by air. In addition to the model experiments with inhibitors, a solution of 139 mM glucose (corresponding to 25 g/L, the approximate glucose concentration in the hydrolysate) was used with 8.0 g of the resin at room temperature for 1 h to determine sugar losses under different conditions. The influence of sulfate ions was studied by the addition of sodium sulfate (23 mM) corresponding to the concentration of sulfate ions determined in the hydrolysate (unpublished data).

Ion-Exchange Resins

The anion exchanger (AG 1-X8, 20–50 mesh, 3.2 meq/g [dry]) and the cation exchanger (AG 50W-X8, 20–50 mesh, 5.1 meq/g [dry]) were obtained from Bio-Rad (Richmond, CA). AG 1-X8, a strong anion-exchange resin, was used in its HO⁻ form. The cation exchanger was changed to sodium form prior to use. XAD-8 (40–60 mesh) was obtained from Supelco (Supelco Park, Bellefonte, PA). The different resins were carefully washed with distilled water. The well-washed resins contained approx 50% water. The concentrations determined in the treated hydrolysate were adjusted for the dilution. The pH was adjusted with NaOH or H_2SO_4.

Detoxification with Ion-Exchange Resins

The mixture of model compounds and the hydrolysate were treated at room temperature for 1 h. The hydrolysate was stirred with the different resins for 1 h and then filtered. The ion exchangers were used in two different quantities, 8.0 or 3.6 g in 50 mL of hydrolysate. The amounts chosen were based on the amount of anion exchanger needed to adjust the pH of the hydrolysate to 10.0 and 5.5, respectively.

Analysis of Hydrolysate Composition

Formic, acetic, lactic, and oxalic acid were determined using a DX 500 high-performance liquid chromatography (HPLC) system (Dionex, Sunnyvale, CA) and an AS11-HC (Dionex) column, eluted with 80% (v/v) water and 20% (v/v) of a mixture consisting of 0.4 mM NaOH and methanol (50% [v/v]) at a flow rate of 1.4 mL/min. Levulinic acid was determined using an AS15 column (Dionex) and isocratic elution with 25 mM NaOH. The samples were directly injected after dilution with water and filtration through a 0.2-µm MFS-25 filter (Advantec MFS, Pleasanton, CA). The amounts of aliphatic acids were determined from external calibration curves using a conductivity detector. The analyzed acids were collectively reported as aliphatic acids.

The furan aldehydes, HMF and furfural, were determined by HPLC using a Gynkotek system 480 equipped with a UVD 340S diode array detector (Gynkotek, Germering, Germany). The furan aldehydes were separated on an ODS-AA column (50 × 4.6 mm, 5-µm particles) from YMC (Waters, Milford, MA). The flow rate was 0.8 mL/min. Elution was performed with a gradient consisting of water containing 5–25% (v/v) acetonitrile and 0.025% (v/v) trifluoroacetic (TFA) acid for 10 min. For quantification, syringic acid was used as an internal standard.

The concentrations of phenolic compounds were estimated by HPLC. The separation was performed with an ODS-AQ column (150 × 3 mm, 3-µm particles) from YMC (Waters). The flow rate was 0.4 mL/min. Elution was performed with a gradient composed of Eluent A and B. Eluent A consisted of water containing 1% (v/v) acetonitrile and 0.025% (v/v) TFA. Eluent B consisted of acetonitrile containing 1% (v/v) water and 0.025% (v/v) TFA. The gradient was formed in three steps for 45 min by going from (1) 100% Eluent A to 86% A and 14% B in 20 min, (2) 86% A and 14% B to 75% A and 25% B in 20 min, and (3) 75% A and 25% B to 100% B in 5 min. The areas for all peaks eluting after furfural (from 15.5 to 45 min) and detected at 280 nm were given as the sum of low molecular weight phenols. The quantifications were made relative to the untreated hydrolysate.

The total concentrations of phenols were also estimated with a spectrophotometric method based on the Folin & Ciocalteaus reagent *(8)*. The samples were diluted 400 times with distilled water and 14 mL was mixed with the Folin & Ciocalteaus reagent (Sigma, Steinheim, Germany), and 5 mL of saturated Na_2CO_3 was added. The absorbance at 725 nm was determined after 30 min at room temperature. The amounts of phenols were determined from a calibration curve based on vanillin, since it was the most abundant phenol in the hydrolysate.

Analyses of Model Compound Experiments

Furan aldehydes, vanillin, guaiacol, and coniferyl aldehyde were separated by HPLC using a reversed-phase Nucleosil 100-5 C18 column (Macherey-Nagel, Düren, Germany). The model compounds were eluted with a gradient of methanol and water, both containing TFA at a concentration of 0.025%. The methanol concentration was increased from 5 to 100% over 35 min at a flow rate of 0.8 mL/min. The amounts were calculated using syringic acid as internal standard.

The concentration of glucose was determined by high-performance anion-exchange chromatography with a Dionex DX 500 chromatography system coupled with pulsed amperometric detection (Dionex ED 40) and using a CarboPac PA-1 column (all from Dionex). The column was first equilibrated with a mixture of 200 mM NaOH and 70 mM sodium acetate for about 5 min. After sample injection, an isocratic elution with pure water and postcolumn addition of 300 mM NaOH was applied. L-Fucose was used as an internal standard.

The concentrations obtained after treatment with the resins were compensated for the dilution caused by the water in the resins. The changes in concentrations are reported as percentages.

Fermentation

Compressed baker's yeast, *Saccharomyces cerevisiae* (Jästbolaget AB, Rotebro, Sweden), was used in all fermentations. The pH of the fermentations was 5.5. The hydrolysate was supplemented with nutrients as previously described *(1)*. Fermentations of a solution with only the nutrients, 35 g/L of glucose, and no hydrolysate were performed for every fermentation and are hereafter referred to as the reference fermentations. All fermentations were carried out with equipment and conditions as described elsewhere *(1)*. Each sample was fermented on at least two separate occasions.

The fermentability was evaluated by comparing the total ethanol yield, calculated as produced ethanol divided by the amount of fermentable sugars (glucose and mannose) present in the hydrolysate (Y_{tot} [g/g]); the ethanol yield on the amount of consumed fermentable sugars (Y_{cons} [g/g]); and the maximum mean volumetric productivity, hereafter referred to as the volumetric productivity, calculated as ethanol produced within the first 6 h of fermentation divided by 6 (Q_{6h} [g·L^{-1}·h^{-1}]), because the maximum mean volumetric productivity was obtained after 6 h in the reference fermentations. Anaerobic growth yield, hereafter referred to as biomass yield (Y_x [g/g]), was calculated as the produced biomass in 24 h (dry wt) divided by the consumed amount of fermentable sugars.

HPLC and Dry Weight Analyses of Fermentations

In all fermentation samples glucose, ethanol, lactic acid, and glycerol were separated with a high-performance liquid chromatograph (Waters) equipped with a refractive index detector (RID-6A; Shimadzu, Kyoto, Japan), using an Aminex HPX-87H column (Bio-Rad) at 45°C with 5 mM H_2SO_4 at a flow rate of 0.6 mL/min as the mobile phase. The system was equipped with a Cation-H Refill Cartridge (Bio-Rad) prior to using the HPX-87H column. Samples were diluted and filtered through 0.2-μm membrane filters (Advantec MFS) prior to analysis. Dry weight was determined at the beginning and the end of the fermentation as described previously *(1)*.

Results

Nine different detoxification treatments were performed (Table 1, methods 2–10). The ethanol yield on consumed sugar (Y_{cons}) ranged from 0.42 to 0.47 g/g and did not deviate much from either the untreated hydrolysate (Y_{cons} = 0.42 ± 0.03 g/g) or the reference fermentation (Y_{cons} = 0.42 ± 0.01 g/g). The highest Y_{cons} was for the sample treated with 3.6 g of XAD-8 at pH 5.5, which, on the other hand, did not give so high a total ethanol yield (Y_{tot}) on fermentable sugars.

The total ethanol yield (Table 1) was 0.06 ± 0.02 for the untreated hydrolysate and 0.42 ± 0.02 for the reference. The highest total ethanol yield was achieved after the anion-exchange treatment at pH 10.0, using either 8.0 or 3.6 g of resin, and after the XAD-8 treatment at pH 10.0. Even though the corresponding treatments at pH 5.5 (Table 1, methods 8 and 6) resulted in considerable improvement in fermentability compared with the control (Table 1, method 1), they could not compare with the treatments performed at pH 10.0. The cation exchanger was the only resin that at pH 10.0 did not result in a total ethanol yield comparable or better than the reference. An increased amount of anion-exchange resin at pH 10.0 did not greatly improve yield, since the yield was high already with 3.6 g, whereas an increased amount of XAD-8 at pH 5.5 raised the Y_{tot} from 0.19 to 0.28 g/g.

The highest biomass yield (Table 1) was obtained after anion exchange with 8.0 g at pH 10.0, and the yield was the same as that of the reference (0.05 ± 0.01 g/g). The anion-exchange treatment with 3.6 g at pH 10.0 resulted in lower biomass yield (0.040 g/g), and treatment with XAD-8 at pH 10.0, which also provided high Y_{tot}, showed even lower biomass yield.

The volumetric ethanol productivity, Q_{6h}, of the untreated hydrolysate was 0.21 ± 0.04 g/(L·h). Treatment at pH 10.0 without any resin present raised the productivity to 0.34 g/(L·h) (Fig. 2). The productivity was for all treatments lower than in the reference fermentation (0.92 ± 0.17 g/[L·h]),

Table 1
Ethanol and Biomass Yield After Detoxification

Detoxification method	Y_{cons} (g/g)	Y_{tot} (g/g)	Y_x (g/g)
1. None (NaOH, pH 5.5)	0.42	0.06	0.000
2. NaOH, pH 10.0	0.45	0.19	0.000
3. Cation exchanger (8.0 g), pH 5.5	0.45	0.17	0.000
4. Cation exchanger (8.0 g), pH 10.0	0.45	0.22	0.000
5. XAD-8 (3.6 g), pH 5.5	0.47	0.19	0.000
6. XAD-8 (8.0 g), pH 5.5	0.46	0.28	0.001
7. XAD-8 (8.0 g), pH 10.0	0.46	0.45	0.004
8. Anion exchanger (3.6 g), pH 5.5	0.42	0.20	0.004
9. Anion exchanger (3.6 g), pH 10.0	0.45	0.45	0.040
10. Anion exchanger (8.0 g), pH 10.0	0.46	0.46	0.050
Reference	0.42	0.42	0.050

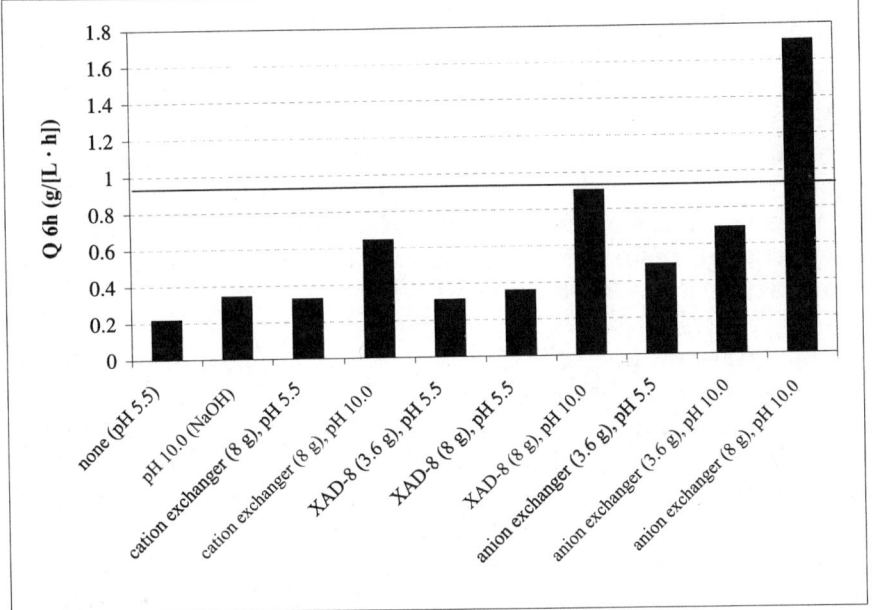

Fig. 2. Volumetric ethanol productivity in a hydrolysate treated with different types of resins. Solid line indicates the productivity in the reference fermentations.

with the exception of the anion-exchange treatment at pH 10.0 (Fig. 2), which resulted in an ethanol productivity of 1.71 g/(L·h). When the amount of anion exchanger used was decreased, the volumetric productivity decreased as well, to 0.69 g/(L·h). When the pH of the anion-exchange treatment was decreased to 5.5, the productivity decreased even further, to 0.49 g/(L·h). The treatment with 8.0 g of XAD-8 at pH 10.0 resulted in the next best ethanol productivity, 0.90 g/(L·h), which is comparable with that of the reference fermentation. At pH 5.5, there was a minor difference in productivity between 3.6 (0.31 g/[L·h]) and 8 g of XAD-8 (0.36 g/[L·h]). Treatment with 8 g of cation exchanger at pH 10.0 gave higher productivity (0.64 g/[L·h]) than at pH 5.5 (0.33 g/[L·h]).

Chemical analyses of the different groups of inhibiting compounds in the hydrolysate were done before and after detoxification with the resins (Table 2). The untreated hydrolysate has been thoroughly analyzed (unpublished data) and the concentrations of monosaccharides, furan aldehydes, aliphatic acids, and 14 phenolic compounds were reported previously *(1)*.

No decrease in fermentable sugars, glucose and mannose, was observed after the different treatments of the hydrolysate (data not shown). Adjustment of the concentrations, to take the dilution into account, is one way in which the experimental approach differs compared to previous detoxification with ion exchange *(1)*. The concentration of aliphatic acids was affected only by the anion exchanger (Table 2). At pH 5.5, almost all the aliphatic acids were removed by the anion exchanger. When the anion exchange was performed at pH 10.0 with the same amount of resin (3.6 g), less than half of the aliphatic acids was removed. However, when the amount of anion exchanger was increased to 8.0 g at pH 10.0, then again the aliphatic acids were efficiently removed. All treatments with resins caused a decrease in the concentrations of HMF and furfural in the hydrolysate (Table 2). The method that most efficiently removed furan aldehydes was anion exchange with 8.0 g at pH 10.0. The cation- and the anion-exchange resin removed roughly equal proportions of HMF and furfural, while the XAD-8 removed a higher percentage of furfural than of HMF.

The total amount of phenols in the hydrolysate was determined spectrophotometrically and by HPLC. More than 50 individual phenols were detected in the hydrolysate by HPLC (unpublished data). The ultraviolet (UV)-absorbing peaks at 280 nm, with retention times and UV spectra making them likely to be phenols, were summarized to a relative value of the content of phenols in the hydrolysate before and after the different treatments. Among the phenols, five were of the Hibbert's ketone type and could be separated and quantified with acceptable certainty. The sum of the Hibbert's ketones was not affected differently from the other phenols when treated with the different resins. At pH 5.5, some of the nonionized phenols were trapped by the cation-exchange resin in spite of its negatively charged groups. The XAD-8 resin with no ionizable groups removed phenols from the hydrolysate and was more efficient at the low pH when few of the phenols were charged. Efficiency increased when the amount of resin was increased. The anion exchanger was most efficient in removing the phenols from the hydrolysate. This ability increased with pH and the amount of anion exchanger used.

The ability of a strong anion exchanger to efficiently extract glucose from a water solution was confirmed in model experiments (Table 3). A solu-

Table 2
Effect of Different Treatments on Chemical Composition of Hydrolysate[a]

Detoxification method	Aliphatic acids[b] (IC) (%)	HMF[b] (HPLC) (%)	Furfural[b] (HPLC) (%)	Phenols[b] (Folin & Ciocalteau) (%)	Phenols[b] (HPLC) (%)	Hibbert's ketones[b] (HPLC) (%)
1. None (NaOH, pH 5.5)	100	100	100	100	100	100
2. NaOH, pH 10.0	100	100	100	100	ND	ND
3. Cation exchanger (8.0 g), pH 5.5	ND	81	76	84	ND	ND
4. Cation exchanger (8.0 g), pH 10.0	100	73	64	97	100	100
5. XAD-8 (3.6 g), pH 5.5	100	93	59	47	ND	ND
6. XAD-8 (8.0 g), pH 5.5	100	80	41	34	ND	ND
7. XAD-8 (8.0 g), pH 10.0	100	58	35	69	82	88
8. Anion exchanger (3.6 g), pH 5.5	9	76	70	43	ND	ND
9. Anion exchanger (3.6 g), pH 10.0	62	70	75	38	36	32
10. Anion exchanger (8.0 g), pH 10.0	4	35	32	21	19	10

[a]Data presented are the percentages remaining in the hydrolysate after each treatment. ND, not determined; IC, ion chromatography; HPLC, high-performance liquid chromatography.
[b]The value 100% corresponds to 7.97 g/L of aliphatic acids (the sum of formic, acetic, lactic, oxalic, and levulinic acids), 5.10 g/L of HMF, 0.82 g/L of furfural, 3.7 g/L of phenols determined as vanillin equivalents with the Folin & Ciocalteau method, 517 au (arbitrary absorbance units at 280 nm) of phenols (HPLC), and 59 au of Hibbert's ketones (HPLC), respectively.

Table 3
Effect of Treating a Glucose Solution with Different Resins[a]

Treatment	Glucose (%)
Cation exchanger (8.0 g), pH 10.0	100
XAD-8 (8.0 g), pH 5.5	97
XAD-8 (8.0 g), pH 10.0	100
Anion exchanger (8.0 g), pH 5.5	83
Anion exchanger (8.0 g), pH 10.0, 30 min[b]	25
Anion exchanger (8.0 g), pH 10.0	25
Anion exchanger (8.0 g), pH 10.0, added SO_4^{2-}	99

[a]Data presented are the percentages of remaining glucose of the initial concentration.
[b]The other samples were incubated for 1 h.

tion of 25 g/L of glucose treated with 8.0 g of anion exchanger at pH 10.0 showed a loss of glucose of 75% (the mean of two experiments giving 79 and 70%, respectively). The pH was maintained at 10.0 by the addition of carbon dioxide, since the release of hydroxide ions tended to increase the pH. At pH 5.5, the anion exchanger trapped 17% of the glucose. Neither the cation exchanger nor the XAD-8 resin affected the concentration of glucose. The ability of the anion exchanger to trap ionizable monosaccharides could be eliminated by the addition of sulfate ions. No significant loss of glucose was observed when 23 mM Na_2SO_4 was added. This result should be compared with the experiment in which the hydrolysate was treated with 8.0 g of anion exchanger at pH 10.0, and no decrease in the concentration of monosaccharides was observed.

The ability of the different resins to remove fermentation-inhibiting furan aldehydes was further investigated in a series of model experiments. The aqueous solution containing 0.2 mM each of the five model compounds was treated with 8 or 3.6 g of the different resins at pH 10.0 and 5.5. Table 4 shows the resulting decrease in concentrations. Control experiments were also done with no added resins. It was observed that part (11%) of the volatile furfural was lost. In addition, a negligible amount of the semi-volatile guaiacol was lost by evaporation. No other losses were observed in the control experiments. The furan aldehydes, furfural and HMF, were partly removed by all the resins tested (Table 4). The XAD-8 resin, with no ionized functional groups, efficiently removed almost half of the HMF and two-thirds of the furfural. XAD-8 was most efficient at pH 10.0, but the difference was very small. The anion and the cation exchangers were less efficient than XAD-8 in removing the furan aldehydes (Table 4). The addition of more resin slightly increased the removal of furan aldehydes for both the anion and the cation resin, but mostly for the XAD-8 resin.

The removal of phenolic compounds in a hydrolysate with an anion exchanger was shown at both high pH and a pH far below the pK_a value of the ionizable phenolic group (Table 4). The cation exchanger was much less

Table 4
Experiments with Model Compounds[a]

Detoxification method	HMF (%)	Furfural[b] (%)	Vanillin (%)	Guaiacol (%)	Coniferyl aldehyde (%)
None (NaOH, pH 5.5)	100	100	100	98	100
NaOH, pH 10.0	100	100	100	97	100
Cation exchanger (3.6 g), pH 5.5	96	89	80	85	66
Cation exchanger (8.0 g), pH 5.5	91	81	66	72	54
Cation exchanger (8.0 g), pH 10.0	87	82	100	82	100
XAD-8 (3.6 g), pH 5.5	78	57	12	13	6
XAD-8 (8.0 g), pH 5.5	56	39	4	4	1
XAD-8 (8.0 g), pH 10.0	54	38	96	7	30
Anion exchanger (3.6 g), pH 5.5	96	89	6	47	12
Anion exchanger (8.0 g), pH 5.5	82	84	1	28	2
Anion exchanger (8.0 g), pH 10.0	78	88	0	4	0
Anion exchanger (8.0 g), added SO_4^{2-}, pH 10.0	77	74	5	19	3

[a]Data presented are percentages of the compound remaining after treatment.
[b]The values are compensated for the losses owing to evaporation, estimated to 11%.

efficient, but still significant removal of phenols was obtained, especially at low pH. At high pH all ionized phenols were left unaffected. The fully ionized vanillin and coniferyl aldehyde were completely excluded. Guaiacol, which is ionized to 50% at pH 10.0, was only partly trapped. At pH 5.5, when almost all the phenols are nonionized, the phenols were trapped to a considerable extent, which showed the active participation of the resin matrix. This was confirmed by the use of a resin without ionized groups, XAD-8. At pH 5.5, all the phenols were almost completely trapped by the XAD-8 matrix. By contrast, only the least acidic phenol, guaiacol, was efficiently removed at pH 10.0. The fully ionized vanillin was excluded, while the somewhat less acidic but more lipophilic coniferyl aldehyde was partly removed. The drastic effect of the addition of sulfate to the anion exchanger at pH 10.0, as shown for glucose, was less pronounced for the phenols, although the removal of phenols decreased slightly. The added SO_4^{2-} ions competed for the positive sites of the anion exchanger. The amount of resin used also influenced the degree of removal, as expected by increased removal with increased amount used.

Discussion

The use of model compounds facilitated the interpretation of the detoxification effects seen with fermentation experiments on a dilute-acid hydrolysate from softwood when combined with chemical analysis of the remaining inhibiting compounds. Anion exchange at pH 10.0 was the most efficient method for increasing the fermentability, in terms of both ethanol yield and productivity, which agrees well with previous comparisons of different detoxification methods (1).

Loss of any fermentable sugar would have a negative influence on the economy of the detoxification step. The decrease in fermentable sugar that was previously observed when anion exchange was used for detoxification (1) was studied in model experiments. The strong alkaline surface of the anion exchanger containing quaternary ammonium groups with hydroxyl ions as counterions was able to ionize and capture the neutral monosaccharide. In the model experiment with glucose, up to 75% could be removed with the anion exchanger. Consequently, the amount of anion exchanger used to achieve detoxification must not be too high. This ability to trap ionizable monosaccharides could be eliminated by the addition of sulfate ions, which have a higher affinity to the quaternary ammonium groups. However, in the hydrolysate the concentrations of ionized aliphatic acids, phenols, and inorganic ions such as sulfate were sufficient to efficiently compete for the sites in the anion-exchange resin. The amount of anion exchanger that can be added without loss of sugars is thus determined by the presence of both inorganic and organic ions at the present pH. Considering the result obtained with the hydrolysate, in which ionized acids, phenolics, and sulfate ions assisted in displacing all the glucose so that no loss occurred, loss of fermentable sugars is unlikely to be a problem in industrial processes.

Even at low pH, the quaternary ammonium groups in the anion exchanger catch the aliphatic acids very efficiently, because the acids occur mostly in their ionized forms at pH 5.5. At the high pH of 10.0, most of the phenolic groups have become ionized, and the phenolates then contribute to that the competition for the cationic sites increases. When the number of compounds competing for the positive sites increased at pH 10.0, the anion exchanger retained less of the aliphatic acids. For the hydrolysate, 3.6 g of the anion exchanger was not enough to fully trap the acids; however, 8.0 g was sufficient.

When comparing results with the reference fermentations, it is noteworthy that the buffer capacity may be quite different compared with that of fermentations with hydrolysate and that the final pH, therefore, also may be different. This is owing to the fact that the hydrolysate contains compounds, e.g., aliphatic acids, that act as buffers at the pH of fermentation. Detoxification of the hydrolysate may also affect the buffering capacity.

The increase in productivity after treatment with the different resins can probably be attributed to the decrease in the concentrations of furan aldehydes. The effect of furfural and HMF on ethanolic fermentation by *S. cerevisiae* has recently been studied *(9–11)*. The furan aldehydes are known to cause a lag phase in ethanol formation, resulting in a substantial impact on the values for volumetric ethanol productivity. The neutral furan aldehydes were removed by the matrixes by another mechanism than ion exchange. The ionized groups of the anion and cation exchangers make them more hydrophilic than the noncharged XAD-8 resin. Consequently, the anion and cation exchangers were less attractive for the furan aldehydes than XAD-8, as shown by the model compound experiments. The interaction between furan aldehydes and the resins was not affected by pH in the model compound experiments. In the experiments with hydrolysate, pH 10.0 was more efficient for removing the furan aldehydes, at least with XAD-8. The high ionic strength in the hydrolysate at high pH is a plausible explanation. Removal also increased when more of the resins was used.

Phenolics are important inhibitors in dilute-acid hydrolysates of spruce. This was shown by enzymatic detoxification of a hydrolysate with laccase, which specifically removed the low molecular weight phenolics and simultaneously caused a major improvement in the fermentability *(1)*. From the experiments with the different resins and the hydrolysate and the model compounds, it can be concluded that the mechanisms for removal of phenols by ion-exchange resins are complex. As expected, the ionized phenols were completely trapped on the anion exchanger and repelled by the cation exchanger, both in the model experiments and in the hydrolysate. At pH 5.5, only a small fraction of the phenols are in their ionized form, but they were still efficiently trapped. The influence of the acidity of the phenolic group was shown: the acidic phenols with conjugated carbonyl groups were completely trapped even at low pH. The dissociation constants for the model compounds are vanillin, 7.40; guaiacol, 9.93; and coniferyl aldehyde, 7.94 *(12)*. The positively charged quaternary ammonium ions form strong ionic bonds with the negatively charged phenols and the negatively charged sulfonic acid groups make the cation exchanger nonattractive for all of the identified inhibiting compounds. When the phenols were ionized, they were completely repelled by the sulfonic acid groups of the cation exchanger. The previously presented detoxification effect of a cation exchanger *(3)* probably can be explained by being performed at a pH allowing lipophilic nonionized phenols to be retained by the resin together with the uncharged furan aldehydes. The importance of the resin matrix was shown by the ability to trap phenols on a cation exchanger and by the most efficient effect on all the model compounds by XAD-8 at low pH. The phenols used in the model experiments were completely removed by the anion exchanger at pH 10.0. With the hydrolysate, the anion exchanger at pH 10.0 was the most efficient way to remove phenols. The group of phenols referred to as Hibbert's ketones, which previously have been suspected to cause inhibition of the ethanolic fermentation *(13)*, were affected in the same way as the other phenols when treated with the different resins, with the possible exception of the anion exchanger at pH 10.0, which removed them almost completely.

It has previously been shown that a cation exchanger can be useful for removing inhibiting metal ions present in a bagasse hydrolysate prepared in a stainless steel vessel *(5)*. Our comparison of cation and anion exchangers indicates that inhibition by metal ions is not a major problem in the hydrolysate used in our study. The minor improvement in fermentability, which we observed after treatment with the cation exchanger, could possibly be attributed to hydrophobic interactions between the resin and organic inhibitors present in the hydrolysate.

In conclusion, treatment with the anion exchanger at pH 10.0 was the most efficient way to remove fermentation-inhibiting compounds from all three groups measured: phenols, furan aldehydes, and aliphatic acids. Treatment with anion-exchange resin at pH 10.0 could even result in a hydrolysate showing higher productivity than a reference fermentation with comparable amounts of fermentable sugars. The presence of counterions, such as sulfate, was found to be important to prevent unwanted losses of fermentable sugars at high pH with the anion exchanger. All three types of resins at least partly removed all groups of fermentation inhibitors investigated, with the exception of the aliphatic acids, which were only removed by the anion exchanger. Regardless of the type of resin used, treatment at pH 10.0 was always better in terms of improved fermentability than treatment at pH 5.5. The positively charged functional groups of the anion exchanger contributed to the detoxification effect. On the other hand, the negatively charged groups of the cation exchanger resulted in a resin providing only a poor detoxification effect. This was attributed to a repulsion effect of anionic inhibitors present in the hydrolysate. Future work will address optimization of the use of the anion exchanger at high pH for detoxification of dilute-acid hydrolysates.

Acknowledgments

We gratefully acknowledge Christer Larsson and Linda Björklund for technical assistance. This work was supported by a grant from the Swedish National Energy Administration.

References

1. Larsson, S., Reimann, A., Nilvebrant, N.-O., and Jönsson, L. J. (1999), *Appl. Biochem. Biotechnol.* **77–79,** 91–103.
2. Olsson, L. and Hahn-Hägerdal, B. (1996), *Enzyme Microb. Technol.* **18,** 312–331.
3. Buchert, J., Niemelä, K., Puls, J., and Poutanen, K. (1990), *Proc. Biochem.* **25,** 176–180.
4. Tran, A. V. and Chambers, R. P. (1986), *Enzyme Microb. Technol.* **8,** 439–444.
5. Watson, N. E., Prior, B. A., Lategan, P. M., and Lussi, M. (1984), *Enzyme Microb. Technol.* **6,** 451–456.
6. Rivard, C. J., Engel, R. E., Hayward, T. K., Nagle, N. J., Hatzis, C., and Philippidis, G. P. (1996), *Appl. Biochem. Biotechnol.* **57/58,** 183–191.

7. Dominguez, J. M., Cao, N., Gong, C. S., and Tsao, G. T. (1997), *Bioresour. Technol.* **61,** 85–90.
8. Singleton, V. L., Orhofer, R., and Lamuela-Raventos, R. M. (1999), *Methods Enzymol.* **299,** 152–178.
9. Larsson, S., Palmqvist, E., Hahn-Hägerdal, B., Tengborg, C., Stenberg, K., Zacchi, G., and Nilvebrant, N.-O. (1999), *Enzyme Microb. Technol.* **24,** 151–159.
10. Taherzadeh, M. J., Gustafsson, L., Niklasson, C., and Lidén, G. (1999), *J. Biosci. Bioeng.* **87,** 169–174.
11. Palmqvist, E., Almeida, J. S., and Hahn-Hägerdal, B. (1999), *Biotechnol. Bioeng.* **62,** 447–454.
12. Ragnar, M., Lindgren, C. T., and Nilvebrant, N.-O. (2000), *J. Wood Chem. Technol.* **20,** 277–305.
13. Clark, T. A. and Mackie, K. L. (1984), *J. Chem. Tech. Biotechnol.* **34B,** 101–110.

Fourier Transform Infrared Quantitative Analysis of Sugars and Lignin in Pretreated Softwood Solid Residues

Melvin P. Tucker,*,[1] Quang A. Nguyen,[1] Fannie P. Eddy,[1] Kiran L. Kadam,[1] Lynn M. Gedvilas,[2] and John D. Webb[2]

*National Renewable Energy Laboratory,
[1]Biotechnology Center for Fuels and Chemicals
and [2]Center for Measurements and Characterization,
1617 Cole Boulevard, Golden, CO 80401,
E-mail: melvin_tucker@nrel.gov*

Abstract

Hydrolysates were obtained from dilute sulfuric acid pretreatment of whole-tree softwood forest thinnings and softwood sawdust. Mid-infrared (IR) spectra were obtained on sample sets of wet washed hydrolysates, and 45°C vacuum-dried washed hydrolysates, using a Fourier transform infrared (FTIR) spectrophotometer equipped with a diamond-composite attenuated total reflectance (ATR) cell. Partial least squares (PLS) analysis of spectra from each sample set was performed. Regression analyses for sugar components and lignin were generated using results obtained from standard wet chemical and high-performance liquid chromatography methods. The correlation coefficients of the predicted and measured values were >0.9. The root mean square standard error of the estimate for each component in the residues was generally within 2 wt% of the measured value except where reported in the tables. The PLS regression analysis of the wet washed solids was similar to the PLS regression analysis on the 45°C vacuum-dried sample set. The FTIR-ATR technique allows mid-IR spectra to be obtained in a few minutes from wet washed or dried washed pretreated biomass solids. The prediction of the solids composition of an unknown washed pretreated solid is very rapid once the PLS method has been calibrated with known standard solid residues.

Index Entries: Fourier transform infrared; biomass; softwood; dilute-acid pretreatment; acid hydrolysis.

*Author to whom all correspondence and reprint requests should be addressed.

Introduction

The conversion of renewable lignocellulosic resources to bioethanol requires a number of process steps. We have used dilute-acid pretreatment with sulfuric acid as one of the early process steps to hydrolyze a significant portion of the hemicellulose in softwood feedstocks consisting of whole-tree forest thinnings or sawdust from sawmill wastes. A second pretreatment at higher temperatures and acid concentrations was used to hydrolyze the remaining hemicellulose and some of the cellulose from the first-stage solid residues. The extent of the pretreatment reactions must be controlled and optimized for each feedstock to maximize the production of bioethanol. In conversion processes requiring enzymatic hydrolysis of the residual cellulose, the pretreatment steps may increase the susceptibility of the remaining cellulose to hydrolysis by cellulase enzymes resulting in increased yields of sugars or ethanol.

First- and second-stage pretreatment of softwood feedstocks with dilute sulfuric acid is usually a short process step involving only a few minutes at high temperature in the reactor. Controlling the reactions requires rapid and accurate methods of analysis for monitoring the reactor. Conventional methods such as wet chemical analysis and high-performance liquid chromatography are time-consuming and too slow to be useful for the control of the reactor. The standard wet chemical method for solids compositional analysis (1) may take several days to complete. Although this method is quite accurate, it is too slow to enable control of the short dilute-acid pretreatment step. However, if a rapid method to monitor and control the pretreatment reactor is available, ethanol production can be increased. Fourier transform infrared (FTIR) spectroscopy is rapid and quantitative, and coupled with advanced high-temperature, high-pressure attenuated total reflectance (ATR) probes the capability of monitoring pretreatment reactors is possible.

FTIR spectroscopic analysis is a rapid and nondestructive technique for the qualitative and quantitative identification of components in solids in the mid-IR region (2). The usefulness of FTIR spectroscopy of solids has increased as sampling device technology has advanced. The FTIR analysis of wet solid residues, such as pretreated whole slurries or washed solids from those slurries, has been severely restricted in the past because of the high IR background absorbance of water. Extremely short pathlength cells (not practical for analyzing high-solids slurries) were needed to obtain transmission spectra (3). However, in the ATR mode (4–7), the high background absorbance caused by water in pretreated slurries and washed solids can be partially controlled by choosing ATR cells incorporating single or multiple reflections within the crystal. This allows attenuation of the incident radiation by the insoluble fiber to give the mid-IR spectra without the high water background absorbance completely obscuring the spectra. Recent developments in ATR cells and probes have extended the application of mid-IR spectroscopy to the qualitative and quantitative analysis of wet and abrasive solid residue (8).

Diffuse reflectance IR Fourier transform spectroscopic methods or transmission spectroscopy utilizing KBr pellets previously has been used to study whole wood, pretreated wood, wood surfaces, lignin, and pulp and paper substrates *(9–17)*. However, the use of FTIR-ATR spectroscopy to study solids was severely limited because of the soft ATR crystals (i.e., ZnSe, ZnS, and KRS-5) then available. Recent availability of ATR cells utilizing very hard crystals of silicon and diamond has overcome this difficulty. These crystals possess considerable hardness and chemical inertness; thus, the samples can be pressed onto the ATR crystal at high pressures, allowing a more uniform penetration by the evanescent radiation *(5)* and a higher degree of reproducibility. The recent development of diamond-composite ATR cells has extended FTIR spectroscopy into areas of research not possible before because the diamond is highly chemically inert, exhibits high hardness and mechanical strength, and is optically transparent in the visible and most of the mid-IR region. Because a diamond surface has the lowest coefficient of friction of all the available crystal materials, few materials stick to this surface *(8)*. Diamond-composite ATR probes are now available that work at temperatures as high as 250°C and at 1500 psi (10 MPa).

In the present study we used a diamond-composite ATR cell and an FTIR spectrometer to obtain mid-IR spectra of washed pretreated softwood solid residues, and washed and dried pretreated softwood solid residues. The partial least squares (PLS) option in the commercially available software package TQ Analyst™ (Nicolet Instruments, Madison, WI) was used to regress the spectra into methods capable of predicting the glucan, mannan, galactan, xylan, and lignin compositions of pretreated softwood solid residues. Using a calibrated method, FTIR-ATR can rapidly determine (in a few minutes) the solids composition of washed residues from a pretreatment reactor.

Materials and Methods

Preparation and Pretreatment of Feedstock

Whole-tree forest thinning was obtained by the Pacific Wood Fuels Company, Redding, CA, from a site near Quincy, CA, and prepared as previously reported *(18)*. Pretreatment of this whole-tree forest thinnings feedstock for first- and second-stage experiments was performed using a 4-L steam explosion reactor (NREL Digester) as described earlier *(19)*. Conditions ranged from 180 to 215°C, 0.35 to 2.5% (w/w) sulfuric acid, and 120-s to 9-min residence times.

Barrels of typical sawdusts from southeastern Alaska were obtained from sawmills of Ketchikan Pulp, by Sealaska of Juneau, Alaska. The barrels of fresh sawdusts were rapidly shipped to NREL to minimize degradation in transit. A mixture of 62% (dry wt basis) hemlock, 26% Sitka spruce, and 12% red cedar sawdusts was prepared by mixing four times on a large tarpaulin using the method of coning and quartering. The composition of mixed sawdusts was chosen to represent the statistical species populations of softwood trees in the forests of southeastern Alaska. Each barrel of sawdust was obtained from a different sawmill conducting a campaign utilizing an individual softwood species for a particular forest product or customer. Samples of this sawdust mixture were pretreated in the 4-L steam explosion reactor. First-stage pretreatment conditions varied from 180 to 190°C, 3 to 4 min, and 0.7% (w/w) H_2SO_4. Second-stage pretreatment conditions varied from 205 to 215°C, 3- to 4-min residence times, and 0.7 to 1% (w/w) H_2SO_4.

Pretreated slurry samples were extensively washed by centrifugation before obtaining mid-IR spectra of the wet solids. Representative samples (10 g) of the pretreated slurries were suspended in 40 mL of deionized-glass distilled water and centrifuged at 6000g for 5 min, and the supernatants were discarded. The washing procedure was repeated a minimum of five times. The percentage of moisture of the washed samples was obtained by standard oven-drying methods at 105°C. Washed and dried solids were prepared from the wet washed solids by drying under vacuum at 45°C for 3 d. Representative samples of the wet washed and dried washed solid samples were ground using an agate mortar and pestle before application to the diamond surface of the ATR cell. Grinding was minimal (usually <30 s on a 0.5-g sample) to prevent the solids from being degraded by the heat of grinding.

IR Spectroscopy

Approximately 50–100 mg of the ground sample was required to fill the sample well of the diamond-composite ATR sample cell. The sample cell was equipped with a spring-loaded anvil to reproducibly press the solid sample uniformly and tightly against the diamond surface. Mid-IR spectra were obtained by averaging 512 scans from 4000 to 400 cm^{-1}, at 2-cm^{-1} resolution using a Nicolet Avatar 360® FTIR spectrometer (Nicolet Instrument) and a six-reflection diamond-composite ASI DurasamplIR™ ATR cell (ASI SensIR Technologies, Danbury, CT).

PLS Method

PLS calibration methods for solid compositions were determined using the PLS regression analysis option in TQ Analyst™ (version 6.0) on 35 wet or 34 dried solid sample spectra. Wet chemical results for percentage of solids, percentage of sugars, and Klason and acid-soluble lignin for each sample were entered into the software package spreadsheet before calibrating the FTIR method. The software automatically calculates the number of factors to use in the regression analysis by calculating the predicted residual error sum of squares (PRESS) before calibrating the method. The number of factors suggested by the PRESS analysis was used in the PLS regression analysis. The mid-IR regions selected for analysis of the sugars and lignin were 3800 cm^{-1} to 2400 m^{-1} and 1846–644 cm^{-1}. The region between

Table 1
Correlation Coefficients (r), r^2, and SEE
for 34 Washed and Vacuum-Dried Pretreated Whole-Tree Forest Thinnings
and Southeast Alaska Softwood Residues Determined
by PLS Regression Analysis

Component	Correlation coefficient (r)	r^2	SEE (wt%)
Glucose	0.9552	0.9125	4.4
Mannose	0.9736	0.9479	[a]
Galactose	0.9110	0.8299	[a]
Xylose	0.9303	0.8655	[a]
Arabinose	0.5333	0.2844	[a]
Lignin	0.9934	0.9868	2.0

[a]SEE values less than the 1.5 wt% variability in wet chemical solids compositional analysis.

1900 and 2200 cm^{-1} was avoided because of the very strong diamond IR absorption. The regression was performed on the entire set of 35 spectra, except one sample (chosen randomly by the software) was not used in the calibration. This sample was used instead as a validation standard for the PLS method. In addition, the regression analysis was cross-validated using the leave-one-out regression analysis option in the software package. The leave-one-out cross-validation regression generally gave poorer correlation of predicted vs measured values; however, the correlation coefficients were >0.9 (e.g., the correlation coefficients in Table 1 for glucose decreased from 0.9552 to 0.9392 and lignin decreased from 0.9934 to 0.9759). The lower correlation for the leave-one-out cross-validation is the result of the limited number of samples for which wet chemical analyses are available. Results from the first PLS regression analysis (not the leave-one-out analysis) were used to plot the FTIR predicted wt% compositional values vs the measured solid compositional analyses.

Results and Discussion

Figure 1 shows the PLS regression analysis for glucose of 34 washed and 45°C vacuum-dried first- and second-stage pretreated whole-tree forest thinnings and southeast Alaska softwood residues. Figure 2 shows the PLS regression analysis for lignin of the same 34 washed and 45°C vacuum-dried samples.

Table 1 lists the correlation coefficients (r), r^2, and root mean square SEE for 34 washed and vacuum-dried pretreated whole-tree forest thinnings and southeast Alaska softwood residues determined by PLS regression analysis. The correlation coefficients for galactose, xylose, and arabinose are low (<0.9) because the compositional values that were measured for each solid residue were low and near the detection limit for our FTIR-ATR instrument and the wet chemical methods. In addition, little variance in the calibration set is expected when the results are less than the wet chemical error of ~1.5% for solids compositional analysis, resulting in

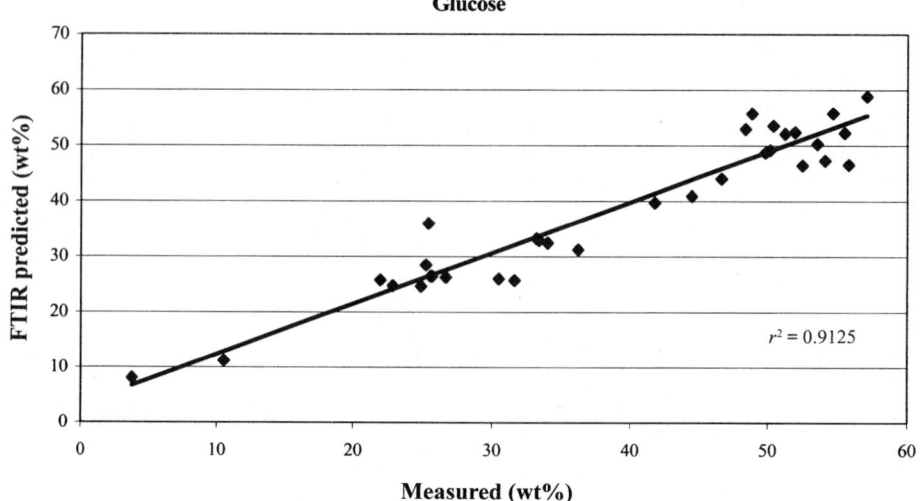

Fig. 1. PLS regression analysis for glucose of 34 washed and 45°C vacuum-dried first- and second-stage pretreated whole-tree forest thinnings and southeast Alaska softwood samples.

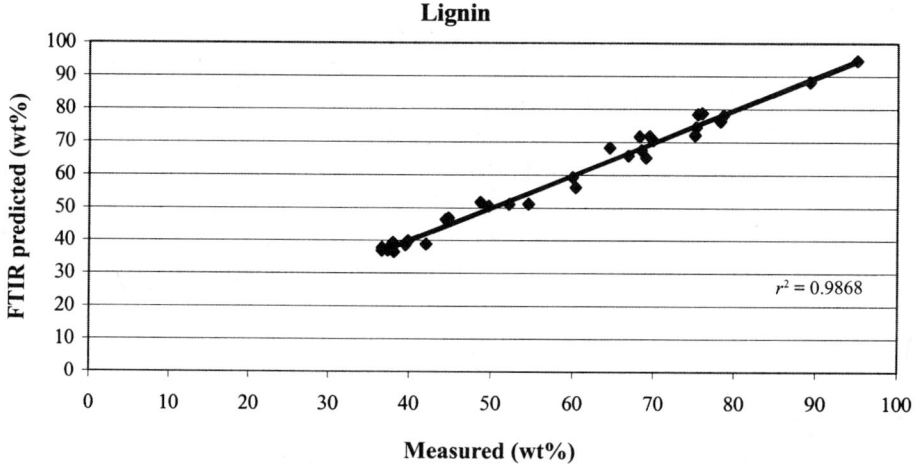

Fig. 2. PLS regression analysis for lignin of 34 washed and 45°C vacuum-dried first- and second-stage pretreated whole-tree forest thinnings and southeast Alaska softwood samples.

little correlation of predicted values to measured values. The wet chemical solid compositional analysis on the 34 dried pretreated samples ranged from 3.77 to 57.09% for glucose, 0.06 to 2.91% for mannose, 0 to 2.87% for galactose, 0 to 7.87% for xylose, 0 to 1.94% for arabinose, and 36.58 to 95.04% for lignin. The number of factors used in the PLS regression analysis shown in Table 1 is two used for glucose, nine for mannose, nine for galactose, nine for xylose, one for arabinose, and five for lignin. The SEE for glucose

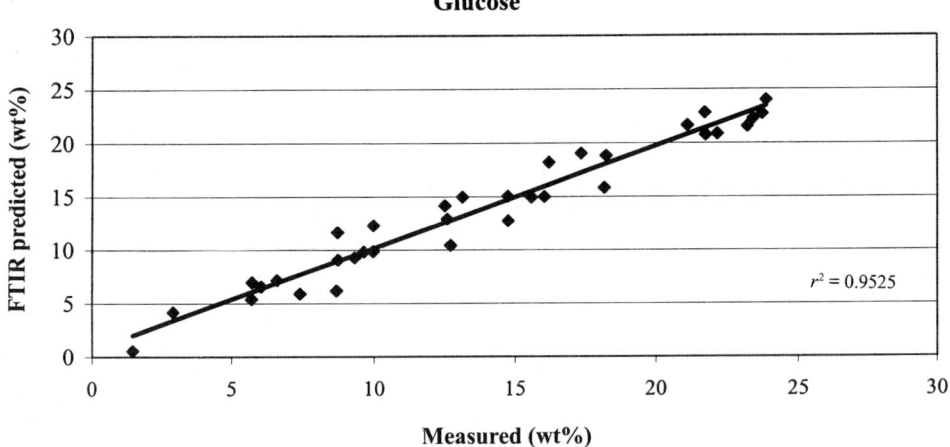

Fig. 3. PLS regression analysis for glucose from 35 wet washed solid residues of pretreated softwood feedstocks.

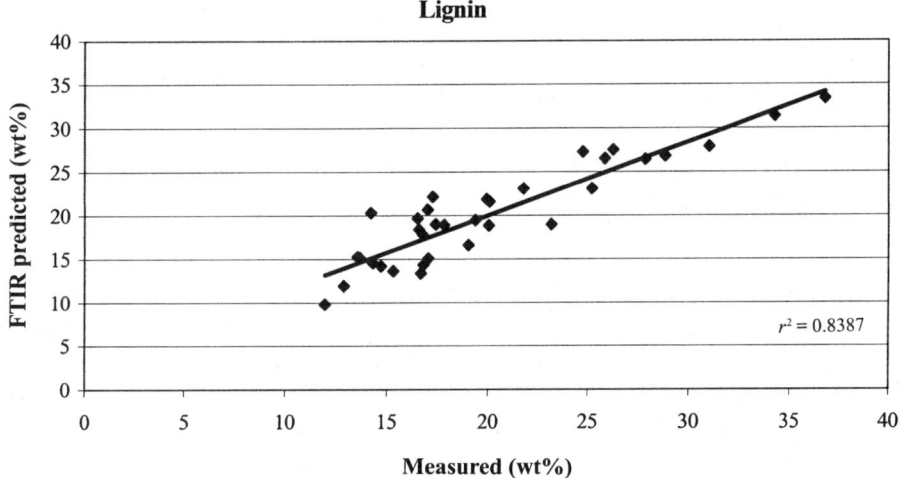

Fig. 4. PLS regression analysis for lignin from 35 wet washed solid residues of pretreated softwood feedstocks.

reported in Table 1 is >2 wt% because the wet chemical compositional analysis for pretreated softwood residues pretreated under severe conditions is inaccurate owing to the large amount of extractives in the feedstock and degradation products produced.

Figure 3 shows the PLS regression analysis for glucose of 35 wet washed first- and second-stage pretreated softwood samples (both whole-tree forest thinnings and southeast Alaska sawdust feedstocks). Figure 4 shows the PLS regression analysis for lignin of the same 35 wet washed samples. Solids compositional data for each sample in the calibration set were adjusted for moisture before entering the data into the PLS spreadsheet.

Table 2
Correlation Coefficients (r), r^2, and SEE
for 35 Washed (Wet) Pretreated Whole-Tree Forest Thinnings
and Southeast Alaska Softwood Residues Determined
by PLS Regression Analysis of FTIR-ATR Calibration Set

Component	Correlation coefficient (r)	r^2	SEE (wt%)
Water	0.9030	0.8154	3.5
Glucose	0.9760	0.9525	1.4
Mannose	0.9706	0.9421	[a]
Galactose	0.7392	0.5464	[a]
Xylose	0.7624	0.5813	[a]
Arabinose	0.6781	0.4598	[a]
Lignin	0.9158	0.8387	2.5

[a]SEE values less than the 1.5 wt% variability in wet chemical solids compositional analysis.

The percentage of moisture was determined in triplicate on each sample by taking aliquots of the wet solids used for collecting the FTIR spectra and drying them in an oven overnight at 105°C.

Table 2 lists the correlation coefficients (r), r^2, and root mean square standard error of the estimate (SEE) for compositional analysis of 35 wet washed pretreated whole-tree forest thinnings and southeast Alaska softwood residues determined by PLS regression analysis of the FTIR-ATR calibration set. The closer the value r^2 is to 1, the higher the probability that the FTIR predicted value (y-axis) is related to the measured solids compositional value for that component (x-axis). The correlation coefficients for galactose, xylose, and arabinose reported in Table 2 are <0.9 because the solid compositional values measured for each solid residue are low and near the detection limit for this FTIR-ATR instrument and wet chemical solids compositional analysis. This is the result of hydrolysis of the hemicellulosic sugars in the feedstock resulting in low values for these three sugars in the residues. In addition, dilution by water in the wet residues further reduces the effective concentration of all components in the residues. The low correlation coefficients (<0.9) for these three sugars may be the result of little variability in the wet chemical solids compositional analysis (~1.5 wt%). The wet chemical compositional analysis on the 35 pretreated samples (corrected for moisture) in Table 2 ranged from 0.02 to 1.28% for mannose, 0 to 1.07% for galactose, 0 to 2.94% for xylose, and 0 to 0.89% for arabinose. The factors calculated from the PRESS analysis used in the PLS regression analysis for the individual components in Table 2 are as follows: one factor was used for water, five for glucose, eight for mannose, three for galactose, four for xylose, four for arabinose, and three for lignin.

Method validation requires that an analytical method have precision and accuracy. An FTIR spectrometer is expected to have a frequency error of approx 0.01 cm^{-1} (2). The intensity (absorbance) error using the Nicolet Avatar 360 FTIR-ATR instrument combination was determined to have a standard deviation of 0.002 absorbance units (au) for eight replica spectra obtained from a typical dry pretreated solid residue for a major absorbance band at 1032 cm^{-1}. These eight replica spectra were obtained following collection of 512 scans each. The sample was not removed from the cell or disturbed in order to test the spectrophotometric and software reproducibility. Each replicate spectrum was baseline corrected by drawing a baseline from 1850 to 915 cm^{-1}. The absorbance for the major absorption band at 1032 cm^{-1} was measured using the peak height tool in the Omnic® (version 5.2) software package. In addition, spectra from three individual aliquots of the same sample were obtained to test the sampling error on pretreated biomass residues. The intensity error found was 0.004 au for the major band at 1032 cm^{-1} between the three spectra. If an obvious inhomogeneity in the ground sample is chosen (e.g., a splinter that was not pretreated), an absorbance error of 0.028 au or greater is found. This demonstrates that the sample must be ground to a homogeneous mixture before placing a sample in the diamond cell. The root-mean-square of the noise of this FTIR-ATR combination was determined to be 0.00054 au in the region between 2800 and 2600 cm^{-1} using a standard software tool in the Omnic software.

The six-reflection diamond-composite cell used in this study has an apparent lower quantitation limit *(20)* of approx 2 wt% for glucan with this FTIR-ATR spectrometer combination if a 10:1 signal-to-noise ratio is used. This limit is directly related to the variance in the wet chemical method (~1.5 wt%) used for the calibration of the PLS regression method and can be improved only if a more accurate method of solid compositional analysis is developed. Models of these cells with higher number of reflection are commercially available and could extend the limits of component detection in the wet solids to lower levels, unless the background IR absorption by water in the wet washed solids increases sufficiently to interfere with the analysis.

Conclusion

In this study, we used FTIR-ATR to obtain mid-IR spectra on washed solid residues from first- and second-stage pretreated whole-tree forest thinnings and softwood sawdusts from southeastern Alaska. Spectra from 35 wet washed pretreated solid samples were subjected to PLS regression analysis, as reported in Table 2. The correlation coefficients for the regression analyses for glucose and lignin were found to be >0.9, with SEE of 1.4 and 2.5 wt%. The correlation coefficients for galactose, xylose, and arabinose were low because little variance in the calibration set is expected when composition is less than the wet chemical error of ~1.5%, resulting in little correlation of predicted values to measured values. The moisture contents of the washed solid residues in this calibration set varied between 50 and 80%. The large amount of moisture in the residues would have made obtaining mid-IR spectra very difficult because of the high background absorbance caused by water; however, the FTIR-ATR technique was able to overcome this difficulty. The method is capable of predicting moisture content of unknown wet washed pretreated softwood residues.

The PLS regression analyses for 34 washed and 45°C vacuum-dried solid residues from first- and second-stage pretreated whole-tree forest thinnings and softwood sawdusts from southeastern Alaska are reported in Table 1. The correlation coefficients for the regression analyses for glucose, mannose, galactose, xylose, and lignin were found to be >0.9, with root-mean-square SEE varying between 2 and 4.4 wt%. The moisture contents of the 45°C vacuum-dried solid residue samples in this calibration set were measured at <1%. The wet washed samples can be generated rapidly by centrifugation, whereas the 45°C vacuum-drying step takes 3 d. Similar results are obtained if 105°C dried residues are used (data not shown). FTIR spectroscopy can be accomplished in a couple of minutes, and solids composition can be predicted rapidly once the method has been calibrated.

The FTIR-ATR technique allows mid-IR spectra to be rapidly obtained on whole slurries, washed solids, and dried washed solids from pretreated biomass. The pretreated softwood slurries used in this study were obtained from a very complex feedstock consisting of chipped whole trees, including limbs, bark, and needles. PLS methods generated using higher temperature and pressure probes in a reactor under process conditions should allow extension of this technology to rapidly monitor and control a pretreatment reactor to maximize product yield. Larger databases of FTIR spectra and solids compositional analysis will increase the accuracy of predicting the composition of pretreated solids with this method. The results presented in this study for SEE are close to the 1.5 wt% errors expected for the standard wet chemical method for solids compositional analysis. However, methods developed using PLS for one feedstock should be redeveloped when changing to other feedstocks.

Acknowledgments

We wish to thank Peter Huberth of Ketchikan Pulp, Alaska, for generously supplying the hemlock, Sitka spruce, and red cedar sawdust samples from various sawmills in southeastern Alaska. This work was funded by the Office of Fuels Development, the US Department of Energy.

References

1. Moore, W. E. and Johnson, D. B. (1967), *Procedures for the Chemical Analysis of Wood and Wood Products*, US Forest Products Laboratory, US Department of Agriculture, Madison, WI.
2. Griffiths, R. P. (1975), *Chemical Infrared Fourier Transform Spectroscopy*, John Wiley, New York.
3. Krishnan, K. and Ferraro, J. R. (1982), in *Fourier Transform Infrared Spectroscopy*, vol. 3, Krishnan, K. and Ferrraro, J. R., eds., Academic, New York, p. 203.
4. Doyle, W. M. (1990), *Appl. Spectrosc.* **44**, 50.
5. Harrick, N. J. (1967), *Internal Reflection Spectroscopy*, John Wiley, New York.

6. Kuehl, D. and Crocombe, R. (1984), *Appl. Spectrosc.* **38(6)**, 907–909.
7. Faix, O. (1988), *Mikrochim. Acta* **1(6)**, 21–25.
8. Milosevic, M., Sting, D., and Rein, A. (1995), *Spectroscopy* **10(4)**, 44–49.
9. Schultz, T. P., Templeton, M. C., and McGinnis, G. D. (1985), *Anal. Chem.* **57(14)**, 2867–2869.
10. Grandmaison, J. L., Thibault, J., Kaliaguine, S., and Chantal, P. D. (1987), *Anal. Chem.* **59(17)**, 2153–2157.
11. Faix, O. and Bottcher, J. H. (1992), *Holz Als Roh-Und Werkstoff* **40(6)**, 221–226.
12. Pandey, K. K. (1999), *J. Appl. Polymer Sci.* **71**, 1969–1975.
13. Zavarin, E., Jones, S. J., and Cool, L. G. (1990), *J. Wood Chem. Technol.* **10(4)**, 495–513.
14. Faix, O. and Bottcher, J. H. (1993), *Holzforschung* **47**, 45–49.
15. Rodrigues, J., Faix, O., and Pereira, H. (1998), *Holzforschung* **52**, 46–50.
16. Heitz, M., Rubio, M., Wu, G., and Khorami, J. (1995), *Anales Quimica* **91S**, 685–689.
17. Schultz, T. P. and Glasser, W. G. (1986), *Holzforschung* **40**, 37–44.
18. Nguyen, Q. A., Tucker, M. P., Keller, F. A., and Eddy, F. P. (2000), *Appl. Biochem. Biotech.* **84–86**, 39–50.
19. Nguyen, Q. A., Tucker, M. P., Boynton, B. L., Keller, F. A., and Schell, D. J. (1998), *Appl. Biochem. Biotechnol.* **70–72**, 77–87.
20. Krull, I. and Swartz, M. (1997), *LC:GC* **15(6)**, 534–538.

Bleachability and Characterization by Fourier Transform Infrared Principal Component Analysis of Acetosolv Pulps Obtained from Sugarcane Bagasse

ADILSON R. GONÇALVES* AND DENISE S. RUZENE

Departamento de Biotecnologia, FAENQUIL, CP 116, CEP 12.600-000 Lorena, SP, Brazil, E-mail: adilson@debiq.faenquil.br

Abstract

Sugarcane bagasse Acetosolv pulps were bleached by xylanase and the pulps classified by using Fourier transform infrared (FTIR) spectroscopy and principal component analysis (PCA). Pulp was treated with xylanase for 4–8 h with stirring at 30°C. Some samples were further extracted with NaOH for 1 h at 65°C. FTIR spectra were recorded directly from the dried pulp samples by using the diffuse reflectance technique. Reduction in kappa number of 69% was obtained after sequence xylanase (4 h)-alkaline extraction. During bleaching the viscosity decreased only 12%. FTIR-PCA showed that the first three principal components (PCs) explained more than 90% of the total variance of the pulp spectra. PC2 × PC1 plot showed that the points related to pulps from sequence xylanase (4 h)-alkaline extraction are different from the other. This group is enlarged by plotting PC3 × PC1 or PC3× PC2 containing all pulps submitted to alkaline extraction. PC2 and PC3 are the principal factor for differentiation of the pulps. These PCs suffer influence of the ester bands (1740 and 1244 cm^{-1}). On the other hand, the pulps bleached only with xylanase could not be differentiated from the nonbleached pulps.

Index Entries: Xylanase bleaching; Fourier transform infrared spectra of pulps; sugarcane bagasse pulps; Acetosolv pulping.

Introduction

Organosolv pulping of agricultural residues has been studied in both acidic and basic conditions *(1)*. The cellulose in acidic pulps is typically more degraded than in basic ones, and after conventional bleaching

---

*Author to whom all correspondence and reprint requests should be addressed.

processes the degradation of the fibers should increase *(2)*. The use of xylanases as a step in the bleaching of acidic pulps can avoid the changes in pH since the pulp is obtained at pH 4.0 to 5.0, corresponding to the maximum activity of xylanases. Treatment of eucalyptus kraft pulps with xylanases has already been investigated and was shown to decrease the kappa number in 11% *(3)*. In the present study we propose the xylanase bleaching of sugarcane bagasse pulp obtained from the Acetosolv process.

Materials and Methods

Pulping and Bleaching

Acetosolv pulping of depithed sugarcane bagasse was carried out as described by Benar *(4)*. After a 2-h cooking time, the pulp was filtered and washed with acetic acid and further with water until reaching pH 4.0. Samples of bagasse Acetosolv pulp (1 g dry) were suspended under agitation in 30.2 mL of water (3.2% consistence) at 30°C for 10 min. Cartazyme HS® (Sandoz) was added at 17.8 U/g of dry pulp. The samples were maintained in a shaker at 30°C for 4–12 h, followed by filtration and washing with 200 mL of distilled water at 30°C for the complete removal of the enzyme. One set of enzymatic bleached pulps was further submitted to alkaline extraction. Samples obtained at different bleaching times (1 g of dry pulp) and original pulp were extracted with 50 mL of 1 mol/L of NaOH at 65°C for 1 h under magnetic stirring. After filtration, the pulps were washed with 50 mL of 1 mol/L of NaOH for 1 h at 65°C and further with distilled water at 65°C until reaching pH 6.0. Chlorite bleaching was also carried out as an example of the classic bleaching process. Dry pulps (1 g) were suspended in 50 mL of water (2% consistence) and heated to 70°C. Sodium chlorite (1.3 mL of 40% aqueous solution) and glacial acetic acid (0.2 mL) were added. The solution was further heated at 70°C for 5 min, and the bleached pulp obtained was exhaustively washed with water. Pulps were oven-dried at 110°C for 15 min and analyzed with respect to kappa number and viscosity by standard methods *(5,6)*.

Hydrolysis of Pulps

One gram of dry pulp (original and bleached) was treated with 10 mL of 72% H_2SO_4 with stirring at 45°C for 7 min. The reaction was interrupted by adding 50 mL of distilled water, the mixture was transferred to a 500-mL Erlenmeyer flask, and the volume reached 275 mL. The flask was autoclaved for 30 min at 1.05 bar for the complete hydrolysis of oligomers. The mixture was filtered and the filtrate (hydrolysate) complete to 500 mL. A 40-mL sample of the hydrolysate was diluted to 50 mL and the pH was adjusted to 2.0 with 2 mol/L of NaOH. After filtration in a Sep-Pak C_{18} cartridge to remove aromatic compounds, the hydrolysate was analyzed in an Aminex HPX-87H column (300 × 7.8 mm) (Bio-Rad) at 45°C by using a Shimadzu chromatograph and refraction-index detector. The mobile phase was 0.005 mol/L of H_2SO_4 at 0.6 mL/min. Sugar concentrations reported as xylan and glucan were determined using calibration curves of pure compounds.

Fourier Transform Infrared Principal Component Analysis of Bleached and Unbleached Pulps

Fourier-transform infrared (FTIR) spectra were obtained directly from the bleached and unbleached refined pulps utilizing the diffuse reflectance technique (DRIFT), under the conditions described in ref. 7. Spectra were recorded (64 scans) in a Nicolet 520 spectrometer. After polygonal baseline correction (7), the spectra were normalized by the absorption at 900 cm^{-1}, which corresponds to the anomeric carbon atom of the group O-C-O in polysaccharides and suffers no influence from other groups (8). Spectra were converted to text files using OMNIC software (Nicolet). The normalized absorbances in the range of 400–4000 cm^{-1} (1866 data points per pulp spectrum) were submitted to principal component analysis (PCA) calculations using the BIOTEC and FAEN programs compiled in FORTRAN, which were written in our laboratory based on the work of Scarminio and Bruns (9). Graphic presentations were easily made with Microsoft EXCEL 5.0.

Results and Discussion

Table 1 gives the viscosity values, kappa number, and yields for the pulps obtained in the xylanase treatment, alkaline extraction, and sequence xylanase-alkaline extraction. For comparison, results from the treatment of the bagasse Acetosolv pulp with sodium chlorite are also given.

Unbleached and nonextracted pulps were not refined before viscosity measures, and the presented results cannot be directly compared with those of extracted pulps. Viscosity values for alkaline-extracted pulps were greater than those for the other pulps, probably owing to losses of lower molecular weight carbohydrate fractions, as can be seen by the yields in Table 1.

Alkaline extraction is responsible for the solubilization of lignin fragments released after pulping or xylanase action. Alkaline extraction without the use of xylanase reduced the kappa number from 28 to 13.4, but the pulp yield was only 85%.

Enzymatic treatment extended to 12 h did not improve the bleaching efficiency, as can be seen in Table 1. Viscosity values were maximum with 4 h of xylanase treatment and the decrease in kappa number from 4 to 12 h was only 8%. Xylanase treatment followed by alkaline extraction was the best sequence. These values cannot be compared with the bleaching results achieved by chlorite treatment because with the same values of viscosity, the kappa number for xylanase-alkaline-extracted pulp was still 28% higher.

The decrease in kappa number followed a pattern along different enzymatic treatment times. Reduction in kappa number from 28 to 8.6 was achieved by using the better experimental condition: 4 h of enzymatic treat-

Table 1
Viscosity, Kappa Number, and Yields of Different Treated Pulps

Sequence[a]	Viscosity (cP)	Kappa number	Yield (%)
Original	10.7	28.0	—
Chlorite bleaching	11.5	6.7	72.5
X (4 h)	9.0	25.4	92.0
XE (4 h)	11.5	8.6	89.2
X (8 h)	8.7	23.9	98.1
XE (8 h)	10.1	9.9	89.0
X (12 h)	7.1	23.4	88.4
XE (12 h)	11.1	10.9	87.5
E	11.0	13.4	85.2

[a]X, xylanase; E, alkaline extraction; XE, xylanase followed by alkaline extraction.

Table 2
Carbohydrate Composition of Pulps and Xylan/Glucan Ratio

Pulp[a]	Cellobiose (%)	Glucose (%)	Total glucan (%)	Xylan (%)	Xylan/glucan ratio	Acetyl groups (%)
Original	1.32	72.20	73.52	9.40	0.13	4.02
X	—	60.74	60.74	8.01	0.13	4.25
XE	4.01	67.20	71.21	6.48	0.09	<1.0
E	—	—	56.90	4.40	0.08	<1.0

[a]X, xylanase treated (4 h); XE, xylanase treated (4 h) followed by alkaline extraction.

ment followed by alkaline extraction. A tendency of the kappa number to increase when the time was increased was observed. The recondensation of the lignin over the fibers can occur, diminishing the action of the alkaline extraction. A decrease of 12% in the viscosity values after 4 h of treatment was also observed.

Carbohydrate quantification of the pulps was performed to evaluate possible selective degradation of pulp constituents. Table 2 gives the results of the quantification of dimers and monomers of cellulose and xylans as well as the xylan/glucan ratio.

The relative amount of xylan was preserved even after xylanase treatment. Only after alkaline extraction was a reduction in xylan detected, which means that xylanase acts over xylans but the fragments formed are not easy released. Alkaline extraction makes solubilization of both lignin and xylan fragments feasible. Removal of glucan and xylan (decreased pulp yield) caused changes in xylan/glucan ratio.

Figure 1 shows the FTIR spectra of the unbleached and three bleached pulps (with chlorite, xylanase, and xylanase + alkaline extraction). The spectra are quite similar and the differences among them were evaluated by PCA.

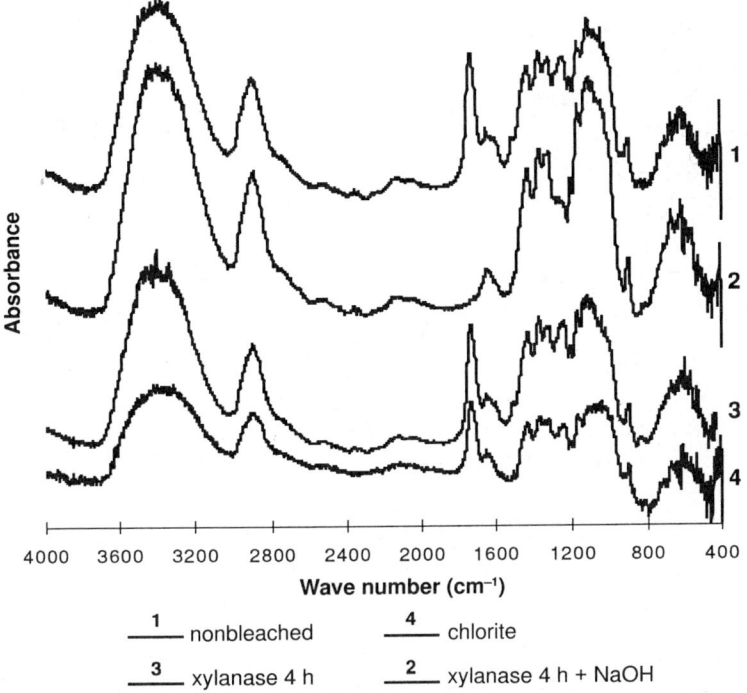

1 ____ nonbleached

4 ____ chlorite

3 ____ xylanase 4 h

2 ____ xylanase 4 h + NaOH

Fig. 1. FTIR spectra of the pulps.

The first three principal components (PCs) explain more than 90% of the total variance of the system. This means that the 1866 properties (data points) of each spectrum can be reduced to only 3 with 90% confidence. The principal components are obtained by the linear combination of the 1866 data, by also discarding the regions that do not change among the samples, and the results are score values. Figures 2–4 show the PC's graphics for the 23 FTIR spectra obtained from 6 different assemblies (bleached and unbleached pulps).

In Fig. 2 (PC2 × PC1), three of the points corresponding to the FTIR of bleached pulps with xylanase (4 h) followed by alkaline extraction are grouped in a different way with respect to the other points (highlighted by the ellipse). In Fig. 3, a larger group is evidenced containing FTIR spectra of bleached pulps with xylanase at different times and followed by alkaline extraction. This group can be differentiated from the others, since the closer the groups, the more similar the corresponding spectra.

Score points corresponding to the unbleached pulp form a group with small dispersity in the PC3 × PC2 plot (Fig. 4), and the pulps extracted with NaOH form a well-defined group. The points corresponding to more drastic treatment (chlorite and xylanase for 8 h) are very different with respect to the unbleached pulp.

This finding suggests that PC2 and PC3 are the principal factors for the differentiation between the pulps' spectra. This is better analyzed by the loading values of each PC (Fig. 5). From Fig. 5 the influence of infrared

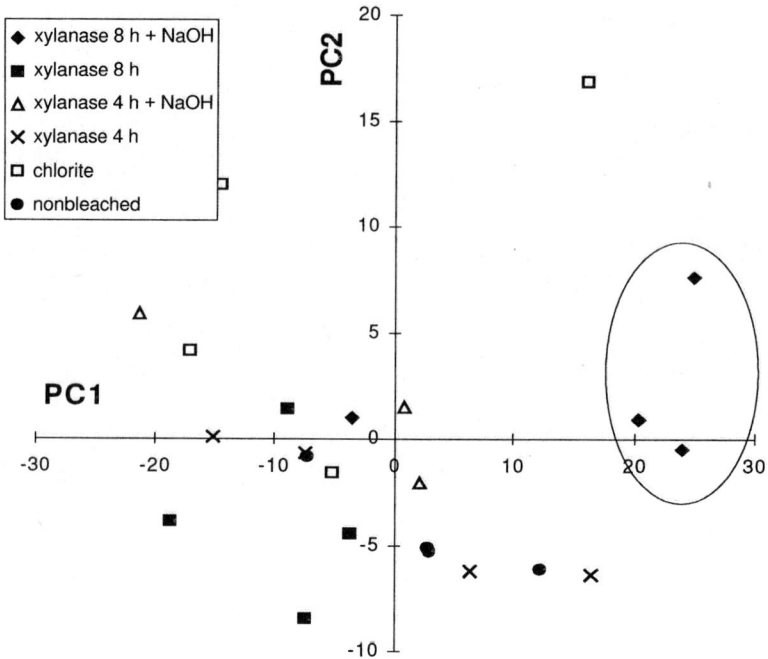

Fig. 2. Score values of PC2 × PC1 from FTIR spectra of bleached and unbleached bagasse pulps.

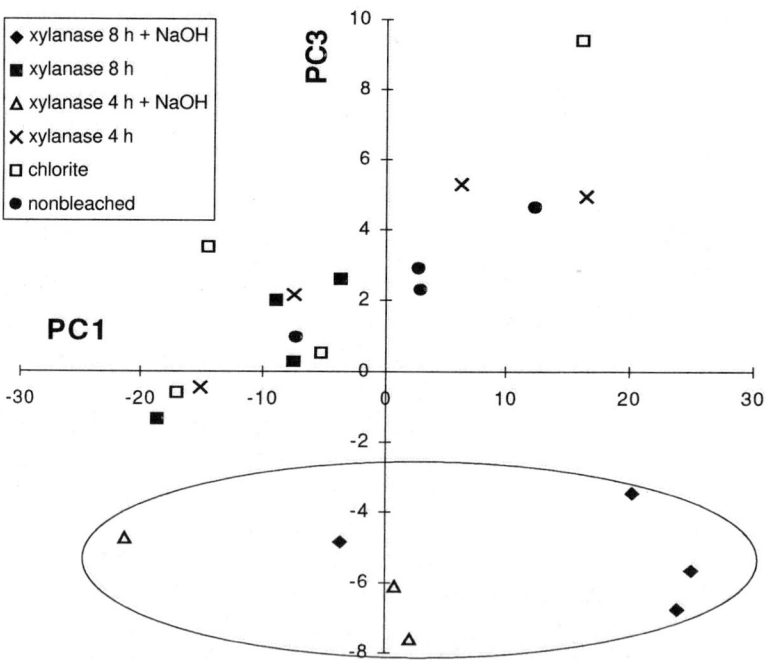

Fig. 3. Score values of PC3 × PC1 from FTIR spectra of bleached and unbleached bagasse pulps.

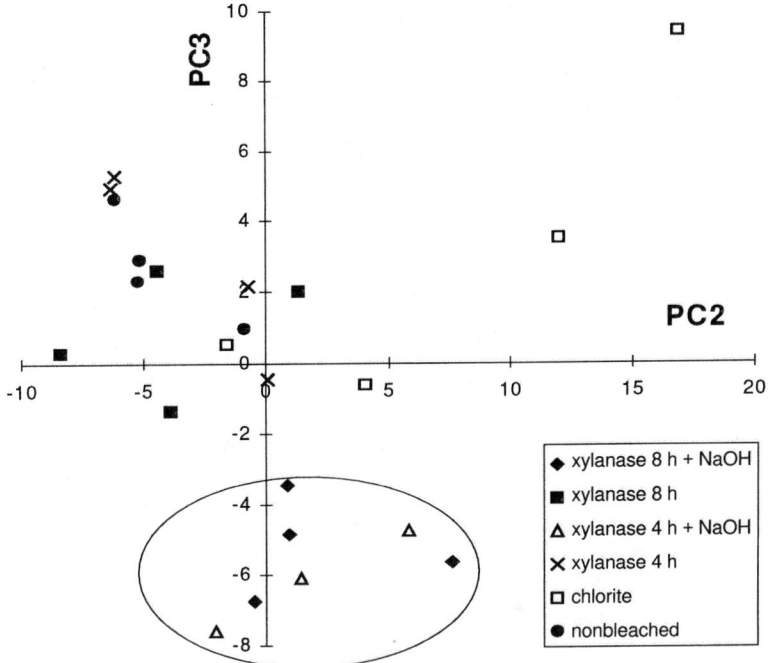

Fig. 4. Score values of PC3 × PC2 from FTIR spectra of bleached and unbleached bagasse pulps.

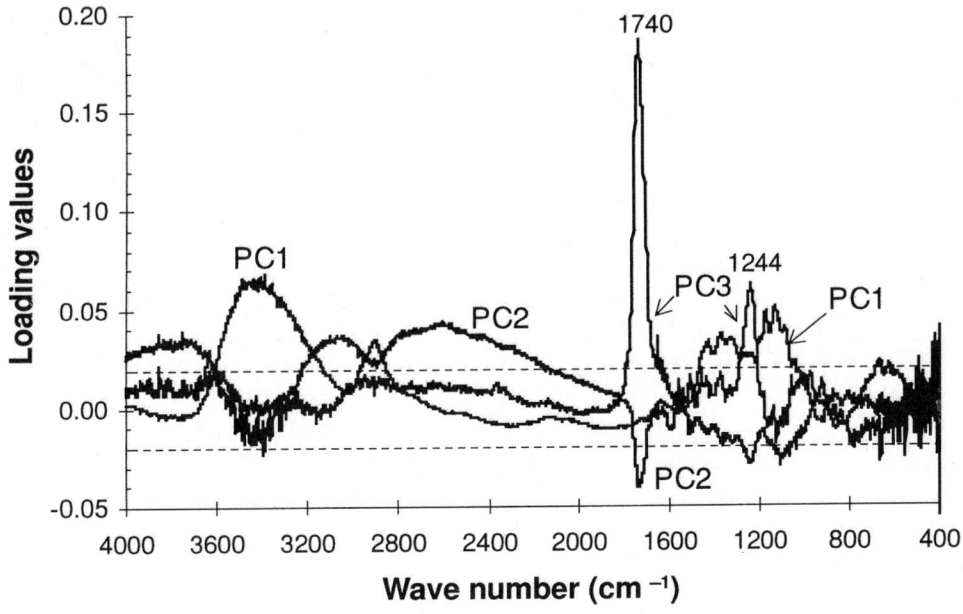

Fig. 5. Loading values of PC1, PC2, and PC3 of FTIR spectra of bleached and unbleached bagasse pulps.

bands on PC scores can be evaluated. The plot of PC3 is influenced by carbonyl (1740 cm^{-1}) and C-O (1244 cm^{-1}) bonds, characteristic of esters. The difference between the pulps can be explained by the acetyl groups (ester) formed during the Acetosolv pulping that are removed during the alkaline extraction. This fact is also corroborated by the chemical analysis previously discussed, with a decrease in the acetyl groups.

Acknowledgments

This work was supported by Conselho Nacional do Desenvolvimento Científico e Tecnológico and Fundação de Amparo à Pesquisa do Estado de São Paulo.

References

1. Raymond, A. Y. and Akhtar, M. (1998), *Environmentally Friendly Technologies for the Pulp and Paper Industry*, Wiley, New York.
2. Minor, J. L. (1996), in *Pulp Bleaching, Principles and Practice*, Dence, C. W. and Reeve, D. W., eds., Tappi, Atlanta, pp. 25–57.
3. Milagres, A. M. F., Erismann, N. M., and Durán, N. (1992), in *Proceedings of 2nd Brazilian Symposium on the Chemistry of Lignin and Other Wood Components*, Durán, N. and Espósito, E., eds., UNICAMP, Campinas-Brazil, pp. 372–376.
4. Benar, P. (1992), MS thesis, Instituto de Química, UNICAMP, Campinas, Brazil.
5. TAPPI, Technical Association of the Pulp and Paper Industry. (1982), *TAPPI Standard Methods*, T 230 om-82.
6. TAPPI, Technical Association of the Pulp and Paper Industry. (1985), *TAPPI Standard Methods*, T 236 cm-85.
7. Faix, O., Böttcher, J. H., and Bertelt, E. (1992), in *8th International Conference in Fourier Transform Spectroscopy Proceedings SPIE 1575*, Heise, M., Korte, E. H., and Siesler, H. W., eds., The International Society for Optical Engineering, San Diego, CA, pp. 428–430.
8. Morohoshi, N. (1991), in *Wood and Cellulosic Chemistry*, Hon, D. N. S. and Shiraishi, N., eds., Marcel Dekker, New York, pp. 331–392.
9. Scarminio, I. S. and Bruns, R. E. (1989), *Trends Anal. Chem.* **8**, 326, 327.

Production of Oxychemicals from Precipitated Hardwood Lignin

QIAN XIANG AND Y. Y. LEE*

Department of Chemical Engineering, Auburn University, Auburn, AL 36849, E-mail: yylee@eng.auburn.edu

Abstract

Lignin is a major byproduct in the biomass-to-ethanol process. The lignin produced from acid treatment of biomass has characteristics suitable for further conversion to organic chemicals. It is free of contaminants and has a relatively low molecular weight. In this study, catalytic oxidative conversion of the acid-soluble lignin precipitated from acid hydrolysates of hardwood was investigated. The process is based on aqueous alkaline oxidation of lignin with dissolved O_2 in the presence of Fe^{3+} and Cu^{2+} catalysts at moderate reaction temperatures (160–180°C). Aromatic aldehydes, ketones, and organic acids are found to be the primary products identifiable on extraction with ether. The combined weight yield of the total ether extractable products is about 20–25% of the initial lignin. The yield of the aldehydes (vanillin + syringaldehyde) is in the vicinity of 15% with an additional 3 to 4% of aromatic ketones. The yields of aldehydes plus ketones observed in this work far exceeded those obtainable from the conventional alkaline air oxidation of spent sulfite liquors. This article also provides comprehensive batch reaction data on conversion and product distribution.

Index Entries: Lignin; oxidation; degradation; oxygen.

Introduction

Saccharification of cellulosic biomass by dilute acid has a much longer history than the enzymatic process. A number of saccharification processes are currently being developed solely on the basis of acid catalysis. They include the total hydrolysis process of the National Renewable Energy Laboratory (NREL) (1), the ethanol plant of BC International (Jennings, LA), and the Arkenol Process (Mission Viejo, CA). The acid-based hydrolysis processes are potential sources of a special grade of lignin. In the acid processes, a large fraction of the lignin is solubilized into the hot acidic medium, as high as 70% in the case of the NREL total hydrolysis process.

*Author to whom all correspondence and reprint requests should be addressed.

About one-third to one-half of this lignin is precipitated after cooling of the hydrolysate and becomes easily recoverable. This precipitated hardwood lignin (PHL) is relatively pure, free of sulfur or other contaminants. It has a low molecular weight, high solubility in various solvents, and high reactivity. It is amenable to further conversion into chemicals and fuels. Because lignin is a major byproduct in the biomass-to-ethanol process, enhancing its value would have a significant impact on the economics of the overall process.

Oxidation in alkaline medium is an established process for the depolymerization of lignins and lignosulfonates. Vanillin, other aromatic aldehydes, and acids are the principal products (2). Among them, only vanillin is presently manufactured on an industrial scale. It is produced from spent sulfite liquors in which the yield is only 5–10% (3). The industrial production of vanillin by direct oxidation of sulfite-spent liquor was practiced in the United States as early as 1936. It has since been replaced by the eugenol-based route. In other countries, especially in Europe, the lignin-based processes are still being employed (4).

Vanillin is one of widely accepted flavor chemicals used in the food industry. It has a broad market for various other applications, such as a ripening agent, an antifoaming agent in lubrication oils, a brightener in zinc-coating baths, an attracting material for insecticides, intermediates for pharmaceuticals such as Aldomet (Merck) and L-Dopa (Monsanto), herbicides, a vulcanization inhibitor, and disinfectants. In addition, it shows bacteriostatic properties (3). Because angiosperm lignin comprises both guaiacylpropane and syringylpropane units, vanillin produced via alkaline oxidation from hardwood lignin is accompanied by substantial quantities of syringaldehyde. Syringaldehyde has a chemical structure and properties similar to those of vanillin and, therefore, has similar applications. Pharmaceutical uses of syringaldehyde have been recently proposed for synthesis of drugs for cancer treatment (5,6).

The overall goal of the present investigation was to assess the technical feasibility of converting the PHL to high-value chemicals through alkaline oxidation by molecular oxygen. Identification of the products and determination of the yields of main products (vanillin, syringaldehyde) are of prime interest. The extent of conversion, its relationship with the reaction severity, and refinement of the reaction conditions were also within the scope of this work.

Materials and Methods

Preparation of Lignin

Yellow poplar wood chips (*Liriodendron tulipifera*) provided by NREL were used as the feedstock for the generation of PHL. The hydrolysis of this lignocellulosic biomass was conducted in a percolation reactor using extremely dilute acid (0.07% sulfuric acid) at 210–220°C. About 70% of the lignin was rendered soluble in the hydrolysate. On cooling, the PHL was precipitated out. It was filtered and dried for oxidation experiments. The details of lignin generation procedures and the characteristics of PHL are described elsewhere *(7)*.

Oxidation of Lignin

Oxidation of lignin was carried out in a 55-mL tubing bomb reactor (SS-316 tubing capped with Swagelok end fitting) connected to a pressurized oxygen source (200 psi). In a typical experiment, 20 mL of an alkaline solution (2 N) was placed into the reactor with 0.8 g of lignin and the required quantity of catalysts ($CuSO_4$ and $FeCl_3$). The reactor was Swagelok-sealed, vacuumed, and filled with oxygen to a desired pressure. The reaction was initiated by immersing the reactor into a preheated sand bath. The heating rate was about 110°C/min. After the reaction, the reactor was quenched in a cold-water bath. When catalysts were used in the reaction, they were removed by filtration from the reaction mixture. The liquid in the reactor was first acidified to pH ≈ 2.0 with an H_2SO_4 solution. The unreacted solid lignin was filtered, washed, and dried. Conversion of lignin to acid-soluble products was calculated by the weight difference.

Severity Factor

When it is deemed appropriate, the experimental conditions are converted into the severity factor, defined as $R_0 = t \exp[(T - 100)/14.75]$, in which t is expressed in minutes and T in degrees Celsius *(8,9)*.

Analytical Methods

Liquid samples (the filtrate obtained after acidification) were first analyzed by high-performance liquid chromatography (HPLC). The remainder of each sample was then divided into two parts. The first part was extracted with diethyl ether until the ether layer appeared colorless. After removal of the ether by evaporation over anhydrous sodium sulfate, the yield of the ether-soluble fraction was determined from one part of the filtrate. Internal standard solution was added to the other part of the filtrate, and extraction and evaporation were ensued. The second part was subjected to gas chromatography (GC) or GC-mass spectrometry (GC-MS) analysis.

HPLC Analysis

The filtrate samples were analyzed for organic acids content by HPLC (Water Associates) equipped with a Bio-Rad Aminex HPX-87H column, a refractive index (RI) and ultraviolet (UV) detectors. The conditions used were as follows: detection, RI and UV (205 nm); eluent, 0.005 M H_2SO_4; flow rate, 0.6 mL/min; column temperature, 65°C.

The organic acids in the oxidized lignin solutions were identified by their retention time. Identification of the acids was confirmed by comparing the retention times with those of pure acids using a second effluent, 10% $CH_3CN/0.01\ N\ H_2SO_4$. External standards were used for quantification.

GC/GC-MS Analysis

The ether extracts were analyzed with a Varian model 3700 GC equipped with a capillary column (DB-5, 30 m × 0.25 mm × 0.25 μm) and flume ionization detector. Helium was used as the carrier gas. Injector and detector temperatures were 200 and 280°C, respectively. The temperature program was set to reach 250°C at a rate of 8°C/min starting from 50°C, with an initial time delay of 2 min and holding at 250°C for 15 min. For quantitative analysis, 4-ethylresorcinol was used as an internal standard. The ether extracts were also analyzed with a GC-MS (VG70E) equipped with an Opus V3.1 and DEC 3000 Alpha Station data system for component identification.

Results and Discussion

PHL produced both vanillin and syringaldehyde on oxidation with oxygen. This is in agreement with chemotaxonomy of plants. Lignins in wood of angiosperms consist of both guaiacyl units and syringyl units derived, respectively, from enzyme-induced dehydrogenative polymerization of coniferyl alcohol and sinapyl alcohol. Because α- and β-aryl ether linkages are the dominant types of linkages in both softwood and hardwood lignins, the cleavage of these bonds is the primary reaction in lignin degradation.

Gierer and Imsgard *(10)* theorized the step-by-step processes involved in the alkaline oxygen degradation of lignin based on their work on model compounds having a β-aryl ether structure. According to them, lignin oxidation in alkaline condition occurs in three main steps. First is the formation of hydroperoxide intermediates through either an electrophilic attack directly by molecular oxygen or nucleophilic addition of hydroperoxide anions to carbonyl and conjugated carbonyl structures. Then the hydroperoxide intermediates undergo several types of reactions, i.e., rearrangement reactions, conversion into quinols, and dehydration or other elimination reactions. Finally, these hydroperoxides or other derivatives are further oxidized and degraded into a series of low molecular weight compounds, such as aldehydes and organic acids.

A series of preliminary experiments were carried out to obtain the profiles of the oxidation reaction. The conditions were varied with different oxygen partial pressures (with system closed or open to constant pressured oxygen source), temperatures, reaction times, and catalysts. Efforts were made to optimize the conditions to raise the yields of identifiable low molecular weight degradation products. Table 1 summarizes the results of the typical preliminary experiments. The data include the amounts of identifiable phenolic aldehydes (vanillin and syringaldehyde) and ketones (acetovanillone and acetosyringone) at various conditions. The amounts produced are expressed as a percentage of the original lignin.

Table 1

Preliminary Study of PHL Oxidation by Oxygen

					Reaction no.			
	1	2	3	4	7	8	9	10
Reaction system	Oxygen fed-batch	Oxygen fed-batch	Closed batch	Closed batch	Closed batch	Closed batch	Closed batch	Closed batch
Oxygen partial pressure	Constant	Constant	Reducing	Reducing	Reducing	Reducing	Reducing	Reducing
Temperature (°C)	170	170	170	170	170	170	170	170
Time (min)	5	10	5	10	10	10	10	20
Log(R_o/min)	2.76	3.06	2.76	3.06	3.06	3.06	3.06	3.36
Input: aqueous solution (mL)	20.0	20.0	20.0	20.0	20.0	20.0	20.0	20.0
Contains:								
Lignin (g)	0.8	0.8	0.8	0.8	0.8	0.8	0.8	0.8
$CuSO_4$ (g)	None	None	None	None	0.04	None	0.04	0.04
$FeCl_3$ (g)	None	None	None	None	None	0.004	0.004	0.004
O_2 (g)	>0.65	>0.65	0.65	0.65	0.65	0.65	0.65	0.65
Conversion (%)[a]	58.3	69.2	46.5	53.4	62.7	58.9	64.3	70.3
Vanillin (%)	2.1	2.2	1.8	2.8	3.5	3.6	3.9	4.7
Syringaldehyde (%)	3.7	1.6	3.4	5.1	6.8	5.3	7.2	8.8
Total aldehyde (%)	5.8	3.8	5.2	7.9	10.3	8.9	11.1	13.5
Acetovanillone (%)	0.4	0.4	0.3	0.6	0.7	0.6	0.7	0.7
Acetosyringone (%)	0.7	0.6	0.7	1.5	2.8	2.2	2.5	2
Total ketones (%)	1.1	1.0	1.0	2.1	3.5	2.8	3.2	2.7

[a]All percentages refer to total lignin (Klason + acid-soluble lignin), which represents 99.7% of the dry brut lignin used.

To study the effect of O_2 partial pressure or the oxidation process, a parallel reaction system was designed. This unit has an open connection to the oxygen tank that allows the lignin to react with oxygen under a constant level of O_2 partial pressure (data represented by reactions no. 1 and 2 in Table 1). By contrast, a normal batch reactor system was initially filled with oxygen to a certain level and closed. The O_2 partial pressure in this unit steadily decreases as the reaction proceeds. The data from these runs are represented by reactions no. 3 and 4 in Table 1. A higher O_2 partial pressure brings about a higher rate of oxygen transfer from the gas phase into the liquid phase and, consequently, a higher dissolved concentration of oxygen in the solution. This results in a higher total conversion of PHL. However, the yields of aldehydes and ketones decreased with time. We think it is because these phenolic compounds easily degrade under a high oxygen pressure. A previous kinetics study on vanillin oxidation *(11)* notes that the reaction rate is first order in dissolved oxygen concentration at pH > 12.0. Under the conditions used in the alkaline oxygen oxidation of lignin, the syringyl structure in lignin is more sensitive to oxygen pressure and more reactive than guaiacyl structure *(12,13)*. To obtain high yields of the phenolic products, it is preferable to keep an initial high pressure of oxygen to rapidly break down the lignin and then gradually lower the oxygen pressure to continue the oxidation of the lignin fragments and reaction intermediates but prevent further degradation of the phenolic products. The normal batch process is therefore a better reactions scheme for our purpose than the oxygen fed-batch process.

The preliminary experiments also addressed the effects of Cu^{2+}, Fe^{3+}, and their mixture on the yields of aldehydes. With no catalysts, selectivity toward aldehydes and ketones was very poor. In commercial production of vanillin via air oxidation of lignosulfonates, $CuSO_4$ is used as the catalyst. The results of our preliminary studies indeed confirmed that $CuSO_4$ is an extremely effective catalyst for the oxidation of PHL. We also found that Fe^{3+} enhances the selectivity for aldehydes. A combination of Cu^{2+} and Fe^{3+}, however, was far more effective in improving the selectivity toward the oxidation reactions. For example, the data in Table 1 show that the use of both Cu^{2+} and Fe^{3+} increases the yield of total aldehydes from 7.9 to 11.1% and the conversion from 53.4 to 64.3% at 170°C and 10 min, in direct comparison with the control experiment without catalyst. It has been suggested *(14)* that Cu^{2+} acts as an electron acceptor accelerating the formation of the phenoxy radical and that Fe^3 can form a new reaction intermediate, O_2-Fe^{3+}-lignin complex, which acts as an oxygen carrier and can attack lignin to form degradation products, thus enhancing the oxidation reaction.

The data also indicate that the optimal conditions are in the range of 160–180°C and 10–30 min in the batch reactor system with combined Cu^{2+}/Fe^{3+}. A series of comprehensive oxidation experiments were therefore conducted near the optimal conditions; Table 2 summarizes the results. Figure 1 shows lignin conversion profiles during the oxidation reaction of

Table 2
Data of PHL Oxidation Near Optimal Oxidation Conditions

	Temperature												
	180°C				170°C					160°C			
Time (min)	5	10	20	30	5	10	20	30	60	10	20	30	60
Log(R_o/min)	3.05	3.36	3.66	3.83	2.76	3.06	3.36	3.54	3.84	2.77	3.07	3.24	3.54
Input: aqueous solution (mL)	20	20	20	20	20	20	20	20	20	20	20	20	20
Contains:													
Lignin (g)	0.8	0.8	0.8	0.8	0.8	0.8	0.8	0.8	0.8	0.8	0.8	0.8	0.8
NaOH (g)	1.6	1.6	1.6	1.6	1.6	1.6	1.6	1.6	1.6	1.6	1.6	1.6	1.6
CuSO$_4$ (g)	0.04	0.04	0.04	0.04	0.04	0.04	0.04	0.04	0.04	0.04	0.04	0.04	0.04
FeCl$_3$ (g)	0.004	0.004	0.004	0.004	0.004	0.004	0.004	0.004	0.004	0.004	0.004	0.004	0.004
O$_2$ (g)	0.65	0.65	0.65	0.65	0.65	0.65	0.65	0.65	0.65	0.65	0.65	0.65	0.65
Conversion (%)[a]	63.2	71.5	77.8	81.2	56.2	64.3	70.3	75.8	81.4	55.7	62.9	68.5	74.6
Vanillin (%)	3.3	4.6	4.3	3.9	3.1	3.8	4.7	4.1	3.8	2.9	3.7	5.1	4.5
Syringaldehyde (%)	6.7	9.8	8.4	7.1	6.3	7.2	8.8	8.2	5.8	6.9	7.7	8.6	7.5
Total aldehyde (%)	10.0	14.4	12.7	11.0	9.4	11.0	13.5	12.3	9.6	9.8	11.4	13.7	12.0
Acetovanillone (%)	0.5	0.7	0.8	0.4	0.3	0.5	0.7	0.7	0.3	0.3	0.5	0.6	0.6
Acetosyringone (%)	2.4	2.6	2.6	1.8	1.5	2.3	2.5	2.6	1.6	1.9	2.4	2.7	1.8
Total ketones (%)	2.9	3.3	3.4	2.2	1.8	2.8	3.2	3.3	1.9	2.2	2.9	3.3	2.4
Oxalic acid (%)	8.4	12.5	16.3	15.4	7.2	8.6	14.1	16.2	16.0	6.7	9.4	13.4	15.3
Formic acid (%)	5.6	6.2	10.3	13.4	4.2	5.2	6.7	9.3	12.5	4.1	6.1	7.2	10.1
Acetic acid (%)	10.7	15.6	27.4	28.6	10.3	12.6	18.5	26.1	31.7	11.4	14.2	19.5	27.5
Malonic acid (%)	1.3	1.6	1.5	1.5	0.8	1.2	1.6	1.6	1.2	0.7	1.1	1.5	1.2
Succinic acid (%)	0.4	0.4	0.5	0.4	0.1	0.2	0.4	0.3	0.3	0.1	0.3	0.3	0.4
Vanillic acid (%)	1.2	1.7	2.0	2.4	0.8	1.1	1.6	2.3	1.9	0.9	1.8	2.4	2.3
Syringic acid (%)	2.8	3.5	3.9	3.8	2.1	2.5	4.2	3.7	2.9	2.2	3.4	3.7	3.7
Total acids (%)	30.4	41.5	61.9	65.5	25.5	31.4	47.1	59.5	66.5	26.1	36.3	48.0	60.5

[a]All percentages refer to total lignin (Klason + acid-soluble lignin), which represents 99.7% of the dry brut lignin used.

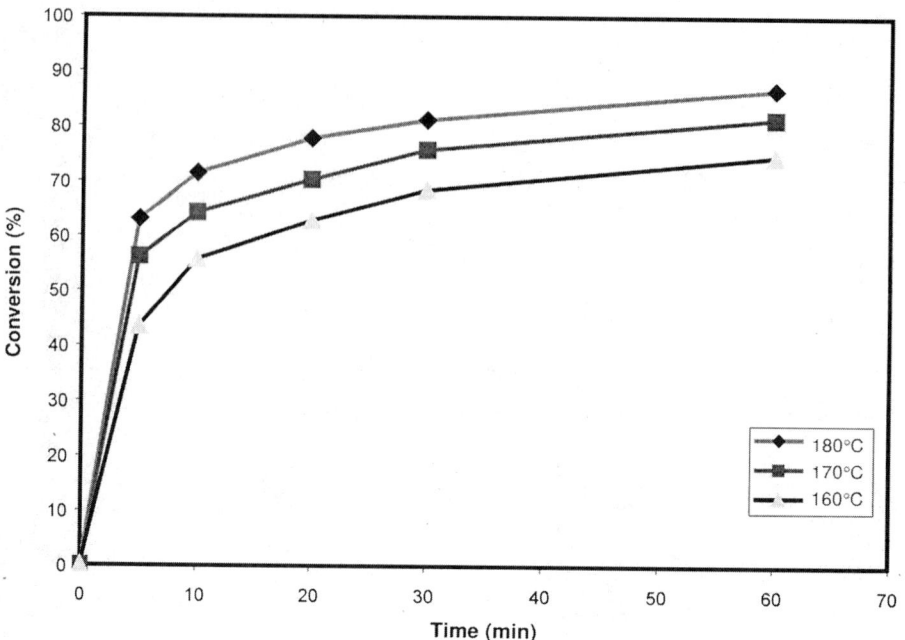

Fig. 1. PHL conversion profiles near the optimal oxidation conditions.

PHL. At a severity index (SI) above 3.36, the PHL conversion reaches above 70%. Even at a lower SI region, more than 50% of lignin conversion is attainable. This indicates that PHL undergoes a rapid initial degradation into liquid products by alkaline oxidation. More severe conditions, although providing higher conversion, did not improve the yields of aldehydes and ketones. Figures 2 and 3 show the yield profiles of aldehydes and ketones under the optimized oxidation conditions. In the best case, the yields reach 14.4% for aldehydes and 3.4% for ketones, which it occurred within the time span of 10–20 min and the SI range of 3.36–3.66. About 5% of aromatic acid (vanillic acid and syringic acid) and a larger amount of low molecular organic acids (mainly acetic, oxalic, and formic acids) were also produced as byproducts. The total yields of acids were in the range of 30–65%.

Conclusion

Degradation of PHL with alkaline oxidation by molecular oxygen was investigated as a means of producing value-added oxychemicals. PHL is readily depolymerized and solubilized by alkaline oxidation to produce aromatic aldehydes, ketones, and organic acids under moderate reaction conditions with the SI in the range of 3.0–3.8. Catalysts can enhance the reaction selectivity toward useful products. A combination of Cu^{2+} and Fe^{3+} catalysts was most effective in raising the yield of ketones and aldehydes. With the implementation of catalysts, the yields of aldehydes and ketones

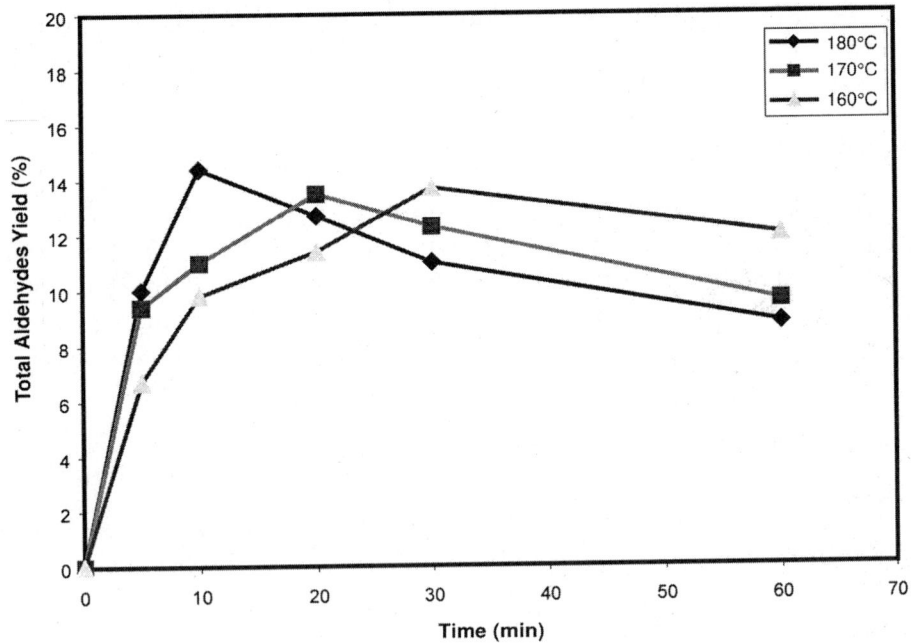

Fig. 2. Total aldehydes yield profiles near the optimal oxidation conditions.

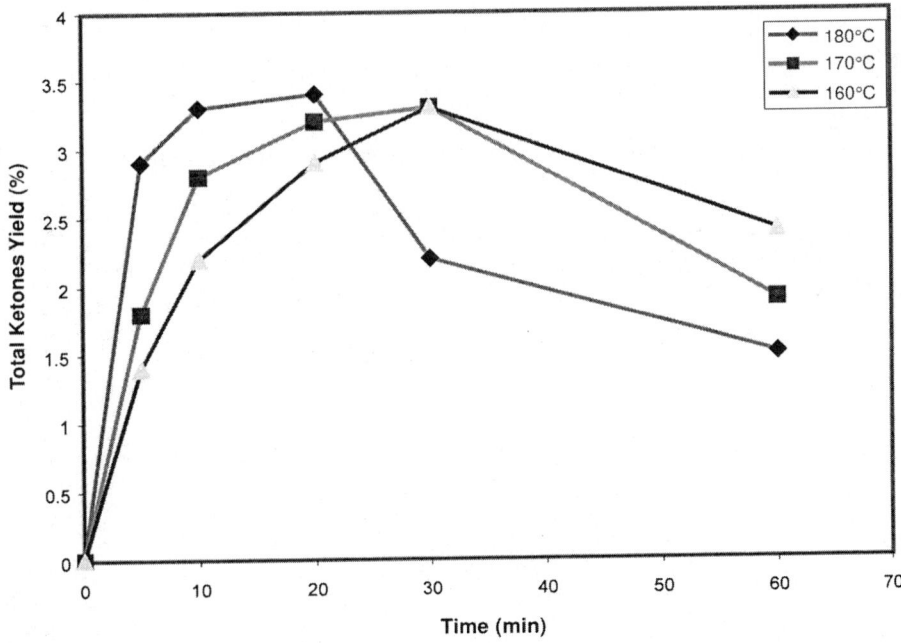

Fig. 3. Total ketones yield profiles near the optimal oxidation conditions.

are improved to the level of 4.6% for vanillin, 9.8% for syringaldehyde, 0.7% for acetovanillone, and 2.6% for acetosyringone. Organic acids are the major byproducts of the oxidation process. The aromatic products are unstable under high oxygen tension, being degraded to acids. A reaction pattern with declining O_2 further improves the yields of aldehydes and ketones.

Acknowledgments

We wish to thank Dr. George Goodloe of the Spectroscopy Center, Auburn University for his service in the GC-MS analysis of the samples. We also gratefully acknowledge the financial support provided for this work by the National Renewable Energy Laboratory (subcontract XGC-7-17041-01) and the Engineering Experiment Station of Auburn University.

References

1. Torget, R. W., Nagel, N., Jennings, E., Ibsen K., and Elander, R. (1999), 21st Symposium on Biotechnology for Fuels and Chemicals, Fort Collins, CO.
2. Chang, H.-M. and Allan, G. G. (1971), in *Lignins Occurrence, Formation, Structure and Reactions*, Sarkanen, K. V. and Ludwig, C. H., eds., Wiley-Interscience, New York, pp. 433–485.
3. Goheen, D. W. (1981), in *Organic Chemicals from Biomass*, Goldstein, I. S., ed., CRC, Boca Raton, FL, pp. 143–161.
4. Hocking, M. B. (1997), *J. Chem. Ed.* **74,** 1055–1059.
5. Pommier, Y., Macdonald, T. L., and Madalengoitia, J. S. (1994), *PCT Int. Appl. WO* **94 10,175.**
6. Cushman, M. S. and Hamel, E. (1993), *PCT Int. Appl. WO* **93 23,357.**
7. Xiang, Q. and Lee, Y. Y. (2000), *Appl. Biochem. Biotechnol.* **84/86,** 153–162.
8. Overend, R. P. and Chornet, E. (1990), *Can. J. Physiol.* **68(9),** 1105–1111.
9. Abatzoglou, N., Chornet, E., Belkacemi, K., and Overend, R. P. (1992), *Chem. Eng. Sci.* **47(5),** 1109–1122.
10. Gierer, J. and Imsgard, F. (1977), *Sven. Papperstidn.* **80(16),** 510–518.
11. Fargues, C., Mathias, A., Silva, J., and Rodrigues, A. (1996), *Chem. Eng. Technol.* **19,** 127–136.
12. Sultanov, V. S. and Wallis, F. A. (1991), *J. Wood Chem. Technol.* **11(3),** 291–305.
13. Wu, G., Heitz, M., and Chornet, E. (1994), *Ind. Eng. Chem. Res.* **33,** 718–723.
14. Wu, G. and Heitz, M. (1995), *J. Wood Chem. Technol.* **15(2),** 189–202.

Effect of Pretreatment Reagent and Hydrogen Peroxide on Enzymatic Hydrolysis of Oak in Percolation Process

SUNG BAE KIM,*,[1] BYUNG HWAN UM,[1] AND SOON CHUL PARK[2]

[1]Division of Applied Chemical Engineering and RIIT, Gyeongsang National University, Chinju, 660-701 South Korea, E-mail: sb_kim@nongae.gsnu.ac.kr; and [2]Biomass Research Team, Korea Institute of Energy Research, Taejon, 305-600 South Korea

Abstract

The effect of pretreatment reagent and hydrogen peroxide on enzymatic digestibility of oak was investigated to compare pretreatment performance. Pretreatment reagents used were ammonia, sulfuric acid, and water. These solutions were used without or in combination with hydrogen peroxide in the percolation reactor. The reaction was carried out at 170°C for the predetermined reaction time. Ammonia treatment showed the highest delignification but the lowest digestibility and hemicellulose removal among the three treatments. Acid treatment proved to be a very effective method in terms of hemicellulose recovery and cellulose digestibility. Hemicellulose recovery was 65–90% and digestibilities were >90% in the range of 0.01–0.2% acid concentration. In both treatments, hydrogen peroxide had some effect on digestibility but decomposed soluble sugars produced during pretreatment. Unlike ammonia and acid treatments, hydrogen peroxide in water treatment had a certain effect on hemicellulose recovery as well as delignification. At 1.6% hydrogen peroxide concentration, both hemicellulose recovery and digestibility were about 90%, which were almost the same as those of 0.2% sulfuric acid treatment. Also, digestibility was investigated as a function of hemicellulose removal or delignification. It was found that digestibility was more directly related to hemicellulose removal rather than delignification.

*Author to whom all correspondence and reprint requests should be addressed.
Current address: Department of Chemical Engineering, Auburn University, Auburn, AL 36849.

Index Entries: Pretreatment; ammonia; acid; water; hydrogen peroxide; enzymatic hydrolysis.

Introduction

Cellulose, a polymer of glucose, can be transformed into fermentable sugars by chemical or biochemical processes. Enzymatic hydrolysis, a method that is designed to produce glucose from cellulosic materials, has several advantages over acid hydrolysis. Enzyme reaction is very specific and does not produce undesirable byproducts. Also it is an energy-saving reaction since it takes place at mild temperatures. But enzymes are expensive and their reactions are slow. Even if the chemical and physical characteristics of native lignocellulosic material vary with the origin of raw material, its enzymatic hydrolysis is hindered by the following substrate-related factors *(1–3)*: cellulose in lignocellulosic biomass contains highly resistant crystalline structure, lignin and hemicellulose surrounding cellulose form a physical barrier, and sites available for enzymatic attacks are limited. Thus, effective pretreatment is an essential prerequisite to enhance the susceptibility of lignocellulosic residues to enzyme action.

The primary goal of any pretreatment process is to maximize digestibility. Various pretreatment methods have been investigated to improve enzymatic hydrolysis of woody biomass. Also, the economics of the bioalcohol process using various pretreatments have been evaluated *(4,5)*. These pretreatments are broadly classified into physical *(3)*, chemical *(6)*, and biologic *(7)* according to the principal mode of action on the substrates. Of the number of studies on the efficiency of enzymatic digestibility, there have been few comparing various pretreatment methods *(8)*. In the present study, ammonia, sulfuric acid, and water were employed as pretreatment reagents of woody biomass. It is generally known that an alkaline solution such as ammonia has a great capability in removing lignin but a relatively low capability in solubilizing hemicellulose, whereas acidic solution such as sulfuric acid has the opposite characteristics in removing the noncellulosic constitutes *(9,10)*. The results of water pretreatment are similar to those of acid. To overcome these problems, hydrogen peroxide was added to each pretreatment. The role of hydrogen peroxide is to promote removal of lignin and break bonds between lignin and carbohydrates *(11,12)*.

The purpose of the present study was to examine the effect of pretreatment reagent and hydrogen peroxide on enzymatic digestibility of oak. In addition, the effect of hemicellulose and lignin removal on enzymatic digestibility was investigated. The reactor used was a percolation reactor, which had a proven performance in our previous studies *(9,10)*. The chief advantage of this reactor is its ability to attain high sugar yield by minimizing sugar decomposition.

Materials and Methods

Materials

The oak chips supplied from Korea Institute of Energy Research (KIER) were ground to an average size of 20–60 mesh (0.25–0.84 mm) using a laboratory knife mill. Oak is one of the hardwood species indigenous to Korea. It is considered a renewable resource suitable for conversion to ethanol.

Pretreatment

The percolation system is composed of an aqueous ammonia reservoir, aqueous hydrogen peroxide reservoir, water reservoir, pump, programmable dry oven, reactor, and liquid holding tank. The details of experimental apparatus were described elsewhere *(13)*. A metering pump (TSP Minipump) supplied various solutions into the percolation reactor through a preheating coil. Ten grams of oak chips was presoaked with reaction solution overnight (over a 12-h period). To carry out the reaction, the oven temperature was initially set at 230°C. Also, nitrogen back-pressure was applied to the reaction system at selected pressure (2.758 MPa for ammonia and 1.379 MPa for sulfuric acid and water) to prevent the vaporization of each solution. After preheating for 20 min, the interior of the reactor reached the reaction temperature (170°C). Each reaction solution was pumped into the reactor while maintained at 170°C at a flow rate of 1 mL/min for 15–60 min. As soon as the reaction was completed, water was pumped into the reactor to remove residual sugars trapped in the biomass. After leaching for 20–50 min, the solid sample was discharged from the reactor and then dried at 105°C overnight for solid analysis. The liquid collected in the liquid-holding tank was measured for sugars after secondary hydrolysis.

Enzyme and Enzymatic Hydrolysis

Enzymatic digestibility was measured on the various extensively washed chemical-treated and control samples (untreated, α-cellulose) using National Renewable Energy Laboratory (NREL) standard Laboratory Analytical Procedure no. 009 *(14)*. Commercial cellulase and β-glucosidase (Novo Nordick, Bagvard, Denmark) supplied from KIER were used. Celluclast (80 IU or international filter paper units [IFPU]/mL, 80 mg of protein/mL) and Novozym 188 (792 cellobiase units [CBU]/mL, 73 mg of protein/mL) were used for cellulose hydrolysis with a volume ratio of 4 IU of celluclast/CBU of Novozym to alleviate end-product inhibition by cellobiose. The amount of washed solids required to give 0.5 g of cellulose in 50 mL was added to a 250-mL flask. The buffer for the digestion was 0.05 *M* citrate, pH 4.8. The cellulase enzyme loading was adjusted to 60 IFPU/g of biomass of cellulose in the flask. The contents of the flask were preheated to 50°C before the enzyme was added. Then the flask was placed on a water shaker bath operating at 50°C and 90 rpm. Using the same method, untreated substrate and α-cellulose were placed in the bath as control substrates. The hydrolysis was carried out for 96 h while removing 1 mL of sample every 24 h. The glucose content of this sample was analyzed using high-pressure liquid chromatography (HPLC), and then its digestibility was calculated.

Analytical Methods

Solid biomass samples were analyzed for moisture, sugars, klason lignin, and ash by NREL standard procedures (Laboratory Analytical Procedures no. 001-005) *(14)*. All experiments were done in duplicate. Sugars and decomposition products were measured by HPLC (Thermo Separation Products) using Bio-Rad Aminex HPX-87C (conditions: 0.6 mL/min, 85°C, and water) and 87H (conditions: 0.6 mL/min, 65°C, and 0.005 M sulfuric acid) columns. Since they do not resolve xylose, mannose, and galactose, the three components were presented as xylose + mannose + galactose (xmg). The quantity of xmg was calculated based on xylose analysis, because xylose consisted of >90% of these three components *(9)*.

Results and Discussion

Solutions of ammonia, sulfuric acid, and water were used singly or in combination with hydrogen peroxide for oak pretreatment. Each pretreatment was carried out at 170°C, using a percolation flow rate of 1 mL/min. The reaction time for each pretreatment was predetermined from our previous studies to maximize hemicellulose recovery and minimize sugar degradation *(9,10)*. Although enzymatic hydrolysis was performed for 96 h, the digestibility at 72 h was shown in Tables 1–3 to compare the pretreatment performance. Since xmg of oak comprises >90% of total hemicellulose, in this study we preferred the terminology of hemicellulose to xmg.

Ammonia Pretreatment

Ammonia has a number of favorable attributes regarding the processing of lignocellulosic biomass. It is very effective in removing lignin and can easily be recovered from aqueous mixtures owing to its high volatility. It also cleaves the lignin-hemicellulose bond and changes cellulose structure. Similar to most alkaline treatment, ammonia treatment does not cause significant loss of carbohydrate *(15)*.

Table 1 indicates the compositions of liquid hydrolysate and solid residue after ammonia treatment and percentage of digestibility at 72 h. When the ammonia concentration was increased from 0.5 to 20% (40 times of increment) as shown in Table 1, hemicellulose recovery (=hemicellulose recovered in liquid phase/initial hemicellulose) was increased from 19 to 30%. Also, it was found that ammonia effectively removed lignin, because delignification (=[initial lignin − residual lignin]/initial lignin) was increased from 34 to 66% as the ammonia concentration was increased from

Table 1
Composition of Ammonia-Treated Oak and Digestibility[a]

Ammonia (wt%)	H$_2$O$_2$ (wt%)	Solid remaining (%)	Glucan (%)		xmg (%)		Klason lignin (%)	Digestibility at 72 h (%)
			Liquid	Solid	Liquid	Solid		
Untreated biomass	—	100.0	—	48.3	—	18.5	22.9	7.2
0.5	0	76.6	0.5	48.0	3.6	13.5	15.0	51.5
5.0	0	72.0	0.6	46.9	4.4	12.7	11.8	55.0
10.0	0	70.9	0.6	46.0	4.8	12.4	10.2	65.4
20.0	0	66.8	0.9	45.4	5.5	12.1	7.9	66.1
10.0	1.7	69.4	0.9	43.5	5.6	11.3	9.7	66.2
10.0	3.3	69.1	1.2	40.1	5.8	10.8	7.5	78.4
10.0	6.6	54.4	1.6	38.6	6.8	8.4	5.4	87.6

[a]All sugar contents are based on the original oven-dry untreated biomass and expressed as glucan, xylan, mannan, and galactan equivalents (reaction conditions: 170°C, 60 min, 1.0 mL/min).

Table 2
Composition of Sulfuric Acid–Treated Oak and Digestibility[a]

H$_2$SO$_4$ (wt%)	H$_2$O$_2$ (wt%)	Solid remaining (%)	Glucan (%)		xmg (%)		Klason lignin (%)	Digestibility at 72 h (%)
			Liquid	Solid	Liquid	Solid		
Untreated biomass	—	100.0	—	48.3	—	18.5	22.9	7.2
0.01	0	68.5	1.7	45.4	12.1	4.5	17.5	90.2
0.1	0	65.9	1.9	45.7	14.9	1.9	17.9	92.2
0.2	0	64.2	2.8	44.4	16.6	0	16.9	93.8
0.1	0.4	64.9	2.3	44.4	15.4	1.6	17.4	90.6
0.1	0.8	63.9	2.9	45.2	14.6	1.7	17.2	92.5
0.1	1.6	61.3	4.1	43.5	13.4	0	15.9	100.0
0.1	3.2	58.2	4.1	41.7	13.6	0	14.8	100.0

[a]All sugar contents are based on the original oven-dry untreated biomass and expressed as glucan, xylan, mannan, and galactan equivalents (reaction conditions: 170°C, 15 min, 1.0 mL/min).

Table 3
Composition of Water-Treated Oak and Digestibility[a]

H_2O_2 (wt%)	Solid remaining (%)	Glucan (%)		xmg (%)		Klason lignin (%)	Digestibility at 72 h (%)
		Liquid	Solid	Liquid	Solid		
Untreated biomass	100.0	—	48.3	—	18.5	22.9	7.2
0	68.4	1.7	45.6	13.4	3.5	17.4	76.5
0.8	60.9	2.2	44.5	14.3	2.2	14.5	87.5
1.6	57.4	3.5	40.4	16.5	1.4	11.7	92.9
3.2	53.9	3.3	38.5	17.0	1.3	10.7	97.2

[a]All sugar contents are based on the original oven-dry untreated biomass and expressed as glucan, xylan, mannan, and galactan equivalents (reaction conditions: 170°C, 60 min, 1.0 mL/min).

0.5 to 20%. Meanwhile, residual cellulose content was not affected much by ammonia concentration, leaving it almost intact. The digestibility of the 20% ammonia–treated sample increased by nine times compared with that of the untreated sample. Thus, the ammonia solution used in the pretreatment could be considered an ideal pretreatment solution regarding delignification and very good sugar accountability (sugar in liquid + sugar in solid). However, ammonia pretreatment was not suitable for hemicellulose recovery because it left 70–80% of hemicellulose in the solid phase. Since hemicellulose and cellulose cannot be fermented simultaneously to alcohol, this low hemicellulose recovery is an undesirable attribute because another pretreatment to remove residual hemicellulose fraction is required for subsequent fermentation.

To alleviate this problem, hydrogen peroxide was added into the ammonia stream as an oxidant. As shown in Table 1, the effect of hydrogen peroxide was investigated at the fixed ammonia concentration, 10 wt%. As the concentration of hydrogen peroxide was increased, cellulose, hemicellulose, and lignin in solid phase were substantially decreased. But the amounts of hemicellulose and cellulose recovered in liquid phase were relatively small. This means that hydrogen peroxide decomposed hemicellulose as well as cellulose. Cellulose digestibility, however, greatly increased as hydrogen peroxide concentration increased.

Acid Pretreatment

Dilute sulfuric acid treatment is a very effective method in terms of hemicellulose recovery and cellulose digestibility. The main drawbacks of this method are sugar decomposition and corrosive action owing to the high temperature and low pH employed. Also, this process needs the neutralization step of residual sulfuric acid with lime, which must be disposed of. In spite of these problems, this pretreatment has been evaluated as the most competitive one for a commercial biomass-to-ethanol process *(8)*.

Table 2 shows the results obtained when oak was treated by sulfuric acid alone or 0.1% sulfuric acid with various hydrogen peroxide concentrations. The reaction time adopted in this treatment was 15 min, which is a quarter of that in the ammonia treatment. Compared to ammonia treatment, dilute-acid treatment has considerably higher hemicellulose removal and lower lignin removal. The hemicellulose recovery at 0.01 and 0.2% acid concentration was 65 and 90%, respectively. However, delignification was only 22% at a concentration of 0.1% acid, which was significantly lower than the values obtained in ammonia treatment. Even when hydrogen peroxide was added in the sulfuric acid stream, delignification was not increased much. In addition, it was found that hydrogen peroxide significantly decomposed the hemicellulose extracted into the liquid phase. By comparing the sugar balances of glucan and hemicellulose in ammonia treatment with those of acid treatment, it is apparent that hydrogen peroxide would degrade a certain sugar preferentially in each treatment; i.e., glucose in ammonia treatment and hemicellulose monomers in acid treatment.

In all cases studied in acid treatment, enzymatic digestibilities were >90%, which were significantly higher than the values obtained in ammonia treatment. This means that acid treatment is superior to ammonia treatment in terms of hemicellulose removal and enzymatic digestibility.

Water Pretreatment

In its natural state, hemicellulose exists in an amorphous form. Thus, it is easy to hydrolyze under mild reaction conditions or even in high-temperature water. The organic acids formed during water hydrolysis are acetic and formic acids derived from the cleavage of acetyl and methoxyl groups from hemicellulose *(16)*. These acids lower the pH to a range of 3.0 to 4.0, which permits the removal of hemicellulose from wood.

Table 3 shows the results of water or water–hydrogen peroxide treatment. The reaction time used in this treatment was 60 min because of low acidity found in hydrolysate. The hemicellulose recovery of water treatment was 72%, which was a little higher than that of 0.01% sulfuric acid. This could be owing to the four times longer reaction time. It is very difficult to achieve >70% of hemicellulose recovery in water treatment, because it is impossible to increase catalytic concentration to a certain degree using only water *(6)*.

Unlike ammonia and sulfuric acid treatments, hydrogen peroxide in water treatment has a certain effect on hemicellulose recovery as well as delignification. As the concentration of hydrogen peroxide increased from 0 to 0.8, 1.6, and 3.2%, hemicellulose recovery and delignification increased from 72 to 77, 89, and 92%, and from 24 to 37, 49, and 53%, respectively. In addition, enzymatic digestibility increased from 76.5 to 87.5, 92.9, and 97.2% in the same concentration change of hydrogen peroxide. This means that if hydrogen peroxide is added in the water stream, all three measured values (i.e., hemicellulose recovery, delignification, and enzymatic digestibility) can be increased significantly. When 1.6% hydrogen peroxide was added, both hemicellulose recovery and enzymatic digestibility were about 90%. These values were almost the same as those of 0.2% sulfuric acid treatment, and its delignification was 23% higher than that of 0.2% acid treatment.

As the concentration of hydrogen peroxide increased, it was observed that glucose degraded significantly, whereas hemicellulose was preserved well, as shown in Table 3. It was found that water–hydrogen peroxide treatment may bring a better result relative to sulfuric acid treatment if the amount of hydrogen peroxide is optimized. Also, this treatment can avoid the neutralization step that is required in dilute-acid treatment.

Digestibility as Function of Time

The enzymatic digestibilities shown in Tables 1–3 are percentage of digestibility at 72 h, which is suggested by NREL standard procedure. However, to investigate the overall trend of enzymatic digestibility,

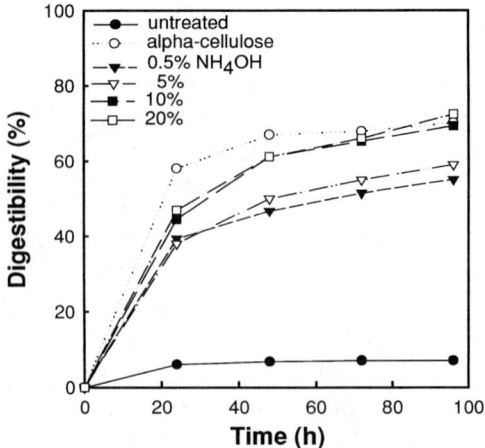

Fig. 1. Effect of ammonia concentration on enzymatic digestibility.

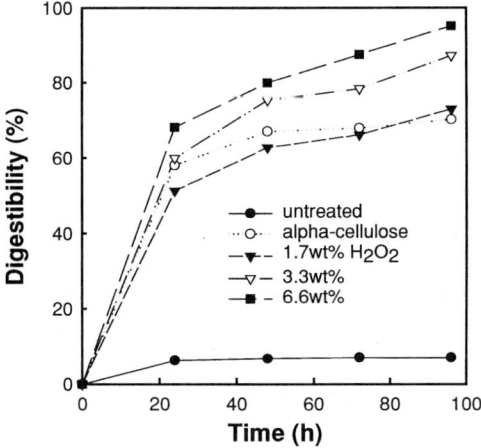

Fig. 2. Effect of H_2O_2 concentration in 10% ammonia solution on enzymatic digestibility.

samples were taken every 24 h for 96 h. The results are illustrated in Figs. 1–5. The untreated substrate and α-cellulose were used as the control substrates. Figure 1 shows the effect of ammonia concentration on digestibility. The digestibility increased as the ammonia concentration increased. The digestibilities from various ammonia treatments were much higher than those of the untreated sample, but they were generally lower than those of α-cellulose. Figure 2 presents the results of digestibilities obtained at various concentrations of hydrogen peroxide in 10% ammonia stream. The digestibility increased significantly as the concentration of hydrogen peroxide increased. Unlike ammonia treatment, ammonia–hydrogen peroxide treatments showed almost equal or higher digestibilities than those of α-cellulose. Note that measured digestibilities showed a gradually growing trend until 96 h in all cases.

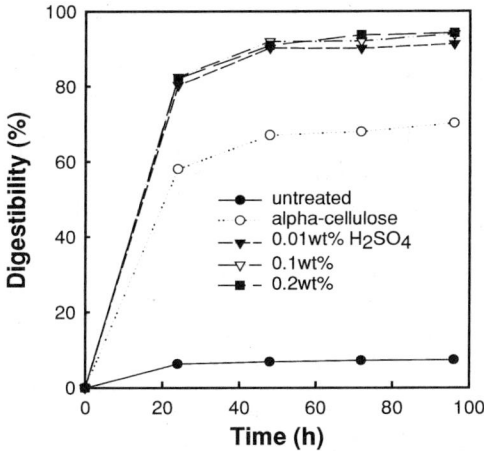

Fig. 3. Effect of acid concentration on enzymatic digestibility.

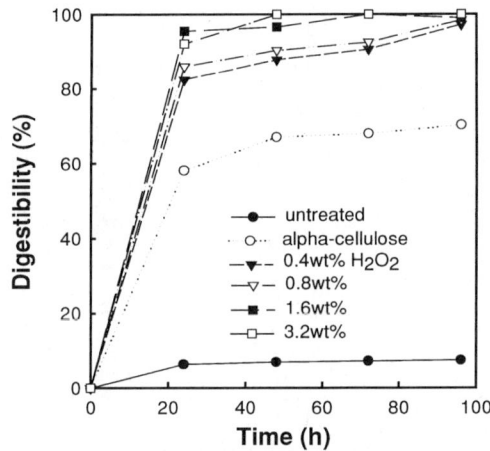

Fig. 4. Effect of H_2O_2 concentration in 0.1% sulfuric acid solution on enzymatic digestibility.

In the sulfuric acid treatment shown in Fig. 3, there was almost no difference in digestibilities as acid concentration was varied. In all cases digestibilities reached about 90% in 48 h and did not increase after that time. Figure 4 shows the effect of hydrogen peroxide on digestibility in 0.1% sulfuric acid stream. Digestibility increased as the concentration of hydrogen peroxide increased. When the 24-h digestibilities of acid–hydrogen peroxide treatments are compared with those of the acid treatment, it is found that initial digestibilities were affected somewhat by hydrogen peroxide, resulting in increasing initial rate. Figure 5 shows the effect of hydrogen peroxide on digestibility in water treatment. The digestibilities obtained from all cases are higher than those of α-cellulose. The digestibilities increased gradually at <0.8% hydrogen peroxide, as in the ammonia–

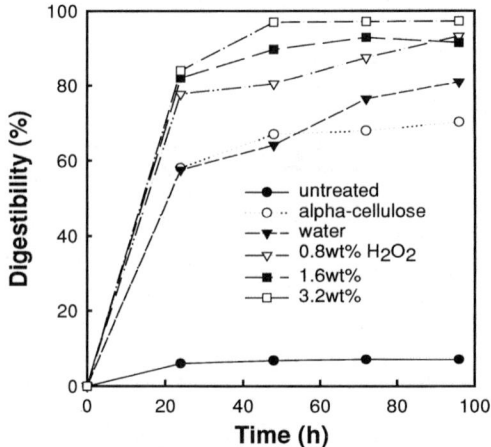

Fig. 5. Effect of H_2O_2 concentration in water on enzymatic digestibility.

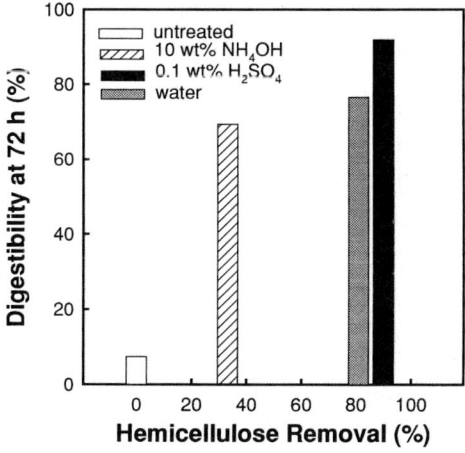

Fig. 6. Enzymatic digestibility vs percentage of hemicellulose removal.

hydrogen peroxide treatment but did not change after 48 h at >1.6% hydrogen peroxide.

Digestibility as Function of Hemicellulose Removal or Delignification

Figure 6 shows 72-h enzymatic digestibility as a function of hemicellulose removal (=[initial hemicellulose – residual hemicellulose]/initial hemicellulose). The data were obtained after treating with ammonia, acid, and water without the addition of hydrogen peroxide. Digestibility increased as hemicellulose removal increased. This can be explained in terms of the substrate-related aspect as follows: the chances of enzyme adsorption onto the cellulose surface are increased as hemicellulose

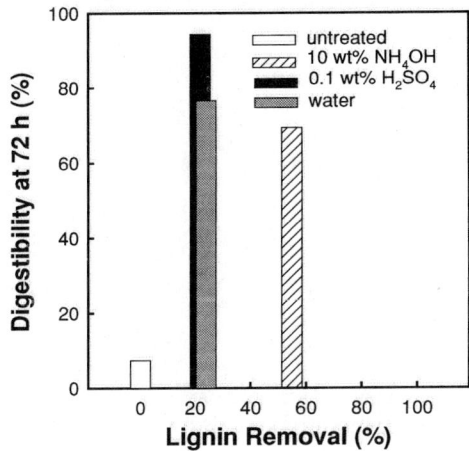

Fig. 7. Enzymatic digestibility vs percentage of lignin removal.

removal is increased. The hemicellulose removal and digestibility of each treatment were as follows: ammonia, 33% (as removal) and 65% (as digestibility); water, 81 and 77%; and acid, 90 and 92%. This means that digestibility is not linearly increased with hemicellulose removal. In the case of ammonia and water treatments, hemicellulose removal in water treatment was about 2.5 times higher than that in ammonia treatment, but the digestibility difference was only 1.2 times higher. Thus, it is assumed that digestibility can be affected by other factors, such as delignification. In spite of relatively low hemicellulose removal, it seems that high digestibility in ammonia treatment is attributed to relatively high delignification.

Digestibility was also affected by delignification, as shown in Fig. 7. In ammonia treatment, digestibility was 65% at 55% delignification. This was about 30% lower than the digestibility in the acid treatment case, whose delignification was only 22%. In the case of water treatment, digestibility was 77% at 24% delignification, which was 12% higher than that of ammonia treatment. This high digestibility at the low delignification can be explained by high hemicellulose removal in both water treatment and acid treatment.

Several studies have reported that hemicellulose (17) or lignin (18,19) hinders enzyme adsorption on cellulose. The results of the present study indicate that digestibility is more directly related to hemicellulose removal than delignification. However, it is not clear which component is more responsible for enzyme hydrolysis because neither hemicellulose nor lignin can be extracted separately without structural change of the other component. Therefore, in the pretreatment of lignocellulosic material, it is important to develop a proper pretreatment method suitable to a specific substrate because the mechanism of pretreatment has not been clearly illustrated and enzyme hydrolysis is affected by many diverse factors (20).

Acknowledgment

We gratefully acknowledge the financial support provided for this work by the Ministry of Commerce, Industry and Energy, South Korea.

References

1. Abraham, M. and Kurup, G. M. (1996), *Appl. Biochem. Biotechnol.* **62,** 201–211.
2. Gregg, D. J. and Saddler, J. N. (1996), *Biotechnol. Bioeng.* **51,** 375–383.
3. Fan, L. T., Gharpuray, M. M., and Lee, Y. H. (1987), *Cellulose Hydrolysis,* Springer-Verlag, Berlin, Germany.
4. Wyman, C. E. (1994), *Bioresour. Technol.* **50,** 3–16.
5. Ballerini, D., Desmarquest, J. P., and Pourquie, J. (1994), *Bioresour. Technol.* **50,** 17–23.
6. Iyer, P. V. (1995), MS thesis, Auburn University, Auburn, AL.
7. Wyman, C. E. (1996), *Handbook on Bioethanol: Production and Utilization,* Taylor & Francis, Washington, DC.
8. McMillan, J. D. (1992), *Processes for Pretreating Lignocellulosic Biomass: A Review,* NREL/TP-421-4978, November, National Renewable Energy Laboratory, Golden, CO.
9. Huh, S. J., Kim, S. B., and Park, S. C. (1999), *Korean J. Biotechnol. Bioeng.* **14(3),** 1–7.
10. Kim, S. B., Yum, D. M., and Park, S. C. (2000), *Biores. Technol.* **72,** 289–294.
11. Gierer, J. (1981), *Proc. Int. Symp. Wood and Pulping Chem.,* Ekman-Days, Stockholm, vol. 2, p. 12.
12. Dence, C. W. (1980), *Chemistry of Delignification with Oxygen, Ozone and Peroxide,* Uni Publishers, Tokyo, Japan, p. 199.
13. Kim, S. B. and Lee, Y. Y. (1996), *Appl. Biochem. Biotechnol.* **57/58,** 147–156.
14. Chemical Analysis and Testing Standard Procedures (1996), National Renewable Energy Laboratory, Golden, CO.
15. Iyer, P. V., Wu, Z., Kim, S. B., and Lee, Y. Y. (1996), *Appl. Biochem. Biotechnol.* **57/58,** 121–132.
16. Weil, J., Sarikaya, A., Rau, S., et al. (1997), *Appl. Biochem. Biotechnol.* **68,** 21–40.
17. Knappert, D., Grethlein, H., and Converse A. (1981), *Biotechnol. Bioeng. Symp.* **11,** 67–77.
18. Ramos, L. P., Breuil, C., and Saddler, J. N. (1992), *Appl. Biochem. Biotechnol.* **34/35,** 37–47.
19. Mooney, C. A., Mansfield, S. D., Touhy, M. G., and Saddler, J. N. (1998), *Biores. Technol.* **64,** 113–119.
20. Mansfield, S. D., Mooney, C., and Saddler, J. N. (1999), *Biotechnol. Prog.* **15,** 804–816.

SESSION 2
Applied Biological Research

SESSION 2
Applied Biological Research

Fingerprinting *Trichoderma reesei* Hydrolases in a Commercial Cellulase Preparation

T. B. Vinzant,[1] W. S. Adney,[1] S. R. Decker,[1] J. O. Baker,[1] M. T. Kinter,[2] N. E. Sherman,[2] J. W. Fox,[2] and M. E. Himmel*,[1]

[1]*Biotechnology Center for Fuels and Chemicals, National Renewable Energy Laboratory, Golden, CO 80401, E-mail: mike_himmel@nrel.gov; and* [2]*Biomolecular Research Facility, University of Virginia Medical School, Charlottesville, VA 22908*

Abstract

Polysaccharide degrading enzymes from commercial *T. reesei* broth have been subjected to "fingerprint" analysis by high-resolution 2-D gel electrophoresis. Forty-five spots from 11×25 cm Pharmacia gels have been analyzed by LC-MS/MS and the resulting peptide sequences were compared to existing databases. Understanding the roles and relationships of component enzymes from the *T. reesei* cellulase system acting on complex substrates is key to the development of efficient artificial cellulase systems for the conversion of lignocellulosic biomass to sugars. These studies suggest follow-on work comparing induced and noninduced *T. reesei* cells at the proteome level, which may elucidate substrate-specific gene regulation and response.

Index Entries: Cellulase; *Trichoderma reesei*; two-dimensional gel electrophoresis; liquid chromatography–mass spectrometry/mass spectrometry.

Introduction

One key technical challenge to enable the production of sugars and ethanol from lost-cost feedstocks remains the reduction in cost of cellulases acting on pretreated biomass *(1–3)*. Our approach to this dilemma is to increase the specific activity of the enzymes in the cellulase complex. Understanding the roles and relationships of cellulase component enzymes acting on specific substrates is vital to the development of an efficient artificial cellulase system for the conversion of cellulosic biomass to sugars. The *Trichoderma reesei* biomass degrading system consists of many glycosyl hydrolases, of which five β-1,4-endoglucanases (EG I–EG V),

*Author to whom all correspondence and reprint requests should be addressed.

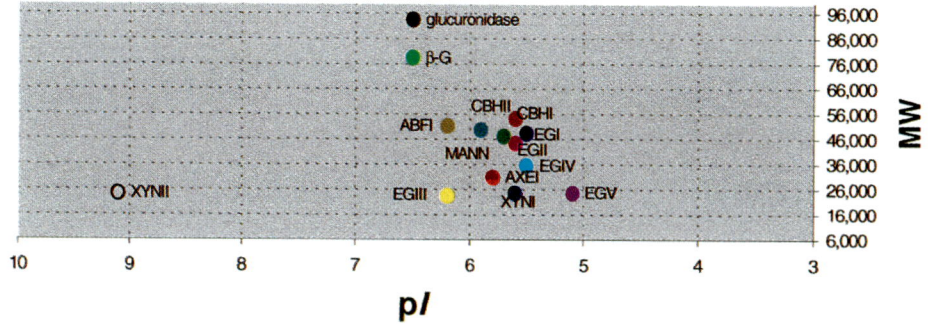

Fig. 1. 2D gel simulation of *T. reesei* hydrolases based on literature values for molecular weight and isoelectric pH.

two β-1,4-exoglucanases (cellobiohydrolase [CBH] I and CBH II), two xylanases (XYN I and XYN II), a β-D-glucosidase, an α-L-arbinoflurano-sidase, an acetyl xylan esterase, a β-mannanase, and an α-glucuronidase have been sequenced *(4)* (Fig. 1, Table 1).

Application of preliminary proteonomics analysis to a standard commercial cellulase product was judged appropriate for demonstrating fingerprinting methodology *(5)*. Although two-dimensional (2D) protein gels have been used on occasion to follow the expression of selected *T. reesei* cellulase components *(6)* and their glycosylated forms *(7,8)*, a systematic display of the entire system of enzymes found in *T. reesei* culture broth has not been reported. We describe here fingerprinting via 2D gel electrophoresis and internal peptide sequence analysis of a commercial cellulase preparation. The Biomolecular Research Facility at the University of Virginia Medical School processed gels prepared at National Renewable Energy Laboratory for spot sequencing and identification. Forty-five spots were identified by liquid chromatography–mass spectrometry/mass spectrometry (LC-MS/MS). The experimental results were compared to ~1100 known glycosyl hydrolases from all species, with positive hits only arising from *T. reesei* proteins. Most of these were of known proteins, but several novel proteins were detected.

Materials and Methods

Enzyme Sample and Reagents

Genencor (Palo Alto, CA) generously provided a sample of the commercial cellulase preparation, Laminex, for testing. Buffers, trypsin, dithiothreitol (DTT), and electrophoresis supplies were obtained from Sigma (St. Louis, MO) and Amersham Pharmacia Biotech (Alameda, CA).

2D Electrophoresis

Characterization of the Laminex cellulase product was performed using sodium dodecyl sulfate polyacrylamide gel electrophoresis (SDS-PAGE) and

Table 1
Critical Values for *T. reesei* Glycosyl Hydrolases[a]

T. reesei enzyme/reference	Molecular weight	Number of amino acids	Calculated pI (experimental)	Average hydrophobicity	Hydrophobic (%)/ hydrophilic (%)	Folded/ extended area (native shape)
EG I	48,209	459	5.50 (4.7)	-3.7	1.15	0.23
EG II	44,228	418	5.60 (5.5)	-1.9	1.02	0.24
EG III	23,481	218	6.20 (7.4)	-3.6	0.91	0.28
EG IV (9)	35,512	344	5.50	-0.7	0.98	0.25
EG V	24,412	242	5.10	-1.1	0.79	0.28
CBH I	54,075	513	5.60	-4.4	1.12	0.22
CBH II	49,655	471	5.90	-1.2	0.93	0.23
XYN I	24,583	229	5.60	-2.8	1.06	0.28
XYN II	24,173	222	9.10	-4.7	1.04	0.28
β-D-Glucosidase I (10)	78,436	744	6.50	-1.6	0.95	0.20
α-L-Arabinofuranosidase I (12)	51,117	500	6.20	-1.1	0.84	0.23
Acetyl xylan esterase (13)	30,755	302	5.80	-1.5	0.95	0.26
β-Mannanase[b]	47,054	437	5.70	-2.6	0.99	0.23
α-Glucuronidase (11)	93,427	847	6.50	-2.5	0.96	0.19

[a]Sequence data are from SwissProt or Genbank and calculated using PepTools (Edmonton, AB, Canada).
[b]Accession no. L25310 NID g506847.

2D gels under denatured conditions. Protein sample preparation was initiated by trichloroacetic acid precipitation followed by acetone washing and drying prior to resolublization in nonionic denaturing loading buffer.

All the 2D gel electrophoreses were performed in a horizontal format utilizing the Pharmacia Multiphor II electrophoresis system and Immobiline Drystrip kit. The first dimension of isoelectric focusing was carried out in an immobilized pH gradient dry strip gel (pH 3.0–10.0 linear). The second dimension was run on 8–18% SDS polyacrylamide gradient gel (ExcelGel) and stained with colloidal blue. The standard methods and procedures were followed directly from the Pharmacia manual.

The precast 180-mm Immobiline Drystrips were hydrated overnight in 8 M urea, 0.5% Triton X-100, Pharmalyte 3-10, and DTT and then were loaded into the running tray and overlaid with oil. Approximately 75–100 µg of total protein was added to the sample buffer, loaded into the loading cups, and pulled into the first dimension gel at 500 V and 1 mA for 5 h and then allowed to run at 3500 V and 1 mA for 14 h (total of 55,250 V-h). After the strips were equilibrated in a Tris-HCl, urea, glycerol, and SDS buffer, they were placed on the second-dimension SDS-PAGE gel (8–18% SDS ExcelGel; precast 245 × 110 mm) at a 90° angle to the electrical field. This dimension was run for 1.5–2 h at 600 V and 50 mA. All subsequent gels were stained with either silver stain or colloidal blue.

Liquid Chromatography–Mass Spectrometry/Mass Spectrometry

The gel piece was precisely cut and transferred to a siliconized tube and washed and destained in 200 µL of 50% methanol overnight. The pieces were then dehydrated in acetonitrile, rehydrated in 30 µL of 10 mM DTT in 0.1 M ammonium bicarbonate, and reduced at room temperature for 0.5 h. The DTT solution was removed and the sample alkylated in 30 µL of 50 mM iodoacetamide in 0.1 M ammonium bicarbonate at room temperature for 0.5 h. The reagent was removed and the gel pieces were dehydrated in 100 µL of acetonitrile. The acetonitrile was removed and the gel pieces were rehydrated in 100 µL of 0.1 M ammonium bicarbonate. The pieces were then dehydrated in 100 µL of acetonitrile, the acetonitrile was removed, and the pieces were completely dried by vacuum centrifugation. The gel pieces were rehydrated in 20 ng/µL of trypsin in 50 mM ammonium bicarbonate on ice for 10 min. Any excess trypsin solution was removed and 20 µL of 50 mM ammonium bicarbonate added. The sample was digested overnight at 37°C and the peptide formed extracted from the polyacrylamide in two 30-µL aliquots of 50% acetonitrile/5% formic acid. These extracts were combined and evaporated to 25 µL for LC-MS analysis.

The liquid chromatography–mass spectrometry system consisted of a Finnigan LCQ ion trap mass spectrometer system with a Protana nanospray ion source interfaced to a self-packed 8 cm × 75 µm id Pnenomenex Jupiter 10-µm C18 reversed-phase capillary column. Volumes of the extract (0.5–2 µL) were injected and the peptides eluted from the column by an acetonitrile/0.1 M acetic acid gradient at a flow rate of 0.25 µL/min. The nanospray ion source was operated at 2.8 kV. The digest was analyzed using the double play capability of the instrument acquiring full-scan mass spectra to determine peptide molecular weights and product ion spectra to determine amino acid sequence in sequential scans. This mode of analysis produces approx 400 collisionally associated desorption (CAD) spectra of ions ranging in abundance over several orders of magnitude. Not all CAD spectra are derived from peptides.

The data were analyzed by database searching using the Sequest search algorithm. Peptides that were not matched by this algorithm were interpreted manually and searched vs the expressed sequence tag databases using the Sequest algorithm. Experimental data were compared to a database of ~1100 known glycosyl hydrolases plus the NCBI database. Proteins were positively identified by 100% identity to two or more peptide fragments per protein.

Results and Discussion

Recent advances in protein chemistry tools including rapid separation techniques such as 2D electrophoresis and LC-MS/MS afford unique opportunities to examine complex protein mixtures, such as the multicomponent mixture of hydrolytic enzymes secreted by *T. reesei*. These techniques permit fingerprinting of mixtures of proteins found in cellulase preparations and their degradation products over a wide range of pHs and molecular weights. The capability of 2D electrophoresis to separate several hundred proteins in a single experiment can ultimately be used to identify products expressed by different genomes as well as to identify unique proteins that may be expressed in response to induction by different biomass substrates. The data generated from the analysis of complex cellulase mixtures not only gives important information about this complex biologic system relevant to biomass conversion, but also may yield information about degradation of proteins in commercial preparations.

2D gel electrophoresis of a commercial *T. reesei* cellulase preparation yielded approx 70 discernible spots on a 245 × 110 mm gel (Figs. 2–4). Fifty-five to 60 gel species were observable in the pH and molecular weight window we considered appropriate for secreted fungal hydrolases (pH 2.5–9 and 20–180 kDa). Thirty-four of these spots were identified as known glycosyl hydrolases by LC-MS/MS. Eleven spots contained unknown proteins.

Although a number of spots were found by LC-MS/MS to contain a single protein component (i.e., EG I, EG IV, EG V, CBH I, CBH II, AXE, and β-D-glucosidase), analysis of many of these spots yielded multiple identities (Table 2). We feel that the sampling and identification conditions applied to these gels were stringent enough to rule out contamination or misidentification of spots. The likely explanations for this observation lie in both the methodology used and the inherent nature of the sample. Some proteins may still be unresolved under the electrophoretic conditions applied (Fig. 3). For example, some improvement in the separation win-

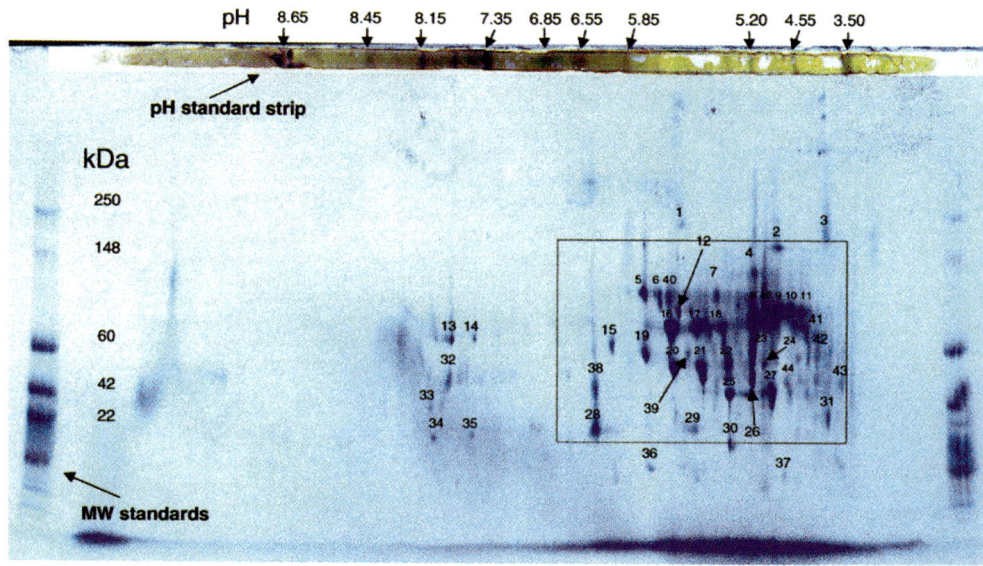

Fig. 2. Pharmacia 2D gel of *T. reesei* cellulase preparation (Laminex; Genencor) visualized with Coomassie blue. MW, molecular weight.

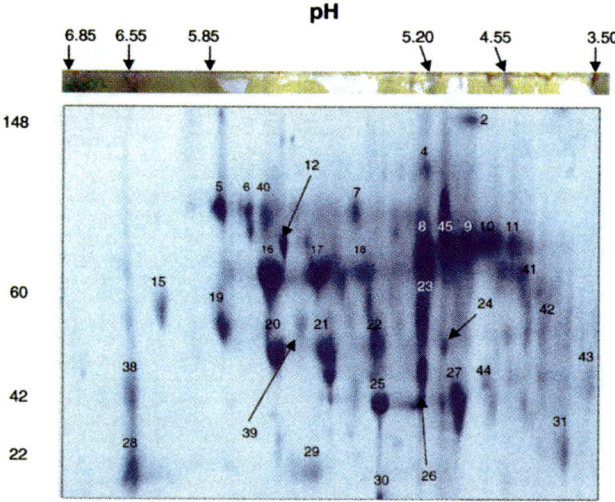

Fig. 3. Pharmacia 2D gel of *T. reesei* cellulase preparation (Laminex; Genencor). Closer view of the gel zone defined by pH 3.5–6.85 and molecular weight (MW) of 20–150 kDa.

dow may be possible with new recently available "narrow" pH range ampholines. Also, isozymes or variably glycosylated forms of the proteins may explain small changes in molecular weight or isoelectric point (p*I*);

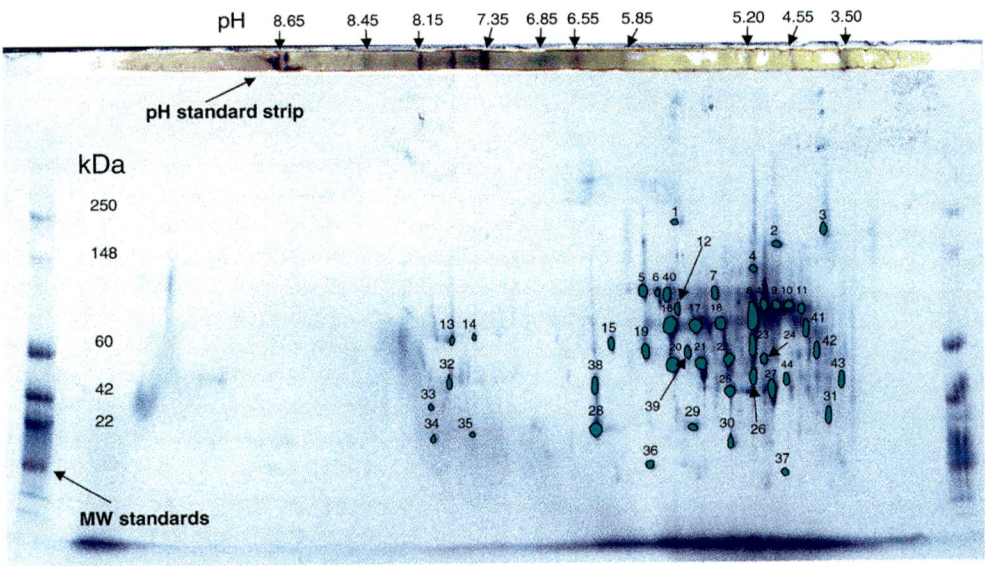

Fig. 4. Pharmacia 2D gel of *T. reesei* cellulase preparation (Laminex; Genencor). Analysis of the 45 spots shown was made by LC-MS/MS. MW, molecular weight.

however, this explanation would not account for the large alterations in the properties observed in the case of EG I, CBH I, and CBH II. Severe proteolysis of some proteins during production, storage, or sample preparation may provide the answer. Except for sample preparation alteration, the presence of these altered proteins raises several interesting possibilities. Can overall system-specific activity be increased through prevention of these altered enzymes? Can purified cellulases be used to reconstitute a cellulase mixture that is more active than native? The phenomenon of multiple electrophoretic forms was most obvious with CBH II. Having many CBH II spots widely distributed on the gel raises two obvious questions. Is CBH II really needed for efficient hydrolysis? and Can the addition of holo-CBH II (the intact enzyme) to complex cellulase mixtures enhance activity?

It is now logical to compare compositions of variably induced cellulase preparations from *T. reesei* by high-resolution 2D electrophoresis and note the differences as a function of the individual enzyme species present and the overall activity of the cellulase complex on pretreated biomass. Special assays such as the diafiltration saccharification assay can be used to assess overall cellulose digestibility. The results of these comparisons may lend yet another perspective to the biochemical interactions of the individual enzyme species within the *T. reesei* cellulase complex needed for the rapid and total saccharification of cellulose in biomass.

Table 2
2D Gel Component Analysis
of *T. reesei* Preparation by LC-MS/MS[a]

Spot no.	Spot assignment
1	CBH II
2	CBH II
3	EG I, CBH II
4	Unknown 1, same as spot 31
5	Unknown 2, same as spot 6
6	Unknown, same as unknown 2
7	EG IV
8	CBH I, CBH II, EG I, EG II
9	CBH I, EG I
10	CBH I, EG I
11	CBH I, EG I
12	Unknown 3
13	β-D-Glucosidase
14	β-D-Glucosidase
15	CBH II, β-D-glucosidase
16	CBH II
17	CBH II
18	CBH II, CBH I
19	CBH II
20	CBH II, CBH I
21	CBH II, CBH I
22	EG II, CBH II, CBH I
23	EG II, CBH II, CBH I
24	EG II, CBH II, CBH I
25	CBH II
26	EG II, CBH II, CBH I
27	EG II
28	Endo
29	Endo, CBH II, EG II
30	Acetyl xylan esterase
31	Unknown, same as unknown 1
32	Unknown 4, same as spot 33
33	Unknown, same as unknown 4
34	Unknown 5, same as spot 35
35	Unknown, same as unknown 5
36	Unknown 6
37	Unknown 7
38	Endo
39	β-D-Mannanase
40	EG IV, CBH II
41	CBH I
42	EG I, CBH I
43	EG V
44	Unknown 8
45	EG I

[a]EG I: 8–11, 42, 45; EG II: 3, 8, 22–24, 26, 27, 29; EG IV: 18, 24, 40; EGV: 43; β-glucosidase: 13–15; CBH I: 8–11, 20–24, 26, 41, 42; CBH II: 1–3, 8, 15–26, 40; AXE I: 30; β-mannanase: 39; novel: 5–7, 30–36, 44.

Acknowledgments

This work was partially funded by the Biochemical Conversion Element within the Biofuels Systems Program of the Office of Fuels Development of the US Department of Energy. This work was also funded by the University of Virginia Pratt Committee.

References

1. Wyman, C. E., Bain, R. L., Hinman, N. D., and Stevens, D. J. (1993), *Renewable Energy: Sources for Fuels and Electricity*, Island Press, Washington, DC.
2. Sheehan, J. J. (1994), in *Enzymatic Conversion of Biomass for Fuels Production*, vol. 566, Himmel, M. E., Baker, J. O., and Overend, R. P., eds., American Chemical Society, Washington, DC, pp. 1–52.
3. Bergeron, P. (1996), in *Handbook on Bioethanol*, Wyman, C. E., ed., Taylor & Francis, Washington, DC, pp. 179–195.
4. Swiss-Prot site <http://www.expasy.ch/sprot/>.
5. Herbert, B. R., Sanchez, J.-C., and Bini, L. (1997), in *Proteonome Research: New Frontiers in Functional Genomics*, Wilkins, M. R., Williams, K. L., Appel, R. D., and Hochstrasser, D. F., eds., Springer Verlag, NY, pp. 13–30.
6. Schmidt, C. S. and Wolf, G. A. (1999), *Eur. J. Plant Pathol.* **105**, 285–295.
7. Pakula, T. M., Uusitalo, J., Saloheimo, M., Salonen, K., Aarts, R. J., and Penttila, M. (2000), *Microbiology-UK* **146**, 223–232.
8. Maras, M., De Bruyn, A., Vervecken, W., Uusitalo, J., Penttila, M., Busson, R., Herdewijn, P., and Contreras, R. (1999), *FEBS Lett.* **452**, 365–370.
9. Saloheimo, M., Nakari-Setala, T., Tenkanen, M., and Penttila, M. (1997), *Eur. J. Biochem.* **249**, 584–591.
10. Barnett, C. C., Berka, R. M., and Fowler, T. (1991), *BioTechnology* **9**, 562–567.
11. Margolles-Clark, E., Saloheimo, M., Siika-aho, M., and Penttila, M. (1996), *Gene* **172**, 171, 172.
12. Margolles-Clark, E., Tenkanen, M., Nakari-Setala, T., and Penttila, M. (1996), *Appl. Environ. Microbiol.* **62**, 3840–3846.
13. Margolles-Clark, E., Tenkanen, M., Soderlund, H., and Penttila, M. (1996), *Eur. J. Biochem.* **237**, 553–560.

Development of High-Performance and Rapid Immunoassay for Model Food Allergen Lysozyme Using Antibody-Conjugated Bacterial Magnetic Particles and Fully Automated System

REIKO SATO, HARUKO TAKEYAMA,
TSUYOSHI TANAKA, AND TADASHI MATSUNAGA*

*Department of Biotechnology, Tokyo University of Agriculture
and Technology, 2-24-16, Naka-cho Koganei, Tokyo 184-8588, Japan,
E-mail: tmatsuna@cc.tuat.ac.jp*

Abstract

A high-performance and rapid chemiluminescence immunoassay for model food allergen lysozyme, one of the major allergenic components in egg white, using antibody-conjugated bacterial magnetic particles and a fully automated system was developed. This system contains a reaction station, tip rack, and an eight-tip pipettor that is able to attach and detach a strong magnet to the tip surface. The immunoreaction time was shortened to 5 min, and the assay was completed within 20 min. The lower detection limit for lysozyme was 10 ng/mL. This system can be used to perform 24 samples in 60 min within 10% coefficient of variation.

Index Entries: Sandwich immunoassay; food allergen; lysozyme; bacterial magnetic particles; fully automated system.

Introduction

A food allergy is defined as an adverse reaction to food triggered by specific allergenic components. Crossed immunoelectrophoresis, radio-immunoelectrophoresis, and IgE inhibition enzyme-linked immuno-sorbent assay (ELISA) have been used to estimate the allergenicity of food allergens (1–3). However, their major disadvantage is the reliance on human serum from allergenic individuals as a reagent. Such human serum is

*Author to whom all correspondence and reprint requests should be addressed.

difficult to obtain, often poorly characterized, and inherently variable *(1)*. Recently, Koppelman et al. *(3)* reported that a sandwich ELISA using IgG purified from immunized animals showed similar results to an IgE inhibition ELISA using human serum in terms of its sensitivity and specificity in detection of hazelnut proteins in food products.

On the other hand, Kushimoto and Aoki *(4)* reported that the allergens obtained from gluten, gliadin, and glutenin by pepsin digestion of wheat yielded the other allergenic fragments that elicit IgE antibodies, and these allergens were different from wheat and gluten allergens obtained by simple extraction. The allergen could arise on digestion of the food. Therefore, the preparation of various IgG antibodies is required for detection of food allergens. However, sandwich ELISA is not appropriate for this purpose because it is time-consuming. A rapid, simple, high-performance assay system needs to be developed.

Magnetic bacteria synthesize intracellular magnetite particles *(5–8)*. The amount of antibody coupling with BMPs is higher than that with artificial magnetite particles of the same size *(9)*, because the bacterial magnetic particles (BMPs) are small (50–100 nm) and are superior to artificial magnetite particles in their dispersion ability as covered with a lipid membrane *(10)*. BMPs permit the development of highly sensitive and rapid chemiluminescence enzyme immunoassays, because they have higher relative surface area and antibodies can be efficiently chemically coupled to their surfaces *(11)*. We have constructed a fully automated, reliable immunoassay system for human serum insulin using antibody and BMP complexes in which 1 ng/mL of insulin was detected in 60 min *(12)*. Furthermore, BMPs have been applied to genotyping of microbes and fish in which target-specific DNA probes were immobilized on BMPs and DNA-BMPs were used for magnetic capture hybridization *(13,14)*.

Two to 8% of children and 1 to 2% of adults are reported to have food allergies. The most common allergenic foods for children are eggs and milk. Lysozyme is one of the major allergenic components in egg white *(15)*. In the present study, lysozyme was targeted as a model allergen. We report on a high-performance and rapid immunoassay system for lysozyme using antibody-conjugated BMPs and a fully automated system.

Materials and Methods

Materials

Rabbit anti-lysozyme IgG fraction (polyclonal) was obtained from Rockland (Gilbertsville, PA). Lysozyme was purchased from Seikagaku (Tokyo, Japan). Alkaline phosphatase (ALP) was purchased from Toyobo (Osaka, Japan). Lumi-phos 530 was obtained from Wako (Osaka, Japan). Sulfo-succinimidyl 6-[3'-(pyridyldithio)-propionamido]hexanoate (Sulfo-LC-SPDP) and sulfo succinimidyl 4-(*N*-maleimidomethyl)cyclohexane-1-carboxylate (Sulfo-SMCC) were purchased from Pierce (Rockford, IL).

Other reagents were commercially available analytical reagents or laboratory-grade materials. Deionized distilled water was used in all procedures.

Preparation of BMPs and Conjugation of Antibody onto BMPs

BMPs were isolated from the magnetic bacterium *Magnetospirillum magneticum* AMB-1 by a method previously described *(11)*. The BMP-antibody conjugation was performed according to Tanaka and Matsunaga *(12)*. Sulfo-SMCC (1 mg) was added to 500 μL of antibody (anti-lysozyme IgG antibody) solution (1 mg/mL of phosphate-buffered saline [PBS]) and incubated for 0.5 h at room temperature. The sample was purified on an NAP-5 column (Amersham Pharmacia Biotech) according to the manufacturer's instructions. Sulfo-LC-SPDP (10 mM) was added to 1 mL of BMP suspension (1 mg/mL) and incubated for 0.5 h at room temperature. After incubation, modified BMPs were isolated magnetically from the reaction mixture using a neodymium-boron (Nd-B) magnet (produced by TDK, Tokyo, Japan) and washed three times with 1 mL of PBS. The modified BMPs were dispersed in 1 mL of 20 mM dithiothreitol in acetate buffer (100 mM sodium acetate; 100 mM NaCl, pH 4.5) and incubated for 0.5 h at room temperature. After washing, BMPs having thiol groups were incubated with the sulfo-SMCC-modified antibody solution for 20 h at 4°C. Antibody-conjugated BMPs (antibody-BMPs) were washed three times with PBS to remove excess antibody.

Preparation of ALP-Conjugated Antibody

A mixture containing 500 μL of ALP from calf intestine and aqueous anti-lysozyme antibody (500 μL; 1 mg/mL) was dialyzed for 12 h against PBS. ALP was crosslinked to antibodies by the addition of 25% (v/v) glutaraldehyde solution (8 μL) and subsequently incubated for 2 h at room temperature. The solution was dialyzed against PBS (1000 mL) for 12 h and then against Tris-HCl buffer, pH 8.0, for 24 h.

Sandwich Immunoassay Using Automated System

The sandwich immunoassay was performed with an automated immunoassay system *(12)*. The automated immunoassay system contains a reaction station, tip rack, and an eight-tip pipettor that is able to attach and detach a strong magnet to the tip surface for use with 96-well microtiter plates mounted in the reaction station. Each reagent is dispensed in each eight-well row of the microtiter plate. Figure 1 illustrates the assay.

Lysozyme solution (50 μL) that was prepared in washing buffer (PBS containing 0.1% bovine serum albumin and 0.05% Tween-20) was added to 50 μL of antibody-conjugated BMPs. The mixture was dispersed by the pipettor and incubated at 30°C (the first immunoreaction). Immunocomplexes were separated magnetically using an Nd-B magnet and then washed by automated resuspension (20 cycles of pipet action) in 100 μL of washing buffer. To the washed complexes, 100 μL of ALP-antibody was added and dispersed with the automated pipet and incubated (the second

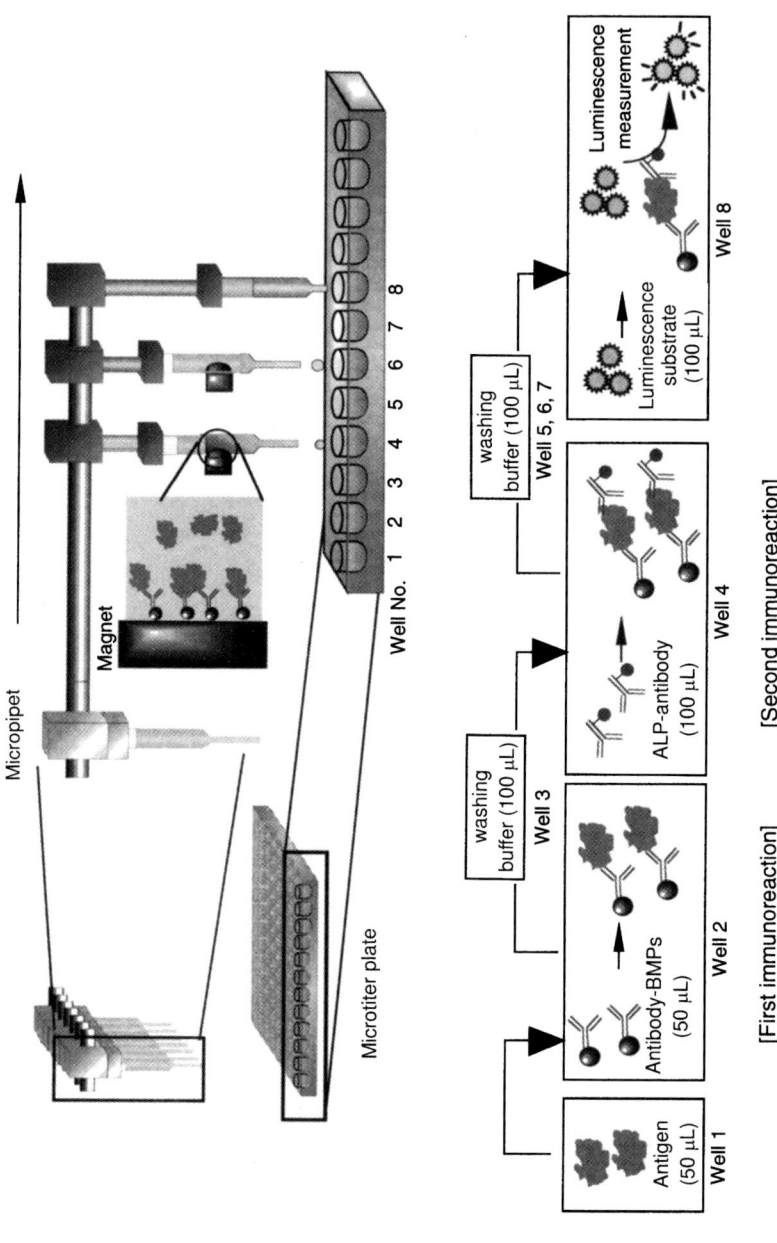

Fig. 1. Fully automated sandwich immunoassay system using antibody-BMPs and ALP-antibody.

immunoreaction). After incubation, antigen-antibody complexes were separated magnetically and washed three times by repeated pipetting in washing buffer. Antigen-antibody complexes were finally suspended in 100 μL of Lumi-phos 530, and the luminescence intensity was measured.

Optimization of the amount of antibody-BMPs added into lysozyme solution was first carried out with the immunoreaction for 20 min and the subsequent immunoreaction with 100 μL of ALP-conjugated antibody solution (10 μg/mL) for 20 min. The amount of ALP-antibody in the second immunoreaction was also optimized after the first immunoreaction between lysozyme and antibody-BMPs for 20 min. The time for the first immunoreaction was fixed under the optimized conditions for amounts of antibody-BMPs and ALP-antibody in which the second immunoreaction was performed for 20 min.

The precision (coefficient of variation [CV]) of this assay method was calculated as follows: standard deviation/mean value × 100.

Results and Discussion

Optimum Assay Conditions
for Rapid Fully Automated Sandwich Immunoassay

Figure 2 shows the relationship between luminescence intensity and BMP concentration in the sandwich immunoassay using the automated system. The luminescence increased with increasing antibody-BMP concentration in the range between 100 and 600 μg of BMP/mL. However, the increase was not proportionate to antibody-BMP concentration. A large number of antibodies normally result in faster immunoassay completion. Excess antibody-BMPs in solution, however, physically block luminescence. This phenomenon in Fig. 2 may be a result of BMPs blocking the luminescence or particle aggregation. The most effective amount of antibody-BMPs was 600 μg/mL in this assay without the influence of BMP shadowing on luminescence at measurement. Therefore, in the immunoassay a concentration of 600 μg/mL of BMPs was used.

ALP-antibody concentration affects the luminescence based on the amount of ALP-antibody bound to antigen-antibody-BMP conjugates. Our previous reports *(11,12)* indicated that higher ALP-antibody concentrations induced high background in immunoassays using BMPs; therefore, sandwich immunoassays using 0–100 μg/mL of ALP-antibody were performed to determine the level of ALP-antibody to employ in the assay. Luminescence increased with increasing ALP-antibody concentrations to 80 μg/mL. Therefore, 80 μg/mL of ALP-antibody was used for further experiments (data not shown).

To optimize the immunoreaction time, the time course of luminescence intensity based on the reaction of lysozyme and antibody-BMPs was investigated. The luminescence intensity in the presence of 10 μg/mL of lysozyme increased up to the reaction time of 20 min and remained constant for times longer than 10 min (Fig. 3). The immunoreaction time was

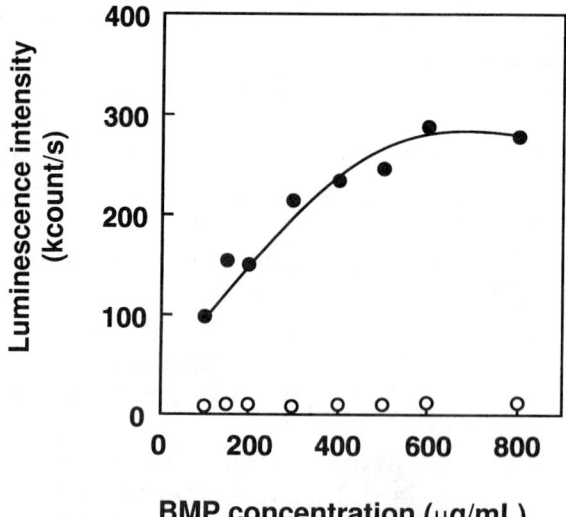

Fig. 2. Correlation between antibody-BMP concentrations and luminescence intensity. Lysozyme solution (50 µL: ○, 0 µg/mL; ●, 10 µg/mL) was mixed with 50 µL of antibody-BMP solution for 20 min. Antigen-antibody complex was collected with an Nd-B magnet and washed with PBS. BMPs were collected and dispersed by pipetting in 100 µL of ALP-conjugated antibody solution (10 µg/mL). Antigen- antibody complexes were dispersed in 100 µL of Lumi-phos 530 and luminescence intensity was measured.

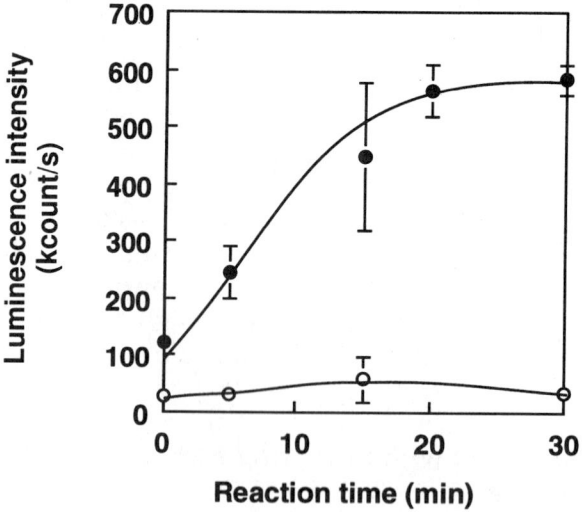

Fig. 3. Time course of the luminescence intensity in immunoreaction based on the lysozyme and antibody-BMP. Lysozyme solution (50 µL: ○, 0 µg/mL; ●, 10 µg/mL) was mixed with antibody-BMPs (50 µL, 600 µg/mL). BMPs were dispersed by pipetting in 100 µL of ALP-conjugated antibody solution (80 µg/mL) for 20 min. Antigen-antibody complexes were dispersed in 100 µL of Lumi-phos 530 and luminescence intensity was measured.

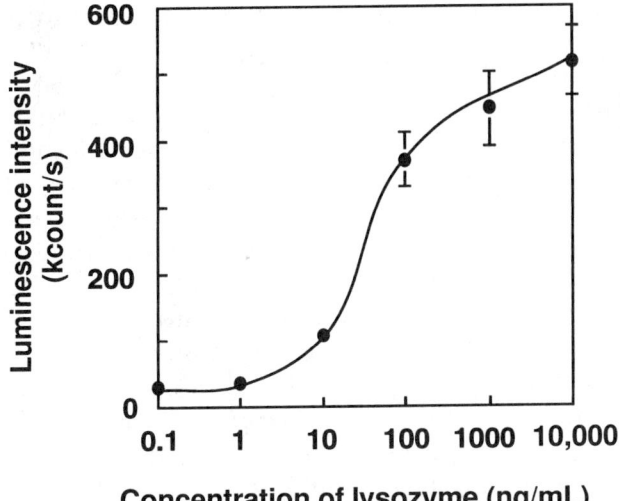

Fig. 4. Relationship between lysozyme concentration and luminescence intensity. The first and second immunoreactions were carried out for 5 min. Lysozyme solution (50 µL) was mixed with antibody-BMPs (50 µL, 600 µg/mL). BMPs were dispersed by pipetting in 100 µL of ALP-conjugated antibody solution (80 µg/mL). Antigen-antibody complexes were dispersed in 100 µL of Lumi-phos 530 and luminescence intensity was measured.

fixed at 5 min in the assay, because the minimum detectable limit obtained with the immunoreaction times at 5 min was not significantly different from that at 20 min (data not shown).

Fully Automated Sandwich Immunoassay for Lysozyme

The relationship between luminescence intensity and concentration of lysozyme was investigated using anti-lysozyme antibody-BMPs and ALP-labeled anti-lysozyme antibody (Fig. 4). Luminescence intensity increased with increasing concentration of lysozyme solution. Luminescence intensity was dependent on lysozyme concentration in the range between 10 ng/mL and 100 µg/mL. The minimum detectable concentration of lysozyme was 10 ng/mL. Luminescence intensity at <1 ng/mL did not show a significant difference from that of a lysozyme concentration of 0 ng/mL. This limit is acceptable for the detection of food allergens *(16)*. In this assay, the second immunoreaction time was fixed for 5 min, because the minimum detectable concentration of lysozyme when using an immunoreaction time of 5 min was equal to that for 20 min.

One assay is completed within 20 min as follows: magnetic separation steps and pipetting steps, 8.5 min; first immunoreaction time, 5 min; second immunoreaction time, 5 min; and luminescence measurement steps, 1.5 min. The fully automated immunoassay system was equipped with an eight-tip pipettor, and, therefore, this system can assay 24 samples within 60 min. The precision (CV) at each lysozyme concentration was within 10%.

Lysozyme, ovalbumin, ovotransferrin, and ovomucoid are major allergens in egg white. In the present study, we constructed an assay system for lysozyme because lysozyme is found in the lowest concentration of the four in egg white and the assay for lysozyme required higher sensitivity than that for the other three allergens. Detection methods for allergens in food must be sufficiently sensitive to detect trace amounts (<10 ppm) of the offered food *(16)*. In this system, the minimum detectable concentration of lysozyme was 10 ng/mL, and it is sufficiently sensitive for the detection of lysozyme in food.

In conclusion, a fully automated chemiluminescence sandwich enzyme immunoassay for a food allergen lysozyme using antibody-BMP complexes was developed. This fully automated system allows rapid detection of lysozyme, within 20 min for eight samples. This system could be applicable to the multiple detection of antigens. Future studies must be directed toward detecting lysozyme in foods using this assay system.

References

1. Ebbehøj, K., Dahl, A. M., Frøkiær, H., Nørggard, A., Poulsen, L. K., and Barkholt, V. (1995), *Allergy* **50**, 133–141.
2. Lehrer, S. B., McCants, M. L., and Salvaggio, J. E. (1985), *Int. Arch. Allergy Appl. Immunol.* **77**, 192–194.
3. Koppelman, S. J., Knulst, A. C., Koers, W. J., Penninks, A. H., Peppelman, H., Vlooswijk, R., Pigmans, I., Duijn, G., and Hessing, M. (1999), *J. Immunol. Methods* **299**, 107–120.
4. Kushimoto, H. and Aoki, T. (1985), *Arch. Dermatol.* **121**, 355–360.
5. Blakemore, R. P. (1975), *Science* **190**, 377–379.
6. Matsunaga, T., Tadokoro, F., and Nakamura, N. (1990), *IEEE Trans. Magnet.* **26**, 1557–1559.
7. Matsunaga, T., Sakaguchi, T., and Tadokoro, F. (1991), *Appl. Microbiol. Biotechnol.* **35**, 651–655.
8. Sakaguchi, T., Burgess, J. G., and Matsunaga, T. (1993), *Nature (Lond.)* **365**, 47–49.
9. Nakamura, N., Hashimoto, K., and Matsunaga, T. (1991), *Anal. Chem.* **63**, 268–272.
10. Balkwill, D. L., Maratea, D., and Blekemore, R. P. (1980), *J. Bacteriol.* **141**, 1399–1408.
11. Matsunaga, T., Kawasaki, M., Yu, X., Tsujimura, N., and Nakamura, N. (1996), *Anal. Chem.* **68**, 3551–35454.
12. Tanaka, T. and Matsunaga, T. (2000), *Anal. Chem.* **72**, 3518–3522.
13. Takeyama, H., Tuzuki, H., Chow, S., Nakayama, H., and Matsunaga, T. (2000), *Marine Biotechnol.* **2**, 309–313.
14. Matsunaga, T., Nakayama, H., Okochi, M., and Takeyama, H. *Biotechnol. Bioeng.*, in press.
15. Holen, E. and Elsayed, S. (1990), *Int. Arch. Allergy Appl. Immunol.* **91**, 136–141.
16. Taylor, S. L. and Nordlee, J. A. (1996), *Food Technol.* **61**, 231–238.

Fermentation Performance Assessment of a Genomically Integrated Xylose-Utilizing Recombinant of *Zymomonas mobilis* 39676

HUGH G. LAWFORD* AND JOYCE D. ROUSSEAU

Bio-engineering Laboratory, Department of Biochemistry, University of Toronto, Toronto, Ontario, Canada M5S 1A8, E-mail: hugh.lawford@utoronto.ca

Abstract

In pH-controlled batch fermentations with pure sugar synthetic hardwood hemicellulose (1% [w/v] glucose and 4% xylose) and corn stover hydrolysate (8% glucose and 3.5% xylose) lacking acetic acid, the xylose-utilizing, tetracycline (Tc)-sensitive, genomically integrated variant of *Zymomonas mobilis* ATCC 39676 (designated strain C25) exhibited growth and fermentation performance that was inferior to National Renewable Energy Laboratory's first-generation, Tc-resistant, plasmid-bearing *Zymomonas* recombinants. With C25, xylose fermentation following glucose exhaustion was markedly slower, and the ethanol yield (based on sugars consumed) was lower, owing primarily to an increase in lactic acid formation. There was an apparent increased sensitivity to acetic acid inhibition with C25 compared with recombinants 39676:pZB4L, CP4:pZB5, and ZM4:pZB5. However, strain C25 performed well in continuous fermentation with nutrient-rich synthetic corn stover medium over the dilution range 0.03–0.06/h, with a maximum process ethanol yield at $D = 0.03$/h of 0.46 g/g and a maximum ethanol productivity of 3 g/(L·h). With 0.35% (w/v) acetic acid in the medium, the process yield at $D = 0.04$/h dropped to 0.32 g/g, and the maximum productivity decreased by 50% to 1.5 g/(L·h). Under the same operating conditions, rec Zm ZM4:pZB5 performed better; however, the medium contained 20 mg/L of Tc to constantly maintain selective pressure. The absence of any need for antibiotics and antibiotic resistance genes makes the chromosomal integrant C25 more compatible with current regulatory specifications for biocatalysts in large-scale commercial operations.

Index Entries: Recombinant *Zymomonas* C25; genomic integrant; xylose; ethanol; biomass hydrolysate; acetate inhibition.

*Author to whom all correspondence and reprint requests should be addressed.

Introduction

For the past two decades our laboratory has been involved in research and development associated with the production of fuel ethanol—more specifically batch and continuous ethanol fermentations using both wild-type and genetically engineered bacteria and a variety of feedstocks *(1–6)*. Recently, we have been investigating the physiologic characteristics of xylose-utilizing recombinants of *Zymomonas mobilis* that were created at National Renewable Energy Laboratory (NREL) *(7,8)* with a view to their efficacy in the production of cellulosic ethanol *(9–16)*. To be economic, the production of ethanol from cellulosic biomass and wastes must involve the rapid and efficient conversion of both hexose and pentose sugars *(17)*. To exploit the superior fermentation characteristics of *Z. mobilis* for the production of cellulosic ethanol, this bacterium has been genetically engineered to broaden its substrate utilization profile to include pentose sugars *(7,18,19)*. In NREL's first-generation strains, the ability to ferment xylose and arabinose was accomplished using a native Zm plasmid vector with inserted *Escherichia coli* genes coding for both pentose assimilation and pentose phosphate pathways together with an antibiotic resistance gene to facilitate selection *(7,20,21)*. In the more recent second-generation constructs *(22)*, genetic stability has been enhanced in the absence of antibiotic selection through the genomic integration of the *E. coli* xylose-fermenting genes. One of the integrants that was derived from *Z. mobilis* ATCC 39676 lacked the tetracycline (Tc) resistance gene and was designated as strain C25 by NREL *(20)*.

The purpose of the present study was to assess the fermentation performance characteristics of a selected xylose-fermenting integrant, strain C25, in both pH-controlled batch and continuous fermentations using pure sugar synthetic biomass hydrolysate media under conditions similar to those previously employed to study plasmid-bearing *Zymomonas* recombinants *(9–16)*.

Materials and Methods

Organisms

The xylose-utilizing, plasmid-bearing, Tc-resistant, recombinant *Z. mobilis* strains 39676:pZB4L, CP4:pZB5, ZM4:pZB5 *(7,8)*, and the Tc-sensitive genomically integrated strain C25 (derived from Zm ATCC 39676) *(19)* were obtained from M. Zhang (NREL, Golden, CO). Stock cultures were stored in glycerol at –70°C and precultures were prepared as described previously *(9)*.

Preparation of Inoculum

A 1-mL aliquot of a glycerol preserved culture was removed from cold storage (freezer) and transferred to about 100 mL of "modified" RM medium (5 g/L of yeast extract and 2 g/L of KH_2PO_4) containing about 20 g/L of xylose and 20 g/L of glucose in 125-mL screw-cap flasks and grown statically overnight at 30°C in an incubator. The medium was supplemented with 10 mg/L Tc when using Tc-resistant strains. This preseed was subcultured into inoculation flasks containing modified RM with 20 g/L of glucose, 20 g/L of xylose, and 10 mg/L Tc when appropriate and grown statically overnight at 30°C in an incubator. This overnight culture was used at a level of ~10% (v/v) to inoculate the batch fermentors. The initial optical density (1-cm light path at 600 nm) was in the range of 0.2–0.25, corresponding to 60–75 mg of dry cell mass (DCM)/L.

Fermentation Medium

The fermentation medium (designated as ZM) *(15)* was prepared with glass-distilled water and contained the following ingredients: 5 g/L of Difco yeast extract (Difco, Detroit, MI), 3.48 g/L of KH_2PO_4, 0.25 g/L of $MgSO_4$, 0.01 g/L of $FeSO_4 \cdot 7H_2O$, 0.21 g/L of citric acid, and 20 mg/L of Tc (for Tc-resistant cultures). The amount of glucose, xylose, and acetic acid added to the medium was variable. The medium and stock sugar solutions were autoclaved separately.

Fermentation Equipment

pH-stat batch fermentations were conducted with about 1500 mL of medium in 2-L bioreactors (model F2000 MultiGen; New Brunswick Scientific, Edison, NJ) fitted with agitation (150 rpm), pH, and temperature control (30°C). Continuous fermentations were conducted with either NBS C30 BioFlo chemostats or 2-L NBS Bioflo 2000 bioreactors. The working volume of these chemostats was about 350 and 1500 mL, respectively. Steady state was assumed only after a minimum of 3 vol had exchanged and when samples of effluent taken on successive days gave similar values for cell mass, and sugar and ethanol concentrations. The pH was monitored using a sterilizable combination pH electrode (Ingold). The standard pH control set point was either 5.75 or 6.0, and the pH was kept constant by automatic titration with 4 N KOH. In some batch fermentations the pH set point was adjusted after 24 h from 5.75 to 6.5. The temperature was controlled at 30°C using a circulating water bath and the agitation was moderate (approx 100–150 rpm). The continuous fermentations were started in the batch mode using ZM medium with concentrations of glucose and xylose as determined by the condition specified for individual experiments. Flow was started 24 h after inoculation (preferably when the residual xylose concentration was <10 g/L) *(16)*.

Analytical Procedures, Growth, and Fermentation Parameters

Growth was measured turbidometrically at 600 nm (1-cm light path). In all cases, the blank cuvet contained distilled water. DCM was determined by microfiltration of an aliquot of culture followed by washing and drying of the filter to constant weight under an infrared heat lamp. Fermentation media and cell-free spent media were compositionally analyzed by

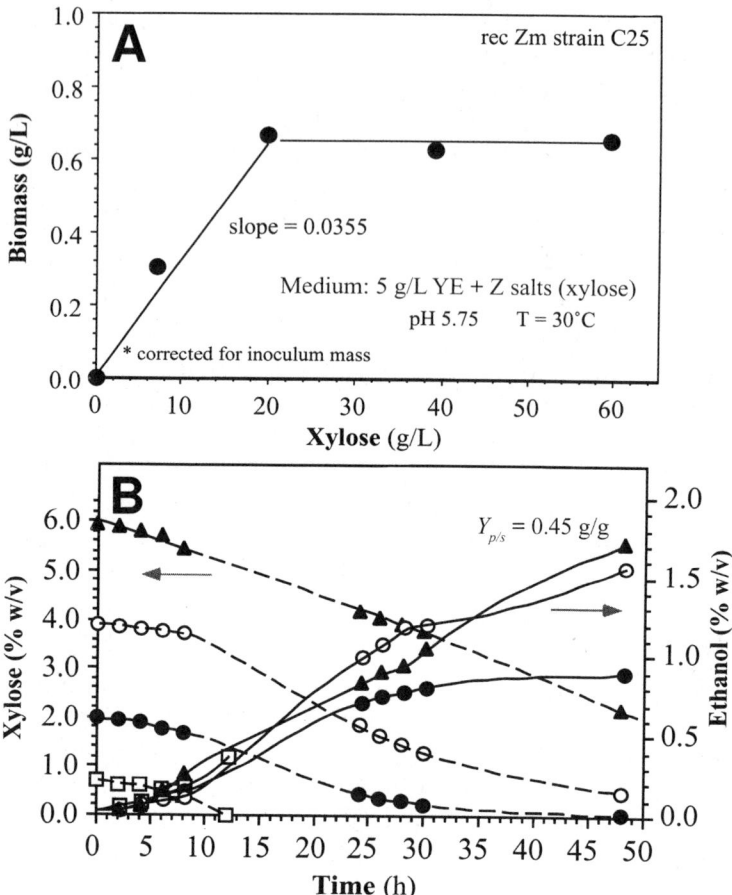

Fig. 1. Growth and fermentation of xylose as the sole sugar by rec Zm strain C25. **(A)** Cell mass as a function of xylose in the medium; **(B)** time course of xylose utilization and ethanol production: (□) 0.8% (w/v) xylose, (●) 2% xylose, (○) 4% xylose, (▲) 6% xylose.

high-performance liquid chromatography as described previously *(9)*. The ethanol yield $(Y_{p/s})$ was calculated as the mass of ethanol produced per mass of sugar consumed; the "process" yield (proc $Y_{p/s}$) was calculated as the mass of ethanol produced per mass of fermentable sugar in the medium (i.e., glucose + xylose).

Results and Discussion

Batch Fermentations

Experiments with Strain C25 and Xylose as Sole Fermentable Sugar

The growth characteristics of strain C25 with xylose as the sole fermentable sugar (Fig. 1) were of interest to us because previously we reported on the growth yield for the plasmid-bearing rec Zm CP4:pZB5 and inferred that the Y_{ATP} was significantly less for xylose than glucose *(15)*. With recombinant CP4:pZB5, at relatively low xylose concentrations (<20 g/L), the final cell mass was proportional to the amount of sugar in the medium *(15)*. Despite the paucity of data, Fig. 1A shows that this also holds for C25. In previous studies with plasmid-bearing recombinants, the maximum cell mass with xylose alone was about 0.75 g of DCM/L *(15)*, and this is also the case with strain C25 (Fig. 1A). The reason for this plateauing of biomass is not yet fully resolved. However, in the context of ethanol production, perhaps the most notable feature of this experiment with strain C25 is that the ethanol yield was only 0.45 g/g (based on sugar used) and the process yield at 48 h was lower with 6% xylose because the fermentation is incomplete (Fig. 1B). Previous work with NREL's xylose-utilizing Zm recombinants by us *(9,11)* and others *(7,23–29)* had shown that these cultures characteristically exhibit a high ethanol yield based on sugar consumed with perhaps xylitol being a byproduct of concern *(24)* because of its putative inhibition of xylulose kinase *(15)*. The reason for the decrease in ethanol yield with C25 relative to the nonintegrated recombinants is primarily that there is increased production of lactic acid (results not shown). Figure 2 illustrates the effect of pH on xylose utilization by strain C25. When the pH set point on the controller was adjusted from 5.75 to 6.5 at 24 h, the final cell density increased from 0.74 to 0.92 g of DCM/L (Fig. 2A), and the process yield for ethanol at 48 h increased from 0.38 to 0.45 g/g (Fig. 2B).

Effect of Glucose on Xylose Utilization

By far the majority of our previous work with rec Zm has been done with synthetic hardwood hemicellulose hydrolysate in which the concentration of xylose was 4% (w/v) and glucose was 0.8–1% (w/v). Figure 2 also shows the growth and fermentation performance of C25 in this pure sugar mixture, which was formulated to mimic the NREL hardwood prehydrolysate. The addition of glucose results in an increase in cell density (Fig. 2A), which probably accounts for the increase in xylose utilization (Fig. 2B). Note also that C25 performs well with equal concentrations of xylose and glucose, although above 3.5% (w/v) of each sugar, xylose utilization is not complete at 48 h (results not shown). It is important to understand that these fermentations were conducted in the absence of acetic acid in the medium and, therefore, are not representative of true fermentation performance in unconditioned biomass hydrolysate medium.

Comparative Effect of Acetic Acid on C25
and Other Xylose-Utilizing rec Zm Strains

Acetic acid is a component of biomass hydrolysates and its inhibitory effects on ethanologenic microorganisms are well documented *(30)*. There are several reports on the effect of acetic acid on rec Zm *(11,14,24)*. In the present study, the pure sugar synthetic hardwood prehydrolysate medium containing 4% xylose, 1% glucose, and varying amounts of acetic acid (HAc) over the range 0–1% (w/v) was used to assess the growth and fermentation performance of different Zm recombinants. We were interested to compare

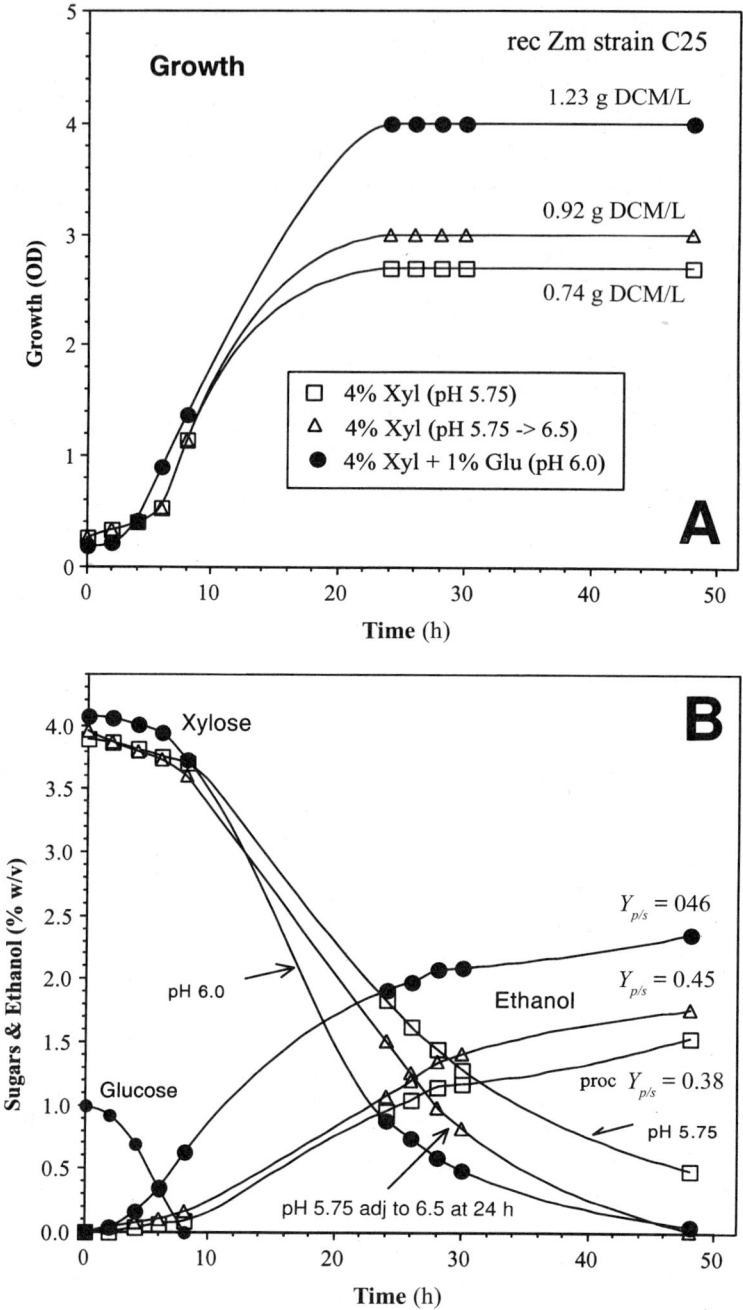

Fig. 2. Effect of pH and glucose on xylose fermentation by rec Zm C25. **(A)** Growth;
(B) sugar utilization and ethanol production.

the effect of acetic acid on strain C25 and a closely related strain, 39676:
pZB4L, and a nonrelated strain, CP4:pZB5. Under these assay conditions,
we observed that C25 behaved quite similarly in terms of growth (as final

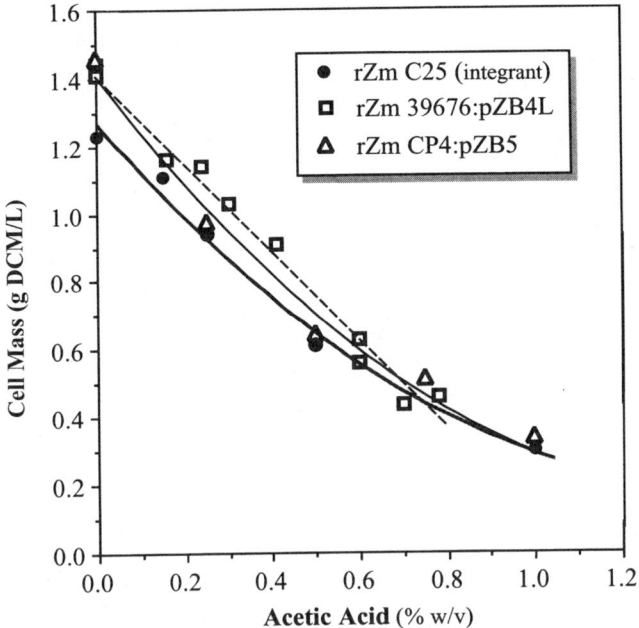

Fig. 3. Comparative relationship between final cell mass concentration and level of acetic acid in the medium for three xylose-utilizing *Zymomonas* recombinants.

cell density) to the plasmid-bearing recombinants 39676:pZB4L and CP4:pZB5 (Fig. 3). In terms of fermentation performance, Fig. 4 shows that strain C25 is more sensitive to inhibition of xylose utilization by acetic acid than the closely related strain 39676:pZB4L. In terms of xylose utilization, recombinant CP4:pZB5 is the most HAc tolerant strain that was tested in this series of experiments with the synthetic hardwood prehydrolysate medium (Fig. 5).

Batch Fermentations with Synthetic Corn Stover Hydrolysate

Various agricultural wastes are being considered for the production of cellulosic ethanol, and among these corn stover is a strong contender. Our work with corn stover fermentations was initiated when Iogen selected this as one of the feedstocks to be processed in its demonstration facility *(31)*. In a separate but closely related study on oat hull hydrolysate, we concluded that rec Zm ZM4:pZB5 was the best of the xylose-utilizing NREL Zm recombinants tested *(32)*. The concentration of glucose and xylose in the Iogen corn stover hydrolysate is 8% glucose and 3.5% xylose and the acetic acid is about 1% (w/v). Proprietary technologies will be employed by Iogen to reduce the acetate content to 0.3% or less (w/v) (Tolan, J., personal communication). The objective of the present study was to compare the effect of acetic acid on the growth and fermentation performance of ZM4:pZB5 and the integrant C25 using pure sugar synthetic corn stover hydrolysate (8% glu + 3.5% xyl) in which the pH was controlled within the range 5.75–6.5. Figure 6 shows the relationship between final cell mass and

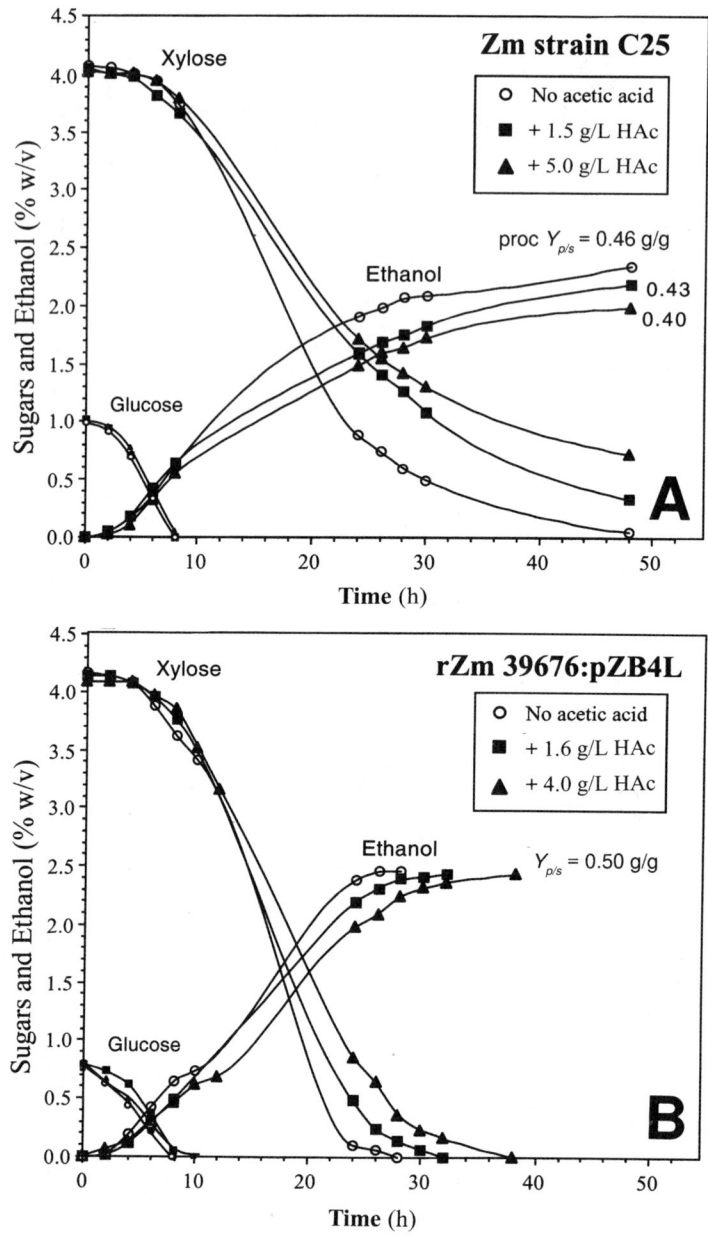

Fig. 4. Sugar utilization and ethanol production using a pure sugar synthetic hardwood prehydrolysate medium. **(A)** Strain C25; **(B)** rec Zm 39676:pZB4L.

the level of acetic acid for ZM4:pZB5 and C25. Over the range of acetate tested, the growth yield for C25 was less than for ZM4:pZB5 (Fig. 6).

Figure 7 shows the effect of acetic acid on the cofermentation performance of these two recombinants. Even in the absence of acetic acid, strain C25 was unable to complete the fermentation within 48 h (Fig. 7A). The rate

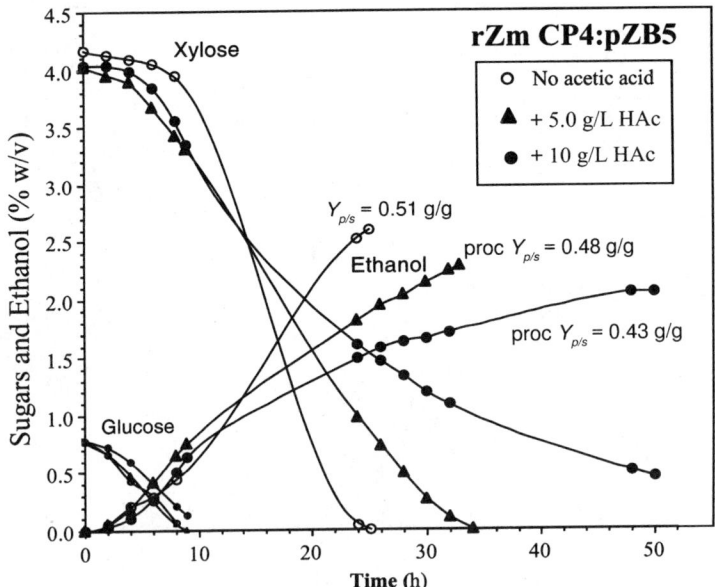

Fig. 5. Sugar utilization and ethanol production by rec Zm CP4:pZB5 using a pure sugar synthetic hardwood prehydrolysate medium. Compare with experiments shown in Fig. 4. Note that higher levels of acetic acid were used in these fermentations with CP4:pZB5.

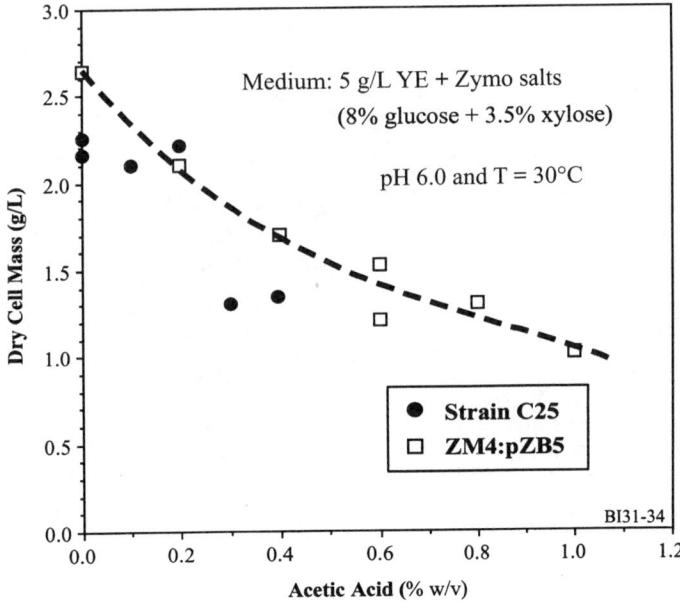

Fig. 6. Comparative relationship between final cell mass concentration and level of acetic acid in a synthetic corn stover hydrolysate medium for *Zymomonas* recombinants ZM4:pZB5 and C25.

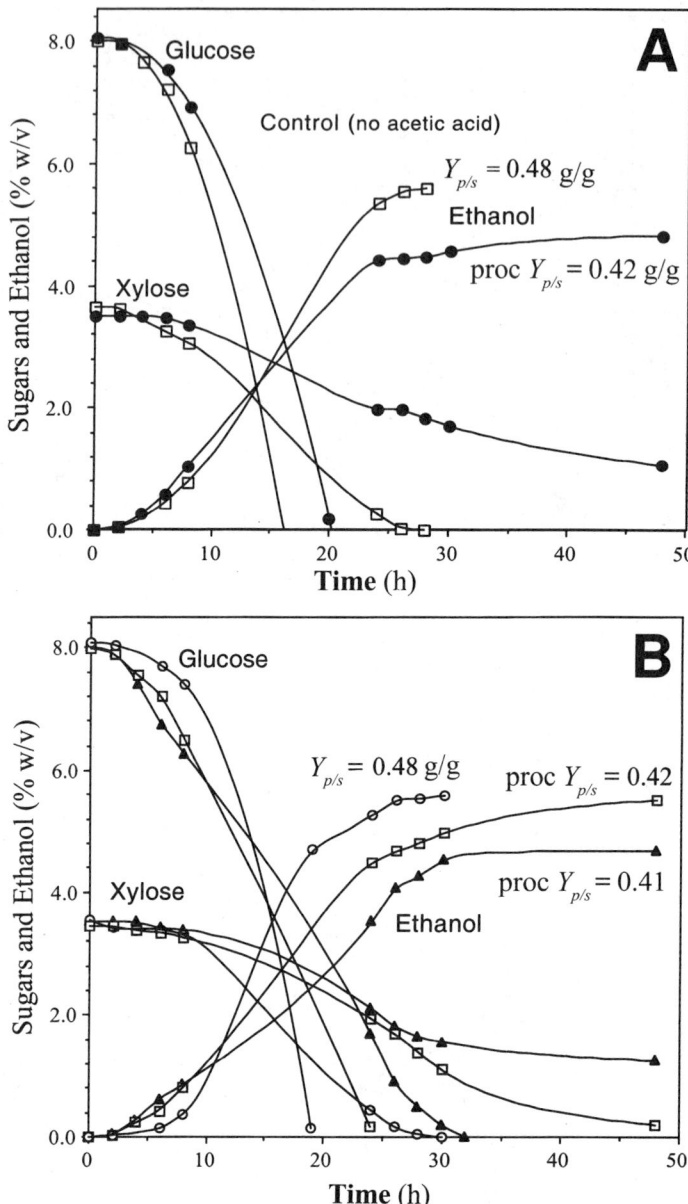

Fig. 7. Time course of sugar utilization and ethanol production using a synthetic corn stover hydrolysate medium and Zm recombinants ZM4:pZB5 and C25. **(A)** Without added acetic acid: (●) strain C25 (no HAc), (□) ZM4:pZB5 (no HAc); **(B)** with added acetic acid (▲) strain C25 (0.4% HAc), (○) ZM4:pZB5 (0.4% HAc), (□) ZM4:pZB5 (0.8% HAc).

of xylose utilization with strain C25 was markedly slower following glucose exhaustion (Fig. 7A). However, ZM4:pZB5 completed the batch fermentation in 24 h with an excellent product yield of 0.48 g/g (Fig. 7A).

At a level of 4 g/L of HAc, it is the rate of glucose utilization that is affected with both strains, whereas the rate of xylose utilization for both strains is relatively unaffected (Fig. 7). With ZM4:pZB5, the fermentation is essentially complete in 48 h even with 8 g/L of acetic acid in the medium (Fig. 7B). For C25 to achieve complete fermentation within 48 h, we recommend that the corn stover hydrolysate be diluted to 85% strength (i.e., 68 g/L of glucose and 30 g/L of xylose). These results point to the superiority of recombinant ZM4:pZB5 for batch fermentations with media that contain acetic acid.

Continuous Fermentations with rec Zm ZM4:pZB5 and C25: Pure Sugar Synthetic Corn Stover Hydrolysate (8% glu + 3.5% xyl)

Since Iogen is contemplating a continuous fermentation process using corn stover hydrolysate *(33)*, it was important to assess the performance of the chromosomal integrated recombinant in the continuous fermentation mode. Because strain C25 lacks the Tc resistance gene, it is more compatible with current regulatory requirements. Figure 8 shows the continuous fermentation of a pure sugar synthetic corn stover hydrolysate using strain C25. The medium was a standard 5 g/L of yeast extract with Zymo salts (ZM medium) *(15)*, the pH was controlled at 6.0, and the temperature was 30°C. Over the 21 d the chemostat was operated, the dilution rate was increased in increments of 0.01 from 0.03 to 0.06/h (Fig. 8). The maximum ethanol productivity was 3 g/(L·h). Figure 8B shows the steady-state levels of both ethanol and xylose; glucose was not detected in the effluent over this time period. At $D = 0.06$/h, approx 25% of the xylose was unfermented, the process yield was 0.434 g/g (85% efficiency), and the experiment was terminated. The effect of starting at a higher D value is not known, but since 0.06/h is probably close to washout with respect to xylose utilization, it would seem prudent to start at a lower D value and progress toward the upper flow rate of 0.06/h.

Figure 9 demonstrates the effect of adding 3.5 g/L of acetic acid to the nutrient-rich synthetic corn stover medium. After operating for about a week at $D = 0.03$/h, the residual xylose was about 3 g/L; however, when the dilution rate was increased to 0.04/h, the level of unfermented xylose increased to 25 g/L (Fig. 9A). The process yield at $D = 0.04$/h was 0.32 g/g and the productivity was 1.5 g/(L·h). Curiously, the level of xylose did not decrease after the dilution rate was lowered to 0.016/h (Fig. 9A); this suggested that the culture was somehow diminished in its capacity to utilize xylose. Steady-state values for effluent ethanol and xylose are shown in Fig. 9B. Glucose was not detected in the effluent. For comparison, Fig. 9B also shows data from a chemostat experiment with rec Zm ZM4:pZB5 under identical operating conditions. Note that with this Tc-resistant recombinant, Tc (20 mg/L) was added to the medium for maintenance of selective pressure. Although previously we have operated chemostats with other plasmid-bearing xylose-utilizing recombinants without Tc, the medium contained a higher proportion of xylose relative to glucose, and under this

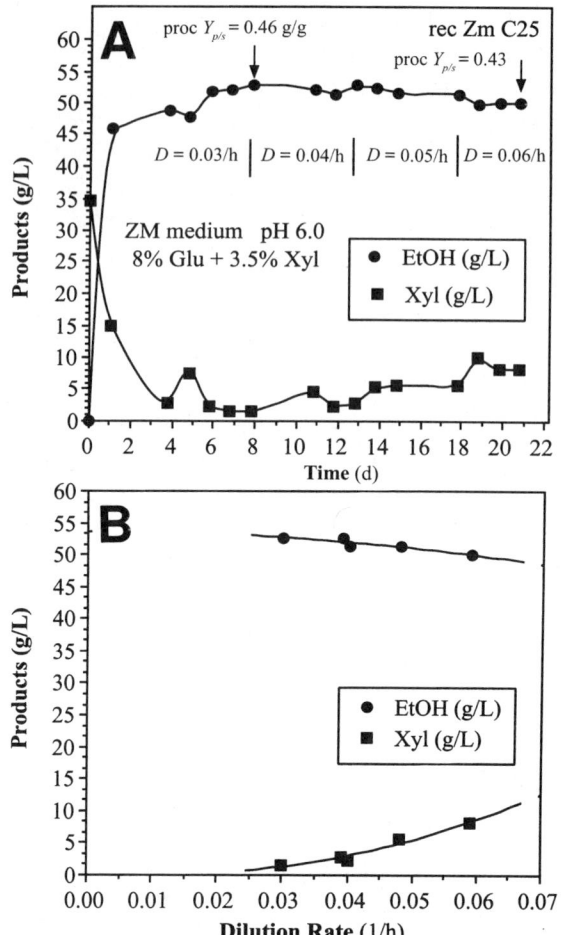

Fig. 8. Continuous fermentation of synthetic corn stover hydrolysate using rec Zm strain C25. **(A)** Time course of ethanol and residual xylose in effluent stream; **(B)** steady-state concentrations of ethanol and xylose as a function of dilution rate.

condition xylose would appear to provide the necessary selective pressure for sustained biocatalyst stability *(14,16)*. Some preliminary chemostat experiments with ZM4:pZB5 in the absence of Tc suggest that perhaps the high glucose-to-xylose ratio in this medium mitigates against the selective effect of xylose. The extent to which Tc is an absolute requirement of stability of ZM4:pZB5 with respect to xylose utilization has not yet been extensively investigated in our laboratory. The absence of any need for antibiotics and antibiotic resistance genes makes the genomic integrant C25 more compatible with current regulatory specifications for biocatalysts in large-scale commercial operations.

These results suggest that C25 could be used in a continuous fermentation process provided that the acetic acid concentration is not >3.5 g/L, the dilution rate is not >0.03/h, and the pH is controlled at about 6.0.

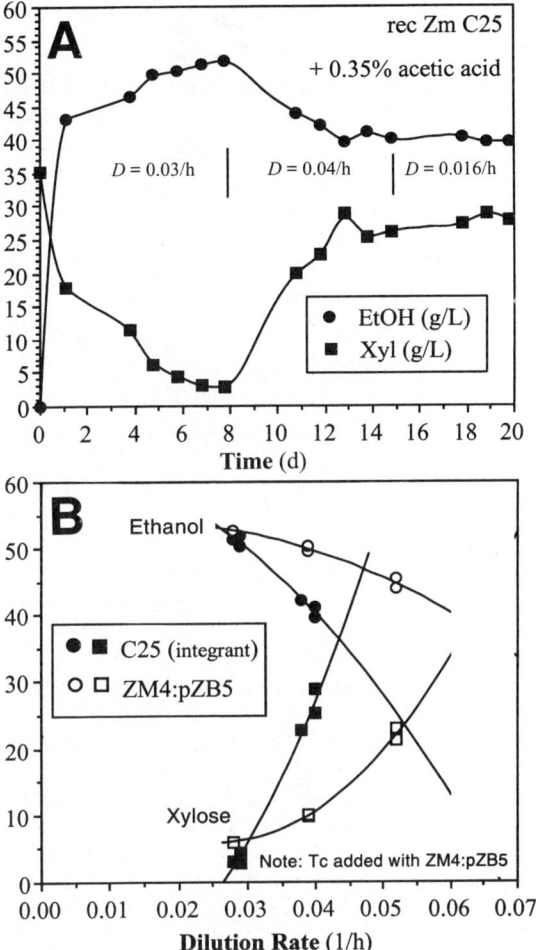

Fig. 9. Continuous fermentation of synthetic corn stover hydrolysate containing 0.35% acetic acid using rec Zm strain C25. **(A)** Time course of ethanol and residual xylose in effluent stream; **(B)** steady-state concentrations of ethanol and xylose as a function of dilution rate. Note that data from a similar experiment with rec Zm ZM4:pZB5 are added to (B) for comparison.

Unfortunately, these conditions are not compatible with current process design parameters being contemplated by Iogen in its new demonstration plant *(33)*.

Conclusion

In making a general conclusion for these collective observations with strain C25, it is important to emphasize that this study focused attention on only one of the many genomic type integrants that have been created *(34)*. The newest generation of integrants contains genes for arabinose as well as xylose fermentation *(35)*, although the fermentation performance of these recombinants has not yet been thoroughly tested. A promising alternative approach to random insertion that has been explored by NREL for integrating the pentose assimilation and metabolism genes into the Zm chromosome involves site-specific gene targeting for insertion *(22)*. Finally, in future genetic developments, it would seem prudent, in light of previous experience, to exploit the superior qualities of Zm ZM4 as a host for chromosomal integration experimentation.

Acknowledgments

We are grateful to Dr. Min Zhang (NREL) for providing the recombinant *Z. mobilis* strains. This work was funded jointly by Iogen (Ottawa, Canada) and the Biochemical Conversion Element of the Office of Fuels Development of the U.S. Department of Energy (NREL subcontract ZDH-9-29009-02).

References

1. Lawford, G. R., Lavers, B. H., Good, D., Charley, R. C., Fein, J. E. and Lawford, H. G. (1982), in *Proceedings of the International Symposium on Ethanol from Biomass*, Duckworth, H., ed., Royal Society of Canada, Ottawa, Canada, pp. 482–507.
2. Lawford, H. G. (1987), US Patent 4,647,534.
3. Lawford, H. G. (1988), in *VIII International Symposium on Alcohol Fuels*, New Energy and Industrial Technology Development Organization, Tokyo, Japan, pp. 21–27.
4. Lawford, H. G. (1988), in *Canadian Power Alcohol Conference*, Candlish, B., ed., Biomass Energy Institute, Winnipeg, Manitoba, Canada, pp. 245–251.
5. Lacis, L. S. and Lawford, H. G. (1989), in *Bioenergy—Proceedings of the 7th Canadian Bioenergy R&D Seminar*, Hogen, E., ed., NRC Canada, Ottawa, pp. 411–416.
6. Lawford, H. G. and Rousseau, J. D. (1991), in *Energy from Biomass and Wastes XV*, Klass, D. L., ed., Institute Gas Technology, Chicago, pp. 583–622.
7. Zhang, M., Eddy, C., Deanda, K., Finkelstein, M., and Picataggio, S. K. (1995), *Science* **267,** 240–243.
8. Picataggio, S., Zhang, M., Eddy, C. K., Deanda, K., and Finkelstein, M. (1996), US Patent 5,514,583.
9. Lawford, H. G., Rousseau, J. D., and McMillan, J. D. (1997), *Appl. Biochem. Biotechnol.* **63–65,** 269–286.
10. Lawford, H. G. and Rousseau, J. D. (1997), *Appl. Biochem. Biotechnol.* **63–65,** 287–304.
11. Lawford, H. G. and Rousseau, J. D. (1998), *Appl. Biochem. Biotechnol.* **70–72,** 161–172.
12. Lawford, H. G., Rousseau, J. D., Mohagheghi, A., and McMillan, J. D. (1998), *Appl. Biochem. Biotechnol.* **70–72,** 353–368.
13. Lawford, H. G. and Rousseau, J. D. (1999), *Appl. Biochem. Biotechnol.* **77–79,** 235–249.
14. Lawford, H. G., Rousseau, J. D., Mohagheghi, A., and McMillan, J. D. (1999), *Appl. Biochem. Biotechnol.* **77–79,** 191–204.
15. Lawford, H. G. and Rousseau, J. D. (2000), *Appl. Biochem. Biotechnol.* **84–86,** 277–294.
16. Lawford, H. G., Rousseau, J. D., Mohagheghi, A., and McMillan, J. D. (2000), *Appl. Biochem. Biotechnol.* **84–86,** 295–310.
17. Hinman, N. D., Wright, J. D., Hoagland, W., and Wyman, C. E. (1989), *Appl. Biochem. Biotechnol.* **20/21,** 391–401.
18. Sprenger, G. A. (1993), *J. Bacteriol.* **27,** 225–237.
19. Feldman, S. D., Sahm, H., and Sprenger, G. A. (1992), *Appl. Microbiol.* **38,** 354–361.
20. Picataggio, S. K., Zhang, M., Eddy, C. K., Deanda, K., and Finkelstein, M. (1998), US Patent 5,726,053.
21. Deanda, K. A., Eddy, C., Zhang, M., and Picataggio, S. (1996), *Appl. Environ. Microbiol.* **62,** 4465–4470.

22. Zhang, M., Chou, Y. C., Lai, X. K., Milstrey, S., Danielson, N., Evans, K., Mohagheghi, A., and Finkelstein, M. (1999), 21st Symposium on Biotechnology for Fuels and Chemicals, Fort Collins, CO (abstract no. 2-16).

23. Rogers, P. L., Joachimsthal, E. L., and Haggett, K. D. (1997), *J. Australasian Biotechnol.* **7,** 304–309.

24. Joachimsthal, E., Haggett, K. D., and Rogers, P. L. (1999), *Appl. Biochem. Biotechnol.* **77–79,** 147–157.

25. Krishnan, M. S., Blanco, M., Shattuck, C. K., Nghiem, N. P., and Davison, B. H. (2000), *Appl. Biochem. Biotechnol.* **84–86,** 525–542.

26. Joachimsthal, E. L. and Rogers, P. L. (2000), *Appl. Biochem. Biotechnol.* **84–86,** 343–356.

27. Dennison, E. and Abbas, C. (2000), 22nd Symposium on Biotechnology for Fuels and Chemicals, Gatlinburg, TN (abstract no. 2-04), Humana, Totowa, NJ.

28. Ngheim, N. P., Krishnan, M. S., Davison, B. H., Jackson, A. N., and Cofer, T. M. (2000), 22nd Symposium on Biotechnology for Fuels and Chemicals, Gatlinburg, TN (abstract no. 3-25), Humana, Totowa, NJ.

29. Dowe, N., Newman, M. M., Mohagheghi, A., and McMillan, J. D. (2000), 22nd Symposium on Biotechnology for Fuels and Chemicals, Gatlinburg, TN (abstract no. 6-20), Humana, Totowa, NJ.

30. McMillan, J. D. (1994), in *Enzymatic Conversion of Biomass for Fuels Production*, Himmel, M. E., Baker, J. O., and Overend, R. A., eds., ACS Symposium Series 566, American Chemical Society, Washington, DC, pp. 411–437.

31. Foody, B. F. (2000), 21st Symposium on Biotechnology for Fuels and Chemicals, Fort Collins, CO (abstract no. 6-01), Humana, Totowa, NJ.

32. Lawford, H. G., Rousseau, J. D., and Tolan, J. S. (2000), 22nd Symposium on Biotechnology for Fuels and Chemicals, Gatlinburg, TN, Humana, Totowa, NJ.

33. Foody, B. F. and Tolan, J. S. (2000), 22nd Symposium on Biotechnology for Fuels and Chemicals, Gatlinburg, TN (abstract no. 6-07), Humana, Totowa, NJ.

34. Zhang, M., Chou, Y. C., Mohagheghi, A., Evans, K., Milstrey, S., Lai, X. K., and Finkelstein, M. (2000), 22nd Symposium on Biotechnology for Fuels and Chemicals, Gatlinburg, TN (abstract no. 2-03), Humana, Totowa, NJ.

35. Zhang, M., Chou, Y.-C., Picataggio, S. K., and Finkelstein, M. (1998), US Patent 5,843,760.

Comparative Ethanol Productivities of Different *Zymomonas* Recombinants Fermenting Oat Hull Hydrolysate

Hugh G. Lawford,[*,1] Joyce D. Rousseau,[1] and Jeffrey S. Tolan[2]

[1]*Bio-engineering Laboratory, Department of Biochemistry, University of Toronto, 5303 Medical Sciences Building, Toronto, Ontario, Canada M5S 1A8, E-mail: hugh.lawford@utoronto.ca; and [2]IOGEN Corporation, 400 Hunt Club Road, Ottawa, Ontario, Canada K1V 1C1*

Abstract

Iogen Corporation of Ottawa, Canada, has recently built a 50 t/d biomass-to-ethanol demonstration plant adjacent to its enzyme production facility. Iogen has partnered with the University of Toronto to test the C6/C5 cofermentation performance characteristics of National Renewable Energy Laboratory's metabolically engineered *Zymomonas mobilis* using its biomass hydrolysates. In this study, the biomass feedstock was an agricultural waste, namely oat hulls, which was hydrolyzed in a proprietary two-stage process involving pretreatment with dilute sulfuric acid at 200–250°C, followed by cellulase hydrolysis. The oat hull hydrolysate (OHH) contained glucose, xylose, and arabinose in a mass ratio of about 8:3:0.5. Fermentation media, prepared from diluted hydrolysate, were nutritionally amended with 2.5 mL/L of corn steep liquor (50% solids) and 1.2 g/L of diammonium phosphate. The estimated cost for large-scale ethanol production using this minimal level of nutrient supplementation was 4.4¢/gal of ethanol. This work examined the growth and fermentation performance of xylose-utilizing, tetracycline-resistant, plasmid-bearing, patented, recombinant *Z. mobilis* cultures: CP4:pZB5, ZM4:pZB5, 39676:pZB4L, and a hardwood prehydrolysate-adapted variant of 39676:pZB4L (designated as the "adapted" strain). In pH-stat batch fermentations with unconditioned OHH containing 6% (w/v) glucose, 3% xylose, and 0.75% acetic acid, rec Zm ZM4:pZB5 gave the best performance with a fermentation time of 30 h, followed by CP4:pZB5 at 48 h, with corresponding volumetric productivities of 1.4 and 0.89 g/(L·h), respectively. Based on the available glucose and xylose, the process ethanol yield for both strains was 0.47 g/g (92% conversion efficiency). At 48 h, the

*Author to whom all correspondence and reprint requests should be addressed.

process yield for rec Zm 39676:pZB4L and the adapted strain was 0.32 and 0.34 g/g, respectively. None of the test strains was able to ferment arabinose. Acetic acid tolerance appeared to be a major determining factor in cofermentation performance.

Index Entries: Recombinant *Zymomonas*; oat hull hydrolysate; xylose; biomass hydrolysate; ethanol yield; acetic acid; productivity.

Introduction

Iogen Corporation of Ottawa, Canada, is a major manufacturer of industrial enzymes. Iogen primarily produces cellulase and hemicellulase enzymes for the textiles, pulp and paper, and poultry feed industries.

In April 2000, Iogen started operations of a 50-t/d biomass-to-ethanol demonstration plant *(1,2)*. The plant will produce ethanol from oat hulls, corn stover, and wheat straw. The plant is located at the site of Iogen's enzyme plant, which offers the advantages that the enzyme can be used without the expenses of stabilization and preservation, and that the process sugars can be used for enzyme production.

The "Iogen Process" for biomass depolymerization consists of a dilute sulfuric acid catalyzed steam explosion at 200–250°C, followed by enzymatic hydrolysis using cellulase enzymes. The process stream contains monomers of glucose, xylose, and arabinose in a mass ratio of about 8:3:0.5, with little sugar oligomers.

Yield and productivity are the key technoeconomic parameters in the production of fuel ethanol from biomass and wastes *(3)*. The fermentation of xylose to ethanol is important in a biomass-to-ethanol process because xylose fermentation has the potential to increase the ethanol yield by up to 50% at little additional cost *(3)*. Several microbes that have been engineered to ferment xylose to ethanol (for a review, *see* refs. *4* and *5*). In a survey of industrial biocatalysts to identify promising host strains for genetic transformation directed to rapid and efficient ethanologenic pentose metabolism, the Gram-negative bacterium *Zymomonas mobilis* met the selection criteria, which were based on several fermentation performance characteristics considered to be essential, as well as a number of secondary traits considered to be desirable, for a commercial biomass-to-ethanol process *(6,7)*. For example, *Z. mobilis* offers the advantage of having generally-regarded-as-safe status; however, wild-type strains are not capable of fermenting the pentose sugars that are produced during the hydrolysis of the hemicellulose component of cellulosic feedstocks. The impediment to exploiting the superior fermentation characteristics of *Zymomonas* was removed by the construction of recombinant strains that expressed *Escherichia coli* genes for pentose metabolism *(8,9)*. At the University of Toronto, we have been conducting a systematic physiological assessment of different National Renewable Energy Laboratory (NREL) recombinant strains using synthetic hydrolysates prepared with pure chemicals, and our observations with both batch and continuous systems have been the subject of several presentations at previous symposia *(10–17)*. In anticipation of this plant start-up, Iogen collaborated with the University of Toronto to determine the capabilities of the NREL recombinant *Zymomonas* strains to ferment xylose from Iogen's sugar stream to ethanol.

The purpose of the present study was to compare the batch fermentation performance characteristics of several different xylose-utilizing recombinants of *Z. mobilis* using appropriately diluted biomass hydrolysate that had been nutritionally amended in a cost-effective fashion. Iogen's demonstration plant is configured for both batch and continuous fermentations, but this laboratory study was carried out using batch culture. This study focused on one of Iogen's feedstocks, oat hulls, as the first that was obtained by Iogen in large quantities.

Materials and Methods

Microorganisms

Xylose-utilizing *Z. mobilis* recombinants 39676:pZB4L *(8)*, a hardwood hydrolysate-adapted variant of 39676:pZB4L (designated as rec Zm "adapted") *(13,15)*, CP4:pZB5 *(8)*, and ZM4:pZB5 (also known as 31821: pZB5) *(18)* were obtained from Min Zhang at NREL. Cryovials of frozen concentrated stock culture were maintained in RM medium (10 g/L of yeast extract and 2 g of KH_2PO_4) *(19)* supplemented with 10 mg/L of tetracycline (Tc) and 15% (w/w) glycerol at –70°C.

Preparation of Inoculum

A 1-mL aliquot of a glycerol preserved culture was removed from cold storage (freezer) and transferred to about 100 mL of modified RM medium (5 g/L of yeast extract and 2 g/L of KH_2PO_4) containing about 20 g/L of xylose and 20 g/L of glucose supplemented with Tc (10 mg/L) in 125-mL screw-cap flasks and grown statically overnight in a 30°C incubator. This preseed was subcultured into inoculation flasks containing modified RM with 20 g/L of glucose, 20 g/L of xylose, and 10 mg/L of Tc and grown overnight in a 30°C shaker. This overnight culture was used at a level of ~10% (v/v) to inoculate the batch fermentors. The initial optical density (1-cm light path at 600 nm) was in the range of 0.2–0.25, corresponding to 60–75 mg of dry cell mass (DCM)/L.

Oat Hull Hydrolysate

The oat hull hydrolysate (OHH) was prepared from oat hulls (Quaker Oat, Peterborough, Ontario, Canada) by the Iogen Process (Iogen, Ottawa, Ontario, Canada) using a combination of steam explosion for hemicellulose disruption and cellulase for cellulose hydrolysis. The relevant composition of the OHH (postevaporator) was 21.65% (w/v) glucose, 10.65% (w/v) xylose, and 2.7% (w/v) acetic acid (HAc).

Fermentation Media

The synthetic, pure-sugar, OHH medium contained Zymo salts (3.48 g/L of KH_2PO_4, 0.25 g/L of $MgSO_4$, 0.01 g/L of $FeSO_4 \cdot 7H_2O$, 0.21 g/L of citric acid) *(15)* and 5 g/L of Difco Yeast Extract (Difco, Detroit, MI) with varying amounts of glucose, xylose, and acetic acid. OHH media were supplemented with Zymo salts and yeast extract or, alternatively, with 0.25% (v/v) whole corn steep liquor (CSL) (approx 50% solids) (Casco, Cardinal, Ontario, Canada) and 1.2 g/L of diammonium phosphate (DAP) *(17)*. All media contained 20 mg/L of Tc. All media were sterilized by autoclaving at 121°C for 30–45 min. Stock sugar solutions were autoclaved separately. Tc was added to the sterilized medium after cooling.

Fermentation Equipment

pH-stat batch fermentations were conducted with about 1500 mL of medium in 2-L bioreactors (model F2000 MultiGen or model Bioflo 2000; New Brunswick, Edison, NJ) and fitted with agitation, pH, and temperature control (30°C). The pH was monitored using a sterilizable combination Ingold pH electrode. The standard pH control set point was either 5.75 or 6.0. In some experiments, the pH set point was adjusted at 24 h from 5.75 to 6.5. pH was maintained by automatic titration with 4 N KOH. Temperature was controlled at 30°C using a circulating water bath and the agitation was moderate (approx 100–150 rpm).

Analytical Procedures, Growth, and Fermentation Parameters

Growth was measured turbidometrically at 600 nm (1-cm light path). In all cases the blank cuvet contained distilled water. DCM was determined by microfiltration of an aliquot of culture followed by washing and drying of the filter to constant weight under an infrared heat lamp. Fermentation media and cell-free spent media were compositionally analyzed by high-performance liquid chromatography as described previously *(10)*. The metabolic ethanol yield ($Y_{p/s}$) was calculated as the mass of ethanol produced per mass of sugar consumed. The process ethanol yield was determined by dividing the ethanol concentration by the total sugar concentration in the feed medium. The volumetric ethanol productivity was calculated by dividing the final ethanol concentration by 48 h or, alternatively, by the time taken to complete the fermentation (i.e., 100% utilization of glucose and xylose).

Results and Discussion

Since the hemicellulose component of lignocellulosic biomass consists of pentosans that are acetylated to varying degrees, all biomass hydrolysates contain acetic acid (for a review *see* ref. *20*). Because of its bacteriostatic properties, acetic acid is a well-known food preservative. The mechanism of acetic acid inhibition of bacterial growth is well understood

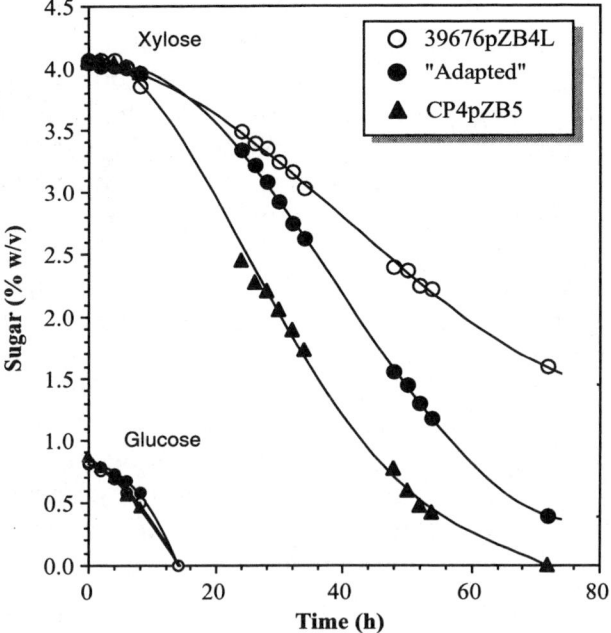

Fig. 1. Effect of 1% (w/v) acetic acid on xylose-utilizing recombinant Z. *mobilis* in a cCSL-based pure-sugar synthetic hardwood hydrolysate medium. The medium contained Zymo salts (*see* Materials and Methods) with 1% (v/v) clarified CSL (cCSL), 20 mg/L of Tc, 4% (w/v) glucose, 0.8% (w/v) xylose, and 1% (w/v) acetic acid. The pH was 6.0 and the temperature was 30°C. The values for maximum DCM concentration, ethanol yield, and productivity are given in Table 1.

(21), and its pH-dependent effect on both wild-type *(10)* and xylose-utilizing recombinant *Zymomonas* has been documented *(12)*. Recently, Joachimsthal et al. *(22)* described the isolation of an acetic acid–tolerant mutant of wild-type *Zymomonas* ZM4.

Several "conditioning" procedures have been described for reducing the toxicity of biomass hydrolysates, thereby making them less recalcitrant to fermentation *(20)*. Among these detoxification procedures, ion-exchange and liquid-liquid extraction are relatively specific for the removal of acetic acid *(23)*. However, there is an added cost associated with hydrolysate conditioning, and in this study we decided to test biocatalyst fermentation performance using unconditioned hydrolysate.

In prior separate studies, we examined the acetic acid sensitivity of different NREL xylose-utilizing recombinant Z. *mobilis* strains using a synthetic hardwood prehydrolysate medium *(10,12)*. At the time the present study was initiated in the spring of 1999, the following three NREL rec Zm strains were in our culture collection: CP4:pZB5, 39676:pZB4L, and the so-called hydrolysate-adapted variant of 39676:pZB4L (designated as adapted). Figure 1 shows the glucose and xylose consumption trajectories for a pH-stat batch fermentation of a synthetic hardwood prehydrolysate using the three rec Zm strains. The nutrient-rich medium contained 1% (w/v) acetic acid and the pH was controlled at 6.0 (Fig. 1). Whereas all three recombinants exhibited identical ethanol yields based on sugar consumed (0.48 g/g), the process yield based on the total initial sugar concentration was 0.32 g/g for recombinant 39676:pZB4L and 0.47 g/g for CP4:pZB5, with the adapted strain exhibiting a yield that was intermediate between the others (Table 1). Under these assay conditions, only rec Zm CP4:pZB5 was able to complete the fermentation in 3 d (Fig. 1). From the results of the experiment illustrated in Fig. 1, we concluded that strain CP4:pZB5 was more tolerant than the other recombinants to acetic acid, and we proceeded to examine its growth and cofermentation performance with unconditioned OHH and pure-sugar synthetic media.

The acetic acid concentration of the evaporated OHH was 2.7% (w/v). When the OHH concentrate was diluted with water to an acetic acid level of 1% (w/v), the concentrations of glucose and xylose were 8 and 3.9%, respectively. Figure 2 compares the cofermentation performance of rec Zm CP4:pZB5 in OHH (1% acetic acid) and a pure-sugar synthetic OHH medium. Both media were supplemented with 5 g/L of yeast extract and Zymo salts (*see* Materials and Methods). The growth and fermentation parameters are summarized in Table 1. The significantly better performance of CP4:pZB5 in the synthetic medium suggests a possible role of inhibitory substances other than acetic acid in the unconditioned hydrolysate, e.g., soluble phenolic compounds derived from lignin decomposition. In addition to acetic acid, ethanol is known to inhibit xylose utilization by rec Zm with ethanol concentrations of 5.5–6% (w/v) causing complete inhibition *(16,24)*. Hence, even with the synthetic medium, ethanol could conceivably be a contributing factor in protracting or, in other experiments not shown, stalling the batch fermentation (Fig. 2).

Figure 3 shows that the combined inhibitory effects of acetic acid and ethanol on CP4:pZB5 cofermentation performance could be reduced by diluting the OHH concentrate such that the acetic acid concentration was <1% (w/v). One of the operational parameters that was included in our biocatalyst performance assessment criteria was that the batch fermentation be essentially completed within 2 d. At an acetic acid concentration of 0.84% (w/v), xylose utilization was incomplete after 48 h (Fig. 3), and the process yield was only 0.44 g/g (Table 1). However, at an acetic acid level of 0.75% (w/v), the fermentation was complete at 48 h (Fig. 3), with a final ethanol concentration of 4.2% (w/v) representing a volumetric productivity of 0.88 g/(L·h) and an ethanol yield of 0.47 g/g or 92% theoretical maximum conversion efficiency (Table 1). Further reduction in acetic acid level to <0.75% (w/v) resulted in an increase in volumetric productivity, but with a proportionately reduced final ethanol concentration (Fig. 3, Table 1). From the results of the experiment shown in Fig. 3, it was concluded that with rec Zm CP4:pZB5 the optimal hydrolysate composition (for complete fermentation within 48 h) with respect to sugars and acetic acid was 6% glucose, 3% xylose, and 0.75% acetic acid.

Table 1
Summary of Growth and Fermentation Parameters[a]

Figure/ experiment rec Zm	Amount of sugar (% [w/v])	Glu (% [w/v])	Xyl (% [w/v])	HAc (% [w/v])	Cell mass (g DCM/L)	Yield(p) (g/g)	Yield(m) (g/g)	Productivity (g/[L·h])	Maximum EtOH (g/L)
Figure 1									
3976pZB4L	4.8	0.8	4	1.0	0.63	0.32	0.48	0.22[b]	15.8
"Adapted"	4.8	0.8	4	1.0	0.72	0.44	0.48	0.29[b]	21.2
CP4pZB5	4.8	0.8	4	1.0	0.74	0.47	0.48	0.33[b]	23.5
Figure 2									
CP4pZB5	12 Pure sugars	8	4	1.0	1.04	0.47	0.48	1.10	58
CP4pZB5	12 OHH	8	4	1.0	1.32	0.36	0.47	0.89	43
Figure 3									
CP4pZB5	6 OHH	4	2	0.5	1.38	0.47	0.47	0.94	29
CP4pZB5	9 OHH	6	3	0.75	1.51	0.47	0.47	0.88	42
CP4pZB5	10.5 OHH	7.0	3.5	0.84	1.36	0.44	0.47	0.96	46
Figure 4									
CP4pZB5	9 Pure sugars	6	3	0.75	1.07	0.48	0.48	0.90	43
CP4pZB5	9 OHH	6	3	0.75	1.51	0.47	0.47	0.88	42
CP4pZB5	9 OHH	6	3	0.75	1.53	0.47	0.48	0.88	42
Figure 5									
CP4pZB5	9 OHH	6	3	0.75	1.53	0.47	0.48	0.89	42
39676	9 OHH	6	3	0.75	1.18	0.32	0.48	0.62	29
"Adapted"	9 OHH	6	3	0.75	1.15	0.34	0.47	0.66	31
ZM4pZB5	9 OHH	6	3	0.75	1.38	0.47	0.48	1.40	42
Figure 7									
CP4pZB5	10.5 OHH	7	3.5	0.9	0.97	0.48	0.44	0.96	46
ZM4pZB5	10.5 OHH	7	3.5	0.9	1.16	0.48	0.47	1.04	50

[a]For fermentations <48 h, the productivity was based on the time required to complete the fermentation; otherwise, the productivity was based on ethanol concentration at 48 h. DCM, dry cell mass; Yield(p), process yield based on available sugar; Yield(m), metabolic yield based on sugar utilized; HAc, acetic acid; OHH, oat hull hydrolysate.
[b]Productivity was based on a fermentation time of 72 h.

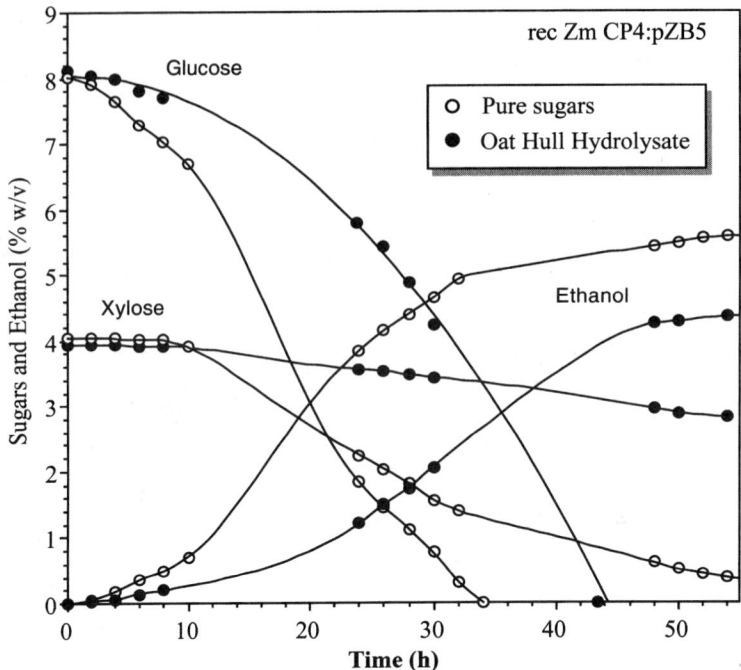

Fig. 2. Time course of pH-stat batch fermentation of OHH and a synthetic pure-sugar hydrolysate by rec Zm CP4:pZB5. The synthetic OHH medium contained Zymo salts, 5 g/L of yeast extract, 20 mg/L of Tc, 8% (w/v) glucose, 3.9% (w/v) xylose, and 1% (w/v) acetic acid. The pH was 5.75 and the temperature was 30°C. The values for maximum DCM concentration, ethanol yield, and productivity are given in Table 1.

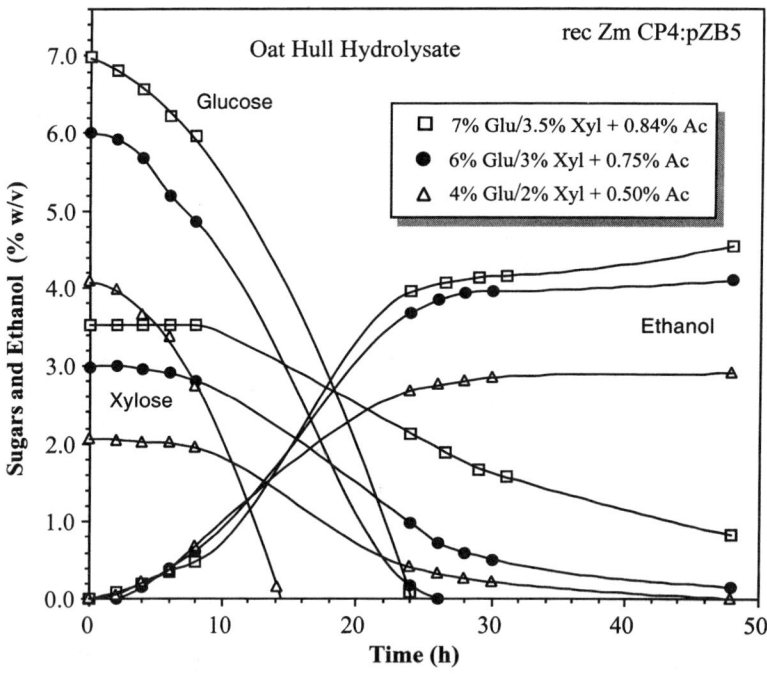

Fig. 3. Effect of dilution of the concentrated OHH on the cofermentation performance of rec Zm CP4:pZB5. The diluted OHH was supplemented with Zymo salts, 5 g/L of yeast extract, and 20 mg/L of Tc. The pH was 5.75 and the temperature was 30°C. The values for maximum DCM concentration, ethanol yield, and productivity are given in Table 1.

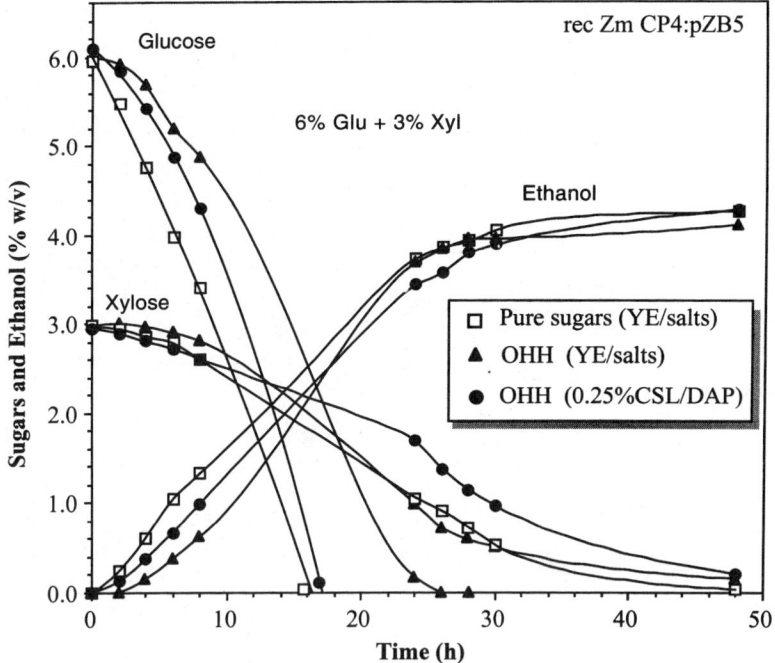

Fig. 4. Comparative cofermentation performance of rec Zm CP4:pZB5 in OHH and synthetic pure-sugar hydrolysate with 9% (w/v) sugar loading and 0.75% (w/v) acetic acid: effect of nutritional supplements. The concentration of glucose and xylose in all media was 6% (w/v) and 3% (w/v), respectively. The nutrients added were either Zymo salts and 5 g/L of yeast extract (YE) or 0.25% (v/v) CSL and 1.2 g/L of DAP. All media contained 20 mg/L of Tc. The pH was 5.75 (which was adjusted to 6.5 at 24 h) and the temperature was 30°C. The values for maximum DCM concentration, ethanol yield, and productivity are given in Table 1.

Cost-Effective Nutritional Amendment of Hydrolysate

The requirement for nutritional supplementation of fermentation media has a significant economic impact on large-scale production of fuel ethanol. Complex supplements such as yeast extract are very costly, and we have previously shown that CSL is a cost-effective substitute for yeast extract *(17)*. At low-level amendment of hydrolysate by CSL, there is a requirement for added nitrogen, which can be supplied in the form of inorganic nitrogen as ammonium salts *(11)*. Figure 4 shows that rec Zm CP4:pZB5 performs comparably when the hydrolysate is amended either with a combination of 0.25% (v/v) whole CSL (50% solids) and 1.2 g/L of DAP or with 5 g/L of yeast extract and Zymo salts *(see* Materials and Methods). Recently, it has been estimated that the economic impact of this nutritional supplementation on large-scale production of ethanol is 4.4¢/US gal of ethanol *(17)*.

Comparative Cofermentation Performance Assessment Using OHH

In September 1999, we received another xylose-utilizing recombinant Zm strain from NREL: ZM4:pZB5 (also known as 31821:pZB5). This strain

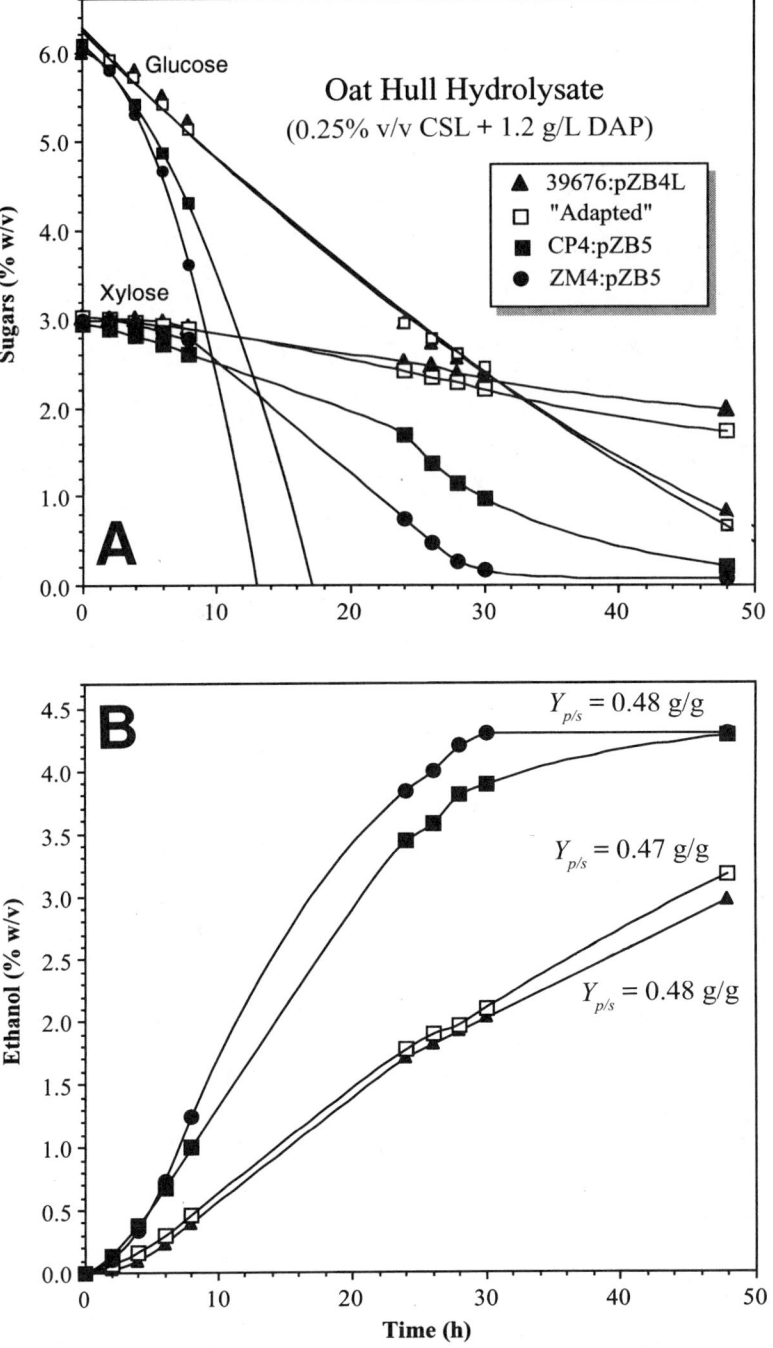

Fig. 5. Comparative cofermentation performance of four different *Z. mobilis* recombinants in oat hull hydrolysate (0.75% [w/v] HAc) with cost-effective minimal nutrient supplementation. **(A)** Sugar utilization; **(B)** ethanol production. The concentration of glucose and xylose in the oat hull hydrolysate was 6% (w/v) and 3% (w/v), respectively. Tc (20 mg/L) was added to the medium. The level of acetic acid was 0.75% (w/v). The nutrients added were 0.25% (v/v) CSL and 1.2 g/L of DAP. The pH was 5.75 (which was adjusted to 6.5 at 24 h), and the temperature was 30°C. The values for maximum DCM concentration, ethanol yield, and productivity are given in Table 1.

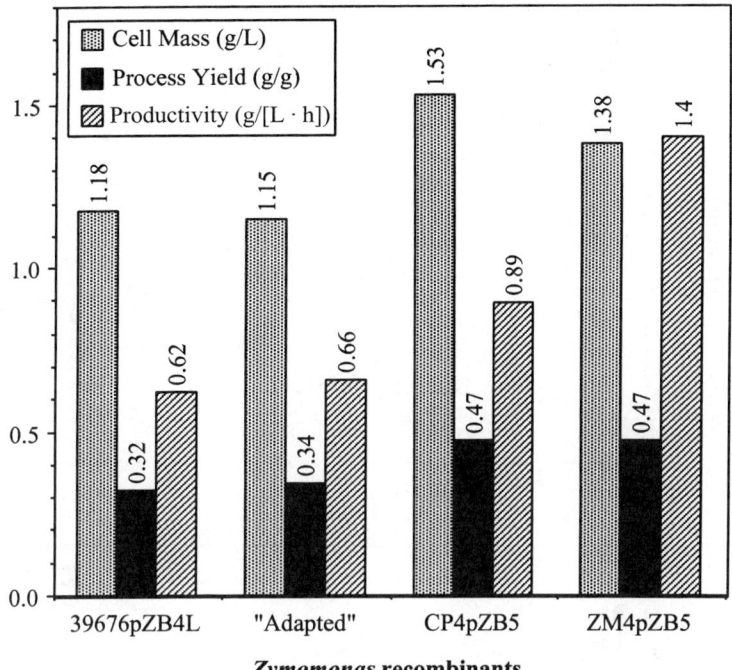

Fig. 6. Comparative growth and fermentation performance profiles for four different *Z. mobilis* recombinants using oat hull hydrolysate. Cell (dry) mass, process ethanol yield, and volumetric ethanol productivity data are taken from the experiment illustrated in Fig. 5 (*see* also Table 1). Ethanol productivity was based on a 48-h batch fermentation except for rec Zm ZM4:pZB5, of which the fermentation was complete in 30 h.

was purported to be superior to CP4:pZB5 in both batch and continuous fermentations *(18)*. Side-by-side pH-stat batch fermentations were conducted with all four xylose-utilizing rec Zm strains using OHH (6% glucose, 3% xylose, and 0.75% acetic acid) supplemented with 0.25% (v/v) CSL and 1.2 g/L of DAP (Fig. 5). The pH was controlled initially at 5.75, but the control set point was adjusted at 24 h to 6.5. Under these test conditions, neither the recombinant 39676:pZB4L nor the adapted variant performed very well (Fig. 5). The improved performance exhibited by CP4:pZB5 under this test condition was consistent with the results of the comparative performance assessment using synthetic hardwood prehydrolysate with 1% acetic acid (Fig. 1). Recombinant CP4:pZB5 completed the fermentation of the unconditioned OHH in 48 h with a yield of 0.47 g/g and a productivity of 0.89 g/(L·h) (Fig. 5, Table 1). However, ZM4:pZB5 outperformed CP4:pZB5 by completing the fermentation in 30 h with a yield of 0.47 g/g and a productivity of 1.4 g/(L·h) (Fig. 5, Table 1). The growth and fermentation parameters for all four recombinants are compared graphically in Fig. 6. The superior performance exhibited by rec Zm ZM4:pZB5 substantiates earlier claims made by Joachimsthal and Rogers *(18)*.

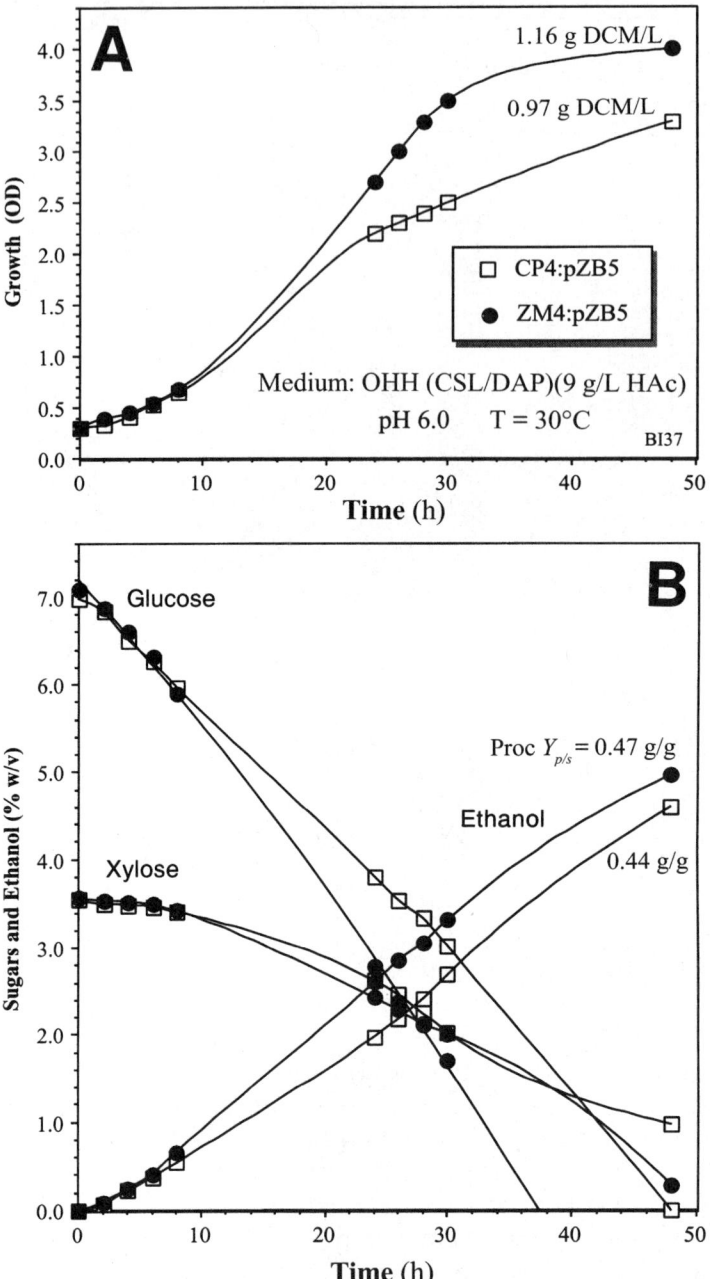

Fig. 7. Comparative cofermentation performance of rec Zm CP4:pZB5 and ZM4:pZB5 in oat hull hydrolysate (OHH) and synthetic pure-sugar hydrolysate with 10% (w/v) sugar loading and 0.9% (w/v) acetic acid. **(A)** Growth; **(B)** sugar utilization and ethanol production. The concentration of glucose and xylose was 7% (w/v) and 3.5% (w/v), respectively. The nutrients added were 0.25% (v/v) CSL and 1.2 g/L of DAP. Tc (20 mg/L) was added to the medium. The pH was 6.0 and the temperature was 30°C. The values for maximum DCM concentration, ethanol yield, and productivity are given in Table 1.

The excellent performance of ZM4:pZB5 prompted us to test its capacity to ferment a higher level of hydrolysate with proportionately high concentrations of glucose, xylose, and acetic acid. Figure 7 shows that ZM4:pZB5 still outperforms CP4:pZB5 at an acetic acid level of 0.9% (w/v). In fact, the CP4pZB5 fermentation appears to be "stuck" at 48 h, and the process yield falls from 0.47 to 0.44 g/g (Fig. 7). For ZM4:pZB5, productivity falls from 1.4 to 1.0 g/(L·h) (Table 1).

From a regulatory perspective, all of the rec Zm strains tested in this study are disadvantaged because they all carry the Tc-resistance gene, which was incorporated into the plasmid carrying the xylose utilization genes for convenience of selection as part of the process of genetic engineering. Also, that all these recombinants are xylose utilizing by virtue of the expression of plasmid-borne genes makes them all susceptible to instability in long-term operations such as either series batch fermentations (draw-and-fill) or continuous fermentations. Finally, none of these recombinants had the ability to utilize arabinose. These issues have been addressed by NREL, and in the future we anticipate testing the new genome integrated C5-utilizing Zm strains developed at NREL that are both devoid of the Tc resistance gene and able to utilize xylose and arabinose *(25)*.

Acknowledgments

We are grateful to Min Zhang at NREL for the recombinant Zm strains used in this study and to Chris Hindle and Sylvia McHugh at Iogen for technical assistance. This work was funded by Iogen Corporation (Ottawa, Canada) and the Biochemical Conversion Element of the Office of Fuels Development of the U.S. Department of Energy. Research conducted at the University of Toronto was part of Subcontract ZDH-9-29009-02 from NREL.

References

1. Foody, B. F. (2000), 21st Symposium on Biotechnology for Fuels and Chemicals, Fort Collins, CO (abstract no. 6-01).
2. Foody, B. F. and Tolan, J. S. (2000), 22nd Symposium on Biotechnology for Fuels and Chemicals, Gatlinburg, TN (abstract no. 6-07).
3. Hinman, N. D., Wright, J. D., Hoagland, W., and Wyman, C. E. (1989), *Appl. Biochem. Biotechnol.* **20/21,** 391–401.
4. McMillan, J. D. (1994), in *Enzymatic Conversion of Biomass for Fuels Production,* Himmel, M. E., Baker, J. O., and Overend, R. A., eds., ACS Symposium Series 566, American Chemical Society, Washington, DC, pp. 411–437.
5. Hahn-Hägerdal, B., Hallborn, J., Jeppsson, H., Olsson, L., Skoog, K., and Walfridsson, M. (1993), in *Bioconversion of Forest and Agricultural Plant Residues,* Saddler, J. N., ed., C.A.B. International, Wallingford, UK, pp. 411–437.
6. Zhang, M., Franden, M. A., Newman, M., McMillan, J., Finkelstein, M., and Picataggio, S. (1995), *Appl. Biochem. Biotechnol.* **51/52,** 527–536.
7. Picataggio, S. K., Zhang, M., and Finkelstein, M. (1994), in *Enzymatic Conversion of Biomass for Fuels Production,* Himmel, M. E., Baker, J. O., and Overend, R. A., eds., ACS Symposium Series 566, American Chemical Society, Washington, DC, pp. 342–362.
8. Zhang, M., Eddy, C., Deanda, K., Finkelstein, M., and Picataggio, S. K. (1995), *Science* **267,** 240–243.

9. Deanda, K. A., Eddy, C., Zhang, M., and Picataggio, S. (1996), *Appl. Environ. Microbiol.* **62,** 4465–4470.
10. Lawford, H. G., Rousseau, J. D., and McMillan, J. D. (1997), *Appl. Biochem. Biotechnol.* **63–65,** 269–286.
11. Lawford, H. G. and Rousseau, J. D. (1997), *Appl. Biochem. Biotechnol.* **63–65,** 287–304.
12. Lawford, H. G. and Rousseau, J. D. (1998), *Appl. Biochem. Biotechnol.* **70–72,** 161–172.
13. Lawford, H. G., Rousseau, J. D., Mohagheghi, A., and McMillan, J. D. (1998), *Appl. Biochem. Biotechnol.* **70–72,** 353–368.
14. Lawford, H. G. and Rousseau, J. D. (1999), *Appl. Biochem. Biotechnol.* **77–79,** 235–249.
15. Lawford, H. G., Rousseau, J. D., Mohagheghi, A., and McMillan, J. D. (1999), *Appl. Biochem. Biotechnol.* **77–79,** 191–204.
16. Lawford, H. G. and Rousseau, J. D. (2000), *Appl. Biochem. Biotechnol.* **84–86,** 277–294.
17. Lawford, H. G., Rousseau, J. D., Mohagheghi, A., and McMillan, J. D. (2000), *Appl. Biochem. Biotechnol.* **84–86,** 295–310.
18. Joachimsthal, E. and Rogers, P. L. (1999), *Appl. Biochem. Biotechnol.* **84–86,** 343–356.
19. Goodman, A. E., Rogers, P. L., and Skotnicki, M. L. (1982), *Appl. Environ. Microbiol.* **44,** 496–498.
20. McMillan, J. D. (1994), in *Enzymatic Conversion of Biomass for Fuels Production,* Himmel, M. E., Baker, J. O., and Overend, R. A., eds., ACS Symposium Series 566, American Chemical Society, Washington, DC, pp. 411–437.
21. Booth, I. R. (1985), *Microbiol. Rev.* **49,** 63–91.
22. Joachimsthal, E., Haggett, K. D., Jang, J.-H., and Rogers, P. L. (1998), *Biotechnol. Lett.* **20,** 137–142.
23. McMillan, J. D., Newman, M. M., Templeton, D. W., and Mohagheghi, A. (1999), *Appl. Biochem. Biotechnol.* **77–79,** 649–665.
24. Rogers, P. L., Joachimsthal, E. L., and Haggett, K. D. (1997), *J. Australasian Biotechnol.* **7,** 304–309.
25. Zhang, M., Chou, Y. C., Mohagheghi, A., Evans, K., Milstrey, S., Lai, X. K., and Finkelstein, M. (2000), 22nd Symposium on Biotechnology for Fuels and Chemicals, Gatlinburg, TN (abstract no. 2-03), Humana, Totowa, NJ.

Isolation of *Magnetospirillum magneticum* AMB-1 Mutants Defective in Bacterial Magnetic Particle Synthesis by Transposon Mutagenesis

ARIS TRI WAHYUDI, HARUKO TAKEYAMA,
AND TADASHI MATSUNAGA*

*Department of Biotechnology,
Tokyo University of Agriculture and Technology,
2-24-16 Naka-Cho, Koganei, Tokyo 184-8588, Japan,
E-mail: tmatsuna@cc.tuat.ac.jp*

Abstract

Nonmagnetic mutants of *Magnetospirillum magneticum* AMB-1 were recovered following mini-Tn5 transposon mutagenesis. Transconjugants with kanamycin resistance were obtained at a frequency of 2.7×10^{-7} per recipient. Of 3327 transconjugants, 62 were defective for bacterial magnetic particle (BMP) synthesis. The frequency of independent transposition events for nonmagnetic mutants was about 1.4% in transconjugants. Further analysis of DNA sequences flanking transposon by inverted polymerase chain reaction allowed isolation of at least 10 genes or DNA sequences involved in BMP synthesis in *M. magneticum* AMB-1.

Index Entries: Transposon mutagenesis; mini-Tn5; *Magnetospirillum magneticum* AMB-1; inverse polymerase chain reaction; bacterial magnetic particles; sequence analysis.

Introduction

Magnetic bacteria are a diverse group of Gram-negative prokaryotes and synthesize membrane-bound intracellular particles of either magnetite (Fe_3O_4) or greigite (Fe_3S_4) that are aligned in chains of 10–30 along the length of the cell. Formation of bacterial magnetic particles (BMPs) is achieved by a biomineralization process regulated at the genetic level. Accumulation of iron and the deposition of specific particle sizes occur within membrane vesicles at specific locations in the cell *(1)*. Studies on

---

*Author to whom all correspondence and reprint requests should be addressed.

BMP synthesis usually use *Magnetospirillum magnetotacticum* MS-1 *(2)* and *M. magneticum* AMB-1 *(3)* as model systems. Little is known about the biosynthetic ability and molecular genetics of BMP formation in these bacteria. *M. magneticum* AMB-1 is the only strain with an established transformation system that can form colonies on laboratory medium. This feature facilitated the genetic analysis of BMP synthesis in *M. magneticum* AMB-1 *(4)*.

Transposon Tn5 is one of the best-studied and most versatile transposons *(5,6)*. Tn5 exhibits a relatively high transposition frequency and is stably inserted into the genome *(7)*. Mini-Tn5 derivatives have proven to be extremely useful tools in bacterial genetics for mutagenesis or insertion of cloned DNA into the chromosome of Gram-negative bacteria *(8–10)*. These derivatives consist of an antibiotic-resistance marker flanked by 19-bp inverted repeats of Tn5 that are essential for transposition *(11)*. The transposase gene is located outside the inverted repeats on the vector. Therefore, insertions produced by mini-Tn5 are generally more stable because of the absence of transposase-mediated secondary transpositions *(8)*. Characterization, cloning, and sequencing of transposon-inserted genes require the isolation of flanking DNA sequences. This is generally achieved by cloning DNA fragments containing the transposon or by inverse polymerase chain reaction (IPCR) *(12)*. The *magA* gene encoding an iron-transporter protein located on BMP membrane and involved in BMP synthesis was isolated from *M. magneticum* AMB-1 by Tn5 transposon mutagenesis *(13)*.

In the present study, we isolated numerous *M. magneticum* AMB-1 mutants that are defective in BMP synthesis by mini-Tn5-mediated mutagenesis and analyzed the DNA sequences interrupted with mini-Tn5.

Materials and Methods

Bacterial Strains, Culture Conditions, and Plasmids

Escherichia coli S17-1 λ *pir (11)* harboring plasmid pUTmini-Tn5Km1 *(5,9)* was grown on Luria Bertani (LB) broth (5.0 g/L of tryptone, 10 g/L of NaCl, 5.0 g/L of yeast extract) containing ampicillin (50 μg/mL) and kanamycin (25 μg/mL) at 37°C. *E. coli* DH5α was grown in LB at 37°C. *M. magneticum* AMB-1 (ATCC 700264) was cultured anaerobically with magnetic spirillum growth medium (MSGM) *(2)* at 25°C. Transconjugants of AMB-1 with mini-Tn5 transposon were anaerobically cultured with MSGM supplemented with kanamycin (5 μg/mL). The pGEM-T Easy vector (Promega, Madison, WI) was used for cloning of IPCR products.

Transposon Mutagenesis of M. magneticum AMB-1

Transposon mini-Tn5Km1 (pUT-miniTn5Km1) was transferred from *E. coli* S17-1 λ *pir* into AMB-1 by conjugation. The donor, *E. coli* S17-1 λ *pir*, was grown in LB medium containing ampicillin and kanamycin. AMB-1 cells responsive for magnetic field were magnetically collected and used for cultivation. AMB-1 and *E. coli* S17-1 λ *pir* cells in logarithmic phase (approx 1×10^8 cells/mL) were centrifuged, washed three times with 0.85% NaCl, and resuspended in MSGM at a concentration of about 10^{11} cells/mL. Fifty microliters of the donor and recipient were mixed in varying ratios of *E. coli* to AMB-1 (1:1, 1:10, and 10:1) and placed on a sterile nitrocellulose filter on MSGM agar, and mating was performed aerobically for 6 h at 25°C. The filters were then transferred into 300 μL of MSGM in a microcentrifuge tube and vortexed. One hundred microliters of each cell suspension was plated on MSGM agar supplemented with kanamycin (5 μg/mL) and incubated for 10–14 d anaerobically. Nonmagnetic mutants were defined by microscopic observation under magnetic fields.

Southern Hybridization Analysis

Isolation of genomic DNA from nonmagnetic mutants of AMB-1 was performed as described by Wilson *(14)*. The genomic DNA digested with *Pst*I or *Eco*RI was used for Southern hybridization analysis using 1.7-kb DNA fragments containing the kanamycin resistance gene from plasmid pUTmini-Tn5Km1 labeled as a probe with a digoxigenin (DIG) chemiluminescence kit (Boehringer Mannheim GmbH Biochemica). Hybridization was performed using a DIG DNA Detection Kit (Boehringer Mannheim).

IPCR Amplification and DNA Sequencing

Figure 1 shows the strategy for IPCR. IPCR was performed against DNA sequences flanking mini-Tn5 in genomic DNA isolated from non-magnetic mutants. The genomic DNA digested with *Eco*RV or *Bst*XI and *Apa*I was circularized and used as templates for IPCR. Amplification was performed using primers designed from the mini-Tn5 sequence near the insertion sequence: primer 1, 5'-ACACTGATGAATGTTCCGTTG-3'; primer 2, 5'-ACCTGCAGGCATGCAAGCTTC-3'. A GeneAmp PCR System 2400 (Perkin-Elmer) was used to amplify DNA by denaturation at 95°C for 2 min, primer annealing at 60°C for 1 min, and primer extension at 72°C for 1 min for 30 cycles and 10 min for the last cycle.

The IPCR products were purified by gel electrophoresis and cloned using pGEM-T Easy vector. Recombinant plasmids were extracted from *E. coli* DH5α by standard methods *(15)* and used for DNA sequencing utilizing an automatic DNA sequencer, DSQ-2000L (Shimadzu, Japan). LASERGENE (DNASTAR, Madison, WI) was used for analysis of DNA homology and for detection of transposon insertion sites in the genome.

Results and Discussion

Isolation of Transposon Mutants Defective in BMP Synthesis

The highest transconjugation frequency was 2.7×10^{-7} when a 10:1 ratio of *E. coli* and AMB-1 was employed (Table 1). AMB-1 with BMPs form brown-black colonies. A total of 3327 transconjugants were screened, of

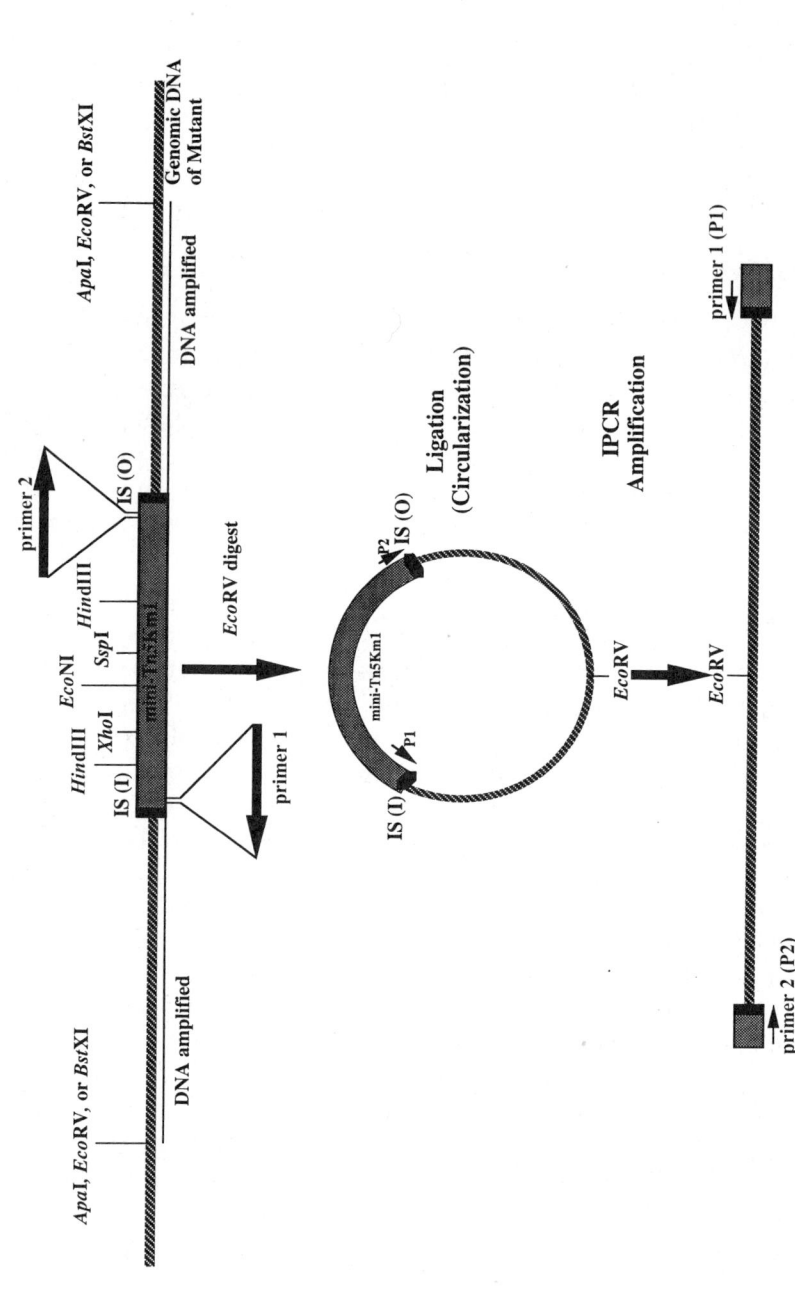

Fig. 1. The strategy for IPCR. Genomic DNA of AMB-1 containing inserted transposon was digested with restriction enzyme *ApaI*, *BstXI*, or *EcoRV*. IS, insertion sequence of the transposon, 19 bp (inner and outer).

Table 1
Frequency of Transconjugation
of Transposon mini-Tn5Km1
from *E. coli* to *M. magneticum* AMB-1

Mating cell ratio of *E. coli* and AMB-1	Frequency[a]
1:10	2.3×10^8
1:1	4.6×10^8
10:1	2.7×10^7

[a]Frequency of transconjugation is calculated per recipient. Values are the mean of 10 experiments.

which 62 colonies were white and thus defective for BMP synthesis. All nonmagnetic mutants grew well in MSGM supplemented with kanamycin under anaerobic conditions, similar to the AMB-1. Because spontaneous mutants with kanamycin resistance did not occur on MSGM plates supplemented with kanamycin, AMB-1 mutants defective in BMP synthesis were owing presumably to the transposon insertion into the genome.

Southern Hybridization Analyses of Nonmagnetic Mutants by IPCR

Thirty-four among 62 nonmagnetic mutants were arbitrarily selected for further analysis. Southern hybridization analysis of nonmagnetic mutants revealed the presence of the transposon in genomic DNA. Furthermore, 26 nonmagnetic mutants were derived from independent transposition events (11 of 34 nonmagnetic mutants are shown in Fig. 2). The frequency of transposition events for nonmagnetic mutants was about 1.9% in total transconjugants (62 nonmagnetic mutants of 3327 transconjugants), and 76% of nonmagnetic mutants analyzed were independent mutants (26 of 34 nonmagnetic mutants). Assuming that AMB-1 has a typical genome size of ~4000 genes like *E. coli*, these data suggest that as many as 60 genes may be involved in BMP synthesis. Biosynthesis of BMPs in AMB-1 is a complex system involving many genes. Jiang et al. *(10)* speculated that 200 genes in *Rhodospirillum centenum* (assuming ~4000 genes in the genome) might be involved in swarm cell response to light because the frequency of transposition events was 5% of the screened 23,000 mutants.

IPCR Amplification of DNA Fragments Flanking Transposon and Their Sequence Analysis

DNA fragments flanking the mini-Tn5 transposon were amplified by IPCR using primers designed from mini-Tn5 sequence. Single bands of 1.3–4.7 kb were amplified from 14 nonmagnetic mutants. Sequence analysis of IPCR products from these nonmagnetic mutants showed that 10 mutants were derived from independent transposition events. This suggested that at least 10 genes or DNA sequences are required for BMP synthesis in AMB-1.

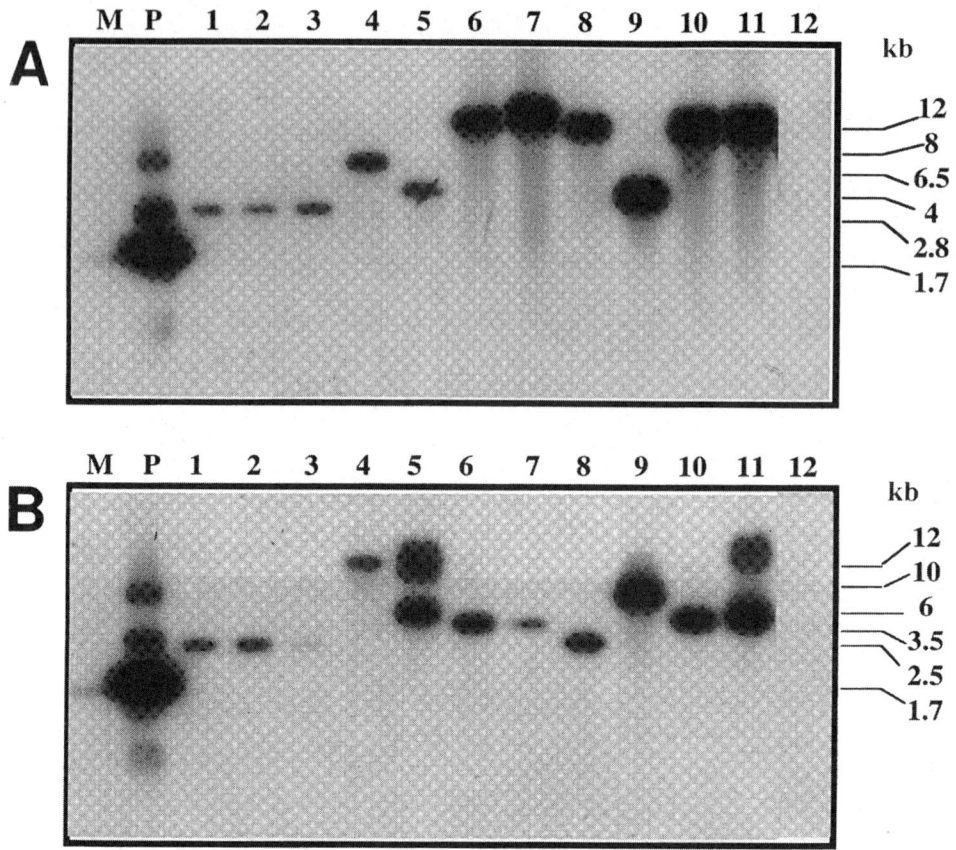

Fig. 2. Southern hybridization analysis of transconjugants using kanamycin-resistant gene as a probe. **(A)** Genomic DNA digested with *Eco*RI, and **(B)** *Pst*I. Lanes 1–11, nonmagnetic mutant genomes no. 1–11; lane 12, wild-type AMB-1 genome (negative control); lane M, 1-kb DNA ladder; lane P, pUTmini-Tn5Km1 digested with *Eco*RI and *Pst*I (positive control).

Table 2 shows the results of homology searches on the IPCR products from these 10 mutants. The amino acid sequence derived from the DNA sequence flanking the inserted mini-Tn5 in mutant 9 had 51% identity with cytochrome C–type SHP in *R. sphaeroides*, which functions as an oxygen-binding heme protein during autooxidation *(16)*.

Further studies are currently being conducted to clarify the role of these genes. Complementation analysis using wild-type genome will help further the understanding of BMP synthesis in *M. magneticum* AMB-1. Elucidation of the remaining mutants will also give more information about the genes involved in BMP synthesis.

Acknowledgment

This work was supported in part by Grant-in-Aid for Scientific Research on Priority Area (A) no. 10145102, The Ministry of Education, Science, Sport and Culture.

Table 2
DNA Sequences Interrupted by Transposon and Homolog Search Analysis

NMA[a] no.	IPCR product (bp)	Insertion site	Homolog		
			Name	Identity/positive (%)	Comment/reference
6=7	2285	GCGC ATCC	Inositol polyphosphate phosphatase	39/45	From *Caenorhabditis elegans* (17)
9	2089	GTCC GGCC	Cytochrome C–type protein SHP	51/62	Oxygen-binding heme protein from Rhodobacter sphaeroides (16)
10	1159	GGGC GGCC	Y4XK protein	30/44	Hypothetical lypoprotein Y4XK precursor from *Rhizobium meliloti*, membrane protein; unknown function (18)
14	1012	GGGC CTCC	Regulatory component of sensor transduction	49/64	From *Synechocystis* sp. (19)
15	1180	GGGG GTCC	Putative mobilization protein	47/49	Uncultured eubacterium; putative role in conjugative DNA replication (20)
17	1000	AGGC ATCC	gra-ORF30	34/44	From *Streptomyces violaceoruber* (21)
19	1100	TGCG GGAC	Flagellar M ring protein	29/51	From *Caulobacter crescentus*; for energy transduction (22)
20=27	1006	GGAT GTTC	Hypothetical protein	38/68	From *Aquifex aeolicus* (23)
33	850	TCAT GTCC	Hypothetical protein	54/60	From *Pseudomonas*; SDSB region (24)
34	890	CATC GTCC	Growth factor receptor–bound protein	27/43	From *Mus musculus* (25)

[a]NMA, non-magnetic mutant of AMB-1.

References

1. Schuler, D. and Frankel, R. B. (1999), *Appl. Microbiol. Biotechnol.* **52,** 464–473.
2. Blackemore, R. P., Maratea, D., and Wolfe, R. S. (1979), *J. Bacteriol.* **140,** 720–729.
3. Matsunaga, T., Sakaguchi, T., and Tadokoro, F. (1991), *Appl. Microbiol. Biotechnol.* **35,** 651–655.
4. Matsunaga, T., Nakamura, C., Burgess, J. G., and Sode, K. (1992), *J. Bacteriol.* **174,** 2748–2753.
5. DeBruijn, F. J. and Lupski, J. R. (1984), *Gene* **27,** 131–149.
6. Simon, R., Preifer, U., and Puhler, A. (1983), *Bio/Technology* **1,** 784–791.
7. Berg, D. E. (1989), in *Mobile DNA*, Berg, D. E and Howe, M. M., eds., American Society for Microbiology, Washington, DC, pp. 185–210.
8. Herrero, M., deLorenzo, V., and Timmis, K. N. (1990), *J. Bacteriol.* **172,** 6557–6567.
9. Holtwick, R., Meinhardt, F., and Keweloh, H. (1997), *Appl. Environ. Microbiol.* **63,** 4292–4297.
10. Jiang, Z. Y., Rushing, B. G., Bai, Y., Gest, H., and Bauer, C. E. (1998), *J. Bacteriol.* **180,** 1248–1255.
11. DeLorenzo, V., Herrero, M., Martinko, J. M., and Parker, J. (1990), *J. Bacteriol.* **172,** 6568–6572.
12. Martin, V. J. J. and Mohn, W. W. (1999), *J. Microbiol. Methods* **35,** 163–166.
13. Nakamura, C., Burgess, J. G., Sode, K., and Matsunaga, T. (1995), *J. Biol. Chem.* **270,** 28,392–28,396.
14. Wilson, K. (1994), in *Current Protocols in Molecular Biology*, vol. 1, Ausubel, F. M., Brent, R., and Kingston, R. E., eds., John Wiley & Sons, New York, pp. 2.4.1–2.4.5.
15. Sambrook, J., Fritsch, E. F., and Maniatis, T. (1989), *Molecular Cloning: A Laboratory Manual*, 2nd ed., Cold Spring Harbor Laboratory Press, New York.
16. Klarskov, K., Van Driessche, G., Barkers, K., Dumortier, C., Meyer, T. E., Tollin, G., Cusanovich, M. A., and Beeumen, J. J. V. (1998), *Biochemistry* **37,** 5995–6002.
17. Wilson, R., Ainscough, R., Anderso, K., et al. (1994), *Nature* **368,** 32–38.
18. Freiberg, C., Fellay, R., Bairoch, A., Broughton, W. J., Rosenthal, A., and Parret, X. (1997), *Nature* **387,** 394–401.
19. Kaneko, T., Sato, S., Kotani, H., et al. (1996), *DNA Res.* **3,** 109–136.
20. Tietze, E. (1998), *Plasmid* **39,** 165–168.
21. Sherman, D. H., Malpartida, F., Bibb, M. J., Kieser, H. M., and Hopwood, D. A. (1989), *EMBO J.* **8,** 2717–2725.
22. Ramakrishnan, G., Zhao, J. L., and Newton, A. (1994), *J. Bacteriol.* **176,** 7587–7600.
23. Deckart, G. and Warren, P. V. (1998), *Nature* **392,** 353–358.
24. Davison, J., Brunel, F., Phanopoulos, A., Prozzi, D., and Terpstra, P. (1992), *Gene* **114,** 19–24.
25. Ooi, J., Yatnik, V., Immanuel, D., Gordon, M., Moskow, J. J., Buchberg, A. M., and Margolis, B. (1995), *Oncogene* **10,** 1621–1630.

Synthesis of Bacterial Magnetic Particles During Cell Cycle of *Magnetospirillum magneticum* AMB-1

CHEN-DONG YANG, HARUKO TAKEYAMA, TSUYOSHI TANAKA,
AKI HASEGAWA, AND TADASHI MATSUNAGA*

*Department of Biotechnology,
Tokyo University of Agriculture and Technology,
2-24-16, Naka-cho, Koganei, Tokyo 184-8588, Japan,
E-mail: tmatsuna@cc.tuat.ac.jp*

Abstract

We investigated the relationship between the synthesis of bacterial magnetic particles (BMPs) and the transcription of *magA* gene–encoding iron transport protein using synchronous culture of *Magnetospirillum magneticum* AMB-1. Synchronously cultured cells were subjected to transmission electron microscopic observation and fluorescence *in situ* hybridization. The average number of BMPs slowly increased in the cell with increasing cell size. A sharp increase in BMPs occurred just before cell division and resulted in maximum BMP production of 30 particles/cell. The transcription of *magA* was regulated immediately before and after cell division.

Index Entries: Synchronous culture; *Magnetospirillum magneticum* AMB-1; bacterial magnetic particle; cell cycle; *magA* gene.

Introduction

Magnetic bacteria have been found in freshwater and marine sediment *(1)*. They produce magnetic particles, which are small (50–100 nm) and covered with a stable lipid membrane *(2)*. A magnetic bacterium, *Magnetospirillum magneticum* AMB-1, has been isolated *(3)* and the bacterial magnetic particles (BMPs) have been used for highly sensitive immunoassays in which enzymes and antibodies immobilized on BMPs were used *(4,5)*. Genetic analysis of BMP synthesis has led to the isolation of the *magA* gene from *M. magneticum* AMB-1, which encodes an iron transport membrane protein on the BMPs *(6)*. Functional proteins have been displayed on BMPs using MagA as an anchor protein through the gene fusion *(7)*. Protein A–

---

*Author to whom all correspondence and reprint requests should be addressed.

displayed BMPs through the fusion with MagA protein has also been used in immunoassays *(8,9)*. These applications emphasize the need to enhance the production of BMPs for biotechnological applications.

To design cell cultivation processes and to achieve efficient production, several factors need to be considered. To overproduce BMPs, fed-batch culture of *M. magneticum* AMB-1 has been performed with continuous addition of nitric acid and iron to the medium *(10,11)*. The cell density during fed-batch cultures of AMB-1 was 10 times higher than that in batch culture; however, the yield of BMPs per cell was decreased. Our recent work shows that the yield of BMPs per cell decreases only after logarithmic growth in a fed-batch culture (unpublished results). The cell-cycle population is one of important factors, because the characteristics of the cells in phases may differ with respect to gene expression. The cell-cycle dependency of rice α-amylase production by a recombinant yeast has been reported *(12)*. Cyclic appearance of intracellular nitrogenase activity was observed in the synchronously grown marine cyanobacterium *Synechococcus* sp., revealing cell-cycle-dependent nitrogenase activity *(13)*. This characteristic has been used for efficient production of hydrogen mediated by nitrogenase in this cyanobacterium *(14)*. We previously described how a synchronous culture of *M. magneticum* AMB-1 was successfully established by repeated cold treatment of cells *(15)*.

In the present study, we investigated the relationship between *magA* expression and BMP synthesis during different cell-cycle phases. The results obtained in this study may contribute to control of cell-cycle phases for efficient BMP production in the strain AMB-1.

Materials and Methods

Bacterial Strain and Growth Medium

M. magneticum AMB-1 (ATCC700264) was cultured with 40 mL of magnetic spirillum growth medium in a 50-mL flask at 25°C *(16)*. The oxygen concentration in the gas phase was reduced to <1 ppm by repeated flushing with argon.

Synchronous Cultivation

Synchronous cultivation of AMB-1 cells was achieved by using repeated cold treatment at 5°C *(15)*. Precultured AMB-1 cells were inoculated into 40 mL of medium to a density of 1×10^5 cells/mL. The culture was incubated for 14 h in a constant 25°C (optimum growth temperature) water bath and then transferred between constant 5°C and 25°C water baths at 5-h intervals for five cold treatments. The cold-treated cells were inoculated into a 40-mL fresh medium to a density of 2×10^6 cells/mL and cultured at 25°C. The cell concentration was determined every hour by direct cell count using a hemocytometer. Cells immediately before division that were longer and had a narrow middle section were considered "doubling cells."

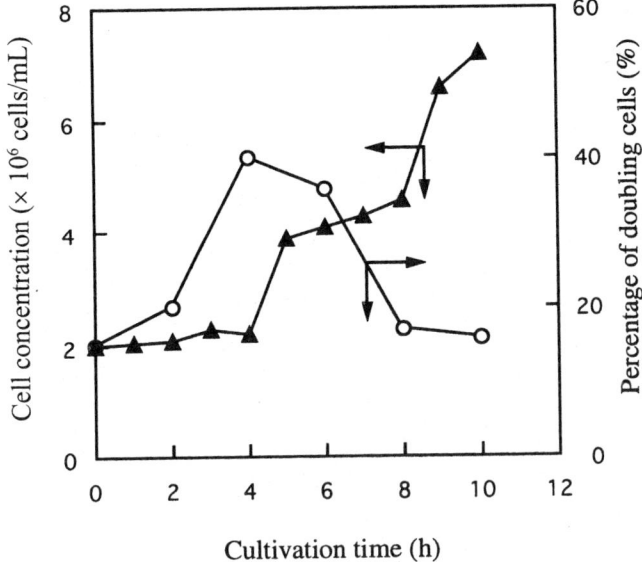

Fig. 1. Relationship between cell growth and percentage of doubling cells in synchronous culture. ▲, synchronous cell growth curve; ○, percentage of doubling cells.

Transmission Electron Microscopic Observation of Synchronous Culture

Synchronously cultivated cells were sampled at 2-h intervals and observed with a transmission electron microscope (Hitachi H700) operating at 150 kV. Electron-dense particles in the cells were counted and an average number of particles/cell were determined.

Fluorescence In Situ Hybridization Analysis

Synchronously cultivated cells were sampled at 2-h intervals and used for fluorescence *in situ* hybridization (FISH) analysis for *magA* expression. FISH was carried out using chemically synthesized fluorescein isothiocyanate (FITC)-labeled *magA* probe (FITC-5'-CAGATCGCGGAACGA ATGGACATG-3') according to the method of Chomczynski *(17)*.

Results and Discussion

Figure 1 shows the cell growth curve of synchronously grown AMB-1 at 25°C. Cell density remained constant for approx 4 h, doubled during a short period of about 1 h, then remained constant for another 4 h, which indicated synchronicity of cells. The percentage of doubling cells in culture increased from 20 to 40% rapidly and then decreased by 17%. This alteration of the proportion of doubling cells also indicated synchronized cell division. However, the percentage of doubling cells at dividing phase (4-h cultivation time) was 40%. In synchronously grown *Synechococcus* sp. using light/dark cycles, higher synchronicity (70–80% of doubling cells)

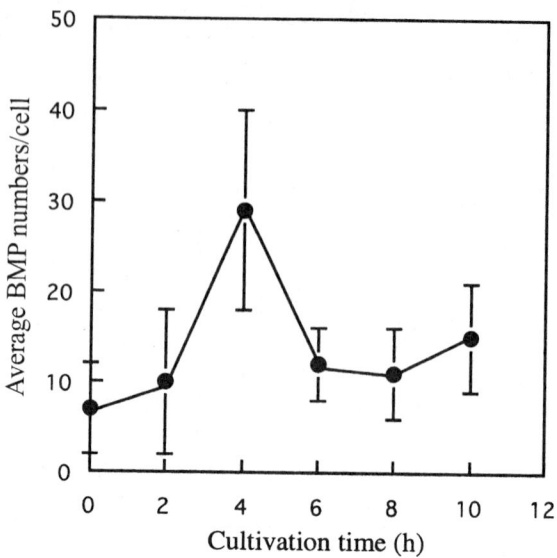

Fig. 2. Average BMPs/cell at various intervals of synchronous culture.

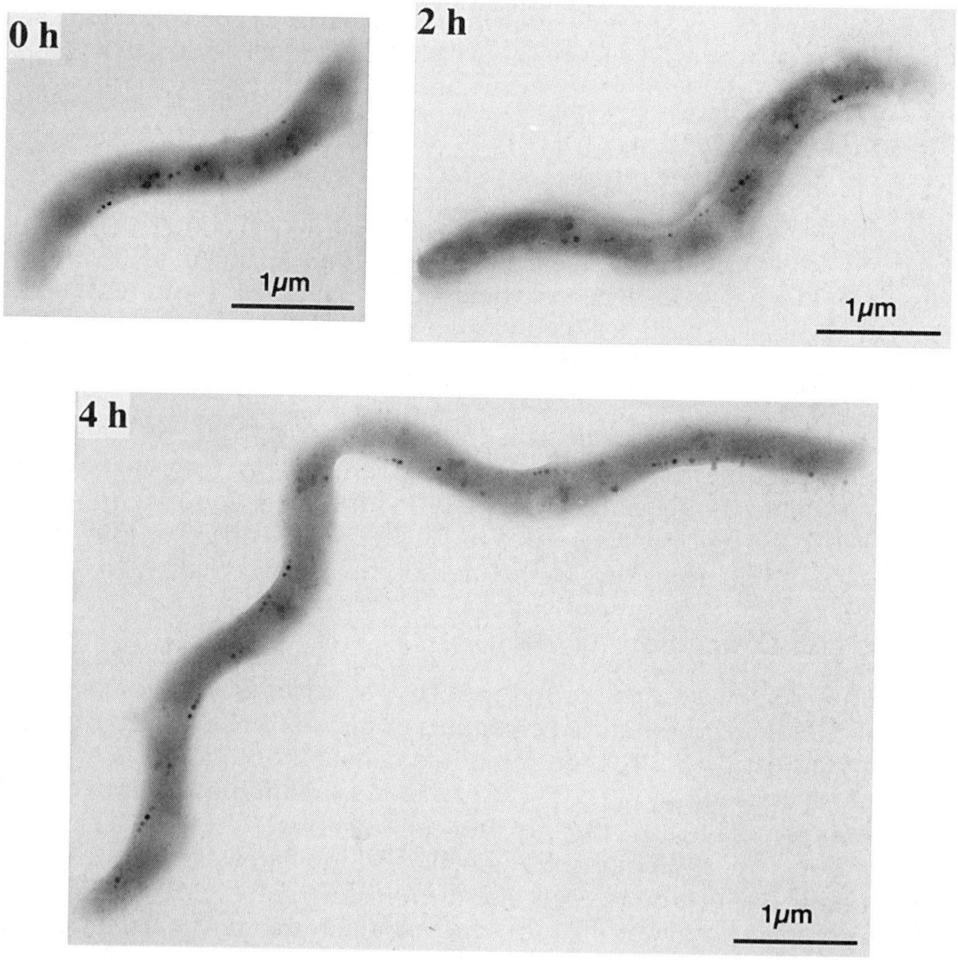

0 h

1μm

2 h

1μm

4 h

1μm

Fig. 3. Transmission electron micrographs of AMB-1 cells sampled at various intervals during synchronous growth. Bars = 1 μm.

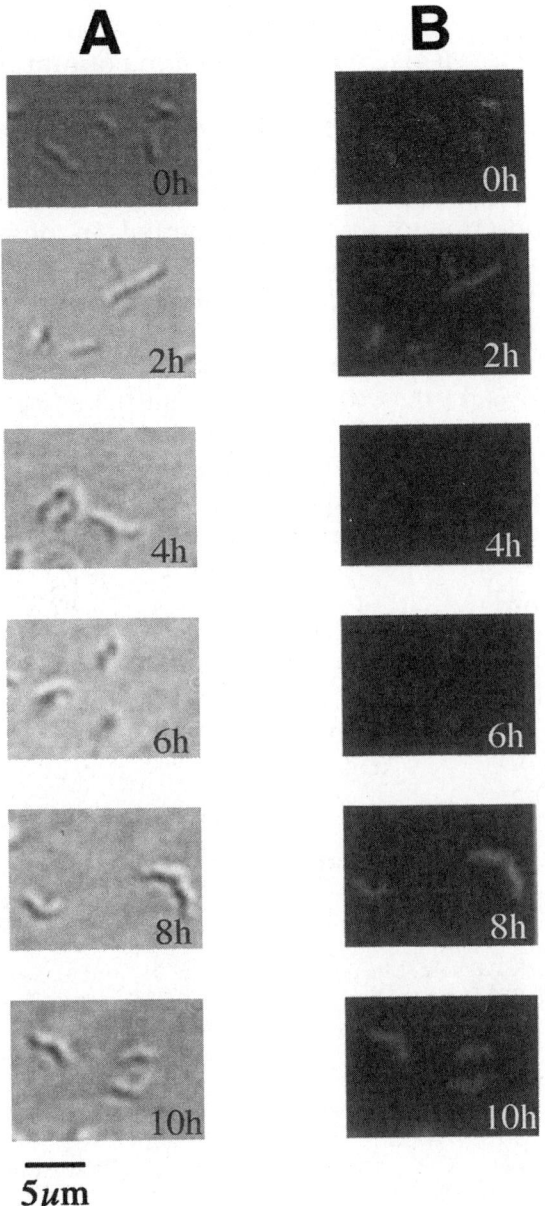

Fig. 4. Detection of *magA* mRNA in synchronous cells by FISH using FITC-labeled *magA* probe. (A) Difference interference contrast image; (B) fluorescence image.

was performed under continuous light conditions *(13)*. By contrast, lower synchronicity in AMB-1 may be caused by taking a time for temperature shift of medium and its shorter generation time than cyanobacteria.

To investigate BMP synthesis in synchronously grown AMB-1, change of the number of BMPs per cell was investigated. BMPs per cell gradually increased with increasing cell size, as shown in Figs. 2 and 3.

However, the number of BMPs sharply increased immediately before cell division in spite of cell size and reached a maximum number of 30 particles/cell at 4 h. The magnetic particles were equally divided to daughter cells. These data suggest that BMP synthesis by AMB-1 occurs just before cell division following a short period.

FISH gave strong fluorescence by FITC in whole cells at 0, 2, 8, and 10 h (Fig. 4). On the other hand, fluorescence signals in the cells were weak at 4 and 6 h. These results show that *magA* was transcribed during most of the cell cycle and repressed just before and after cell division. The repression of *magA* transcription at the phase of active BMP synthesis occurred, which may be caused by fluctuation of iron concentration in the cells. The transcription of *magA* is repressed during iron-sufficient conditions.

The BMP synthesis was found to rapidly increase just before cell division. The control of cell-cycle phase will contribute to enhanced BMP production in AMB-1. In future studies, we will design a new fed-batch culture system by controlling cell cycle for efficient BMP production.

Acknowledgment

This work was partially supported by Grant-in-Aid for Scientific Research on Priority Areas (A) no. 10145102, The Ministry of Education, Science, Sports, and Culture, Japan.

References

1. Blakemore, R. P. (1982), *Annu. Rev. Microbiol.* **36**, 217–238.
2. Balkwill, D. L., Maratea, D., and Blakemore, R. P. (1980), *J. Bacteriol.* **141**, 1399–1480.
3. Matsunaga, T., Sakaguchi, T., and Tadokoro, F. (1991), *Appl. Microbiol. Biotechnol.* **35**, 651–655.
4. Matsunaga, T., Kawasaki, M., Yu, X., Tsujimura, N., and Nakamura, N. (1996), *Anal. Chem.* **68**, 3551–3554.
5. Nakamura, N., Hashimoto, K., and Matsunaga, T. (1991), *Anal. Chem.* **63**, 268–272.
6. Nakamura, C., Burgess, J. G., and Matsunaga, T. (1995), *J. Biol. Chem.* **270**, 28,392–28,396.
7. Nakamura, C., Kikuchi, T., Burgess, J. G., and Matsunaga, T. (1995), *J. Biochem.* **118**, 23–27.
8. Matsunaga, T., Sato, R., Kamiya, S., Tanaka, T., and Takeyama, H. (1999), *J. Magn. Magn. Mater.* **194**, 126–131.
9. Tanaka, T. and Matsunaga, T. (2000), *Anal. Chem.* **72**, 3518–3522.
10. Matsunaga, T., Tsujimura, N., and Kamiya, S. (1996), *Biotechnol. Techniques* **10**, 495–500.
11. Matsunaga, T., Togo, H., Kikuchi, T., and Tanaka, T. *Biotechnol. Bioeng.*, in press.
12. Uchiyama, K. and Shioya, S. (1999), *J. Biotechnol.* **71**, 133–141.
13. Mitsui, A., Kumazawa, S., Takahashi, A., Ikemoto, H., Cao, S., and Arai, T. (1986), *Nature* **323**, 720–722.
14. Campbell, C., Takeyama, H., and Mitsui, A. (1994), *J. Mar. Biotechnol.* **2**, 39–43.
15. Sato, R., Miyagi, T., Kamiya, S., Sagaguchi, T., Thornhill, R. H., and Matsunaga, T. (1995), *FEMS Microbiol. Lett.* **128**, 15–20.
16. Blakemore, R. P., Maratea, D., and Wolfe, S. (1979), *J. Bacteriol.* **140**, 720–729.
17. Chomczynski, P. (1987), in *Current Protocols in Molecular Biology*, vol. 3, Ausuble, F. M., ed., Wiley, New York, pp. 14.1.1–14.7.11.

Overexpression
of Glucose-6-Phosphate Dehydrogenase
in Genetically Modified
Saccharomyces cerevisiae

FERNANDO H. LOJUDICE,[1] DANIEL P. SILVA,[2]
NILSON I. T. ZANCHIN,[1] CARLA C. OLIVEIRA,[1]
AND ADALBERTO PESSOA, JR.*[,2]

*[1]Department of Biochemistry, Chemistry Institute
and [2]Biochemical and Pharmaceutical Technology Department/FCF/
University of São Paulo, Av. Prof. Lineu Prestes, 580/B16, 05508-900,
São Paulo, SP, Brazil, E-mail: pessoajr@usp.br*

Abstract

Glucose-6-phosphate dehydrogenase (G6PD) (EC 1.1.1.49) is an abundant enzyme in *Saccharomyces cerevisiae*. This enzyme is of great interest as an analytical reagent because it is used in a large number of quantitative assays. A strain of *S. cerevisiae* was genetically modified to improve G6PD production during aerobic culture. The modifications are based on cloning the G6PD sequence under the control of promoters that are upregulated by the carbon source used for yeast growth. The results showed that *S. cerevisiae* acquired from a commercial source and the same strain produced by aerobic cultivation under controlled conditions provided very similar G6PD. However, G6PD production by genetically modified *S. cerevisiae* produced very high enzyme activity and showed to be the most effective procedure to obtain glucose-6-phosphate dehydrogenase. As a consequence, the cost of producing G6PD can be significantly reduced by using strains that contain levels of G6PD up to 14-fold higher than the level of G6PD found in commercially available strains.

Index Entries: Glucose-6-phosphate dehydrogenase; *Saccharomyces cerevisiae*; aerobic culture; molecular biology.

Introduction

Glucose 6-phosphate dehydrogenase (G6PD) (EC 1.1.1.49), which is an abundant enzyme in *Saccharomyces cerevisiae*, is the first enzyme of the

*Author to whom all correspondence and reprint requests should be addressed.

pentose phosphate pathway and shows a wide distribution in nature, being found in almost all animal tissues and microorganisms. This enzyme presents great interest as an analytical reagent because it is used in many quantitative assays, including the measurement of hexokinase and creatine kinase activities, adenosine triphosphate and hexose concentration, and as a marker for enzyme immunoassays *(1)*. The use of G6PD for measuring glucose in the presence of fructose constitutes an important tool for detecting illegal sugar addition in the final products of the wine and fruit juice industries. Thus, studies related to G6PD production by *S. cerevisiae* should become an important matter. This yeast has been demonstrated to be a very useful organism for the expression of various genes for protein purification and analysis, and for studies of control of gene expression.

As a eukaryotic organism, *S. cerevisiae* is a suitable host for high-level production of both secreted and soluble cytosolic proteins. Most vectors used to overexpress proteins in yeast are based on the multicopy 2μ natural plasmid, but vectors containing centromeric regions that determine the presence of a single copy per cell have also been developed. These plasmids contain sequences for replication and selection in *Escherichia coli* and *S. cerevisiae*, as well as yeast promoters and terminators for transcription by RNA polymerase II (RNA pol II) *(2)*. The *S. cerevisiae* phosphoglycerate kinase 1 (PGK1) gene encodes one of the most abundant mRNA and protein species in the cell, accounting for between 1 and 5% of the total cellular mRNA and protein *(3)*. Therefore, the PGK1 promoter is an attractive option for obtaining high levels of protein expression. Sequences encoding proteins of interest have also been cloned under the control of the GAL1 promoter, which is one of the most powerful tightly regulated promoters of *S. cerevisiae (4)*. Expression of the yeast GAL1 gene is undetectable in cells grown in the absence of galactose and is induced by more than 1000-fold when galactose becomes available *(5,6)*. Fusion of different promoters has also been successfully used for expression of proteins in yeast, such as the GAL1-PGK1 fusion *(7,8)*.

The present work is a comparative study of the overexpression of *S. cerevisiae* G6PD using genetically modified strains, which carry plasmid constructs containing the G6PD coding sequence under the control of the PGK1 and GAL1-PGK1 (GPF) promoters.

Materials and Methods

DNA Analysis Methods and Plasmid Construction

DNA cloning and electrophoresis analysis were performed as described by Sambrook et al. *(9)*. Both the G6PD coding sequence and the PGK1 promoter were amplified from yeast genomic DNA by using polymerase chain reaction (PCR). Oligodeoxynucleotides that served as primers for PCR reactions were purchased from Bio-Synthesis (Lewisville, TX). The primers used to PCR-amplify the G6PD coding sequence were

5'ggct<u>ggatcc</u>acagaaagagtaa3' (GPD-Bam5'), and 5'gtga<u>tctaga</u>cgataaatgaatg3' (GPD-Xba3'), which contain sites for the restriction enzymes *Bam*HI and *Xba*I, respectively (underlined). The primers used to PCR-amplify the PGK1 promoter were 5'cacc<u>ctcgag</u>ctattatcagggc3' (PGK-Xho5'), and 5'aatt<u>ggatcc</u> ttgatgatctgta3' (PGK-Bam3'), which contain sites for the restriction enzymes *Xho*I and *Bam*HI, respectively. PCR amplification reactions were performed according to instructions provided by the supplier (Gibco-BRL). Vectors YCpSUPEX1 *(8)* and YEplac181 *(10)* were used to construct plasmids YCpGPF-G6PD and YEpPGK-G6PD, which were used for expression of G6PD in yeast.

Media, Yeast Strains, and Genetic Techniques

Different carbon sources were added to yeast extract/peptone medium (YP) and synthetic medium (YNB). YPD and YNB-Glu contained 2% of glucose and YNB-Gal contained 2% of galactose as the carbon source. Yeast strains were incubated at 30°C as described by Sherman et al. *(11)*. *S. cerevisiae* W303-1a (MATa ade2-1 leu2-3,112 his3-11,15 trp1-1 ura3-1 can1-100) was used as the host strain for the plasmids YCpGPF-G6PD and YEpPGK-G6PD. YCpGPF-G6PD contains the URA3 gene, which is a genetic marker for yeast. Cells carrying this plasmid (named W303/YCpGPF-G6PD) were grown in YNB-Glu/Gal medium, containing 2% glucose; 2% galactose; 20 µg/mL of adenine, histidine, and tryptophan; and 30 µg/mL of leucine. YEpPGK-G6PD plasmid contains the LEU2 genetic marker, allowing for the cells (W303/YEpPGK-G6PD) to grow in YNB-Glu medium, containing 20 µg/mL of adenine, uracil, histidine, and tryptophan. Plasmids were transformed into yeast cells by using the lithium acetate method, as described previously *(11)*.

RNA Isolation

Total yeast RNA was isolated by using the hot phenol method *(12)* from 50-mL cultures. Briefly, cells were harvested by centrifugation, washed with cold water, and resuspended in 500 µL of 50 m*M* sodium acetate buffer, pH 5.0. Cells were lysed by vortexing, following the addition of 100 µL of 10% sodium dodecyl sulfate (SDS), 0.4 g of glass beads (0.4-mm diameter), and 1.0 mL of prewarmed (65°C) phenol. The aqueous phase was extracted with 1.0 mL of phenol-chloroform and precipitated with 1.5 mL of ethanol. RNA pellets were resuspended in 100 µL of diethylpyrocarbonate-treated water, and RNA concentration was determined by measuring the optical density (OD) at 260 nm.

Northern Hybridization

RNA electrophoresis was performed as described by Sambrook et al. *(9)*. Typically, a 30-µL sample was prepared by mixing 20 µg of total yeast RNA (up to 15 µL) with 6 *M* glyoxal (6 µL); DMSO (8 µL); and 0.1 *M* sodium phosphate buffer, pH 7.0 (3 µL). Samples were incubated at 50°C for 1 h and chilled on ice. Subsequently, 4 µL of stop solution (30% glycerol, 0.25% bromophenol blue, and 0.25% xylene cyanolFF) was added to the RNA samples, which were loaded onto a 1.3% agarose gel prepared with 10 mM sodium phosphate buffer, pH 7.0. Following electrophoresis, the RNA was blotted onto nylon membranes (Hybond-N; Amersham) and crosslinked by ultraviolet irradiation for 2 min (Ultra-Lum Electronic Dual Light Transillumination). DNA probes were labeled by random priming with ^{32}P-α-dATP by using the Gibco-BRL labeling system. Hybridization was performed at 42°C using a hybridization solution containing 5X SSPE, 5X Denhardt's solution, 50% formamide, 0.1% SDS, and 50 µg/mL of herring sperm DNA *(9)*. The DNA fragments used as probes corresponded to the yeast G6PD and ACT1 (actin 1) coding sequences. Radioactive bands were quantitated using a Molecular Dynamics Phosphorimager.

Aerobic Culture Process

To prepare the inoculum, *S. cerevisiae* (isolated from pressed yeast cake) was maintained in agar slant tubes containing 23 g/L of nutrient-agar (Difco, Detroit, MI) and 1.0 g/L of glucose, at 4°C. The cells were transferred to 250-mL Erlenmeyer flasks containing 50 mL of growth medium (15 g/L of glucose; 2 g/L of sucrose; 5.0 g/L of peptone; and 3.0 g/L of yeast extract, pH 4.5) and incubated for 10 h at 35°C on a rotary shaker (175 rpm) (New Brunswick). A volume of 0.45 L of inoculum (4.7 g$_{dry cell}$/L) was poured into a 5-L NBS-MF 105-New Brunswick bench fermentor (coupled with NBS dissolved oxygen controller, DO-81) containing 2.55 L of culture medium. The composition of the culture medium was 20 g/L of glucose, 5.0 g/L of peptone, 3.0 g/L of yeast extract, 2.4 g/L of Na$_2$HPO$_4$·12H$_2$O, 0.075 g/L of MgSO$_4$·7H$_2$O, and 5.1 g/L of (NH$_4$)$_2$SO$_4$ at pH 4.0. The culture was grown batchwise at 35 ± 0.5°C and pH 4.0 ± 0.1. Foaming was controlled by the addition of a mixture containing silicone emulsion (10%) and water (90%) (Thomas Scientific, Swedesboro, NJ) (addition dropwise when needed). Agitation and aeration rates were, respectively, 800 rpm and 2.3 vvm (oxygen transfer volumetric coefficient, K_La = 230 h^{-1}). Dissolved oxygen tension was measured by polarographic electrode (Ingold). Airflow was measured by an in-line rotatometer and was set using a needle valve. The pH of the medium during the aerobic culture was measured by combination electrode (Ingold) at the desired value by automatic addition of 0.5 M NaOH and 0.5 M H$_2$SO$_4$.

Measurement of Cell Concentration and Protein Determination

The cell concentration values of the cultures were obtained by using a calibration curve to correlate OD with dry weight (grams/liter). The amount of total protein was determined according to the Coomassie blue method described by Bradford *(13)* using bovine serum albumin as a protein concentration standard.

G6PD Activity Assays

Precultures of control strains and of strains containing either plasmid YCpGPF-G6PD or YEpPGK-G6PD were incubated for 18 h at 30°C in 2 mL of YNB or YPD medium. Cells from the preculture were transferred to a 50-mL culture of the same medium and incubated for 18 h at 30°C up to the OD$_{600}$ of approx 1.0. Subsequently, cells were harvested, immediately frozen, and stored at –20°C. For extract isolation, cells were resuspended in 50 mM Tris-HCl buffer, pH 7.5; 5 mM MgCl$_2$; 0.2 mM EDTA; 10 mM β-mercaptoethanol; 2 mM aminocaproic acid; and 1 mM phenylmethylsulfonyl fluoride. The cells were disrupted by submitting to a vortex (PHOENIX AT56) in the presence of glass beads (0.5-mm diameter). Cell debris and glass beads were removed by centrifugation (2880g for 10 min at 1°C), and the G6PD activity of the supernatant was measured by spectrophotometric quantitation of reduced NADP at 30°C, as described by Bergmeyer *(1)*. One G6PD unit was defined as the amount of enzyme catalyzing the reduction of 1 μmol of NADP/min under the assay conditions.

Results and Discussion

Construction of Plasmids YCpGPF-G6PD and YEpPGK-G6PD

One of the aims of this work was to obtain yeast strains expressing G6PD protein at levels higher than the level of G6PD of the commercially available wild-type strains. Therefore, plasmids were constructed in which the G6PD coding region is under the control of two strong RNA pol II transcription promoters: the PGK1 promoter and the GPF promoter that was constructed by making a fusion between the GAL1 and PGK1 promoters *(8)*. The GPF promoter has the advantage of maintaining the features of the inducible GAL1 promoter and of having a unique transcription start site, which is determined by the PGK1 portion of GPF *(8)*. The yeast G6PD coding sequence was isolated by PCR amplification using genomic DNA as template and a pair of primers complementary to its 5' and 3' ends. The PCR reaction yielded a product showing an electrophoretic mobility of 1.5 kb pairs on agarose gel, which is the expected size of the G6PD coding sequence. This DNA fragment was submitted to restriction digestion with the enzymes *Bam*HI and *Xba*I and ligated with plasmid YCpSUPEX1, which had been digested with the same restriction enzymes. The ligated DNA fragments were transformed into *E. coli* cells. Plasmid DNA was isolated from the transformants and screened for clones containing the G6PD insert by means of restriction digestion with the enzymes *Hind*III, *Xba*I, and *Bam*HI. The resulting plasmid was named YCpGPF-G6PD (Fig. 1A). The PGK1 promoter was PCR-amplified from genomic DNA using a pair of primers complementary to its 5' and 3' ends that contain sites for the restriction enzymes *Xho*I and *Bam*HI, respectively. The DNA fragment obtained showed the correct size (0.6 kb) and was submitted to restriction digestion

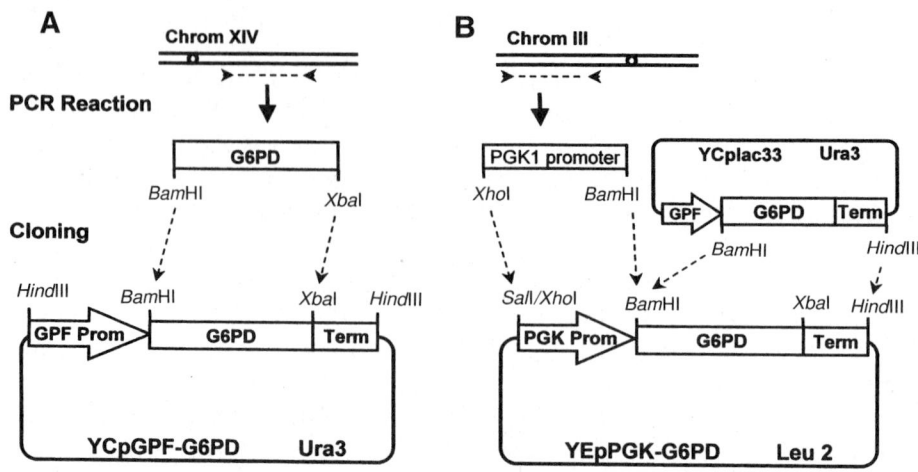

Fig. 1. Construction of plasmids YCpGPF-G6PD and YEpPGK-G6PD. **(A)** The G6PD coding sequence was PCR-amplified from genomic DNA and inserted into the *Bam*HI and *Xba*I sites of vector YCpSUPEX1 to construct plasmid YCpGPF-G6PD. **(B)** The PGK1 promoter was amplified by using PCR from genomic DNA. A 1.8-kb *Bam*HI-*Hind*III fragment containing the G6PD coding sequence and the PGK1 transcription terminator was isolated from YCpGPF-G6PD. This fragment was ligated to PGK1 promoter digested with *Xho*I and *Bam*HI, and plasmid YEplac181 digested with *Sal*I and *Hind*III. Tha *Sal*I and *Xho*I sites are compatible for ligation, although none of them is regenerated. The resulting plasmid was named YEpPKG-G6PD.

with *Xho*I and *Bam*HI. This fragment was cloned into the vector YEplac181 *(10)* at the same time as the G6PD coding sequence and the PGK1 transcription terminator, which were isolated from the YCpGPF-G6PD plasmid as a *Bam*HI-*Hind*III 1.8-kb fragment (Fig. 1B). The new plasmid was named YEpPGK-G6PD.

Quantitation of G6PD mRNA

The gene copy number per cell and the strength of the transcription promoter are major factors that determine the level of a given protein in the cell, provided that there is no posttranscriptional regulatory mechanism. In this work, we used both strong promoters and, in the case of YEpPGK-G6PD, a high copy vector. This should lead to an increase in both G6PD mRNA and protein levels in the cell. We performed Northern analysis to determine the steady-state level of G6PD mRNA in control cells and in cells carrying plasmids YCpGPF-G6PD and YEpPGK-G6PD (Fig. 2). Actin 1 (ACT1) encodes a constitutively expressed protein that has been widely used as an internal control. Interestingly, W303-1a cells grown in synthetic medium (YNB) showed an increased amount of both G6PD and ACT1 mRNAs compared to W303-1a cells incubated in rich medium (YPD). Quantitation of G6PD mRNA revealed a fourfold increase in G6PD mRNA in cells containing the YCpGPF-G6PD plasmid, when incubated in media

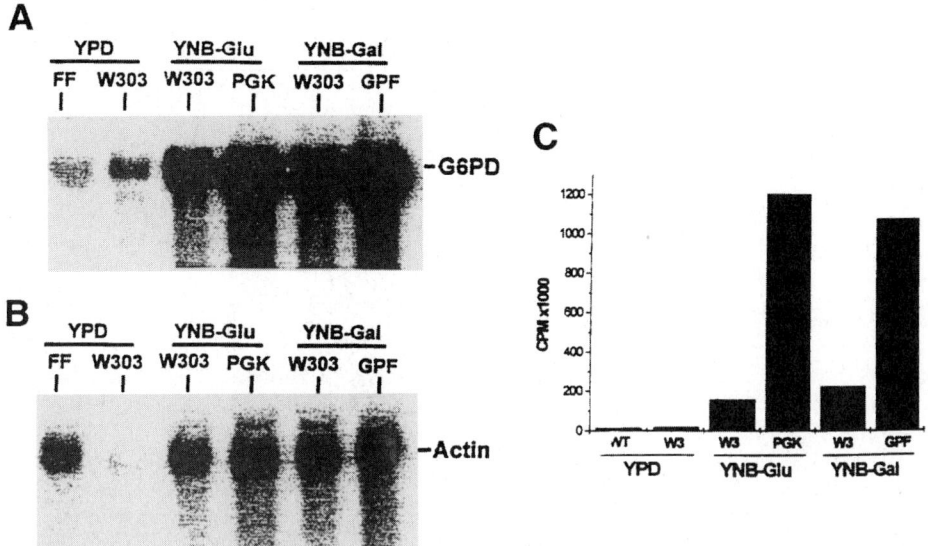

Fig. 2. Quantitation of G6PD mRNA steady-state level. **(A,B)** Northern blot analysis of G6PD and actin mRNAs, respectively; **(C)** graph showing quantitation of Northern blots (G6PD/actin ratio). WT, commercially available strain; W303, W303-1a; PGK, W303-1a cells carrying plasmid YEpPKG-G6PD; GPF, W303-1a cells carrying plasmid YCpGPF-G6PD. YPD, YNB-Glu, and YNB-Gal indicate the medium in which cells were incubated.

containing galactose, compared to the W303-1a parental strain (Fig. 2C). In cells carrying plasmid YEpPGK-G6PD, the level of G6PD mRNA was increased fivefold relative to the W303-1a strain.

Although the GAL1 promoter has been demonstrated to be stronger than the PGK1 promoter, the results obtained in this work can be explained because the GPF-G6PD construct is in a single-copy plasmid, whereas the PGK-G6PD clone is present in a multicopy plasmid. The advantage of plasmid YEpPGK-G6PD is that expression can be performed in medium containing glucose as the carbon source, which is a reagent less expensive than galactose.

G6PD Production

Table 1 shows the G6PD activities obtained from the different *S. cerevisiae* strains. The genetically modified strains produced higher (up to ~40 times $U/mg_{protein}$ and up to ~108 times U/g_{cell}) enzyme-specific activities compared with those of the wild types. Specifically, the strains carrying plasmids YCpGPF-G6PD and YEpPGK-G6PD showed a 6- and 14-fold increase in G6PD activity, respectively, compared with the parental strains grown under the same conditions. Consistent with the results obtained from G6PD RNA quantitation, cells carrying vector YEpPGK-G6PD showed a higher G6PD activity than cells carrying vector YCpGPF-

Table 1
G6PD Activities Obtained from Different Sources

Enzyme source		G6PD activity	
		U/mg_{prot}	U/g_{cell}
Baker's yeast (commercial source)		0.22	50
S. cerevisiae produced by aerobic cultivation		0.30	49
Strain[a]	Medium		
W303	YPD	0.36	191.8
W303GAL	SC-Gal	0.28	140.8
GAL-G6PD[b]	SC-Gal	1.72	1437.9
W303GLU	SC-Glu	0.61	211.7
PGK-G6PD[b]	SC-Glu	8.90	5303.8

[a]W303GAL, W303 grown in galactose medium; GAL-G6PD, W303/YCpGPF-G6PD in galactose medium; W303GLU, W303 grown in glucose medium; PGK-G6PD, W303/UEpPGK-G6PD in glucose medium.

[b]Genetically modified *S. cerevisiae*

G6PD. This results clearly show that the technique employed in this work to improve G6PD production is efficient and promising.

Conclusion

S. cerevisiae acquired from a commercial source and the same strain produced by aerobic cultivation under controlled conditions provided very similar G6PD productivities. However, G6PD production by *S. cerevisiae* can be improved by using the techniques of recombinant DNA. Its production can be influenced by substituting the native transcription promoter by promoters that are upregulated specifically by the carbon source of the culture medium used to grow the cells. Therefore, the genetically modified *S. cerevisiae* produced very high enzyme activity and proved to be the most effective procedure to obtain G6PD.

The results presented here show that the cost of producing G6PD can be significantly reduced by using strains that contain levels of G6PD up to 14-fold higher than the level of G6PD found in commercially available strains.

Acknowledgments

We acknowledge receipt of fellowships from Fundacao de Amparo a Pesquisa do Estado de Sao Paulo (FAPESP)/Brazil. We also acknowledge the financial support of FAPESP and Conselho Nacional de Desenvolvimento Científico e Tecnologico (CNPq)/Brazil and Prof. Dr. José Abrahão-Neto.

References

1. Bergmeyer, H. U. (1984), in *Methods of Enzymatic Analysis*, vol. 2, 3rd ed., Bergmeyer, H. V., Bergmeyer, J., and Grasl, M., eds., *Verlag Chemie*, Weinheim, Germany, pp. 222, 223.
2. Romanos, M. A., Scorer, C. A., and Clare, J. J. (1992), *Yeast* **8,** 423–488.
3. Ogden, J. E., Stanway, C., Kim, S., Mellor, J., Kingsman, A. J., and Kingsman, S. M. (1986), *Mol. Cell. Biol.* **6,** 4335–4343.
4. West, R. W. Jr., Yocum, R. R., and Ptashne, M. (1984), *Mol. Cell. Biol.* **4,** 2467–2478.
5. Johnston, M. (1987), *Microbiol. Rev.* **51,** 458–476.
6. Axelrod, J. D., Reagan, M. S., and Majors, J. (1993), *Genes Dev.* **7,** 857–869.
7. Kingsman, S. M., Cousens, D., Stanway, C. A., Chambers, A., Wilson, M., and Kingsman, A. J. (1990), *Methods Enzymol.* **185,** 329–341.
8. Oliveira, C. C., van den Heuvel, J. J., and McCarthy, J. E. G. (1993), *Mol. Microbiol.* **9,** 521–532.
9. Sambrook, J., Fritsch, E. F., and Maniatis, T. (1989), *Molecular Cloning—A Laboratory Manual*, 2nd ed., Cold Spring Harbor Laboratory Press, Cold Spring Harbor, NY.
10. Gietz, R. D. and Sugino, A. (1988), *Gene* **74,** 527–534.
11. Sherman, F., Fink, G. R., and Hicks, J. B. (1986), *Laboratory Course Manual for Methods in Yeast Genetics*, Cold Spring Harbor Laboratory, Cold Spring Harbor, NY.
12. Köhrer, K. and Domdey, H. (1991), *Methods Enzymol.* **194,** 398–405.
13. Bradford, M. A. (1976), *Analyt. Biochem.* **72,** 248–254.

Biosorption of Heavy Metals
by Bacteria Isolated from Activated Sludge

WA C. LEUNG,[1] HONG CHUA,[2] AND WAIHUNG LO*,[1]

*Departments of [1]Applied Biology and Chemical Technology
and Open Laboratory of Chiral Technology
and [2]Civil and Structural Engineering, The Hong Kong Polytechnic University,
Hung Hom, Hong Kong SAR, China, E-mail: bctlo@polyu.edu.hk*

Abstract

Twelve aerobic bacteria from activated sludge were isolated and identi-fied. These included both Gram-positive (e.g., *Bacillus*) and Gram-negative (e.g., *Pseudomonas*) bacteria. The biosorption capacity of these strains for three different heavy metals (copper, nickel, and lead) was determined at pH 5.0 and initial metal concentration of 100 mg/L. Among these 12 isolates, *Pseudomonas pseudoalcaligenes* was selected for further investigation owing to its high metal biosorption capacity. The lead and copper biosorption of this strain followed the Langmuir isotherm model quite well with maximum biosorption capacity (q_{max}) reaching 271.7 mg of Pb^{2+}/g of dry cell and 46.8 mg of Cu^{2+}/g of dry cell at pH 5.0. Study of the effect of pH on lead and copper removal indicated that the metal biosorption increased with increasing pH from 2.0 to 7.0. A mutual inhibitory effect was observed in the lead-copper system because the presence of either ion affected the sorption capacity of the other. Unequal inhibitions were observed in all the nickel binary systems. The increasing order of affinity of the three metals toward *P. pseudoalcaligenes* was Ni < Cu < Pb. The metal biosorptive potential of these isolates, especially *P. pseudoalcaligenes*, may have possible applications in the removal and recovery of metals from industrial effluents.

Index Entries: Activated sludge; biosorption; copper adsorption; lead removal; bioremediation.

Introduction

Over the past decade, the consumption of metals and chemicals in the process industries has increased dramatically. Industrial uses of metals such as metal plating and tanning, as well as industrial processes utilizing metal as catalysts, have generated large amounts of aqueous effluents that contain high levels of heavy metals. These heavy metals include cadmium,

*Author to whom all correspondence and reprint requests should be addressed.

chromium, cobalt, copper, manganese, mercury, nickel, silver, and zinc. Metal-polluted industrial effluents discharged into sewage treatment plants could lead to high metal concentrations in the activated sludge. Microbial populations in metal-polluted environments contain micro-organisms that have adapted to the toxic concentrations of heavy metals and become "metal resistant" *(1)*.

At present, metal-polluted industrial effluents are mostly treated by chemical methods, such as chemical precipitation, electrochemical treatment, and ion exchange. These methods provide only partially effective treatment and are costly to implement and use, especially when the metal concentration is low. The alternative use of microbe-based biosorbents for the removal and recovery of toxic metals from industrial effluents can be an economical and effective method for metal removal. The metal-removing ability of microorganisms including bacteria *(2,3)*, microalgae *(4,5)*, and fungi *(6)* has been studied extensively. Their capacity for heavy metal removal is apparently higher than for conventional methods and the uptake of heavy metals can be selective *(7,8)*. Microbial cells can also be supplied inexpensively as waste from industrial fermentation processes as well as biologic wastewater treatment plants *(9,10)*.

The present study was conducted to characterize the metal biosorption behavior of bacteria in an activated sludge process treating both municipal and metal-contaminated industrial wastewater. Heavy metals studied included copper, nickel, and lead. One bacterial strain, which was effective in removing metals, was selected for further investigation. The effects of initial metal concentration and pH on the metal biosorption capacity of the selected strain were studied extensively. The metal biosorption isotherms of the selected strain were also compared with activated sludge.

Materials and Methods

Isolation Procedures and Identification

Fresh activated sludge was collected from the return sludge channel at the Shatin Sewage Treatment Works in Hong Kong. The activated sludge was serially diluted with distilled and deionized water. Aliquots (0.1 mL) were spread on nutrient agar and cultivated in a Shell Lab model 2020 incubator at 30°C for 3 d. The isolated colonies were identified by Sherlock Microbial Identification System (MIDI, Newark, DE) and API 20NE as well as 20E tests (BioMerieux, France). The API tests employ a series of biochemical tests, such as amino acid decarboxylation and carbohydrates fermentation, for identifying the bacteria.

Cultivation of Biomass

The bacterial cells of each isolate were grown in 1-L conical flasks containing 200 mL of nutrient broth on an orbital shaker at 200 rpm and 30°C. The 72-h cultivated cells were harvested by centrifugation (Beckman

Model J2-21) at 9000g force for 30 min. After two rinses with distilled and deionized water, the cells were suspended in a designated volume of distilled and deionized water for preparing the biomass stock solution. The concentration of the biomass stock solution was determined gravimetrically by withdrawing 5 mL of the solution and oven drying it at 105°C for 24 h.

Screening Tests for Biosorption

Enough bacterial cells to create a final concentration of 1 to 2 g of cell/L was suspended in 100 mL of solution containing 100 mg/L of respective heavy metals (Cu, Ni, Pb) in Nalgene propylene bottles, which were gently agitated at 25°C. The pH of all the metal solution was adjusted to 5.0 by adding 0.1 M NaOH and 0.1 M HNO$_3$ just before the experiments and at 21 h during the experiments. Samples (5 mL) were taken from the solution at 3 and 24 h and were subsequently centrifuged at 9000g force for 10 min (Beckman Model J2-21). The concentration of each heavy metal in the supernatants was determined using a model 100 Perkin-Elmer atomic absorption spectrophotometer. The biomass concentration was determined after oven drying at 105°C for 24 h.

Kinetics of Biosorption

The kinetics of metal biosorption by a selected isolate, *Pseudomonas pseudoalcaligenes*, was investigated for two metals: lead and copper. One hundred milliliters of a metal solution containing 50 ppm of Cu or 500 ppm of Pb was mixed with about 1.0 g/L of biomass and incubated on a shaker at 200 rpm and 25°C. Samples (3 mL) were withdrawn during the 24 h of incubation at predetermined time intervals and centrifuged immediately (Sigma® 201m) at 15,000g force. The metal concentration of the supernatant was analyzed by atomic absorption spectrophotometry.

Adsorption Isotherms

One of the isolated bacteria, *P. pseudoalcaligenes*, had outstanding metal-biosorption ability and was selected for further studies. The selected isolate and activated sludge (1 to 2 g of cell/L) were suspended in solutions containing different heavy metal concentrations. The pH of the metal solution was adjusted to 5.0 by adding 0.1 M NaOH and 0.1 M HNO$_3$ just before experiments and at 21 h during experiments. After 24 h of incubation at 25°C, 5-mL samples were taken from the solutions, and the metal concentrations in the supernatants were measured by atomic absorption spectrophotometry.

Effect of pH on Biosorption

The initial metal concentration of the experiment was 100 mg/L for copper and 200 mg/L for lead. The pH of the metal and biomass suspension was adjusted to the appropriate value by adding 0.1 M NaOH and

0.1 M HNO$_3$ just before experiments and at 21 h during experiments. After 24 h of incubation at 25°C, 5-mL samples were taken from the solutions and centrifuged, and the metal concentrations in the supernatants were measured.

Binary Metal Biosorption System

Three metals—lead, copper, and nickel—were considered for binary biosorption studies; two of the three metals were considered each time. The initial metal concentration of metal one was changed from 0.2 to 2.0 mmol/L, and the initial concentration of metal two was held constant at 0.2, 0.5, 1.0, and 2.0 mmol/L. The other conditions were the same as in the single metal biosorption experiments.

Results and Discussion

Isolation and Identification

In nonselective isolation, a dilution of activated sludge was streaked on nutrient agar plates and 12 colonies were isolated. Isolation of microbial strains from activated sludge or other waste streams for biosorption of heavy metals has been studied extensively *(11–13)*. It is impossible to isolate all the bacteria that might be present in the sludge; only the dominant cultivable species were isolated in the nonselective isolation. Table 1 gives the identification results by the MIDI Sherlock Microbial Identification system as well as API 20 NE and 20E systems. Species A–M were nonselectively isolated bacteria.

The heterotrophic isolates from the activated sludge belonged to a wide variety of species including Gram negative and Gram positive. Most were rods and a few bacteria were filamentous or coccus. *Pseudomonas*, *Bacillus*, and *Aeromonas* species are commonly found in activated sludge *(1)*.

Biosorption of Heavy Metals

Batch biosorption experiments were conducted to investigate the metal-removing ability of each bacterium isolated. The metals chosen for tests were copper, nickel, and lead; Figure 1 presents the results.

For copper biosorption, species L and M had a Q_e (equilibrium biosorption capacity, mg of metal Cu^{2+}/g of dry biomass) >35 mg/g among the 12 isolates with an initial concentration of 100 mg/L (Fig. 1). Only three species had a Q_e >10 mg/g among all the isolates in nickel biosorption test, and no outstanding species for nickel removal were found. Under the conditions of initial concentration of 100 mg of Pb^{2+}/L and pH 5.0, the Q_e values of species H, L, and M were about 60 mg/g and about five species had a Q_e of about 50 mg/g.

Significant differences were observed in biosorption of metal ions by the various bacterial strains examined. Because metal biosorption from solution is predominantly owing to physicochemical interactions between

Table 1
Identification of Species

Species	Nonselectively isolated
A	*Bacillus pumilus* GC subgroup B
B	*Neisseria sicca*
C	*Aeromonas hydrophila*
D	*Pseudomonas* sp.
E	*Xanthomonas maltophilia*
F	*Bacillus lentimorbus*
G	*Pseudomonas putida*
H	*Bacillus subtilis*
J	*Gordona bronchialis*
K	*Kocuria varians*
L	*Pseudomonas pseudoalcaligenes*
M	*Micrococcus luteus* GC subgroup B

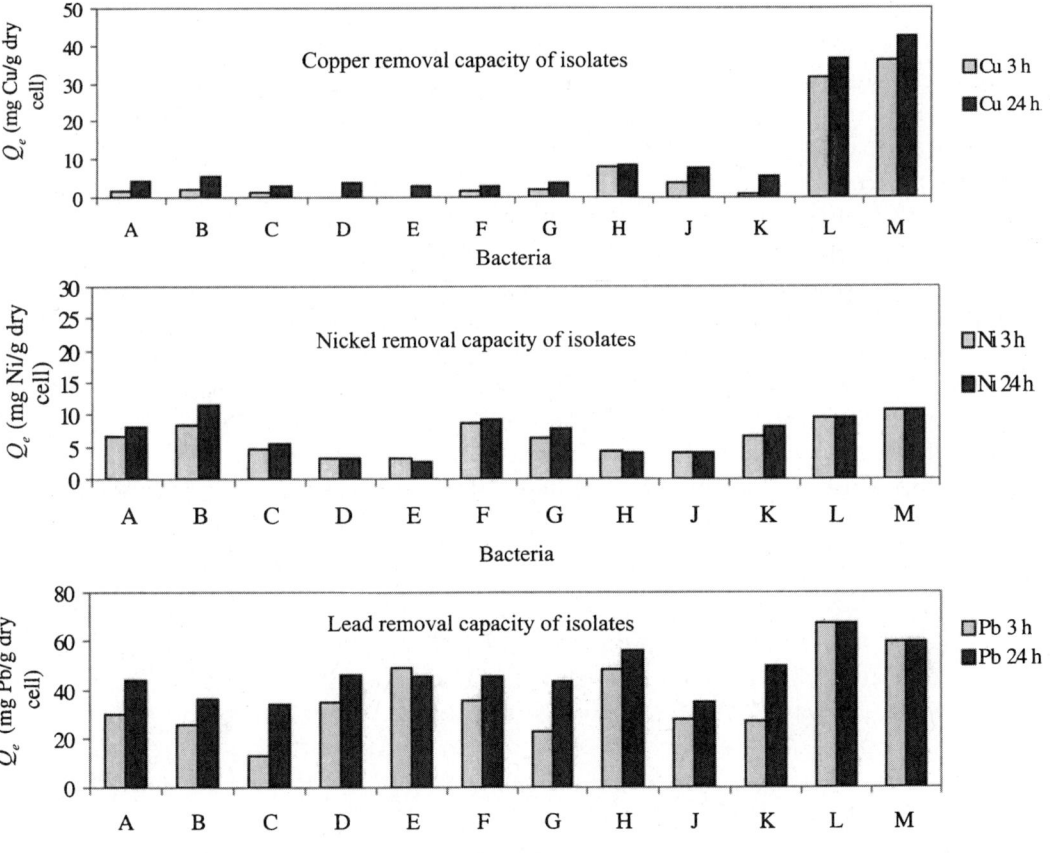

Fig. 1. Metal removal capacity of isolates with retention times of 3 and 24 h on 100 ppm initial concentration at pH 5.0.

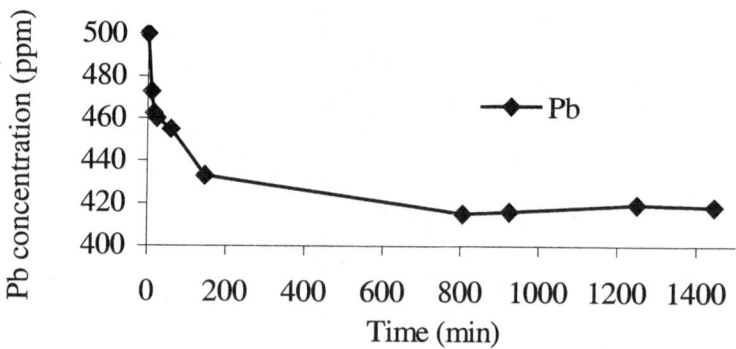

Fig. 2. Kinetics of lead sorption by *Pseudomonas pseudoalcaligenes* at pH 5.0.

the biomass and metal in solution, morphologic differences existing within the biomass can greatly influence the biosorption process. The stereochemical differences in the structures of the cell envelope can make a significant difference in the acceptance of metallic ions by these structures *(14)*. Cell wall is the most important structure that may form the cell envelope, but capsules, S-layers, and sheaths are commonly found superimposed on the wall *(15)*. The Gram-negative bacteria (e.g., *Pseudomonas*) possess cell walls that are chemically and structurally more complex than the Gram-positive bacteria (e.g., *Bacillus*), resulting in different metal biosorption capacities. Many species of bacteria isolated from activated sludge have been shown to produce extracellular polymers, which provide surface sites for adsorbing and complexing heavy metals. Increased production of extracellular polymers may enhance metal binding *(16)*.

Most studies have reported that copper and lead are removed more efficiently than many other metals, and nickel has one of the lowest removal efficiencies associated with it *(17)*. In the present study, similar results were obtained. Copper and lead were significantly removed whereas nickel was poorly removed. Among all the isolates, species L (*P. pseudoalcaligenes*) was one of the most effective bacteria in removing copper and lead from the aqueous solutions. Hence, this species was selected for further investigation on the kinetics and equilibrium of biosorption and effect of pH and competing cations.

Kinetics of Biosorption

The kinetics of metal biosorption by *P. pseudoalcaligenes* is shown in Figs. 2 and 3. Biosorption was rapid, since the initial decrease in metal concentration was very fast. Half of the total metal adsorption occurred within 10 min. For copper, the equilibrium concentration was attained at about 500 min (Fig. 3), and 800 min was required for lead adsorption to reach equilibrium (Fig. 2).

Metal Biosorption Isotherms

The Langmuir and Freundlich isotherms were used to simulate the biosorption of lead and copper by species L and activated sludge. Table 2

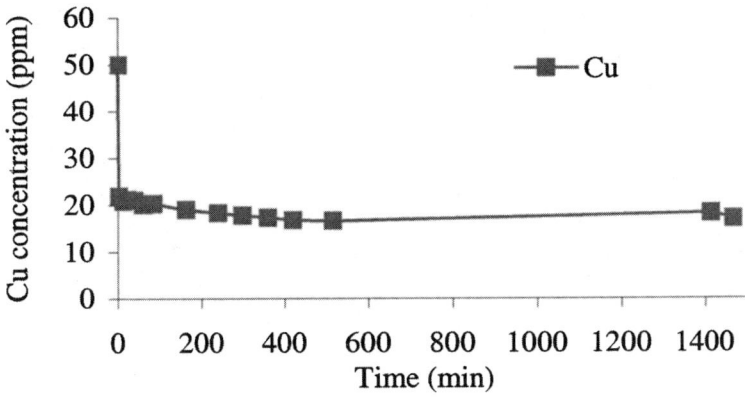

Fig. 3. Kinetics of copper sorption by *P. pseudoalcaligenes* at pH 5.0.

Table 2
Parameters for Langmuir and Freundlich Isotherms

Parameter	Definition
Q_e or Q	Equilibrium adsorption capacity (mg or mmol of metal/g of dry cell)
C_e	Equilibrium metal concentration (mg/L or mmol/L)
Q_{max}	Maximum biosorption capacity (mg of metal/g of dry cell)
b	Langmuir isotherm constant (L/mg metal)—measures effectiveness of biosorption at low metal concentrations
k	Freundlich coefficient (L/g of dry cell)—represents amount of metal adsorbed when concentration of solution in equilibrium is unity
n	Freundlich isotherm constant—measures impact on biosorption of change in residual solution concentration from unity

gives the parameters used in the Langmuir and Freundlich isotherms. Figures 4 and 5 show the copper biosorption isotherm and linearized Langmuir isotherm for both *P. pseudoalcaligenes* and activated sludge.

Figures 6 and 7 show the lead biosorption isotherms for both *P. pseudoalcaligenes* and activated sludge. Table 3 gives the calculated parameters for the Langmuir and Freundlich isotherms. The copper and lead biosorption data were well described by both the Langmuir and Freundlich isotherms with good linear relation (Table 3). Q_{max} represents the saturation level of sorbed metal at high metal concentrations. Based on the Q_{max} values, species L has higher maximum biosorption capacity for both copper and lead than activated sludge. Species L also has higher b values for both copper and lead compared to activated sludge. Since the parameter b measures the effectiveness of biosorption at low metal concentrations, species L has higher adsorption ability compared to activated sludge at low metal concentrations. In other words, species L has higher affinity than activated sludge.

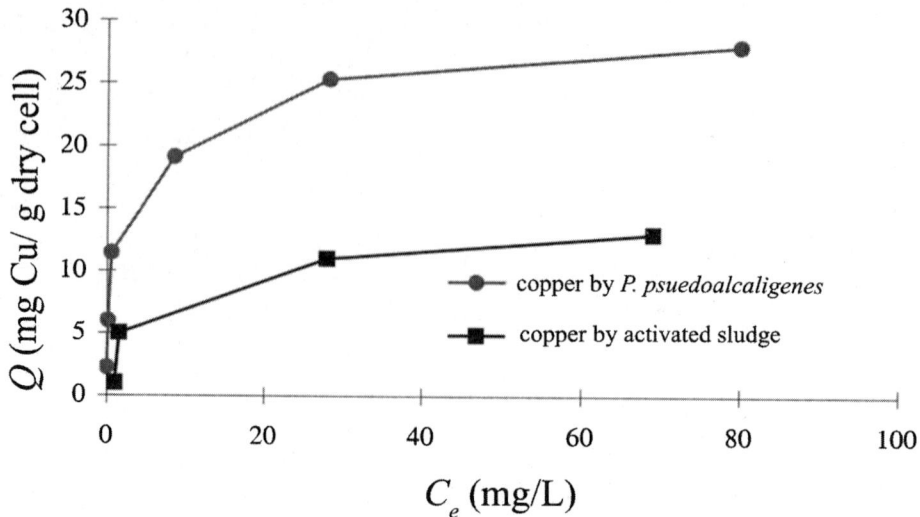

Fig. 4. Equilibrium isotherm for copper biosorption by *P. pseudoalcaligenes* and activated sludge.

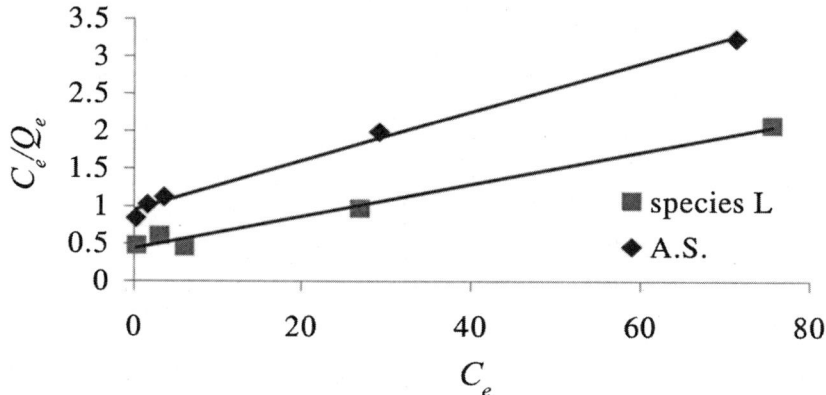

Fig. 5. Linearized Langmuir isotherm plot for copper biosorption by *P. pseudo-alcaligenes* and activated sludge (A.S.).

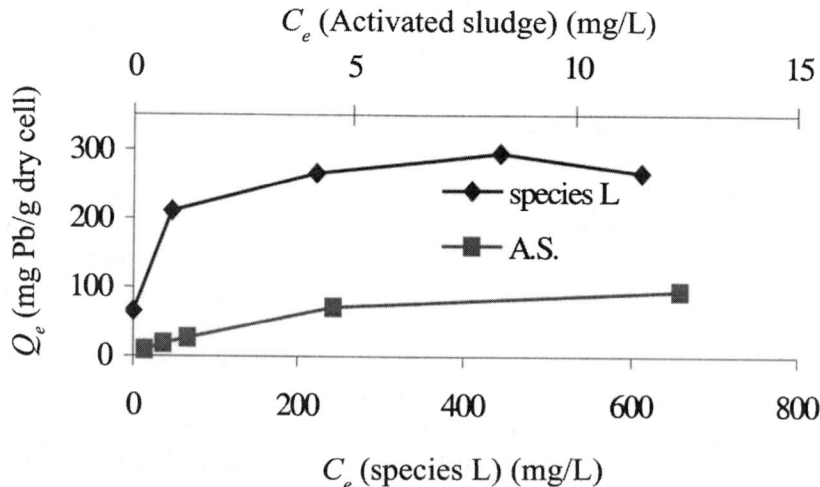

Fig. 6. Equilibrium isotherm for lead biosorption by *P. pseudoalcaligenes* and activated sludge (A.S.).

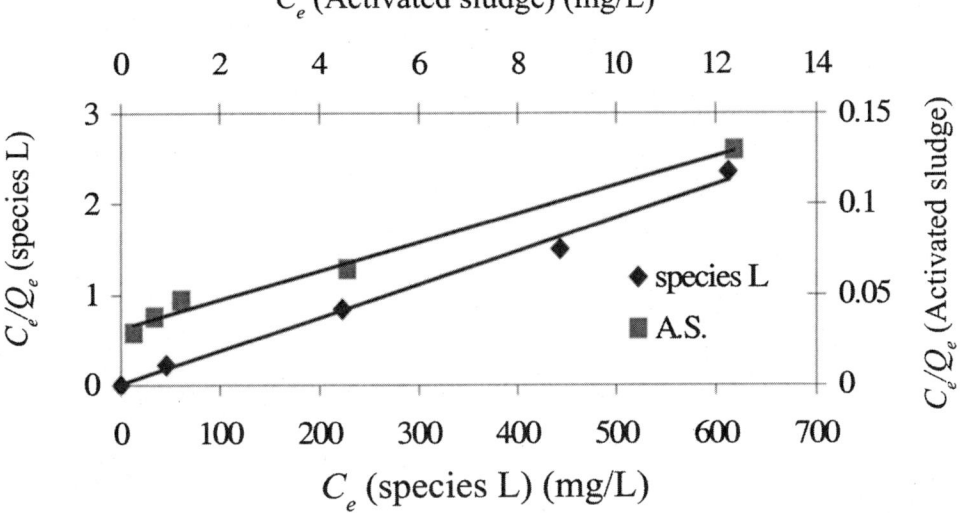

Fig. 7. Linearized Langmuir isotherm plot for lead biosorption by *P. pseudoalcaligenes* and activated sludge (A.S.).

Table 3
Langmuir and Freundlich Parameters for Copper and Lead Biosorption

	Copper		Lead	
Langmuir	Species L	Activated sludge	Species L	Activated sludge
Q_{max} (mg/g of dry cell)	46.8	30.7	271.7	126.4
b (L/mg)	0.048	0.034	0.46	0.25
r^2	0.984	0.994	0.992	0.989
Freundlich				
n	1.33	1.30	8.96	1.58
k	2.08	1.01	140.0	22.6
r^2	0.952	0.992	0.802	0.983

The Freundlich coefficient, k, represents the amount of metal adsorbed when the concentration of the solution in equilibrium is unity. The value of k for *Pseudomonas* is higher than that for activated sludge for both metals. This is consistent with the Q_{max} values in the Langmuir isotherm model. On the other hand, n measures the impact on biosorption of a change in residual solution concentration from unity. A low value of n implies a relatively large change in sorbed metal when residual concentration deviates from unity, either above or below.

Effect of pH

The lead and copper biosorption by *P. pseudoalcaligenes* was strongly affected by solution pH, as indicated in Fig. 8. Metal uptake was negligible

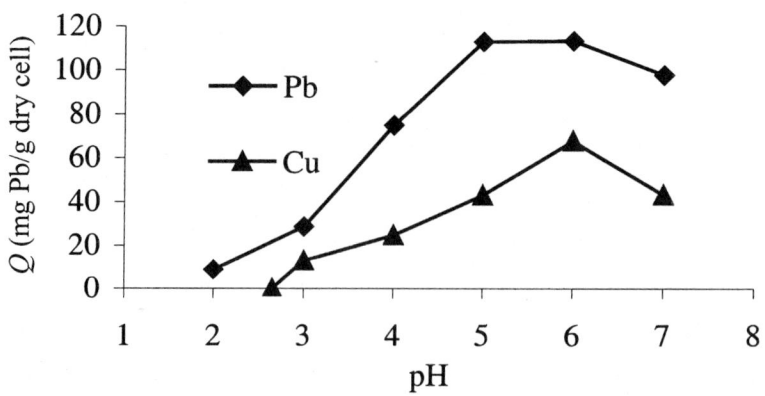

Fig. 8. Effect of pH on lead and copper biosorption by *P. pseudoalcaligenes*.

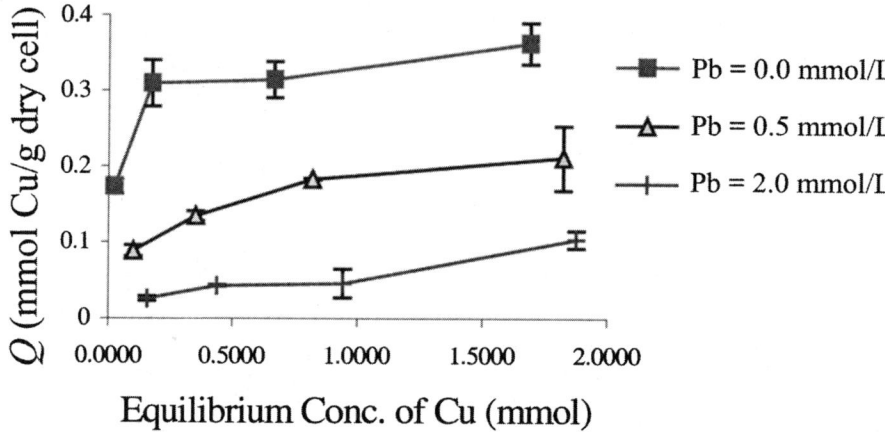

Fig. 9. Effect of lead on biosorption of copper by *P. pseudoalcaligenes* at pH 5.0.

at pH 2.0 and then increased rapidly with increasing pH. These results could be explained by the competition between hydrogen ions and metal ions for the sorption sites of cells *(18)*. At very low pH values, metal cations and protons compete for binding sites on the cell walls, which results in lower uptake of metal. As pH levels are increased, more ligands with negative charge would be exposed with a subsequent increase in attraction for positively charged metals ions.

Binary Metal Biosorption System

The adsorption isotherms of copper ions in the absence and presence of lead or nickel ions are shown in Figs. 9 and 10. The equilibrium copper uptake increased with increasing initial copper concentration up to a certain concentration at which the equilibrium uptake reached a steady level. The presence of lead ions (Fig. 9) decreased the amount of copper adsorbed at equilibrium. Figure 10 shows that the biosorption of copper is not signifi-

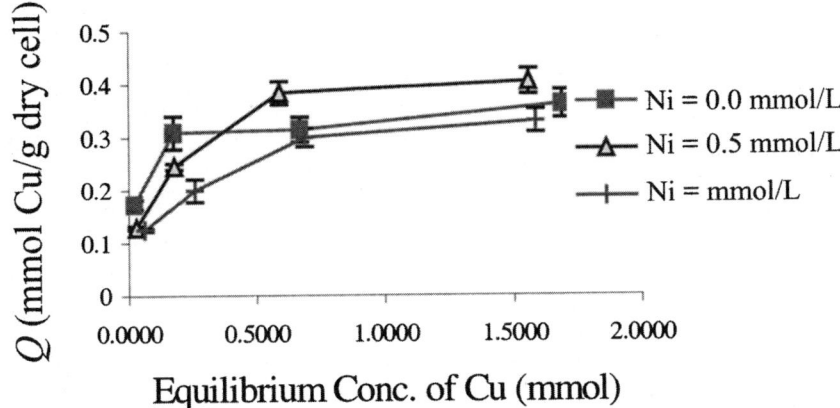

Fig. 10. Effect of nickel on biosorption of copper by *P. pseudoalcaligenes* at pH 5.0.

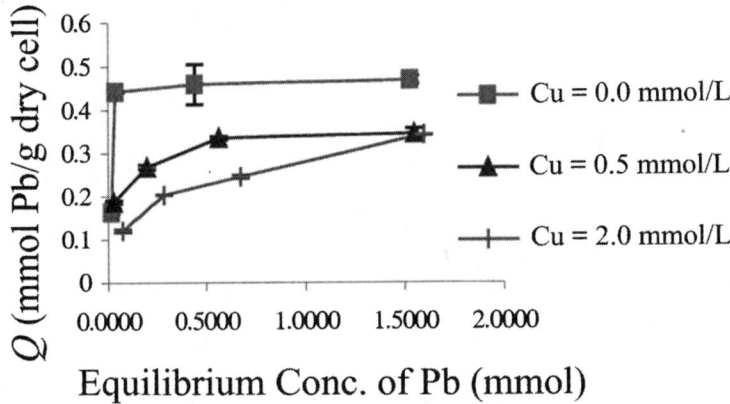

Fig. 11. Effect of copper on biosorption of lead by *P. pseudoalcaligenes* at pH 5.0.

cantly decreased by the presence of nickel. It was obvious that the effect of lead on copper sorption was greater than that of nickel.

Figures 11 and 12 depict the variations in lead uptakes at equilibrium with increasing initial concentration of copper and nickel. The presence of copper ion reduced moderately the uptake of lead (Fig. 11). The reduction was about 40% when copper was present at 2.0 mmol/L after 3 h. The presence of nickel ion also reduced lead biosorption, but to a much lesser extent (Fig. 12).

The presence of lead or copper affected significantly the nickel biosorption by *P. pseudoalcaligenes*. The nickel uptake was reduced by 80% in the presence of 2.0 mmol/L of either one of the ions (Figs. 13 and 14). Furthermore, the extent of inhibition was enhanced at increasing competing ion concentrations. This progressive interference in nickel biosorption by lead and copper ions indicated the possibility of overlap in the sorption sites. In the case of copper inhibition (Fig. 13), a decrease of about 40% uptake was observed at an initial copper concentration of 0.05 mmol/L, whereas

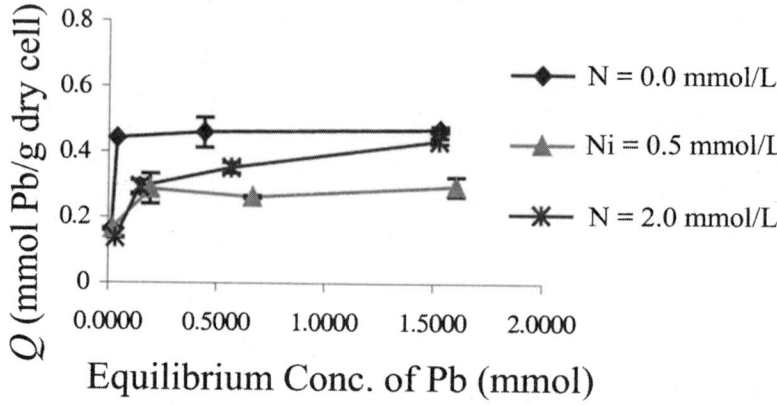

Fig. 12. Effect of nickel on biosorption of lead by *P. pseudoalcaligenes* at pH 5.0.

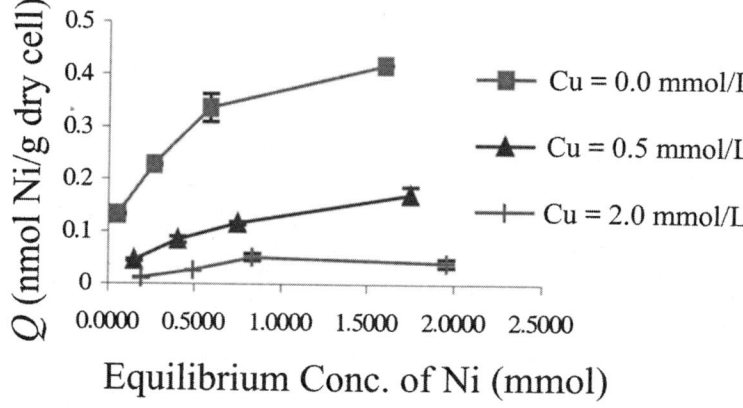

Fig. 13. Effect of copper on biosorption of nickel by *P. pseudoalcaligenes* at pH 5.0.

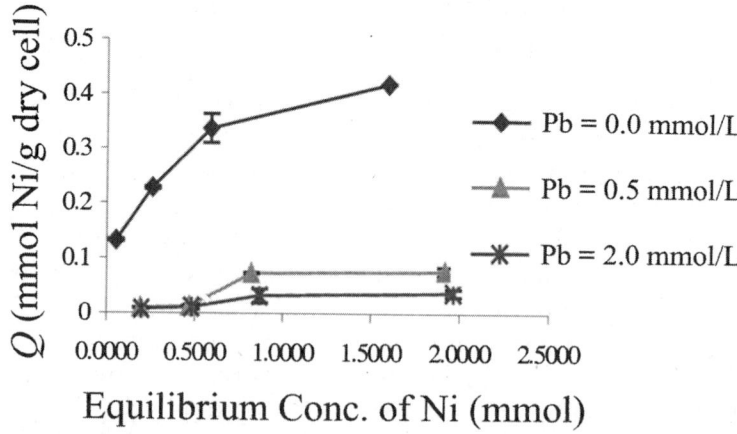

Fig. 14. Effect of lead on biosorption of nickel by *P. pseudoalcaligenes* at pH 5.0.

a decrease of about 60% was observed in lead biosorption isotherms (Fig. 14). Thus, the order of increasing inhibition for nickel biosorption was Cu < Pb.

In the present studies, a mutual inhibitory effect was observed in the lead-copper system because the presence of either one of the ions affected the biosorption capacity of the other. However, unequal inhibitions were observed in all the nickel binary systems, including the nickel-copper and nickel-lead systems. The presence of either lead or copper reduced significantly the biosorption capacities of nickel, whereas the uptakes of lead and copper were not greatly affected by nickel. Similar unequal inhibitions were reported in the zinc-cadmium biosorption system by *Chlorella vulgaris* (*19*). Zhang et al. (*18*) reported that lead biosorption by *Rhizopus nigricans* was depressed in the presence of zinc or iron whereas it was not inhibited by the presence of manganese. Wong et al. (*20*) observed that the biosorption of copper by *Pseudomonas putida* II-11 was reduced in the presence of lead, whereas it was not affected by zinc and nickel. Tobin et al. (*21*) have proposed various mechanisms of metal uptake from binary metal systems. These include direct competition for the binding sites and preferential binding to different sites. In the present study, lead seems to be preferentially bound to the binding sites in addition to the capability of competing for the binding sites. The increasing order of affinity of the three metals toward *P. pseudoalcaligenes* was Ni < Cu < Pb.

Conclusion

Twelve bacteria were isolated and identified from an activated sludge process treating both metal-contaminated industrial effluents and municipal wastewaters. These isolates included both Gram-positive (e.g., *Bacillus*) and Gram-negative (e.g., *Pseudomonas*) bacteria. The biosorption capacity of these strains for three different heavy metals—copper, nickel, and lead—was determined at pH 5.0 with an initial metal concentration of 100 mg/L. Among these isolates, *P. pseudoalcaligenes* was selected for further investigation owing to its high metal biosorption capacity. Both Langmuir and Freundlich adsorption isotherms represent adequately the distribution of copper and lead for this bacterium. Study of the effect of pH on both copper and lead removal by the selected strain indicated that biosorption increased with increasing pH from 2.0 to 6.0. A mutual inhibitory effect was observed in the lead-copper system because the presence of either ion affected the biosorption capacity of the other. Unequal inhibitions were observed in all the nickel binary systems. The increasing order of affinity of the three metals toward *P. pseudoalcaligenes* was Ni < Cu < Pb. The metal biosorptive potential of these isolates, especially *P. pseudoalcaligenes*, may have possible applications in the removal and recovery of metals from industrial effluents.

Acknowledgments

We thank Dr. K. C. Cheung and Louisa Fok for isolating the microbial strains. We also gratefully acknowledge financial support from the Hong

Kong Polytechnic University Research Committee and the Hong Kong Research Grants Council (grant no. PolyU 5001/00E).

References

1. Kasan, H. C. and Baecker, A. A. W. (1989), *Water Sci. Technol.* **21,** 297–303.
2. Chang, J. S., Law, R., and Chang. C. C. (1997), *Water Res.* **31(7),** 1651–1658.
3. Leung, W. C., Wong, M. F., Chua, H., Lo, W., Yu, P. H. F., and Leung, C. K. (2000), *Water Sci. Technol.* **41(12),** 233–240.
4. Kratochvil, D. and Volesky, B. (1998), *Trends Biotechnol.* **16,** 291–300.
5. Volesky, B. and Holan, Z. R. (1995), *Biotechnol. Prog.* **11,** 235–250.
6. Lo, W. H., Chua, H., Lam, K. H., and Bi, S. P. (1999), *Chemosphere* **39(15),** 2723–2736.
7. Unz, R. F. and Shuttleworth, K. L. (1996), *Curr. Opin. Biotechnol.* **7(3),** 307–310.
8. Loaec, M., Olier, R., and Guezennec, J. (1997), *Water Res.* **31(5),** 1171–1179.
9. Volesky, B. and May-Phillips, H. A. (1995), *Appl. Microbiol. Biotechnol.* **42,** 797–806.
10. Atkinson, B. W., Bux, F., and Kasan, H. C. (1996), *Water Sci. Technol.* **43(9),** 9–15.
11. Goddard, P. A. and Bull, A. T. (1989), *Appl. Microbiol. Biotechnol.* **31(3),** 308–313.
12. Gourdon, R., Bhende, S., Rus, E., and Sofer, S. S. (1990), *Biotechnol. Lett.* **12(11),** 839–842.
13. Pumpel, T., Pernfuss, B., Pigher, B., Diels, L., and Schinner, F. (1995), *J. Ind. Microbiol.* **14(3–4),** 213–217.
14. Thompson, J. B. and Beveridge, T. J. (1993), in *Particulate Matter and Aquatic Contaminants,* Rao, S. S., ed., Lewis Publishers, Chelsea, MI, pp. 65–104.
15. Beveridge, T. J. (1993), *J. Appl. Bacteriol.* **74,** S143–S153.
16. Norberg, A. B. and Enfors, S. O. (1982), *Appl. Environ. Microbiol.* **44,** 1231–1237.
17. Brown, M. J. and Lester, J. N. (1979), *Water Res.* **8,** 817–837.
18. Zhang, L., Zhao, L., Yu, Y., and Chen, C. (1998), *Water Res.* **32(5),** 1437–1444.
19. Ting, Y. P., Lawson, F., and Prince, I. G., (1990), *Aust. J. Biotechnol.* **4,** 197–200.
20. Wong, P. K., Lam K. C., and So, C. M., (1993), *Appl. Microbiol. Biotechnol.* **39,** 127–131.
21. Tobin, J. M., Cooper, D. G., and Neufeld, R. J., (1988), *Biotechnol. Bioeng.* **31,** 282–286.

Utilization of Cyanobacteria in Photobioreactors for Orthophosphate Removal from Water

ALEXEA M. GAFFNEY, SERGEI A. MARKOV,** AND M. GUNASEKARAN*

*Department of Biology, Fisk University, Nashville, TN 37208,
E-mail: markov@marshall.edu*

Abstract

The effectiveness of photosynthetic free-living and polyurethane foam (PU) immobilized *Anabaena variabilis* cells for removal of orthophosphate (P) from water in batch cultures and in a photobioreactor was studied. Immobilization in PU foams was found to have a positive effect on P uptake by cyanobacteria in batch cultures. The efficiency of P uptake by immobilized cells was higher than by free-living cells. A laboratory scale photobioreactor was constructed for removal of P from water by the immobilized cyanobacteria. The photobioreactor was designed so that the growth medium (water) from a reservoir was pumped through a photobioreactor column with immobilized cyanobacteria and back to the reservoir. This created a closed system in which it was possible to measure P uptake. No leakage of cells into the photobioreactor medium reservoir was observed during the operation. The immobilized cells incorporated into a photobioreactor column removed P continuously for about 15 d. No measurable uptake was demonstrated after this period. Orthophosphate uptake efficiency of 88–92% was achieved by the photobioreactor.

Index Entries: Orthophosphate; water clean-up; immobilization; cyanobacteria; photobioreactor.

Introduction

Inorganic phosphorus (orthophosphate and phosphate) is an essential element for the growth of plants and animals, but it could be harmful for the environment. High levels of phosphorus in rivers and lakes owing to pollution can cause a negative environmental impact through the process

---

*Author to whom all correspondence and reprint requests should be addressed.
**Current address: Division of Biological Sciences, Marshall University, 400 Hal Greer Boulevard, Huntington, WV 25755-2510.

of eutrophication *(1)*. Eutrophication results when an excess of inorganic phosphorus enters a waterway. Algae and aquatic plants grow fast, choking waterways and consuming large amounts of dissolved oxygen *(2)*. The rapid and uncontrollable growth of aquatic organisms will cause decay and, eventually, destruction of the aquatic ecosystem. Removal of phosphorus from municipal and industrial wastewater is required to protect water quality. According to federal government standards, phosphate levels in water should not exceed 0.01–0.1 mg/L *(3)*. There are several chemical and physical methods for removing phosphorus from water, such as distillation *(4)*. Treatment plants have been designed to remove phosphorus, often by using chemicals *(4)*. Chemical and physical phosphorus-removing methods require a great deal of energy to operate efficiently and are high-maintenance systems. Studies have shown that cyanobacteria are good candidates for use in removal of phosphorus from water *(5,6)*. They can use orthophosphate for their growth (formation of photosynthetic adenosine triphosphate) with solar light as their energy source.

The goal of the present study was to compare the effectiveness of photosynthetic free-living and polyurethane (PU) foam–immobilized *Anabaena variabilis* cells for the removal of orthophosphate from various sources of water in batch cultures and in a photobioreactor. Sources of water included a cyanobacterial standard medium, municipal tap water, and water from a local lake. Capabilities of orthophosphate uptake by ammonia excreting a mutant and a wild-type cyanobacterium were compared as well. The ultimate goal of this study was to develop and operate a photobioreactor with PU foam–immobilized cyanobacterial cells for the removal of orthophosphate from polluted water. One of the important advantages of the immobilized cells is the very large surface-to-volume ratio, which enhances mass removal of orthophosphate by the cells. In addition, cyanobacteria, when immobilized in matrices such as agar, cotton, polyurethane, or polyvinyl foams, stabilize and increase their physiologic functions *(7)*.

A. variabilis was chosen for the current study of orthophosphate removal from water because higher efficiencies of inorganic nitrogen and phosphate removal by this cyanobacterium in a hollow-fiber photobioreactor were observed in preliminary studies *(8)*. Because hollow-fiber photobioreactors might currently be too expensive for water treatment, other options for a photobioreactor design (PU foam immobilization) were considered. Immobilization of *A. variabilis* on different substrates has been studied thoroughly *(9)*. The immobilization of *A. variabilis* on hollow fibers led to stabilization of H_2 photoproduction for several months *(9)*.

Materials and Methods

Chemicals

Chemicals were purchased from Fisher (Atlanta, GA) and Sigma (St. Louis, MO).

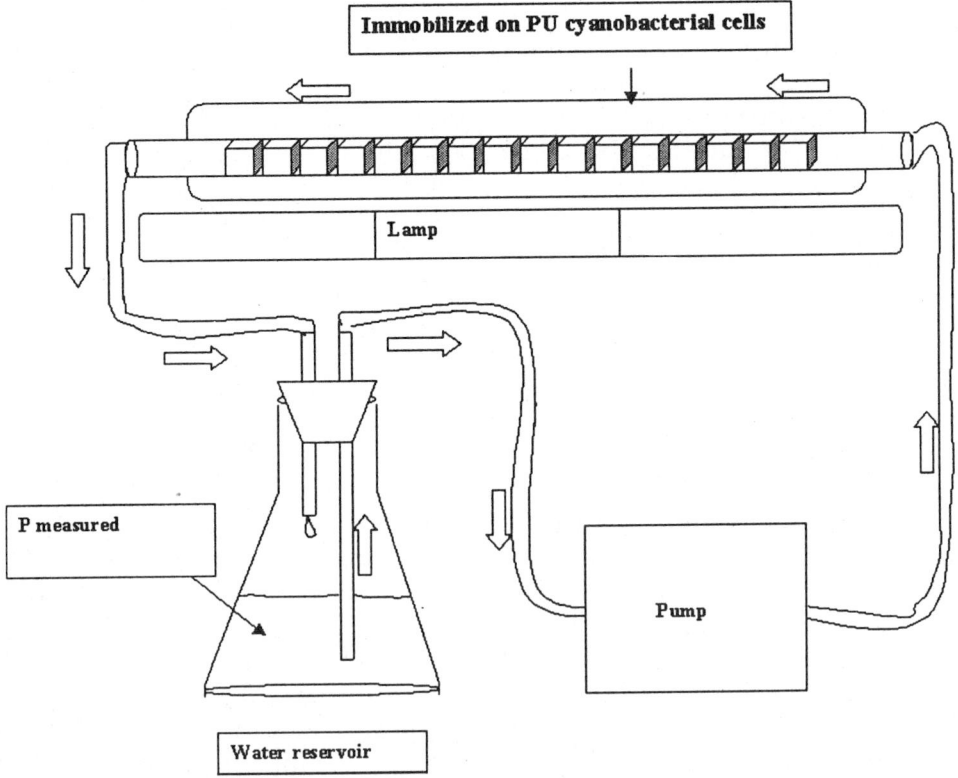

Fig. 1. Schematic diagram of the photobioreactor with PU foam–immobilized cyanobacteria for removal of orthophosphate from water.

Culture Growth

Cyanobacterium *A. variabilis* (wild-type SA-0 and ammonia-excreting mutant SA-1) was obtained from Dr. K. T. Shanmugam (University of Florida). Batch cultures of the cells were grown in the medium of Allen and Arnon *(10)*, municipal tap water, or water from J. Percy Priest Lake, Nashville, TN, at 26–28°C. Continuous light was provided by cool white fluorescent lamps (30–50 ft candles [15–25 μmol/m²s] on the surface of the culture) in an incubator as described previously *(11)*. Cell concentration for batch cultures during inoculation was 0.65 mg of cell dry wt/mL and increased because of cyanobacterial growth.

Photobioreactor Design

A laboratory-scale photobioreactor was constructed for the removal of orthophosphate from water (Allen-Arnon medium) by *A. variabilis* SA-0 cells (Fig. 1). The photobioreactor glass column (19 cm long × 7 cm diameter) was filled with PU foams. Cyanobacterial cells (0.65 mg of cell dry wt/mL) were added to the photobioreactor column under sterile conditions. Cyanobacterial growth medium was continuously returned to a water reservoir that created a closed system in which it was possible to measure the uptake of orthophosphate. The column was maintained at room temperature and illuminated continuously with a cool white fluorescent lamp at the bottom. The irradiance above the column was measured at approx 25 ft candles and below the column at approx 30 ft candles (15 μmol/m^{2}s).

Cell Immobilization

Cyanobacterial cells were immobilized by adsorption on PU foam (1-cm cubes). Before inoculation with cells, foams were washed with distilled water. Seven pieces of PU foam were added to each flask. The flasks were autoclaved (121°C for 15 min) and cooled to room temperature. Then cyanobacterial cells were added.

Orthophosphate Assay

Orthophosphate content was measured regularly using a modified ascorbic method from *Standard Methods for Examination of Water and Wastewater (12)*. Samples (0.5 mL) were obtained from each flask and the reservoir of the photobioreactor. Samples were then diluted in 100 mL of distilled water. Eight milliliters of the combined reagent was added to half of each sample. The combined reagent was mixed as follows: 50 mL of H_2SO_4, 5 mL of potassium antimony tartrate solution, 15 mL of ammonium molybdate solution, and 30 mL of ascorbic acid solution. The mixture was allowed to stand for at least 10 min, but no more than 30 min. The samples were read on a Hitachi U-2000 Spectrophotometer at 880 nm, providing a 1-cm light path.

Orthophosphate Uptake in Batch Cultures

For each experiment, water samples were taken from 20 flasks (10 flasks for each cyanobacterial strain and 5 flasks each for free-living or immobilized cells) and analyzed for orthophosphate content by the ascorbic acid method just described.

Orthophosphate Uptake in Photobioreactor

Water samples were collected from the water reservoir of the photobioreactor and analyzed for orthophosphate content by the ascorbic acid method just described.

Calculation of Orthophosphate Uptake Efficiency

Orthophosphate uptake efficiency (E) was defined as: $E = [(I - F)/I] \times 100\%$, in which I and F are the initial and final concentrations of orthophosphate, respectively *(5)*. An efficiency value of 100% was obtained when no orthophosphate appeared in the water (i.e., $F = 0$).

Biomass

Biomass was determined after filtering and drying the cell suspension at 90°C to a constant weight.

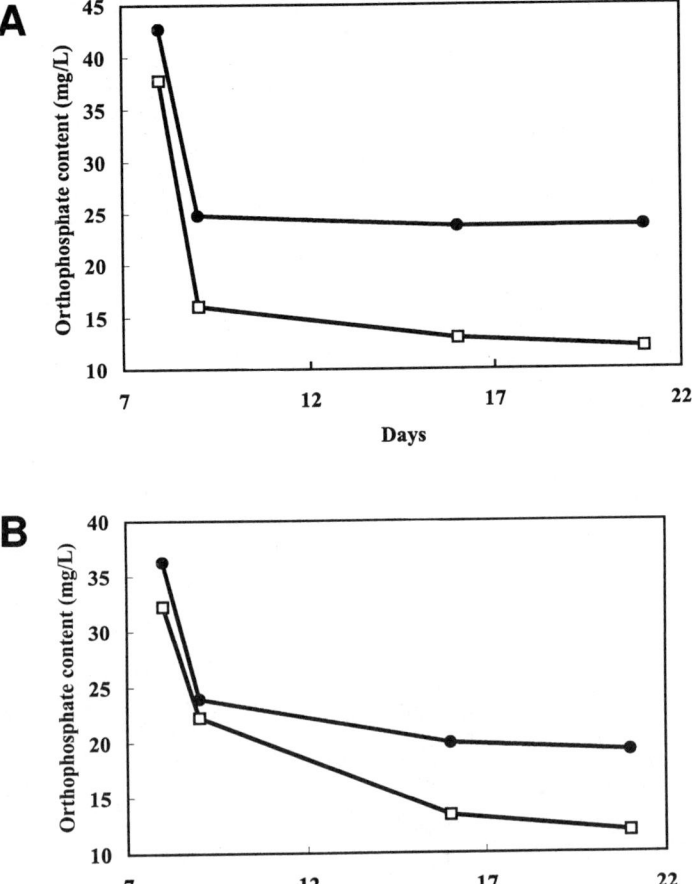

Fig. 2. Orthophosphate uptake from standard cyanobacterium medium by free-living (●) or PU foam–immobilized (□) cyanobacterium *A. variabilis* wild type SA-0 **(A)** or *A. variabilis* mutant SA-1 **(B)** in batch cultures.

Results

Orthophosphate Uptake in Batch Cultures

Both immobilized and free-living cells absorbed orthophosphate from a variety of water sources used in experiments (Figs. 2–5). In general, PU foam–immobilized cells absorbed orthophosphate faster than free-living cells during the experiment, regardless of the water source. There was practically no difference in orthophosphate uptake from water by cells of *A. variabilis* wild strain (SA-0) or ammonia-excreting mutant *A. variabilis* SA-1.

Orthophosphate Uptake from Standard Cyanobacterium Medium

Orthophosphate uptake by free-living and immobilized cells of *A. variabilis* SA-0 and SA-1 is presented in Fig. 2. Initially, orthophosphate

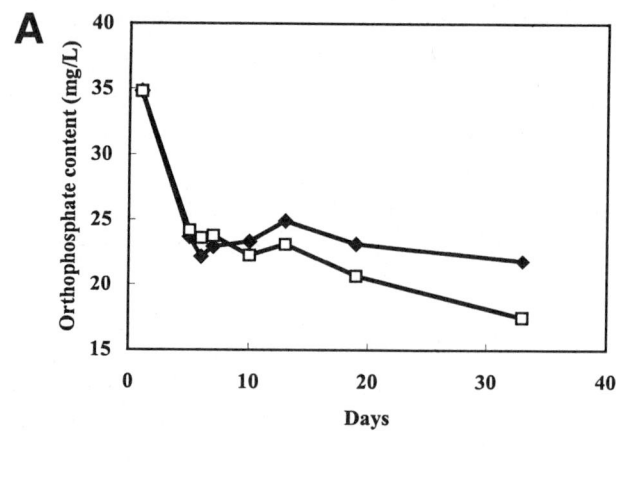

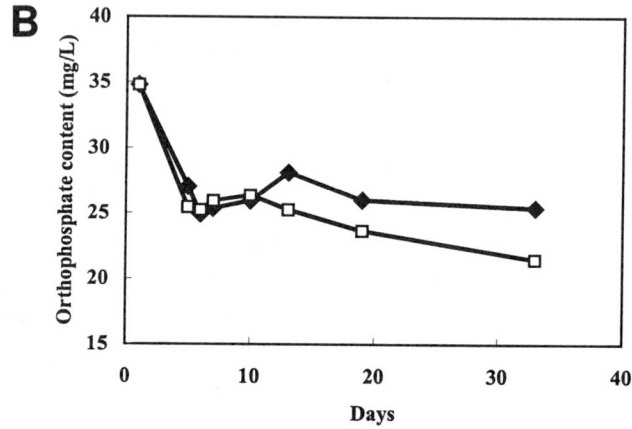

Fig. 3. Orthophosphate uptake from lake water by free-living (◆) or PU foam–immobilized (□) cyanobacterium *A. variabilis* SA-0 **(A)** and SA-1 **(B)** in batch cultures.

uptake by cyanobacterial cells was faster, but it subsequently decreased to a lower steady uptake. The decrease in the orthophosphate uptake down to the steady uptake in batch cultures was apparently owing to the age of the cyanobacterial culture and lack of nutrients other than orthophosphate. The efficiency of orthophosphate uptake by immobilized cells was higher than by free-living cells (up to 30% higher).

Orthophosphate Uptake from Lake Water

Inorganic phosphorus was not found in water from J. Percy Priest Lake at the time of sample collection (October 1999). For experiments, orthophosphate was added at concentrations similar to those for a standard cyanobacterial medium. Cyanobacterial cells, both free living and immobilized, were found to remove orthophosphate better when they were added to the standard cyanobacterial medium rather than to lake water, probably owing to the difference in ion compositions (Fig. 3). The efficiency

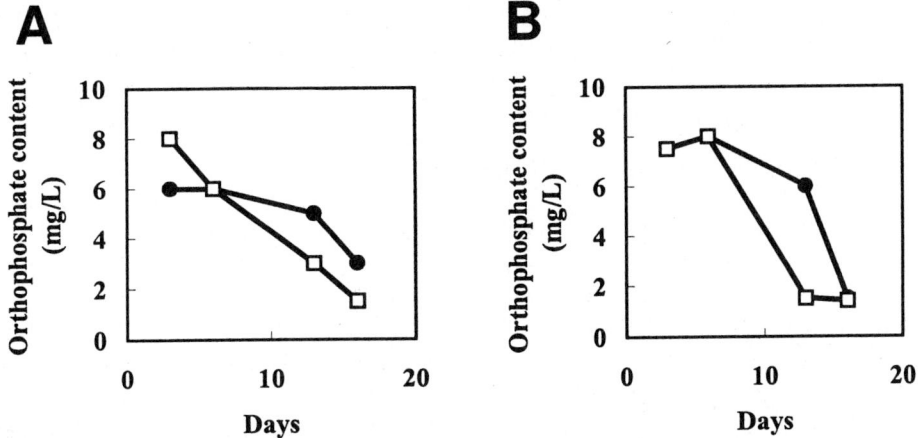

Fig. 4. Orthophosphate uptake from municipal tap water by free-living (●) or PU foam–immobilized (□) cyanobacterium *A. variabilis* SA-0 **(A)** and SA-1 **(B)** in batch cultures.

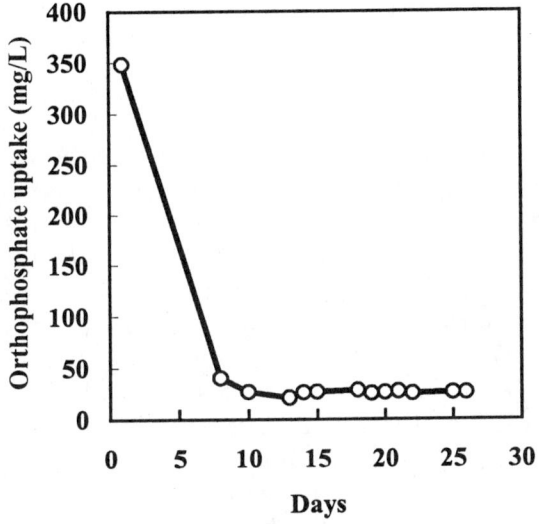

Fig. 5. Orthophosphate uptake by *A. variabilis* SA-0 immobilized on PU foam in a photobioreactor.

of orthophosphate uptake from lake water by immobilized cells was higher than by free-living cells (up to 12% higher).

Orthophosphate Uptake from Municipal Tap Water

Orthophosphate was found in municipal tap water (up to 8 mg/L). The concentration of orthophosphate in tap water was not stable and varied from day to day. However, this orthophosphate content was lower as compared to standard cyanobacterial medium. At low initial orthophosphate concentrations, cyanobacterial cells were able to almost completely utilize orthophosphate from tap water (Fig. 4). At these concentrations, there was no big difference in orthophosphate uptake by free-living or PU foam–immobilized cyanobacteria.

Orthophosphate Uptake in Photobioreactor

The batch experimental results, which showed higher orthophosphate uptake by PU foam–immobilized cells compared to free-living cells, allowed us to select PU foam–immobilized *A. variabilis* for photobioreactor studies. The photobioreactor was run continuously for 26 d after inoculation with a cyanobacterial suspension. Orthophosphate uptake was measured, as shown in Fig. 5. Orthophosphate uptake in a photobioreactor was very similar, with batch cultures showing a higher initial orthophosphate uptake rate and a slower steady uptake days later. The efficiency of orthophosphate uptake was calculated. After 8 d of photobioreactor operation, the efficiency of orthophosphate uptake was 88%. After 26 d of orthophosphate uptake in the photobioreactor, the efficiency had reached 92%. After 27 d, no significant improvement in the efficiency of orthophosphate uptake was demonstrated.

Discussion

The results presented herein demonstrate that there is potential for the use of PU foam–immobilized cyanobacteria in photobioreactors for removing excess inorganic phosphorus from water to prevent eutrophication. It was found that immobilized cyanobacterial cells take up orthophosphate faster than free-living cells in batch culture during the experimental period, regardless of the water source. Each batch culture was inoculated with the same amount of cyanobacterial cells, so any increases in biomass are owing to cyanobacterial growth. There was no statistical difference in biomass amount (biocatalyst amount) during cyanobacterial growth between free-living and immobilized cells. Thus, it is cell immobilization, not biocatalyst amount, that increases the orthophosphate uptake. Similar results were found by Garbisu et al. *(4)*, in which polyvinyl foam–immobilized cyanobacteria took up phosphate from cyanobacterial medium faster than free-living cells during the incubation period. At low orthophosphate concentrations (which were found in tap water), cyanobacterial cells were able to completely consume orthophosphate from water in a batch culture (100% efficiency). A photobioreactor with a continuous flow of water, described herein, was also operated with higher efficiencies of orthophosphate uptake by PU foam–immobilized cyanobacteria (88–92%), despite a higher concentration of orthophosphate. These efficiencies were similar to previously reported efficiencies for phosphate uptake by chitosan-immobilized *Phormidium* in continuous cultures *(13)*. Efficiencies of orthophosphate uptake in a photobioreactor with PU foam–immobilized cyanobacteria and the low cost of PU foam make this photobioreactor suitable for future practical applications. In addition, we observed other advantages of immobilized cells compared to free-living cells during the operation of the photobioreactor. When the immobilized cyanobacterial cells were used in the photobioreactor to remove orthophosphate, no significant leakage of cells into the photobioreactor effluent was seen. This eliminates any problem of recovering the microorganisms from the treated effluent during the operation with PU foam–immobilized photobioreactors.

Acknowledgments

We wish to acknowledge support from NASA (Grant NAG 2-6015) and the Howard Hughes Medical Institute (Grant 71194-527-802). Many thanks go to Christel Hall for technical assistance.

References

1. Sirenko, L. A. and Gavrilenko, M. Y. (1978), *Water Blooms and Eutrophication*, Naukova Dumka, Kiev, Russia.
2. Hallegraeff, G. M. (1993), *Phycologia* **32,** 79–99.
3. Environmental Protection Agency (1991), National primary drinking water regulations; final rule. 40 CFR Parts 141, 142, and 143. *Federal Register* 56, 20, 3526-97.
4. Garbisu, C., Hall, D. O., and Serra, L. (1993), *J. Chem. Tech. Biotechnol.* **57,** 181–189.
5. Garbisu, C., Gil, J. M., Bazin, M. J., Hall, D. O., and Serra, J. L. (1991), *J. Appl. Psychol.* **3,** 221–234.
6. Hall, D. O., Markov, S. A., Watanabe, Y., and Rao, K. K. (1995), *Photosynthesis Res.* **46,** 159–167.
7. Hall, D. O. and Rao, K. K. (1989), *Chimicaoggi.* **7,** 40–47.
8. Markov, S. A. (1998), in *20th Symposium on Biotechnology for Fuels and Chemicals, Program and Abstracts*, Gatlinburg, TN, ORNL, Oak Ridge, TN.
9. Markov, S. A., Lichtl, R. R., Rao, K. K., and Hall, D. O. (1993), *Int. J. Hydrogen Energy* **18,** 901–906.
10. Allen, M. and Arnon, D. I. (1955), *Plant Physiol.* **30,** 366–372.
11. Spiller, H. and Gunasekaran, M. (1991), *Appl. Microbiol. Biotechnol.* **35,** 798–804.
12. (1985), *Standard Methods for Examination of Water and Wastewater*, American Public Health Association.
13. de la Noue, J. and Proulx, D. (1988), in *Algal Biotechnology*, Stadler, T., Mollion, J., Verdus, M. C., Karamanos, W., Morvan, H., and Christiaen, D., eds, Elsevier Applied Science, Barking, Essex, England, UK, 159–168.

Enhancement of the Conversion of Toluene by *Pseudomonas putida* F-1 Using Organic Cosolvents

Miguel Rodriguez, Jr., K. Thomas Klasson,* and Brian H. Davison

*Bioprocessing Research and Development Center,
Oak Ridge National Laboratory, PO Box 2008-6226,
Oak Ridge, TN 37831-6226, E-mail: klassonkt@ornl.gov*

Abstract

Pseudomonas putida F-1 (ATCC 700007) was used as a model organism in stirred tank reactors to study conversion enhancement of poorly soluble substrates by organic cosolvents. After a literature study, silicone oil was used as a solvent system to enhance the mass transfer rate. To study the benefits of the organic solvent addition, batch experiments were conducted in two side-by-side fermentation vessels (experimental and control) at three different levels of silicone oil (10, 30, and 50%). Results showed that the presence of silicone oil resulted in a 100% increase in the toluene mass transfer compared to the control. Experiments in continuous stirred-tank reactors showed that improved conversion could be obtained at higher agitation rates.

Index Entries: *Pseudomonas putida* F-1; toluene; organic solvents; silicone oil.

Introduction

To promote higher bioconversion of poorly water-soluble components, cosolvents and surfactants are often added to the fermentation broth. The logarithm of the partition coefficient (log P or log P_{ow}) of an organic solvent in a standard octanol-water two-phase system is a useful parameter to predict what solvent would be most suitable for a bioconversion *(1,2)*. The partition coefficients for some common organic solvents are listed in Table 1. The relationship between log P and bioactivity is based on the assumption that the octanol-water system provides a sufficient description of hydrophobic and transport interactions when it is introduced into a biologic system *(3,4)*. In general, organic solvents with a log P value between 1 and 5 are toxic to microorganisms *(5)*.

*Author to whom all correspondence and reprint requests should be addressed.

Table 1
Log *P* Values for Selected Solvents *(2)*

Solvent	Log *P*	Solvent	Log *P*
Ethanol	−0.24	1-Decanol	4.0
1-Heptanol	2.4	Dodecanol	5.0
Toluene	2.5	Decane	5.6
1-Octanol	2.9	Dodecane	6.6
Hexane	3.5	Oleyl alcohol	7.5
Heptane	4.0	Hexadecane	8.8

In general, Gram-negative bacteria appear to have a higher solvent tolerance than Gram-positive bacteria, and species within a genus sometimes show a range of tolerances *(6–9)*. It has been suggested that the difference in solvent tolerance is caused by the presence of the outer membrane in Gram-negative bacteria containing lipopolysaccharides, which protect the cells against hydrophobic compounds. The most resistant Gram-negative species have been reported in the genus *Pseudomonas (10–12)*. One of the key processes in the adaptation of some *Pseudomonas* strains enabling them to tolerate organic solvents appears to be the isomerization of *cis-* into *trans-*unsaturated fatty acids *(13)*.

Silicone oil has been used for the biological elimination of alkanes from gases using biotrickling filters. One study reported the use of silicone oil with an aqueous medium in a 1:1 ratio being recirculated in a biotrickling filter to remove hexane *(14)*. An 89% elimination efficiency of hexane was achieved. The oil was reused after separation by natural gravity or centrifugation. Column experiments were performed with intermittent replacement of nutrients. In these studies, a control column was not used, making it difficult to positively prove the benefit of the oil addition.

In a study by Budwill and Coleman *(15)*, silicone oil was used as an additive in peat-based biofilters for the removal of hexane. Peat was coated with 20% (v/v) silicone oil and loaded into the biofilter columns. An average 60%, or 16 g/(m³·h), hexane removal was reported in the column containing silicone oil compared to 24%, or 8.2 g/(m³·h), in the untreated control. These investigators speculated that the presence of silicone oil increased the mass transfer of hexane from the gas to the liquid phase by increasing the contact of microorganisms with the dissolved gas at the water–silicone oil interface.

A group of French researchers have been investigating the applications of silicone oil as an organic solvent for the degradation of poorly water-soluble xenobiotic compounds such as xylene, butyl acetate, 2,4,6-trichlorophenol, and styrene *(16–19)*. One of their studies reported that microorganisms were able to grow more in a two-phase system (70% medium and 30% silicone oil [v/v]) on xylene and butyl acetate (70%/30% [w/w]) than in a one-phase system *(16)*. The two-phase system resulted in an increase in optical density at 540 nm for several substrate concentrations. No appreciable growth was noted in the one-phase system.

Other studies performed by the French group have also used a biphasic aqueous-silicone oil system with 20% oil (v/v) in a continuous stirred-tank reactor (CSTR). One of these studies demonstrated a more efficient degradation of 2,4,6-trichlorophenol in the biphasic system when compared to a monophasic aqueous system *(17)*. As the dilution rate was changed from 0.033 to 0.22/h, the volumetric conversion rate increased from 21.3 to 85.8 g/(m³·h) in the biphasic system compared with an increase from 13.9 to 40.2 g/(m³·h) in the monophasic system. Ascon-Cabrera and Lebeault *(18)* demonstrated the effect of silicone oil in the degradation of chlorinated and nonchlorinated mixed compounds. They found that the specific growth rate of the microorganisms used in the study was about two times higher in the biphasic system (0.48/h) than in the monophasic system (0.27/h). Statistical analysis showed that the biphasic system was more efficient in the degradation process when compared to the monophasic system. Finally, the French group showed in another study that this type of biphasic system was effective in the degradation of styrene by *Pseudomonas aeruginosa (19)*. Without silicone oil, the microorganisms were unable to oxidize styrene, but in the presence of silicone oil, the lethal dose of styrene in aqueous medium (70 mg/L) was avoided.

The French studies were performed by mixing an aqueous phase and silicone oil laden with organic contaminants, and they were able to demonstrate the benefit of silicone oil as a "reservoir" for the contaminant. By doing this, the toxic levels of some of the compounds in the aqueous phase could be lowered and toxicity avoided. By contrast, our scope was to study the benefit of using silicone oil for enhancement of mass transfer of dilute gaseous organics (not necessarily toxic) to microorganisms in the aqueous phase when conducting batch or continuous fermentations.

Materials and Methods

Microbial Culture and Media

Pseudomonas putida F-1 strain ATCC 700007 was obtained from the American Type Culture Collection (Manassas, VA). Seed cultures were grown aerobically at 30°C in a mineral salts medium consisting of the following ingredients: 0.4 g/L of KH_2PO_4, 0.5 g/L of K_2HPO_4, 0.5 g/L of $MgSO_4 \cdot 7H_2O$, 0.04 g/L of $CaCl_2$, 0.5 g/L of NH_4Cl, 0.5 g/L of KNO_3, and 1 mL of Pfennig trace metals *(20)*. The medium was distributed in 50-mL aliquots in 125-mL serum bottles and sealed with butyl rubber stoppers and aluminum crimps (Wheaton, Millville, NJ). Toluene was added (20 µL) as a liquid to each culture as a carbon source.

Reactor Experiments

Experiments were conducted in 1-L nominal volume, stirred batch reactors (SBRs) and in a CSTR (Virtis, Gardiner, NY). Silicone oil (DC 200,

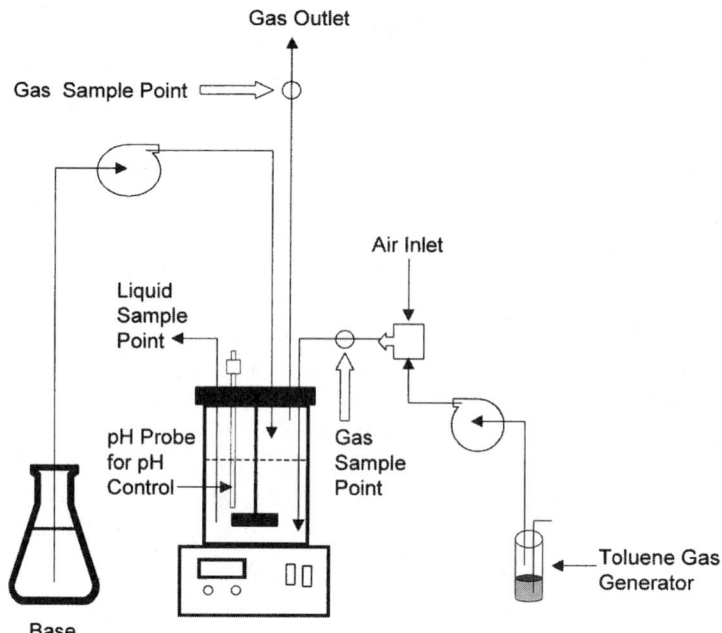

Fig. 1. Schematic of experimental setup for SBRs.

~53 mPa·s, polydimethylsiloxane) (Sigma/Fluka) was selected as an organic solvent to increase the conversion of toluene. A schematic of the batch reactor configuration is shown in Fig. 1. Modifications to the reactor for continuous feed included the addition of two feed pumps (for medium and oil). Both gas and liquid exited the continuous reactor through a tube through the top of the reactor positioned at the interface (Fig. 2).

The batch experiments were conducted in two side-by-side fermentation vessels (experimental and control) with a total liquid volume in each reactor of 1 L. The SBRs were operated at 30°C and an agitation of 300 rpm. The pH was controlled at 7.0. The aeration was set at 1 L/min, and the air was prefiltered using a Gelman Sciences Acrodisk (ACRO 50 APT, 0.2 µm polytetrafluoroethylene). Toluene was pumped as a gas into the inlet air at a flow rate of 10 mL/min, resulting in concentration of approx 35 ppmv (1.4×10^{-6} mol/L). After an initial growth phase without silicone oil, the contents of both reactors were mixed and a portion of the broth was returned to the fermentation vessels. Then, either silicone oil or water was added to a total volume of 1 L. Thus, the same population of viable cells was present in each reactor when measurements began. Silicone oil was tested at three different concentrations: 10, 30, and 50% (v/v).

The continuous experiments were conducted with a total liquid volume in the reactor of 1 L. The CSTR was operated at 30°C and an agitation of 300 rpm. The pH, aeration, and toluene addition was the same as in the SBRs. After an initial growth phase without silicone oil, the oil and medium feed streams were started at a total flow rate of 0.45 mL/min (15% silicone oil).

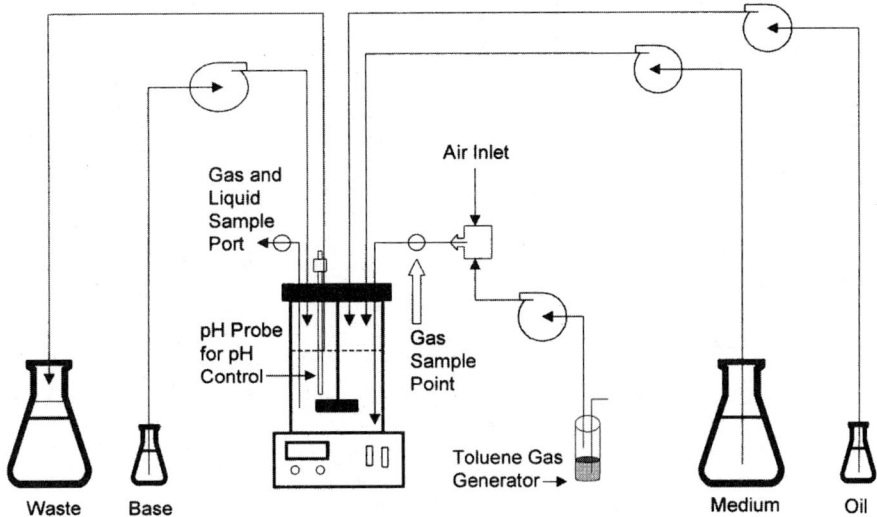

Fig. 2. Schematic of experimental setup for CSTR.

Analytical Techniques

Gas samples were collected from the inlet gas and the reactor headspace in gastight syringes, and 100 µL was injected into a gas chromatograph (Hewlett Packard HP 5890 Series II) equipped with an HP WAT (cross-linked polyethylene glycol) capillary column (30 m × 0.53 mm) with a 1.0-µm film thickness. Temperatures of the column, injection port, and flame ionization detector were 40, 175, and 200°C, respectively. Helium was used as a carrier gas. The calibration was based on 8.66 and 86.6 µg/L of toluene standards in hexane. Liquid samples were collected from the aqueous phase, which was allowed to settle by temporarily turning off the agitation. To measure toluene in the aqueous phase, samples were centrifuged for 5 min at 14,200*g* and 2 µL of the aqueous phase was injected into the gas chromatograph. The column temperature program was initially 35°C followed by ramping to 50°C at 25°C/min with a 2.0-min hold, then followed by ramping to 150°C at 20°C/min with a 0-min hold. Temperatures of the injection port and the flame ionization detector were 245 and 265°C, respectively.

The increase in dry cell weight (DCW) was measured by optical density (OD) at 600 nm after hexane had been used to extract the remaining silicone oil present in each sample. No emulsion was observed in the samples after extraction. The calibration curve was prepared from samples with known cell concentration. The potential interference of hexane in the procedure was determined by comparing OD measurements of hexane-extracted and nonhexane-extracted samples.

Results and Discussion

The addition of silicone oil enhanced the conversion of gaseous toluene for all conditions studied in the batch experiments. In Fig. 3, the conver-

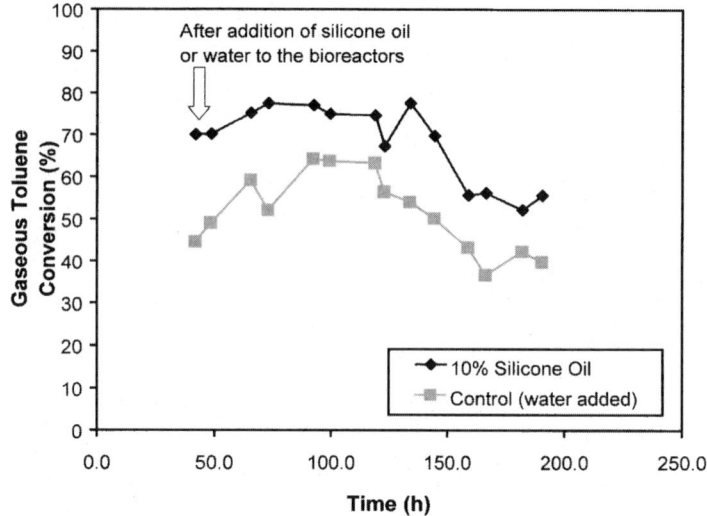

Fig. 3. Toluene conversion in SBRs with or without the addition of 10% (v/v) silicone oil.

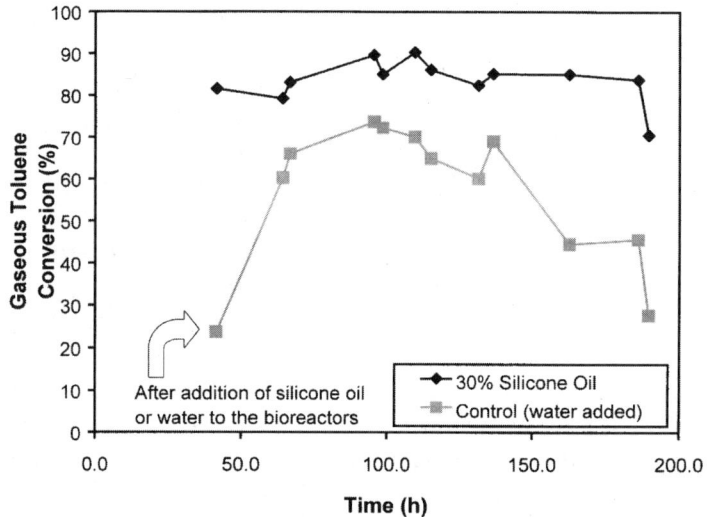

Fig. 4. Toluene conversion in SBRs with or without the addition of 30% (v/v) silicone oil.

sion of toluene has been plotted as a function of fermentation time when 10% (v/v) silicone oil was present in one of the reactors. Data prior to the addition of the oil has not been plotted, and it should be noted that the contents of the reactors were mixed just before the addition of the silicone oil. The toluene conversion was substantially higher when silicone oil was present. At the end of the batch fermentation, the conversion dropped

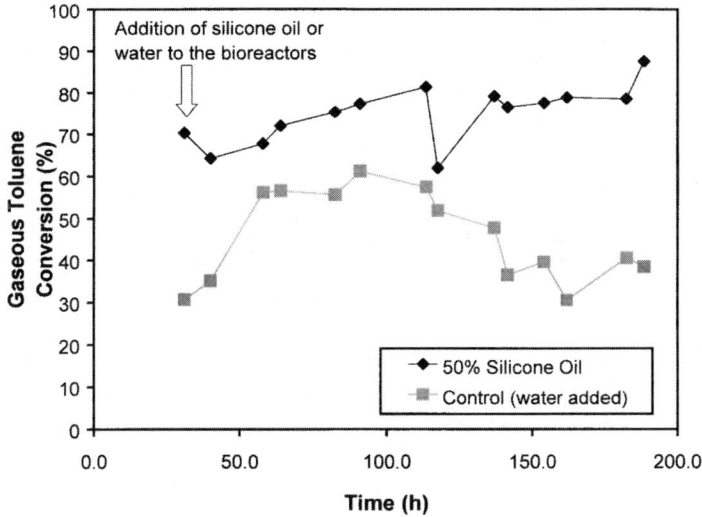

Fig. 5. Toluene conversion in SBRs with or without the addition of 50% (v/v) silicone oil.

in both cases, presumably because of nutrient limitations. A short lag phase was noted in most cases after the mixing/initiation of the experiment. Figures 4 and 5 show results from the addition of 30 and 50% silicone oil. In both cases, the conversion of toluene was higher in the reactor containing silicone oil. Toluene concentration in the aqueous phase remained below detection (data not shown) in these experiments, indicating gas mass transfer limiting conditions.

The cell growth in the study in which 30% silicone oil was used is shown in Fig. 6. The increase in cell concentration is dramatic in the case in which silicone oil was used. This, of course, can be attributed to the higher conversion obtained in this reactor by the apparent improved mass transfer rate of the toluene from the gas to the aqueous phase. It is clear from the results that the silicone oil was not toxic to the cells. The log P value for silicone oil used in these experiments with a molecular mass of 3000 g/mol was estimated to be 2.93 *(21)*.

The mass transfer coefficient ($K_L a$) was calculated (Eq. 1) from the measured consumption of toluene, the composition of the headspace in the reactors, and the assumption that mass transfer–limited conditions were present.

$$\text{rate of toluene conversion} = (K_L a / H)(p_{\text{toluene}})(V_{\text{liq}}) \tag{1}$$

in which p_{toluene} is the partial pressure of toluene in the headspace and V_{liq} is the total liquid volume in the reactor. The results showed that the mass transfer of toluene from the gas increased by a factor of 2 in the presence of silicone oil (Table 2). No trend was found between the mass transfer rate and amount of silicone oil added to the reactor.

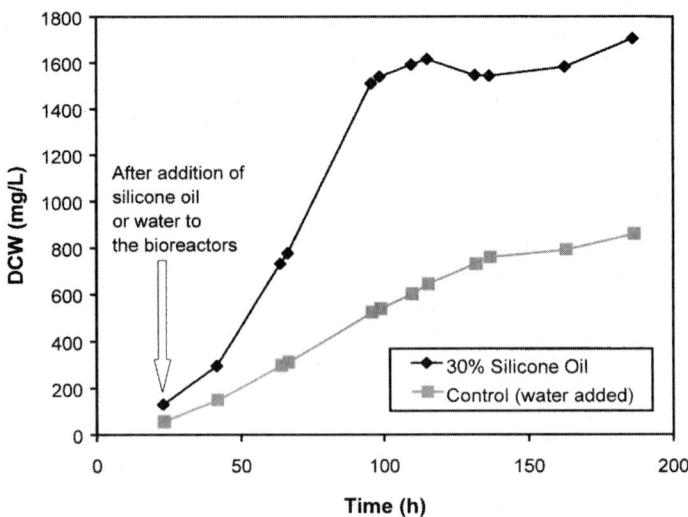

Fig. 6. Cell growth in SBRs with or without the addition of 30% (v/v) silicone oil.

Table 2
Average K_La' for Conversion of Toluene
by *P. putida* F-1 at Different Concentrations of Silicone Oil[a]

Silicone oil (%)	Experimental average K_La' ±SD	Control average K_La' ±SD
10	7.9 ± 1.6	3.7 ± 1.1
30	16.7 ± 6.6	5.3 ± 2.1
50	8.2 ± 3.0	3.1 ± 1.0

[a]Average K_La' was determined from samples taken in an interval of approx 40–140 h during the course of the experiment. The unit for the mass transfer coefficient ($K_La' = K_La/H$) is mol/(h·L·atm) and includes the Henry's law constant (H).

To confirm the enhancement of the mass transfer rate by the addition of silicone oil, the control and experimental reactors were switched so that the reactor normally used as control became the experimental reactor and vice versa. Within experimental error, the results were the same as before (data not shown).

Since this was a non-steady-state condition, it is important to estimate the amount of toluene that initially may accumulate in the silicone oil. An overestimated microbial uptake rate may be calculated if the accumulation is significant. Using a Henry's law constant of 0.071 atm/(L·mol) *(22)* for the toluene/silicone oil system, we can calculate that it would take 15–76 min to saturate the silicone oil if no microbial conversion existed. This time is considerably shorter than the fermentation time with silicone oil, which lasted approx 150 h.

The CSTR was operated for 35 d in a study conducted to investigate the effect of agitation rate. After an initial batch growth without silicone oil,

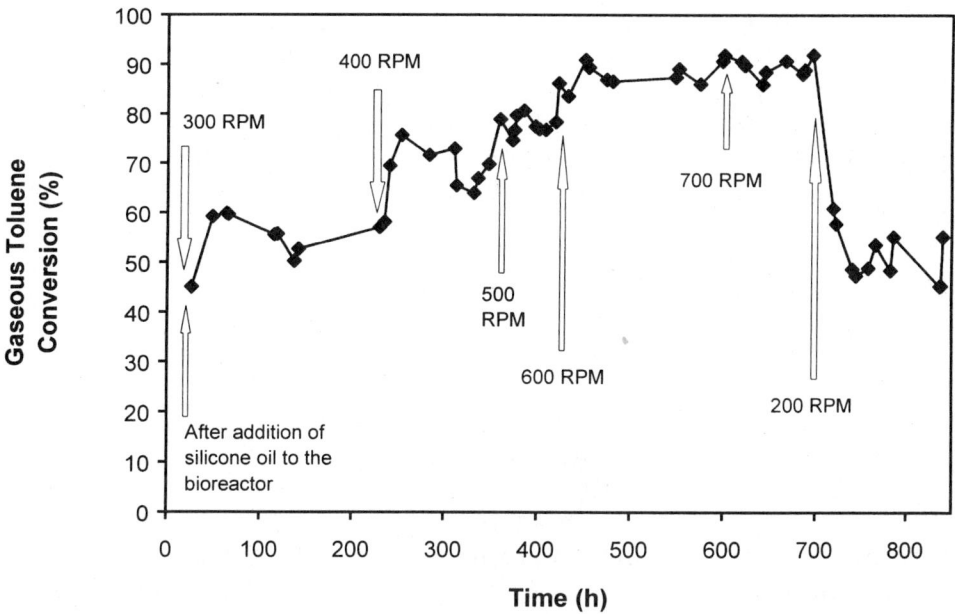

Fig. 7. Effect of agitation on toluene conversion in CSTR with 15% (v/v) silicone oil. Arrows indicate agitation changes (in revolutions per minute).

the liquid feeds were started and oil was added to the reactor. As expected, the conversion of toluene was higher at higher agitation rates, although the dependence was more apparent at a low agitation rate (Fig. 7). In these studies we also confirmed that it was possible to recycle the silicone oil after a gravity separation and filtration through a filter paper (Whatman, Clifton, NJ). Calculations using a simple mass balance over the CSTR showed that the maximum toluene loss in the silicone oil exiting the reactor was 3.5% of the gaseous toluene exiting the reactor.

Conclusion

Silicone oil is an organic cosolvent that efficiently enhances the conversion of toluene by *P. putida* F-1. We speculate that the increases seen may be attributed to an increase in effective transfer area between the toluene-rich and toluene-poor phases. This is more apparent at lower agitation rates where the gas holdup in the liquid is low. Toluene is absorbed into the silicone oil, and the silicone oil disperses the toluene into the aqueous medium. Subsequently, the availability of toluene increases, thus increasing the consumption of the gas by the microorganism. In our studies with low levels of toluene (average of 15 ppmv) in the headspace air, the calculated concentration of oxygen in the aqueous phase (2.5×10^{-4} mol of O_2/L) is 70 times greater than the calculated toluene concentration (3.5×10^{-6} mol of toluene/L) under equilibrium conditions *(23,24)*. Thus, the concentration of O_2 is eight times higher than is needed to completely oxidize the toluene, assuming a theoretic oxygen-to-toluene molar ratio of 9. It is therefore safe to assume that the limiting reactant is toluene.

This process would be applicable for biologic conversion of other poorly water-soluble gases such as nitric oxide or synthesis gases. Since the silicone oil can be reused, it minimizes the generation of waste and the capital cost. Further research needs to be conducted to expand this process for industrial applications.

Acknowledgments

This work was supported by the Advanced Research and Technology Development Program of the Office of Fossil Energy. Oak Ridge National Laboratory is managed and operated for the U.S. Department of Energy by UT-Battelle, LLC under contract DE-AC05-00OR22725. This article was authored by a contractor of the U.S. Government under contract DE-AC05-00OR22725. Accordingly, the U.S. Government retains a nonexclusive royalty-free license to publish or reproduce the published form of the contribution, or allow others to do so, for U.S. Government purposes.

References

1. Inoue, A. and Horikoshi, K. (1991), *J. Fermentation Bioeng.* **71,** 194–196.
2. Laane, C., Boeren, S., Vos, K., and Veeger, C. (1987), *Biotechnol. Bioeng.* **30,** 81–87.
3. Laane, C. and Tramper, J. (1990), *CHEMTECH* **20,** 502–506.
4. Aono, R. and Inoue, A. (1998), in *Extremophiles: Microbial Life in Extreme Environments,* Horikoshi, K. and Grant, W. D., eds., John Wiley & Sons, New York, pp. 287–310.
5. Isken, S. and de Bont, J. A. M. (1998), *Extremophiles* **2,** 229–238.
6. Harrop, A. J., Hocknull, M. D., and Lilly, M. D. (1989), *Biotechnol. Lett.* **11,** 807–810.
7. Rajagopal, A. N. (1996), *Enzyme Microb. Technol.* **19,** 606–613.
8. Vermue, M., Sikkema, J., Verhuel, A., Bakker, R., and Tramper, J. (1993), *Biotechnol. Bioeng.* **42,** 747–758.
9. Weber, F. J. and de Bont, J. A. M. (1996), *Biochim. Biophys. Acta* **1286,** 225–245.
10. Cruden, D. L., Wolfram, J. H., Rogers, R. D., and Gibson, D. T. (1992), *Appl. Environ. Microbiol.* **58,** 2723–2729.
11. Harbron, S., Smith, B. W., and Lilly, M. D. (1986), *Enzyme Microb. Technol.* **8,** 85–88.
12. Inoue, A. and Horikoshi, K. (1989), *Nature* **338,** 264–266.
13. Weber, F. J., Isken, S., and de Bont, J. A. (1994), *Microbiology* **140,** 2013–2017.
14. van Groenestijin, J. W. and Lake, M. E. (1998), Paper presented at the Air & Water Management Association's 91st Annual Meeting & Exhibition, San Diego, CA.
15. Budwill, K. and Coleman, R. N. (1997), *Med. Fac. Landbouww. Univ. Gent* **62/4b,** 1521–1528.
16. Gardin, H., Lebeault, J. M., and Pauss, A. (1999), *Biodegradation* **10,** 193–200.
17. Ascon-Cabrera, M. A. and Lebeault, J.-M. (1995), *J. Fermentation Bioeng.* **80,** 270–275.
18. Ascon-Cabrera, M. and Lebeault, J.-M. (1993), *Appl. Environ. Microbiol.* **59,** 1717–1724.
19. El Aalam, S., Pauss, A., and Lebeault, J.-M. (1993), *Appl. Microbiol. Biotechnol.* **39,** 696–699.
20. McInerney, M. J., Bryant, M. P., and Pfennig, N. (1979), *Arch. Microbiol.* **122,** 129–135.
21. Watanabe, N., Nakamura, T., Watanabe, E., Sato, E., and Ose, Y. (1984), *Sci. Total Environ.* **38,** 167–172.
22. Poddar, T. K., Majumdar, S., and Sikar, K. K. (1996), *AIChE J.* **42,** 3267–3282.
23. Lide, D. R. (1999), *CRC Handbook of Chemistry & Physics,* 80th ed., CRC Press, Boca Raton, FL, pp. 8–87.
24. Robbins, G. A., Wang, S., and Stuart, J. D. (1993), *Anal. Chem.* **65,** 3113–3118.

Effect of Temperature on Biofiltration of Nitric Oxide

K. Thomas Klasson* and Brian H. Davison

*Bioprocessing Research & Development Center,
Oak Ridge National Laboratory, PO Box 2008,
MS-6226, Oak Ridge, TN 37831-6226,
E-mail: klassonkt@ornl.gov*

Index Entries: Nitric oxide; *Paracoccus denitrificans*; denitrification; trickle-bed biofilter; mass transfer.

Introduction

In 1997, a record level of 3.13 trillion kilowatt-hours (kWh) of electricity was generated in the United States. Coal-fired power generation accounted for 57% of this production, generating an estimated 7.2 million t of nitrogen oxide (NO_x) compounds. Generally, >95% of the NO_x compounds in combustion gas streams will be in the form of NO. Titles I and IV of the 1990 Clean Air Act Amendments list an overall goal of reducing total NO_x emissions to 697–775 kg/million kWh *(1)*.

The NO_x compounds consist of nitric oxide (NO) and nitrogen dioxide (NO_2). NO and NO_2 contribute to photochemical smog, acid rain, visibility degradation, and fine particulates in the atmosphere *(1)*. NO_x compounds tend to be local and regional concerns, because of their relatively short lifetime in the atmosphere. The other NO_x, nitrous oxide (N_2O), has a long life span in the atmosphere and is a strong absorber of infrared radiation in the troposphere. It is claimed that it is a major component of global warming and depletion of ozone *(1)*. Because N_2O is stable in the troposphere, it is transported to the stratosphere, where it is the largest source of stratospheric NO. By being the major natural chemical sink, N_2O helps establish the stratospheric ozone concentration.

Recent scientific articles report daily increases in particle air pollution. These are associated with daily increases in harmful human health effects,

---

*Author to whom all correspondence and reprint requests should be addressed.

The submitted manuscript has been authored by a contractor of the U.S. government under contract DE-AC05-00OR22725. Accordingly, the U.S. government retains a non-exclusive royalty-free license to publish or reproduce the published form of the contribution, or allow others to do so, for U.S. government purposes.

primarily among the elderly and children with increased hospitalizations for respiratory diseases *(2)*. NO is considered a signaling molecule that acts primarily in the nervous and cardiovascular systems, and it diffuses freely across cell boundaries to activate nearby "target" cells *(3)*. Moreover, NO has cytotoxic properties and is implicated as the causative factor for neuron degeneration associated with Parkinson disease, autoimmune deficiency syndrome, dementia, and stroke. NO is produced by a variety of cell types, including neurons in the central and peripheral nervous systems, endothelial cells, platelets, and certain activated cells of the immune system *(3)*.

One possible alternative for NO_x treatment is microbial denitrification. Denitrification is the process in which nitrate is reduced to nitrogen through nitrite, NO, and N_2O intermediates via the following reaction *(4)*:

$$NO^{3-} \Rightarrow NO^{2-} \Rightarrow NO \Rightarrow N_2O \Rightarrow N_2 \tag{1}$$

The goal of the present study was to examine the effect on operating temperature in a trickle-bed biofilter when using the bacterium *Paracoccus denitrificans* (renamed from *Thiosphaera pantotropha*) to carry out the last two steps in the aforementioned denitrification process with a simulated combustion gas containing NO.

Materials and Methods

The bacterium *P. denitrificans* (ATCC 35512) was obtained from the American Type Culture Collection (Manassas, VA) and was maintained on a *Thiosphaera* mineral medium consisting of 3.0 g of KNO_3, 2.7 g of $CH_3COONa \cdot 3H_2O$, 0.8 g of K_2HPO_4, 0.3 g of KH_2PO_4, 0.4 g of NH_4Cl, 0.4 g of $MgSO_4 \cdot 7H_2O$, and 2 mL of trace metal solution *(5)* per liter of medium. The pH of the medium was adjusted to 8.0 with NaOH before autoclaving at 121°C for 30 min. The cells were aerobically grown in the medium for 2 d at 30°C before they were placed in a refrigerator (4°C) for later use in the reactor. To inoculate the trickle-bed reactor, 90 mL of refrigerated culture was added to the liquid recycle loop, and the system was used in full recycle with an air purge for 2 d, after which the continuous feed medium was started at a flow rate of 0.2 mL/min. After 7 d the medium was replaced with a medium with sodium acetate and KNO_3 concentrations of 8.25 and 1.5 g/L, respectively. After an additional 13 d, the medium was replaced with a medium without nitrate and a sodium acetate concentration of 8.25 g/L. The gas feed was also changed to a mixture of ~500 ppmv of NO and 15.5% CO_2 in N_2. Finally, after 10 d the medium was again changed to the *Thiosphaera* medium, less the KNO_3. By changing the medium in this manner, we were able to establish rapid cell growth on the packing under aerobic conditions and slowly transition the system to a nitrate-free medium under anaerobic conditions. The total start-up period was 30 d, and the data presented were collected over a 72-d period.

The trickle-bed biofilter consisted of a jacketed glass column (51 mm id, 600 mm high), packed with porous foam material (Fluval Foam; Rolf C.

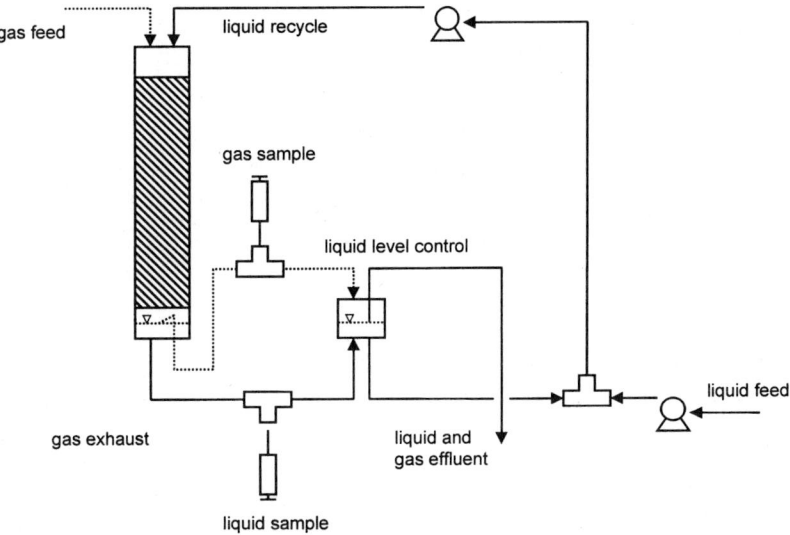

Fig. 1. Schematic of trickle-bed bioreactor. Fresh gas feed entered the top of the reactor and passed over plastic foam packing acting as a solid support for the biofilm. NO is removed by the microbial consortium as it is transferred into the liquid phase. Trickling mineral medium containing acetate as the carbon and energy source is recirculated through the reactor.

Hagen, Montreal, P.Q., Canada) to a height of 525 mm. Both the gas and the liquid recycle streams entered from the top and were separated below the packing. Figure 1 presents a schematic of the system.

The liquid recycle was kept at 200 mL/min and the fresh medium feed rate was 0.2 mL/min. The temperature of the reactor was maintained by circulating heated water through the jacket surrounding the column. The gas flow rate was varied with a calibrated rotameter control valve.

The gas analysis ($NO/O_2/CO_2$) was performed by collecting approx 2 L of gas in a Tedlar PVF gas sample bag and immediately analyzing the content of the bag using a Micro Emissions Analyzer (Model Enerac 400 EMS; Energy Efficiency Systems, Westbury, NY). Because of the already high concentration of N_2 in the inlet gas (80–85%), N_2 produced by the conversion of NO could not be measured.

Results and Discussion

Figure 2 depicts conversion of gaseous NO within the trickle bed during the course of the work. As noted, the removal of NO was 100% at gas flow rates corresponding to an empty bed residence time of approx 2 min in the packed section of the bed. Interestingly, there was no apparent temperature effect over the interval studied. A few of the data points collected at 42°C were below expected removal efficiency (Fig. 2); however, these points were collected immediately after a few days of interrupted gas supply (thus, the electron acceptor, NO, was not present). Also note that the

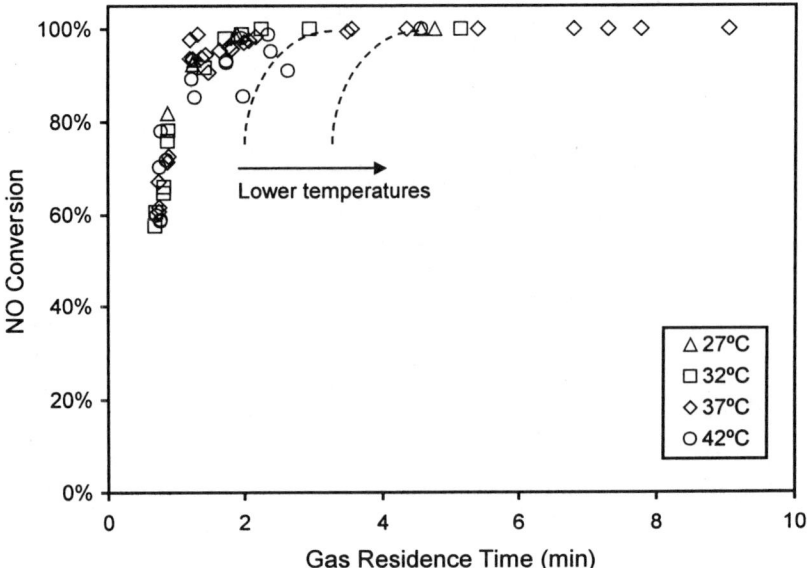

Fig. 2. Typical NO removal efficiency profiles as function of gas residence time and temperature.

temperature was changed in the order of 37° → 32° → 27° → 42°, and that gas flow rates were altered between high and low values in a random pattern. Each flow rate was allowed to stabilize for at least 30 gas residence times.

Production of the intermediate N_2O was not measured in the experiment. However, in batch serum bottle experiments with a gas phase of He and NO, complete conversion of NO to N_2 was noted without accumulation of any intermediates (data not shown). If the system were operating in a kinetic limited region, it would be expected to perform worse at lower temperatures, because the culture becomes less active. Dashed lines in Fig. 2 have been constructed to illustrate this point. Because the system was unaffected by temperature, we may assume that the conversion of NO was mass transfer limited. We have discussed in detail the use of temperature control to investigate kinetic and mass transfer operational limits in previous articles *(6,7)*. (Note that in ref. 7 K_La and K_La_{eff} incorporate ε_L.)

To estimate the mass transfer coefficient, we followed the approach taken by Cowger et al. *(8)* and Barton et al. *(9)* by focusing on the data collected at <100% NO conversion (Fig. 3). (Note that Eq. 3 in ref. 9 is incorrect; the left-hand side should be $\ln([y_i]_{outlet}/[y_i]_{inlet})$.) The slope of the line in Fig. 3 is proportional to the mass transfer coefficient. Using a value of Henry's law constant of 455 atm·L/mol for NO (in water at 34.5°C) *(10)*, we can calculate a value of approx 2100 h^{-1} for the mass transfer coefficient $(K_La\varepsilon_L)$ in the experiments. This value is six times higher than those given by Bredwell et al. *(11)*, who list mass transfer coefficients of 36–360 h^{-1} for trickle-bed reactors with conventional packing materials. (Note that in ref. *11* the K_La values listed in Table 1 incorporate ε_L.)

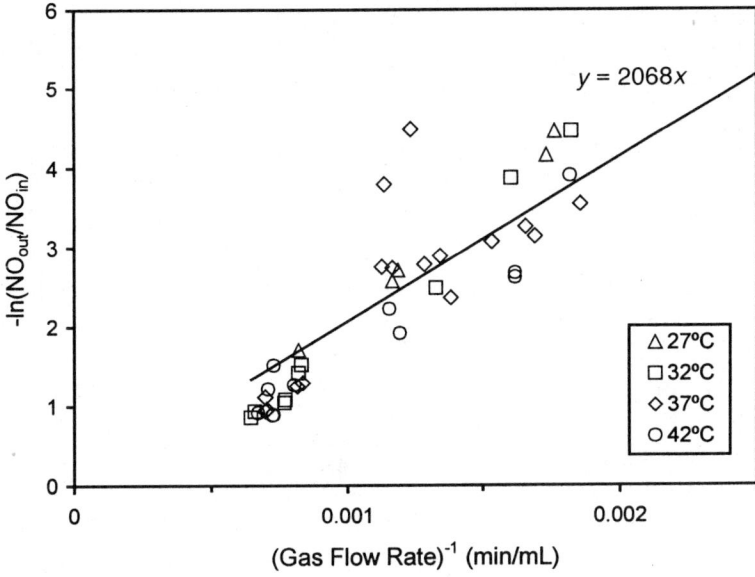

Fig. 3. Determination of mass transfer in a trickle-bed system *(8,9)*.

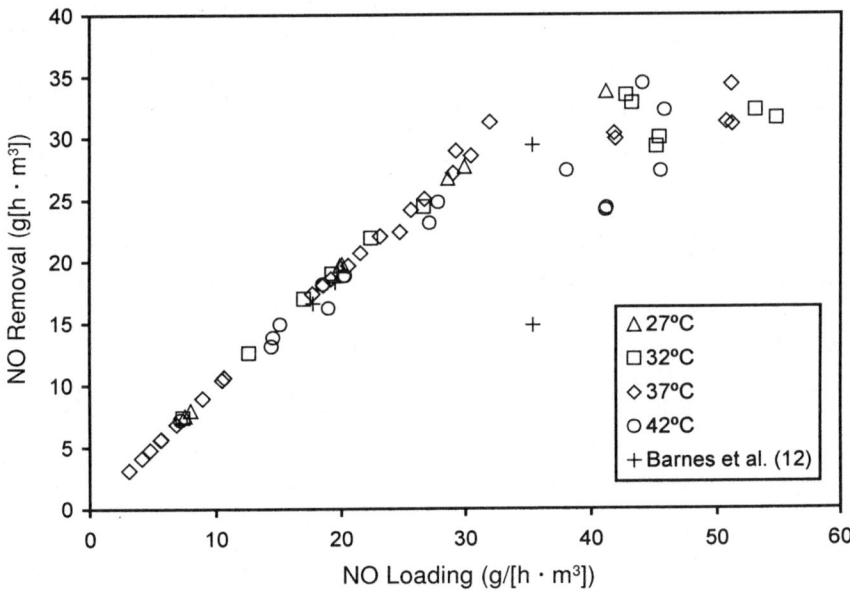

Fig. 4. Volumetric removal rate as a function of loading rate.

The reactor removal rate is displayed in Fig. 4, where it has been plotted as a function of the loading based on empty packed-bed volume. The obtained removal rate in our current studies compares well with the maximum rates obtained by Barnes et al. *(12)* in an anaerobic compost biofilter augmented with dextrose or lactate. These results have been incorporated into Fig. 4 for comparison. The maximum NO removal at loading rates above 35 g/(h·m³) was approx 32 g/(h·m³). This indicates that the system capacity is reached, corresponding to mass transfer limiting condition in our case.

The normally anaerobic trickle bed was also operated under aerobic conditions by mixing air with the gas stream before entering the reactor. The aerobic studies were conducted at 27°C and with a gas residence time of approx 7.5 min. The oxygen concentration was varied between 0.5 and 4.5% with the system responding with a near linear decrease in NO conversion from 75 to 45%. This corresponds to a decrease in reactor removal rate of 3.2–1.4 g/(h·m³) for a loading of 4.2–3.1 g/(h·m³). Under anaerobic conditions, NO conversion would be 100% at 7.5-min residence time (*see* Fig. 2); less was achieved because of the presence of O_2. The removal rate compared favorably to results presented by du Plessis et al. *(13)*, who found the NO removal rate to be 0.53–0.73 for a loading of 0.75 g/(h·m³) in an aerobic biofilter. Nagase et al. *(14)* reported an NO removal rate of 0.82 for a loading of 1.4 g/(h·m³) in an aerobic bubble column inoculated with *Dunaliella tertiolecta*. In an aerobic soil biofilter, Okuno et al. *(15)* observed a removal rate of 0.018 at a loading rate of 0.12 g/(h·m³). NO can arguably be converted to NO_2 in the presence of oxygen; however, this is a slow reaction at low NO concentrations *(14)*. Because this reaction rate is proportional to the oxygen concentration, the abiotic conversion of NO should increase with increasing oxygen levels. This was not observed in our biotic studies with oxygen in which the NO conversion decreased as the oxygen concentration increased. This indicates, indirectly, that the abiotic conversion was small in our aerobic studies.

The formation of N_2 from the conversion of NO (*see* Eq. 1) could not be confirmed in the trickle bed because of the low concentration of NO in the system. However, in separate batch experiments (not shown), conducted with elevated concentrations of NO in a helium atmosphere, accumulation of N_2 was noted to increase with time. Also, at the end of experimentation, the temperature of the reactor was increased to 50°C, which inactivated the culture. The abiotic reactor stopped converting NO at that point, and the effluent gas composition was the same as the inlet gas composition, within experimental error (10–15% for a single measurement).

Acknowledgment

This work was supported by the U.S. Department of Energy's Office of Fossil Energy Advanced Research and Technology Program. Oak Ridge National Laboratory is operated by UT-Battelle, LLC for the U.S. Department of Energy under contract DE-AC05-00OR22725.

References

1. Muzio, L. J. and Quartucy, G. C. (1997), *Prog. Energy Combust. Sci.* **23**, 233–266.
2. Ball, J. C., Hurley, M. D., Straccia, A. M., and Gierczak, C. A. (1999), *Environ. Sci. Technol.* **33**, 1175–1178.

3. Stepanov, A. L. and Korpela, T. K. (1997), *Biotechnol. Appl. Biochem.* **25,** 97–104.
4. Sakurai, N. and Sakurai, T. (1997), *Biochemistry* **36,** 13,809–13,815.
5. Barford, C. L., Montoya, J. P., Altabet, M. A., and Mitchel, R. (1999), *Appl. Environ. Microbiol.* **65,** 989–994.
6. Barton, J. W., Davison, B. H., Klasson, K. T., and Gable, C. C., III (1999), *Environ. Prog.* **18,** 87–92.
7. Barton, J. W., Hartz, S. M., Klasson, K. T., and Davison, B. H. (1998), *J. Chem. Technol. Biotechnol.* **72,** 93–98.
8. Cowger, J. P., Klasson, K. T., Ackerson, M. D., Clausen, E. C., and Gaddy, J. L. (1992), *Appl. Biochem. Biotechnol.* **34/35,** 613–624.
9. Barton, J. W., Klasson, K. T., Koran, L. J., Jr., and Davison, B. H. (1997), *Biotechnol. Prog.* **13,** 814–821.
10. Lide, D. R. and Frederikese, H. P. R., eds. (1995), *CRC Handbook of Chemistry and Physics*, 76th ed., CRC Press, Boca Raton, FL.
11. Bredwell, M. D., Srivastava, P., and Worden, R. M. (1999), *Biotechnol. Prog.* **15,** 834–844.
12. Barnes, J. M., Apel, W. A., and Barrett, K. B. (1995), *J. Hazard Mater.* **41,** 315–326.
13. du Plessis, C. A., Kinney, K. A., Schroeder, E. D., Chang, D. P. Y., and Scrow, K. M. (1998), *Biotechnol. Bioeng.* **58,** 408–415.
14. Nagase, H., Yoshihara, K.-I., Eguchi, K., Yokota, Y., Matsui, R., Hirata, K., and Miyamoto, K. (1997), *J. Ferment. Bioeng.* **83,** 461–465.
15. Okuno, K., Hirai, M., Sugiyama, M., Haruta, K., and Shoda, M. (2000), *Biotechnol. Lett.* **22,** 77–79.

Biodegradation of Formaldehyde by a Formaldehyde-Resistant Bacterium Isolated from Seawater

Tomohioko Yamazaki, Wakako Tsugawa, and Koji Sode*

*Department of Biotechnology,
Tokyo University of Agriculture and Technology, 2-24-16, Nakamachi,
Koganei, Tokyo, 184-8588, Japan, E-mail: sode@cc.tuat.ac.jp*

Abstract

A formaldehyde-tolerant bacterium designated as a DM-2 strain was used to biodegrade formaldehyde. The cells, precultivated in the presence of 400 ppm of formaldehyde, were able to degrade formaldehyde in a minimal medium supplemented with up to 400 ppm of formaldehyde in the presence of 3% NaCl. The rate of formaldehyde degradation achieved in this study was 45 ppm/h when the DM-2 culture's optical density at 660 nm was 1.2.

Index Entries: Biodegradation; formaldehyde; marine bacterium; screening.

Introduction

Formaldehyde, a volatile organic compound, is widely used in medicine, agriculture, and industrial processes as a disinfectant for killing bacteria and fungi. Moreover, formaldehyde is contained in materials such as pesticides, plastics, and adhesives. However, considering its high cytotoxicity toward human health and the environment, the removal of formaldehyde from soil, water, and air has become a necessity.

Regarding the biodegradation of formaldehyde from a polluted environment, a number of studies have been conducted on the isolation of bacteria resistant to high concentrations of formaldehyde. *Pseudomonas* sp. *(1,2)*, *Escherichia coli (3)* and *Halomonas* sp. *(4)*, and *Trichosporon* sp. *(5)* have been isolated and characterized. These bacteria were isolated from soil or river water near a chemical plant that used formaldehyde. The biodegradation of formaldehyde under anaerobic conditions was also reported *(6,7)*.

In the marine environment, wastewaters containing formaldehyde are being discharged from rivers. In addition, formaldehyde has also been used as a fungicide for fishing implements such as nets. Formaldehyde has

*Author to whom all correspondence and reprint requests should be addressed.

been detected in seawater, and the removal of formaldehyde has also been required. However, the isolation of formaldehyde-degrading bacterium and the biodegradation of formaldehyde in seawater have not been reported.

Recently, we succeeded in isolating a formaldehyde-tolerant bacterium from a marine environment. Since this bacterium can use formaldehyde as a carbon source, its application for the bioremediation of formaldehyde in a marine environment is expected.

In this article, we describe the application of the formaldehyde-tolerant bacterium for the degradation of formaldehyde.

Materials and Methods

Chemical

All chemicals used in this study were reagent grade. Formaldehyde (36% solution) was purchased from Kanto (Tokyo, Japan). Iron (III) chloride and 3-methyl-2-benzothiazolinone hydrazone hydrochloride (MBTH) were obtained from Wako Pure Chemical (Osaka, Japan). Yeast extract and trypton were from Difco (Detroit, MI).

Bacteria and Culture Conditions

A formaldehyde-resistant bacterium, DM-2 strain, used in this study was isolated from coastal seawater in Japan. It was cultured aerobically at 28°C with shaking in a modified DM-2 medium (30 g of NaCl, 5 g of trypton, 5 g of yeast extract, 6 g of Na_2HPO_4, 3 g of KH_2PO_4, 1 g of NH_4Cl, 1.2 g of $MgSO_4$, and 111 mg of $CaCl_2$ in 1 L, pH 6.8). Formaldehyde was added to the medium as indicated in each experiment. Cell density was monitored by measuring optical density at 660 nm (OD_{660}). Cells cultivated in the medium in either the presence or the absence of formaldehyde were collected by centrifugation at 4000g for 10 min at 4°C. The pellet was washed two times with 10 mM potassium phosphate buffer (pH 7.0) containing 3% NaCl and used for biodegradation experiments.

Biodegradation of Formaldehyde

Washed cells of the DM-2 strain that had been grown until the late log growth phase with or without 200 or 400 ppm of formaldehyde were resuspended in 50 mL of M9S medium (30 g of NaCl, 6 g of Na_2HPO_4, 3 g of KH_2PO_4, 1 g of NH_4Cl, 1.2 g of $MgSO_4$, and 111 mg of $CaCl_2$ in 1 L) in 100-mL Erlenmeyer flasks. Formaldehyde was added to the medium at the concentration indicated in each experiment. Cell density in the medium was adjusted by measuring OD_{660}. The medium containing cells was incubated at 28°C with shaking (130 strokes/min). One milliliter of sample was removed from the medium, and the sample was centrifuged at 4000g for 3 min to remove cells. Supernatant of the medium was collected, and the formaldehyde concentration was analyzed.

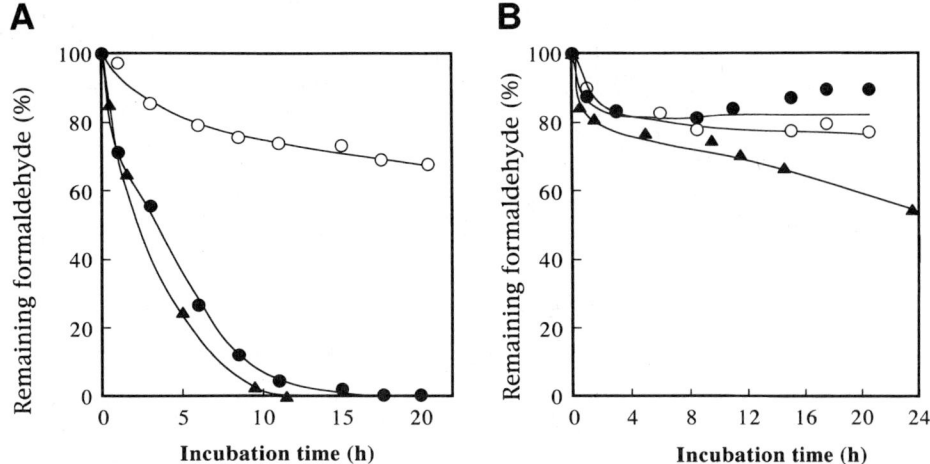

Fig. 1. Time course of formaldehyde degradation by a formaldehyde-tolerant bacterium, DM-2. Formaldehyde concentration in the minimal medium was 200 **(A)** and 400 ppm **(B)**, respectively. DM-2 strains were precultivated in modified DM-2 medium in the absence of formaldehyde (○), in the presence of 200 ppm of formaldehyde (●), or in the presence of 400 ppm of formaldehyde (▲). Cell density in the medium was adjusted to 0.6 OD$_{660}$.

Analysis of Formaldehyde Concentration

Formaldehyde concentration in the culture supernatant was determined using a colorimetric reaction with FeCl$_3$ and MBTH as reported previously *(8)*. To 200 μL of sample containing up to 10 ppm of formaldehyde, 40 μL of 0.4% MBTH in 2 N HCl was added. After 20 min of incubation at room temperature, 10 μL of 1% (w/v) FeCl$_3$ was added. After 10 min of incubation at room temperature, 250 μL of acetone was added. The final solution was incubated for 20 min at room temperature for the color to stabilize. The absorbance at 670 nm was measured using a spectrophotometer (UV1200, Shimazu, Kyoto, Japan).

Results and Discussion

Figure 1 shows the time course of formaldehyde degradation in minimal medium containing 3% NaCl. DM-2 strain cells were precultivated until the late logarithmic growth phase in either the presence or the absence of 200 or 400 ppm of formaldehyde and were used for the biodegradation experiments. Cells were added to M9S minimal medium containing either 200 ppm (Fig. 1A) or 400 ppm (Fig. 1B) of formaldehyde as the sole carbon source.

DM-2 was able to degrade formaldehyde in M9S medium containing 200 ppm of formaldehyde, as shown Fig. 1A, and formaldehyde was completely degraded within 15 h. DM-2, which was precultivated in the absence of formaldehyde, also showed the degradation of formaldehyde; however, >70% of formaldehyde remained even after 20 h.

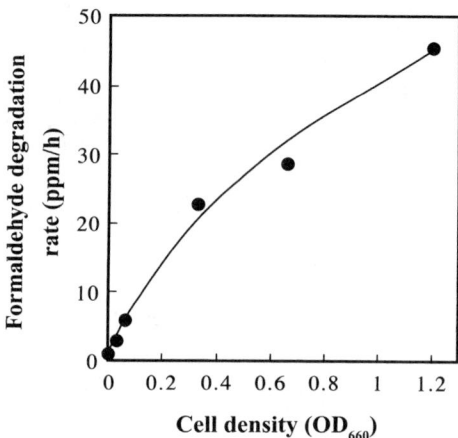

Fig. 2. Effect of cell density on formaldehyde degradation rate. Cell density was indicated as the absorbance at 660 nm (OD_{660}). The formaldehyde concentration in the M9S minimum medium was 200 ppm.

The biodegradation of 400 ppm of formaldehyde was also attempted. DM-2 cells were able to degrade formaldehyde; therefore, the cells were tolerant of 400 ppm of formaldehyde. However, the ability of formaldehyde degradation was dependent on the precultivation conditions. The cells precultivated in the presence or absence of 200 ppm of formaldehyde could degrade formaldehyde for 4 h (Fig. 1B). After 4 h, these cells did not show formaldehyde degradation. On the other hand, the cell precultivated in the presence of 400 ppm of formaldehyde degraded formaldehyde continuously and 45% of formaldehyde in the medium was degraded within 24 h. Therefore, the DM-2 strain is classified as a high formaldehyde-tolerant microorganism. It can degrade and tolerate up to 400 ppm of formaldehyde, a concentration at which most, if not all, formaldehyde-tolerant bacteria, such as *E. coli (3)* and *Halomonas* sp. *(4)*, are unable to survive. The DM-2 strain can be used for the construction of a formaldehyde degradation system that may be adapted to a higher formaldehyde concentration environment.

It has been reported that formaldehyde is metabolized in a reaction catalyzed by the combination of formaldehyde dehydrogenase and formate dehydrogenase in *Pseudomonas* sp. *(1,2)*, *E. coli (3)*, and *Halomonas* sp. *(4)*. Induction of glutathione-dependent formaldehyde dehydrogenase activity in *E. coli* and *Haemophilus influenza* was reported *(3)*. Production of formaldehyde dehydrogenase may be induced by formaldehyde during the growth phase; consequently, precultivated DM-2 strain in the presence of formaldehyde showed high formaldehyde degradation activity.

Figure 2 shows the effect of cell concentration in the medium on the formaldehyde degradation rate. In minimal medium, 200 ppm of formaldehyde was added. As seen from Fig. 2, increased cell concentrations resulted in increased rates of formaldehyde degradation. The formaldehyde degradation rate was dependent on the cell density. The highest formaldehyde degradation rate achieved was 45 ppm/h (45 mg of formaldehyde/(L·h)) at a DM-2 strain concentration corresponding with an OD_{660} of 1.2. Lu and Hegemann *(6)* reported that under 200 or 400 ppm of formaldehyde concentration, >90% of formaldehyde was degraded by incubation with anaerobic sludge for 20 d *(6)*. This study demonstrated, for the first time, the potential application of marine bacteria for the biodegradation of formaldehyde.

Conclusion

We demonstrated formaldehyde degradation by using a novel formaldehyde-tolerant bacterium isolated from a marine environment, the DM-2 strain. The bacterium could degrade high concentrations of formaldehyde up to 400 ppm in M9 minimal medium containing 3% NaCl. This bacterium can be used as the basis of a bioremediation system that can remove formaldehyde from marine environments.

Acknowledgment

This work was financially supported by the Sasakawa Scientific Research Grant from The Japan Science Society.

References

1. Kato, N., Kobayashi, H., Shimao, M., and Sakazawa, C. (1984), *Agric. Biol. Chem.* **48,** 2017–2023.
2. Yanase, H., Noda, H., Kiyotaka, A., Kita, K., and Kato, N. (1995), *Biosci. Biotechnol. Biochem.* **59,** 197–202.
3. Gutheil, W. G., Kasimoglu, E., and Nicholson, P. C. (1997), *Biochem. Biophys. Res. Commun.* **238,** 693–696.
4. Azachi, M., Henis, Y., Oren, A., Gurevich, P., and Sarig, S. (1995), *Can. J. Microbiol.* **41,** 548–553.
5. Kaneko, Y., Ito, M., Ogura, Y., and Ishikawa, J. (1985), US patent no. 4,505,821.
6. Lu, Z. and Hegemann, W. (1998), *Water Res.* **32,** 209–215.
7. Omil, F., Méndez, D., Vidal, G., Méndez, R., and Lema, J. M. (1999), *Enzyme Microb. Technol.* **24,** 255–262.
8. Eberhardt, M. A. and Sieburth, J. M. (1985), *Mar. Chem.* **17,** 199–212.

Dissemination of Catabolic Plasmids Among Desiccation-Tolerant Bacteria in Soil Microcosms

FREDERIC WEEKERS,*,[1] CHRISTIAN RODRIGUEZ,[1]
PHILIPPE JACQUES,[2] MAXIMILIEN MERGEAY,[3] AND PHILIPPE THONART[1,2]

[1]*Walloon Center for Industrial Biology, University of Liege, B40,
4000 Liege, Belgium, E-mail: p.thonart@ulg.ac.be;*
[2]*Walloon Center for Industrial Biology,
Agricultural University of Gembloux, Passage des Deportes,
2-5030 Gembloux, Belgium; and [3]Belgian Nuclear Research Center,
Boeretang 200, B-2400 Mol, Belgium*

Abstract

The dissemination of catabolic plasmids was compared to bioaugmentation by strain inoculation in microcosm experiments. When *Rhodococcus erythropolis* strain T902, bearing a plasmid with trichloroethene and isopropylbenzene degradation pathways, was used as the inoculum, no transconjugant was isolated but the strain remained in the soil. This plasmid had a narrow host range. *Pseudomonas putida* strain C8S3 was used as the inoculum in a second approach. It bore a broad host range conjugative plasmid harboring a natural transposon, RP4::Tn4371, responsible for biphenyl and 4-chlorobiphenyl degradation pathways. The inoculating population slowly decreased from its original level (10^6 colony-forming units [CFU]/g of dry soil) to approx 3×10^2 CFU/g of dry soil after 3 wk. Transconjugant populations degrading biphenyl appeared in constant humidity soil (up to 2×10^3 CFU/g) and desiccating soil (up to 10^4 CFU/g). The feasibility of plasmid dissemination as a bioaugmentation technique was demonstrated in desiccating soils. The ecologic significance of desiccation in bioaugmentation was demonstrated: it upset the microbial ecology and the development of transconjugants.

Index Entries: Bioaugmentation; drought tolerance; conjugation; plasmid dissemination; microcosm; isopropylbenzene.

Introduction

Bioaugmentation, the addition of microorganisms to enhance a specific activity, has been used in several areas such as agriculture (1) and

*Author to whom all correspondence and reprint requests should be addressed.

wastewater treatment *(2)*. However, it is not considered an efficient technique for the remediation of soils polluted with hydrocarbons. In many cases, the advantages of increasing the biocatalyst activity do not offset the advantages of niche fitness demonstrated by indigenous microorganisms *(3,4)*. Therefore, bioaugmentation should be considered only when the intrinsic catabolic activity is not present in the soil. The bioaugmentation product should combine the advantages of good catalytic activity and niche fitness.

A way to fulfill these two criteria is the dissemination of catabolic plasmids to indigenous bacterial strains adapted to the environment. Then, the inoculating microorganism could be a plasmid-delivering strain chosen according to technologic and ecologic selective criteria:

1. A degrading activity toward the more recalcitrant hydrocarbons, borne on plasmids or not.
2. A high productivity in bioreactors preferentially with inexpensive growth media.
3. A good tolerance to desiccation in order to prepare a starter culture in a desiccated form.
4. A good shelf-life with full maintenance of metabolic activity on rehydration *(5)*.
5. Adaptation to soil physicochemical conditions, to survive long enough to transmit its plasmid or to achieve pollution removal itself.
6. Compatibility with autochthonous bacterial strains, which could be measured by the ability of the inoculating strain to disseminate its catabolic plasmid to the indigenous flora.

For microorganisms to meet these criteria, drought-tolerant bacteria were selected from desiccated polluted soils *(6)*. Desiccation tolerance is relevant for the production of a dry bioaugmentation product, but it also has ecologic significance in soils submitted to daily or seasonal variations of hydration *(7)*.

The introduction of new catabolic genes borne on plasmids into the "technologic strains" was achieved by means of natural conjugation in order to improve their catabolic activity *(8)*. Many catabolic genes are located on plasmids that are self-transmissible and have a broad host range *(9)*. The newly obtained strains could be used in microcosm experiments either as bioaugmentation strains with good catalytic activity able to remove the pollution or as plasmid delivery systems. They should exhibit a high catalytic activity and be adapted to the soil biotope.

This article compares the catabolic gene dissemination strategy and bioaugmentation by strain inoculation. Some studies have dealt with catabolic plasmid dissemination *(10–14)*, but none have addressed the ecologic importance of desiccation or used the concept of technologic strains as tools to produce and deliver the plasmid to the autochthonous microflora.

In the present study, two strains were used as model systems. In the first experiment, a *Rhodococcus erythropolis* strain was used as the starter strain and as the plasmid delivery system. These Gram-positive nocardioform *Actinomycete* eubacteria should survive and develop in polluted soils. The *R. erythropolis* strain carries a conjugative plasmid responsible for mercury resistance, trichloroethene (TCE) and isopropylbenzene (IPB) degradation that could disseminate. It has a narrow host range, but in vivo experiments have shown that unexpected transfers could occur in field conditions *(15,16)*. This strain is desiccation tolerant and exhibits good technologic properties. The fate of this strain, the dissemination of the plasmid, and the appearance of IPB-degrading autochthonous microflora were monitored in IPB-treated and untreated soil microcosms.

In the second approach, a *Pseudomonas putida* strain was used as the inoculum. It carried a broad host range plasmid responsible for biphenyl (BP) and 4-chlorobiphenyl pathways. This strain did not exhibit technologic properties but was adapted to the soil biotope since it was first isolated from the soil in which it served as the inoculum. Its survival in BP-treated and untreated soil microcosms and the dissemination of the catabolic plasmid were monitored. The BP concentration was measured.

When desiccation stress was imposed on the ecosystems, the appearance of drought-tolerant strains expressing the plasmid was monitored. The behavior of desiccating systems was compared to constant water activity systems and the ecologic significance of desiccation was evaluated.

Materials and Methods

Bacterial Strains and Plasmids

In the first microcosm experiment, *R. erythopolis* strain T902.1 was used as the inoculum. It was first isolated from a desiccated polluted soil *(6)*. It was mated with another *R. erythropolis* strain harboring pBD2, a conjugative megaplasmid responsible for mercury resistance and TCE and IPB degradation, in order to broaden its metabolic activity *(8)*. This plasmid has a narrow host range *(17)* as measured in vitro. A rifampicin mutant of this strain was used as the inoculum in the microcosm experiments.

The second microcosm experiment was conducted with an IncP, broad host range conjugative plasmid harboring a natural transposon RP4::Tn4371 responsible for BP and 4-chlorobiphenyl pathways *(18)*. It also confers resistance to ampicillin, tetracycline, and kanamycin. This plasmid was originally hosted in *Escherichia coli* strain CM844, but it was transferred by conjugation to *P. putida* strain C8S3 prior to soil inoculation.

Transfer of Plasmids to Soil Isolates

The *E. coli* strain CM844 harboring the RP4::Tn4371 plasmid is a polyauxothroph and is not adapted to soil conditions. To inoculate the soil with a strain adapted to it and harboring the plasmid, a filter mating was done between *E. coli* strain CM844 and a soil microbial extraction obtained by adding 5 g of soil to 45 mL of $MgSO_4$ solution (10 mM) with 0.1% of

Tween-80 and shaking for 1 h at 30°C. The transconjugant was selected on minimal medium 284 *(5)* with tetracycline (50 mg/L) and glucose (2 g/L) as the sole source of carbon. It did not express the biphenyl (BP) degradation capability. This property enabled good counterselection of the donor. A rifampicin mutant of this strain, C8S3, was used as donor throughout this study. C8S3 was identified with BIOLOG Fingerprinting (BIOLOG, Hayward, CA) as *P. putida*.

Soil Microcosms

Two types of soil were used in the first microcosm experiment: The first was a silt soil collected from woods with a pH (H_2O) of 3.9, a very low organic matter content, and a field capacity moisture content of 51% on dry soil. The second was a clay soil collected from a field plot with a pH (H_2O) of 7.0, an organic matter content of about 2%, and a field capacity moisture content of 50% on dry soil. Only clay soil was used in the second experiment. Soil samples were moistened to 75% of their water-holding capacities and preincubated at 20 ± 2°C for 3 wk prior to the start of the experiment. The soil microcosms were prepared by adding 100 g of soil to 250-mL tightly closed glass jars for the IPB-amended soil and to opened jars for the BP-amended soils.

Four different experimental conditions were set up: the first set of microcosms was inoculated with the plasmid donor strain (10^6 colony-forming units [CFU]/g of dry soil) and amended with 1000 ppm of either IPB or BP, the second was not inoculated but amended with the pollutant, the third was inoculated but not amended, and the fourth was untreated and served as the control.

Strains used as inoculant were grown overnight at 30°C in selective 869 based [21] rich media ($869 + 5 \mu M$ $HgCl_2 + 150$ mg/L of ripampicin for T902.1 and $869 + 50$ mg/L of tetracycline + 150 mg/L of ripampicin for C8S3), centrifuged at 10,000g, washed in $MgSO_4$ solution (10 mM), and resuspended in $MgSO_4$ solution (10 mM). A Bürker's chamber count was made to determine the required volume of this suspension to reach 10^6 CFU/g of dry soil in the microcosms.

The total heterotrophic count of the soil was determined by plating 0.1 mL of serial 10-fold dilutions of soil in $MgSO_4$ solution (10 mM) onto 869 rich medium containing cycloheximide (150 mg/L). Strain T902.1, the inoculating strain in the first experiment, was enumerated on 284 minimal medium containing cycloheximide (150 mg/L) and rifampicin (150 mg/L) with IPB as the sole source of carbon. Transconjugants with strain T902.1 were sought on minimal selective medium with cycloheximide (150 mg/L), $HgCl_2$ (5 μM), and IPB as the sole source of carbon. No counterselection of strain T902.1 could be made. Strain C8S3, the inoculating strain in the second experiment, was enumerated on 869 rich medium with cycloheximide (150 mg/L), tetracycline (50 mg/L), and rifampicin (150 mg/L). Transconjugants with strain C8S3 were sought by plating the dilutions onto 284 minimal medium with cycloheximide (150 mg/L), tetracycline (50 mg/L), and BP as the sole source of carbon. The donor strain could not grow on this medium.

Owing to IPB volatility, IPB-amended microcosms were incubated in closed jars. No variation in water content could be made. Therefore, desiccation-tolerant transconjugants were sought by plating the dilution of freeze-dried microcosm soil samples onto the minimal selective medium. Because BPs are not volatile, microcosm jars could be left open and two experimental conditions could be set up. One set of microcosms was kept at 75% of the water-holding capacity by daily weighed additions of water. The other set of microcosms was submitted to cycles of desiccation and rehydration in order to impose a water depletion stress to the biotope. The desiccation-tolerant transconjugants were sought by plating dilutions from the desiccating microcosms onto the 284 selective medium.

BP Analysis of Soil Samples

Concentrations of BPs in the soil were determined as follows. Five-gram samples were collected in triplicate from the microcosms and were extracted with 45 mL of cyclohexane by mixing the soil and the solvent in a blender for 1 min. The supernatant was collected and filtered, and the absorbance at 268 nm was measured and compared to a standard curve.

Identification of Strains

A few strains exhibiting transconjugant properties were identified to the species level with BIOLOG fingerprinting. The procedure was realized according to the manufacturer's instructions (BIOLOG).

Plasmid DNA Isolation

Individual colonies of transconjugants were isolated by streaking them onto selective agar. They were cultured in 869 selective broth. DNA isolation was performed as described previously *(8)*, and the plasmids were resolved on a 0.8% agarose gel essentially as described.

Drought-Tolerance Measurements

The strains were cultivated overnight in 869 broth, centrifuged, washed in $MgSO_4$ solution (10 m*M*), and resuspended in the same volume of solution. Then they were freeze-dried as described previously *(19)*. Soil samples were freeze-dried under the same conditions.

Results and Discussion

Bioaugmentation in IPB-Amended Soil Microcosms

The first microcosm experiment was conducted with *R. erythropolis* strain T902.1 used as the inoculum. This strain harbored a plasmid responsible for IPB and TCE degradation as well as mercury resistance. In vitro experiments have shown that this linear plasmid has a narrow host range

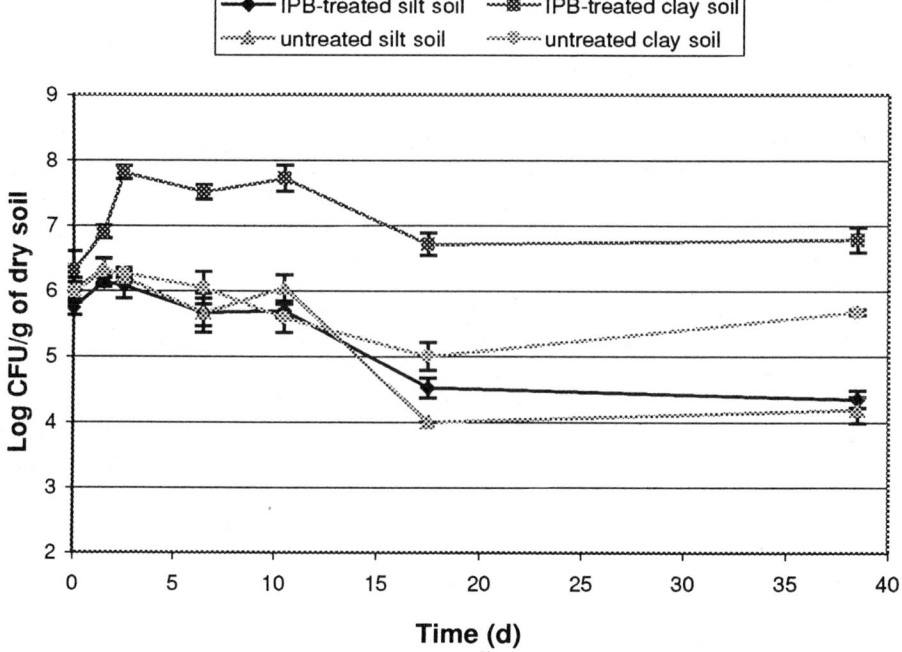

Fig. 1. Fate of strain T902.1 in treated and untreated silt soil and clay soil.

(17). However, in vivo observations may differ from what occurs in vitro *(15,16)*. This strain is drought tolerant and adapted to the technologic production and conditioning constraints. It could either deliver the plasmid and disappear or develop and do the remediation itself.

Two types of soil were used : a forest acid silt soil and an agricultural clay soil. The fate of the starter culture was monitored in both soils (Fig. 1). The behavior is different according to the nature of the soil.

In IPB-amended clay soil, strain T902.1 increased 30-fold after only 2 d. Selective pressure owing to the presence of IPB gave an advantage to strain T902.1. In nonamended soil, T902.1 maintained its count to its original level (10^6 CFU/g of dry soil) during the time of the experiment (40 d). Strain T902.1 was well adapted to these soil conditions. By maintaining its population in the soil, the starter culture guaranteed the presence of IPB degradation activity in the soil. This presence reinforced the standardization of the bioaugmentation treatment whether or not plasmid dissemination occurred and independently of the nature of the indigenous microflora. No transconjugants could be isolated from either of the two soils. This was probably owing to the narrow host range nature of the pBD2 plasmid. Although no plasmid dissemination occurred, the strain bioaugmentation approach was successful because the inoculated strain developed in the clay soil.

In the acid silt soil, the T902.1 population declined slowly from its original level (10^6 CFU/g of dry soil) to 10^4 CFU/g in amended or

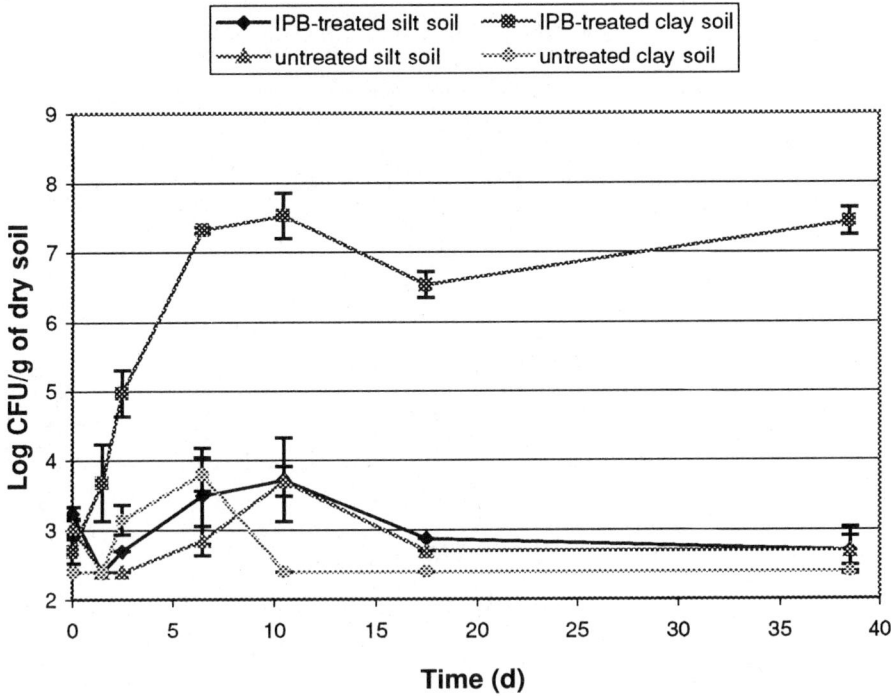

Fig. 2. Evolution of indigenous IPB-degrading populations in silt soil and clay soil.

nonamended soil. The starter maintained the presence of the catabolic activity in the soil.

The autochthonous population of the uninoculated clay soil adapted rapidly to IPB pollution and reached 2×10^7 CFU/g of dry soil (Fig. 2). IPB was not recalcitrant and intrinsic degrading activity was present and could develop. Two different strains were isolated among this population: a *Pseudomonas pseudoalcaligenes* strain and an *R. erythropolis* strain. They both exhibited IPB 2,3-dioxygenase activity. Gram-negative *Pseudomonas* strains are known to decompose a large variety of aromatic compounds *(20)*. A high homology was found between IPB dioxygenase from *Rhodococci* and dioxygenase of other aromatic compounds found in *Pseudomonads (21)*, which explained how they had similar activities in IPB-amended soils.

Desiccation of microcosm samples was achieved in order to select drought-tolerant IPB-degrading microorganisms (Fig. 3) and to monitor the ecologic significance of desiccation in soils. The sensitivity to desiccation of the IPB-degrading population increased as this population appeared and developed. Before IPB treatment, the desiccation-tolerant population was of the same order of magnitude as the overall population. After 7 d, only 0.5% of the global population was still desiccation tolerant. The microorganisms responsible for IPB degradation were more drought sensitive than the original average population of the soil. IPB amendment disturbed the soil ecology, resulting in a change in the drought tolerance of the

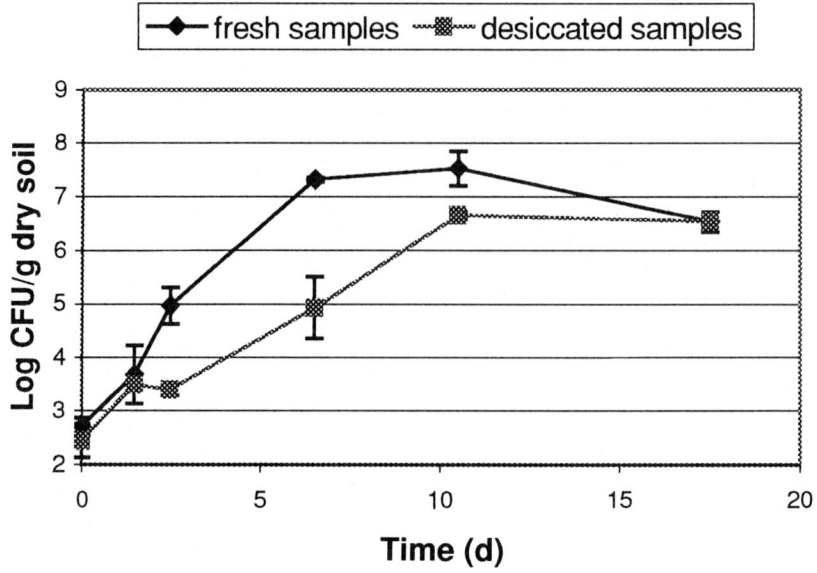

Fig. 3. Evolution of indigenous IPB-degrading population in clay soil in fresh samples and after desiccation.

autochthonous microflora. Because the microcosm itself was not submitted to desiccation stress, it did not exert a selective pressure on the IPB-degrading population. In the second experimental setup, some microcosms were submitted to desiccation in order to measure the ecologic significance of desiccation as a selective agent.

Plasmid Dissemination in BP-Treated Soil Microcosms

In the second approach, a broad host range IncP plasmid harboring a natural catabolic transposon (RP4::Tn4371) was used to study plasmid dissemination as an alternative bioaugmentation approach. This plasmid was previously transferred in Gram-positive *Arthrobacter* bacteria *(16)*, a microorganism usually found in soils *(22)*. The plasmid was originally hosted in *E. coli* strain CM844. This strain was polyauxothrophic and not adapted to the soil. To inoculate the microcosms with an adapted strain, a filter mating was achieved between *E. coli* strain CM844 and an extract of the soil to be treated. Because of its polyauxothrophic character, the donor strain was easily counterselected on selective minimal medium. A transconjugant was isolated on medium 284+tetracycline (50 mg/L) with glucose as the sole source of carbon. This transconjugant was also resistant to kanamycin and ampicillin but did not express the BP degradation property. This enabled easy counterselection of the inoculum in microcosm isolates. The presence of the plasmid in the strain was confirmed by extraction and isolation by agarose gel electrophoresis. The transconjugant was identified by BIOLOG fingerprinting as *P. putida*. A rifampicin mutant of this strain was used to inoculate the microcosms. The clay soil was used in this experiment.

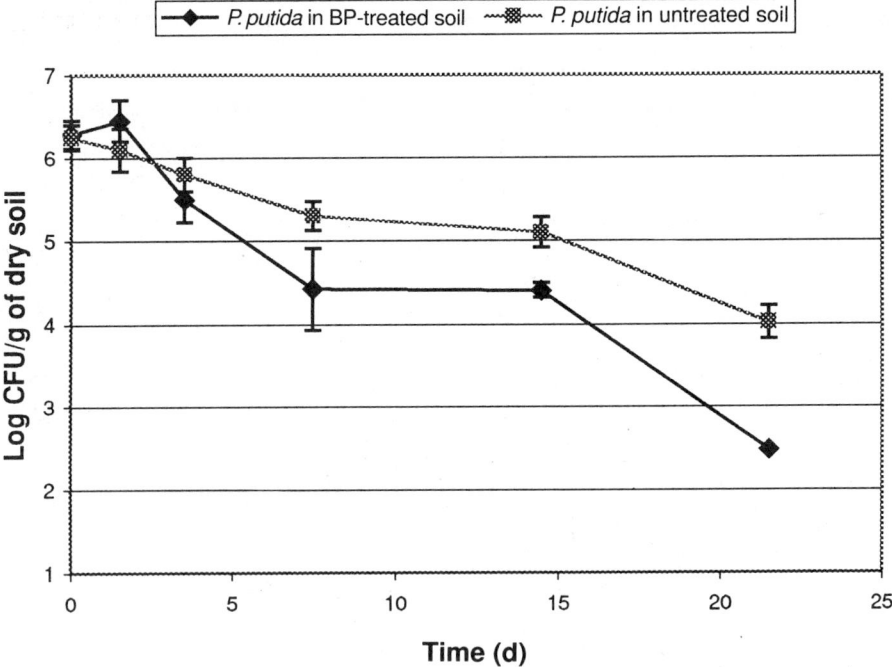

Fig. 4. Fate of the inoculum (*P. putida*) in BP-treated and untreated microcosms under constant-humidity conditions.

To measure the influence of desiccation on microbial ecology in polluted soils, desiccating soil microcosms were set up in parallel with constant humidity systems.

The fate of *P. putida* was monitored in constant-humidity soil (Fig. 4). Its population slowly declined from its original level (2×10^6 CFU/g) to 10^4 CFU/g in untreated soil and to 3×10^2 CFU/g in BP-treated soil after 21 d. This inoculating strain could not maintain its population in the soil. Furthermore, the decline was faster in the BP-amended soil than in the untreated control, which means that it was sensitive to the presence of BP. A BP-degrading population appeared after 12 d in the constant-humidity soil (Fig. 5). Individual colonies of this population were collected and further analyzed. They exhibited resistance to tetracycline, kanamycin, and ampicillin, which is a good indication that these strains are transconjugants. They were identified by BIOLOG fingerprinting as *Pseudomonas fluorescens*, *Acinetobacter johnsonii*, and *Sphingobacterium* sp. RP4 is an IncP plasmid. These plasmids are well expressed in γ-proteobacteria *(23)*, to which the isolated strains belonged.

In the inoculated soil in which transconjugants were found, a decrease of 20% in the BP concentration was measured after 30 d as compared with the uninoculated soil in which BP concentration remained unchanged after the same time. The lack of any addition of nutrients could have limited the biomass development and blocked the BP degradation in this experiment.

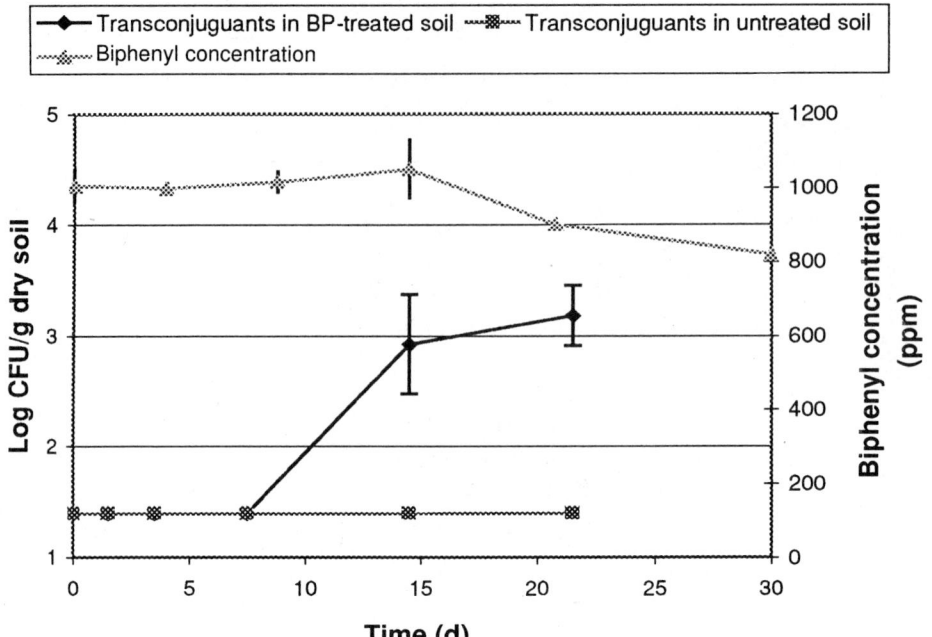

Fig. 5. Fate of transconjugants in BP-treated and untreated soil under constant-humidity conditions and BP concentrations.

In parallel, microcosms were submitted to desiccation. The *P. putida* population was monitored in BP-treated and nontreated desiccating microcosms (Fig. 6). Soil water content was measured daily.

The overall behavior of the inoculum was the same as in constant-humidity soil. Its population declined with time and the rate of mortality was higher in BP-treated soil than in untreated soil. The rate of disappearance of the starter culture over the duration of the experiment was higher in the desiccating microcosm than in the constant-humidity microcosm. However, as the soil began to desiccate, the population of *P. putida* rose 10-fold. This could be owing to the decrease in competition under these conditions. When the soil was rehydrated, the competition became more important again and the decline in the inoculum accelerated. This reflected the upset effect of desiccation on the microbial ecology of the soil.

Transconjugants in the inoculated BP-treated soil appeared after 5 d, but desiccation of the soil slowed the rate of growth of this population (Fig. 7). It increased again when the soil was rehydrated. The second cycle of desiccation made this population disappear under the limit of detection and it reappeared on rehydration. Desiccation upset the development of transconjugants. Transconjugants developed faster in desiccating soil than in constant-humidity soil, also likely owing to decreased competition.

In both soil conditions (desiccating or constant humidity), no BP-degrading population was observed in noninoculated soil. As opposed to IPB-dioxygenase activity, no intrinsic BP-degrading activity was present in

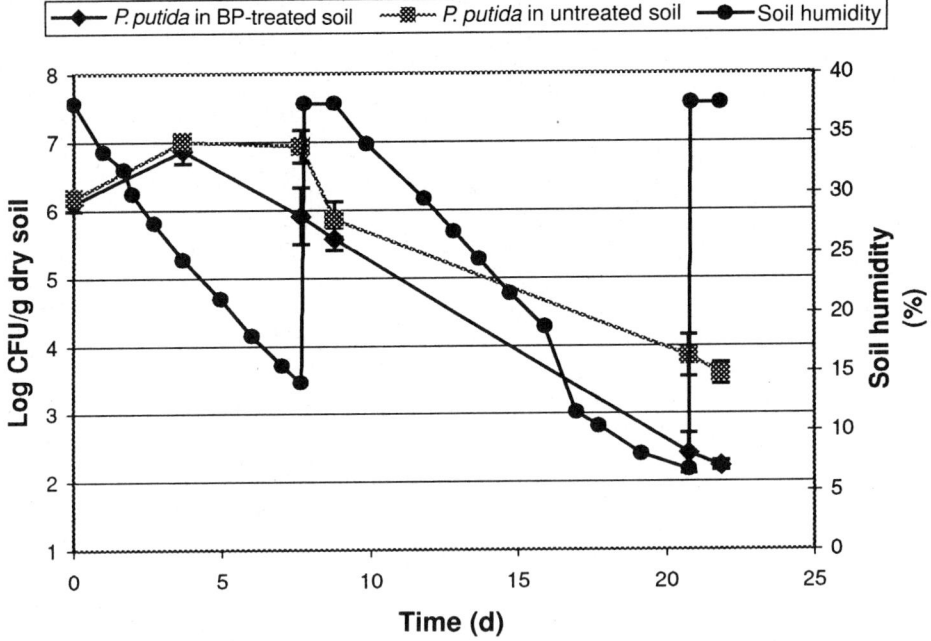

Fig. 6. Fate of the inoculum in desiccating BP-treated and untreated soil microcosms and soil humidity.

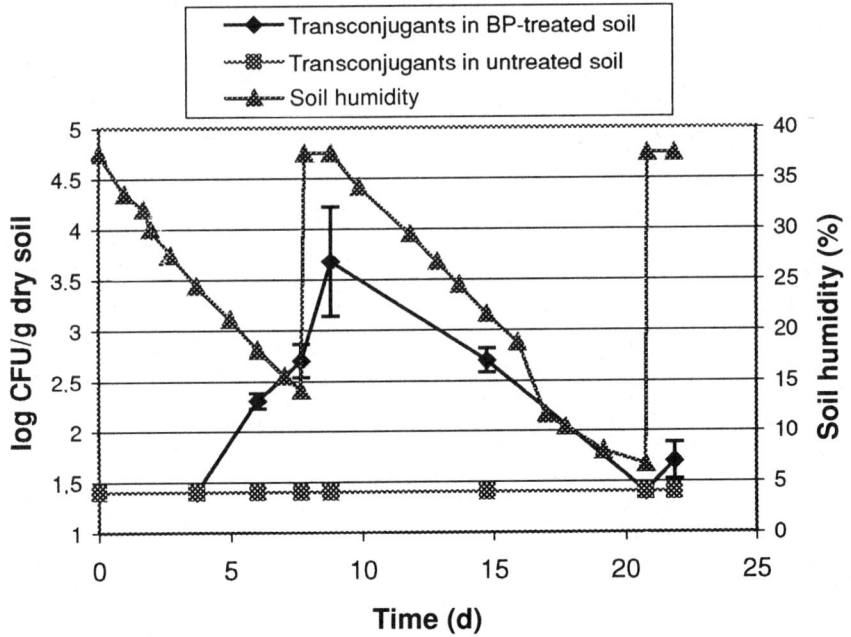

Fig. 7. Fate of transconjugants in desiccating BP-treated and untreated soil microcosms.

Table 1
Drought Tolerance of Transconjugants

Strain	Origin	Survival to desiccation (%)	Gram stain
Escherichia coli CM844	Plasmid collection	0.005	–
Pseudomonas putida C8S3	Inoculum	<0.002	–
Pseudomonas fluorescens	Inoculated BP-treated microcosm under constant humidity	0.0005	–
Sphingobacterium sp.	Inoculated BP-treated microcosm under constant humidity	ND[a]	–
Acinetobacter johnsonii	Inoculated BP-treated microcosm under constant humidity	0.4	–
Alcaligenes denitrificans	Desiccating inoculated BP-treated microcosm	1	–
ND[a]	Desiccating inoculated BP-treated microcosm	3	–

[a]ND, not determined.

the soil. With this pollution, inoculation was necessary. The plasmid dissemination strategy yielded catalytically active, adapted indigenous microorganisms responsible for partial removal of pollution.

The transconjugants from the desiccating microcosms were isolated and further analyzed. They all exhibited resistance to tetracycline, kanamycin, and ampicillin. They were identified with the BIOLOG system as *Alcaligenes denitrificans*. Another transconjugant strain could not be identified. The drought tolerance of these strains was determined to compare with the tolerance of the strains isolated from a constant-humidity microcosm (Table 1). It was 10- to 1000-fold higher than the drought tolerance of the transconjugants isolated from the constant-humidity microcosm. The nature and phenotype of the transconjugants were fundamentally different owing to the desiccation of the soil.

Although the strains from the desiccating soil were more resistant to desiccation than the strains from constant-humidity soil, their survival was still low (about 1%). All isolated transconjugants were Gram-negative strains expressing the RP4 plasmid. Although Gram-positive strains could express this plasmid *(16)* and are more resistant to desiccation *(19)*, no desiccation-tolerant Gram-positive transconjugants could be selected under these conditions.

Conclusion

When an *R. erythropolis* strain was used as the inoculum, it maintained its population in the polluted soil but could not disseminate its plasmid to indigenous soil microorganisms. The pollution removal activity could be guaranteed by the presence of the inoculating strain in the soil.

When a broad host range plasmid was introduced in the soil by means of a soil isolate, transconjugants appeared rapidly and BP concentration decreased as compared to the uninoculated control. However, some limiting factor blocked the pollution removal process before it was completed. The addition of nutrient could solve this problem. When water availability decreased, the ecology of the system was upset. The transconjugant nature was different and their development was compromised, limiting the biodegradation of the pollutant.

The dissemination of plasmids from technologic strains as remediation strategy should be used with broad host range plasmids. The desiccation tolerance of the inoculum could be an important factor when the treated soil is submitted to hydration variations. This property could guarantee the maintenance of the catalytic activity in the soil when desiccation-sensitive inocula or transconjugants could disappear, jeopardizing the pollution removal procedure.

Acknowledgment

This work was supported by Fonds pour la Formation à la Recherche dans l'Industrie et l'Agriculture grant.

References

1. Jasper, D. (1994), *Aust. J. Soil Res.* **32**, 1301–1319.
2. Rittmann, B. and Whiteman, R. (1994), *Water Qual. Int.* **1**, 12–16.
3. Möller, J., Steckel, T., Wedebye, E., and Werstermann, P. (1995), *Bull. Environ. Contam. Toxicol.* **54**, 913–918.
4. Vogel, T. (1996), *Curr. Opin. Biotechnol.* **7**, 311–316.
5. Weekers, F., Jacques, P., Springael, D., Mergeay, M., Diels, L., and Thonart, P. (1998), *Appl. Biochem. Biotechnol.* **70–72**, 311–322.
6. Weekers, F., Jacques, P., Springael, D., Mergeay, M., Diels, L., and Thonart, P. (1996), *Med. Fac. Landbouww. Univ. Gent* **61/4b**, 2161–2164.
7. Potts, M. (1994), *Microbiol. Rev.* **58**, 755–805.
8. Weekers, F., Jacques, P., Springael, D., Mergeay, M., Diels, L., and Thonart, P. (1999), *Appl. Biochem. Biotechnol.* **77–79**, 251–266.
9. Sayler, G., Hooper, S., Layton, A., and King, J. (1990), in *Microbial Ecology*, Fietcher, M., ed., Springer-Verlag, New York.
10. Brokamp, A. and Schmidt, F. (1991), *Curr. Microbiol.* **22**, 299–306.
11. Daane, J., Molina, J., Berry, E., and Sadowski, M. (1996), *Appl. Environ. Microbiol.* **62**, 515–521.
12. De Rore, H., Demolder, K., De Wilde, K., Top, E., Houwen, F., and Verstraete, W. (1994), *FEMS Microbiol. Ecol.* **15**, 71–78.
13. Di Giovanni, G., Neilson, J., Pepper, I., and Sinclair, N. (1996), *Appl. Environ. Microbiol.* **62**, 2521–2526.
14. Top, E., Van Daele, P., De Saeyer, N., and Forney, L. (1998), *Antonie Van Leeuwenhoek* **73**, 87–94.
15. Dong, Q., Springael, D., Schoters, J., Nuyts, G., Mergeay, M., and Diels, L. (1998), *Water Sci. Technol.* **37**, 465–468.
16. Margesin, R. and Schinner, F. (1997), *J. Bas. Microbiol.* **37**, 217–227.

17. Dabrock, B., Kesseler, M., Averhoff, B., and Gottschalk, G. (1994), *Appl. Environ. Microbiol.* **60,** 853–860.
18. Springael, D., Kreps, S., and Mergeay, M. (1993), *J. Bacteriol.* **175,** 1674–1681.
19. Weekers, F., Jacques, P., Springael, D., Mergeay, M., Diels, L., and Thonart, P. (2000), in *Focus on Biotechnology*, Hofman, M. and Anne, J., eds., Kluwer Academic, Amsterdam.
20. Galli, E., Barbieri, P., and Bestteti, G. (1992), in *Pseudomonas: Molecular Biology and Biotechnology*, Galli, E., Silver, S., and Withold, B., eds., American Society for Microbiology, Washington, DC.
21. Kesseler, M., Dabbs, E., Averhoff, B., and Gottschalk, G. (1996), *Microbiology* **142,** 3241–3251.
22. Paul, E. and Clark, F. (1988), *Soil Microbiology and Biochemistry*, Academic Press, London.
23. Mergeay, M. and Springael, D. (1996), in *Bioremediation Protocols*, Sheehan, D., ed., Humana Press, Totowa, NJ.

SESSION 3
Bioprocessing Research

Bioprocessing Research

Thomas R. Hanley[1] and David N. Thompson[2]

[1]University of Louisville, Louisville, KY and [2]Idaho National Engineering and Environmental Laboratory, Idaho Falls, ID

Millions of tons of renewable lignocellulosic biomass are produced each year worldwide, along with enormous quantities of other renewables such as food processing effluents. Production of chemicals and materials from this biomass is limited in large part by the cost of existing chemical and bioprocessing technologies. The close association of lignin with cellulose and hemicellulose in lignocellulosic biomass, for example, necessitates the use of costly pretreatments to allow near-quantitative enzymatic conversion of the polysaccharides to fermentable carbohydrates. This adversely affects the economics of lignocellulose utilization, severely limiting its economic competitiveness with existing technologies to produce the desired fuels and chemicals.

Considerable research has led to great strides in the development of pretreatment, fermentation, and product recovery technologies for renewable biomass. For example, genetic engineering of yeasts to ferment xylose to ethanol was an important step in increasing yields of ethanol from pretreated lignocellulose. Further improvements in the methods of production and harvest of biomass in the energy, chemical, and enzyme inputs to the processes, in the production and recovery of processing enzymes, and in product recovery can reduce costs and improve the overall economics. In the end, advances will be necessary in each of these areas to make bioprocessing of lignocellulose and other renewables to fuels and chemicals economically competitive.

Seven research papers and over forty related posters were presented in this session, focusing on many of the steps in the bioprocessing of lignocellulose and other renewable biomass sources. Many authors addressed production of fuels and chemicals from pulp effluents and other lignocellulosic and food-processing effluents. Several authors presented research on the development of pretreatment technologies, while others focused on product and/or enzyme recovery using a wide range of techniques. Finally, several authors addressed bacterial and fungal bioremediation of lignocellulose-derived effluents. Each target area above was addressed in this session, illustrating many important advances in the bioprocessing of renewable biomass.

Ethanol Production
from Lignocellulosic Byproducts
of Olive Oil Extraction

IGNACIO BALLESTEROS, JOSE MIGUEL OLIVA, FELICIA SAEZ,
AND MERCEDES BALLESTEROS*

CIEMAT, Renewable Energies Department, Av. Complutense,
No. 22, 28040-Madrid, Spain, E-mail: m.ballesteros@ciemat.es

Abstract

The recent implementation of a new two-step centrifugation process for extracting olive oil in Spain has substantially reduced water consumption, thereby eliminating oil mill wastewater. However, a new high sugar content residue is still generated. In this work the two fractions present in the residue (olive pulp and fragmented stones) were assayed as substrate for ethanol production by the simultaneous saccharification and fermentation (SSF) process. Pretreatment of fragmented olive stones by sulfuric acid–catalyzed steam explosion was the most effective treatment for increasing enzymatic digestibility; however, a pretreatment step was not necessary to bioconvert the olive pulp into ethanol. The olive pulp and fragmented olive stones were tested by the SSF process using a fed-batch procedure. By adding the pulp three times at 24-h intervals, 76% of the theoretical SSF yield was obtained. Experiments with fed-batch pretreated olive stones provided SSF yields significantly lower than those obtained at standard SSF procedure. The preferred SSF conditions to obtain ethanol from olives stones (61% of theoretical yield) were 10% substrate and addition of cellulases at 15 filter paper units/g of substrate.

Index Entries: Ethanol; olive oil extraction byproducts; pretreatment; enzymatic saccharification; fermentation.

Introduction

The olive oil industry represents one of the most important economic agro-food sectors in Spain. On a worldwide scale, Spain is the main producer and exporter of olive oil, with an average of 590,000 t/yr, which makes up >33% of the world's production (1).

*Author to whom all correspondence and reprint requests should be addressed.

The extraction process of olive oil yields a highly contaminating residue that causes serious environmental concerns, still unsolved in the olive oil–producing countries. The production of oil mill wastewater in the Mediterranean area is estimated to be approx 1.2 million t/yr *(2)*. Traditionally, a three-step centrifugation process performs the continuous extraction of olive oil, in which the addition of water in a proportion of 1:1 is needed, that large amounts of highly contaminated oil mill wastewater are generated *(3)*.

Currently, a new two-step centrifugation process (without exogenous addition of water) to extract the olive oil has been developed that dramatically reduces the oil mill wastewater *(4)*. Implementation of the two-phase centrifugation process by many olive oil–producing Spanish industries has reduced water consumption and, consequently, oil mill wastewater. However, in this new process, a residue is still generated (estimated to be 800 kg/t of olive), containing the pulp, the water content of the olive, and portions of the seed husks and olive stones. This new residue, called pommace, consists of 70% water, 21% organic substances, and 9% mineral substances, respectively *(5)*. The organic fraction comprises 25% sugars, making it a potentially attractive, low-cost feed material for biologic conversion. Upgrading this residue by fractionation, with the possibility of obtaining ethanol from the free fermentable sugars and the cellulose present in the residue, would mean an improvement in the management of the byproducts of the olive oil industry. Currently, the extraction industries that use the two-phase centrifugation process separate the residue into two fractions: the pulp, formed by the olive pulp itself and the vegetation liquors engaged in nutritive and growth function, and the woodlike portion, comprising fragments of the olive stones.

In the present study, the two fractions contained in the pommace (olive pulp and olive stone fragments) were evaluated as a carbon source for the production of ethanol. This article reports the analysis (carbohydrates, lignin, ash, and other components) of these materials. This composition has been used in the calculation of conversion efficiencies. Steam explosion was tested as a pretreatment to enhance enzymatic hydrolysis. Raw materials were hydrolyzed with cellulase enzyme at different substrate and enzyme loadings. Finally, the simultaneous saccharification and fermentation (SSF) bioconversion process of olive pulp and olive stones was tested.

Materials and Methods

Substrates

Olive pulp and olive stone fragments, generated as a residue from the two-phase centrifugation olive oil production process, were supplied by Oleicoa El Tejar S.C.L. (Córdoba, Spain). The substrates were chemically analyzed according to the following standard methods: ASTM D-1348 *(6)* for moisture content, ASTM D-1102-84 *(7)* for ash content, ASTM D-1111-84 *(8)* for hot water extracts, and ASTM D-1107-87 *(9)* for ethanol/toluene extracts. The sample was first hydrolyzed using 72% H_2SO_4 for 60 min and then hydrolyzed a second time with 4% H_2SO_4 for 60 min at 120°C. This analysis gave the hemicellulosic sugars content (expressed as the sum of the xylose + arabinose + galactose + mannose), cellulose (expressed as glucose), and klason lignin.

Pretreatment

Washed feedstocks were pretreated in a steam explosion pilot unit, operated by batches and equipped with a 2-L reaction vessel. The plant description and working methodology were described in a previous article *(10)*. The temperature (210°C) and residence time (4 min) conditions of the biomass pretreatment were selected with regard to the maximum glucose recovery after 72 h of enzymatic hydrolysis. The reactor was filled with 200 g of feedstock per batch and was directly heated to the desired temperature with saturated steam. After the explosion, the material was recovered in a cyclone. The wet material was cooled to about 40°C and then filtered for solid recovery. The water-insoluble fraction was analyzed for xylans, glucans, and lignin content. The carbohydrate content of the filtrate was also analyzed. A variation of the described pretreatment procedure using olive stones was carried out in the presence of an acid catalyst. Two hundred grams of dry material was soaked in 1 L of acid solution (0.5% [w/v] H_2SO_4), and then a vacuum was applied to extract the air from the sample and leave it well impregnated. The vacuum was then released and allowed to stand for 1 h. It was next filtered through a Buchner funnel and rinsed with plenty of water to remove any remaining acid.

Microorganisms and Growth Conditions

Kluyveromyces marxianus CECT 10895, a thermotolerant mutant yeast strain obtained in our laboratory *(11)*, was used in fermentation and SSF experiments. Active cultures for inoculation were prepared by growing the organism on a rotary shaker at 180 rpm for 16 h at 42°C in a growth medium containing the following: 5 g/L of yeast extract (Difco), 5 g/L of peptone (Oxoid), 2 g/L of NH_4Cl, 1 g/L of KH_2PO_4, 0.3 g/L of $MgSO_4 \cdot 7H_2O$, and 30 g/L of glucose.

Enzymatic Hydrolysis and SSF Tests

The washed feedstocks were enzymatically hydrolyzed to determine sugar yield. Enzymatic hydrolysis was performed in 0.1 M sodium acetate buffer (pH 4.8), and substrate concentrations and enzyme loadings were tested at 50°C for 72 h. The enzyme preparation Celluclast 1.5L was a gift from Novo Nordisk (Bagsvaerd, Denmark).

SSF experiments were carried out in 100-mL Erlenmeyer flasks that contained 50 mL of the growth medium as described above and then agitated at 150 rpm. Glucose was substituted by the lignocellulose biomass at different substrate concentrations; the cellulolytic complex (Celluclast

1.5L), at different enzyme loadings, was also added. Enzymatic hydrolysis and SSF assays were conducted at 42°C for 72 h.

In the SSF experiments, flasks were inoculated with 10% (v/v) yeast cultures and periodically checked during the tests for ethanol and glucose. A fed-batch variation of the described SSF process was performed, in which olive pulp and pretreated stones were fed in discrete and successive charges. Fed-batch SSF experiments were initiated as described above, using 15 and 10% olive pulp and olive stone substrate concentrations, respectively, and 15 filter paper units (FPU)/g substrate enzyme loading. A charge of fresh substrate was added 24 h after the onset of SSF. The mixture was incubated for an additional 24 h. Then, a new charge of fresh substrate was added again and the mixture incubated for a further 24 h. Supplementation of enzyme to maintain the initial enzyme loading of 15 FPU/g of substrate was performed at the same time as fresh substrate was added.

Various fed-batch tests using the previously described variation of SSF process were performed. Substrate charges were carried out at 24-h intervals as follows:

1. 15% pulp as an initial substrate + 5% pulp + 5% pulp.
2. 15% pulp as an initial substrate + 7.5% pulp + 7.5% pulp.
3. 15% pulp as an initial substrate + 10% pulp + 5% olive stones pretreated by steam explosion.
4. 15% pulp as an initial substrate + 10% pulp + 5% olive stone pretreated by steam explosion in the presence of acid.
5. 10% olive stone pretreated by steam explosion as an initial substrate + 15% pulp + 5% pulp.
6. 10% of olive stone pretreated by steam explosion in the presence of acid as an initial substrate + 15% pulp + 5% pulp.
7. 10% olive stone as an initial substrate + 10% olive stone + 10% olive stone, all pretreated by steam explosion.
8. 10% olive stone as an initial substrate + 10% olive stone + 10% olive stone, all pretreated by steam explosion in the presence of acid.

Analytical Procedures

Composition of feedstocks and water-insoluble fraction after pretreatment has been determined by total hydrolysis with H_2SO_4 *(12)*. Enzymatic activities (filter paper and β-glucosidase) were measured according to the methods described by Ghose *(13)*.

Sugars were quantified by high-performance liquid chromatography in a 1081B Hewlett Packard (HP) apparatus with differential refractometer detector under the following conditions: column, AMINEX HPX-87P (Bio-Rad, Hercules, CA); temperature, 85°C; eluent, water at 0.1 mL/min.

Ethanol was measured by gas chromatography, using an HP 5890 Series II apparatus, with flame ionization detector and a column of Carbowax 20 M (2 m × 0.3175 cm) at 95°C. Both injector and detector temperatures were 150°C.

Table 1
Composition of Pulp and Olive Stone Fractions Generated
as Residues in Olive Oil Extraction
by the Two-Phase Centrifugation Process

Composition	Olive pulp (%)	Olive stone (%)
Extracts	53.3	19.2
Free sugars	6.4	0.5
Hemicellulosic sugars	13.4	23.5
Xylose	8.4	20.6
Arabinose	2.5	1.5
Galactose	1.0	1.0
Mannose	1.5	0.4
Cellulose as glucose	15.8	27.6
Klason lignin	15.2	29.1
Ash	2.3	0.6

Results

Feedstock Composition

Table 1 shows the compositions of the two fractions (pulp and olive stone fragments) that make up the residue from the olive oil extraction by a two-phase centrifugation process. Water extraction dissolved 31.6% of the pulp, of which 6.4% corresponds to free sugars; mainly, glucose. The process using organic solvents extracted 21.7%. The hemicellulose fraction comprises 13.4% of the pulp, xylose being the main sugar (8.4%). Cellulose and lignin represent 15.8 and 15.2%, respectively.

The olive stones basically consist of a strongly lignified secondary wall (29.1%), rich in cellulose (27.6%) and hemicelluloses (23.5%), with low salt content (0.6%). A percentage of 56% (dry wt) of the olive stone consists of carbohydrates, with a cellulose content >27%.

Steam Explosion Pretreatment

The results of the steam explosion pretreatment at 210°C and 4 min for the pulp and the olive stone fractions (with no catalyst and with impregnation by sulfuric acid) are shown in Table 2.

The majority of the pulp was solubilized during pretreatment, and, therefore, only 25% of the initial material was recovered as solids. Although 72% of the hemicellulose and 67% of the cellulose present in the pulp dissolved during pretreatment, practically no free sugars were detected in the filtrate, indicating almost total degradation of the dissolved sugars. Lignin and ash were also solubilized during pretreatment of the pulp by 15 and 56%, respectively.

In the pretreatment of the olive stone fraction as a substrate, about 60% of the material was recovered as solids. Analysis of the results of the com-

Table 2
Composition of Filtrate (g/100 g Raw Material) and Water-Insoluble Fiber (%) of Olive Pulp and Olive Stone Residues at 210°C and 4-min Steam Explosion Pretreatment Conditions[a]

| | Olive pulp | | Olive stone | | | |
| | | | No catalyst | | 0.5% (w/v) H_2SO_4 | |
Composition	Filtrate	Water-insoluble fiber	Filtrate	Water-insoluble fiber	Filtrate	Water-insoluble fiber
Glucose	0.8	20.8 (5.2)	0.21	40.5 (24.3)	0.28	42.3 (23.3)
Hemicellulosic sugars	0	14.7 (3.7)	14.90	9.8 (5.9)	15.9	7.7 (4.2)
Xylose	0	11.9 (3.0)	13.40	9.8 (5.9)	14.2	7.7 (4.2)
Galactose	0	0.9 (0.3)	0.6	0 (0)	0.7	0 (0)
Arabinose	0	1.5 (0.4)	0.9	0 (0)	1.0	0 (0)
Mannose	0	0.4 (0.1)	ND	ND	ND	ND
Lignin	—	39.6 (9.9)	—	45.7 (27.4)	—	44.0 (24.2)
Ash		4.1 (1.0)		0.6 (0.4)		0.5 (0.3)
Solubilization (%)		75		40		45

[a]Data are expressed in parentheses as a percentage based on dry wt of raw material. ND, No determined.

Table 3
Enzymatic Hydrolysis of Olive Pulp
at Different Substrate and Enzyme Loadings[a]

Substrate (%)	Enzyme loading (FPU/g substrate)	Potential glucose (g/L)	Free glucose (g/L)	Enzymatic hydrolysis yield (%)
5	7.5	7.9	3.0	38.0
	15		3.0	38.0
	30		3.6	45.6
10	7.5	15.8	6.2	39.2
	15		6.4	40.5
	30		7.7	48.7
15	7.5	23.7	9.6	40.5
	15		9.8	41.3
	30		11.6	48.9
20	7.5	31.6	12.2	38.6
	15		12.5	39.6
	30		13.7	43.3

[a]Yield is expressed as glucose obtained in the enzymatic hydrolysis divided by potential glucose in the raw material.

position of the water-insoluble fraction of the olive stones showed that 75% of the hemicellulose and 12% of the cellulose dissolved during pretreatment. When sulfuric acid was used as a catalyst, a slight increase in the fraction of carbohydrate that was solubilized (82 and 16% for the hemicellulose and cellulose, respectively) was observed. Almost all the dissolved glucose was degraded to other products during pretreatment with and without catalyst (94%), and, therefore, no free glucose was found in the liquid fraction. Regarding the hemicellulosic sugars, 24% of the dissolved sugars was degraded during treatment with no impregnation and 26% when sulfuric acid was used. The lignin content of the water-insoluble fraction obtained after pretreatment of the olive stone fragments without a catalyst remained unchanged, whereas some dissolution of the lignin resulted (12.3%) in the sulfuric-acid-catalyzed steam explosion.

Enzymatic Hydrolysis Tests

To establish the effect of steam explosion pretreatment on the susceptibility to enzymatic attack of the cellulose from the pulp and olive stone fragments, samples of these feedstocks, with no pretreatment and exploded, were submitted to enzymatic hydrolysis tests using the commercial cellulolytic complex Celluclast 1.5L.

Enzymatic hydrolysis yields of the exploded pulp decreased when compared with the sample as received (data not shown). Consequently, tests to evaluate the effect of the initial concentrations of the substrate and enzyme (Table 3) were carried out using pulp, which had not been previously pretreated. Enzymatic hydrolysis yields were in the range of 38–49%

Table 4
Enzymatic Hydrolysis of Steam-Exploded Olive Stones
at 10% (w/v) Substrate and 15 and 30 FPU/g Substrate Enzyme Loading

Pretreatment conditions (210°C, 4 min)	Enzyme loading (FPU/g substrate)	Potential glucose (g/L)	Free glucose (g/L)	Enzymatic hydrolysis yield (%)
No catalyst	15	40.5	16.4	40.5
	30		22.4	55.3
0.5% (w/v) H$_2$SO$_4$	15	42.3	22.3	52.7
	30		25.8	61.0

for all conditions tested. As can be seen, these values did not vary significantly when the concentration of the substrate was increased. At different enzyme loadings, no variations were observed at 7.5 and 15 FPU/g of substrate. However, an enzyme loading of 30 FPU/g of substrate caused a significant increase in enzymatic yield. Owing to the relatively low cellulose content of the pulp, little glucose was released during enzymatic hydrolysis; thus, a high initial substrate loading (above 15%) was needed to obtain glucose concentrations >10 g/L.

Concerning the stone fraction, there was no enzymatic hydrolysis when the olive stone residue was not pretreated (data not shown). The enzymatic hydrolysis results for the pretreated olive stone fraction at 10% substrate concentration (w/v) and 15 and 30 FPU/g of substrate enzyme loads are shown in Table 4. When steam-exploded olive stone fragments without a catalyst were used as substrate, enzymatic hydrolysis yields of 40 and 55% were obtained at enzyme loadings of 15 and 30 FPU/g of substrate, respectively. These yields increased up to 53 and 61%, respectively, when sulfuric acid was used as a catalyst during pretreatment.

SSF Tests

Results of the SSF tests for the untreated pulp, using different initial substrate concentrations, are shown in Table 5. Yields of the SSF process decreased as the initial substrate concentration increased. Initial substrate loading >25% (w/v) could not be used because of the difficulty of keeping solids in suspension in the SSF media (poor mixing). The best ethanol yield (67% of theoretical) was obtained at 15% substrate concentration.

The results of the SSF experiments, using the pretreated olive stone fraction as substrate, are shown in Table 6. As in the case of the pulp, the ethanol yield of the process decreased as the initial substrate concentration increased. No ethanol production was obtained at 25% initial substrate loading (data not shown). The best ethanol yields in the SSF process (59% of theoretical) were obtained at 10% olive stone steamed with sulfuric acid as a catalyst.

Results of the fed-batch experiments by adding substrate (olive pulp and pretreated olive stones) three times at 24-h intervals are shown in

Table 5
Effect of Initial Substrate Concentration on SSF Yield
from Untreated Olive Pulp (enzyme loading of 15 FPU/g of substrate)

Substrate (%)	Ethanol (g/L)	Potential glucose (g/L)	SSF yield	Theoretical yield (%)
15	8.1	23.7	0.34	66.6
20	10.5	31.6	0.33	64.7
25	11.8	39.5	0.30	58.9

Table 6
Effect of Initial Substrate Concentration on SSF
from Pretreated Olive Stones (enzyme loading of 15 FPU/g of substrate)

Pretreatment conditions (210°C, 4 min)	Substrate (%)	Potential glucose (g/L)	Ethanol (g/L)	SSF yield	Theoretical yield (%)
No catalyst	10	40.5	7.8	0.19	37.2
	20	81.0	14.5	0.18	35.3
0.5% (w/v) H_2SO_4	10	42.3	12.9	0.30	58.8
	20	84.6	17.9	0.21	41.2

Table 7. For assays with pulp, the SSF process is better when fed-batch of pulp is used (76.5% of theoretical) than when the substrate is added all at once (58.9% of theoretical). No significant improvement in SSF yield was achieved when pulp and stones are combined. Moreover, a high free glucose concentration in the media is observed after 72 h from the onset of SSF, indicating continuance of cellulosic activity and cessation of ethanol production. Experiments with fed-batch, pretreated olive stones provide SSF yields significantly lower than those obtained at standard SSF procedure; however, high residual glucose concentration remains unmetabolized in the medium.

For comparing the ethanol production rate of different fed-batch experiments, the ethanol production kinetics in the SSF process are shown in Fig. 1. Ethanol production of fed-batch experiments by adding pulp (Fig. 1A) increased after each addition of substrate, reaching values of 15.5 and 18.6 g/L for 25 and 30% final substrate concentration, respectively. In the SSF assays using pulp and olive stone for the second addition (Fig. 1B), no increase in ethanol production was obtained from 48 to 72 h. The SSF assay with pretreated olive stones as initial substrate combined with a fed batch of pulp (Fig. 1C) gives ethanol concentrations in the range of 15.6–18.5 g/L. For experiments using olive stones as the sole substrate, at fed-batch mode of operation (Fig. 1C), no ethanol production was observed during the last 24 h, when 30% total substrate loading was reached.

Table 7
SSF for Fed-Batch Mode Operation Experiments[a]

Substrate	Substrate loading (%)	Ethanol (g/L)	Potential glucose (g/L)	Residual glucose (g/L)	SSF yield	Theoretical yield (%)
Pulp	25 (15 + 5 + 5)	15.5	39.5	2.1	0.39	76.5
Pulp	30 (15 + 7.5 + 7.5)	18.6	47.4	6.3	0.39	76.5
Pulp + stone	30 (15 + 10 + 5)	13.6	59.7	16.1	0.23	45.1
Pulp + stone[b]	30 (15 + 10 + 5)	15.0	60.6	16.4	0.25	49.0
Stone + pulp	25 (10 + 10 + 5)	15.6	64.2	12.3	0.24	47.1
Stone[b] + pulp	25 (10 + 10 + 5)	18.5	66.0	12.9	0.28	54.9
Stone	30 (10 + 10 + 10)	17.5	121.5	56.2	0.14	28.2
Stone[b]	30 (10 + 10 + 10)	19.7	126.9	63.2	0.16	31.4

[a]Supplementation of enzyme to maintain the initial 15 FPU/g of substrate was performed at the same time as fresh substrate was added.
[b]Olive stones impregnated with 0.5% H_2SO_4 prior to steam explosion pretreatment.

Discussion

Upgrading of the lignocellulosic residue generated in the production of olive oil by the two-phase centrifugation process (pulp and olive stone fragments) would yield a significant advance in the management of byproducts originating from the olive oil extraction industry. In the present study, the possibility of obtaining ethanol from the fermentable sugars and cellulose present in the residue was investigated.

The results obtained from analysis of the composition of the raw material showed that the main hemicellulosic sugar, both in the pulp and in the olive stone fractions of olives, was xylose, which indicated that the predominant hemicelluloses in these kinds of materials was xylan. However, the xylose content of the olive stone fraction represents nearly 88% of the hemicellulosic sugars, whereas in the pulp it represents about 63%. The lower amount of xylose in the hemicellulosic sugars in the pulp compared with those obtained for stones resulted from the fact that this fraction contained the olive seed, which had an appreciable quantity of hemicelluloses of the arabinane and glucomanane type (14). Although fermentable sugars content of both feedstocks was lower than that from typical lignocellulosic biomass (15–17), the cellulose content in these materials was high enough to be considered as potential substrate for ethanol production.

The percentage of material dissolved during both the aqueous and organic extraction was much greater in the pulp than in the fragmented stone fractions. According to Brenes et al. (18,19), the olive pulp fraction, composed of the rest of the olive skin, the pulp, and the seed, contains a low quantity of true lignin and presents a series of substances, including low molecular weight phenols, which can be extracted using organic solvents. This lignin structure would explain the high percentage value for dissolution (20.1%) during organic extraction of the pulp.

During the pretreatment, a large part of the fraction determined as acid-insoluble lignin contained in the pulp was dissolved. This fact differs from that of the klason lignin present in woody substrates, which basically remains unchanged during steam explosion pretreatment, or even increases, owing to the formation of pseudolignin as a result of the repolymerization of the decomposition products of hemicellulose and lignin (20,21). This different performance of acid-insoluble lignin of olive pulp can be explained by the fact that the olive pulp and seed do not contain true lignin, but highly polymerized phenolic glycosides (22).

The steam explosion process was not suitable as pretreatment for olive pulp, because it produced a high cellulose solubilization and did not increase enzymatic hydrolysis. On the other hand, the nonpretreated pulp was an appropriate substrate for enzymatic hydrolysis, because, under all initial substrate concentrations and 7.5 and 15 FPU/g enzyme loadings, hydrolysis yields in the range of 38–41.5% were obtained. These yields can be considered to be in the range of values obtained for other types of materials with high hemicellulose content. Higher enzymatic hydrolysis

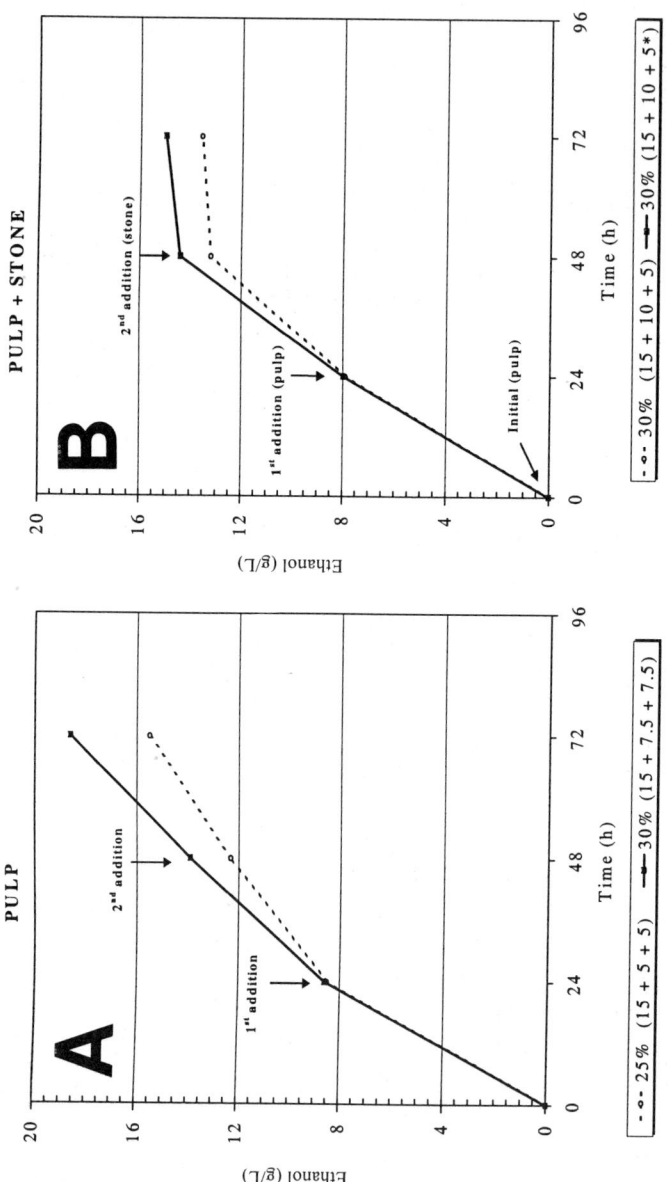

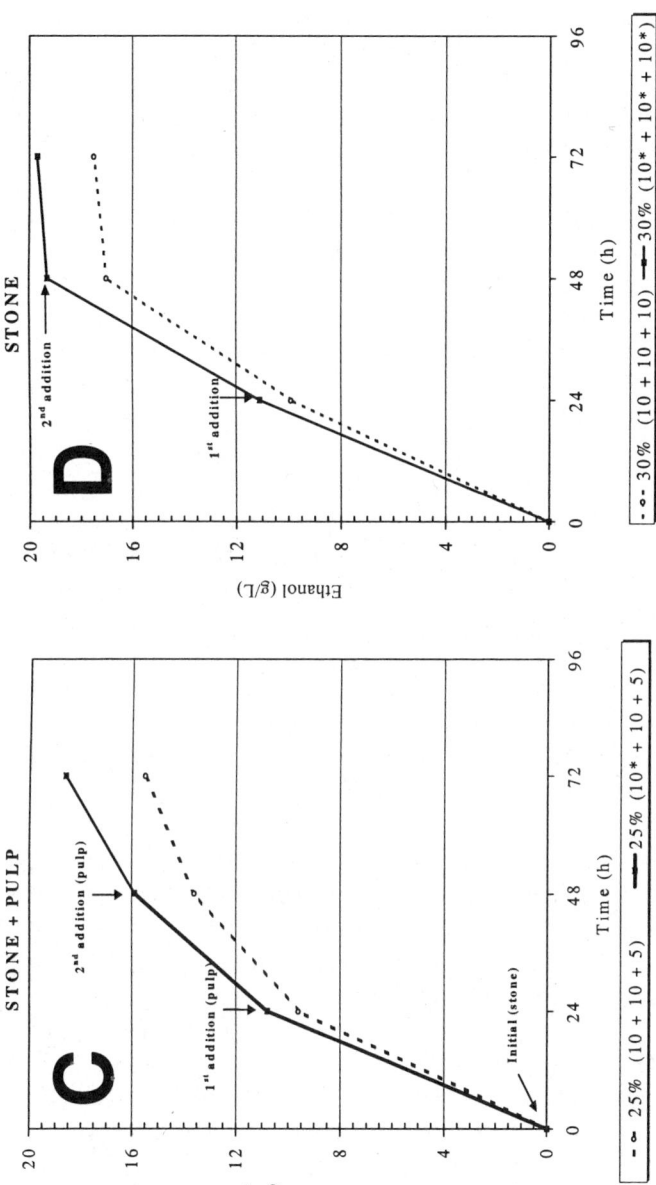

Fig. 1. Ethanol concentration in fed-batch SSF process as a function of time. Supplementation of enzyme to maintain the initial 15 FPU/g substrate was performed at the same time as fresh substrate was added. **(A)** Untreated pulp; **(B)** untreated pulp plus steam-exploded stones; **(C)** steam-exploded stones plus untreated pulp; **(D)** steam-exploded stones plus untreated pulp. *Olive stone impregnated with 0.5% (w/v) prior to steam explosion pretreatment.

yields were achieved at 30 FPU/g of substrate, but this enzyme loading can be considered too costly.

Enzymatic hydrolysis of the olive stone fragment fraction was favored by sulfuric acid impregnation prior to pretreatment. Under these conditions, with an enzyme loading of 15 FPU/g of substrate, a hydrolysis yield of 53% of theoretical was obtained. In the tests without a catalyst, olive stone portions remained intact after pretreatment, reducing the accessibility of the enzyme to the cellulose of the olive stone. Penetration of the acid inside the structure of the olive stone brings about a better hydrolysis of the acetyl groups of the xylanes *(23)*. During autohydrolysis more acetic acid is formed, which favors the cleavage of a series of bonds. This ensures the cohesion of the material, and during decompression and expansion of the steam, the material is subsequently reduced to smaller particles.

Yields obtained in the SSF tests of nonpretreated pulp (Table 5) were in the range of those obtained from other lignocellulosic materials using Celluclast 15L and *K. marxianus* CECT 10895 *(24,25)*. SSF yields decreased as initial substrate concentration increased. This decrease is owing to lower yields in the fermentation stage and not in enzymatic hydrolysis step, because, as previously stated, hydrolysis yields were not affected by an increased substrate concentration (Table 3). This inhibition may be owing to the presence of phenolic substances in pulp and seed, which at higher concentrations are toxic for microorganisms *(26,27)*. It has been reported *(28,29)* that oil and pulp from the two-phase centrifugation extraction process contain a higher amount of phenolic compounds, mainly in the form of glycosides and esters, than those present when the conventional extraction process is used.

In the SSF tests using the olive stone fraction as substrate, the best yield (59% of theoretical) was obtained using 10% of the pretreated material in the presence of sulfuric acid catalyst. Initial steam-exploded olive stone concentrations of 25% inhibited ethanol production. This shows that toxic compounds present in steam-exploded olive stones have a stronger inhibitory effect on the microorganisms than those present in unpretreated pulp, since initial pulp loads of 25% were feasible for the SSF process. Although the characteristics of olive stone lignin are similar to that of hardwood (syryngil-guayacil type), recently, glucosides of tyrosol and hydroxityrosol have been identified as components of olive stone *(30)*. These polyphenols are well known as part of olive pulp, and they have not been described in any lignocellulosic material. They account for the characteristics of olive stones such as hardness and high reluctance to acid and enzymatic attacks, and, during steam explosion pretreatment, chemical changes (probably hydrolysis) in these compounds can occur that increase the toxicity. This suggestion is under research in our laboratory.

As in pulp, lower initial substrate loads are needed to obtain higher SSF yields, but it means low sugar concentration for fermentation. Consequently, it is necessary to establish SSF conditions that make an adequate substrate concentration in the media (providing a reasonable sugar concentration for fermentation) compatible with high SSF yields. It would then be possible to obtain ethanol concentrations in the SSF process, which would allow later distillation of the ethanol from the fermentation medium to be economically viable. The variation of the SSF process, by fed-batch operation mode, allows achievement of high SSF yields, together with ethanol concentrations suitable for fermentation.

In pulp assays, for the same substrate loading, the SSF process is better when the fed-batch procedure is used (76% of theoretical yield) than when substrate is added all at once (59%). Under these conditions not only were concentrations of ethanol in the medium >18 g/L achieved, but also a significant increase in the SSF yield was observed. The addition of pulp by pulses every 24 h allowed the microorganism to maintain the capacity to ferment at high substrate concentration, thus reducing the inhibitory effect of toxic compounds (phenols and tannins) present in the pulp.

The addition of pretreated olive stones in a sequential model did not permit use of substrate loadings above 20% as occurred in standard SSF. Fermentation ceases when substrate concentration increases up to 30%, confirming the formation of unknown toxic compounds during steam explosion pretreatment.

Considering carbohydrate composition of raw materials and yields for the different SSF conditions studied, it can be concluded that lignocellulosic byproducts from olive oil extraction by the two-phase centrifugation process are suitable feedstocks for ethanol production.

The most suitable scheme to obtain ethanol by the SSF process from pulp would be a fed-batch operation mode using untreated pulp at 15% + 7.5% + 7.5% addition each 24 h. At this condition, 10 kg of pulp will yield 1 L of ethanol.

Owing to olive stone fraction, sulfuric acid impregnation prior to steam explosion pretreatment improves SSF yields. Substrate loading above 20% was not suitable because of strong inhibition of ethanol production. The preferred SSF conditions would be 10% acid steam-exploded olive stones and 15 FPU/g enzyme loading. In these conditions, SSF yields of about 59% of theoretical yield can be obtained; thus, by using 6 kg of acid-exploded stones, 1 L of ethanol could be obtained.

Acknowledgment

We wish to acknowledge CICYT (Spanish Interministerialist Commission of Technology and Science) for its financial support.

References

1. Civantos, L. (1995), *Olivae* **59,** 18–21.
2. Netti, S. and Wlassics, I. (1995), *Riv. Ital. Sostanze Grasse.* **72,** 119–125.
3. Giovanchino, DI L. (1991), *Olivae* **36,** 14–41.
4. Alba, J. (1994), *Fruticultura Profesional* **62,** 85–95.
5. Espínola, F. (1996), *Alimentación, Equipos y Tecnología* **4,** 1–11.

6. ASTM D-1348 (1995), *Annual Book of ASTM Standards*, vol. 04. 10, American Society for Testing Materials, Philadelphia.
7. ASTM D-1102-84 (1995), *Annual Book of ASTM Standards*, vol. 04. 10, American Society for Testing Materials, Philadelphia.
8. ASTM D-1111-84 (1995), *Annual Book of ASTM Standards*, vol. 04. 10, American Society for Testing Materials, Philadelphia.
9. ASTM D-1107-87 (1995), *Annual Book of ASTM Standards*, vol. 04. 10, American Society for Testing Materials, Philadelphia.
10. Carrasco, J. E., Martínez, J. M., Negro, M. J., Manero, J., Mazón, P., Sáez, F., and Martín, C. (1989), in *Biomass for Energy and Industry, 5th Conference*, vol. 2, Grassi, G., Gosse, G., and Dos Santos, G., eds., Elsevier Applied Science, Essex, England, pp. 38–44.
11. Ballesteros, I., Oliva, J. M., Ballesteros, M., and Carrasco, J. (1993), *Appl. Biochem. Biotechnol.* **39/40**, 201–211.
12. Puls, J., Poutanen, K., Körner, H. V., and Viikari, L. (1985). *Appl. Microbiol. Biotechnol.* **22**, 416–423.
13. Ghose, T. K. (1987), *Pure Appl. Chem.* **59:2**, 257–268.
14. Heredia, A., Guillén, R., Fernández Bolaños, J., and Rivas, M. (1987), *Biomass* **14**, 143–148.
15. Holtzapple, M. T. (1993), in *Encyclopedia of Food Science, Food Technology, and Nutrition*, Macrae, R., Robinson, R. K., and Sadler, M. J., eds., Academic, London, pp. 758–767.
16. Holtzapple, M. T. (1993), in *Encyclopedia of Food Science, Food Technology, and Nutrition*, Macrae, R., Robinson, R. K., and Sadler, M. J., eds., Academic, London, pp. 2324–2334.
17. Holtzapple, M. T. (1993), in *Encyclopedia of Food Science, Food Technology, and Nutrition*, Macrae, R., Robinson, R. K., and Sadler, M. J., eds., Academic, London, pp. 2731–2738.
18. Brenes, M., García, P., and Garrido, A. (1993), *J. Food Sci.* **58**, 347–350.
19. Brenes, M., Rejano, L., García, P., Sánchez, A. H., and Garrido, A. (1995), *J. Agric. Food. Chem.* **43**, 2702–2706.
20. Chua, M. G. S. and Wyman, M. (1979), *Can. J. Chem.* **57**, 1141–1150.
21. Moniruzzaman, M. (1996), *Appl. Biochem. Biotechnol.* **59**, 283–297.
22. Maestro Durán, R., Borja Padilla, R., Mnartín Martín, A., Fiestas Ros de Ursinos, J. A., and Alba Mendoza, J. (1991), *Grasas Aceites* **42:4**, 271–276.
23. Ropars, M., Marchal, R., Pourquié, J., and Vandecasteele, J. P. (1992), *Biores. Technol.* **42**, 197–204.
24. Ballesteros, I., Ballesteros, M., Oliva, J. M., and Carrasco, J. E. (1994), in *Biomass for Energy, Environ. Agric. Ind.*, Chartier, Ph., Beenackers, A. A. C. M., and Grassi, G., eds., Elsevier Science, Oxford, UK, pp. 1953–1958.
25. Ballesteros, I., Oliva, J. M., Carrasco, J., Cabañas, A., Navarro, A. A., and Ballesteros, M. (1998), *Appl. Biochem. Biotechnol.* **70–72**, 369–381.
26. Sorlino, C., Andreoni, V., Ferrari, A., and Ranall, G. (1986), in *International Symposium on Olive Byproducts Valorization*, Sevilla, Spain, pp. 81–88.
27. Rodríguez, M., Pérez, J., Ramos Cormazana, A., and Martínez, J. (1988), *J. Appl. Bacteriol.* **25**, 219–226.
28. Alba, J., Ruiz, M. A., Hidalgo, F., Martínez, F., and Moyano, M. J. (1993), *Dossier Oleo* **2**, 40–59.
29. Uceda, M., Hermoso, M., and González, J. (1995), *Alimentación Equipos Tecnología* **5**, 93–98.
30. Felizón, B. P. (1997), PhD Thesis, Facultad de Farmacia, Universidad de Sevilla, Spain.

Continuous Countercurrent Extraction of Hemicellulose from Pretreated Wood Residues

KYOUNG HEON KIM, MELVIN P. TUCKER, FRED A. KELLER,
ANDY ADEN, AND QUANG A. NGUYEN*

*National Renewable Energy Laboratory, 1617 Cole Boulevard,
Golden, CO, 80401-3393, E-mail: quang_nguyen@nrel.gov*

Abstract

Two-stage dilute acid pretreatment followed by enzymatic cellulose hydrolysis is an effective method for obtaining high sugar yields from wood residues such as softwood forest thinnings. In the first-stage hydrolysis step, most of the hemicellulose is solubilized using relatively mild conditions. The soluble hemicellulosic sugars are recovered from the hydrolysate slurry by washing with water. The washed solids are then subjected to more severe hydrolysis conditions to hydrolyze approx 50% of the cellulose to glucose. The remaining cellulose can further be hydrolyzed with cellulase enzyme. Our process simulation indicates that the amount of water used in the hemicellulose recovery step has a significant impact on the cost of ethanol production. It is important to keep water usage as low as possible while maintaining relatively high recovery of soluble sugars. To achieve this objective, a prototype pilot-scale continuous countercurrent screw extractor was evaluated for the recovery of hemicellulose from pretreated forest thinnings. Using the 274-cm (9-ft) long extractor, solubles recoveries of 98, 91, and 77% were obtained with liquid-to-insoluble solids (L/IS) ratios of 5.6, 3.4, and 2.1, respectively. An empirical equation was developed to predict the performance of the screw extractor. This equation predicts that soluble sugar recovery above 95% can be obtained with an L/IS ratio as low as 3.0.

Index Entries: Extraction; hemicellulose; softwood; pretreatment; acid hydrolysis.

Introduction

In a previous study *(1)*, we concluded that two-stage dilute sulfuric acid pretreatment of softwood forest thinnings gave higher sugar yields than single-stage pretreatment. Figure 1 shows a simplified block-flow

---

*Author to whom all correspondence and reprint requests should be addressed.

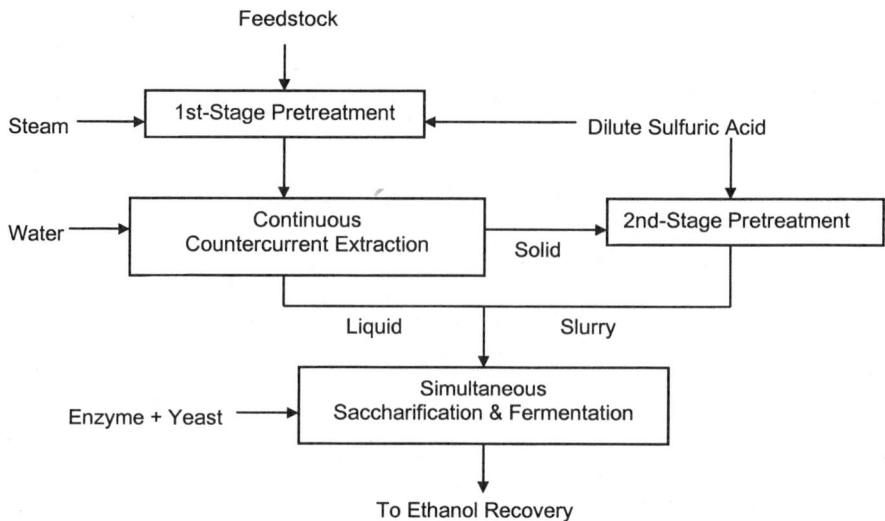

Fig. 1. Two-stage dilute sulfuric acid pretreatment of softwood forest thinnings.

diagram of a two-stage pretreatment process. In this process, high recovery of soluble sugars in the first-stage pretreated material (or hydrolysate) is essential because sugars remaining in the first-stage hydrolysate would be destroyed in the second-stage hydrolysis. Furthermore, the amount of wash water required to achieve high sugar recovery has a significant impact on the process economics. Too much wash water would dilute the sugar stream and increase the cost of fermentation and ethanol distillation. In general, countercurrent washing of the pretreated biomass is required to achieve adequate sugar recovery and high sugar concentration in the extract while maintaining a reasonable water usage requirement. One common term used to describe the amount of water used for extracting solubles from material is the liquid-to-solids (L/S) ratio, which is the ratio of water used in the extraction process over the dry wt of total solids (insoluble and soluble) in the feed. Because the content of soluble solids varies with pretreated materials and diminishes as the solids are being washed, the liquid-to-insoluble solids (L/IS) ratio is also used. The L/IS ratio is essentially constant throughout a countercurrent extractor at steady state, whereas the L/S ratio increases as the soluble solids are removed. We used the L/IS ratio to ensure consistent comparison of extraction characteristics for different pretreated biomass materials.

Figure 2 shows the effect of the L/IS ratio on soluble sugar recovery from first-stage hydrolysate and cost of ethanol production for a 2000 dry t/d softwood-to-ethanol plant using two-stage dilute acid hydrolysis (2). The sugar recovery values were based on the results of three-stage, stagewise, countercurrent extraction experiments (3). The process simulation in Fig. 2 implies that reducing the amount of water used in extracting sugars from first-stage hydrolysate from an L/IS ratio of 5.0 to 3.0 would lower the cost of ethanol production from $0.34/L ($1.29/gal) to $0.31/L

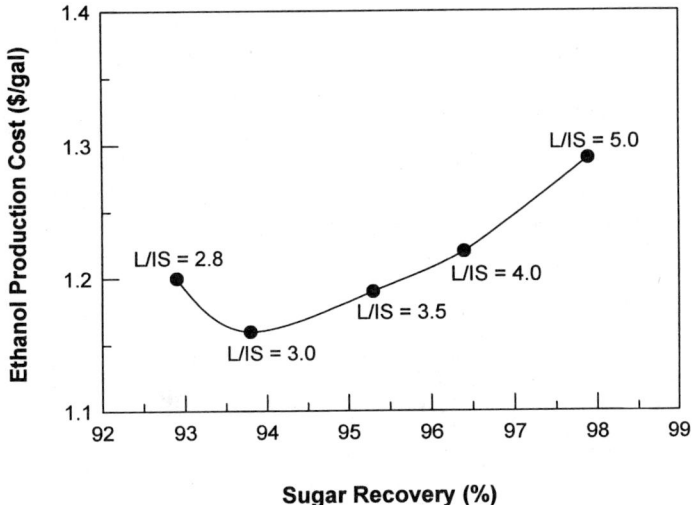

Fig. 2. Effect of L/IS ratio on soluble sugar recovery and production cost of ethanol (2000 dry t/d, $25/dry t softwood residues, two-stage dilute acid hydrolysis process, without enzyme addition).

($1.16/gal), even though the sugar recovery would be lowered from 97.9 to 93.8%. This reduction in production cost is owing mainly to the lower energy requirements to recover ethanol resulting from fermentation of a more concentrated sugar stream. Further reduction of L/IS to 2.8 raises the cost of ethanol production because the negative impact of sugar loss is greater than the savings from lower water usage. If a greater number of extraction stages or a continuous countercurrent extraction device is used, the L/IS ratio can probably be reduced to <3.0 while achieving high sugar recovery and further reducing the cost of ethanol production.

Countercurrent extraction of solubles from biomass materials (such as pulp, sugarcane, fruits, seeds, and pretreated lignocellulose) can be accomplished in a variety of commercial equipment (4,5). The main criteria for selecting countercurrent washing of pretreated biomass to recover soluble sugars include high sugar recovery, high sugar concentration in the extract (i.e., low L/IS ratio), and low capital and operating costs. These criteria are generally the same as those used in the food-processing industry for extraction of sugars and other soluble solids from a variety of feedstock. By comparison, most stagewise washers used in the pulp industry are designed primarily for thorough washing of fibers and not necessarily for obtaining a high concentration of solutes in the wash water. Therefore, our focus is on continuous countercurrent extraction equipment used in the food-processing industry because these systems are most effective in reducing water requirements. Screw conveyors (single or twin) and screw towers are commonly used for extraction of sugar from sugarcane, sugar beet, and fruits (4,6). Pilot-scale screw extractors were used to extract soluble components from sweet sorghum silage (7) and steam-pretreated lignocellulosic materials *(8)*. Twin-screw extractors provide better liquid/solid contact and, thus, are generally more efficient than single-screw extractors. However, twin-screw extractors are generally more expensive. The design of inter-mittent-reversing of screw rotation direction was reported to overcome the compaction problem and inefficient liquid/solid contact in single-screw extractors *(9,10)*. We installed mixing paddles on the auger of our single-screw extractor to improve liquid/solid contact.

Good liquid/solid contact in screw extractors also depends on the drainage characteristics of the pretreated biomass. The particle size of bio-mass may be important in continuous countercurrent extraction because very fine particles tend to compact and cause liquid to channel or block liquid flow completely. Water temperature may also have an effect on the extraction of solubles from the pretreated biomass. The impact of these parameters and the L/IS ratio on soluble sugar recovery from pretreated softwood forest thinnings was explored in a series of experiments using small column percolators and a pilot-scale single-screw extractor.

Materials and Methods

Pretreated Biomass

Whole-tree chips (passing through a 0.5-in. screen) from California softwood forest thinnings were soaked in 0.66% (w/w) sulfuric acid solution. The acid-impregnated chips were air-dried to 43% (w/w) solids (the acid concentration of liquid in air-dried chips was 1.08% w/w), then pretreated at 185°C for 4 min using a 4-L steam explosion reactor described previously *(1)*. At these conditions, approx 85% of the hemicellulose was solubilized. The water-insoluble fraction of the pretreated material was 72.9% on a dry wt basis. Table 1 gives the feedstock composition and the theoretical component yields after pretreatment. As seen in Table 1, the conversion yield of cellulose (i.e., glucan) was lower than that of hemicel-lulose (i.e., mannan, galactan, xylan, and arabinan) owing to the mild pre-treatment condition aimed at maximizing hemicellulose hydrolysis. The total soluble solids concentration of the liquid fraction of the pretreated material was 99.8 g/L, and the sugar composition of the liquid fraction is given in Table 2.

Yellow poplar sawdust was pretreated at 0.3% (w/w) sulfuric acid and 195°C for 5 min using a Sunds™ Hydrolyzer installed at the Process Development Unit of the National Renewable Energy Laboratory (NREL) in Golden, CO. Pretreatment of yellow poplar sawdust using the Sunds Hydrolyzer was reported previously *(11)*. Yellow poplar chips (pulp chip size) were pretreated at 0.55% sulfuric acid and 170°C for 15 min using a Sunds Hydrolyzer installed at the Tennessee Valley Authority pilot plant (Muscel Shoals, AL). The water-insoluble fraction of the yellow poplar chips was 71.8% on a dry wt basis, and the soluble solids concentration of the liquid fraction of the pretreated material was 144.6 g/L.

Table 1
Feedstock Composition and Theoretical Component Yields
of Pretreated Softwood

Component	Feedstock composition (%)	Theoretical yield after pretreatment (%)
Glucan	43.2	
Unconverted		90.5
To monomeric glucose		10.9
To oligomeric glucose		1.1
To HMF[a]		0.3
Unaccounted for		−2.7
Mannan	11.5	
Unconverted		9.8
To monomeric mannose		74.0
To oligomeric mannose		12.3
Mannan to HMF[a]		2.2
Unaccounted for		+1.7
Galactan	4.3	
Unconverted		29.3
To monomeric galactose		61.3
To oligomeric galactose		10.1
Unaccounted for		−0.7
Xylan	7.7	
Unconverted		15.2
To monomeric xylose		76.4
To oligomeric xylose		8.9
To furfural		5.3
Unaccounted for		−5.9
Arabinan	2.2	
Unconverted		9.4
To monomeric arabinose		96.1
To oligomeric arabinose		9.9
Unaccounted for		−15.3

[a]5-hydroxymethyl-2-furaldehyde.

Effect of Water Temperature on Extraction of Pretreated Softwood

To determine the effect of water temperature on the extraction of soluble solids from the pretreated softwood, stagewise batch extraction was carried out in a glass beaker. For a single-batch extraction, 50 g (wet wt) of pretreated softwood chips (68.3% moisture content) was mixed with 280 mL of deionized water at 25, 40, 60, and 80°C. The slurry was stirred for 2 min, then filtered using a vacuum Buchner filter to separate the liquid from insoluble solids. The extracted solids were dried in a 105°C oven overnight. The soluble solids recovery was determined by subtracting the dry wt of the extracted solids from the dry wt of the starting material. Data were also collected for multiple (as many as five), consecutive batch extrac-

Table 2
Sugar Composition of Liquid Fraction
of Starting Pretreated Softwood and Liquid Extract from Continuous Countercurrent Extraction (g/L)

Liquid	Cellobiose[a]	Glucose	Xylose	Galactose	Arabinose	Mannose
Liquid fraction of pretreated softwood	1.9	17.4	22.6	9.4	7.7	32.4
Extract from L/IS = 2.1	ND	13.3	16.4	7.8	6.2	27.8
Extract from L/IS = 3.4	ND	7.6	10.9	5.2	4.3	17.1
Extract from L/IS = 5.6	ND	5.4	7.3	4.2	2.0	10.2

[a]ND, not detected by HPLC.

tions. For a two-batch extraction, the filtered solids obtained from the first extraction were reslurried with 280 mL of deionized water, stirred, and filtered. This was repeated as many times as the number of extraction batches before the extracted solids were dried in the 105°C oven. The weight ratio of added water to dry insoluble solids (L/IS) was 24 for the single-batch extraction, 47 for the two-batch extraction, and 118 for the five-batch extraction. At 80°C, the resulting soluble solids concentrations were 1.2, 0.25, 0.11, 0.05, and 0.02% (w/w) corresponding to the number of extraction of 1, 2, 3, 4, and 5, respectively.

Drainage Rate of Pretreated Biomass

The continuous countercurrent screw extractor used in the present study relies on percolation of water by gravity through the pretreated biomass. If the pretreated biomass has poor water drainage properties (i.e., very slow drainage rate), channeling or blockage may occur inside the extractor, which can result in low sugar recovery or low throughput. Therefore, bench-scale percolation tests were performed using silicone columns to compare the water drainage rates for three pretreated materials: softwood chips, yellow poplar sawdust, and yellow poplar chips.

Pretreated wood residues (15.6 g on a dry wt basis) were placed in a 2.5-cm (1-in.) diameter × 30.5-cm (12-in.) high silicone column, which was fitted with a filter at the bottom. Hot water at 60°C was added to the top of the column. To keep the total slurry concentration in the column in the first percolation batch (15.2% on a dry wt basis) constant per pretreated material, the weight ratio of added water to total dry solids (L/S) was varied depending on pretreated materials. The L/S ratios in the first batch were 3.4, 4.4, and 3.9 for the pretreated softwood, yellow poplar sawdust, and yellow poplar chips, respectively. Seven consecutive percolations, each with the same amount of water, were performed on the same column to determine whether the drainage rate changed as the amount of water used increased. The time period required for the liquid to completely drain from the column was recorded. The average drainage rate was calculated by dividing the mass of liquid collected by the draining time.

Countercurrent Extraction of Pretreated Biomass

Figure 3 shows a schematic diagram of the pilot-scale continuous countercurrent extractor designed by NREL. A 10-cm (4-in.) diameter × 305-cm (10-ft) long, U-trough screw conveyor, driven by a Link-Belt® drive (Rexnord, Philadelphia, PA), was purchased from FMC (Tupelo, MS) and modified for the purpose of this process. The helical screw has 36 short-pitch flights. Half-inch holes were drilled into the flights and mixing paddles were installed between the flights to improve liquid/solid contact. Because the solids discharge opening was 31 cm (1 ft) from the top of the conveyor, the effective length of the extraction zone was only 274 cm (9 ft). The total working volume of the extractor was 28 L. To minimize channel-

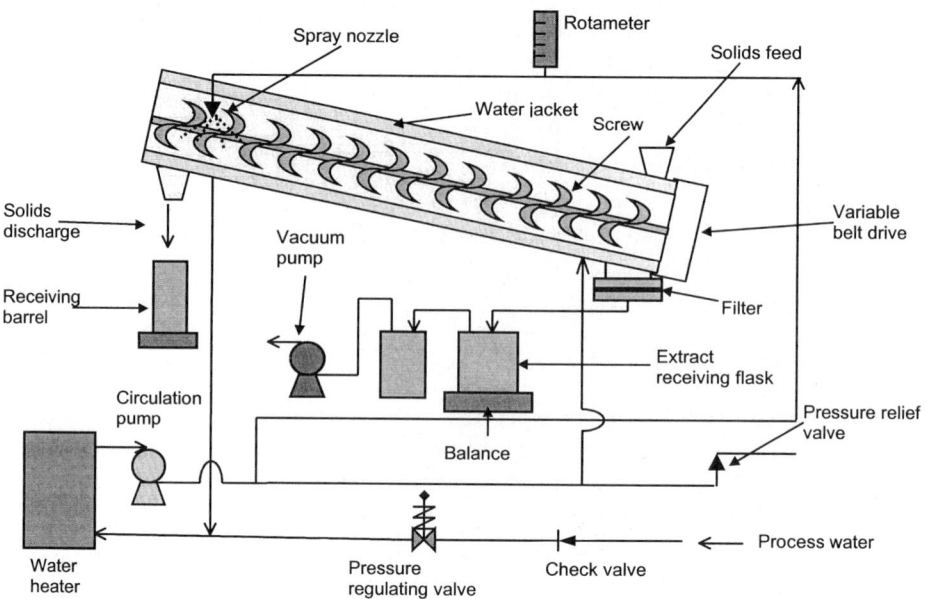

Fig. 3. Schematic of the pilot-scale continuous countercurrent extractor.

ing of water along the bottom of the trough, the screw extractor was mounted with an inclined angle of 50° from horizontal.

At the beginning of a run, a batch of pretreated softwood was loaded into a constant volumetric feeder (Acrison® feeder BDFM; Acrison, Moonachie, NJ). Process water was heated in the water heater to 60°C, then circulated through the screw conveyor jacket. When the return water temperature reached a steady-state value of about 57°C, the feeder was then switched on and set at a predetermined feed rate to begin to introduce pretreated wood into the bottom of the extractor. The conveyor drive was then activated and set at a predetermined forward speed such that the flights were less than 50% filled with pretreated wood. At about the same time, a split stream of hot water was metered through a rotameter and sprayed on top of the pretreated material through a spray nozzle installed approx 31 cm (1 ft) upstream of the solids discharge opening. The extracted liquid passed through coarse filters at the bottom of the extractor and was collected every 5 min into a flask connected to a vacuum pump. The extracted solids were discharged into a barrel via the solids discharge chute at the top of the extractor. The Acrison feeder, extract receiving flask, and barrel for receiving discharged solids were placed on electronic balances, and their weight changes were recorded every 5 min. Each extraction run lasted 110–120 min. Steady state (i.e., no appreciable change in pH of the extracted solids) was obtained after approx 60 min. At the end of each run, samples of solids inside the extractor were taken at 2-ft intervals and analyzed for insoluble solids and percentage of soluble solids extracted.

Analysis of Extracted Solids and Liquid

The concentration of monomeric sugars in the extract was analyzed by high-performance liquid chromatography (HPLC) using the same method for hydrolysate liquor analysis as described previously *(12)*. The fraction of insoluble solids (FIS) of feed materials and extracted solids were determined to find the amount of solubles extracted from pretreated wood by the screw extractor. The FIS is defined as the dry weight ratio of water-insoluble solids over the unwashed solids.

To determine the FIS, approx 7 L of tap water at 40°C was added to solids with known total weight and solid content (approx 180 g of total dry solids or 120 g of insoluble solids for pretreated softwood) in a container. The resulting slurry was mixed vigorously with a portable mixer for 5 min and left standing for 5 min. The approximate L/S and L/IS ratios of the slurry were 44 and 58, respectively. The slurry was filtered with a 24-cm diameter glass-fiber filter (1.5-μm particle retention, Whatman grade 934-AH; Whatman, Maidstone, England) under a vacuum and resuspended with tap water for the next washing. These procedures were repeated at least three more times or until the pH of the slurry was higher than 6.0, and the solids were then washed once more with 40°C deionized water. The total L/S and L/IS ratios used in determining the FIS of pretreated softwood were 220 and 290, respectively. All the washed solids were recovered, mixed, and weighed. Three representative samples were dried overnight in a 105°C oven to determine the solid content of the washed solids. The FIS values for pretreated biomass are generally in the 0.65–0.80 range. We define 100% soluble solids (or solubles) recovery as $100 \times (1 - \text{FIS})$. The percentage of solubles recovery yields of partially extracted solids were calculated according to the following equation:

$$\text{Percentage of solubles recovery} = [(1 - \text{FIS}_i)/(1 - \text{FIS}_0)] \times 100 \qquad (1)$$

in which FIS_i is the FIS of the partially extracted material, and FIS_0 is the FIS of the starting material.

Fitting Empirical Equation to Experimental Data

The following empirical equation was fitted to the experimental data to predict the recovery yield of solubles against the extractor length and the L/IS ratio:

$$R = 100 - a \exp(-bL) \qquad (2)$$

in which R is the recovery yield of solubles (%), a and b are constants, and L is either the length of extractor or L/IS ratio. Nonlinear regression was carried out using the scientific graphing software SigmaPlot® (Jandel, Chicago, IL) to determine constants a and b with different operating parameters.

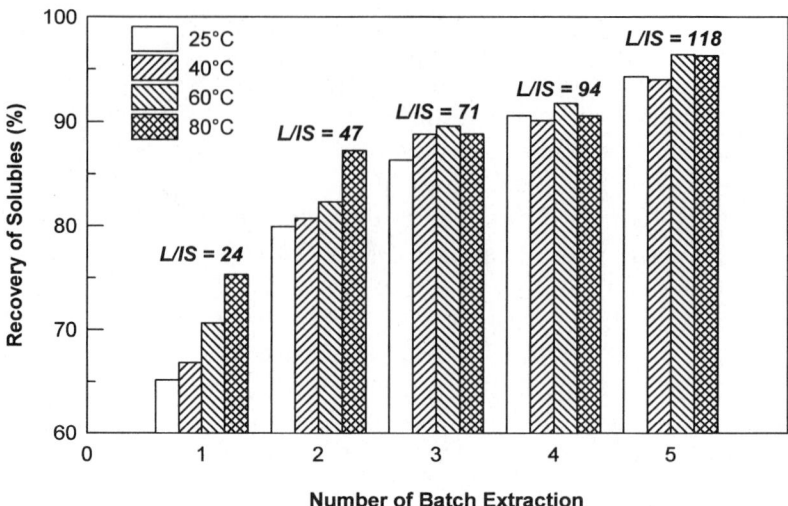

Fig. 4. Effect of wash water temperature on the extraction of solubles from pre-treated softwood.

Results and Discussion

Effect of Water Temperature on Extraction of Pretreated Softwood

As shown in Fig. 4, the recovery of solubles from the pretreated soft-wood forest thinnings increased significantly as the wash water temperature was raised from 25 to 80°C. This effect of water temperature was especially pronounced when low amounts of water were used such as in single- and double-batch extractions, which were carried out with an L/IS ratio of 24 and 47, respectively. Therefore, it is important to use hot water in countercurrent extraction in which a low L/IS ratio is maintained to obtain high extraction efficiency. In the pilot-scale continuous extraction experiments, 57°C water was sprayed on the pretreated biomass and 60°C water was circulated through the screw conveyor jacket.

Comparison of Drainage Rates

Pretreated biomass has different particle sizes depending on the particle size of the starting materials and the pretreatment conditions. If the particle size is too small to give an adequate drainage rate, channeling and blockage of liquid flow may become serious problems for countercurrent screw extractors. In the present study, drainage tests were performed on three different pretreated wood residues; Figure 5 shows the results. The drainage rate through pretreated softwood chips was significantly higher than those obtained with pretreated yellow poplar sawdust and chips, which contain a large amount of fines. This higher drainage rate of the pretreated softwood was most likely attributable to its large particle sizes in comparison to the pretreated yellow poplar materials. The increase in

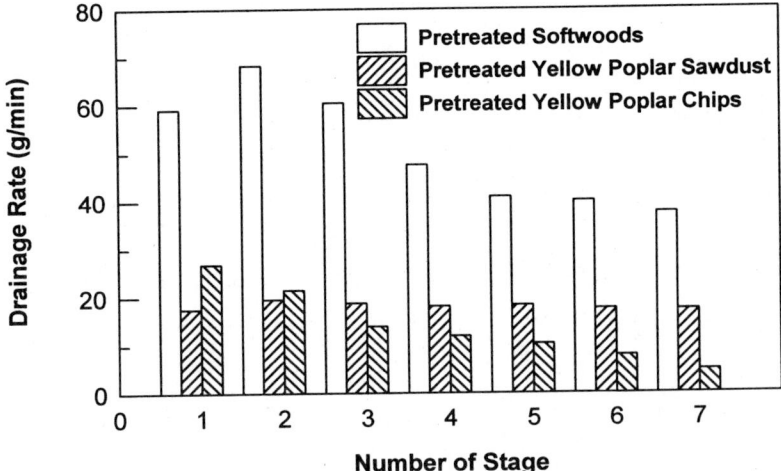

Fig. 5. Comparison of water drainage rates for different pretreated materials using a 1-in. (2.5-cm) diameter × 12-in. (30.5-cm) high column.

drainage resistance probably was caused by the compaction of the bed of washed materials. The bed heights of both the pretreated yellow poplar residues shrank approx 40% after four consecutive extraction batches, whereas that of the pretreated softwood shrank about 25%. A similar packing problem in a column extractor was also reported in oil extraction from fine soybean flour *(13)*. The drainage rates of pretreated wood residues generally decreased as the number of percolation batches increased and later on remained constant. This was probably because water-absorbed wood matrix hampered the water drain in the packed column. Basing our selection on the results of the drainage tests, we chose pretreated softwood for running extraction experiments with the countercurrent extractor. The two pretreated yellow poplar materials with lower water drainage rates will be considered in future studies.

Countercurrent Extraction of Pretreated Softwood Forest Thinnings

Countercurrent extraction of pretreated softwood forest thinnings was performed at three different L/IS ratios and a fixed solid feed rate of approx 220 g/min (228, 209, and 234 g/min for L/IS = 2.1, 3.4, and 5.6, respectively). The L/IS ratio was determined by dividing the amount of water in the extract by the amount of insoluble solids in solids feed. The wash water flow rate was varied for different L/IS ratios (175, 240, and 400 g/min for L/IS = 2.1, 3.4, and 5.6, respectively) and the solids feed rate was kept constant for all runs. Because the screw rotation speed was fixed at 20 rpm for the entire extraction runs, the average solids residence times in the extractor were essentially the same (approx 20 min) for all runs.

Figure 6 shows the concentrations of solubles and pH of extracts recovered during the steady-state operation of the continuous countercurrent extractor at various L/IS ratios. The concentration of solubles in the

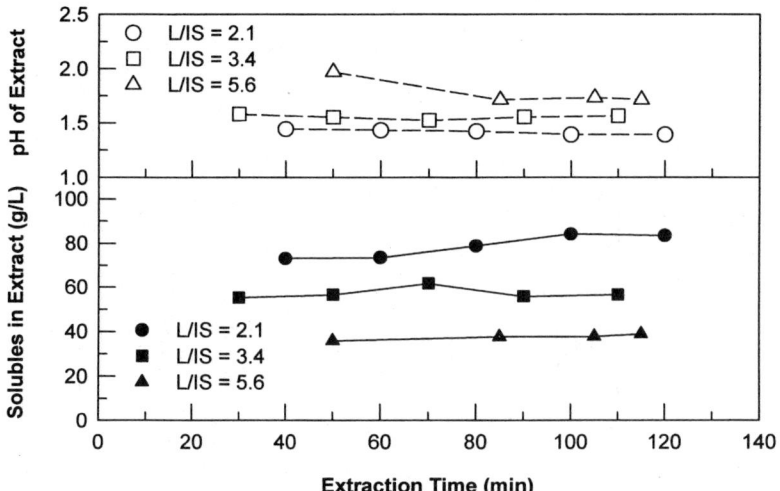

Fig. 6. Extract pH and concentration of solubles in extract recovered from the bottom of the extractor in continuous countercurrent extraction of pretreated softwood at different L/IS ratios.

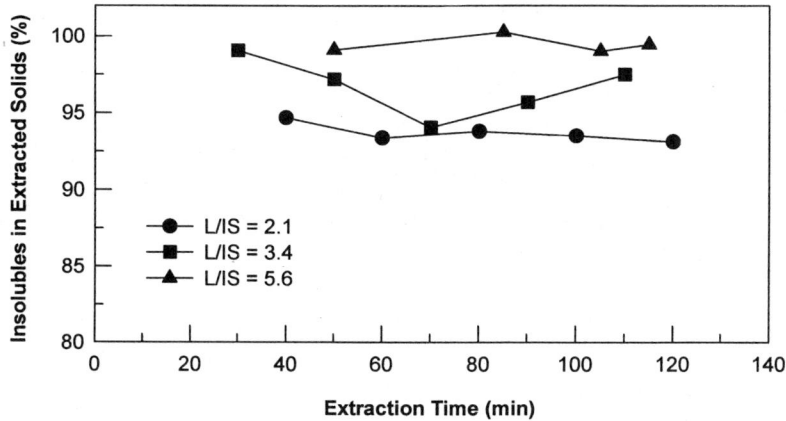

Fig. 7. Percentage of insolubles in extracted solids discharged from the top of the extractor in continuous countercurrent extraction of pretreated softwood at different L/IS ratios.

extract decreased as the L/IS increased because of dilution of the sugar solution contained in the pretreated wood residues at higher water flow rates. Therefore, it is necessary to keep the L/IS ratio low to achieve high solute concentrations in the extract as long as the extraction efficiency is maintained at an adequate level. The increase in extract pH with the L/IS ratio also indicates the dilution effect at higher water flow rates. Table 2 lists the sugar composition of the liquid fraction of the starting pretreated material (i.e., before extraction) and extracts collected from the countercurrent extraction runs at the different L/IS ratios. As expected, the sugar concentration decreased as the L/IS ratio was raised.

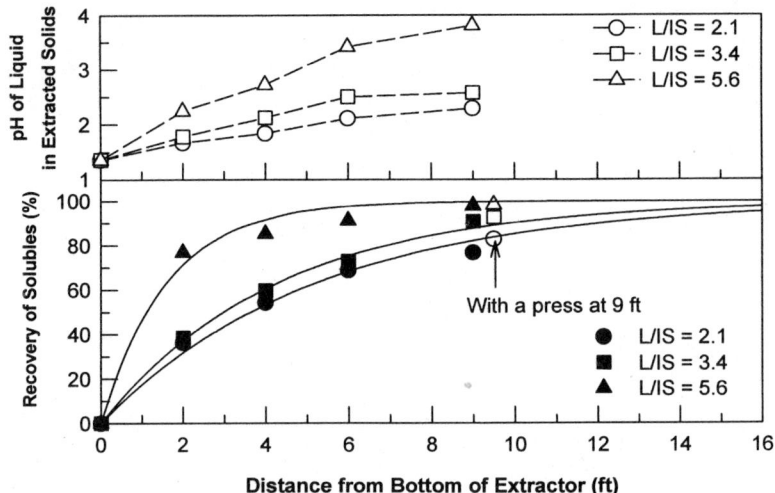

Fig. 8. Solubles recovery and pH of liquid in extracted solids at different locations in the continuous countercurrent extractor operated at steady state after 120 min. Closed symbols indicate experimental data and lines represent the predicted curves.

Figure 7 illustrates the percentage of insolubles in extracted solids discharged during the steady-state operation of the countercurrent extractor. Higher FIS (i.e., insoluble fraction in extracted solids) values were obtained at higher L/IS ratios. This result indicates that higher amounts of soluble solids were extracted at higher L/IS ratios. The lower value of percentage of insolubles in extracted solids at 70-min extraction time from L/IS = 3.4 can be attributed to a disruption in steady-state operation when the clogged filters were replaced. The gradual decline in FIS value for the run of L/IS = 3.4 was most likely caused by plugging of the filters at the bottom of the extractor. After the filter was replaced at 70 min, the extraction efficiency improved, as indicated by the rising FIS trend. We installed a different type of filter for the other two runs and did not observe severe plugging problems.

Operating Line

To establish the operating line of the recovery of solubles with respect to location in the extractor, extracted solids remaining in the extractor were recovered at the end of each extraction run and analyzed for percentage of solubles extracted. The experimental data are indicated by closed symbols in Fig. 8. The solid lines represent the best-fitted operating lines to the experimental data by the empirical equation (Eq. 2).

The percentage of solubles recovery increased as the distance from the bottom of the extractor increased. This was confirmed by an increase in pH of the liquid in the extracted solids. At 274 cm (9 ft) from the bottom of the extractor, where solids were discharged, the solubles recoveries for L/IS ratios of 5.6, 3.4, and 2.1 were 98, 91, and 77%, respectively. In plant opera-

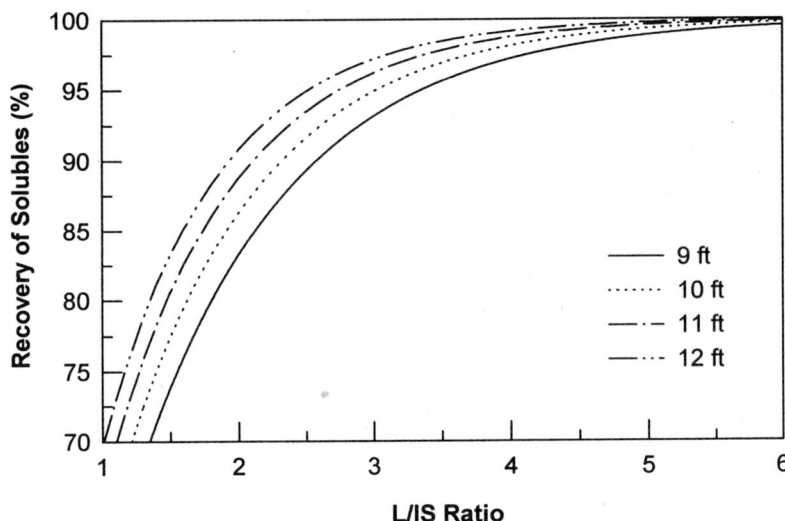

Fig. 9. Prediction of soluble recovery with respect to L/IS ratio for continuous countercurrent extraction of the pretreated softwood with different extractor lengths.

tion, assuming that the extracted solids would be pressed to approx 45% (w/w) solid content to recover the entrained solubles and the obtained liquid returned to the extractor, one would expect a slight enhancement in the solubles recovery, as represented by the open symbols in Fig. 8. The soluble recovery was higher at a higher L/IS ratio for the same extractor length. The values of constants a and b for the empirical equation used (Eq. 2) for plotting the operating lines in Fig. 8 were estimated to be 99.9120 and 0.1910, 99.9938 and 0.2327, and 99.9710 and 0.6251 for L/IS = 2.1, 3.4, and 5.6, respectively.

Figure 9 presents the prediction of soluble recoveries when the length of the extractor is extended to increase the extraction stages. The estimated values of constants a and b for Eq. 1 used in plotting the predicted curves in Fig. 9 were 99.9950 and 0.8937, 99.9969 and 0.9926, 99.9981 and 1.0920, and 99.9988 and 1.1916 for extractor lengths of 274.3, 304.8, 335.3, and 365.8 cm, respectively (9, 10, 11, and 12 ft, respectively). A 12-ft extractor with the same configuration as the current extractor used in this work is expected to achieve >95% recovery of solubles at an L/IS ratio of 3.0. This predicted performance is better than the predicted 93.8% soluble recovery for the three-stage stagewise countercurrent washer mentioned earlier (Fig. 2).

Conclusion

We have demonstrated that continuous countercurrent extraction of hemicellulosic sugars from pretreated softwood residues using a pilot-scale screw extractor can be effectively achieved. The soluble recovery yield decreased as L/IS ratio was reduced. The empirical equation predicts that adequate recovery of soluble sugars can be obtained in the low-range L/IS ratio of 2.5–3.0, if the length of the extractor is extended to about 366 cm (12 ft).

Acknowledgment

This work was funded by the U.S. Department of Energy, Office of Fuels Development.

References

1. Nguyen, Q. A., Tucker, M. P., Keller, F. A., and Eddy, F. P. (2000), *Appl. Biochem. Biotechnol.* **84–86,** 561–576.
2. Nguyen, Q. A. and Aden, A. (1999), Report no. 4083, National Renewable Energy Laboratory, Golden, CO.
3. Schell, D. (1997), Report no. 4115, National Renewable Energy Laboratory, Golden, CO.
4. Schwartzberg, H. G. (1980), *Chem. Eng. Prog.* **76,** 67–85.
5. Smook, G. A. (1992), *Handbook for Pulp and Paper Technologies,* 2nd ed., Angus Wilde, Vancouver, BC.
6. Rundle, K. W. (1989), US patent 4,873,095.
7. Noah, K. S. and Linden, J. C. (1989), *Trans. ASAE* **32,** 1419–1425.
8. Nguyen, Q. A. (1993), Canadian patent 1,322,366.
9. Brinkley, C. R. and Wiley, R. C. (1978), *J. Food Sci.* **43,** 1019–1023.
10. Gunasekaran, S., Fisher, R. J., and Casmir, D. J. (1989), *J. Food Sci.* **54,** 1261–1265.
11. Tucker, M. P., Farmer, J. D., Keller, F. A., Schell, D. J., and Nguyen, Q. A. (1998), *Appl. Biochem. Biotechnol.* **70–72,** 25–35.
12. Nguyen, Q. A., Tucker, M. P., Boynton, B. L., Keller, F. A., and Schell, D. J. (1998), *Appl. Biochem. Biotechnol.* **70–72,** 77–87.
13. Nieh, C. D. and Snyder, H. E. (1991), *J. Am. Oil Chem. Soc.* **68,** 246–249.

Enzymatic Hydrolysis
of Ammonia-Treated Sugar Beet Pulp

Brian L. Foster,[1] Bruce E. Dale,[2]
and Joy B. Doran-Peterson*,[1]

[1]Department of Biology, 127 Brooks Hall, Central Michigan University,
Mt. Pleasant, MI 48859, E-mail: joy.doran@cmich.edu;
and [2]Department of Chemical Engineering,
A 202 Engineering Building, Michigan State University,
East Lansing, MI 48824

Abstract

Sugar beet pulp is a carbohydrate-rich coproduct generated by the table sugar industry. Beet pulp has shown promise as a feedstock for ethanol production using enzymes to hydrolyze polymeric carbohydrates and engineered bacteria to ferment sugars to ethanol. In this study, sugar beet pulp underwent an ammonia pressurization depressurization (APD) pretreatment in which the pulp was exploded by the sudden evaporation of ammonia in a reactor vessel. APD was found to substantially increase hydrolysis efficiency of the cellulose component, but when hemicellulose- and pectin-degrading enzymes were added, treated pulp hydrolysis was no better than the untreated control.

Index Entries: Biomass; sugar beet pulp; enzyme hydrolysis; ammonia pretreatment; fuel alcohol.

Introduction

Ethanol is used as an alternative fuel or an additive to currently used petroleum-based fuels and significantly reduces automobile emissions, decreases the dependence of the United States on foreign oil, and requires few alterations in current vehicle technology to be implemented. Currently, the United States has a production capacity of about 1.7 billion gal of ethanol/yr (1). Much of this is used as a gasoline additive, in the form of 10% ethanol-blended fuels (2). Almost all the ethanol produced today in the United States is from corn-based fermentations by yeast. This industry has been able to function with the help of residual coproducts and government subsidies, even though production costs are not currently competitive

*Author to whom all correspondence and reprint requests should be addressed.

with petroleum *(3)*. Ethanol from corn is being used effectively as a fuel additive but this industry is limited in its growth potential because corn is an important human and animal food source, experiencing price fluctuations related to these food markets. Even if all the corn produced in the United States were devoted to ethanol production alone, only 15% of annual gasoline consumption could be replaced *(3)*. This fact underscores the importance of turning to alternative feedstocks for the production of ethanol. Many categories of feedstocks show potential for this application, including biomass energy crops, agricultural crop residues, woody biomass, and even municipal solid wastes *(3–5)*.

Sugar beet pulp is a coproduct of the table sugar (sucrose) industry and is the remaining plant fiber after the majority of the sucrose has been removed by processing. Sugar beets are farmed throughout the world in temperate climates; however, in the United States, sugar beet farming is concentrated in northern states west of the Mississippi, and also in Michigan and Ohio. Sucrose is also extracted from sugarcane, which is grown in tropical and subtropical areas of the world. Worldwide, about 35% of sucrose is derived from sugar beets, while the other 65% comes from sugarcane. However, in the United States approximately equal amounts of sucrose come from beet and cane sources, each contributing 3 to 4 million t of sugar annually *(6)*. In 1998, US sugar beet production topped 32 million t *(7)*. For a typical sugar beet processing plant, 250 kg of pressed beet pulp (75% [w/w] moisture) remain after the removal of sucrose from 1 t of sugar beets, equivalent to about 62.5 kg of dry matter beet pulp material *(8)*. Sugar beet pulp remains quite carbohydrate rich after processing (with about three-fourths of the dry matter weight being sugars), primarily owing to the sugars that compose the structural polymers, cellulose, hemicelluloses, and pectin *(8)*. Sugar beet pulp consists of 20–24% cellulose, 25–36% hemicellulose, 20–25% pectin or uronic acids, 1 to 2% lignin, and 7 to 8% protein, all expressed as a percentage of dry wt of total solids *(9,10)*. Sugar beet pulp has been used successfully for some time as a low-cost cattle feed, providing sucrose-processing plants a means for disposing of this residual material *(6)*. Taking into consideration that beet pulp is carbohydrate-rich, that sugar beet farming is a widespread and already mature industry, and that beet pulp is abundant and of relatively low value, this coproduct has potential for use as a renewable biomass feedstock for microbial fermentations to ethanol.

Successful conversion of beet pulp to ethanol has been accomplished by simultaneous saccharification and fermentation (SSF) using an engineered ethanologenic bacterium, *Escherichia coli* strain KO11 *(11)*. However, because this bacterium does not produce the cadre of enzymes necessary to degrade cellulose, hemicellulose, and pectin to their simple sugar units, exogenous enzymes are necessary. The level of commercially available fungal enzymes required to increase ethanol yields to distillable levels may be too high, because the enzymes are quite expensive. A variety of chemical pretreatments may be employed prior to SSF to render the sugar beet pulp more easily hydrolyzed, thus reducing the amount of enzymes that must be added to achieve significant ethanol yields.

Cellulosic biomass materials typically require some sort of pretreatment to make the cellulose and other carbohydrate components reactive, or more susceptible, to enzymatic attack and use by microorganisms *(12–14)*. A successful pretreatment must economically promote the efficient conversion of carbohydrates to soluble sugars and should do so without formation of inhibitory products. The desired result of such a pretreatment is therefore to increase the effectiveness of the enzyme hydrolysis of biomass, producing more simple sugars than untreated biomass hydrolyzed under the same conditions. The ammonia pressurization depressurization (APD) treatment is a process for treating biomass with liquid ammonia at elevated pressures, followed by a rapid depressurization and violent evaporation of the ammonia. This process is a modification of the earlier ammonia fiber explosion (AFEX) process, which is well established as a successful biomass pretreatment. In contrast to AFEX, grinding of the materials is not employed and final reactor temperatures are well above freezing in APD *(15,16)*. AFEX has significantly increased the efficiency of hydrolysis of numerous forms of biomass, including wheat straw, rice straw, barley straw, corn stover, and alfalfa *(17)*; coastal bermudagrass and sugarcane bagasse *(18,19)*; switchgrass and rye straw *(20)*; and corn fiber *(21–23)*. APD works by two probable mechanisms. The rapidly evaporating ammonia rips apart the biomass fibers as it exits, exposing more surface area for enzymatic attack. Additionally, ammonia has the effect of decrystallizing the highly ordered cellulose strands found in lignocellulosic materials, allowing the cellulose to be more easily degraded during enzyme hydrolysis. Some delignification and hemicellulose hydrolysis effects of the APD treatment may also contribute to increased efficiency of hydrolysis *(17)*. The cost of ammonia pretreatment is estimated to be $20–$40 per dry t of biomass treated *(24)*.

In this study, we evaluate the effectiveness of APD in treating sugar beet pulp. This substrate has not previously been treated using this process, and in fact no such pectin-rich substrate has been APD treated to our knowledge. Therefore, two APD conditions were varied to establish the best APD operating conditions for sugar beet pulp. Hydrolyses with added commercially available fungal enzymes were done to determine the ease of hydrolysis of treated vs untreated pulp. Statistical analysis was performed to indicate the significance of the various treatment effects.

Materials and Methods

Sugar Beet Pulp

Beet pulp from the common sugar beet *Beta vulgaris* was provided by Monitor Sugar (Bay City, MI) in a pressed form containing approx 25% (w/v) solids. Beet pulp was stored at –20°C when not in use. Prior to APD processing, beet pulp was dried by baking at 45°C for 24 h and was then rehydrated

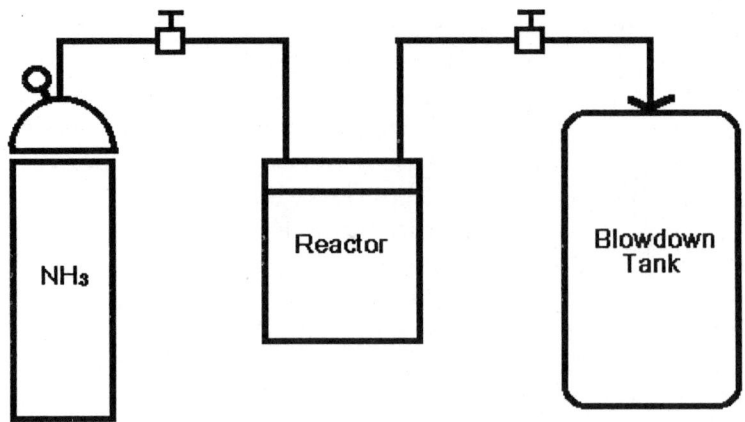

Fig. 1. Diagram of the APD reactor apparatus.

to the desired moisture content levels for APD treatment. Beet pulp fibers were used at the original size at which they emerged from processing, with no further grinding or milling. Fibers were in the size range of several millimeters in width and length.

Ammonia Pressurization Depressurization

The APD reactor apparatus (Fig. 1) was a modified steel pressure vessel attached to an ammonia source and a blowdown tank. In this study, two APD treatment conditions were varied: the beet pulp moisture content and the added ammonia load. Sugar beet pulp was treated at moisture contents of 50, 66, and 75% (w/w), and the ratio of ammonia load to beet pulp was set at levels of 0.5:1, 0.75:1, and 1:1 for each moisture level. Other APD treatment conditions were held constant, with a temperature of 80°C and a treatment time of 5 min for all samples. Following APD treatment, residual ammonia in samples was allowed to evaporate for 24 h in a fume hood, and the treated material was then stored at –20°C until used.

Enzyme Hydrolyses

Beet pulp underwent enzymatic hydrolysis at 40°C with a 5% (w/w) solids load, in a 0.05 M citrate buffer solution (pH 4.8). Hydrolysis proceeded for 48 h, and periodic samples taken for analysis were placed in a boiling water bath for 15 min to deactivate enzymes, which were then stored at –20°C. Sodium azide (0.15% [w/v]) was added to prevent microbial contamination. Three commercially available enzyme mixtures provided by Novo Nordisk (Franklinton, NC) were used for pulp saccharification. Celluclast 1.5L FG contains about 1500 Novo Cellulase U/mL (Novo Nordisk assay), or 102 filter paper units (FPU) of activity/mL (8). Novozym 431 contains approx 250 cellobiase units (CBU)/mL (Novo Nordisk assay). Viscozyme L contains about 32 hemicellulase units (HU) and 2300 polygalacturonase units (PGU)/mL *(8)*. Two sets of hydrolysis experiments were performed; the first set used only cellulase and cellobiase, and the second set added the hemicellulase/polygalacturonase. Enzyme loadings were kept constant at 4.2 FPU, 28.4 CBU, 0.85 HU, and 60.2 PGU/g of dry wt of sugar beet pulp.

Analytical Methods

Scanning electron microscopy (SEM) was used to examine the microscopic appearance of treated and untreated sugar beet pulp fibers. Beet fibers were fixed in 5% glutaraldehyde, dehydrated in a graded ethanol series (35, 50, 85, 95, 100, 100, and 100%), critical point dried, and sputter coated with 30 nm of gold. Multiple samples were observed over a range of magnifications on an Amray AMR 1200 microscope.

A modification of the dinitrosalicylic acid (DNS) assay *(25)* was used to measure the amount of soluble reducing sugars released during hydrolysis. Samples were filtered through a 0.2-µm syringe filter prior to reaction with DNS reagent (containing 1.0 g/L of 3,5-DNS and 300 g/L of sodium potassium tartrate in a 0.4 M NaOH solution). Sugars were reported in reducing sugar units using glucose, arabinose, and galacturonic acid to create the standard calibration curve. A factorial design analysis of variance (ANOVA) test was performed on the glucose equivalent yields (only from hydrolyses with hemicellulase/pectinase) to determine statistical differences from treatment effects, and a Tukey test for multiple comparisons was done to ascertain where differences lay. The Minitab statistical package (version 11) was used for all statistical applications.

High-performance liquid chromatography (HPLC) was performed on selected samples to measure specific sugar contents and organic acids. Samples analyzed for sugars by HPLC were eluted for 60 min with a Milli-Q water mobile phase through a Bio-Rad HPX 87-P column heated at 85°C, and sugars were detected by a refractive index detector. For detection of organic acids, a Bio-Rad HPX-87H column was heated at 65°C, with a mobile phase of 10 mM H_2SO_4 at 0.6 mL/min. Quantification was achieved by chromatogram peak area measurement with external standards.

Samples were submitted to an independent testing lab (Agri-King, Fulton, IL) for compositional analysis according to Official Methods of Analysis of AOAC International.

Results and Discussion

Effect of APD on Sugar Beet Pulp Structure and Composition

The APD treatment had an obvious physical effect on sugar beet pulp fibers. The size of the particles was somewhat reduced and the color of treated beet fibers was substantially darkened. This dark color persisted through hydrolysis experiments, giving the hydrolysis slurry a much darker tint than untreated controls.

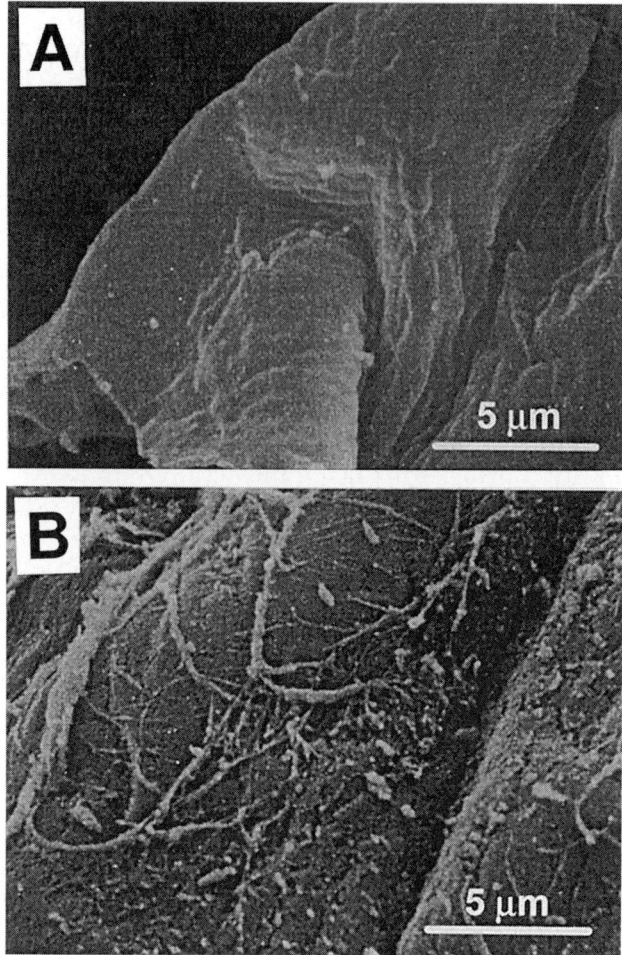

Fig. 2. Scanning electron micrographs of **(A)** untreated and **(B)** APD-treated sugar beet pulp.

Using SEM, the microscopic surfaces of treated and untreated pulp fibers were compared. Examination at an original magnification of ×5000 or higher showed great differences between treated and untreated fibers. As seen in Fig. 2, untreated pulp fibers had a smooth and intact surface, very little debris, and well-defined edges and folds. By contrast, APD-treated pulp fibers exhibited a disrupted surface with large amounts of stringy and fibrous debris, as well as small tears and holes in the pulp surface. The use of SEM with other ammonia-treated biomass types has previously shown a similar disruption of fiber but typically on a larger scale, with splitting fibers parallel to the longitudinal axis *(17)*.

Analysis of samples after treatment indicated that there were no major changes in the cellulose component or in the levels of many minerals. The crude protein content increased three- to fourfold based on the Kjeldahl

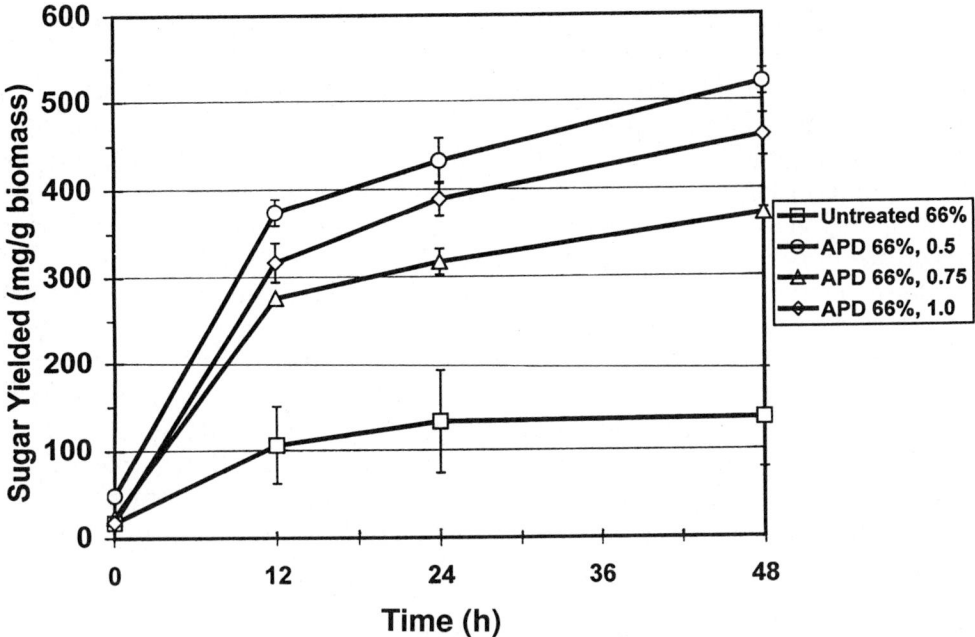

Fig. 3. Reducing sugars yielded from the enzymatic hydrolysis of APD-treated and untreated sugar beet pulp (66% moisture condition) using cellulase and celllobiase.

method, a large increase that has been observed in the previous ammonia treatment of other types of biomass *(15,19)*. At the lowest ammonia load (0.5:1), hemicellulose and pectin contents were approximately equal to those of the untreated controls. However, as ammonia levels increased to 0.75:1 and 1:1, a concomitant decrease in both hemicellulose and pectin was observed. At these higher ammonia loads, 50% of the hemicellulose and 20% of the pectin was destroyed or removed during processing. Under all APD conditions examined, lignin concentrations were found to be threefold higher than for the untreated controls.

Hydrolyses with Cellulase

The first objective was to evaluate the effectiveness of enzymatic hydrolysis of APD-treated and untreated sugar beet pulp. In previous studies, APD has been evaluated by hydrolyses employing cellulase and cellobiase enzymes to determine the effectiveness of cellulose degradation *(16,18,20,26–28)*. Therefore, hydrolysis experiments on treated and untreated SBP were performed with additions of Celluclast 1.5L FG and Novozym 431. Figure 3 shows the reducing sugars yielded from hydrolysis of APD-treated pulp with a moisture content of 66% and varying ammonia loads. As expected, APD substantially increased hydrolysis yields and rates over those of untreated controls. APD-treated pulp produced reducing sugar yields two- to threefold higher than untreated pulp. Previous studies of ammonia-treated agricultural residues reported four- to fivefold sugar yields increases over untreated materials *(17,20)*. Volumetric sugar yields from the first 12 h of hydrolysis were also higher in APD-treated sugar beet pulp, showing a 2.5- to 3.5-fold increase over untreated material. The lowest ammonia loading resulted in the greatest release of reducing sugars. APD-treated material reached about 85% of its total sugar yield after 24 h of enzyme hydrolysis in these experiments. Similar results were observed in cellulase hydrolyses of APD-treated and untreated pulp with moisture contents of 50 and 75% (not shown). Note that all APD-treated hydrolyses resulted in sugar yields higher than the maximum theoretical yield, when only cellulose was hydrolyzed. Complete hydrolysis of the cellulose component would yield approx 240 mg of reducing sugar/g of biomass. It is clear from these hydrolyses, coupled with the SEM analysis, that APD treatment disrupts the fibrous structure and facilitates enzymatic hydrolysis with cellulases. Additional hemicellulose or pectin degradation must also occur to obtain higher yields than expected from cellulose hydrolysis alone.

Hydrolyses with Cellulase and Added Hemicellulase/Pectinase

Susceptibility to enzyme hydrolysis was increased in cellulase hydrolyses of ammonia-treated beet pulp; however, cellulose makes up only about one-third of the total carbohydrates in sugar beet pulp. The cellulase hydrolyses neglect to consider the hydrolysis of the pectin and hemicellulose components, which make up roughly the other two-thirds of the carbohydrates in SBP. For the overall process efficiency of sugar beet pulp hydrolysis to be acceptable, all three large carbohydrate components must be considered. Therefore, further hydrolysis experiments were carried out with the addition of Viscozyme L, an enzyme mixture possessing hemicellulase, pectinase, and polygalacturonase activities. All sets of treated and untreated samples were hydrolyzed under identical conditions with fungal enzymes Celluclast 1.5L FG, Novozym 431, and Viscozyme L.

The results of these sets of hydrolyses were quite different from the previous cellulase- and cellobiase-only experiments, indicating that treated beet pulp consistently produced lower reducing sugar yields than untreated controls, regardless of APD parameters used. Table 1 summarizes the average reducing sugar yields from treated and untreated sugar beet pulp. Untreated beet pulp outproduced treated pulp in reducing sugar yields by about 100 mg/g of biomass, or greater. Among APD-treated samples, those with the lower ammonia load of 0.5:1 were associated with higher sugar yields on average, although differences were relatively small (50 mg/g of biomass) (Fig. 4). APD-treated pulp at 75% moisture produced lower average sugar yields as ammonia loads increased, suggesting an association between higher ammonia loads and poor hydrolysis.

A factorial design ANOVA (two-way ANOVA with replication) was performed on hydrolysis DNS data to determine whether differences

Table 1
Average Reducing Sugars Yielded
from APD-Treated and Untreated Sugar Beet Pulp During Hydrolysis
with Cellulase, Cellobiase, and Pectinase (± Standard Error Term)

Factor 1: SBP moisture	Factor 2: Ammonia load			
	Untreated	0.5:1	0.75:1	1:1
50%	786 ± 35 mg/g	608 ± 32	546 ± 37	548 ± 37
66%	700 ± 85	605 ± 52	554 ± 40	565 ± 19
75%	743 ± 6	561 ± 37	505 ± 26	350 ± 127

between APD-treated and untreated pulp sugar yields were significant, and what effect moisture and ammonia load had on effectiveness of hydrolysis. Table 2 summarizes the statistical results. Moisture content was not found to be a significant factor, with a p value of 0.131. Therefore, the three moisture levels, 50, 66, and 75%, produced statistically equivalent results and are, for all practical purposes, the same. Ammonia load was found to be a significant factor and had a p-value of <0.0005. No interaction was found to exist between the moisture content and ammonia load, because the interaction term had a high p value of 0.434. A Tukey test for multiple comparisons was performed to determine where the differences lay among the levels of ammonia load. The Tukey test revealed that the untreated sugar beet pulp yielded significantly higher reducing sugar yields than any of the treated pulp material and also that no difference existed among the APD-treated pulp samples owing to ammonia load. In other words, the three levels of ammonia load were found to have essentially equivalent effects on hydrolysis effectiveness.

When compared with the untreated control, the decrease in total reducing sugars at the two highest ammonia loads can be explained by the decrease in hemicellulose and pectin content. There is simply less carbohydrate available for enzymatic hydrolysis. However, the decrease in total reducing sugars at the lowest ammonia loading cannot be explained by a difference in the starting carbohydrate content. Cellulose, hemicellulose, and pectin content were essentially identical between the 0.5:1 ammonia load and the untreated controls. To elucidate the profile of carbohydrates released during enzymatic hydrolysis, HPLC analysis was performed.

HPLC analysis provided detection and quantification of specific sugars (cellobiose, glucose, arabinose, and galacturonic acid) throughout enzymatic hydrolysis, making it possible to measure differences in the release of simple sugars from APD-treated and untreated sugar beet pulp. During the first 12 h of enzymatic hydrolysis, the carbohydrate profiles of untreated and 0.5:1 ammonia treated pulp were very similar (Fig. 5). After 12 h more galacturonic acid and arabinose were released from the untreated samples. Approximately 66 mg/g more galacturonic acid and 55 mg/g

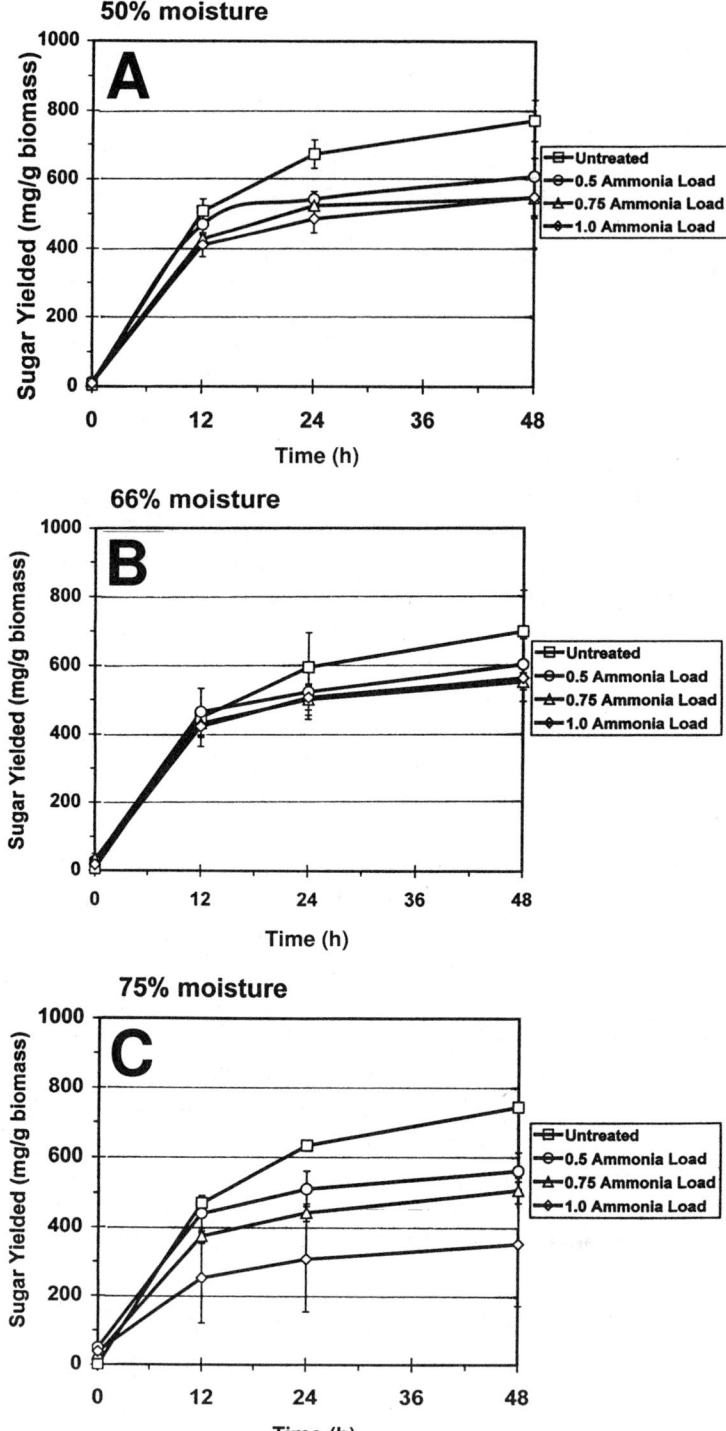

Fig. 4. Reducing sugars yielded from the enzymatic hydrolysis of APD-treated and untreated sugar beet pulp at **(A)** 50%, **(B)** 66%, and **(C)** 75% moisture contents, using cellulase, cellobiase, and pectinase enzymes.

Table 2
ANOVA Table Summarizing Factorial ANOVA Results for the Factors
of Moisture Content and Ammonia Load,
and Their Interaction for APD Treatment of Sugar Beet Pulp

Source	DF	Adjusted MS	F-Ratio	p-Value
Moisture content	2	20,142	2.22	0.131
Ammonia load	3	100,440	11.06	0.000
Moisture × ammonia	6	9291	1.02	0.434
Error	24	9082		

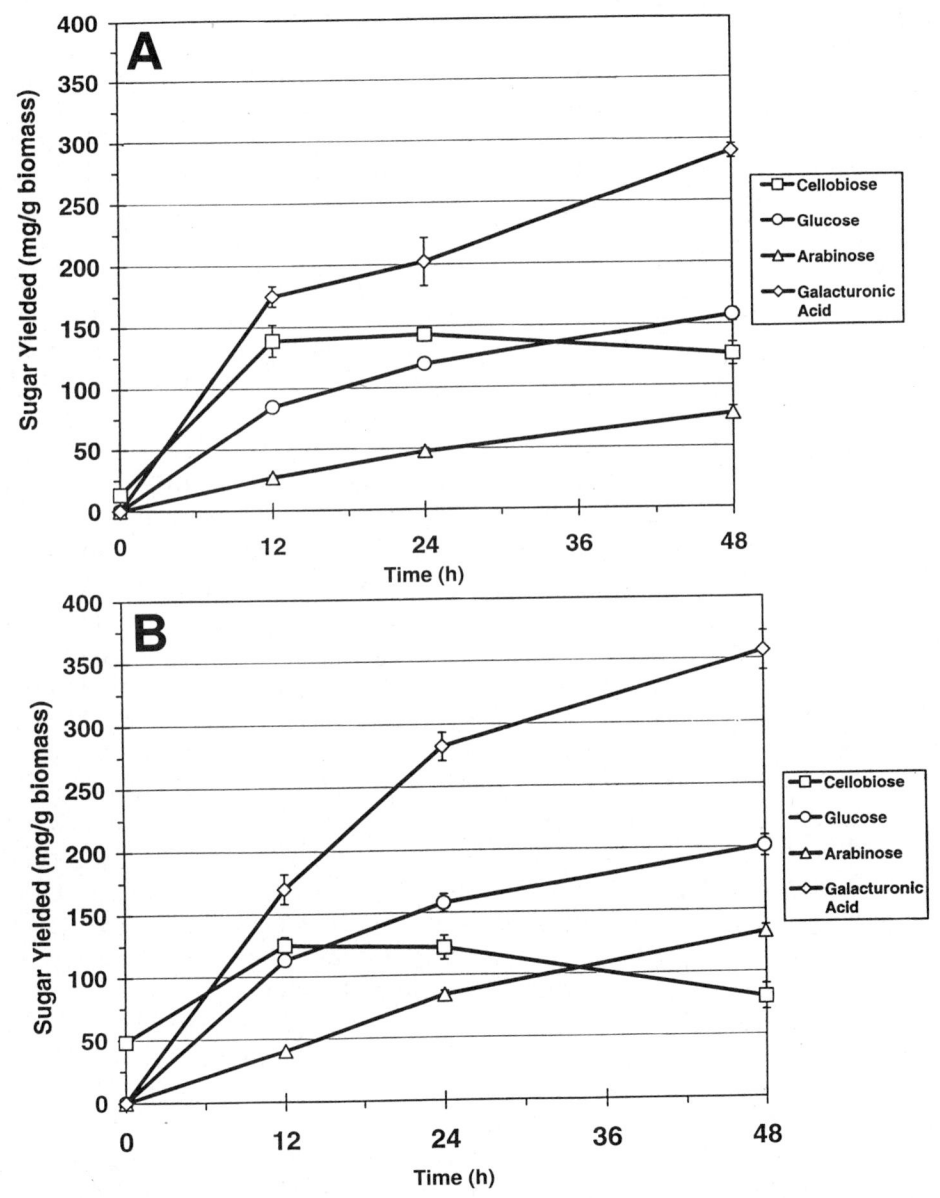

Fig. 5. Sugars yielded from the enzymatic hydrolysis of (A) APD-treated and (B) untreated sugar beet pulp using cellulase, cellobiase, and pectinase, as determined by HPLC (± standard error term).

more arabinose were released by 48 h. The total amount of carbohydrates quantified by HPLC was 650 mg/g for the treated pulp and 770 mg/g for the untreated pulp, ranges within those obtained by the DNS analysis as well (Fig. 4).

Conclusion

In summary, the APD processing of sugar beet pulp appears to have disrupted the structure as evidenced by SEM analysis and significantly enhanced the enzymatic hydrolysis when cellulases and cellobiase were used. Total cellulose content remained unchanged during APD treatment and lignin content increased threefold for all ammonia levels examined. When a third enzyme mixture of hemicellulases and pectinases was added, carbohydrate yields for untreated pulp were higher than for any of the ammonia treatments. Because hemicellulose and pectin contents of the lowest ammonia loading and untreated pulp were essentially identical, APD appears to have altered the structure of pectin and/or hemicellulose, thereby decreasing effective hydrolysis by the enzymes used in this study. Perhaps at even lower ammonia loads, the physical structure of the pulp could be altered without reducing degradation by pectinase enzymes. At ammonia loads of 0.75:1 or higher, hemicellulose or pectin appear to be degraded or destroyed.

Studies on composition of sugar beet-cell walls indicate that the separation of lignocellulosic components is achieved by solubilization of pectin, which in turn is a result of arabinan hydrolysis (29,30). Because galacturonic acid exclusively originates from pectin, an increase in galacturonic acid indicates better hydrolysis of the pectin component. Higher concentrations of both galacturonic acid and arabinose were found in enzymatic hydrolysates from untreated pulp, indicating more effective hydrolysis of the pectin component. Sugar beet pectin is a heteropolysaccharide composed of a "smooth" backbone of polygalacturonan and resistant "hairy" regions consisting of arabinan and arabinogalactan. These "hairy" regions are bound to the backbone by galactopyranosil-rhamnopyranosil or arabinofuranosil-rhamnopyranosil linkages (30). Pectin solubilization is obtained when the glycosidic linkages are broken in the arabinan hairy region. In sugar beets, ferulic acid is esterified to the arabinan and (galacto) arabinan side chains, making it possible to crosslink pectins (31). Such crosslinking of arabinan and galactan side chains by ferulate dehydro-dimers is responsible for decreased digestibility by ruminants (32). Further studies are needed to determine whether the degree of crosslinking of side chains is increased in APD-treated pulp, thus providing one explanation for the decrease in enzymatic hydrolysis of APD-treated beet pulp when compared with untreated pulp. However, under the conditions examined, APD treatment only enhances degradation of cellulose and offers no advantage for enzymatic hydrolysis of the pectin and hemicellulose fractions.

Acknowledgments

We thank Dr. Michael Byers and Guy Stone for technical input and use of the APD reactor, Nathan Krueger for technical assistance, and Dr. Badal Saha for HPLC analysis. This research was supported by the Council of Great Lakes Governors, Inc. and the US Department of Energy (DOE) Grant no. DE-FG45-93R530280; however, any opinions, findings, conclusions, or recommendations expressed herein are those of the authors and do not necessarily reflect the views of the DOE or the Council of Great Lakes Governors, Inc. Additional support was provided by Central Michigan University Department of Biology and Graduate Studies Program.

References

1. Voorhies, M. Biofuels for sustainable transportation and www. ott. doe. gov/ofd/ biomass. html.
2. Miller, R. (1999), *Renewable Fuel Association's Ethanol Industry Outlook*. www. ethanolrfa.org/outlook 99/99 industryoutlook.html, RFA publisher.
3. Keeney, D. and DeLuca, T. (1992), *Am. J. Altern. Agric.* **7,** 137–144.
4. Ingram, L. O., Aldrich, H., Borges, A., Causey, T., Martinez, A., Morales, F., Underwood, S., Yomano, L., York, S., Zaldivar, J., and Zhou, S. (1999), *Biotechnol. Prog.* **15,** 855–866.
5. Lynd, L. R., Cushman, J. H., Nichols, R. J., and Wyman, C. E. (1991), *Science* **251,** 1318–1323.
6. Clark, M. A. and Edye, L. A. (1996), in *Agricultural Materials as Renewable Resources,* Fuller, G., McKeon, T., and Bills, D. D., eds., ACS Symposium Series 647, American Chemical Society, Washington, DC, pp. 228–247.
7. US Dept. of Agriculture. National Agricultural Statistics Service. (1999), *Agricultural Statistics*, United States Government Printing Office, Washington, DC.
8. Spagnuolo, M., Crecchio C., Pizzigallo, M., and Ruggiero, P. (1997), *Biores. Technol.* **60,** 215–222.
9. Micard, V., Renard, C. M., and Thibault, J. F. (1996), *Enzyme Microb. Technol.* **19,** 162–170.
10. Michel, F., Thibault, J. F., and Barry, J. L. (1988), *J. Sci. Food. Agric.* **42,** 77–85.
11. Doran, J. B., Cripe, J., Sutton, M., and Foster, B. (2000), *Appl. Biochem. Biotechnol.* **84–86,** 141–162.
12. McMillan, J. D. (1994), in *Enzymatic Conversion of Biomass for Fuels Production,* Himmel, M. E., Baker, J. O., and Overend, R. P., eds., ACS Symposium Series 566, American Chemical Society, Washington, DC, pp. 373–390.
13. Grohmann, K., in *Bioconversion of Forest and Agricultural Plant Residues,* Saddler, J. N., ed., CAB International, Wallingfor, UK, pp. 183–210.
14. Ramos, L. P. and Saddler, J. N. (1994), in *Enzymatic Conversion of Biomass for Fuels Production,* vol. 566, Himmel, M. E., Baker, J. O., and Overend, R., eds., ACS Press, Washington, DC, pp. 325–341.
15. Ferrer, A., Byers, F. M., Sulbaran de Ferrer, B., Dale, B. E., and Ricke, S. C. (1999), *J. Sci. Food Agric.* **79,** 828–832.
16. Ferrer, A., Byers, F. M., Sulbaran de Ferrer, B., Dale, B. E., and Aiello, C. (2000), *Appl. Biochem. Biotechnol.* **84–86,** 163–179.
17. Dale, B. E., Henk, L. L., and Shiang, M. (1985), *Dev. Ind. Microbiol.* **26,** 223–233.
18. Holtzapple, M. T., Jun, J., Ashok, G., Patibandla, S. L., and Dale, B. E. (1991), *Appl. Biochem. Biotechnol.* **28/29,** 59–74.
19. Holtzapple, M. T., Ripley, E. P., and Nikolaou, M. (1994), *Biotechnol. Bioeng.* **44,** 1122–1131.

20. Dale, B. E., Leong, C. K., Pham, T. K., Esquivel, V. M., Rios, I., and Latimer, V. M. (1996), *Biores. Technol.* **56,** 111–116.
21. Moniruzzaman, M., Dien, B. S., Ferrer, B., Hespell, R. B., Dale, B. E., Ingram, L. O., and Bothast, R. J. (1996), *Biotechnol. Lett.* **18,** 985–990.
22. Hespell, R. B., O'Bryan, P. J., Moniruzzaman, M., and Bothast, R. J. (1997), *Appl. Biochem. Biotechnol.* **62,** 87–97.
23. Moniruzzaman, M., Dale, B. E., Hespell, R. B., and Bothast, R. J. (1997), *Appl. Biochem. Biotechnol.* **67,** 113–126.
24. Wang, L., Dale, B. E., Yurttas, L., and Goldwasser, I. (1998), *Appl. Biochem. Biotechnol.* **70–72,** 51–66.
25. Wood, T. M. and Bhat, K. M. (1988), in *Methods in Enzymology,* vol. 160, Wood, W. A. and Kellogg, S. T., eds., Academic, San Diego, pp. 87–112.
26. Dale, B. E. and Moreira, M. J. (1982), *Biotechnol. Bioeng. Symp.* **12,** 31–43.
27. Mes-Hartree, M., Dale, B. E., and Craig, W. K. (1988), *Appl. Microbiol. Biotechnol.* **29,** 462–468.
28. Vlasenko, E. Yu., Ding, H., Labavitch, J. M., and Shoemaker, S. P. (1997), *Biores. Technol.* **59,** 109–119.
29. Sakamoto, T. and Sakai, T. (1995), *Phytochemistry* **39,** 821–823.
30. Spagnuolo, M., Crecchio, C., Pizzigallo, M. D. R., and Ruggiero, P. (1999), *Biotechnol. Bioeng.* **64,** 685–691.
31. Rombouts, F. M. and Thibault, J. F. (1986), in *Chemistry and Function of Pectins,* vol. 310, Fishman, M. L. and Jen, J. J., eds., ACS, Washington, DC, pp. 49–60.
32. Graham, H. and Aman, P. (1984), *Anim. Feed Sci. Technol.* **10,** 199–211.

Ethanol Production
in a Membrane Bioreactor

Pilot-Scale Trials in a Corn Wet Mill

Jose M. Escobar,[1] Kishore D. Rane,[2] and Munir Cheryan*,[2]

[1]Universidad de los Andes, Departamento de Ingenieria Quimica,
Bogotá, Colombia, S.A.; and [2]University of Illinois,
Agricultural Bioprocess Laboratory, 1302 W. Pennsylvania Avenue,
Urbana, IL 61801, E-mail: mcheryan@uiuc.edu

Abstract

Pilot plant trials were conducted in a corn wet mill with a 7000-L membrane recycle bioreactor (MRB) that integrated ceramic microfiltration membranes in a semi–closed loop configuration with a stirred-tank reactor. Residence times of 7.5–10 h with ethanol outputs of 10–11.5% (v/v) were obtained when the cell concentration was 60–100 g/L dry wt of yeast, equivalent to about 10^9–10^{10} cells/mL. The performance of the membrane was dependent on the startup mode and pressure management techniques. A steady flux of 70 L/(m^2·h) could be maintained for several days before cleaning was necessary. The benefits of the MRB include better productivity; a clear product stream containing no particulates or yeast cells, which should improve subsequent stripping and distillation operations; and substantially reduced stillage handling. The capital cost of the MRB is \$21–\$34/(m^3·yr) (\$0.08–\$0.13/[gal·yr]) of ethanol capacity. Operating cost, including depreciation, energy, membrane replacement, maintenance, labor, and cleaning, is \$4.5–9/m^3 (\$0.017–\$0.034/gal) of ethanol.

Index Entries: Ethanol; membrane; corn wet mill; yeast.

Introduction

Environmental concerns about reducing automobile-polluting emissions, the clean air legislation in the United States, and the impending ban on methyl-tertiary butyl ether as an oxygenate have improved the potential market for ethanol production as a gasoline additive. Little has been done to improve the productivity of the traditional batch process for producing ethanol. Large increases in productivity have been recorded for continuous culture fermentations (50–300%) and immobilized whole-cell bioreactors

---

*Author to whom all correspondence and reprint requests should be addressed.

(10–20 times higher productivity) *(1–4)*. However, these techniques present some drawbacks: washout for the continuous culture approach and gas holdup and high-pressure drops for immobilized cells.

A promising alternative is the membrane recycle bioreactor (MRB), which couples a continuous stirred-tank bioreactor to membrane modules to operate as a cell-recycle system. The main advantage of the MRB is that higher dilution rates can be achieved without washout because the cells are retained by the membrane and recycled back to the fermentor *(1–8)*. In addition, cell concentrations are typically 100–150 g/L on a cell dry wt basis, which is equivalent to cell counts of 10^9–10^{11}/mL. This is much higher than batch fermentations, which typically start with 15–25 g/L of yeast. The combination of these factors results in ethanol productivities that may be 5–20 times higher than for batch fermentors at ethanol concentrations of 75–100 g/L *(1–8)*. The MRB concept has been demonstrated successfully mostly on small laboratory-scale systems of 0.25–5 L *(3–8)*.

Recent advances in membrane technology and a decrease in membrane costs owing to increased competition has improved the prospects for the MRB. We had earlier demonstrated a 1500-L MRB for starch hydrolysis on-site at an ethanol plant *(9)*. In this article, we describe trials with a 7000-L MRB utilizing ceramic microfiltration membranes for ethanol production by fermentation. This was tested on-site at one of the largest corn wet-milling ethanol plants in the world. To bring it to this stage, the project went through three phases in three locations:

1. At our laboratories at the University of Illinois, where research into this concept was conducted to understand the kinetics and microbiology of the fermentation. Several membrane systems were screened and evaluated, the MRB was scaled up from 0.5 to 10 L, and the design of the 7000-L MRB was finalized and industrial partners were selected.
2. At Hudson, WI, where the MRB was fabricated by the engineering contractor (Niro Filtration) based on Phase 1 design and specifications.
3. At Pekin Energy (now Williams Energy), Pekin, IL. The MRB was installed and evaluated side-by-side with the present fermentation process. Further optimization of the MRB continued during this phase.

Because of space restrictions, this article presents only a few of the runs and experiments performed over the 15-mo period of the project at the company's site. More details are available in a project report available from the authors.

Materials and Methods

Membrane Recycle Bioreactor

Figure 1 shows a schematic of the MRB system constructed by Niro Filtration. A jacketed stainless steel tank with a volume of 7000 L (1830 US gal) was used as the fermentation vessel. The system valves, pressure

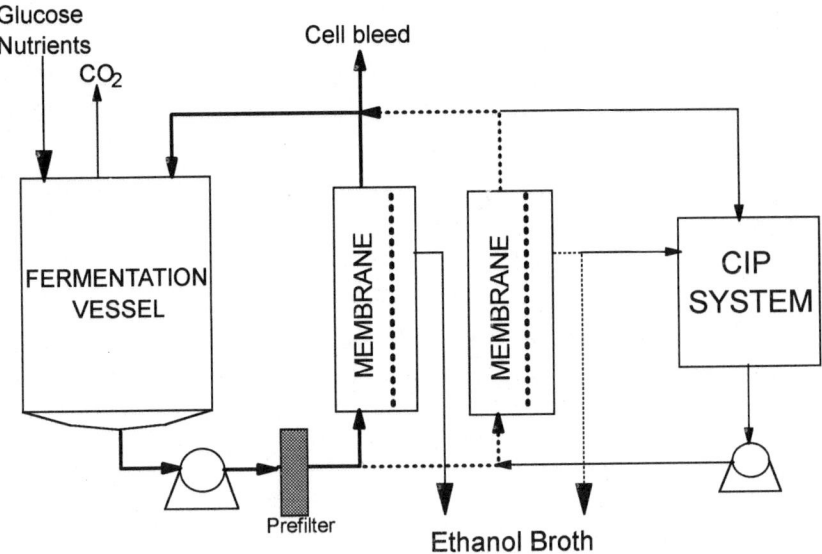

Fig. 1. Schematic of the membrane recycle bioreactor.

gages, and sensors were connected to the programmable logic controller (PLC) (Allen-Bradley, Milwaukee, WI), which had several programs to control the flux by manipulating crossflow rate and transmembrane pressure as needed. Pressure transducers, located at the top and bottom of the tank, were used as a level indicator and controller. This level controller was attached via the PLC to the feed supply valve, which opened and closed depending on the level in the tank.

The fermentation broth was pumped to one of two membrane modules in parallel. All MRB runs reported herein were done with Membralox 19P19-40 modules (U.S. Filter, Warrendale, PA), each containing 3.8 m² (40.7 ft²) of membrane area. Pressure and crossflow velocity to the membrane modules were provided with a centrifugal pump (Fristam, Middleton, WI) rated at 90 m³/h and 700 kPa. This pump was connected to two in-line horizontal prefilters with 400-μ stainless steel screens to protect the membrane modules.

The membrane modules and the fermentation vessel were connected to a separate clean-in-place (CIP) system that could isolate individual components and automatically clean them when needed. Membrane cleaning was done using the following procedure. The membrane loop to be cleaned was "sweetened off" with water to recover the broth in the loop. Then, tap water at 50°C was flushed through the membrane loop, discarding the permeate and retentate streams. Next, cleaning solution containing 5 g/L of Ultrasil-11 (an NaOH-based cleaner from Ecolab-Klenzade, St. Paul, MN) and 200 ppm of chlorine was recirculated through the loop at 50°C for 30 min. Finally, cleaning solution was flushed out with tap water, and the water flux was measured. If it had not reached the original water flux, cleaning was repeated.

Fermentation

Yeast (*S. cerevisiae*) was obtained from the company's seed fermentors or the main fermentors during their growth phase. The seed culture broth, typically containing 20–25 g of yeast/L, was pumped into the MRB fermentation vessel and concentrated to the required cell density using the membrane modules. After the desired yeast concentration was reached, the glucose feed was started and the process was run continuously. Residence time was maintained by adjusting tank volume and flux.

The glucose feed (corn starch hydrolysate, typically 93–95 dextrose equivalents) and corn steep liquor (CSL) were obtained from the company's supply to their regular fermentation system. Samples of permeate (the MRB outlet) and the fermentation broth from the vessel were analyzed for ethanol and residual sugars by high-performance liquid chromatography and for cell dry wt and total plate count by standard methods *(7)*.

Results and Discussion

Tubular ceramic membranes were selected for this phase of the project because of several advantages: they permit flux management techniques such as backwashing, back-pulsing, and UTP/CPF; they operate in highly turbulent flow conditions; they are made of relatively hydrophilic materials, which minimizes fouling; and they can be cleaned with aggressive chemicals if needed (*[2]*; Filson, J., personal communication; Keefe, R., personal communication). Performance of the membrane is expressed as flux, J (L/[m²·h]):

$$J = AP_T$$

in which P_T is the transmembrane pressure (kPa) and A is the permeability coefficient (L/[m²·h]/kPa).

Preconcentration of Yeast

The first task in a run was to grow and preconcentrate the cells to the required cell density. Prior work in our laboratory *(1–3,7)* and elsewhere *(4–6,8)* had indicated that a yeast concentration of about 100 g dry wt/L was necessary to maximize sugar utilization and minimize residence time. Figure 2 shows a concentration run with fermentation broth from the company's main fermentors. A volume of 5600 L with an initial cell concentration (as measured by plate count) of 1.48×10^8 viable cells/mL, equivalent to about 23 g dry wt/L, was pumped into the fermentation vessel. The system was operated at constant transmembrane pressure (140 kPa) and constant flow rate (1350 L/min, equivalent to a crossflow velocity of 5 m/s) with a pressure drop (ΔP) of 140 kPa. Over a period of about 16 h, the flux gradually decreased from 70 to 60 L/(m²·h). This was a small drop in flux, considering that the fermentation broth was concentrated 4.6X and this fermentation broth contained CSL, which has been shown in previous experiments to cause fouling of the module.

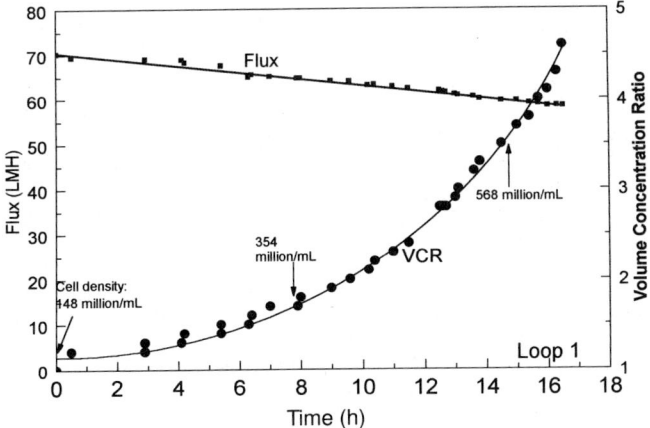

Fig. 2. Preconcentration of yeast cells (run 729). Initial volume, 5600 L; initial cells, 1.48×10^8/mL; transmembrane pressure, 140 kPa; pressure drop (ΔP), 140 kPa; recirculation flow rate, 1350 L/min; temperature, 30°C; LMH, L/(m^2·h), VCR, volume concentration ratio.

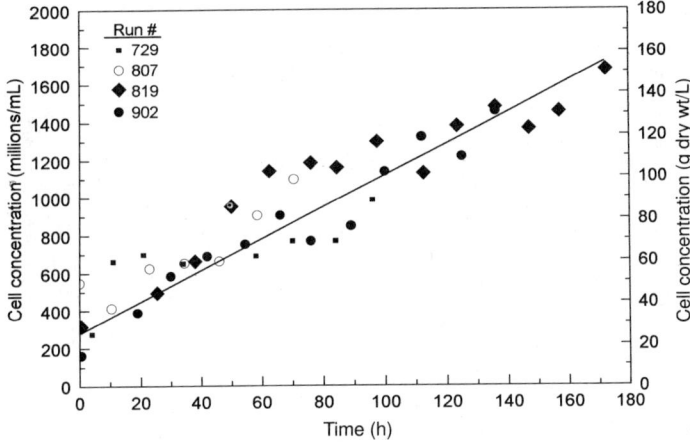

Fig. 3. Cell concentration during MRB runs. Run 819 is expressed as grams/liter, all others as viable cell count in millions/milliliter.

Cell Concentration

Cell concentration in various fermentation runs is shown in Fig. 3. The number of viable yeast cells was 13.5–16.2 $\times$ 10^9 cells/g of dry yeast, with higher cell concentrations tending to decrease the cell dry wt slightly *(10)*. Because of continuous cell growth, it was necessary to bleed cells to control the viscosity of the fermentation broth and fouling of the module. Cell concentration was maintained below 120 g/L to avoid serious disturbances within the system.

Yeast cell viability, as measured by the methylene blue test, averaged 50–70%. Viability was affected by sugar concentration, initial number of viable cells, and the quantity and quality of nutrients, which should be

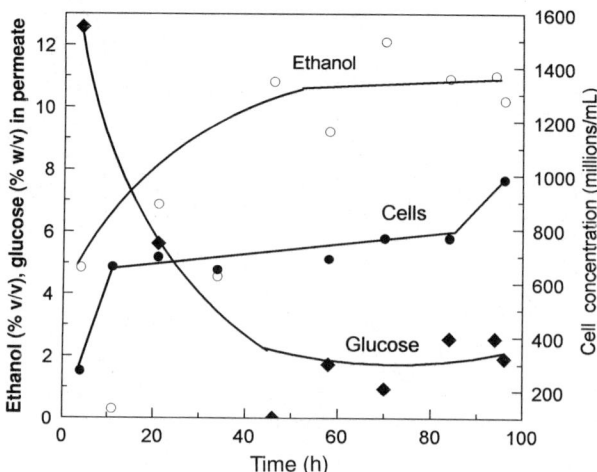

Fig. 4. Fermentation parameters during run 729. Inlet glucose concentration, 210 g/L; residence time, 10 h; temperature, 30°C.

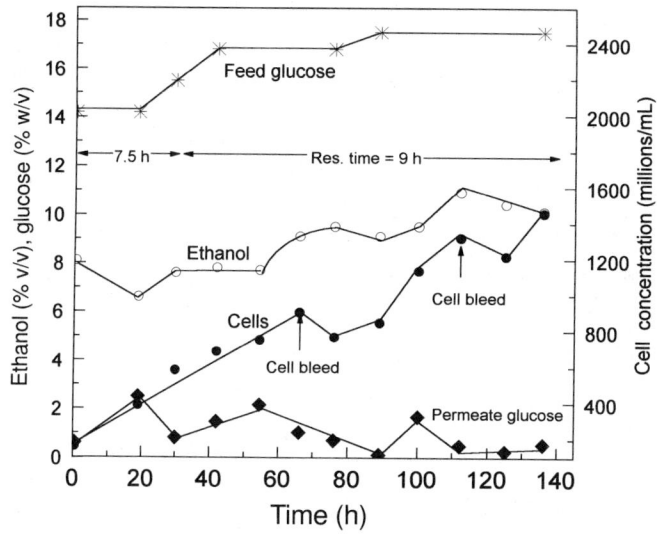

Fig. 5. Fermentation parameters (run 902). Temperature, 31°C.

adjusted for the high cell density. Viability also could have been affected by shear stress from the pumping. With a feed glucose concentration of about 180 g/L, cells reached 6×10^8 viable cells/mL at the end of the precon-centration phase. The higher this value, the better the system performed.

Ethanol Concentration

Ethanol production was directly dependent on glucose concentration in the feed, cell concentration, and residence time, as shown in Figs. 4 and 5. A cell density of about 100 g/L ($1.2–1.5 \times 10^9$ cells/mL) was needed for

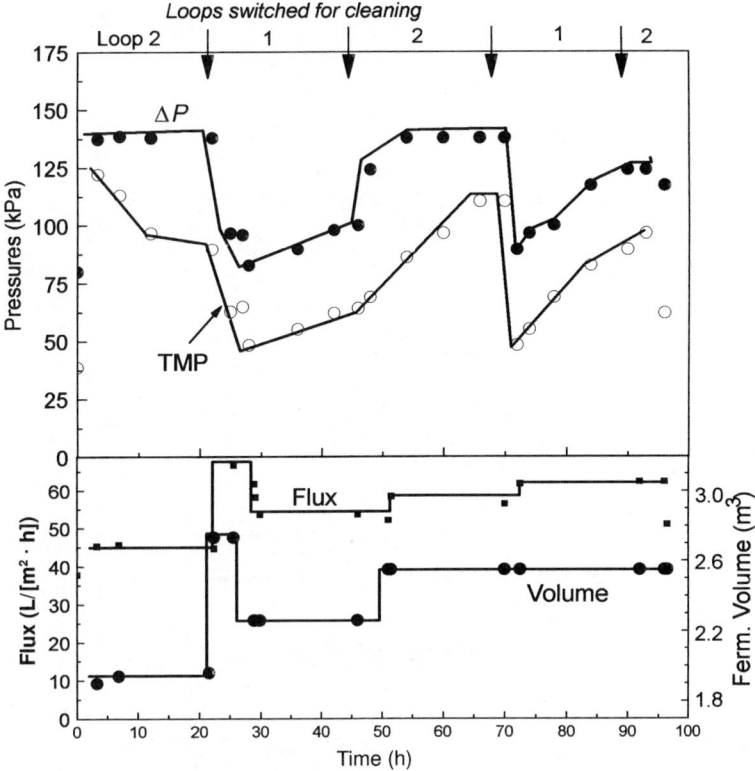

Fig. 6. Membrane parameters during run 729. Recirculation rate for ΔP of kPa, 1350 L/min; residence time, 10 h; temperature, 30°C. TMP, transmembrane pressure.

nearly complete glucose utilization of a feed containing 175 g of glucose/L, resulting in 10.8% (v/v) ethanol and a yield of 0.48 g of ethanol/g of glucose consumed, about 94% of the maximum theoretical value (Fig. 5). Ethanol concentrations varied between 7.5 and 12% (v/v) for feed glucose concentrations of 142–210 g/L.

Membrane Performance

Figures 6 and 7 show selected membrane parameters for the two runs shown in Figs. 4 and 5. Initial experiments were designed to determine those conditions that would provide steady conditions of flux and residence time, which is defined as follows:

Residence time (h) = Volume of fermentation broth in system (L)/
Flow rate through system (L/h)

Flow rate (L/h) = Flux (L/[m²·h]) × membrane area (m²)

If flux declined during operation owing to fouling and the pressures could not be adjusted further, the volume was adjusted to keep the residence time at the required value. When the module was clean and brought

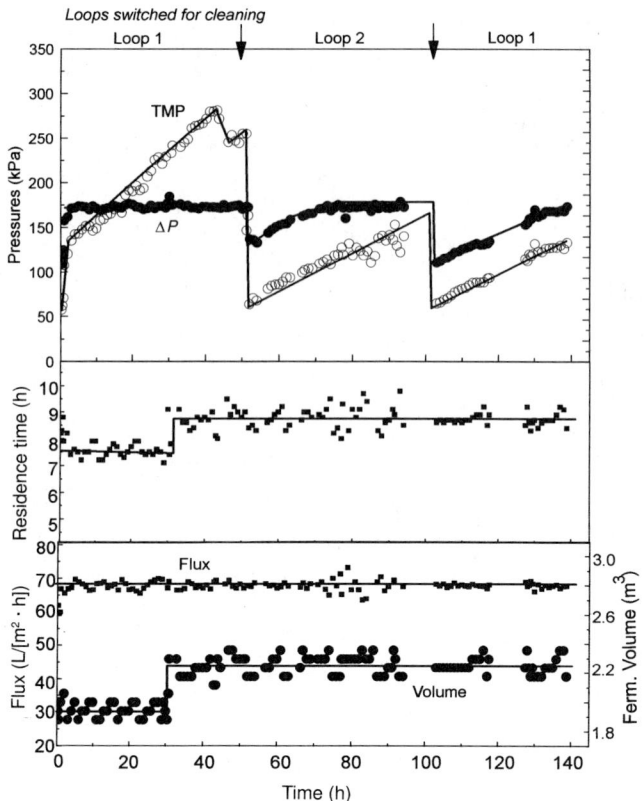

Fig. 7. Membrane parameters (run 902). Recirculation rate, 1450 L/min at ΔP = 172 kPa; temperature, 31°C.

on-line initially, the system transmembrane pressure or crossflow rate (the latter controlled by pressure drop, ΔP) was set very low by the PLC. As the module fouled, these two parameters were gradually increased to keep the flux constant. It was observed that at low fluxes (e.g., below 70 L/[m²·h]) the system could be maintained for as much as 110 h (~4.5 d) before the transmembrane pressure increased to such an extent (>275 kPa) that cleaning was necessary (Fig. 8). Higher fluxes would require cleanings more often. When cleaning was required, the system switched to the other loop containing the second membrane module.

The stability of the MRB was also affected by gas (CO_2) production. When CO_2 production reached levels above 9.2 g of $CO_2/(L\cdot h)$, the pump experienced large cavitation problems. However, it also appeared that the gas produced was sometimes beneficial in reducing membrane fouling, acting as a scrubber to remove some of the fouling material.

A more serious problem is the presence of antifoaming compounds, some of which can cause serious membrane fouling *(2)*. In one run (no. 819, shown in Fig. 8), some of the antifoam used in the plant's main fermentors inadvertently slipped into the feed stream during the initial stages of the

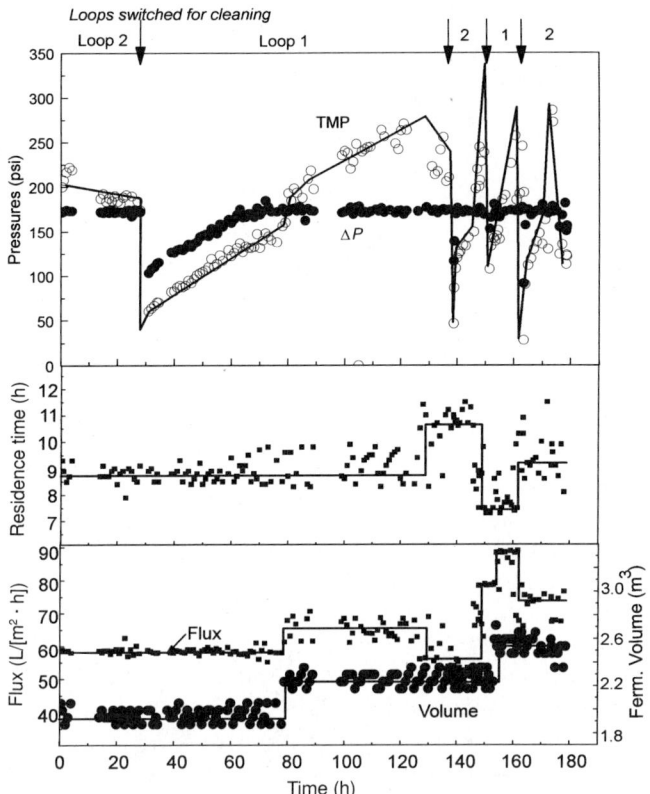

Fig. 8. Membrane parameters for run 819. Recirculation rate, 1450 L/min at $\Delta P =$ 172 kPa; temperature, 31°C.

run using loop 2. The normal cleaning cycle was inadequate to remove the antifoam, so that when loop 1 was called back into service after 135 h of operation, the membrane permeability rapidly declined, forcing a steep increase in the rate at which the pressure was increased (30 vs 2 to 3 kPa/h in normal operation).

A separate experiment was performed to evaluate the effect of the antifoam on fouling of the membrane. Many commercial antifoaming agents (e.g., polyoxyethylene polyoxypropylene oleyl ether, polyglycols, silicone oils) severely foul membranes *(2)*. At the time of this study, the company was using a Polaxamer antifoam compound. A sample of the antifoam compound was brought back to the university pilot plant, and a controlled experiment with a model fermentation broth was performed to gage the effect of the antifoam. As shown in Fig. 9, the effect of even a small concentration of antifoam compound on flux was quite pronounced. It is necessary to use nonchemical means of foam control or to use antifoams that have cloud points above the process temperature. Otherwise, the membrane would require frequent and aggressive cleanings.

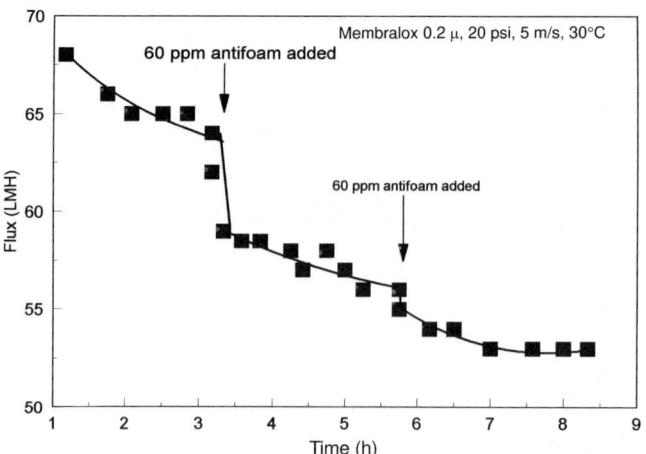

Fig. 9. Effect of Polaxamer antifoam (hydroxy-polyoxyethylene-polyoxypropylene block copolymer) on membrane fouling. Membralox ceramic membrane (1P19-40, 0.2 μ) was used with a model ethanol fermentation broth. LMH, L/(m²·h).

Capital Cost

Table 1 summarizes calculations for capital and operating costs for this application. It was assumed that the same 0.2-μ membrane would be used. However, the newer 60P37-30 (1020 mm long) modules, which contain 60 elements in a single housing, were not available at the time of the trials reported herein. They provide a larger area (21 m²) for the same cost and are available with smaller-diameter channels—3 instead of 4 mm. This means that for the same crossflow velocity, the volumetric flow rate would be lower with only a slight increase in pressure drop. Based on our data, a crossflow velocity of 5 m/s would require 283 m³/h of crossflow per module with a pressure drop of about 205 kPa.

Because the MRB probably will be one of the largest ceramic membrane plants in the world, unit membrane cost should be lower than standard catalog prices. The membrane cost (less housing) for this plant was estimated at $1000/m², and the system (including the valves, pipes, fittings, control system, and first set of membranes) was $1600/m². The CIP system would cost an additional $50,000 for the whole plant, regardless of size (*[2]*; Filson, J., personal communication; Keefe, R., personal communication).

An ethanol plant with a nominal capacity of 3.8×10^6 m³ (100 million gal)/yr would actually be producing 3.6×10^6 m³/yr (at the 5% denaturant level). Assuming an 8000-h operating year and ethanol concentration in the permeate outlet of the MRB is 11% (v/v), the flow rate through the MRB, which is permeate flow rate, is

$$\text{Flow rate} = \frac{3.6 \times 10^6 \text{ m}^3 \text{ ethanol/yr}}{8000 \text{ h/yr} \times 60 \text{ min/h} \times 0.11 \text{ m}^3 \text{ ethanol/m}^3 \text{ broth}}$$

$$= 408{,}608 \text{ L/h}$$

Table 1
Cost Analysis of MRB Fermentation System for Ethanol Production

Item	Base case, using data from in-plant trials	Potential with flux management techniques
Fermentation broth volume @ 11% (v/v) ethanol (m³/h)	408.6	408.6
Volume of tanks needed for 9-h fermentation (m³)	3677	3677
Cost of tanks at $2/gal ($)	1,943,182	1,943,182
Membrane flux (L/[m²·h])	70	150
Area for MRB (m²)	5837	2724
Total area (+10% for cleaning) (m²)	6421	2996
Cost of membranes ($/m²)[a,b]	1000	1000
Cost of membrane system ($/m²)[a,b]	1600	1800
Cost of membrane system ($)	10,273,571	5,393,625
Total capital cost ($)	12,216,753	7,336,807
Capital cost ($/[m³·yr] capacity)	32.2	19.3
Number of 60P37-30 modules	306	143
Flow rate per module at 5 m/s (m³/h)	283	283
Pressure drop per module (kPa)	205	205
Number of parallel flow paths	77	37
Energy consumption (kW/m²)[a]	1.118	1.118
Energy cost/yr ($/yr)[a]	2,010,017	938,008
Capital charge ($/yr)	275,185	171,226
Membrane replacement ($/yr)	642,098	299,646
Maintenance, labor ($/yr)	366,503	220,104
Cleaning cost ($/yr)	44,947	89,894
Operating cost ($/yr)	3,338,749	1,718,878
Operating cost		
($/m³ ethanol)	9	4.5
($/gal ethanol)	0.034	0.017

[a]Ref. 2.
[b]Filson, J., personal communication; Keefe, R., personal communication.

Assuming a flux of 70 L/(m²·h), membrane area = 408,608 L/(h·70 L/ [m²·h]) = 5837 m². With 10% extra needed for cleaning, this comes to 6421 m². The membrane system cost = $1600/m² × $6421 m² = $10.27 million.

Assuming a 9-h residence time, the total volume of tanks required would be 408,608 L/h × 9 h = 3677 m³. At a cost of $528/m³ ($2/gal), the cost of fermentation tanks is $1.94 million (Keefe, R., personal communication; Gadomski, R. T., personal communication). Thus, the total MRB system cost is $10.27 million + $1.94 million + $50,000 for the CIP system = $12.22 million. This is equivalent to $32/(m³·yr) of ethanol capacity ($0.122/[gal·yr]). Since a typical corn wet-milling plant of this capacity would cost at least $250 million to construct (Gadomski, R. T., personal communication), the MRB would be 5% or less of the plant cost.

Operating Costs

Operating costs are based on energy (electric power for recirculating pumps), membrane replacement, depreciation, cleaning, and labor/maintenance. Under the operating conditions of the MRB, the Membralox modules have a unit energy consumption of 1.118 kW/m^2 (2). Assuming a power cost of $0.035/kWh, energy cost per year = 1.118 kW/m^2 × 6421 m^2 × 8000 h/yr × $0.035/kWh = $2.01 million/yr.

The capital charge, which provides for depreciation, insurance, and so forth, was calculated based on a depreciation over a 14-yr period. Membrane costs are not included in this category and are considered a separate operating expense. In this case, system cost less membranes = $1600 – $1000 = $600/m^2. Using a 14-yr straight-line method of depreciation

$$\text{Depreciation} = \$600 \times \$6421/14 = \$275,185/\text{yr}$$

Membrane replacement costs are based on a 10-yr life. Membralox membranes have been in commercial service since 1984, and 10-yr or more lifetimes are now being recorded for such relatively nonaggressive applications (Filson, J., personal communication). Thus, membrane replacement cost = $1000/m^2 × $6421 m^2/10 yr = $642,098/yr. Labor and maintenance are charged at a nominal rate of 3% of installed cost, which in this case is $366,503/yr.

For cleaning cost, conventional detergents and sanitizers can be used. A detailed analysis at the plant site indicated that it would be about $7/(m^2·yr) = $44,947/yr. This includes the additional power needed during the 1-h cleaning cycle.

The total operating cost is $3.34 million/yr. This is equivalent to $9/m^3 of ethanol produced in the plant ($0.034/gal). The highest costs are power requirements (60% of operating costs) and membrane replacement (20%).

Potential to Reduce Costs

Capital costs can be reduced by using polymeric membranes. However, their lower capital cost has to be balanced against the lower flux and more frequent replacement of membranes. Increasing flux will reduce the membrane area as well as capital and operating costs. This can be done in the following ways. First, the flux we used for our calculations (70 L/[m^2·h]) is based on run 902 (*see* Fig. 7). Higher transmembrane pressure will increase the flux. However, it will increase the cleaning frequency from every 4 to 5 d to perhaps a daily cleaning. Considering that the cleaning cost right now is so low, the cost-benefit ratio would favor more frequent cleanings if the flux can be increased. Second, overall flux could also be increased by using back-pulsing or the CPF/UTP mode (2,11,12). Back-pulse devices typically add about 9–18% to the cost of a module. Although these devices and techniques need to be confirmed for this large-scale application, there are several successful industrial applications that use these techniques (Filson, J., personal communication; Keefe, R., personal communication).

Incorporating these factors could increase flux to 150 L/(m²·h). The added cost for the back-pulse devices increases capital cost to $1800/m², but the total capital cost is reduced to $7.34 million and operating cost is reduced by 50% (Table 1). If cheaper membrane modules are used and assuming the flux remains the same with fluid management and back-pulse techniques, the capital cost could decrease even further with an additional reduction in operating cost.

Conclusion

The on-site trials demonstrated that the MRB is a practical concept from a technical viewpoint. One additional advantage of this type of fermentor that became apparent in the plant was that the product stream (the permeate) was clear and contained no suspended matter. This could essentially eliminate the centrifuges needed to process the "stillage" and improve the heat transfer in the beer still and distillation columns. There should be a reduction in energy consumption and in waste treatment costs. None of these factors were specifically included in the cost estimates. The MRB will bring considerable benefits to ethanol production and other fermentations that need higher cell concentrations and productivities.

Acknowledgments

We are especially grateful to Jack Huggins (president), Jacob Duke (plant microbiologist), Stan Janson (quality control manager), Asif Malik (ethanol area coordinator), and Charles Sauder (director of production) at Pekin Energy for their cooperation and contributions during this work. We appreciate the information on equipment and construction costs provided by James L. Filson of US Filter, Richard T. Gadomski of PSI-Lurgi and Robert J. Keefe of Niro. We thank the Universidad de Los Andes, Bogota, Colombia, for permission and leave of absence for one of the authors (J.M.E) to work on this project. This work was supported by the Illinois Department of Commerce and Community Affairs, Illinois Corn Marketing Board, Council of Great Lakes Governors, U.S. Filter, Pekin Energy, and the Illinois Agricultural Experiment Station.

References

1. Cheryan, M. and Mehaia, M. A. (1986), *CHEMTECH* **16**, 676–681.
2. Cheryan, M. (1998), *Ultrafiltration and Microfiltration Handbook*, Technomic, Lancaster, PA.
3. Cheryan, M. and Mehaia, M. A. (1984), *Process Biochem.* **19(6)**, 204–208.
4. Chang, H. N. and Furusaki, S. (1991), *Adv. Biochem. Eng.* **44**, 1074–1110.
5. Lee, S. S., Burt, A., Russotti, G., and Buckland, B. (1995), *Biotechnol. Bioeng.* **48**, 386–400.
6. Tanaka, T., Kamimura, R., Itoh, K., and Nakanishi, K. (1993), *Biotechnol. Bioeng.* **41**, 617–624.
7. Mehaia, M. A. and Cheryan, M. (1991), *Enzyme Microb. Technol.* **13**, 257–261.

8. Groot, W. J., Sikkenk, C. M., Waldram, R. H., van der Lans, R. G. J. M., and Luyben, K. C. A. M. (1992), *Bioproc. Eng.* **8,** 39–47.
9. Cheryan, M. and Escobar, J. M. (1993), in *Proceedings, First Biomass Conference of the Americas, Volume II,* National Renewable Energy Laboratory, Golden, CO, pp. 1068–1077.
10. Patel, P. N., Mehaia, M. A., and Cheryan, M. (1987), *J. Biotechnol.* **5,** 1–16.
11. Saglam, N. (1995), PhD thesis, University of Illinois, Urbana.
12. Pafylias, I., Cheryan, M., Mehaia, M. A., and Saglam, N. (1996), *Food Res. Int.* **29,** 141–146.

Cellulase Recovery via Membrane Filtration

Wendy D. Mores, Jeffrey S. Knutsen, and Robert H. Davis*

Department of Chemical Engineering, University of Colorado, Boulder, CO 80309-0424, E-mail: robert.davis@colorado.edu

Abstract

A combined sedimentation and membrane filtration process was investigated for recycling cellulase enzymes in the biomass-to-ethanol process. In the first stage, lignocellulose particles longer than approx 50 µm were removed by means of sedimentation in an inclined settler. Microfiltration was then utilized to remove the remaining suspended solids. Finally, the soluble cellulase enzymes were recovered by ultrafiltration. The permeate fluxes obtained in microfiltration and ultrafiltration were approx 400 and 80 L/(m²·h), respectively. A preliminary economic analysis shows that the cost benefit of enzyme recycling may be as much as 18 cents/gal of ethanol produced, provided that 75% of the enzyme is recycled in active form.

Index Entries: Sedimentation; microfiltration; ultrafiltration; cellulase enzyme.

Introduction

As traditional energy resources become more depleted, the drive to develop new, nontraditional energy sources grows. One area that has been of interest for decades is the conversion of biomass to ethanol (*1–3*). Although cellulosic biomass can be converted to ethanol in existing processes, this conversion is not currently employed on a large scale because of its expense. One way to reduce the process cost is to recycle the cellulase enzyme used to hydrolyze cellulose. The enzyme used in this process is extremely expensive, representing approx 20% of the total ethanol cost (*4*). Because much of it remains active after hydrolysis, recycling this enzyme could considerably decrease operating costs.

One possible way to separate and recover cellulase enzyme is through the use of sedimentation followed by microfiltration or ultrafiltration. In the sedimentation step, the larger particles are removed so as not to block the tubing or membrane filter in the subsequent filtration step (*5*).

---

*Author to whom all correspondence and reprint requests should be addressed.

Particle removal may be accomplished efficiently using sedimentation vessels with inclined walls *(6–8)*. This simple and inexpensive technique involves pumping feed into an inclined vessel that has a large surface area but small spacing between the inclined surfaces. As the fluid moves up the channel, the particles settle owing to gravity onto the upward-facing surfaces and then slide down to the bottom of the settler, where they can be either removed separately or returned to the feed material. Inclined vessels provide the advantages of short settling distances and large surface areas for sedimentation *(9)*.

After sedimentation, the suspension may be further clarified using microfiltration. In this step, suspended solids >0.2 µm in diameter are removed, but the cellulase enzyme (mol wt = 60–90 kDa) passes through to the permeate side *(10)*. Finally, in the ultrafiltration step, these enzymes are retained by the membrane while the water, sugars, ethanol, and other small molecules pass through the membrane for further processing. The retained cellulase may then be reused for further hydrolysis. The proposed separation strategy may be used either after simultaneous saccharification and fermentation or between separate hydrolysis and fermentation steps. For the latter, lignocellulosic particles and cellulase enzyme would be retained in the hydrolysis reactor while inhibitory sugars pass through the membranes to the fermentation vessel.

In this article, we present the results of initial laboratory studies on sedimentation, microfiltration, and ultrafiltration experiments with lignocellulosic particles and cellulase enzyme. We also present a preliminary cost-benefit analysis.

Materials and Methods

Sedimentation

Feed was drawn through an inclined settler to remove the larger particles from the feed stream prior to microfiltration. A schematic of the inclined settler is shown in Fig. 1. A rectangular glass sedimentation channel of 30-cm length (L), 5-cm width (w), and 0.5-cm depth (b) was used as the settler. A peristaltic pump (Millipore) was used to draw fluid from the feed tank at a prescribed rate (Q_o) between 0.01 and 1.4 mL/s. Particles then settled onto the upward-facing wall of the settler and slid back into the feed tank, producing a clarified overflow. The angle of inclination from vertical, θ, was varied from 15 to 60°. Experiments were conducted using aqueous suspensions of lignocellulosic particles (ground yellow poplar) obtained from the National Renewable Energy Laboratory (NREL) after pretreatment in its pilot plant in Golden, CO. These particles contain approx 40% lignin and 60% cellulose, plus acetic acid and other residual components. The particles are oblong, with a typical length-to-diameter ratio of 2:1, and range in length from 1 to 1000 µm (with the nominal length about 100 µm). The pH of the lignocellulose feed was 3.5.

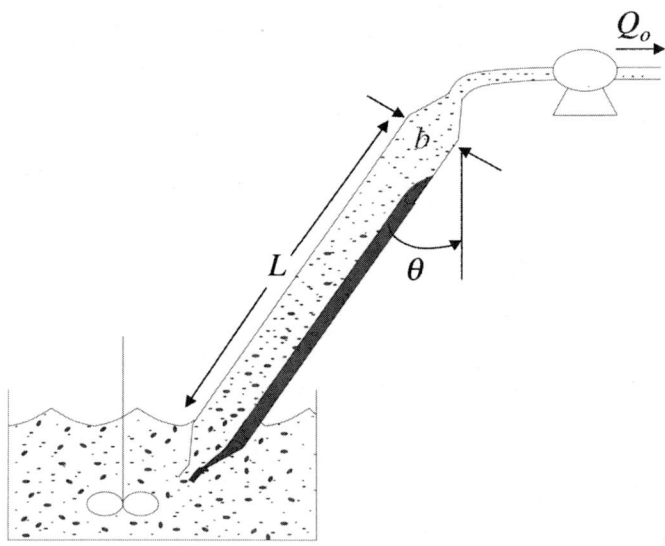

Fig. 1. Schematic of the inclined settler used to remove larger particles from the process stream prior to membrane filtration.

Microfiltration

Dead-End Filtration

Preliminary membrane testing was carried out in a Nuclepore dead-end filtration apparatus (Scientific Products, McGaw Park, IL). The unstirred cell holds up to 70 mL of feed and contains a membrane disk that is 4.7 cm in diameter. In each experiment, about 60 mL of solution was filtered in a single pass through the membrane. The cell is pressurized with nitrogen.

Microfiltration membranes were used to separate cellulase enzymes from lignocellulosic particles. All microfiltration membranes were purchased from Micron Separations (Westborough, MA) and had a nominal pore size of 0.22 μm, including membranes made of cellulose acetate (cat. no. A02SP04700), polysulfone (cat. no. S02SP04700), and nylon (cat. no. N02SP04700). The experiments were run at a lignocellulose concentration of 5% (w/v) in water on a dry weight basis. ABS cellulase (cat. no. C-8546), produced by *Trichoderma reesei*, was obtained from American Biosystems (Roanoke, VA). Its concentration in the permeate was determined using a BCA protein assay kit from Pierce, after calibration. The experiments were run at a cellulase concentration of 0.3% (w/v) in water. All microfiltration experiments were run at a transmembrane pressure of 10 psi and room temperature (22–24°C). The cellulose acetate membranes were used in these short-time experiments only for comparison purposes, because they are degraded by cellulase in long-time experiments.

Backwashing was performed on the microfiltration membranes to determine the effectiveness of cleaning. Backwashing entails reversing the

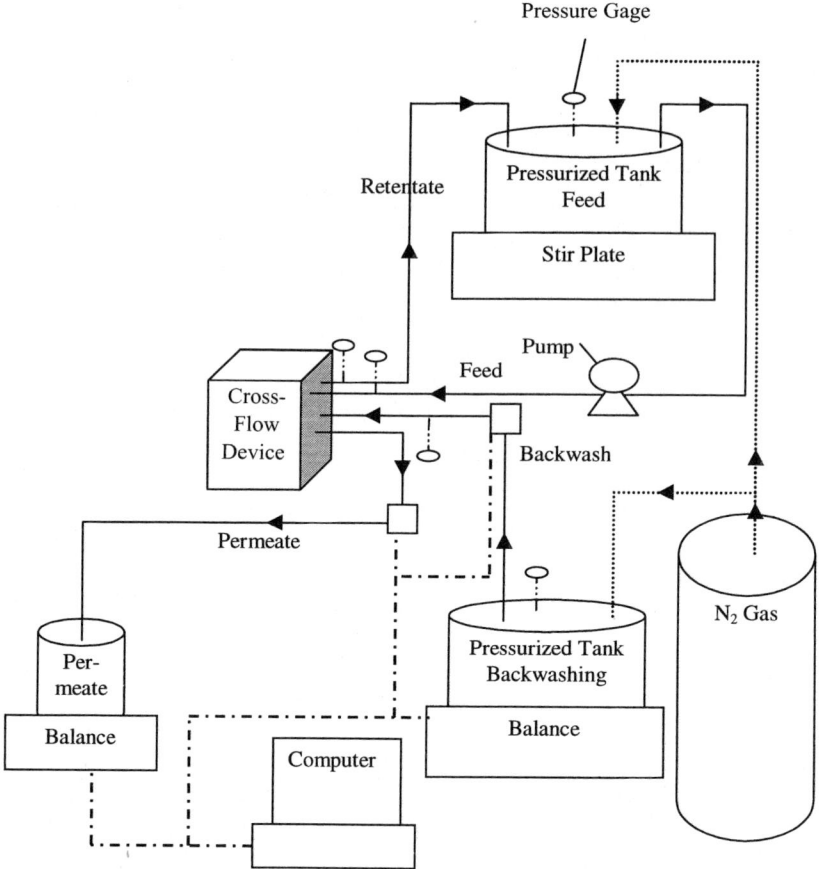

Fig. 2. Schematic of the crossflow membrane filtration apparatus. Solid lines are liquid streams, dotted lines are gas streams, and dashed-dotted lines are electrical connections.

transmembrane pressure to remove the foulant cake from the membrane surface. Membranes were backwashed for 5 min at a reverse transmembrane pressure of 10 psi. Afterward, clean water was filtered through the membrane to determine the flux recovery.

Crossflow Filtration

Lignocellulosic particles (ground yellow poplar, pretreated) were filtered with a Minitan (Millipore) flat-sheet device. This device has three parallel channels, each 7 mm wide by 5 cm long by 3 mm high. A polysulfone membrane with a 0.22-μm nominal pore size (cat. no. S02SP00010) from Micron Separations was employed. A schematic of the crossflow filtration device is shown in Fig. 2. Nitrogen gas is used to provide 10 psi of transmembrane pressure. A peristaltic pump (Millipore) is used to feed the suspension into the filter. The permeate mass is measured by an electronic microbalance (Mettler PG5002) interfaced with a computer. Microfiltration membranes were cleaned by means of backwashing for 5 min at a reverse transmembrane pressure of 10 psi. As in dead-end filtration, clean water was subsequently filtered through the membrane to determine the flux recovery.

Ultrafiltration

Dead-End Filtration

Ultrafiltration membranes were used to separate glucose sugar from the cellulase enzymes. Experiments were conducted using the same apparatus and procedure as used for microfiltration dead-end filtration. Molecular weight cutoffs of 10,000, 30,000, 50,000, and 100,000 Daltons were chosen for the ultrafiltration membranes to test their ability to retain the cellulase (60,000–90,000 Daltons) while passing the glucose (181 Daltons). The polymeric membranes were made of polysulfone or polyethersulfone. Polysulfone membranes were purchased from Sartorius, and the polyethersulfone membranes were purchased from Millipore. The glucose was purchased as D-(+)-glucose anhydrous from Sigma-Aldrich and used at a concentration of 8% (w/v) in water. The glucose concentration in the permeate was determined by refractive index measurements. The cellulase was the same as that used in the microfiltration experiments, and again, had a concentration of 0.3% (w/v). Cellulase concentrations were measured by absorption at 290 to 296-nm wavelength in a Hewlett Packard 8452A Diode Array Spectrophotometer, after calibration. All ultrafiltration stir-cell experiments were performed at a transmembrane pressure of 46 psi and room temperature (22–24°C).

Crossflow Filtration

Crossflow ultrafiltration experiments were done using a setup similar to that used for crossflow microfiltration, as shown in Fig. 2. Experiments were performed with the Mid-Gee and Xampler hollow-fiber cartridges manufactured by A/G Technology. The hollow fibers have inside diameters of 1 mm and lengths of 27 cm. Feed is passed through the fiber lumens, and permeate is collected on the shell side. The Mid-Gee contains 2 fibers (16-cm² total filtration area), and the Xampler contains 13 fibers (110-cm² total filtration area). Polysulfone hollow-fiber membranes (30-kDa mol wt cutoff) were used in all crossflow ultrafiltration experiments.

Results and Discussion

Sedimentation

Guided by the theory of Davis et al. (11), the flow rate through the settler and the angle of inclination were varied to give a desired concentration and particle size in the clarified overflow stream. As shown in Fig. 3, particle concentration in the overflow (θ_o/θ_f) increased with increasing overflow rate, Q_o, and decreasing angle of inclination, θ. As Q_o increased,

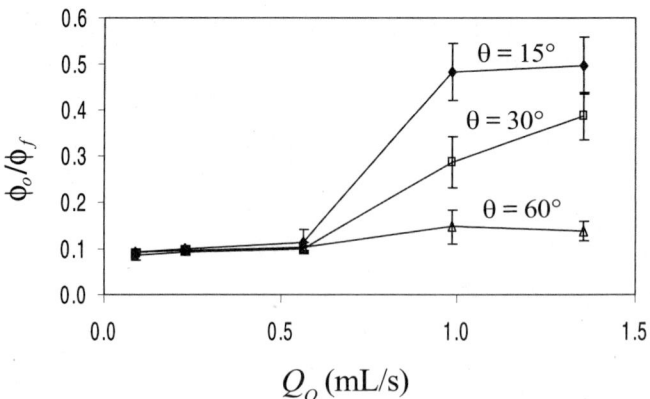

Fig. 3. Overflow particle concentration (θ_o) divided by feed particle concentration (θ_f) vs overflow rate, Q_o, for inclined settling. Values are given for angles of inclination of 15°, 30°, and 60° from vertical. Lignocellulose particles were used at room temperature (22–24°C) at a concentration of 1.25% (w/v) dry weight.

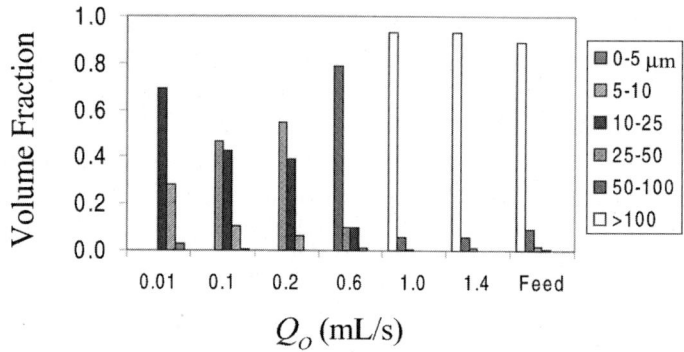

Fig. 4. Overflow particle size distributions by volume for various overflow rates, Q_o, for inclined settling. Particles are grouped according to their length. Lignocellulose particles were used at room temperature (22–24°C) at a concentration of 1.3% (w/v) dry weight. LMH, L/(m²·h).

the particles had less time to settle down the column. As θ decreased, the horizontal projected area for settling decreased, and greater concentrations of particles were drawn into the overflow.

The particle distribution in the sedimentation overflow was also determined as a function of the overflow rate. Figure 4 depicts the volume fraction of particles in different ranges of length (0–5, 5–10 μm, and so on) at various overflow rates. The volume fraction of particles was determined assuming cylindrical particles with a length-to-diameter ratio of 2:1. Particle lengths were measured visually using a Nikon microscope. For overflow rates of 1.0 mL/s and higher, the overflow particle distribution was virtually identical to that of the feed, indicating that no particles settled out during the short holdup time at high flow rates. As the overflow rate

Table 1
Final Flux (L/[m²·h]) During Dead-End Microfiltration
of 60 mL of Solution at 10-psi Transmembrane Pressure Through
a Membrane Disk with 4.7 cm Diameter and 0.2-μm Nominal Pore Size[a]

Membrane	Water	Lignocellulose	Cellulase	Mixture
Cellulose acetate	7300 ± 900	2000 ± 1000	1800 ± 600	400 ± 300
Polysulfone	13,300 ± 600	1500 ± 100	300 ± 200	100
Nylon	2900	800	NA	NA

[a]With ±1 SD provided for repeated experiments with different membranes. NA, not available.

decreased, so did the average particle length in the overflow, owing to more particles settling during the longer holdup times. For an overflow rate of 0.01 mL/s, there were virtually no particles in the overflow >25 μm long. An angle of inclination of 30° from the vertical and a flow rate through the settler of 0.2 mL/s were selected as providing the desired performance; under these conditions, the clarified overflow had a particle concentration of <10% of that in the feed, and nearly all the particles reaching the overflow were <50 μm long.

Microfiltration

Dead-End Filtration

The results of the microfiltration dead-end tests are summarized in Table 1. The clean water flux (defined as the volume of permeate collected per time per surface area of the membrane) was highest for the polysulfone membrane, but the flux in the presence of foulants (lignocellulose and cellulase) was higher for the cellulose acetate membrane. Unfortunately, cellulase gradually degrades cellulosic membranes, and thus cellulose acetate is not an acceptable material for the present application. The nylon membrane gave the lowest flux, so tests with it were discontinued. The flux was significantly lower when lignocellulose and cellulase were mixed together than when filtered separately, possibly owing to the lignocellulose particles becoming sticky when enzyme bound to them. However, the flux improved dramatically by backwashing, increasing from the fouled flux of 100 L/(m²·h) to a cleaned flux of 11,200 L/(m²·h). The flux recovery by backwashing indicates that periodic cleaning by reverse filtration may be effective in maintaining high average fluxes. During forward filtration, the enzyme recovery in the permeate was high (>80% transmission), as desired. Considerable variation between different membranes from the same lot was observed.

Crossflow Filtration

Figure 5 shows the permeate flux vs time for typical crossflow microfiltration experiments with feeds of lignocellulosic particles, cellulase enzyme, and a mixture of particles and enzyme. The lignocellulosic

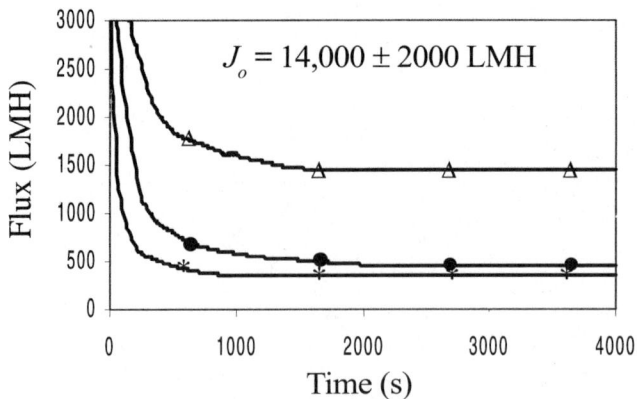

Fig. 5. Flux vs time for crossflow microfiltration with 0.2-μm polysulfone membranes for a feed with 0.2% (w/v) lignocellulose (△), 0.3% (w/v) cellulase (●), and a mixture of both (*). Experiments were conducted at room temperature. The clean water flux, J_o, was 14,000 ± 2000 L/(m²·h) (LMH).

Table 2
Preliminary Crossflow Microfiltration Fluxes[a]

	J_o (L/[m²·h])	J_s (L/[m²·h])	J_r (L/[m²·h])
Lignocellulose	14,000 ± 2000	1300 ± 200	4000 ± 500
Cellulase	15,000	400	2300
Mixture	16,000	360	11,300

[a]J_o is the initial water flux, J_s is the long-term flux after fouling, and J_r is the recovered water flux after a 5-min backwash. For lignocellulose, the results are shown as the average ±1 SD for three experiments.

particle concentration in the feed to the settler was 2.5% (w/v), which resulted in a settler overflow particle concentration of 0.2% (w/v) at the overflow rate of 0.2 mL/s. The cellulase enzyme concentration of 0.3% (w/v) was the same in both the feed to and the overflow from the settler. This settler overflow was used directly as the microfiltration feed. The recirculation rate of feed to the filter was 6 mL/s, which corresponds to a wall shear rate of 190 s⁻¹. In all three cases, the permeate fluxes started at the initial flux of J_o = 14,000 ± 2000 L/(m²·h) and then declined rapidly owing to particle deposition on the membrane surface, before reaching nearly steady values after approx 1000 s. As in the dead-end filtration experiments, the most dramatic fouling was found in the mixed system, in which the flux declined to a value of J_s = 360 L/(m²·h).

In all cases, membranes were backwashed with water for 5 min at 10 psi of reverse transmembrane pressure after 4000 s of filtration, to determine cleaning effectiveness. The results are shown in Table 2. The cleaning resulted in a flux recovery in the lignocellulose system from J_s = 1300 ± 200 L/(m²·h) to J_r = 4000 ± 500 L/(m²·h), reported as the mean ± 1 SD for three repeats. The cellulase system showed less improvement, with a

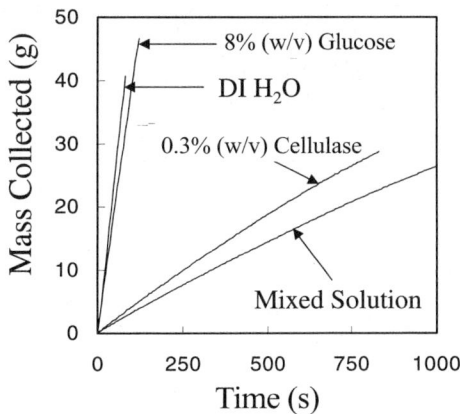

Fig. 6. Permeate mass vs time for dead-end ultrafiltration experiments using a 30-kDa polysulfone membrane and 46-psi transmembrane pressure. DI, deionized.

recovered membrane flux of J_r = 2300 L/(m²·h). However, the mixture of cellulase and lignocellulosic particles showed a very high flux recovery with backwashing, possibly because the cake layer of lignocellulosic particles effectively serves as a secondary membrane that prevents cellulase aggregates and very small particles from fouling the primary membrane *(12)*. These potential foulants are removed with the lignocellulose during backwashing, increasing the effectiveness of cleaning.

Ultrafiltration

Dead-End Filtration

Figure 6 shows the mass of permeate collected vs time for dead-end filtration using a 30-kDa polysulfone membrane. A high rate of water collection was observed, which was reduced by about 30% when glucose was added (owing to the higher viscosity of the glucose solution). When cellulase was added, a much lower collection rate resulted (owing to membrane fouling by the rejected cellulase).

The cellulase concentration in the permeate for the different polysulfone and polyethersulfone membranes was monitored using a feed containing 0.3% (w/v) cellulase and 8% (w/v) glucose. As expected, the cellulase transmission (defined as concentration in the permeate divided by that in the feed) increased with increasing molecular weight cutoff, from 2 to 5 to 40% for polysulfone membranes with 10-, 30-, and 100-kDa molecular weight cutoffs, respectively, and from 2 to 6% for polyethersulfone membranes with 30- and 50-kDa molecular weight cutoffs, respectively. Fortunately, the cellulase transmission was very low (<10%) for all but the 100-kDa polysulfone membrane. Moreover, the glucose showed essentially 100% transmission for all membranes, as desired. Thirty-kilodalton polysulfone, 30-kDa polyethersulfone, and 50-kDa polyethersulfone membranes proved to be the best candidates for crossflow testing, with average

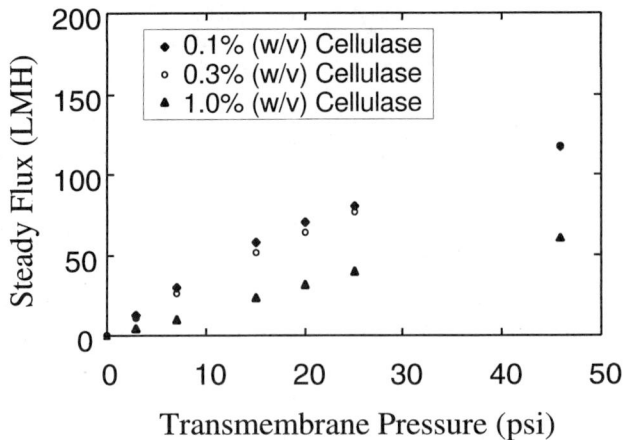

Fig. 7. Long-term flux vs transmembrane pressure for crossflow ultrafiltration of cellulase solutions using hollow-fiber polysulfone membranes with a 30-kDa molecular weight cutoff. LMH, L/(m²·h).

fluxes of 53 ± 6, 49 ± 1, and 64 ± 0 L/(m²·h), respectively, shown with ±1 SD for two experiments. These membranes are stable, give relatively high flux, and reject cellulase while passing sugar and other small molecules.

Crossflow Filtration

Figure 7 shows preliminary results with the Mid-Gee hollow-fiber cartridge containing 30-kDa polysulfone membranes. For these experiments, the wall shear rate was 1500 s⁻¹. Cellulase concentrations of 0.1, 0.3, and 1.0% (w/v) were used, and the transmembrane pressure varied from 0 to 46 psi. The long-term flux increased with increasing transmembrane pressure, indicating that the limiting flux was not reached over the ranges of concentration and pressure studied. At low cellulase concentrations, a steady flux of about 115 L/(m²·h) was obtained at the highest transmembrane pressure (46 psi) investigated. Essentially 100% enzyme retention was achieved with the 30 kDa polysulfone membranes, as desired.

Economic Analysis

Micro- and ultrafiltration fluxes have been used in a preliminary economic analysis of the cost of implementing a membrane-based enzyme recycling system. The total annualized cost of each membrane separation step will depend on the membrane area required and, hence, on the permeate flux achieved. By using an economic model for membrane filtration of fermentation broth described by Kuberkar et al. *(13)*, researchers have made preliminary estimates that indicate that the reduction in enzyme costs when enzyme recovery and recycle are used will significantly outweigh the added cost of employing membrane separation with values of the permeate flux comparable with those obtained in our preliminary tests.

Table 3 shows the cost basis employed and the calculated costs of one membrane separation (micro- or ultrafiltration) for two different permeate

Table 3
Economic Analysis for Enzyme Recovery by Membrane Separation[a]

Category	Cost basis	Cost (cents/gal EtOH) (flux = 10 L/[m²·h])	Cost (cents/gal EtOH) (flux = 50 L/[m²·h])
Capital	$200/(m²·yr)	14.6	2.9
Membranes	$175/(m²·yr)	12.7	2.5
Power	$28/(m²·yr)	2.0	0.4
Cleaning	$25/(m²·yr)	1.8	0.4
Maintenance	$20/(m²·yr)	1.5	0.3
Labor	0.2 person/d	0.01	0.01
Total cost	$450/(m²·yr)	32.6	6.5
Savings with 75% enzyme recycled	—	24.0	24.0

[a]The cost basis is taken from Kuberkar et al. *(13)* and is based on a biotechnology plant capacity of 40,000 L/d. A linear scale-up with membrane area is assumed.

flux values. In both cases, the largest costs are for capital equipment and membranes, with power, labor, maintenance, and cleaning representing only a small fraction of the total. The cost of membrane separation (per gallon of ethanol produced) decreases with increasing flux, owing to the reduced membrane area and equipment size required. Also shown is the cost savings that would occur by recovering and reusing 75% of the cellulase enzyme. The amount of cellulase that can be recycled will depend strongly on its adhesion to solid particles and its deactivation. Both of these factors must be investigated in future work. Fortunately, previous work by Roseiro et al. *(14)* showed no deactivation of cellulase during batch concentration by ultrafiltration for 8 h. The data on the enzyme cost ($0.32/gal of EtOH) and the volume ethanol produced per total volume processed (0.05 gal of EtOH/gal total) are taken from Wooley et al. *(4)*. These data are for a simultaneous saccharification and cofermentation (SSCF) process. Although the proposed membrane system is more directly applied to a process with sequential steps, it may also be adapted to an SSCF process.

When using a cost/benefit analysis, a critical flux must be exceeded so that the cost required for adding membrane separation is more than offset by the cost savings of reducing the total enzyme requirements by recovery and recycle. The tradeoff is illustrated in Fig. 8, where the separation cost for one membrane step is plotted vs the average permeate flux. The total cost of implementing a membrane system is the sum of the individual micro- and ultrafiltration costs, each of which is shown in Fig. 8. Assuming that 75% of the cellulase enzyme is recovered in active form by membrane separation, an average flux of 14 L/(m²·h) must be achieved for a single membrane separation step, and 30 L/(m²·h) must be achieved when two membrane separation steps with equal flux are required. Fortunately, the

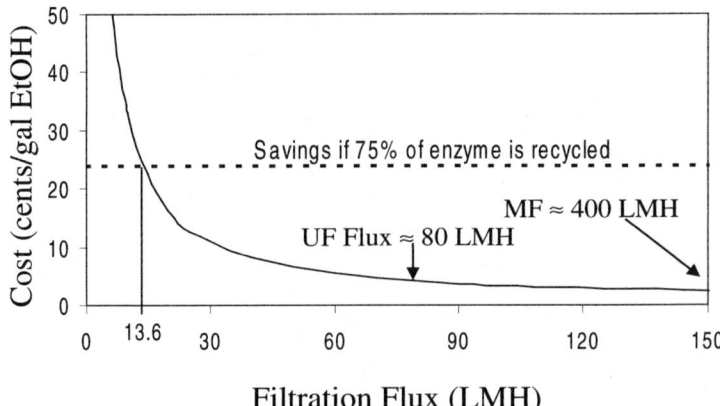

Fig. 8. Cost of one membrane separation step vs average flux during membrane filtration. MF, microfiltration; UF, ultrafiltration; LMH, L/(m²·h).

preliminary fluxes presented herein significantly exceed the critical flux values. Indeed, using typical values of 80 L/(m²·h) for ultrafiltration and 400 L/(m²·h) for microfiltration, the combined cost of the two membrane separation steps is only $0.06/gal of EtOH, whereas $0.24/gal of EtOH would be saved if 75% of the enzyme were reused. The added cost of inclined settling is expected to be small compared with the filtration cost. Although this cost estimate is preliminary, it does encourage further investigation into implementing a membrane system for cellulase recovery.

Acknowledgments

We wish to thank Jim McMillan and Bob Wooley of the NREL for providing advice and materials. We also gratefully acknowledge the Department of Energy for funding the project. W. Mores was supported by the Department of Education's Graduate Assistantships in Areas of National Need program.

References

1. Lynd, L. R., Wyman, C. E., and Gerngross, T. U. (1999), *Biotechnol. Prog.* **15,** 777–793.
2. McCoy, M. (1998), *C&EN* **12,** 29–32.
3. Lee, J. (1997), *J. Biotechnol.* **56,** 1–24.
4. Wooley, R., Ruth, M., Sheehan, J., Ibsen, K., Majdeski, H., and Galvez, A. (1999), NREL/TP-580-26157. National Technical Information Service, Springfield, VA.
5. Nguyen, O. A., Keller, F. A., Tucker, M. P., et al. (1999), *Appl. Biochem. Biotechnol.* **77/79,** 455–472.
6. Hill, W. D., Rothfus, R. R., and Li, K. (1977), *Int. J. Multiphase Flow* **3,** 561–583.
7. Acrivos, A. and Herbolzheimer, E. (1979), *J. Fluid Mech.* **92,** 435–457.
8. Davis, R. H. and Gecol, H. (1996), *Int. J. Multiphase Flow* **22,** 563–574.
9. Davis, R. H. and Acrivos, A. (1985), *Annu. Rev. Fluid Mech.* **17,** 91–118.

10. Kroner, K. H., Schutte, H., Hustedt, H., and Kula, M. R. (1984), *Process Biochem.* **April,** 67–74.
11. Davis, R. H., Zhang, X., and Agarwala, J. P. (1989), *Ind. Eng. Chem. Res.* **28,** 785–793.
12. Kuberkar, V. T. and Davis, R. H. (2000), *J. Membr. Sci.* **168,** 245–260.
13. Kuberkar, V. T., Czekaj, P., and Davis, R. H. (1998), *Biotech. Bioeng.* **60,** 70–87.
14. Roseiro, J. C., Conceição, A. C., and Amaral-Collaço, M. T. (1993) *Bioresour. Technol.* **43,** 155–160.

Effect of Yeast Extract on Growth Kinetics of *Monascus purpureus*

DANIELA GEREVINI PEREIRA AND BEATRIZ VAHAN KILIKIAN*

Departamento de Engenharia Química, Escola Politécnica de São Paulo, CP 61548, CEP 05424-970, São Paulo-SP, Brazil, E-mail: kilikian@usp.br

Abstract

Growth kinetics and red pigment production of *Monascus purpureus* CCT 3802 was studied. A reproducible inoculum with extremely dispersed hyphae for bioreactor runs was obtained through a two-step cultivation in a shaker. First, the spores were cultivated in a complex medium rendering a suspension of vegetative cells. In the second step these cells were grown in a semisynthetic medium. Two types of media were employed in the bioreactor runs: a semisynthetic (glucose, salts, and yeast extract), and a synthetic, without yeast extract. The inclusion of yeast extract, caused an increase in cell yield on glucose ($Y_{x/s}$) as high as 40%. Also, yeast extract probably yielded a higher proportion of red pigment associated with the cell, relative to the synthetic medium. On the other hand, cells grown on the synthetic medium were slightly higher producers of red soluble pigments.

Index Entries: *Monascus purpureus*; yeast extract; red pigment; kinetics.

Introduction

Interest in natural food coloring has increased in recent years. Patents on natural pigments are outnumbering the synthetic ones by five to one *(1)*. The number of permitted synthetic colorings has decreased because some, such as azorubin and tartrazin, have been shown to cause allergies *(2)*. Therefore, the development of processes for the production of microbial organic pigments has become important.

Monascus spp. are millinery used in the East Asian countries. Cultivation is done in semisolid medium (rice) *(3)*, and the whole medium has been used for centuries as a medicinal agent and as a food coloring *(4)*.

Interest in the red pigments of *Monascus* spp. relies on the substitution of the nitrite and nitrate in cured meats, for they are significantly less toxic. Other additives also can be substituted such as cochinila carmim *(5)*. In addition, the fungi synthesize mevinolin, which is a cholesterol-reducing medicine *(6)*.

*Author to whom all correspondence and reprint requests should be addressed.

Growth kinetics and pigment production in the submerged cultivation of *Monascus* spp. have been discussed in several articles (7–10). However, the influence of yeast extract on growth and production activities of *Monascus* spp. has not been determined.

In the present work, two parts of the *Monascus purpureus* culture process were studied: (1) the standardization of the inoculum cultivation procedure in order to get vegetative cells with extremely dispersed hyphae, and (2) the growth kinetics and red pigment production through batch runs on a semisynthetic medium (composed of glucose, salts, and yeast extract) and on a synthetic medium (without yeast extract).

Materials and Methods

Microorganism and Storage

M. purpureus CCT 3802 was obtained from the Centro de Culturas Tropicais of Fundação Tropical de Pesquisas André Tosello, Campinas, SP, Brazil. This strain corresponds to ATCC 36928. A standardized suspension of spores in glycerol (15% [v/v]) was stored at –20°C.

Inoculum for Bioreactor Runs

The inoculum was precultivated in two phases. First, the spores were activated through the cultivation in 500-mL shake flasks (10^6 spores/mL) containing 100 mL of a complex medium (described subsequently) at pH 5.5. The flasks were incubated for 48 h at 30°C and 300 rpm.

Second, 20 mL of the first preculture was used to inoculate 80 mL of a semisynthetic medium (described subsequently). This second preculture was incubated in 500-mL shake flasks for 30 h at 30°C and 200 rpm.

Cultivation Medium

Three types of media were utilized: complex, semisynthetic, and synthetic. The complex medium contained (in distilled water) 10 g/L of glucose, 3.0 g/L of meat extract, and 5.0 g/L of peptone. The semisynthetic medium contained (in distilled water) 20 g/L of glucose, 4.8 g/L of $MgSO_4 \cdot 7H_2O$, 1.5 g/L of KH_2PO_4, 1.5 g/L of K_2HPO_4, 0.01 g/L of $ZnSO_4 \cdot 7H_2O$, 7.6 g/L of monosodium glutamate, 0.4 g/L of NaCl, 0.1 g/L $FeSO_4$, and yeast extract. The semisynthetic medium without yeast extract was called synthetic medium. pH values were adjusted to 5.5. The media were sterilized in an autoclave at 120°C for 20 min.

A solution of yeast extract and the salts (except $MgSO_4$) was sterilized inside the bioreactor (or in an Erlenmeyer flask for the shaker runs). Another solution, composed of glucose and $MgSO_4$, was sterilized separately at pH 4.0, to avoid undesirable reactions.

Bioreactor Cultivation

The two experiments were done in a 5-L bioreactor Bio Flow III (New Brunswick Scientific, Edison, NY).

Temperature was kept at 30°C and agitation was kept at 500 rpm. pH was controlled at 5.0 ± 0.3 through automatic addition of an NaOH solution (2 *N*) or an HCl solution (2 *N*). Airflow rate was kept at 1.0 L/(L·min). Four liters of the media described previously was inoculated with 400 mL of the inoculum.

Analytical Methods

Samples were periodically taken from the bioreactor cultivations in order to analyze cell concentration (*X*) as dry cell weight, glucose concentration, and pigment production (*Abs*). The assessment of *X* was done after samples were vacuum filtrated through a 1.2-μm membrane followed by drying the pellet in a microwave (180 W, 15 min). Glucose concentration was determined by the glucose-oxidase method (Merck, Darmstadt, Germany). Red soluble pigment production was evaluated through the absorbancy measurement at λ = 500 nm, using the filtrate samples.

Ethanol and acetate concentrations were determined by a Waters 600E high-performance liquid chromatograph (Waters, Milford, MA) equipped with a Waters 410 refractometer (35°C) at its outlet. The Shodex Ionpak KC-811 column (Shodex, Japan) at a working temperature of 40°C was employed with a flow of 1 mL/min of H_3PO_4 (0.1%) as the mobile phase. Samples injected to the high-performance liquid chromatograph were filtered through a Waters NH_2 filter, to remove the pigments, which could damage the column.

Results and Discussion

The procedure for inoculum culture was established through shaker runs, whose properties are described on Table 1. According to Table 1, complex media were more appropriate to induce spore germination, because the lag phase became shorter while the media became complex (runs I–III).

The initial spore concentration was standardized on 10^6 spores/mL, in order to not render pellets of germinated cells. To obtain a reproducible inoculum for the bioreactor runs, the two-step procedure of run V was established, which is the same as described in Materials and Methods. Through this procedure, the shorter lag phase was verified in the spore germination and a suspension of extremely dispersed hyphae was obtained, which rendered no lag phase in the subsequent bioreactor cultivation. This is an important feature for industrial purposes because it shortens the growth phase, also reducing the possibility of contaminant growth.

From the shaker runs, it was visually verified that the addition of yeast extract to the medium yielded a higher pigment production associated with the cells relative to the synthetic medium, in addition to higher values of cell concentration. Taking into account the significant influence of the yeast extract on process performance, batch runs R1 (semisynthetic

Table 1
Properties of Runs

Run	Medium	Inoculum	Lag phase (h)
I	Synthetic	Spores	40
II	Semisynthetic	Spores	24
III	Complex	Spores	20
IV	Synthetic	Vegetative[a]	0
V	Semisynthetic	Vegetative[a]	0

[a]Inoculum deriving from spores cultivation in complex medium (III).

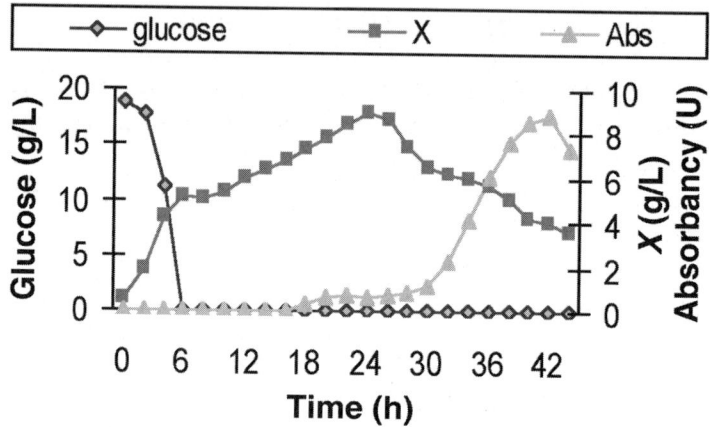

Fig. 1. Cell concentration (*X*); red pigment (*Abs*) and glucose concentration in R1, semisynthetic medium.

medium) and R2 (synthetic medium) were done in the bioreactor, in order to study the effect of yeast extract in well-controlled conditions (Figs. 1 and 2).

In both experiments, R1 and R2, a clear separation between growth and production phases was verified, a typical behavior for secondary metabolites. Red soluble pigment production began when cell growth almost finished. Table 2 summarizes growth features for both runs.

Growth activity was significantly more effective in the medium with yeast extract relative to the synthetic one, regarding the increase of 40% on cell yield on glucose, reported in Table 2. Also, the medium with yeast extract allowed the maintenance of cell viability when glucose was exhausted, opposing the behavior verified with the synthetic medium, in which a sharp decrease in cell concentration was verified before growth on carbon sources other than glucose was established. However, the closer specific growth rate values in both runs indicate that the rates of the metabolic pathways of the growth were not limited in the synthetic medium. Therefore, yeast extract acts only as an additional nutrient supplier.

Monosodium glutamate probably was consumed as carbon and energy source, in addition to nitrogen source, after glucose exhaustion,

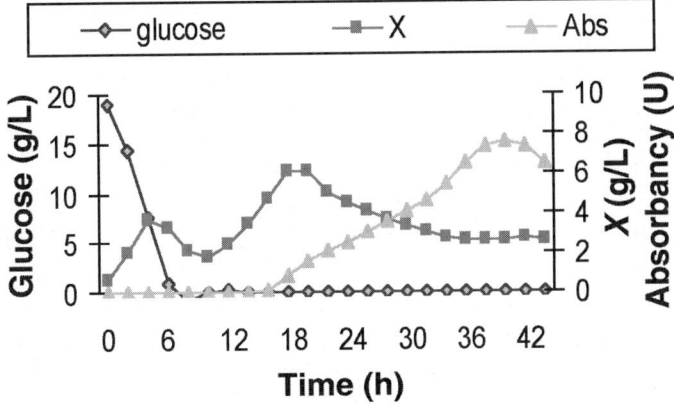

Fig. 2. Cell concentration (*X*); red pigment (*Abs*) and glucose concentration in R2, synthetic medium.

Table 2
Growth Features of Runs R1 and R2[a]

Run	X_{max} (g/L)	$Y_{x/s}$	μ_6 (h^{-1})	μ_{40} (h^{-1})
R1	9.0	0.42	0.44	0.23
R2	6.1	0.30	0.40	0.28

[a]X_{max}, cell concentration at the end of the growth phase; $Y_{x/s}$, cell yield on glucose, at the end of the growth phase; μ_t, specific growth rate at time t.

Table 3
Production Features of Runs R1 and R2[a]

Run	Abs_{max} (500 nm)	*Pec* (*Abs*/[g·L])	*Pep* (*Abs*/[g·L])	*Pp* (*Abs*/h)
R1	8.8	0.08	2.2	0.21
R2	7.8	0.13	2.7	0.19

[a]Abs_{max}, maximum absorbancy at $\lambda = 500$ nm; *Pec* and *Pep*, specific production (relative to cell concentration) at the end of the growth and production phase, respectively; *Pp*, productivity.

in both runs. Although *Monascus* spp. are reported as ethanol and acetate producers *(11,12)*, which also could be consumed, concentrations of these molecules in the range of 10–40 mg/L by the end of the growth on glucose does not explain the increase in cell concentration of about 4 g/L. Table 3 summarizes features of the red pigment production of the runs.

Cells cultivated in the synthetic medium showed a higher level of red soluble pigment production relative to that of the medium enriched with yeast extract (see *Pec* and *Pep* values in Table 3). Regarding that productivity (*Pp*) was almost the same for both runs, the synthetic medium could be the better one. However, the production of red insoluble pigments (not measured) was probably higher in the semisynthetic medium, which can lead to a higher level of total production (soluble plus insoluble red pigments). Further studies of the process including measurements of total red pigment production on both media must be conducted.

Conclusion

A reproducible inoculum with extremely dispersed hyphae for the bioreactor runs was obtained through a two-step cultivation in a shaker. First, the spores were cultivated in a complex medium rendering a suspension of vegetative cells. Second, the culture was subsequently grown in a semisynthetic medium, also utilized in the bioreactor. Moreover, no lag phase was observed in the bioreactor through this procedure.

Cells grown on the synthetic medium were higher producers (23%) of red soluble pigments, relative to cells in semisynthetic medium.

On the other hand, cells grown in the semisynthetic medium (with 1 g/L of yeast extract added), showed a more effective growth (increase of 40% on cell yield on glucose, $Y_{x/s}$) and probably a higher proportion of red pigment associated with the cells.

Further studies on the influence of medium composition on red metabolite solubility must be conducted.

References

1. Francis, F. J. (1989), *Crit. Rev. Food Sci. Nutr.* **28(4),** 273–314.
2. Multon, J. L. (1992), *Additifs et auxiliares de Fabrication dans les Industries Agro-alimentares,* Tec. et Doc. Lavoisier, ed., Apria.
3. Rosenblitt, A., Agosin, E., Delgado, J., and Pérez-Correa, R. (2000), *Biotechnol. Prog.* **16,** 152–162.
4. Went, F. A. F. C. (1985), *Ann. Sci. Nat. Bot. Ser.* **8,** 111–148.
5. Fabre, C. E., Santerre, A. L., Baberian, R., Pareilleux, A., Goma, G., and Blanc, P. J. (1993), *J. Food Sci.* **58(5),** 1099–1110.
6. Wong, H. C. and Bau, Y. S. (1977), *Plant Physiol.* **60,** 578–581.
7. Lin, T. F. and Demain, A. L. (1991), *Appl. Microbiol. Biotechnol.* **36,** 70–75.
8. Juzlová, P., Martinková, L., and Kren, V. (1996), *J. Ind. Microbiol.* **16,** 163–170.
9. Pastrana, L., Blanc, P. J., Santerre, A. L., Loret, M., and Goma, G. (1995), *Process Biochem.* **30(4),** 333–341.
10. Hamdi, M., Blanc, P. J., and Goma, G. (1997), *Bioprocess Eng.* **17(2),** 75–79.
11. Hamdi, M., Blanc, P. J., and Goma, G. (1996), *Process Biochem.* **31(6),** 543–547.
12. Pastrana, L. and Goma, G. (1995), *Process Biochem.* **30(7),** 607–613.

Severity Function Describing
the Hydrolysis of Xylan Using Carbonic Acid

G. Peter van Walsum

*Department of Environmental Studies
and Glasscock Energy Research Center, Baylor University, PO Box 97266,
Waco TX 76798-7266, E-mail: gpeter_van_walsum@baylor.edu*

Abstract

Beech wood derived xylan to hydrolyzed to predominantly xylose mono-mer units after exposure to hot, compressed liquid water saturated with carbon dioxide. Similar treatment without CO_2 saturation resulted in only minor hydrolysis and a smaller fraction of monomers among the hydrolysis products. Severity of the hydrolysis reaction was correlated to reaction time, temperature, and carbon dioxide partial pressure and followed a function similar to those used to characterize mineral acid systems. Results from parallel hydrolysis experiments with an aqueous system and a very dilute sulfuric acid system allowed an approximation of the dissociation constant of carbonic acid in the temperature range of 170–230°C. Results suggest that carbonic acid may be a viable reagent for promoting hydrolysis with-out mineral acids, especially in the case of a bioprocessing plant that pro-duces carbon dioxide.

Index Entries: Xylan; carbonic acid; severity; hydrolysis; biomass conver-sion; pretreatment; carbon dioxide.

Introduction

Rapid use of the world's oil reserves has prompted much research into finding alternative sources for fuels and chemicals. The most likely renew-able resource is biomass (i.e., plant materials such as agriculture and for-estry wastes). The limiting factor in using biomass is hydrolyzing the raw material into fermentable sugars, which can then be biologically converted into a myriad of fuels and chemicals (1).

Two hydrolysis approaches are commonly used: acid and enzymati-cally catalyzed. In the current state of the art, both methods require some acid, although the enzymatic method requires far less (2–4). Eliminating all need for mineral acids (such as sulfuric) from the hydrolysis process would be highly valuable to reduce the operating, construction, and environmen-tal costs associated with a commercial biomass conversion (5,6).

The aim of the present study was to evaluate the potential of carbonic acid (dissolved carbon dioxide) as a hydrolysis agent for xylan, which may be present as a residue in biomass-derived hydrolysates *(7)*. There is increasing interest in the use of carbonic acid in industrial processes *(8,9)*; however, little is known about its fundamental behavior in the range of 150–250°C. Whether carbonic acid can serve the function of a xylan hydrolysis agent is not clear because of its weak acidity and poorly understood behavior at elevated temperatures and pressures. Puri and Mamers *(10)* reported that the addition of carbon dioxide to a steam explosion reactor increased the degree of "in vitro organic matter digestibility." There is at least one report of carbonic acid being capable of hydrolyzing cellulose, but to my knowledge this result has not been replicated *(11,12)*. If carbonic acid is effective, it could prove preferable to stronger acids because of its reduced corrosion and neutralization requirements. In addition, carbonic acid is likely to be available at no cost, because carbon dioxide is a byproduct of the fermentation processes used for conversion of biomass to other products *(10–12)*.

The objectives of the present study were to characterize the hydrolyzing effect of carbonic acid on xylan and to determine a severity function to describe the action of carbonic acid under the conditions tested. Severity was investigated by quantifying the extent of hydrolysis to monomer sugars.

Materials and Methods

Materials

Xylan was derived from beech wood (X-4252; Sigma-Aldrich). This xylan is alkali extracted, which effectively strips it of its acetyl groups (Jeffries, T., personal communication). Sulfuric acid was standard reagent grade, and sodium hydroxide was high-performance liquid chromatography quality (Fisher). Gasses used were regular grade CO_2 and ultra-high-purity He for the high-performance anion-exchange chromatography (HPAE).

Hydrolysis

Hydrolysis experiments were performed either in a 50-mL glass serum vial with crimp seal or in a simple reactor constructed of 1/2-in. 316 stainless steel tubing. The volume of the stainless steel reactor was 15 mL and it was filled and emptied by removing a swage connection on one end. For experiments involving pressurization with CO_2, a 1/8-in. stainless steel tubing connection and valve were fitted to the 15-mL reactor to allow for charging the CO_2 from a gas cylinder (Fig. 1).

In all experiments reported here, the initial concentration of xylan was 1.0 g/L. Temperature control of the hydrolysis reactor was achieved by quickly immersing the reactor in a fluidized sand bath (model SBL 2D;

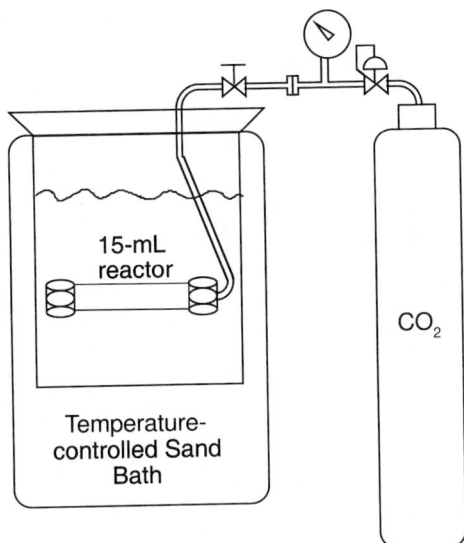

Fig. 1. Schematic diagram of hydrolysis apparatus. Construction of reactor components is 316 stainless steel.

Techhne, Oxford UK) with temperature controller (Techne model TC-8D)
that maintained temperature in the bath to ±1°C. Reaction temperatures
were set to values ranging from 121 to 270°C. Reactions were allowed to
continue for between 2 and 30 min before removing the reactor from the
sand bath and either air cooling (glass vials) or quenching in a cold water
bath (stainless steel reactor).

Carbon dioxide pressure in the reactor was initially set at room
temperature using the gas cylinder regulator (0–800 psig). A standard
correlation between reactor pressures at room temperature and pressures
at elevated temperature was determined by fitting a high-temperature
stainless steel Bourdon tube pressure gage onto the reactor and immersing the entire assembly into the sand bath. To simplify the experimental
apparatus, most hydrolysis experiments did not use the attached pressure gage, and their reaction pressure was inferred from this standard
correlation.

Analysis

The hydrolysate was analyzed using HPAE (models GP50 and AS3500;
Dionex, Sunnyvale CA). A sodium hydroxide eluent was run at a concentration of 25 mM/L for 10 min prior to injection, a gradient from 25 to
500 mM/L NaOH over 30 min, and a final hold at 500 mM/L for a final
15 min. The column used was the Dionex Carbopac PAX-100 column.
Detection was done with pulsed amperometric detection (Dionex model
ED40). Data were collected using the Dionex Peak One software package
running on a Pentium-quality PC.

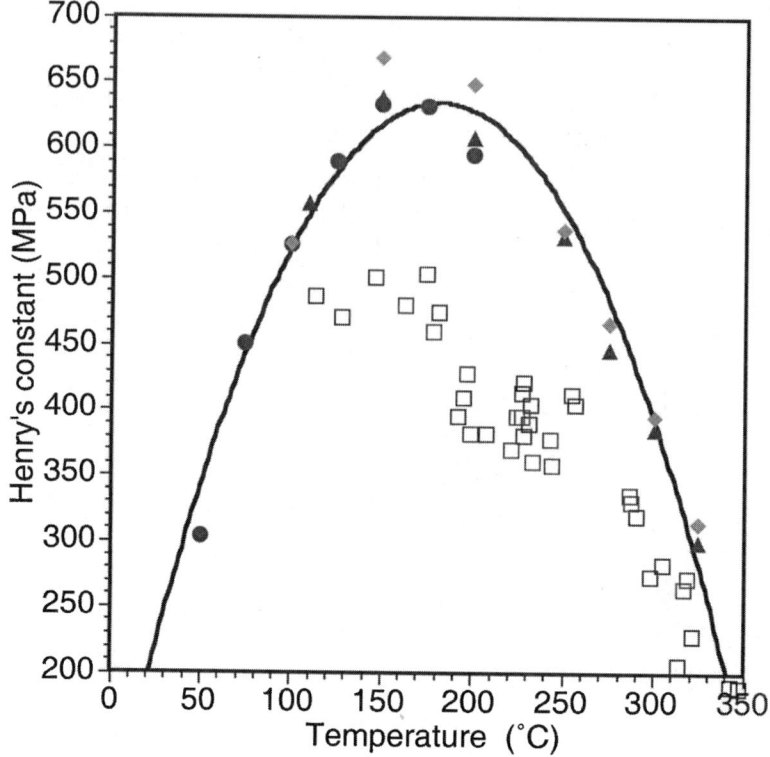

Fig. 2. Literature values for Henry's constant describing CO_2 solubility in water at elevated temperatures. □, *(13)*; ▲, *(14)*; ◆, *(21)*; ●, *(22)*. Curve fit to all data except ref. *13*.

Theory

Solubility and Dissociation Correlations

The chemical engineering literature has relatively little to report on the behavior of carbonic acid in the vicinity of 200°C and elevated pressures. Data are available from studies reported in the geochemistry literature, however, and these are reported here.

The solubility of CO_2 in water achieves a minimum near 160–170°C. Typically, literature reports of solubility relate Henry's constant to temperature. Figure 2 presents values taken from several studies. A curve was fit through the data from the three more recent studies. Points from an earlier study by Ellis *(13)* were not included in this fit because the data do not appear to have been replicated, even by his own later work with Golding *(14)*. The equation was found to be

$$H(T) = -0.017037 \times T^2 + 6.1553 \times T + 78.227 \qquad (1)$$

in which T is expressed in °C and H in MPa, which defines the partial pressure of gaseous CO_2 in equilibrium with a mole fraction of dissolved CO_2.

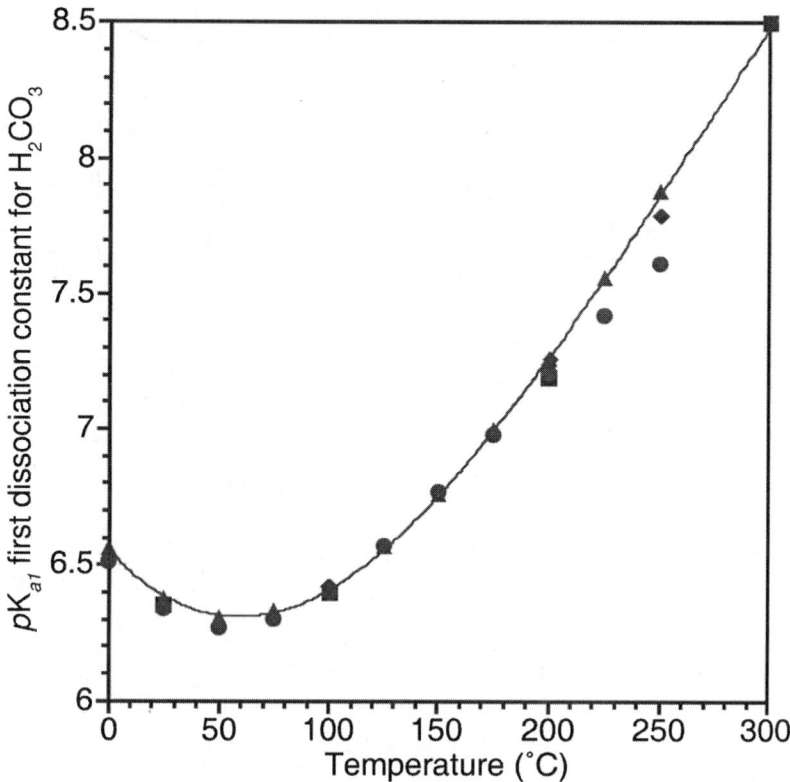

Fig. 3. Literature values for the first dissociation constant of H_2CO_3 in water at elevated temperatures. ■, CRC handbook; ▲, *(15)*; ◆, *(23)*; ●, *(24)*.

The dissociation of H_2CO_3 in water is also highly temperature dependent. Literature values for the pK_{a1} of H_2CO_3 are given in Fig. 3. The correlation of pK_{a1} to temperature proposed by Ryzhenko *(15)* closely matched those of several other investigators and has been adopted for this study:

$$pK_{a1} = (2382.3/T) - 8.153 + 0.02194 \times T \qquad (2)$$

in which T is expressed in degrees kelvin.

Prediction of pH of Carbonic Acid

The pH of carbonic acid was estimated from the dissociation equation for CO_2 in water:

$$K_{a1} = \frac{[H^+][HCO_3^-]}{[CO_{2(aq)}]} \qquad (3)$$

The H^+ ions present originate almost entirely from the dissociation of carbonic acid, because at the temperatures considered, the dissociation constant of water is still relatively low (neutral pH $\cong 5.65$ *[16]*). Thus, as a first

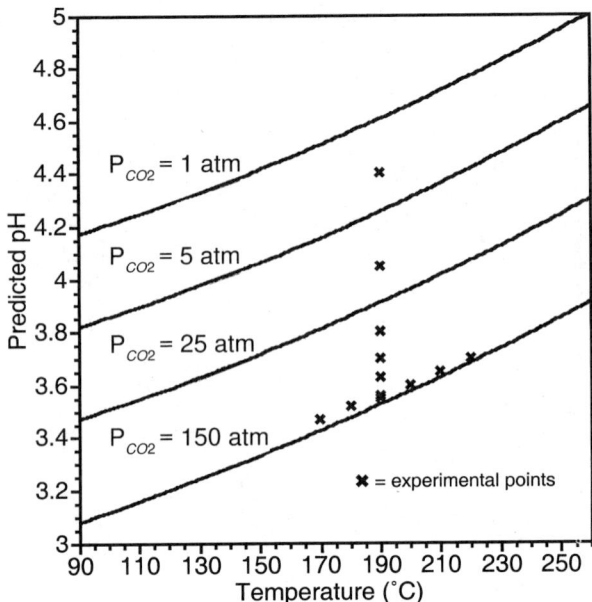

Fig. 4. Calculated values of temperature sensitivity of pH of carbonic acid at different partial pressures of CO_2. P_{CO_2} = 1, 5, 25, or 150 atm as indicated. ×, Experimental conditions of points investigated.

approximation, it is assumed that the concentrations of bicarbonate and hydronium ions are roughly the same:

$$[H^+] \approx [HCO_3^-] \tag{4}$$

At higher temperatures, this assumption no longer holds. Because the solubility of CO_2 in water is affected by the dissociation of carbonic acid in water, the total CO_2 dissolved is determined as a molar balance on carbon:

$$CO_{2(solubilized)} = CO_{2(aq)} + H_2CO_3 + HCO_3^- \tag{5}$$

By assuming a partial pressure of CO_2 in contact with water at a given temperature, the H^+ ion concentration can be determined using Eqs. 1–5 and the steam tables. This involves making an assumption about the contribution of dissolved CO_2 to the liquid volume. In this case, because of the low solubility of CO_2 at the temperatures considered, it was simply assumed that dissolved CO_2 contributed equally, on a molar basis, to the volume of the liquid phase. Figure 4 shows predicted pH values vs temperature and partial pressures of CO_2 (P_{CO_2}). It is clearly seen that pH is a strong function of both temperature and pressure. Eq. 6 was fit to these generated points and expresses the expected pH of the binary CO_2-H_2O system as a function of temperature and pressure in the range of 100–250°C and up to a CO_2 partial pressure of 150 atm:

$$pH = 8.00 \times 10^{-6} \times T^2 + 0.00209 \times T - 0.216 \times \ln(P_{CO_2}) + 3.92 \tag{6}$$

in which T is expressed in °C and P_{CO_2} is in atms. Figure 4 also illustrates the approximate conditions of experiments conducted for this study.

Severity Function

The effectiveness of biomass pretreatment is often correlated using a severity function, which combines the effects of time and temperature into one function. Overend and Chornet *(17)* have defined the following function to quantify the severity of a biomass hydrolysis system in batch or plug-flow reaction modes:

$$R_o = t \times \exp\left[\frac{(T-100)}{14.75}\right] \tag{7}$$

in which t is time in minutes and T is expressed in °C. The effect of acid concentration can also be included by using a combined severity function, which has been shown by Chum et al. *(18)* to be effectively represented as

$$\text{combined severity} = \log R_o - \text{pH} \tag{8}$$

Thus, Eqs. 6 and 7 can be inserted into Eq. 8 to give the combined severity of the carbonic acid system as a function of reaction time, temperature, and P_{CO_2}:

$$\text{combined severity} = \log\left\{t \times \exp\left[\frac{(T-100)}{14.75}\right]\right\} - \tag{9}$$
$$8.00 \times 10^{-6} \times T^2 + 0.00209 \times T - 0.216 \times \ln(P_{CO_2}) + 3.92$$

Quantifying Severity

For studies conducted on raw biomass, severity is often correlated to solubilization of substrate *(18–20)*. However, because purified xylan was completely solubilized under conditions of even the mildest severity, in this study the severity is correlated to the accumulation of xylose monomers, which offered response to severity over a greater range of reaction times.

Results

Effect of Pressure

The effect of varying P_{CO_2} on the hydrolysis system was tested by subjecting a suspension of xylan to varying CO_2 pressures at constant temperature (190°C) and reaction time (16 min). Figure 5 shows HPAE traces from several runs conducted with different CO_2 pressures. The marked pressures refer to the initial pressure of the CO_2 when the reactor was charged at room temperature. It can be seen that at higher pressures, more xylose is released and the low-DP (degree of polymerization) oligomers are also more plentiful. This shows clearly that adding carbonic acid hydro-

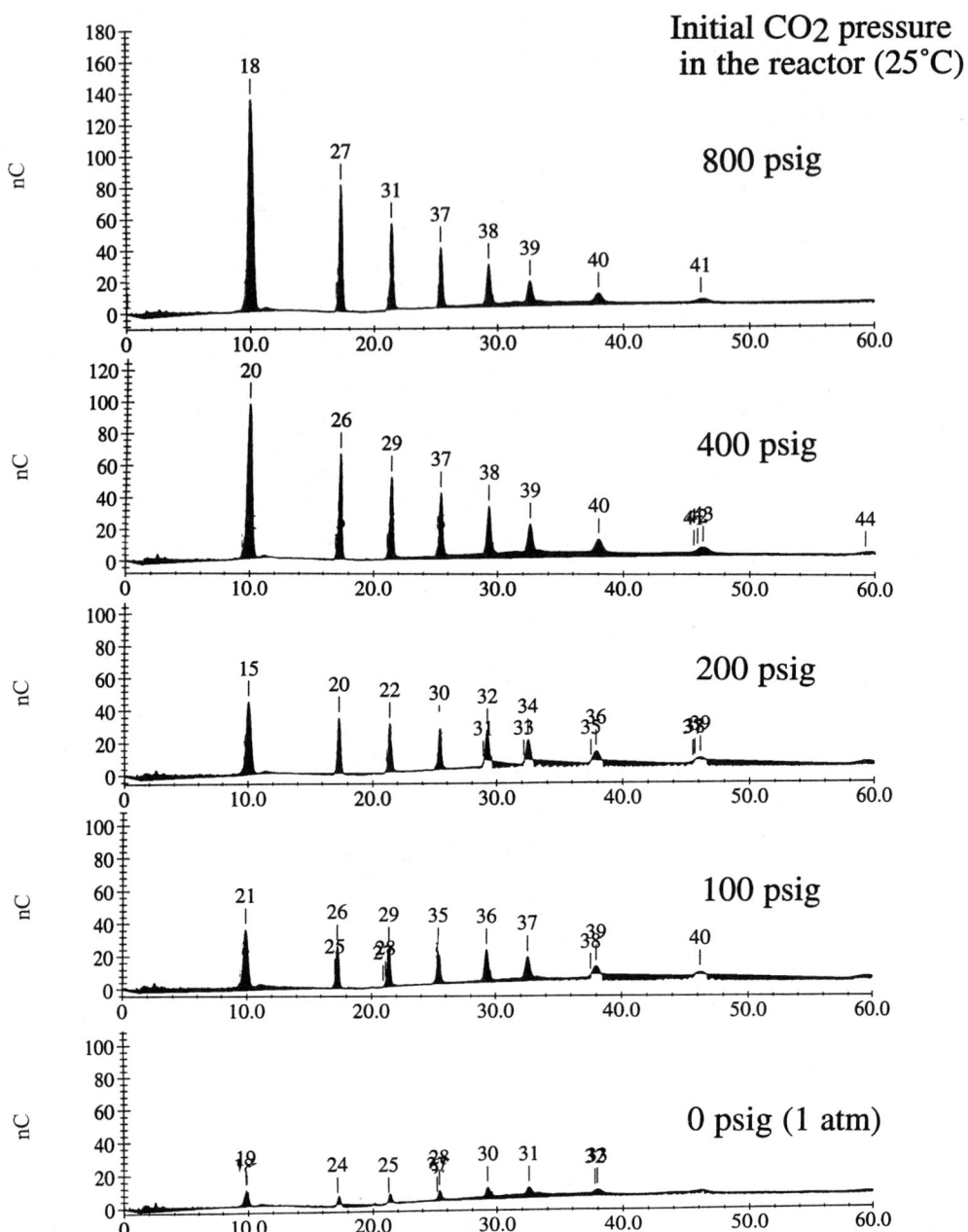

Fig. 5. Output from HPAE-PAD indicating varying degrees of hydrolysis with differing initial reactor pressures of CO_2. Sequential peaks represent oligomers of increasing degree of polymerization.

lyzes xylan. Figure 6 plots the amount of xylose released vs the estimated P_{CO_2} in the reactor at reaction temperature. The trend of increased xylose release with increased pressure is clearly discernible.

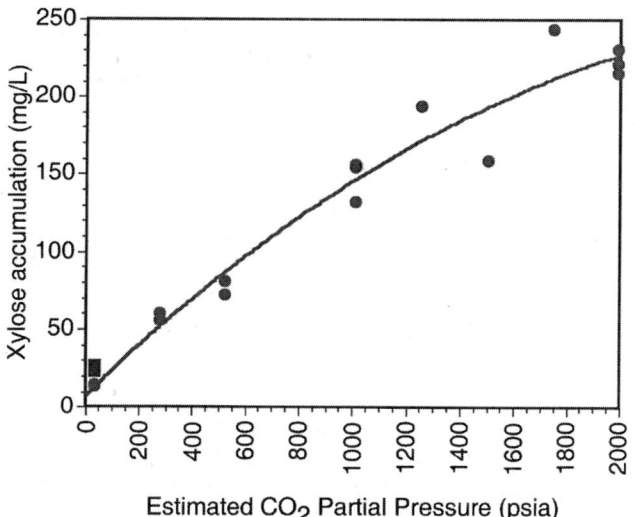

Fig. 6. Xylose accumulation vs P_{CO_2}. Reaction temperature, 190°C; duration, 16 min; ■, no supplemental CO_2; ●, CO_2 added.

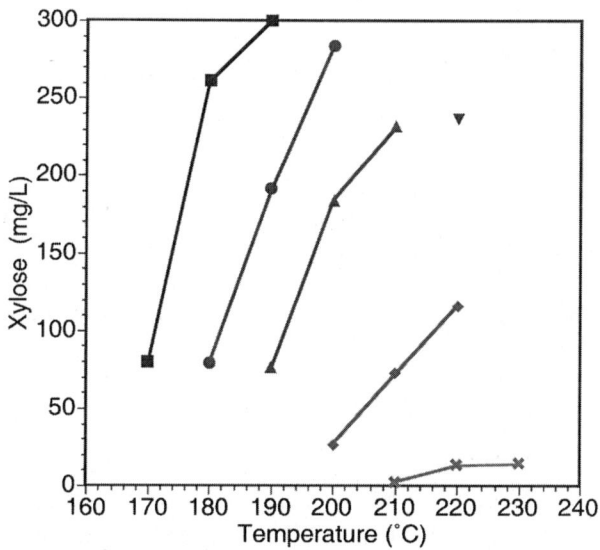

Fig. 7. Xylose release resulting from varying reaction temperatures and durations. Initial CO_2 pressure for all points was 800 psig. ■, 28.5 min; ●, 14.5 min; ▲, 6.5 min; ▼, 5.5 min; ◆, 2.5 min; ×, 0.5 min.

Effect of Temperature and Reaction Time

Figure 7 shows the results of a series of experiments that varied the duration and temperature of the hydrolysis reaction. Immersion times for the reactor varied from 2 to 30 min. Actual reaction times were taken as 1.5 min shorter duration than the immersion time to allow the reactor to

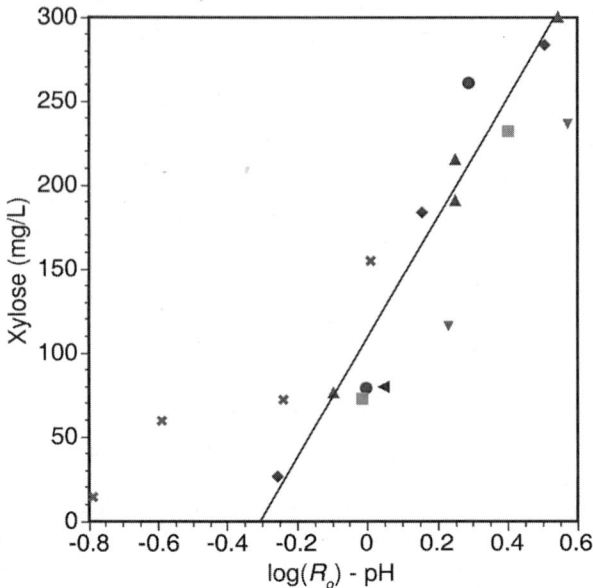

Fig. 8. Collected data plotted vs combined severity factor $\log(R_o)$ – pH. For solid symbols, initial CO_2 pressure 800 psig. ◄, 170°C at 28.5 min; ●, 180°C at 28.5 and 14.5 min; ▲, 190°C at 6.5, 14.5, and 28.5 min; ◆, 200°C at 14.5, 6.5, and 2.5 min; ■, 210°C at 6.5 and 2.5 min; ▼, 220°C at 5.5 and 2.5 min. For × symbols, initial CO_2 pressure varies: × = 190°C, 14.5 min at 400, 200, 100, and 0 psig initial CO_2 pressure.

heat up. All reactions were performed with an initial CO_2 pressure of 800 psig. The effect of higher temperature is clearly visible, as is the effect of increased reaction duration. The reduction in xylose production at higher temperatures suggests that temperature may have a less potent effect on reaction rate than would normally be expected from a system using strong acids.

Combined Severity of Reaction

The applicability of the combined severity function defined in Eq. 9 is shown in Fig. 8, which combines the data presented in Figs. 6 and 7. The effects of pressure and temperature on the reaction severity seem to be adequately expressed by the severity correlation at higher pressures, but the low-pressure data stray from the correlation.

Comparison to Sulfuric Acid and pH Confirmation

Because the determination of pH through Eq. 6 is entirely theoretical and involves a number of simplifying assumptions, an experiment was undertaken to estimate the effective pH of the carbonic acid system by comparing it to a very dilute sulfuric acid system. Hydrolysis of 1.0 g/L of xylan was performed with various concentrations of sulfuric acid, giving a spectrum of pH conditions from 2.9 to 3.7. The pH of the sulfuric acid was

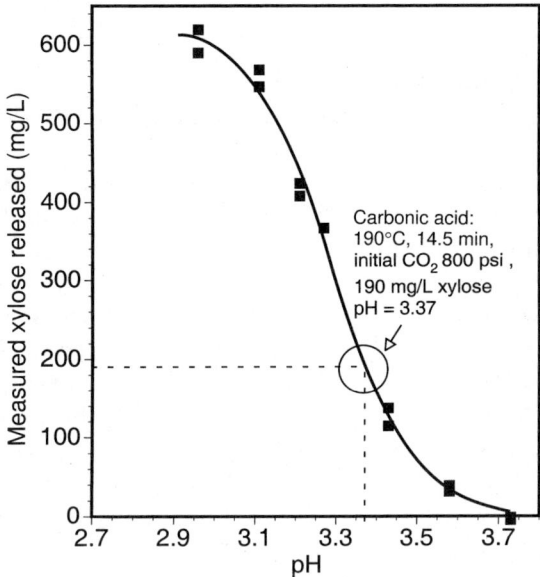

Fig. 9. Comparison of carbonic acid performance to a sulfuric acid system. Data points represent concentration of xylose released from 1.0 g/L xylan slurry reacted for 14.5 min at 190°C in dilute sulfuric acid at the pH indicated.

estimated by assuming complete first dissociation and a formation constant of $K° = 25617.5$ for the second dissociation *(25)*. Under reaction conditions of 190°C and 16-min immersion (14.5-min reaction) time, conversion of xylan to xylose ranged from 0 to 60%. Figure 9 shows the results of this experiment. Also shown is the performance of the corresponding carbonic acid system (1.0 g/L of xylan, 16-min immersion time, 190°C), operating at a P_{CO_2} of 2000 psia. It is apparent that even with the high pressure of CO_2, carbonic acid under the severity conditions tested achieved a lower conversion of xylan to xylose than was possible with dilute sulfuric acid. For the carbonic acid system, the theoretically estimated pH was 3.56; however, it appears that the hydrolysis performance achieved is comparable to a sulfuric acid system with a pH on the order ~3.37. Thus, there appears to be a discrepancy between the theoretical prediction and observed reactivity of the carbonic acid system.

Discussion

In the chemical engineering literature, there is relatively little information about the characteristics of carbonic acid under the conditions tested in the present study. Much of the fundamental work undergirding the theory presented herein originated in the geochemistry field. Yet, it appears that carbonic acid may have utility for some applications requiring mildly acidic conditions under high temperature and pressure. From this investigation, it is clear that xylan hydrolysis is promoted by CO_2 in the tempera-

ture range of 170–230°C, provided that the CO_2 partial pressure is sufficiently high. This may mean that carbonic acid is useful for enhancing
aqueous pretreatment systems, particularly for improving recovery of
monomeric xylose from aqueous pretreatment hydrolysates.

The possible use of CO_2 effluent from a fermentor to assist catalytic
pretreatment has many advantages. If carbonic acid could replace sulfuric
acid for biomass pretreatment, the many disadvantages of sulfuric acid
(purchase cost, high corrosion activity, need for neutralization, separation
of calcium sulfate) could be eliminated. Fermentation-derived CO_2 would
provide a low-cost source of catalyst, and although it is not sequestered by
the process *per se*, it does use an otherwise valueless waste stream.

However, before a clear application can be detailed, some areas of
question remain to be answered. First, although the proposed combined
severity function did incorporate temperature and pressure as pH-moder-
ating influences, and the function could predict the general reactivity trend,
the theory of pH prediction should be verifiable with experimental
results—which has not yet been shown. Second, the question of how carbonic acid will behave in the presence of actual lignocellulosic feedstock is
not yet understood. For example, will the carbonate system act as an acidi-
fying or buffering system in the presence of organic acids liberated from the
biomass? Thus, further investigation is warranted.

Conclusion

Experiments with alkali-extracted xylan show that carbonic acid is
able to catalyze hydrolysis at temperatures from 170 to 230°C. A proposed
combined severity function incorporates the influence of temperature and
P_{CO_2} on the pH of the system, and thus the overall severity of reaction.
However, the pH predictions based on theory have yet to be validated
experimentally. Should it prove a viable catalyst on a realistic substrate,
carbonic acid would offer many process and design advantages over sulfuric acid.

Acknowledgments

This research was supported by the Department of Environmental
Studies and Glasscock Energy Research Center and the University Research
Council, both at Baylor University.

References

1. Lynd, L. R., Cushman, J. H., Nichols, R. J., and Wyman, C. E. (1991), *Science* **251(15),**
   1318–1323.
2. Torget, R. and Hsu, T. (1994), *Appl. Biochem. Biotechnol.* **45/46,** 5–22.
3. Torget, R., Hatzis, C., Hayward, T. K., Hsu, T., and Phillipidis, G. P. (1996), *Appl.
   Biochem. Biotechnol.* **57/58,** 85–101.
4. Lee, Y. Y., Wu, Z., and Torget, R. W. (2000), *Bioresour. Technol.* **71(1),** 29–38.
5. Lynd, L. R., Elander, R. T., and Wyman, C. E. (1996), *Appl. Biochem. Biotechnol.* **57/58,**
   741–761.

6. van Walsum, G. P., Allen, S. G., Spencer, M. J., Laser, M. S., Antal, M. J., and Lynd, L. R. (1996), *Appl. Biochem. Biotechnol.* **54/55,** 157–170.
7. Shevchenko, S. M., Chang, K., Robinson, J., and Saddler, J. N. (2000), *Bioresour. Technol.* **72,** 207–211.
8. Duranceau, S. J., Anderson, R. K., and Teegarden, R. D. (1999), *J. Am. Water Works Assoc.* **91(5),** 85–96.
9. Minkova, V., Marinov, S. P., Zanzi, R., et al. (2000), *Fuel Process. Technol.* **62,** 45–52.
10. Puri, V. P. and Mamers, H. (1983), *Biotechnol. Bioeng.* **25,** 3149–3161.
11. Pavilon, S. J. (1990), US Patent no. 4952504.
12. Pavilon, S. J. (1992), US Patent no. 5135861.
13. Ellis, A. J. (1959), *Am. J. Sci.* **257,** 217–234.
14. Ellis, A. J. and Golding, R. M. (1963), *Am. J. Sci.* **261,** 47–60.
15. Ryzhenko, B. N. (1963), *Geochemistry* **2,** 151–164.
16. Sweeton, F. H., Mesmer, R. E., and Baes, C. F. Jr. (1974), *J. Solut. Chem.* **3(3),** 191–214.
17. Overend, R. P. and Chornet, E. (1987), *Phil. Trans. R. Soc. Lond.* **A321,** 523–536.
18. Chum, H. L., Johnson, D. K., Black, S. K., and Overend, R. P. (1990), *Appl. Biochem. Biotechnol.* **24/25,** 1–14.
19. Abatzoglou, N., Chornet, E., and Overend, R. P. (1992), *Chem. Eng. Sci.* **47(5),** 1109–1122.
20. Bouchard, J., Nguyen, T. S., Chornet, E., and Overend, R. P. (1991), *Bioresour. Technol.* **36,** 121–131.
21. Takenouchi, S. and Kennedy, G. C. (1964), *Am. J. Sci.* **262,** 1055–1074.
22. Zawasza, A. and Malesinska, B. Y. (1981), *J. Chem. Eng. Data* **26,** 388–391.
23. Read, A. J. (1975), *J. Solut. Chem.* **4(1),** 53–70.
24. Park, S. N., Kim, C. S., Kim, M. H., Lee, I.-J., and Kim, K. (1998), *J. Chem. Soc. Faraday Trans.* **94,** 1421–1425.
25. Bilal, B. A. and Müller, E. (1993), *Z. Naturforsch.* **48a,** 1073–1080.

Cellulose Hydrolysis Under Extremely Low Sulfuric Acid and High-Temperature Conditions

Jun Seok Kim,[1] Y. Y. Lee,*,[1] and Robert W. Torget[2]

[1]*Department of Chemical Engineering,*
230 Ross Hall, Auburn University, Auburn, AL 36849,
E-mail: yylee@eng.auburn.edu;
and [2]National Renewable Energy Laboratory, Golden, CO, 80401

Abstract

The kinetics of cellulose hydrolysis under extremely low acid (ELA) conditions (0.07 wt%) and at temperatures >200°C was investigated using batch reactors and bed-shrinking flow-through (BSFT) reactors. The maximum yield of glucose obtained from batch reactor experiments was about 60% for α-cellulose, which occurred at 205 and 220°C. The maximum glucose yields from yellow poplar feedstocks were substantially lower, falling in the range of 26–50%. With yellow poplar feedstocks, a large amount of glucose was unaccounted for at the latter phase of the batch reactions. It appears that a substantial amount of released glucose condenses with nonglucosidic substances in liquid. The rate of glucan hydrolysis under ELA was relatively insensitive to temperature in batch experiments for all three substrates. This contradicts the traditional concept of cellulose hydrolysis and implies that additional factors influence the hydrolysis of glucan under ELA. In experiments using BSFT reactors, the glucose yields of 87.5, 90.3, and 90.8% were obtained for yellow poplar feedstocks at 205, 220, and 235°C, respectively. The hydrolysis rate for glucan was about three times higher with the BSFT than with the batch reactors. The difference of observed kinetics and performance data between the BSFT and the batch reactors was far above that predicted by the reactor theory.

Index Entries: Yellow poplar; cellulose hydrolysis; bed-shrinking flow-through reactor; kinetics.

Introduction

The acid-based treatment of biomass is gaining its position as a viable saccharification process. Among the notable indicators of such is the

---

*Author to whom all correspondence and reprint requests should be addressed.

emergence of the BCI biomass-to-ethanol plant (Jennings, LA) and the Total Hydrolysis Process of National Renewable Energy Laboratory (NREL), Golden, CO *(1)*. The latter is built on three unique technical elements: employing extremely low acid (<0.1%) and high reaction temperature, applying a countercurrent moving bed scheme in the reactor design, and utilizing the bed-shrinking phenomena as a means to improve the reactor performance.

There are distinct advantages of using extremely low acid (ELA) conditions for hydrolysis of lignocellulosic biomass. The low acidity minimizes the gypsum production, if any. The corrosion characteristics of ELA are close to those of neutral aqueous reaction so that standard-grade stainless steel equipment can be used instead of high nickel alloy. ELA gives a significant cost advantage in the equipment. A process using ELA also qualifies as a "green technology" because it has a minimal environmental effect. The recent advancement made in this technology has brought the acid hydrolysis process to a position where it can compete with the enzymatic hydrolysis process in the overall process economics. This breakthrough technology has been proven on an ideally behaving bench-scale kinetic reactor system. An upscale experimental investigation using a continuous countercurrent reactor is being conducted at NREL.

The ELA reaction conditions are beyond the region normally explored in the conventional acid hydrolysis processes. Recent findings at NREL have proven that yields in the vicinity of 90% are attainable under ELA conditions. The NREL data also suggest that the reaction mechanism may be quite different in this region from those found in conventional processes. The present study was undertaken to provide further insights and kinetic data on the reactions taking place under ELA conditions.

Materials and Methods

Material

Yellow poplar sawdust feedstock was provided by NREL. The chemical composition of a representative sample was 45.2% glucan, 15.8% xylan, and 18.4% Klason lignin. It was milled to pass through a 2-mm screen before use. The composition of prehydrolyzed yellow poplar was 68.9% glucan and 26.9% Klason lignin. The prehydrolysis conditions were 174 and 204°C/10 min using 0.07 wt% sulfuric acid by percolation reactor *(2)*.

Batch Kinetic Experiments

All batch reactor experiments were performed using sealed tubular reactors. The reactors (13.5 cm^3 of internal volume) were constructed out of Hastelloy C-276 tubing (0.5 in. [1.27 cm]). Both ends of the reactor were capped with Swagelok end caps measuring 0.5 in. (1.27 cm) wide and 6 in. (15.24 cm) long. The reactors were packed with 0.8 g of solid substrate and 8 mL of acid solution to achieve a solid-to-liquid ratio of 1:10. The sulfuric acid concentration was 0.07 wt% (pH 2.2). The reaction temperatures were

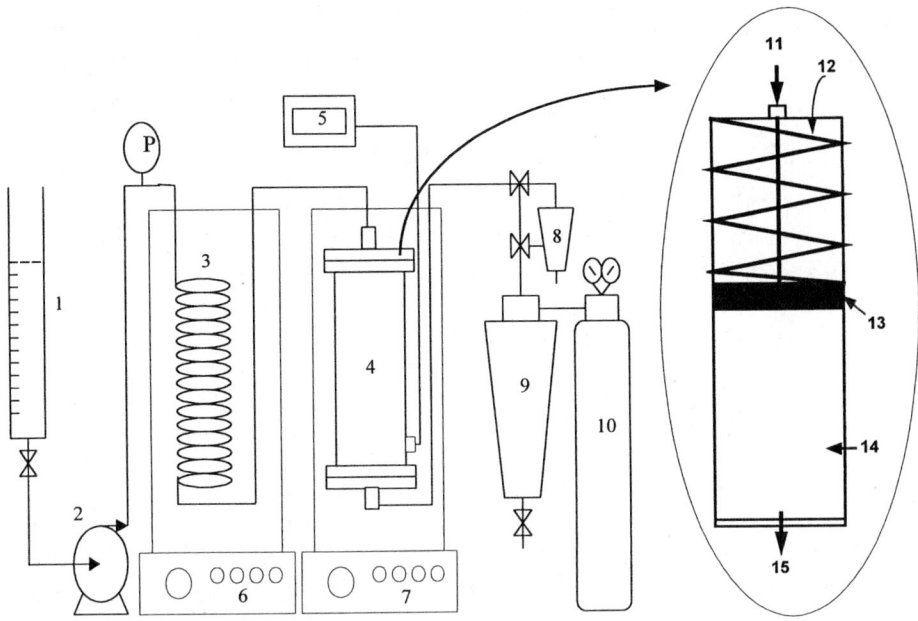

Fig. 1. Laboratory setup for BSFT reactor system: 1, liquid tank; 2, metering pump; 3, preheating coil; 4, bed-shrinking reactor; 5, thermometer; 6 and 7, temperature programmable sand bath; 8, sampling port; 9, pressure holding tank; 10, N_2 gas; 11, acid fluid inlet; 12, spring; 13, movable end; 14, compressed solid biomass; 15, liquid outlet.

controlled in sand baths. The reactors were first submerged into a sand bath set at 50°C above the desired reaction temperature for rapid preheating. The reactors were then quickly transferred into another sand bath set at the precise desired reaction temperature. The reactor temperature was monitored by a thermocouple inserted into the reactor. Reaction temperatures of 205, 220, and 235°C were applied. After the desired reaction time, the reaction was quenched in an ice bath. The contents of the reactor were separated into liquid and solid by filtration and subjected to analyses.

Bed-Shrinking Flow-Through Kinetic Experiments

The bed-shrinking flow-through (BSFT) reactor system invented by NREL is described in Fig. 1. The main body of the reactor is Hastelloy C276 tubing (2 in. [5.08 cm]). The internal volume was 294.4 cm³. The Hastelloy C276 tubing (1/8 in. [1.6 mm] od × 0.03 in. [0.8 mm] id) was used to connect the reactor with other components of the system as well as for the preheating coil. The reactor, ancillary tubing, pump assembly, and collection system were connected and pressurized to 400 psig with N_2 gas. The flow rate of the BSFT runs was kept at 30 mL/min. The amount of initial biomass was 60 g. The reactor is equipped with an internal spring to compress the bed in the reactor as hydrolysis occurs. When the reaction reached the desired time, the flow was stopped and the reactor was quenched in cold water. The liquid sample was collected from the liquid holding tank, and the remaining solid was taken from the reactor for further analysis of composition.

Analytical Methods

The sugars were determined by high-performance liquid chromatography using Bio-Rad Aminex, HPX-87P columns *(3,4)*. A refractive index detector was used. The compositional analysis of all biomass solid samples was carried out by the NREL standard methods *(5)*. The sugars in the liquid sample were determined after being subjected to a secondary acid hydrolysis. The conditions of secondary hydrolysis were 4 wt% sulfuric acid, 121°C, and 1 h.

Results and Discussion

The ELA conditions have been applied mostly for hemicellulose hydrolysis, primarily as a method of pretreatment for the enzymatic hydrolysis. For the past several years, however, it has been investigated from a different angle and with a different purpose at NREL—as a means of cellulose hydrolysis. This work has produced remarkable results in that unusually high glucose yields have been achieved. Yields were particularly high when the experiments were conducted with a BSFT reactor. The observed yields are far above the level projected by the known kinetics, often exceeding 90%. The high yields could not be explained solely from the reactor analysis. In our opinion, there are factors in the kinetics of hydrolysis unique to ELA yet to be identified. Consequently, we became interested in reconfirming the NREL experiments and in seeking explanations for these findings.

Batch Reactions

The reaction kinetics under ELA is far from being established. The literature data on acid hydrolysis of glucan under the ELA conditions are currently limited to those of NREL. In the initial experiments, we conducted a series of batch runs for α-cellulose using 0.07 wt% of sulfuric acid, at varying temperatures of 205, 220, and 235°C. The reaction progress is summarized in Fig. 2. The results are shown in terms of the percentage of glucan remaining in solid and the percentage of glucose released in liquid. The maximum yield of glucose obtained from the α-cellulose was about 60% for 205 and 220°C. However, the maximum yield at 235°C was actually <40%. This is contrary to the conventional concept of cellulose hydrolysis, in which higher yields are obtained at higher temperature, because the activation energy for hydrolysis is higher than that of the decomposition reaction.

The same batch experiments were also conducted using yellow poplar as the feedstock. The overall reaction profiles on the percentage of glucan remaining and the percentage of glucose released were similar to those of α-cellulose runs (Fig. 3). The yield of glucose released increased slightly as

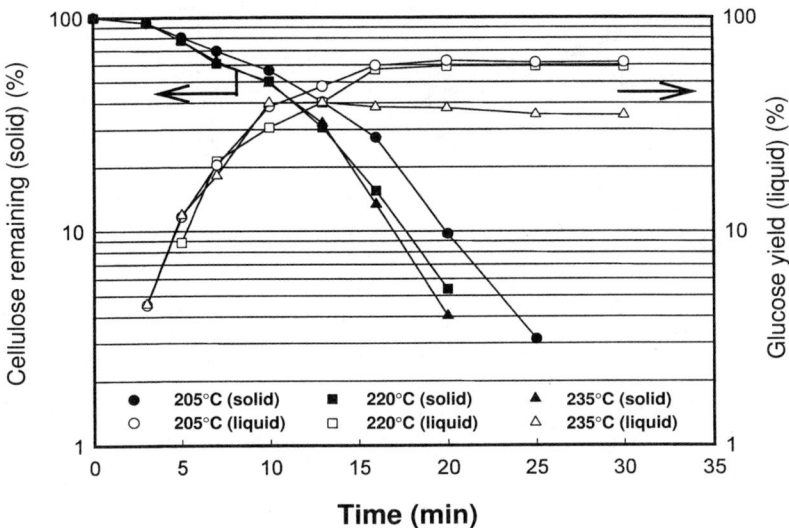

Fig. 2. Semilog plot of cellulose remaining and glucose yield in batch reaction of α-cellulose (0.07 wt% H_2SO_4).

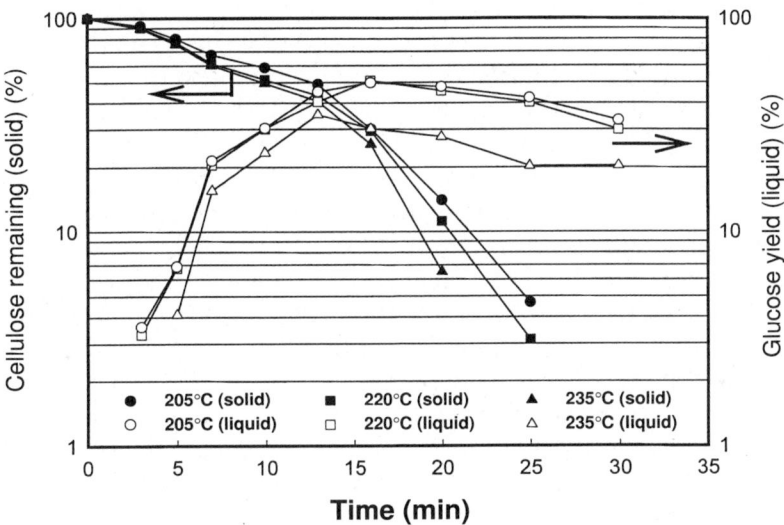

Fig. 3. Semilog plot of glucan remaining and glucose yield in batch reaction of untreated yellow poplar (0.07 wt% H_2SO_4).

the temperature was raised. However, the maximum yield of glucose from yellow poplar was much lower than that from α-cellulose for all three temperatures. The difference was most significant at 235°C, at which the maximum observed batch yield was only 35.2% for yellow poplar feedstock, a drastic departure from the 59.2% observed for α-cellulose. Although there was a similar tendency in the reaction profiles, the batch glucose yields obtained with prehydrolyzed (xylan-free) yellow poplar feedstocks were

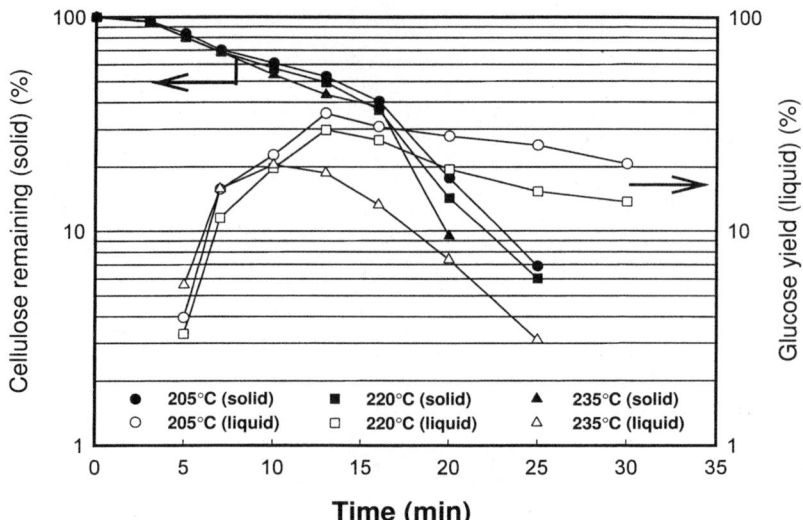

Fig. 4. Semilog plot of glucan remaining and glucose yield in batch reaction of pretreated yellow poplar (0.07 wt% H_2SO_4).

substantially lower than those of the untreated feedstocks for all temperatures (Fig. 4). This is not unique to the ELA conditions; it has been observed in hydrolysis with a higher acid level. The pretreated biomass had a higher fraction of crystalline cellulose, since the easily hydrolyzable glucan was removed during the prehydrolysis process. Prehydrolysis therefore makes the feedstock more difficult to hydrolyze when it comes to acid hydrolysis.

The batch data clearly indicate that the glucose yields from lignocellulosic feedstock (yellow poplar) are substantially lower than those from α-cellulose. One of the potential reasons is that the extraneous materials in the lignocellulosic biomass have a certain degree of buffering capacity for acids *(6)*. The acidity of the liquid, and thus the reactivity, might have been affected. The data in Figs. 2–4, however, preclude this because the decay curves of glucan are essentially the same for α-cellulose and yellow poplar. The hydrolysis reactivity is obviously not affected significantly by the presence of the extraneous materials. Furthermore, the solid-to-liquid ratio applied in the batch experiments was high enough (1:10) to mimic the buffering effects of the biomass.

Another interesting point seen from the batch hydrolysis experiments is that the glucan hydrolysis in solid was relatively insensitive to temperature for all three substrates. The difference in overall slope, although not straight lines, was <10% over the temperature span of 40°C. If one calculates the activation energy for the glucan hydrolysis with the data obtained in this study (ELA), it is approx 1 kcal/mol, an order of magnitude lower than those reported for conventional cellulose hydrolysis. This again contradicts the traditional cellulose hydrolysis and further indicates that there are additional factors and reactions in the hydrolysis under ELA. The estimated rate of glucose decomposition in the traditional sense (glucose to

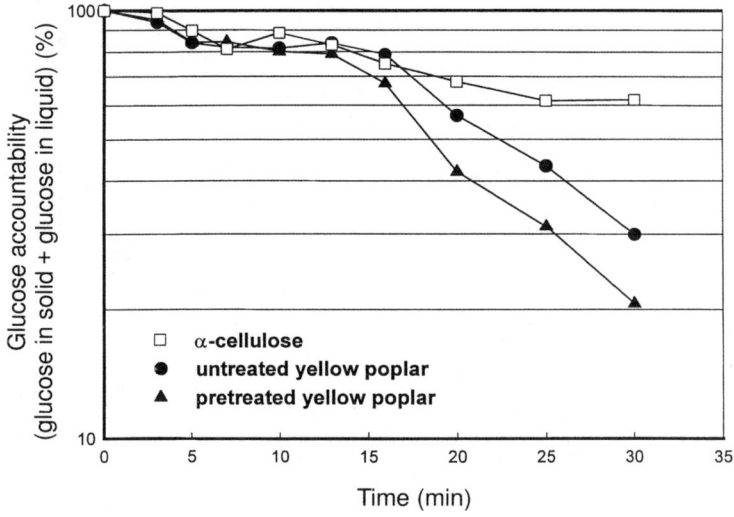

Fig. 5. Semilog plot of glucose accountability for various substrates at 220°C.

hydroxymethyl furfural [HMF]) under the ELA conditions does not fully account for the low yields observed with the yellow poplar feedstocks. What happens to the released glucose under the ELA conditions is uncertain. Some of it would decompose to HMF *(7–9)*, some may recondense with remaining cellulose or lignin (soluble and insoluble), and some may repolymerize as suggested by Conner et al. *(10)*. To verify this point, we prepared a separate plot from the batch data. In this plot, the accountability of glucose, defined as glucan in solid plus the glucose in liquid, is plotted against time for both α-cellulose and yellow poplar (Fig. 5). The difference in the accountability is relatively small between the prehydrolyzed and untreated yellow poplar (lower two curves). However, there is a substantial difference between α-cellulose and yellow poplar. The difference of the accountability occurs only at the latter phase of the reaction, at which the soluble lignin and extraneous compounds tend to accumulate and the released glucose is at a high level. We also note that the true decomposition of glucose under acidic conditions is independent of the feedstock. These findings collectively indicate that the low yields for the lignocellulosic feedstock are primarily owing to the interaction of the released glucose with the nonglucosidic compounds in the liquid. The unaccounted glucose would most likely exist in a condensed form with nonglucosidic substances, which includes the solubilized lignins. However, the existence of the condensed products is yet to be proven. In any event, there is much to be learned regarding the fundamental aspects of the reactions occurring under the ELA conditions.

BSFT Reactor

A series of experiments were conducted using the BSFT reactor invented by NREL. This reactor was used in its original design without any

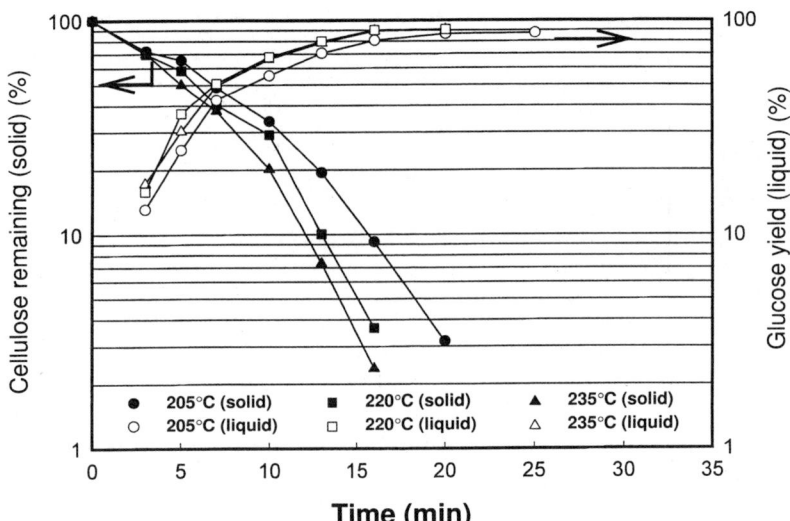

Fig. 6. Semilog plot of cellulose remaining and glucose yield for BSFT reactor, untreated yellow poplar (0.07 wt% H_2SO_4).

Table 1
Maximum Glucose Yields of Batch and BSFT Reactor

Reactor/feedstock	Maximum yield (%)/ reaction time (min)		
	205°C	220°C	235°C
Batch (alpha-cellulose)	61.77/30	59.23/25	40.17/16
Batch (pretreated yellow poplar)	26.62/16	35.45/13	20.43/10
Batch (untreated yellow poplar)	49.82/16	50.98/16	35.22/13
Bed-shrinking (untreated yellow poplar)	87.54/25	90.32/20	90.78/20

modification. The experiments were conducted at 205, 220, and 235°C using 0.07 wt% H_2SO_4. The feedstock was untreated yellow poplar. The shrinking-bed reactor operation was stopped at a desired point in the reaction time; therefore, each run provided only one data point. For each reaction condition, the experiments were repeated to obtain nine data points over the reaction time. The results are summarized in Fig. 6 and Table 1.

The results we obtained are indeed astonishing in that the glucose yields of 87.5, 90.3, and 90.8% were obtained at respective temperatures of 205, 220, and 235°C. The concentrations of glucose in these runs were 2.25, 2.37, and 2.47 wt%, respectively—certainly in a usable range. We have essentially reproduced the results obtained at NREL involving BSFT reactors that can achieve yields in the vicinity of 90%. In addition, we expanded the temperature range down to 205°C. Although the yield is somewhat lower at 87.5%, the low-temperature (205°C) run offers economic benefits

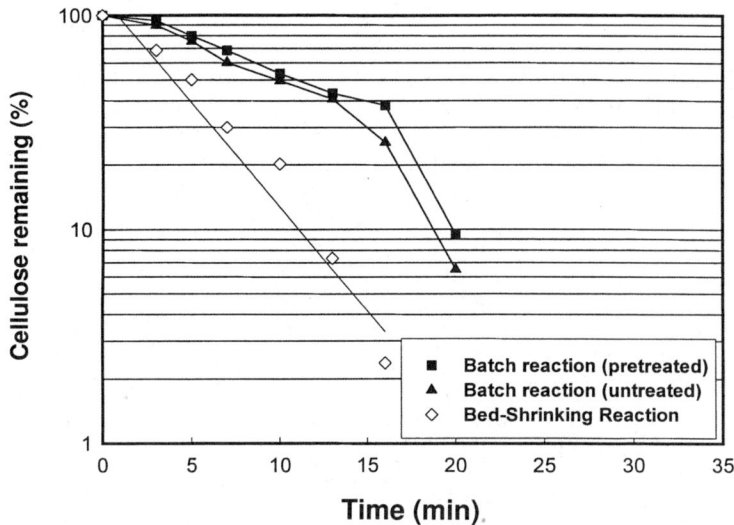

Fig. 7. Hydrolysis profiles of remaining glucan for batch and BSFT reactors at 235°C.

in other areas (lower equipment cost and energy input). It may prove to be a desirable operating condition.

We have previously conducted a modeling investigation that ascertains the positive effect of a bed-shrinking reactor *(11)*. However, the difference in observed kinetics and performance data between the BSFT and batch reactors is far above that predictable by the reactor theory in lieu of the solid-liquid contact pattern. One should also realize that the simplistic approach of representing the acid hydrolysis of cellulose as one set of serial-parallel reaction patterns is grossly inadequate under the ELA conditions.

The drastic difference in reactor performance between the batch and BSFT reactors is reaffirmed in Fig. 7. Here, the semilog plots for the remaining glucan are shown for the batch and BSFT reactors. It is clearly seen that the hydrolysis rate for glucan (estimated from the initial slopes) is about three times higher with the BSFT reactor than with the batch reactors. What causes this difference remains a mystery. For it to be fully understood, the detailed reaction mechanism of this heterogeneous catalytic reaction must be verified, and, therefore, further research is necessary.

Acknowledgments

We wish to thank Nick Nagle of NREL for helpful guidance in the setup and operation of bed-shrinking kinetic reactors. We also gratefully acknowledge the financial support of the US Department of Energy (DE-FC36-99GO10475) and partial support by NREL (subcontract XGC-7-17041-01).

References

 1. Torget, R., Hayward, T. K., and Elander, R. (1997), 19th Symposium on Biotechnology for Fuels and Chemicals, Colorado Springs, CO.
 2. Torget, R., Hatzis, C., Hayward, T. K., Hsu, T., and Philippidis, G. P. (1996), *Appl. Biochem. Biotechnol.* **57/58,** 85–101.
 3. Grohmann, K., Himmel, M. E., Rivard, C., Tucker, M., Baker, J., and Torget, R. (1984), *Biotechnol. Bioeng. Symp.* **14,** 137–157.
 4. Yoon, H. H., Wu, Z., and Lee, Y. Y. (1995), *Appl. Biochem. Biotechnol.* **51/52,** 5–19.
 5. Vinzant, T. B., Ponfick, L., Nagle, N., Ehrman, C. I., Reynolds, J. B., and Himmel, M. E. (1994), *Appl. Biochem. Biotechnol.* **45/46,** 611–626.
 6. Kim, S. B. and Lee, Y. Y. (1986), *Biotechnol. Bioeng. Symp.* **17,** 71–84.
 7. Church, J. and Wooldridge, D. (1981), *Ind. Eng. Chem. Prod. Res. Dev.* **20,** 371–378.
 8. Brenner, W. and Rugg, B. (1985), Report to Environmental Protection Agency, EPA/600/S2-85/137, Washington, D.C.
 9. Bobleter, O., Schwald, W., Concin, R., and Binder, H. (1986), *J. Carbohydr. Chem.* **5(3),** 387–399.
10. Conner, A. H., Wood, B. F., and Hill, C. G. (1985), *J. Wood Chem. Technol.* **5,** 461–489.
11. Chen, R. W., Wu, Z. W., and Lee, Y. Y. (1998), *Appl. Biochem. Biotechnol.* **70/72,** 37–49.

A Hybrid Neural Network Algorithm for On-Line State Inference That Accounts for Differences in Inoculum of *Cephalosporium acremonium* in Fed-Batch Fermentors

ROSINEIDE G. SILVA, ANTONIO J. G. CRUZ, CARLOS O. HOKKA,
RAQUEL L. C. GIORDANO, AND ROBERTO C. GIORDANO*

*Departamento de Engenharia Química,
Universidade Federal de São Carlos, C.P. 676, CEP 13565-905,
São Carlos, SP, Brazil, E-mail: roberto@deq.ufscar.br*

Abstract

One serious difficulty in modeling a fermentative process is the forecasting of the duration of the lag phase. The usual approach to model biochemical reactors relies on first-principles, unstructured mathematical models. These models are not able to take into account changes in the process response caused by different incubation times or by repeated fedbatches. To overcome this problem, we have proposed a hybrid neural network algorithm. Feedforward neural networks were used to estimate rates of cell growth, substrate consumption, and product formation from on-line measurements during cephalosporin C production. These rates were included in the mass balance equations to estimate key process variables: concentrations of cells, substrate, and product. Data from fed-batch fermentation runs in a stirred aerated bioreactor employing the microorganism *Cephalosporium acremonium* ATCC 48272 were used. On-line measurements strongly related to the mass and activity of the cells used. They include carbon dioxide and oxygen concentrations in the exhausted gas. Good results were obtained using this approach.

Index Entries: Neural networks; hybrid model; cephalosporin C production; state inference.

Introduction

Cephalosporin C is a natural β-lactam antibiotic produced by strains of a strictly aerobic fungus, *Cephalosporium acremonium*. Industrial production

---

*Author to whom all correspondence and reprint requests should be addressed.

of this antibiotic is carried out in aerated, stirred bioreactors using submerged cultures of mutant strains. This antibiotic may be hydrolyzed to manufacture 7-amino cephalosporanic acid (7-ACA), which is employed to produce the commercially available, semisynthetic cephalosporins (1).

Many researchers have proposed mechanistic models to simulate the cephalosporin C production process (2–5). These models were based on the overall biosynthetic mechanisms, and although simplifying assumptions were made, the problem of estimating a large number of parameters still remained. Furthermore, changes in strain and small differences in the reactivation of the microorganism may cause appreciable modifications on the system response.

In biochemical processes, it is often difficult to establish quantitative relationships between microbial growth and production rates. This difficulty is even more pronounced when the product is a secondary metabolite. On the other hand, the lack of reliable sensors to measure key process variables makes their inference a necessity for the on-line monitoring and control of the bioprocess.

Artificial neural networks (ANNs) emerged as a very useful tool for process modeling. The ability of ANNs to represent complex nonlinear relationships without prior knowledge of any model structure makes them a promising alternative in many practical situations. One way to use ANNs, called the black box approach, is to replace the complete first-principles model with the artificial net. Applications of this approach to bioprocesses have been reported in the literature (6–11).

An alternative to this technique is called gray box modeling strategy, or hybrid modeling (12). In this case, the mass and energy balances are used in conjunction with ANNs, which serve as an estimator of unmeasured parameters or variables. Psichogios and Ungar (12) reported the use of this hybrid model to study the dynamic behavior of a fed-batch stirred bioreactor. Following this approach, mass and energy conservation would not be violated by the model, which is not ensured when the ANN is a black box. According to van Can et al. (13), another advantage of the hybrid model might be its better extrapolation properties.

In this article, a hybrid neural network model is proposed to infer state variables of the cephalosporin C production process. The hybrid model includes three neural networks, combined with mass balance equations. The first network estimates the specific rate of cell growth from selected on-line measurements and initial conditions. The second and third ones predict glucose consumption and specific antibiotic production, respectively, using the specific growth rate predicted by the previous network. The solution of the mass balances provides the concentrations of cells, substrate, and product, using the output of the three ANNs. Most important, the proposed model was able to forecast the main features of the system response when the inoculation followed a nonstandard procedure.

Modeling Cephalosporin C Production

Production of cephalosporin C by *C. acremonium* has usually been modeled using variations on Monod kinetics. Morphologic differentiation of the microorganism and catabolic repression by glucose are also studied *(2–5)*. Matsumura et al. *(2)* analyzed the induction of cephalosporin C by endogenous methionine, while Chu and Constantinides *(3)* and Basak et al. *(4)* focused on the role of enzymes synthesized by the cells in a complex enzymatic pathway. The complexity of the internal cellular relations makes it very difficult to quantify the production of cephalosporin from the metabolic reactions. All the available information does allow a grasp of the main features of the process, useful for modeling and optimization purposes. Nevertheless, it is virtually impossible that a pure, first-principles ("white box") model will be generic enough to take into account all the consequences of the performance of a biochemical reactor when some random alteration occurs in the inoculum. In this case, all kinetic parameters previously estimated from experimental data might be inaccurate, and systematic deviations from the predicted reactor trajectory would occur.

ANNs in Biochemical Process Simulation

ANNs may be used to circumvent the difficulties described beforehand. ANNs consist of interconnected units, called nodes or neurons, that are organized in layers. A very common type of neural net is the three-layer (input, hidden, and output) feedforward neural (FNN) network. The number of nodes in the input and output layers is equal to the number of input and output variables in the process under investigation. The activity of the input units reflects the raw information flux through the ANN. The optimal number of nodes in the hidden layers depends on the type and complexity of the task the ANN is designed to perform. Usually, this number is determined by trial and error. Each interneuron connection has a weight associated with it, emulating the synapses between actual neurons. These weights are the internal parameters of the network, whose fitting allows the ANN to "learn" information about the system to be modeled. This procedure is called the ANN training.

There are several training algorithms in the literature. Among them, one of the most popular is the backpropagation routine *(14)*, a modified gradient descendent method that minimizes an objective function by redistributing the output error back through the network, appropriately modifying the weights of the nodes' connections. The behavior of the ANNs depends mainly on the weights and the transfer function (typically nonlinear) that is specified for the nodes. The most popular architecture uses sigmoidal functions *(15)* for this task.

FNNs are frequently applied to model (bio)chemical reactors. Thibault et al. *(6)* used an FNN to predict key fermentation variables. Di Massimo et al. *(7)* utilized FNNs to model the production of penicillin. Warnes et al.

(11) tested FNNs and radial-basis function networks to estimate the concentration of biomass and of a recombinant protein produced by *Escherichia coli*—variables that are usually measured off-line. Cruz et al. *(10)* modeled cephalosporin C production using two FNNs, the first one estimating cell concentration and the second inferring antibiotic concentration. They used black box models. The neural network replicates the input-output behavior of the whole system.

With a hybrid model, it is possible to include prior knowledge of the system, even if it is not complete. The ANN would assume the difficult task of emulating aspects that could not be reasonably modeled within the field of a conventional approach. Hybrid models are especially indicated for biochemical processes, in which key variables are not measured on-line and the data training sets are sparse. The ANN would serve as an estimator of unknown parameters or variables of the phenomenologic model of the process. Psichogios and Ungar *(12)*, Thompson and Kramer *(16)*, and van Can et al. *(13,17)* have provided examples of successful applications of the gray box approach to biochemical/enzymatic reactors.

Materials and Methods

Microorganism

C. acremonium ATCC 48272 was used throughout this work. This strain is able to produce more than 1000 mg of cephalosporin C/L, when grown in a synthetic medium *(18)*. It was kept on cryotubes in the presence of glycerol at –50°C.

Culture Media

For inoculum preparation a synthetic medium *(19)* was utilized containing (the following): 30.0 g/L of glucose, 8.8 g/L of ammonium acetate, 5.0 g/L of DL-methionine, 1.5 g/L of oleic acid, 2.3 g/L of KH_2PO_4, 5.8 g/L of K_2HPO_4, 0.16 g/L of $Fe(NH_4)_2(SO_4)_2 \cdot 6H_2O$, and 2.0 g/L of $CaCO_3$. The pH was 7.0 ± 0.1. Micronutrients were provided by a salt solution (50 mL/L) whose composition was 16.2 g/L of Na_2SO_4, 7.68 g/L of $MgSO_4 \cdot 7H_2O$, 1.6 g/L of $CaCl_2 \cdot 2H_2O$, 0.64 g/L of $MnSO_4 \cdot H_2O$, 0.64 g/L of $ZnSO_4 \cdot 7H_2O$, and 0.04 g/L of $CuSO_4 \cdot 5H_2O$. The composition of the production medium was the same as that of the inoculum medium, except for glucose, whose concentration was 27.0 g/L.

Experimental Procedure

The standard procedure for fermentation started with the transfer of 15 mL of cell spores to 135 mL of culture medium. One thousand milliliter flasks were incubated (250 rpm at 26°C) for 48 h (run 1) and 72 h (runs 2 and 3) in a shaker (model G25, New Brunswick Scientific, Edison, NJ). Runs 4 and 5 were repeated batches. The cultivated cells were then used to inoculate the fermentation broth. The inoculum seed was 10% in volume.

The experiments were carried out in a stirred aerated bioreactor with a 5-L working volume (Bioflo II-C; New Brunswick Scientific) at 26°C. Fermentation runs were carried out for about 120 h and samples were taken periodically. The experiments started in batch mode and after depletion of glucose, supplementary medium was added to the fermentor at suitable flow rates. The supplementary medium contained the same components as the production medium, but instead of glucose, hydrolyzed sucrose (glucose + fructose) was used as a carbon source *(20)*. Cell mass and the concentrations of glucose and cephalosporin C were measured off-line.

A paramagnetic oxygen analyzer (model 755; Rosemount Analytical) and an infrared carbon dioxide analyzer (model 880A; Rosemount Analytical) were used to monitor the effluent gas. The analyzers were calibrated with standard gases (5% CO_2/95% N_2, and 99.99% N_2).

A Supervisory Data Acquisition and Control System (Unisoft, São Paulo, Brazil) was used. It includes a Programmable Logic Controller (model 340; GE Fanuc) supervised by a personal computer. The manipulated variables were airflow rate, stirring, cooled water flowrate, and feed rate of supplementary medium. Temperature and dissolved oxygen were the controlled variables. The carbonate buffer was able to maintain the pH.

Analytical Procedures

Cell mass concentration was evaluated in terms of dry weight at 105°C, in grams/liter. Glucose concentration was measured utilizing the enzymatic GOD-PAP method *(21)*. Cephalosporin C concentration was determined by high-performance liquid chromatography (Waters), with a Nova Pak® C-18 column, isocratic operation; eluent; phosphate buffer (1.36% KH_2PO_4, pH 6.0); and 100/3 (v/v) KOH/CH_3CN. Cephalosporin C standards were kindly provided by Laboratório de Tecnologia Enzimática do Instituto de Catalisis y Petróleo Química do CSIC, Madrid, Spain.

Database for Training and Validation of Hybrid Model

Five experimental runs were used for training and validation of the hybrid model. The experiments were conducted in fed-batch and repeated fed-batch modes. On-line measurements included carbon dioxide and oxygen concentrations in the exhausted gas. Table 1 gives the experimental conditions of each data set (runs 1–5). The data-sampling interval was 10 s, but for training purposes a 10-min period was used. The output variables of the ANNs were compared with "experimental" rates during the training procedure. These rates were obtained after smoothing the concentration time course of the state variables. A mechanistic model was used for this purpose *(20)*.

Network Architecture

Two distinct phases are present during fed-batch cephalosporin C production. During the trophophase there is a rapid consumption of substrate to produce biomass. The synthesis of antibiotic occurs during the

Table 1
Experimental Conditions of Bioassays

Run no.	Incubation time in shaker (h)	Inoculum seed (% [v/v])	Flow rate of supplementary medium (mL/h)	Feed starting time (h)	$Y_{x/s}$ (g cell/ g glucose)	pH range
1[b]	48	10	10.00	66.50	0.41	6.3–7.5
2[c]	72	20	10.00	45.00	0.58	6.5–6.9
3[c]	72	20	10.75	44.66	0.58	6.1–6.8
4[b,d]	—	15	13.25	22.75	0.58	6.3–6.6
5[c,d]	—	15	12.00	21.25	0.51	6.3–8.3

[a]T = 26°C; Q_{air} = 3.0 SLPM (standard liter per minute), dissolved oxygen controlled at 40% of saturation, manipulating the stirring speed.
[b]Runs used as validation data set.
[c]Runs used in training data set.
[d]Runs 4 and 5 were sequential repeated fed batch of run 3.

idiophase, when cell growth becomes insignificant. During this phase, glucose should be fed at low flow rates. An increase in carbon dioxide concentration and a decrease in oxygen concentration in the exhausted gas (22) accompany the rapid biomass formation in the trophophase. Therefore, these two on-line variables were chosen as inputs for the ANN. Our strategy also included the specific growth rate at five previous sampling times as additional inputs to the net, as suggested by Hernández and Arkun (23). Hence, the topology of the first ANN included 17 inputs, 1 hidden layer with 5 nodes and 1 output, the specific growth rate at the present time (Fig. 1A).

The second network inferred the specific glucose consumption rate at the present time. Inputs to this network were concentrations of carbon dioxide and oxygen in the exhaust gas, specific growth rate (provided by the first ANN), and specific glucose consumption in five previous time intervals (Fig. 1B).

The third network provided the specific production rate at the present time. The inputs to this network were specific growth rate (from first network) and specific production rate in five previous time intervals (Fig. 1C). The three FNNs had five nodes in the hidden layer. The topology of the FNNs was defined by trial and error.

This architecture follows the "modularization" concept of Di Massimo et al. (7). The information provided by one network (the specific growth rate) was used by another two ANNs (which predicted the specific glucose consumption and antibiotic production rates). When we tried a single network to predict the three rates simultaneously, the results were not promising.

The neuron activation function was sigmoidal (24). The objective function during the training phase was the sum of the squares of the deviation between the neural network outputs and experimental outputs, for all data in the set. The training algorithm was the classic backpropagation of

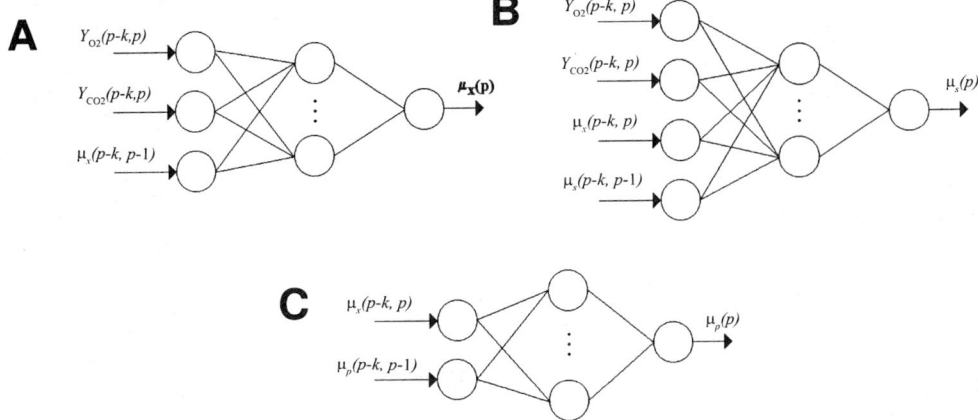

Fig. 1. FNNs used in this study. **(A)** First FNN, to estimate the specific growth rate; **(B)** second FNN, to infer the specific glucose consumption rate; and **(C)** third FNN, for specific production rate. p, present time; k is set to five previous instants; $Y_{O_2}(p-k,p)$, oxygen mole fractions in the exhausted gas; $Y_{CO_2}(p-k,p)$, carbon dioxide mole fractions in the exhausted gas; t, time (h); $\mu_x(p-k, p-1)$, specific growth rates (h^{-1}); $\mu_s(p-k, p-1)$, specific glucose consumption rates (h^{-1}); $\mu_p(p-k, p-1)$, specific product formation rates (h^{-1}).

Rumelhart and McClelland *(14)*. The mean square error (MSE) between the neural network output and the learning data was the objective function for the training algorithm. After 300 presentations of the data set, the MSE did not decrease significantly and the training was interrupted, to avoid overfitting.

Hybrid Model

The mass balance equations of the main components are as follows:

$$\frac{dC_x}{dt} = \mu_x \times C_x \tag{1}$$

$$\frac{dC_s}{dt} = -\mu_s \times C_x \tag{2}$$

$$\frac{dC_p}{dt} = \mu_p \times C_x \tag{3}$$

in which C_p is cephalosporin C concentration (g/L), C_s is glucose concentration (g/L), C_x is cell concentration (g/L), μ_s is specific substrate consumption rate (h^{-1}), μ_x is specific growth rate (h^{-1}), and μ_p is specific production rate (h^{-1}).

The neural networks illustrated in Fig. 2 provided the three specific rates. It is important to stress that μ_s, μ_x, and μ_p are apparent rates, which

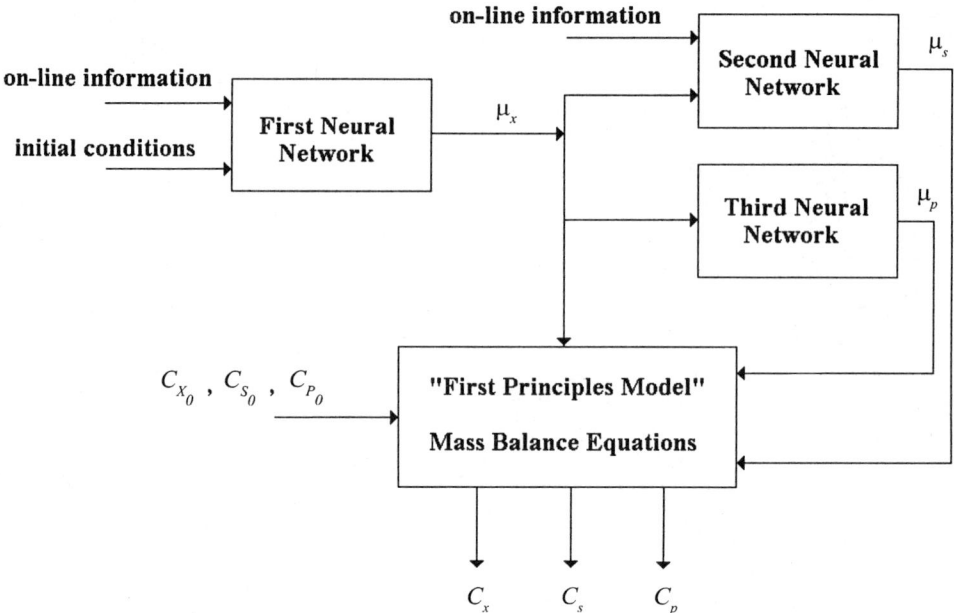

Fig. 2. Schematic hybrid model. Three ANNs estimate three specific rates (cellular growth, glucose consumption, and cephalosporin C production), that are used in the mass balance equations. The rates at the five initial sampling times (initial conditions of the first FNN) were equal to zero.

take into account the process kinetics and the dilution effect caused by the fed-batch operation. This is a characteristic feature of the hybrid approach; the ANNs were trained to lump these two phenomena in one single rate.

Results and Discussions

The network results for μ_x and μ_s for the runs used for training are illustrated in Fig. 3. The two networks provided fairly good estimates at the three different conditions. The experimental rates were obtained by smoothing the raw data. These results showed that the ANNs captured the essential characteristics of the nonlinear process response.

Figure 4 exhibits the performance of the third neural network, inferring the specific production rate for the three training data sets. Figure 5 shows the inferred variables for the training set. The model follows accurately the concentration time course for all the calculated variables.

Run 1 has some distinct features (refer to Table 1) that make it very appropriate to validate our inference algorithm. During this run the lag phase was considerably longer (approx 24 h), owing to the different inoculation procedure. Initial glucose concentration was also higher. And, most important, the overall yield for run 1 was $Y_{X/S} = 0.41$ (g of cell/g of glucose), whereas for all other runs it was in the range $Y_{X/S} = 0.51$–0.58. Therefore, using this specific run to validate the hybrid model is a very strong test of its extrapolation capabilities.

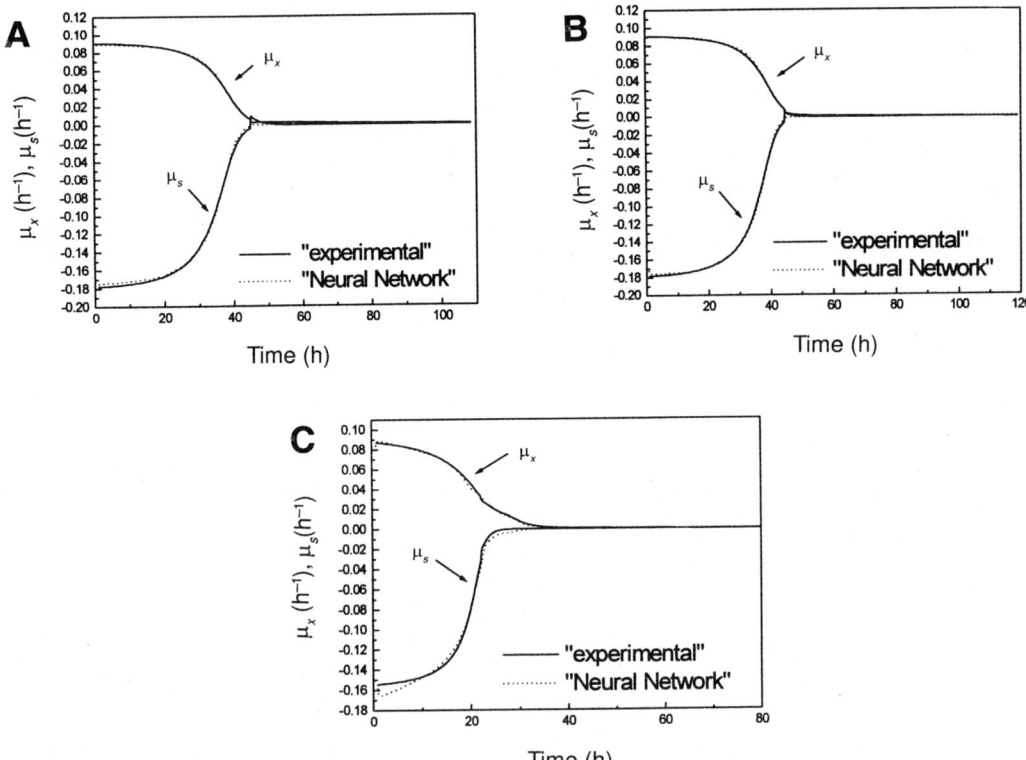

Fig. 3. Specific growth and consumption rates obtained by neural networks during the training phase. **(A)** Run 2, fed batch with standard inoculum preparation; **(B)** run 3, fed batch with standard inoculum preparation; **(C)** run 5, repeated fed batch.

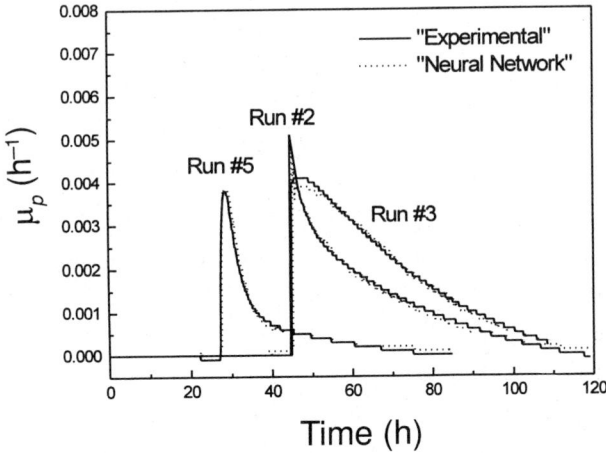

Fig. 4. Specific production rate time course. Run 2, fed batch with standard inoculum preparation; run 3, fed batch with standard inoculum preparation; run 5, repeated fed batch.

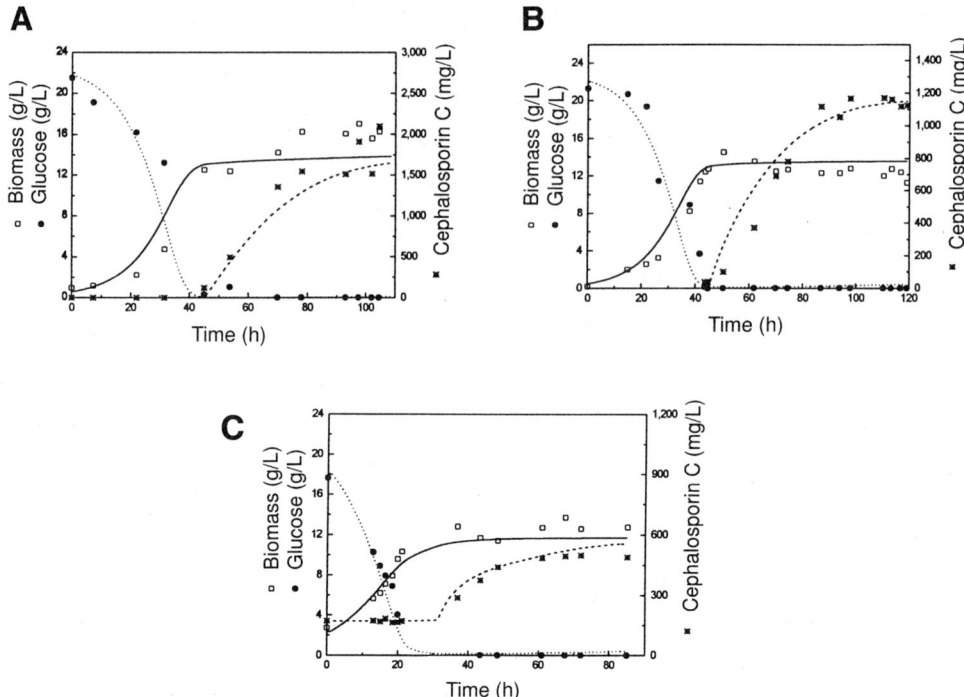

Fig. 5. Hybrid model fitting to the training data sets. **(A)** Run 2, fed batch with standard inoculum preparation; **(B)** run 3, fed batch with standard inoculum preparation; **(C)** run 5, repeated fed batch.

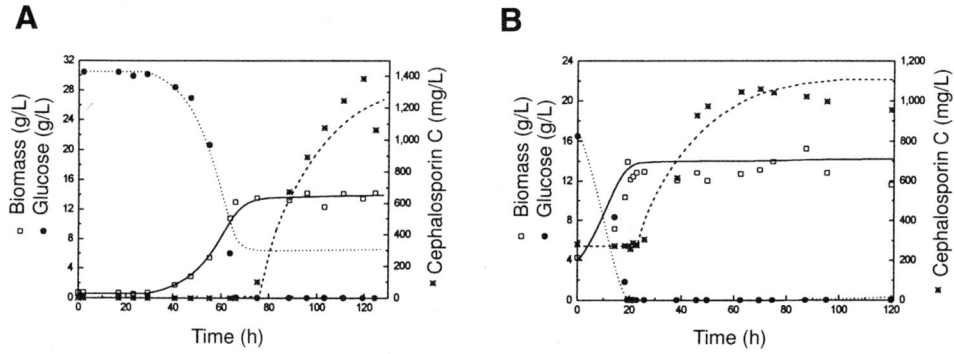

Fig. 6. Hybrid model validation test. **(A)** Run 1, fed batch with cell growth delay; **(B)** run 4, repeated fed batch.

Figure 6 displays the results of the validation tests. Run 4 was not used for the FNN training, and therefore it may be used as a validation set. The inference of cell mass and product concentration was very accurate in both cases. As for glucose concentration, an offset was observed for run 1 after 65 h. Actually, this is an expected result, since the FNN that describes μ_s had no previous information about the possibility of such a different yield coefficient occurring. Nevertheless, the overall performance of the hybrid model was very good with respect to cell and product concentrations. These are the key state variables to support the decision of interrupting the fermentation and are not measured on-line. Run 4, displayed a very good fitting for all variables.

Despite the offset in final glucose concentration observed in run 1, the hybrid model is superior to the white box approach. When using a first-principles model, such as the one described by Cruz et al. *(20)*, it would be necessary to reset the parameter $Y_{x/s}$ and to establish a time condition to trigger the beginning of the cellular growth. In other words, a conventional, white box model whose parameters were estimated from runs 2, 3, and 5 would not be able to simulate the trend of the bioreactor during run 1. It is important to note that the predictive capabilities of the hybrid model might be further enhanced if run 1 were added to the training set.

Conclusion

ANNs were used to estimate reaction rates from on-line data, which were then used in the conservation equations. Our results indicate that, despite the different lag phase duration that occurred for run 1, a hybrid neural network model was able to capture the complex dynamics of the cephalosporin C production bioprocess.

The hybrid model could predict with accuracy the depletion of glucose at the end of the trophophase and, therefore, may be used to trigger the beginning of the supplementary feed.

Our architecture was based on the modularization concept, in which one ANN provides information to the next one. A single ANN was not able to simultaneously infer cell growth, glucose consumption, and product formation rates.

Acknowledgments

We gratefully acknowledge the financial support of CAPES and CNPq and a scholarship from FAPESP (São Paulo State Foundation, Brazil).

References

1. Savidge, T. A. (1984), in *Biotechnology of Industrial Antibiotics*, vol. 22, Vandamme, E. J., Marcel Dekker, NY, pp. 171–224.
2. Matsumura, M., Imanaka, T., Yoshida, T., and Taguchi, H. (1981), *J. Ferment. Technol.* **59(2),** 115–123.
3. Chu, W. B. Z. and Constantinides, A. (1988), *Biotechnol. Bioeng.* **32,** 277–288.
4. Basak, S., Velayudhan, A., and Ladisch, M. R. (1995), *Biotechnol. Prog.* **11,** 626–631.
5. Araujo, M. L. G. C., Oliveira, R. P., Giordano, R. C., and Hokka, C. O. (1996), *Chem. Eng. Sci.* **51(11),** 2835–2840.
6. Thibault, J., Breusegem, V. V., and Chéruy, A. (1990), *Biotechnol. Bioeng.* **36,** 1041–1048.
7. Di Massimo, C., Montague, G. A., Willis, M. J., Tham, M. T., and Morris, A. J. (1992), *Comput. Chem. Eng.* **16(4),** 283–291.

8. Karim, M. N. and Rivera, S. L. (1992), *Comput. Chem. Eng. Suppl.* S369–S377.
9. Syu, M.-J. and Tsao, G. T. (1993), *Biotechnol. Bioeng.* **42,** 376–380.
10. Cruz, A. J. G., Araujo, M. L. G. C., Giordano, R. C., and Hokka, C. O. (1998), *Appl. Biochem. Biotecnol.* **70–72,** 579–592.
11. Warnes, M. R., Glassey, J., Montague, G. A., and Kara, B. (1998), *Neurocomputing* **20,** 67–82.
12. Psichogios, D. C. and Ungar, L. H. (1992), *AIChE J.* **38(10),** 1499–1506.
13. van Can, H. J. L., te Breake, H. A. B., Hellinga, C., Luyben, K. C. A. M., and Heijnen, J. J. (1997), *Biotechnol. Bioeng.* **54(6),** 549–566.
14. Rumelhart, D. E. and McClelland, J. L. (1986), in *Parallel Distributed Processing,* vol. 1, Massachusetts Institute of Technology, Cambridge, pp. 318–362.
15. Ruck, D. W., Rogers, S. K., Kabrisky, M., Maybeck, P. S., and Oxley, M. E. (1992), *IEEE Trans. Pattern Anal. Machine Intell.* **14(6),** 686–691.
16. Thompson, M. L. and Kramer, M. A. (1994), *AIChE J.* **40(8),** 1328–1340.
17. van Can, H. J. L., te Breake, H. A. B., Bijman, A., Hellinga, C., Luyben, K. C. A. M., and Heijnen, J. J. (1999), *Biotechnol. Bioeng.* **62(6),** 666–680.
18. Shen, Y.-Q., Wolfe, S., and Demain, A. L. (1986), *Bio/Technology* **4,** 61–63.
19. Demain, A. L., Newkirk, J. F., and Hendlin, D. (1963), *J. Bacteriol.* **85,** 339–344.
20. Cruz, A. J. G., Silva, A. S., Araujo, M. L. G. C., Giordano, R. C., and Hokka, C. O. (1999), *Chem. Eng. Sci.* **54,** 3137–3142.
21. Trinder, P. (1969), *Ann. Clin. Biochem.* **6,** 24.
22. Silva, A. S., Cruz, A. J. G., Araujo, M. L. G. C., and Hokka, C. O. (1998), *Braz. J. Chem. Eng.* **15(4),** 320–325.
23. Hernández, E. and Arkun, Y. (1992), *Comput. Chem. Eng.* **16(4),** 227–240.
24. Bhat, N. and McAvoy, T. J. (1990), *Comput. Chem. Eng.* **14(4/5),** 573–583.

Kinetics of Ethanol Fermentation with High Biomass Concentration Considering the Effect of Temperature

Daniel I. P. Atala,[1] Aline C. Costa,*[,2] Rubens Maciel,[2] and Francisco Maugeri[1]

[1]DEA/FEA/UNICAMP, Cx. Postal 6121, Campinas, SP, Brazil 13081-970;
and [2]DPQ/FEQ/UNICAMP, Cx. Postal 6066, Campinas, SP, Brazil 13081-970,
E-mail: accosta@feq.unicamp.br

Abstract

A model of ethanol fermentation considering the effect of temperature was developed and validated. Experiments were performed in a temperature range from 28 to 40°C in continuous mode with total cell recycling using a tangential microfiltration system. The developed model considered substrate, product and biomass inhibition, as well as an active cell phase (viable) and an inactive (dead) phase. The kinetic parameters were described as functions of temperature.

Index Entries: Ethanol fermentation; high biomass concentration; temperature.

Introduction

Brazil is one of the greatest ethanol producers in the world as the result of a political strategy initiated in 1975 by the government to cope with the sharp increase in oil prices. Programs in the United States in 1978 and, more recently in Canada, followed this strategy *(1)*. Because of the stabilization of petroleum prices at a low level most of the incentives to the alcohol industries were withdrawn and there was a great interest in the optimization of all the stages of the ethanol production process. Now, a new increase in petroleum prices and improvements in productivity attained in the industries have made alcohol production costs lower than that of gasoline, and, again, there is a good outlook for the alcohol industries in Brazil *(2)*. In addition, alcohol is a clean fuel, producing less harm to the environment.

Operation of the alcoholic fermentation process in a continuous mode is desirable, since higher productivity, improved yields, and better process

*Author to whom all correspondence and reprint requests should be addressed.

control are attained *(3)*. However, industrial implementation of a continuous process requires previous study of the process behavior and its use in the development of an efficient control strategy. The influence of temperature on the kinetic parameters must be considered, because it is usually difficult to support a constant temperature during large-scale alcoholic fermentation. The process is exothermic and small deviations in temperature (2–4°C) can dislocate the process from optimal operational conditions. Also, knowledge about the effect of temperature on ethanol fermentation kinetics can be useful in strategies for process optimization *(4)*.

One way of improving productivity in an ethanol production process is to increase cell concentration. Several methods have been proposed to obtain high cell densities in continuous cultures. Cell immobilization *(5)* and cell recycling are common methods for this purpose. Cell recycling by settling *(6)* requires flocculent yeast strains. Despite being the method used in Brazilian industries *(1)*, cell recycling by centrifugation has some disadvantages because aseptic conditions are difficult to achieve and the process is usually too expensive *(7)*. Membrane filter systems have been studied in recent years and seem to be a good alternative to cell recycling. The membrane module can be located inside *(8)* or outside the reactor *(7,9–11)*. The location outside the reactor facilitates washing, cleaning, and regeneration of membranes without interfering with fermentation *(10)*.

In the present study, a mathematical model was developed to describe the alcoholic fermentation process kinetics taking into account the effect of temperature on kinetic parameters. To achieve this goal, experiments were performed at temperatures between 28 and 40°C in a system with cell recycling by tangential microfiltration. The substrate was sugarcane molasses.

Materials and Methods

Microorganism

The yeast used was *Saccharomyces cerevisiae* cultivated in the Bioprocess Engineering Laboratory in the Faculty of Food Engineering/UNICAMP and obtained from an industrial fermentation plant.

Culture Medium

The growth medium for the inoculum consisted of 50 kg/m^3 of glucose, 5 kg/m^3 of KH_2PO_4, 1.5 kg/m^3 of NH_4Cl, 0.7 kg/m^3 of $MgSO_4 \cdot 7H_2O$, 1.2 kg/m^3 of KCl, and 5 kg/m^3 of yeast extract. The production medium was diluted sugarcane molasses to which 1 kg/m^3 of yeast extract and 2.4 kg/m^3 of $(NH_4)_2SO_4$ were added. Sterilization was done at 121°C for 30 min.

Fermentation System and Operation

The experimental equipment is shown in Fig. 1. The tangential microfiltration system (Ceraflow model; Millipore) was coupled to a

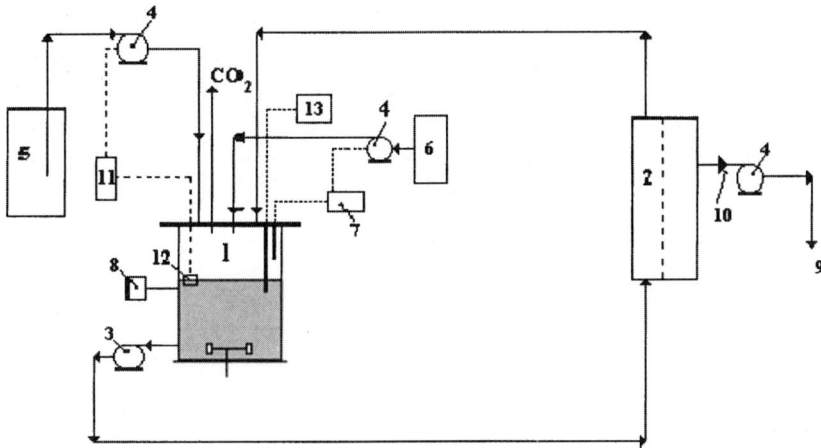

Fig. 1. Fermentation system: 1, fermentor; 2, tangential microfilter; 3, helical pump; 4, peristaltic pumps; 5, feed tank; 6, antifoam tank; 7, foam sensor; 8, temperature indicator; 9, filtrate outlet; 10, reduction pressure valve; 11, level controller; 12, level sensor; 13, turbidimeter.

0.0025-m^3 bioreactor (Bioflow III System; New Brunswick Scientific). Cell broth was recirculated into the microfilter by a helical pump (Netsch 2.NEL.20 A). A reduction pressure valve was installed at the filtrate outlet to avoid outlet rate fluctuation and a turbidimeter was used to monitor cell growth.

Sterilization

Filters were washed with 2 *N* NaOH solution at 40°C for 4 h and then rinsed with water for 4 h. Sterilization was done by circulating saturated steam into the complete system for 2 h.

System Operation

After inoculation, fermentation was operated in batch mode until the end of the exponential phase, monitored by CO_2 production. At this point, the continuous fermentation with total cell recycling was started, by adjusting a peristaltic pump to remove the filtrate in a flow rate corresponding to a dilution rate of 0.1 h^{-1}. Fresh medium was continuously supplied to the fermentor by a peristaltic pump connected to an on–off level controller, so that the fermentor volume was maintained constant. The experiment was finished when the filtrate flow rate could not be maintained constant owing to the decrease in the filtration capacity of the system.

Analytical Methods

Dry cell mass was determined gravimetrically after centrifuging, washing, and drying the cells at 105°C. Viable cells were counted with the methylene blue staining technique *(12)*. Total reducing sugar (TRS) and ethanol concentrations were determined by high-performance liquid chromatography (Varian 9010 model). A column SHODEX KS 801 at 70°C was used. Ultrapure Milli-Q water was used as the eluent at a flow rate of 0.5 mL/min. The standards were mixed solutions of sucrose, fructose, and ethanol at concentrations from 0.1 to 40 kg/m³. The software Millennium® v.2.1 was used for integration and quantification.

Results

Experiments

The substrate was fed at a high rate in all the experiments (about 25 kg/[m³·h] of TRS), so that high substrate and ethanol concentrations were maintained in the fermentor. Then, the kinetics of the fermentation was studied in highly stressing conditions for the microorganism, which enabled the study of inhibition by biomass, substrate, and product.

Figures 2 to 6 show the results of the experiments performed at 28, 31, 34, 37, and 40°C, respectively. They all show the results of the mathematical model developed in the next session. Concentrations of total and viable biomass, substrate (TRS) and ethanol are plotted against time during the operation with total cell recycling. The substrate and ethanol concentrations remained at about a constant value almost since the beginning of the operation. The biomass concentration (total and viable), however, increased with time and, at the end, tended to stabilize near a fixed value. This finding is in agreement with that of other researchers (10), who observed that a pseudo-steady-state biomass concentration was reached in cultures with total cell recycling. Melzoch et al. (10) showed that this is not caused by product inhibition, because even at a low product concentration a constant biomass concentration was achieved.

Comparison of the results from the different experiments is shown in Table 1. Productivity showed a maximum at 31°C and then decreased with temperature; conversion (defined as $100 \times [S_F - S]/S_F$) always decreases with temperature. The highest biomass concentrations were obtained at the lowest temperatures (28 and 31°C) and then diminished with temperature, presenting a very low value at 40°C. Final viability (defined as X_V/X_T) remained constant at about 60% at the three first temperatures studied and then increased at 37 and 40°C. This increase in viability is probably owing to the lower biomass and ethanol concentrations when compared to the experiments at lower temperatures.

The biomass concentrations achieved were low when compared with the values obtained by other investigators. Lafforgue et al. (9) obtained 300 kg/m³ in a continuous fermentation with cells recycling by tangential microfiltration. Lee and Chang (7) reached 210 kg/m³ using a cell recycle system with a hollow-fiber membrane filter. The biomass concentration attained by Melzoch et al. (10) was nearer to that obtained in the present work (60 kg/m³ of biomass using a separation module composed of ultrafiltration tubular membranes). One of the reasons for the low biomass con-

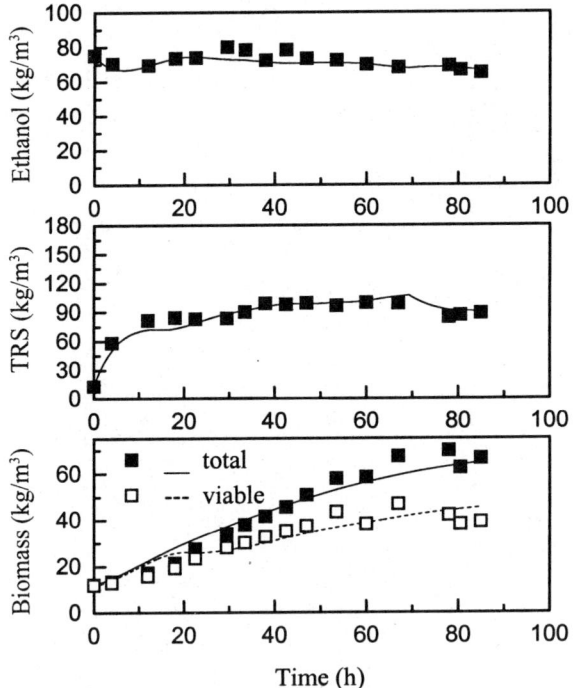

Fig. 2. Experimental (■, □) and modeling (——; – – –)results at 28°C.

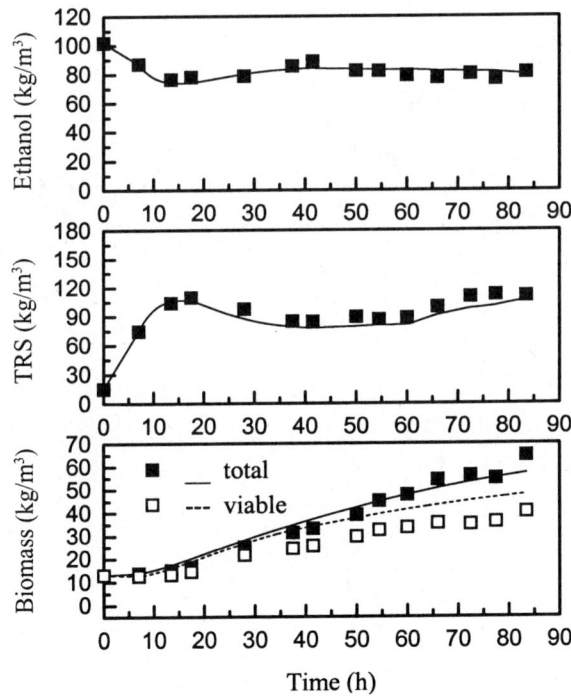

Fig. 3. Experimental (■, □) and modeling (——; – – –) results at 31°C.

Atala et al.

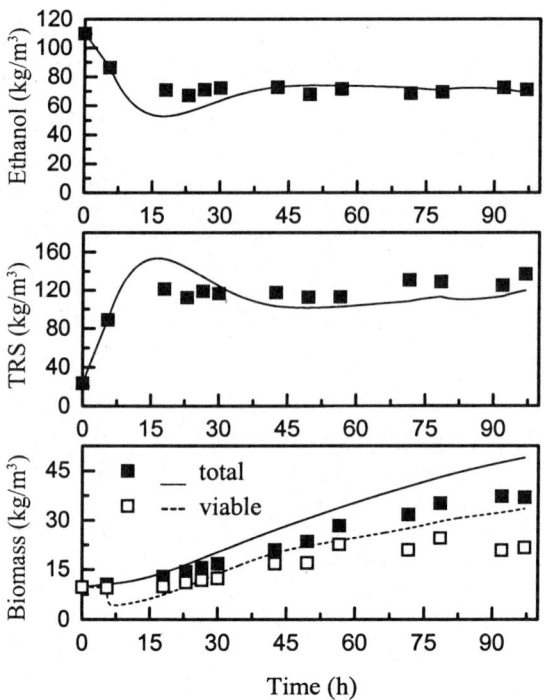

Fig. 4. Experimental (■, □) and modeling (——; − − −) results at 34°C.

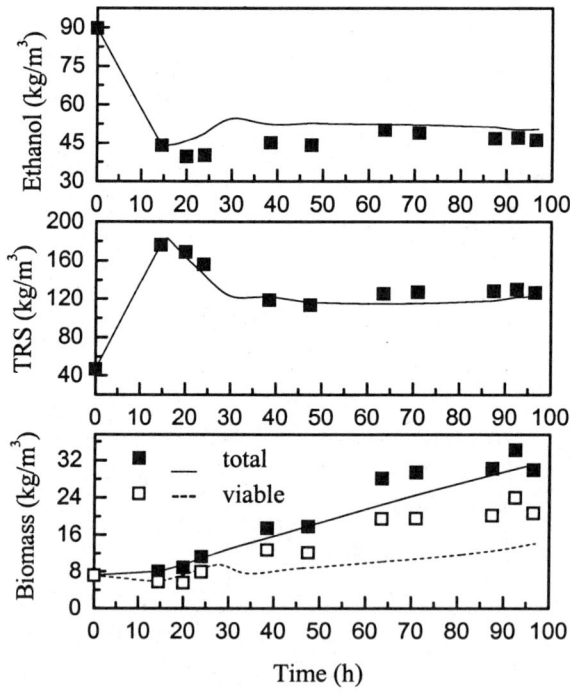

Fig. 5. Experimental (■, □) and modeling (——; − − −) results at 37°C.

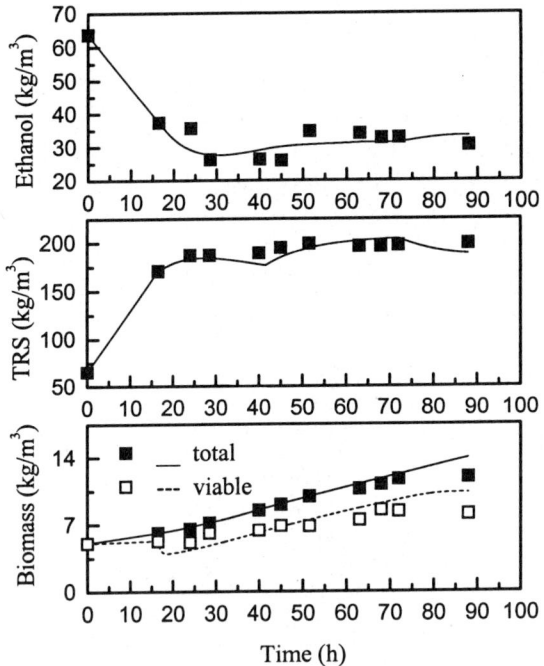

Fig. 6. Experimental (■, □) and modeling (——; – – –) results at 40°C.

Table 1
Comparison of Experimental Results for Different Temperatures

	28°C	31°C	34°C	37°C	40°C
Conversion (%)	60–75	59–67	49–58	29–50	23–30
Productivity (kg/[m³·h])	6.5–7.5	7.1–8.4	6.5–7.2	4.1–4.7	2.5–3.5
Final total biomass (kg/m³)	65–70	65	37	30	8
Viability (%)	60	62	59	70	67

centrations obtained could be the use of sugarcane molasses as substrate, since all the other works cited used synthetic media. It is well known that sugarcane molasses, although a good substrate, has some components that can act as inhibitors for ethanol fermentation. In this work, the objective was to study fermentation kinetics under industrial conditions, so only sugarcane molasses was used as substrate.

Mathematical Modeling

The following approximations were made for the development of an unstructured mathematical model for the alcoholic fermentation process: the fermentor volume is constant, the bubble volume is negligible, and the microfilter operation is ideal (i.e., the filtrate stream contains no cells).

Monbouquette *(13)* showed that the use of models developed from low biomass concentration cultures leads to errors if they are extrapolated to high cell concentrations. He emphasized the need to use intrinsic models (models that take into account cell volume fraction) if the volume fraction occupied by biomass is >10%.

In agreement with the results obtained by many researchers *(9,14)*, the experimental data showed that there is a loss of cell viability during fermentation. Taking this into account, it is assumed that the total biomass comprises a viable (active) phase, X_v, and an inactive (dead) phase, X_d.

The mass balance equations using the intrinsic model are as follows:

$$\frac{d(X_v V)}{dt} = V r_x - V r_d \tag{1}$$

$$\frac{d(X_d V)}{dt} = V r_d \tag{2}$$

substrate:
$$\frac{d\left[\left(1 - \frac{X_t}{\rho}\right) S V\right]}{dt} = F(S_F - S) - V r_s \tag{3}$$

product:
$$\frac{d\left[\left(1 - \frac{X_t}{\rho}\right) P V + \frac{X_t}{\rho} \gamma P V\right]}{dt} = V r_p - F P \tag{4}$$

In Eqs. 3 and 4, ρ is the ratio of dry cell weight per wet cell volume and γ is the ratio of concentration of intracellular to extracellular ethanol. Assuming constant volume and simplifying, the following equations are obtained:

viable cells:
$$\frac{dX_v}{dt} = r_x - r_d \tag{5}$$

dead cells:
$$\frac{dX_d}{dt} = r_d \tag{6}$$

substrate:
$$\frac{dS}{dt} = \frac{D(S_F - S) - r_s + (r_x/\rho)/S}{1 - (X_t/\rho)} \tag{7}$$

product:
$$\frac{dP}{dt} = \frac{r_p - dP + \frac{(1-\gamma)}{\rho} r_x P}{1 - (X_t/\rho) + (\gamma/\rho)/X_t} \tag{8}$$

Table 2
Kinetic Parameters as Functions of Temperature (°C)

Parameter	Expression or value
μ_{max}	$1.57 \exp(\dfrac{-41.47}{T}) - 1.29 \times 10^4 \exp(\dfrac{-431.4}{T})$
X_{max}	$-0.3279 \times T^2 + 18.484 \times T - 191.06$
P_{max}	$-0.4421 \times T^2 + 26.41 \times T - 279.75$
Y_x	$2.704 \exp(-0.1225 \times T)$
Y_{px}	$0.2556 \exp(0.1086 \times T)$
K_s	4.1
K_i	$1.393 \times 10^{-4} \exp(0.1004 \times T)$
m_p	0.1
m_x	0.2
m	1
n	1.5
K_{dP}	$7.421 \times 10^{-3} \times T^2 - 0.4654 \times T + 7.69$
K_{dT}	$4.10^{13} \exp\left[-\dfrac{41{,}947}{1.987 \times (T + 273.15)} \right]$
ρ	390
γ	0.78

Many different models have been proposed to describe the kinetic rates of ethanol fermentation. Herein, the growth rate was described by

$$r_x = \mu_{max} \frac{S}{K_s + S} \exp(-K_i S) \left(1 - \frac{X_t}{X_{max}}\right)^m \left(1 - \frac{P}{P_{max}}\right)^n X_v \qquad (9)$$

which is similar to the equation proposed by Jarzebski et al. *(14)* plus a term to describe substrate inhibition.

The kinetic rates of death, ethanol formation, and substrate consumption are as follows:

$$r_d = [K_{dT} \exp(K_{dP}P)] X_v \qquad (10)$$

$$r_p = Y_{px} r_x + m_p X_v \qquad (11)$$

$$r_s = (r_x / Y_x) + m_x X_v \qquad (12)$$

The kinetic parameters were adjusted as functions of the temperature from the experimental data and are shown in Table 2. Equations 5–12 were solved using a Fortran program with integration by an algorithm based on the fourth-order Runge-Kutta method. A comparison of the results of the mathematical model and the experimental is presented in Figs. 2–6. The proposed model described well the dynamic behavior of the alcoholic fermentation.

Table 3
RSD of the Average of Experimental Values

Temperature (°C)	RSD X_t (%)	RSD X_V (%)	RSD ART (%)	RSD ethanol (%)
28	9.4	13.3	6.6	4.5
31	8.8	26.6	8.3	3.5
34	27.9	·34.0	15.6	10.0
37	13.3	43.9	5.6	10.7
40	7.5	14.8	3.9	9.0

The quality of the prediction of the model can be characterized using the residual standard deviation (RSD), Eq. 13, which provides an indication of the accuracy of the prediction, as suggested by Cleran et al. *(15)*:

$$RSD = \frac{\sqrt{\sum_{i=1}^{n} (y_i - y_{pi})^2}}{n} \qquad (13)$$

in which y_i is the experimental value, y_{pi} is the value predicted by the mathematical model, and n is the number of experimental points.

Because the magnitude of the RSD will vary depending on the magnitude of the variable to be predicted, it is easier to analyze the RSD written as a percentage of the average of the experimental values $\bar{y}_i$:

$$RSD(\%) = \frac{RSD}{\bar{y}_i} \times 100 \qquad (14)$$

The results are shown in Table 3. The deviations are from 3.5% to more than 40%. In general, the model does not predict well the process at 34°C and the data for X_V at all the temperatures. The bad results for X_V are mainly owing to the difficulty in finding a function to describe parameter K_{dP} in all of the studied temperature range. Deviations below 10% can be considered acceptable regarding bioprocess engineering. Therefore, the model can be used to predict system performance and to design process controllers.

Discussion

As reported by Jarzebski et al. *(14)*, at high yeast cell concentrations, conditions for growth are less favorable owing to hindered access to nutrients, space limitations, and cell interaction. These facts, together with the prolonged residence time of cells, lead to a kinetic behavior different from that usually encountered, and an adequate mathematical model must take these factors into account. A model was developed by taking into account the loss of viability during fermentation and the volume fraction occupied by biomass, as proposed by Monbouquette *(13)*.

The experimental data have shown that the alcoholic fermentation kinetics is strongly dependent on temperature. It was shown that the increase in temperature decreases productivity, conversion, and final biomass concentration. These values at 40°C are extremely low: conversion is 23–30%, productivity is 2.5–3.5 kg/(m³·h), and final biomass concentration is about 8 kg/m³. The effect of temperature on viability is not so clear, because it increased at 37 and 40°C when it was expected that the increase in temperature would make it decrease. This increase is probably owing to the low biomass and ethanol concentrations attained in the experiments performed at these temperatures.

Because the experimental data show that temperature has a strong influence on the kinetics of the process, in the mathematical model developed in this work the kinetic parameters are described as functions of temperature. The proposed model was shown to describe the experimental data well.

Cell viability decreased greatly with an increase in fermentation time, reaching values of about 60 and 70% at the end of the experiments. This can be explained by the highly stressing conditions to the microorganism in the experiments, owing to high substrate, ethanol, and biomass concentrations and to the friction forces in the recirculation pump and filters. Higher viability could be obtained if, instead of total cell recycling, partial recycling were used. The use of a purge permits cell renovation and the withdrawal of secondary products accumulated in the fermentor (which could be toxic to the microorganism).

The biomass concentrations were not as high as those reported by some other researchers *(7,9)*, probably because molasses was used as the substrate. Although the experiments were interrupted owing to the decrease in the system's filtration capacity, experimental data have shown that in all the cases the biomass concentration had stabilized near a constant value, which suggests that even if higher-capacity filters were used the attained biomass concentration would not increase.

Dilution rate and feed substrate concentration are important parameters in ethanol fermentation. Their influence on the final biomass concentration should be studied in order to determine the possibility of increasing biomass concentration, conversion, and yields. This study can be made experimentally or by computer simulation using the mathematical model developed herein.

Acknowledgments

We acknowledge Fundação de Amparo à Pesquisa do Estado de São Paulo (process number 98/09198-6) and Conselho Nacional de Desenvolvimento Científico e Tecnológico for financial support.

Nomenclature

$D = F/V$ = dilution rate (h^{-1})
F = substrate feed flow rate (m^3/h)
K_{dP} = coefficient of death by ethanol (m^3/kg)
K_{dT} = coefficient of death by temperature (h^{-1})
K_i = substrate inhibition coefficient (m^3/kg)
K_s = substrate saturation constant (kg/m^3)
m = constant in Eq. 5
m_p = ethanol production associated with growth (kg/[kg·h])
m_x = maintenance coefficient (kg/[kg·h])
n = constant in Eq. 5
P = product concentration (kg/m^3)
P_{max} = product concentration when cell growth ceases (kg/m^3)
r_d = kinetic rate of death (kg/[m^3·h])
r_p = kinetic rate of ethanol formation (kg/[m^3·h])
r_s = kinetic rate of substrate consumption (kg/[m^3·h])
r_x = kinetic rate of growth (kg/[m^3·h])
S = substrate concentration (kg/m^3)
S_F = feed substrate concentration (kg/m^3)
V = reactor volume (m^3)
X_d = dead biomass concentration (kg/m^3)
X_{max} = biomass concentration when cell growth ceases (kg/m^3)
$X_t = X_v + X_d$ = total biomass concentration (kg/m^3)
X_v = viable biomass concentration (kg/m^3)
Y_{px} = yield of product based on cell growth (kg/kg)
Y_x = limit cellular yield (kg/kg)
γ = ratio of concentration of intracellular to extracellular ethanol
μ_{max} = maximum growth rate (h^{-1})
ρ = ratio of dry cell weight per wet cell volume (kg/m^3)

References

1. Wheals, A. E., Basso, L. C., Alves, D. M. G., and Amorim, D. M. G. (1999), *Tibtech.* **17,** 482–487.
2. Furtado, M. R. (1999), *Química Derivados* 8–17.
3. Kargupta, K., Datta, S., and Sanyal, S. K. (1998), *Biochem. Eng. J.* **1,** 31–37.
4. Kalil, S. J., Maugeri, F., and Rodrigues, M. I. (2000), *Process Biochem.* **35,** 539–550.
5. Dale, M. C., Chen, C., and Okos, M. R. (1990), *Biotechnol. Bioeng.* **36,** 983–992.
6. Ghose, T. K. and Tyagi, R. D. (1979), *Biotechnol. Bioeng.* **21,** 1401–1420.
7. Lee, C. W. and Chang, H. N. (1987), *Biotechnol. Bioeng.* **29,** 1105–1112.
8. Chang, H. N., Lee, W. G., and Kim, B. S. (1993), *Biotechnol. Bioeng.* **41,** 677–681.
9. Lafforgue, C., Malinowski, J., and Goma, G. (1987), *Biotechnol. Lett.* **9,** 347–352.
10. Melzoch, K., Rychtera, M., Markivichov, N. S., Pospíchalová, V., Basařová, G., and Manakov, M. N. (1991), *Appl. Microbiol. Biotechnol.* **34,** 469–472.
11. Lafforgue-Delorme, C., Delorme, P., and Goma, G. (1994), *Biotechnol. Lett.* **16,** 741–746.
12. Lee, S. S., Robinson, F. M., and Wang, H. Y. (1981), *Biotechnol. Bioeng. Symp.* **11,** 641–649.

13. Monbouquette, H. G. (1992), *Biotechnol. Bioeng.* **39,** 498–503.
14. Jarzebski, A. B., Malinowski, J. J., and Goma, G. (1989), *Biotechnol. Bioeng.* **34,** 1225–1230.
15. Cleran, Y., Thibault, J., Chéruy, A., and Corrieu, G. (1991), *J. Ferment. Bioeng.* **71,** 356–362.

Fermentation of Xylose into Acetic Acid by *Clostridium thermoaceticum*

Niru Balasubramanian, Jun Seok Kim, and Y. Y. Lee*

Department of Chemical Engineering, 230 Ross Hall, Auburn University, Auburn, AL 36849, E-mail: yylee@eng.auburn.edu

Abstract

For optimum fermentation, fermenting xylose into acetic acid by *Clostridium thermoaceticum* (ATCC 49707) requires adaptation of the strain to xylose medium. Exposed to a mixture of glucose and xylose, it preferentially consumes xylose over glucose. The initial concentration of xylose in the medium affects the final concentration and the yield of acetic acid. Batch fermentation of 20 g/L of xylose with 5 g/L of yeast extract as the nitrogen source results in a maximum acetate concentration of 15.2 g/L and yield of 0.76 g of acid/g of xylose. Corn steep liquor (CLS) is a good substitute for yeast extract and results in similar fermentation profiles. The organism consumes fructose, xylose, and glucose from a mixture of sugars in batch fermentation. Arabinose, mannose, and galactose are consumed only slightly. This organism loses viability on fed-batch operation, even with supplementation of all the required nutrients. In fed-batch fermentation with CSL supplementation, D-xylulose (an intermediate in the xylose metabolic pathway) accumulates in large quantities.

Index Entries: Xylose; fermentation; *Clostridium thermoaceticum*; acetic acid.

Introduction

Acetic acid is an important feedstock for many chemicals including vinyl acetate polymer, cellulose acetate, terephthalic acid/dimethyl terephthalate, acetic acid esters, acetic anhydride, and calcium magnesium acetate. At present these products are made from petroleum-derived acetic acid (1). Fermentation is potentially a cost-effective alternative for acetic acid production. Production of acetic acid via fermentation using renewable biomass feedstock has been studied extensively since the late 1970s (2–4). The cellulose and hemicellulose in lignocellulosic biomass are the two most abundant renewable sources of carbon for fermentation to industrially

*Author to whom all correspondence and reprint requests should be addressed.

useful chemicals. Efficient utilization of these constituents is vital to the development of economically viable bioconversion processes. Homoacetate anaerobic organisms such as *Clostridium thermoaceticum* convert glucose and xylose to acetic acid with a theoretical weight yield of 100% (5–7).

Fermentation of glucose to acetic acid by the modified *C. thermoaceticum* strain (ATCC 49707) has been extensively studied (8,9). However, its ability to ferment xylose into acetic acid is unknown. In this article, we report on the characteristics of acetic acid production by this organism using xylose as the carbon source.

Materials and Methods

Microorganism and Growth Media

A modified/mutant strain of *C. thermoaceticum* registered as ATCC 49707 (10) and renamed *Moorella thermoacetica* was used. The culture was grown in Difco Reinforced Clostridial medium at 59°C. It was maintained in the active state by transferring it alternately between this medium and medium containing 3% sodium acetate. To acclimatize this strain to using xylose as the principal sugar source, it was transferred to the fermentation medium described in the next section for 48 h for three generations alternating with growth in Clostridial broth. This adapted strain was stored at 4°C for future fermentation runs using xylose as the principal sugar source.

Fermentation Medium

The fermentation medium, similar to that used by Ljungdahl (2), contained the following: 1.0 g/L of $(NH_4)_2SO_4$, 0.25 g/L of $MgSO_4 \cdot 7H_2O$, 0.04 g/L of $Fe(NH_4)_2(SO_4)_2 \cdot 6H_2O$, 0.00024 g/L of $NiCl_2 \cdot 6H_2O$, 0.00029 g/L of $ZnSO_4 \cdot 7H_2O$, 0.000017 g/L of Na_2SeO_3, 0.25 g/L of cysteine·$HCl \cdot H_2O$, 5 g/L of yeast extract or corn steep liquor (CSL) (variable), 7.5 g/L of KH_2PO_4, 4.4 g/L of K_2HPO_4, 0.415 g/L of NaOH, 5 g/L of $NaHCO_3$, and xylose (variable) with resazurin to detect trace amounts of oxygen.

The medium was prepared in five parts and sterilized separately at 121°C for 20 min:

1. Xylose.
2. Yeast extract or CSL.
3. Cysteine·HCl.
4. Mineral solution with resazurin.
5. Buffer (KH_2PO_4, K_2HPO_4, NaOH, $NaHCO_3$).

Fermentation

All fermentation experiments were conducted in a New Brunswick Bioflo model C-30 bioreactor at 59°C with a working volume of 400 mL of the previously described fermentation medium. Anaerobic environment was achieved by sparging filtered CO_2 until the oxygen indicator resazurin changed from pink to colorless and then was maintained by supplying CO_2

in the headspace of the reactor. The pH was maintained at 6.7–6.9 with 8 N NaOH so that there would be no appreciable change in working liquid volume. Fermentation was initiated by transferring 7 mL of 24-h xylose adapted inoculum to the reactor medium. Xylose concentration and amount of CSL added were varied in batch and fed-batch experiments. The yield of acetic acid to xylose was based on consumed xylose.

Analytical Methods

The fermentation samples were analyzed for sugar and acetic acid by high-performance liquid chromatography (HPLC) (Water Associates) equipped with an RI detector. Bio-Rad's HPX-87H column was used at 65°C with 0.005 M H_2SO_4 as the mobile phase at a flow rate of 0.6 mL/min. For mixed sugars, the substrate profiles were analyzed using a Bio-Rad HPX-87-P column operated at 85°C with deionized water as the mobile phase and a flow rate of 0.55 mL/min.

The cell density of the fermentation medium was measured by a turbidimeter (Hach Model 2100N) *(11)*. The data on nephelometric turbidity units were calibrated with four different Formazin standards prior to use.

Results and Discussion

The original strain of *C. thermoaceticum* ATCC 49707 was maintained through growth in medium with glucose as the only carbon source. On transfer to a medium containing a mixture of glucose and xylose, it consumed xylose first before consuming glucose. However, xylose uptake was very slow. With subsequent transfers in xylose medium, the rate of utilization increased. This culture was stored at 4°C and used in all fermentation experiments.

Profiles in Xylose Fermentation

Figure 1 shows typical batch fermentation profiles of cell growth, xylose utilization, and acetic acid production through fermentation at pH 6.9 and 59°C by *M. thermoacetica* (ATCC 49707). A lag phase was observed for the first 20 h, after which an exponential growth phase occurred for about 60 h, as depicted in Fig. 2. Almost all the xylose consumption and acetic acid production occurred during the log phase, indicating a growth-associated acid production. Cell numbers decreased after the log phase of growth, indicating that there was an autolytic decay of cells.

Batch fermentation experiments were conducted over a range of initial xylose concentrations to find the optimal initial sugar concentration for acetic acid production. The data indicated that a concentration of 15 g/L resulted in a maximum yield of acetic acid at 0.84 (g of acetic acid/g of xylose consumed). The maximum concentration of product was 15.2 g/L, which occurred with a 20 g/L xylose concentration with a yield of 76%. With increases in xylose concentration, the amount of unconsumed xylose

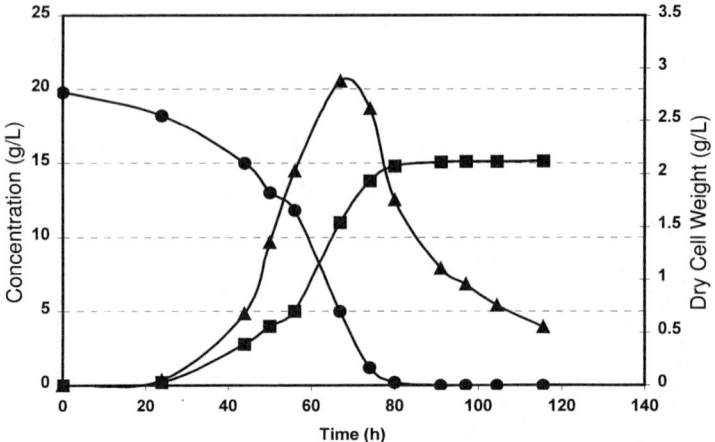

Fig. 1. Batch fermentation profile of *C. thermoaceticum*. Fermentation conditions: 59°C, pH 6.8. (—●—), xylose; (—■—), acetate; (—▲—), cell dry weight.

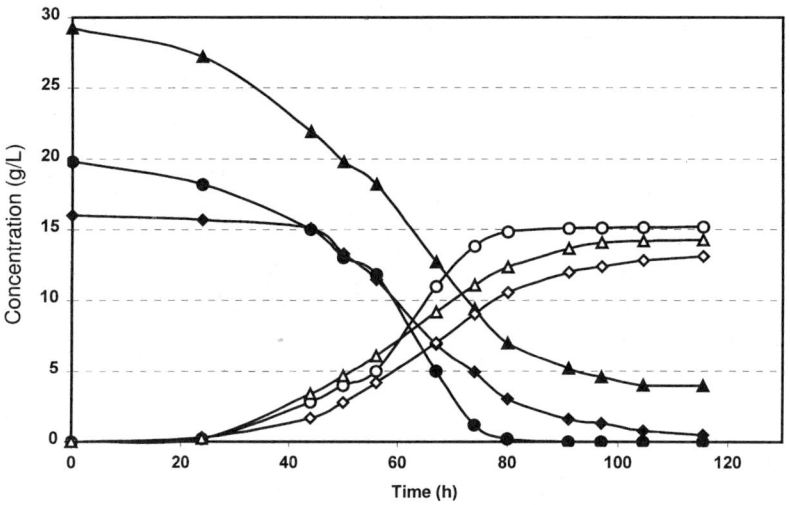

Fig. 2. Effect of initial xylose loading on the acetic acid yield. Fermentation conditions: 59°C, pH 6.8. (—◇—), acetic acid (1.5% xylose); (—○—), acetic acid (2% xylose); (—△—), acetic acid (3% xylose); (—◆—), 1.5% xylose; (—●—), 2% xylose; (—▲—), 3% xylose.

in the medium increased, which decreased the yield. The effects of initial xylose are summarized in Table 1 and Fig. 2. Subsequent fermentation experiments were conducted using an initial xylose concentration of 15–20 g/L.

Effect of CSL as Nitrogen Source

One obstacle to successfully commercializing this bioconversion process is the high cost of nutrients, such as yeast extract, required by *C. thermo-*

Table 1
Batch Fermentation of Xylose into Acetic Acid by *C. thermoaceticum* ATCC 49707 at 59°C, pH 6.8

| Initial loading (g/L) | Xylose consumed (%) | Acetic acid | | Nutrient | | Fermentation time (h) |
		Productivity (g/[L·h])	Yield (g acid/g xylose)	Yeast extract (g/L)	CSL (g/L)	
16	95	12.78	0.84	5	—	100
20	100	15.13	0.76	5	—	100
29.23	84.3	14.16	0.574	5	—	100
19.4	90	10	0.557	—	10	100
19.4	97.5	14.94	0.79	—	20	100
19.4	100	16.57	0.85	—	25	100

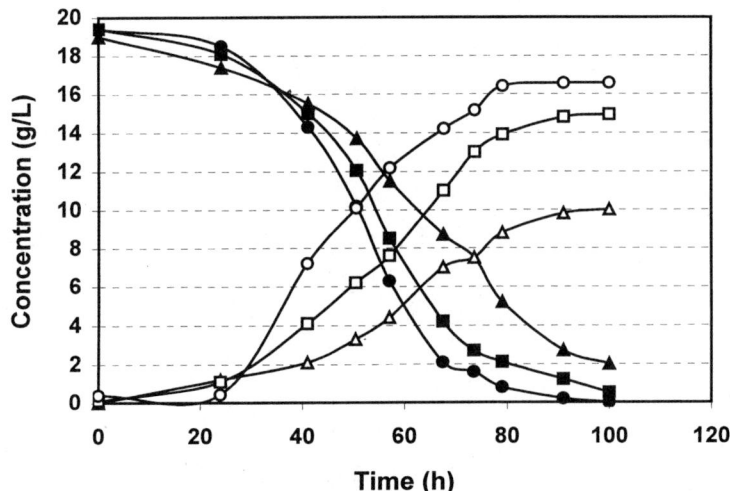

Fig. 3. Effect of concentration of CSL on xylose utilization. Fermentation conditions: 59°C, pH 6.8. (—△—), acetic acid (10 g/L CSL); (—□—), acetic acid (20 g/L CSL); (—○—), acetic acid (25 g/L CSL); (—▲—), 10 g/L CSL; (—■—), 20 g/L CSL; (—●—), 25 g/L CSL.

aceticum. CSL *(12)* has been identified as an inexpensive nitrogen-rich nutrient source *(4,13)*.

CSL, a byproduct of wet milling of corn, is a rich source of amino acids, minerals, and vitamins. It also contains other nitrogen compounds useful for microbial growth *(14)*. CSL has been used as medium for industrial production of penicillin *(15)*. Using this nutrient source, instead of yeast extract, can reduce the fermentation cost significantly *(16)*. Several experiments were performed to estimate the amount of CSL that would be required to obtain a yield of acetic acid comparable with that obtained using yeast extract. The results in Fig. 3 show that the fermentation profile with CSL is similar to that with yeast extract. When the concentration of CSL was 25 g/L and initial xylose loading was 20 g/L, the final concentration and yield of acetic acid were 16.57 g/L and 0.84 (g of acetic acid/g of xylose consumed), respectively. The profiles with varying CSL also are presented in Table 1.

Consumption of Mixture of Sugars

In a medium containing a mixture of glucose, xylose, galactose, fructose, arabinose, and mannose, *C. thermoaceticum* consumes fructose first and then xylose, as shown in Fig. 4. However, the rate of fructose consumption was faster than the rate of xylose consumption. Glucose was the third sugar to be utilized as the carbon source. When xylose, fructose, and glucose were completely consumed, the organism appeared to utilize arabinose, mannose, and galactose in that order, but at an extremely slow rate.

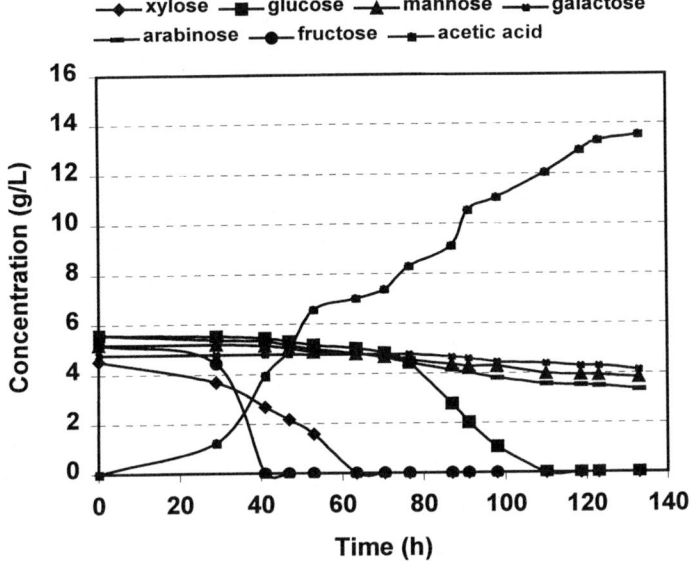

Fig. 4. Batch fermentation of a mixture of sugars into acetic acid. Fermentation conditions: 57°C, pH 6.8.

Fed-Batch Operation

Considering that the yield is higher at low sugar and high nitrogen source, a fed-batch mode of operation was perceived as a way to enhance yield by maintaining the optimal conditions in the reactor. Therefore, three sets of experiments were performed in which the xylose concentration was maintained at 15–20 g/L. The results are given in Table 2 and Fig. 5.

In experiment 1, adding 8 g of xylose led to slightly increased cell viability after the initial log phase. After 126 h of fermentation, however, there was no further uptake of xylose. In experiment 2, the same trend was observed despite adding 8 g of CSL to the 400 mL of medium. It is suspected that cell death could have resulted owing to lack of mineral supplementation. In experiment 3, mineral solution and cysteine-HCl were therefore added in addition to 8 g of CSL and 6 g of xylose. This strategy worked only up to 95 h of fermentation, a slight improvement over the previous set of experiments, in which growth ceased after 80 h. The organism produced acetic acid at the same rate as in the initial log phase of growth. However, further addition of sugar or nutrients did not increase the growth rate, sugar consumption, or acid yield. The organism could not be revitalized after the initial 80 to 90-h period. Accumulation of an intermediate product was also detected in experiment 3. This substance was identified to be D-xylulose using a pure standard in HPLC. It is an intermediate in the xylose metabolism of most bacterial systems formed by the action of xylose isomerase on xylose.

Table 2
Fed-Batch Fermentation of Xylose into Acetic Acid by C. thermoaceticum ATCC 49707 at 59°C, pH 6.8

| Volume | | Xylose | | | Acetic acid | | | |
Initial (mL)	Final (mL)	Initial (g/L)	Total (g/L)	Consumed (%)	Productivity (g/[L·h])	Yield (g/g xylose)	Time (h)	Additional nutrients
400	405	18.9	15.56	57	19.6	0.7	120	Xylose
400	420	18	15.2	52	17.48	0.7	120	Xylose + CSL
400	460	19.7	29.88	74	17.66	0.36	120	Xylose + CSL + minerals

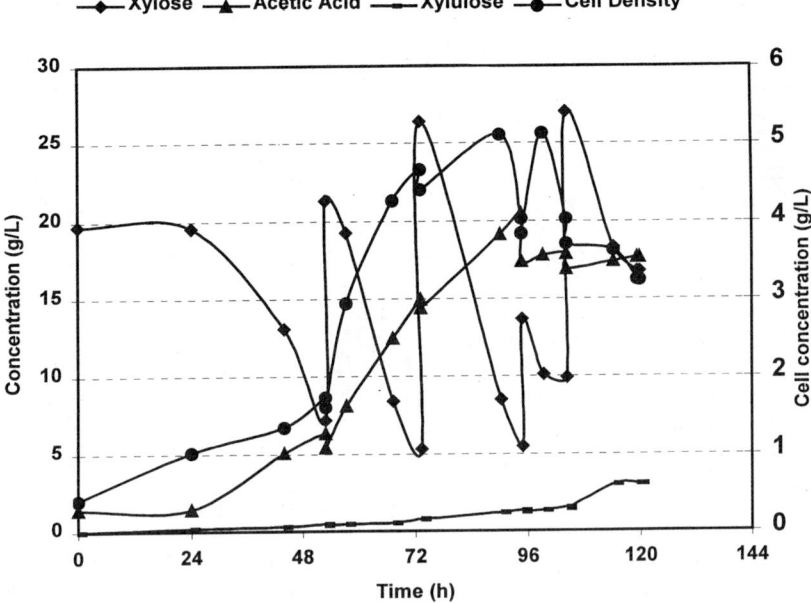

Fig. 5. Fermentation profile under the fed-batch mode of operation. Fermentation conditions: 59°C, pH 6.8. Experiment 3: total addition of 32 g of xylose, 16 g of CSL, and trace salts.

Conclusion

C. thermoaceticum needs to be acclimatized to a xylose environment to obtain high yields of acetic acid. It preferentially consumes xylose over glucose when grown in a medium containing a mixture of glucose and xylose. To maintain viability for xylose fermentation, it is necessary to grow the organism in xylose and glucose medium alternately. In a 20 g/L xylose medium containing 5 g/L of yeast extract, fermentation to acetic acid occurs within 80 h, resulting in a final acetate concentration of 15.2 g/L and yield of 0.76. CSL is efficiently used by this strain as its nitrogen source. With an initial CSL loading of 25 and 20 g/L of xylose, 0.86 yield and 6.6 g/L of final acetic acid concentration are attained. Replacing yeast extract with CSL can significantly reduce production cost. The organism consumes arabinose, mannose, and galactose only when each of these is present with xylose in the medium. In a batch fermentation of a mixture of sugars, the extent of consumption of mannose, arabinose, and galactose is <20% in 130 h. Fed-batch operation did not result in increased yield of acetic acid, because the organism lost viability after a certain period and was not revived by adding extra nutrients or trace elements. This proves to be a major drawback for acetate production from this strain using xylose as the carbon source. Accumulation of D-xylulose is detected in fed-batch fermentation of xylose with CSL as the nitrogen source.

References

1. Johnson, K. L. (1994), *Cryotech Deicing Technologies*, Fort Madison, IA.
2. Ljungdahl, L. G. (1983), *Formation of Acetate Using Homoacetate Fermenting Anaerobic Bacteria in Organic Chemicals from Biomass*, Menlo Park, CA.
3. Sugaya, K. and Jones, J. L. (1986), *Biotechnol. Bioeng.* **28,** 678–683.
4. Wijitra, K. (1994), MS thesis, University of Illinois, Urbana.
5. Fontaine, F. E., Peterson, W. H., McCoy, E., and Johnson, M. J. (1942), *J. Bacteriol.* **43,** 701–715.
6. Andreesen, J. R., Schaupp, A., Neurauter, C., Brown, A., and Ljundahl, L. G. (1973), *J. Bacteriol.* **114,** 743–751.
7. Brumm, P. J. (1988), *Biotechnol. Bioeng.* **32,** 444–450.
8. Parekh, S. R. and Cheryan, M. (1990), *Process Biochem. Int.* **25,** 117–121.
9. Parekh, S. R. and Cheryan, M. (1990), *Biotechnol. Lett.* **16(2),** 139–142.
10. Parekh, S. R. and Cheryan, M. (1990), *Appl. Microbiol. Biotechnol.* **36,** 384–387.
11. Stephanopoulous, G. and San, K. Y. (1985), *Biotechnol. Prog.* **1(4),** 250–259.
12. Liggett, R. W. and Koffler, H. (1948), *Bacteriol. Rev.* **12,** 297–311.
13. Shah, M. M. and Cheryan, M. (1995), *J. Ind. Microbiol.* **15,** 424–428.
14. Bock, S. A., Fox, S. L., and Gibbons, W. R. (1997), *Biotechnol. Appl. Biochem.* **25,** 117–125.
15. Sikyta, B. (1983), *Methods in Ind. Microbiol,* Wiley, New York.
16. Larsson, S., Palmquist, E., and Nilvebrant, N. (1999), *Enzyme Microbiol. Technol.* **24(3/4),** 151–159.

Heat Transfer Considerations in Design of a Batch Tube Reactor for Biomass Hydrolysis

Sigrid E. Jacobsen and Charles E. Wyman*

Chemical and Biochemical Engineering, Thayer School of Engineering, Dartmouth College, Hanover, NH 03755, E-mail: charles.wyman@dartmouth.edu

Abstract

Biologic conversion of inexpensive and abundant sources of cellulosic biomass offers a low-cost route to production of fuels and commodity chemicals that can provide unparalleled environmental, economic, and strategic benefits. However, low-cost, high-yield technologies are needed to recover sugars from the hemicellulose fraction of biomass and to prepare the remaining cellulose fraction for subsequent hydrolysis. Uncatalyzed hemicellulose hydrolysis in flow-through systems offers a number of important advantages for removal of hemicellulose sugars, and it is believed that oligomers could play an important role in explaining why the performance of flow-through systems differs from uncatalyzed steam explosion approaches. Thus, an effort is under way to study oligomer formation kinetics, and a small batch reactor is being applied to capture these important intermediates in a closed system that facilitates material balance closure for varying reaction conditions. In this article, heat transfer for batch tubes is analyzed to derive temperature profiles for different tube diameters and assess the impact on xylan conversion. It was found that the tube diameter must be <0.5 in. for xylan hydrolysis to follow the kinetics expected for a uniform temperature system at typical operating conditions.

Index Entries: Reactors; heat transfer; hydrolysis; kinetics; pretreatment.

Introduction

Biomass provides a unique, vast resource for the production of organic fuels and chemicals on a sustainable basis, and biologic conversion routes offer great promise for low-cost production to support large-scale use *(1)*. However, biomass must be pretreated prior to biologic conversion to achieve the high yields essential for economic success *(2,3)*. Currently,

---

*Author to whom all correspondence and reprint requests should be addressed.

pretreatment represents the most expensive single step in biomass processing and also greatly influences the cost of biologic steps that collectively comprise the largest total costs (4). Thus, it is important to optimize and improve pretreatment if we are to reduce the cost of products from biomass to be competitive with traditional fossil sources.

The use of dilute acid and uncatalyzed pretreatment technologies is favored by many because of the good enzymatic digestibility realized for pretreated cellulose (5). In addition, dilute-acid catalyzed hydrolysis can achieve high yields of sugars from hemicellulose (5–7). However, there are important inconsistencies in the kinetics of hemicellulose hydrolysis as studied to date (8). Therefore, we are interested in exploring hemicellulose hydrolysis with the goal of improving our knowledge of important mechanisms involved and developing models that will more accurately predict reaction performance. Such information will help optimize current systems, support scale-up, and provide insight for advancing the technology.

In our initial research, we chose to employ a closed-batch system to facilitate material balances and simplify the overall experimental design. However, interpretation of the results is more easily accomplished if we can ensure a constant temperature. This is particularly important because kinetics tend to follow an Arrhenius behavior that makes rates change substantially with changes in temperature. Furthermore, the yields of some thermal biomass reactions, such as cellulose hydrolysis, are very sensitive to the temperature applied (9). Thus, we needed to analyze heat transfer in a batch reactor tube to guide the design and be sure the temperature profile is reasonably uniform.

For these reasons, a model was developed that incorporated heat transfer and traditional reaction kinetics to predict the temperature history for a tube batch reactor system and the possible impact on biomass conversion. Such analysis may also indicate whether variations in experimental results reported previously by others could be explained by variable temperature profiles owing to heat transfer effects. Additionally, the model can be used to determine whether temperature transients could be reduced by submerging the reactor tubes initially in a hotter bath before transferring them to one maintained at the target temperature of the reaction.

Batch Tube System and Model

The experimental plan is based on mixing biomass with water to achieve target moisture levels and adding weighed amounts of these materials to several cylindrical metal tubes. The tubes are chosen to have a relatively high length-to-diameter ratio to ensure that the temperature depends primarily on tube radius. The tubes are then sealed at each end and submerged in a fluidized sand bath that is held at a set temperature. If transient times need to be reduced, the tubes can first be inserted in a bath held at a temperature above that targeted for the reaction to speed heat up and then be transferred to a second sand bath that is held at the temperature of interest as soon as the target temperature is attained. One or more of the tubes can then be withdrawn from the bath at selected times and quenched in a water or ice bath to stop the reaction. Measurement of the liquid and solids sugar content and weights can then be used to determine the conversion and yield profiles at each time and temperature.

As designed, the tubes lend themselves to classic heat transfer modeling. First, it is assumed that the initial temperature is equal to room temperature and uniform over the tube length and radius. Second, it is assumed that heat transfer is only a function of the tube radius because the temperature does not change with angular position and the length-to-diameter ratio for the tubes is high. Third, because fluidized sand baths have high heat transfer coefficients, the outside temperature of the tube wall is taken to be essentially equal to that of the sand bath *(10)*. Fourth, it is assumed that the thermal conductivity of the metal tube wall is much higher than that of the biomass slurry and, as a result, that the temperature of the wall is uniform over its radius.

Based on the experimental approach and its geometric representation, we can develop initial and boundary conditions for the system. First, a new temperature variable, v, is defined through the relationship

$$v = [(T - T_0)/(T_1 - T_0)]V \qquad (1)$$

in which T is the temperature as a function of the radius, T_0 is the initial tube temperature, and T_1 is the temperature of the bath. Thus, the following initial and boundary conditions result:

$$v = 0 \text{ at time} <0 \text{ for all } r \qquad (2)$$

$$v = V \text{ at } r = R \text{ for } t > 0 \qquad (3)$$

in which R is the inside radius of the tube.

Because the temperature is assumed to be only a function of radius and time, we can apply the unsteady equation of heat conduction in cylindrical coordinates with only dependence on radius to give *(11)*

$$\frac{\partial v}{\partial t} = k\left(\frac{\partial^2 v}{\partial r^2} + \frac{1}{r}\frac{\partial v}{\partial r}\right) \quad \text{where } k = \frac{K}{\rho C_p} \qquad (4)$$

in which r is the tube radius, k is the thermal diffusivity, K is the thermal conductivity, ρ is the density, and C_p is the heat capacity. Note that this model only includes conduction, and in reality, heating rates would be higher for low solids concentrations because the effects of convection would be significant.

Based on the initial and boundary conditions, Eq. 4 can be solved for the temperature function v:

$$v = AJ_0(\alpha r)e^{-k\alpha^2 t} \qquad (5)$$

in which $J_0(x)$ is the Bessel function of order zero of the first kind *(11)*. Satisfying the boundary conditions requires that α be a root of $J_0(a\alpha) = 0$, giving

$$v = \sum_{n=1}^{\infty} A_n J_0(\alpha_n r) e^{-k\alpha_n^2 t} \tag{6}$$

After integration of the Bessel function and rearranging, keeping in mind constant initial temperature $v = 0$ and a surface maintained at V, we arrive at

$$v = V - \frac{2V}{a} \sum_{n=1}^{\infty} e^{-k\alpha_n^2 t} \frac{J_0(r\alpha_n)}{\alpha_n J_1(a\alpha_n)} \tag{7}$$

Calculation of Temperature and Conversion Profiles

The temperature profile in the tube was calculated as a function of time and radius. To support later calculation of the effect of temperature variations on conversion, the crosssection of the tube was divided into 10 concentric rings of equal thickness, and the time-dependent temperature history in each ring was determined from Eq. 7 by a Matlab program. The solution was developed for each of the 10 rings in 1–5 time intervals to derive 10 terms over a total time of 1000 s. Thus, 1000 temperature points were calculated in each of the 10 rings. The values of these first five roots can be found in Appendix IV of Carslaw and Jaeger *(11)*, and for this simulation only the first five roots of the Bessel functions were needed because further roots made minor contributions.

For the purpose of estimating the effect of temperature on the hydrolysis of hemicellulose to sugars, the following kinetic equations were applied to each ring based on published information *(12)*:

$$X_{r,i} = X_{0,i} e^{-kt} \tag{8}$$

$$\text{where } k = k_{i0} e^{-(E/RT)} \tag{9}$$

in which $X_{r,i}$ is the xylan remaining at time t in any ring i, and $X_{0,i}$ is the amount of xylan in ring i initially. The value of E, the activation energy, and k_{i0}, a kinetic constant, were taken from Estaglalian et al. *(12)*.

The mass of xylan remaining in each ring at each time was calculated based on reported kinetics but using the temperature predicted by the heat transfer analysis. First, the fraction of total original xylan, $X_{0,i}$, in each ring i was calculated by determining the area of each ring and dividing by the total cross-sectional area of the tube. Second, the fraction of xylan remaining, $X_{r,i}$ in each ring was calculated using the time and corresponding temperature from the solution to the conduction Eq. 7 in Eqs. 8 and 9. An average of the temperatures at the inner and outer radius for each ring in Eq. 9 was employed by the Matlab program to determine the kinetic rate constant for each time interval.

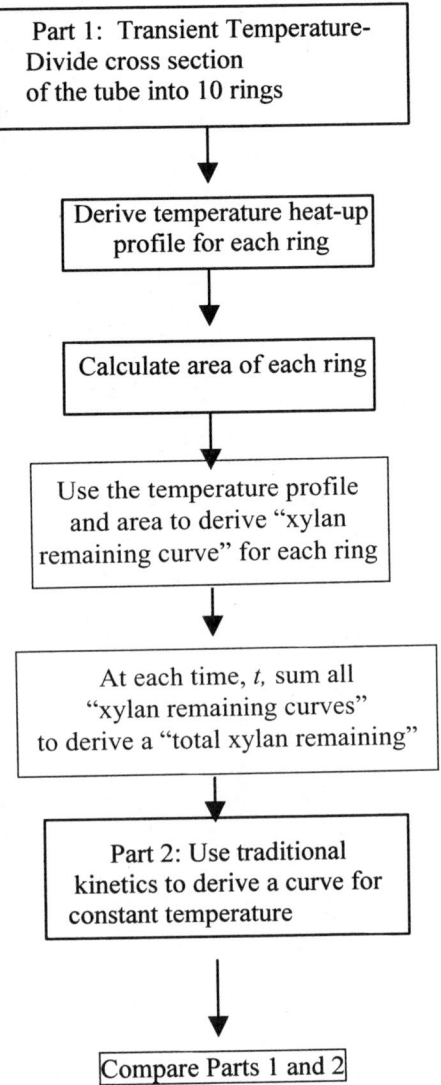

The approach outlined was applied to estimate the temperature and xylan-remaining profiles by the algorithm shown above.

Simulation Results

Figure 1 presents the predicted temperature profiles in each of the 10 radial intervals for reactor diameters of 0.5-in. (radius = 0.64 cm) and 1.0 in. (radius = 1.27 cm) with a 160°C sand bath temperature. Thus, we see that it takes some time for the temperature to approach the wall temperature over the entire tube radius. Furthermore, the model predicts that it will take 3 min to heat the center of the 0.5-in. tube to 140°C vs 14 min for the 1-in. tube.

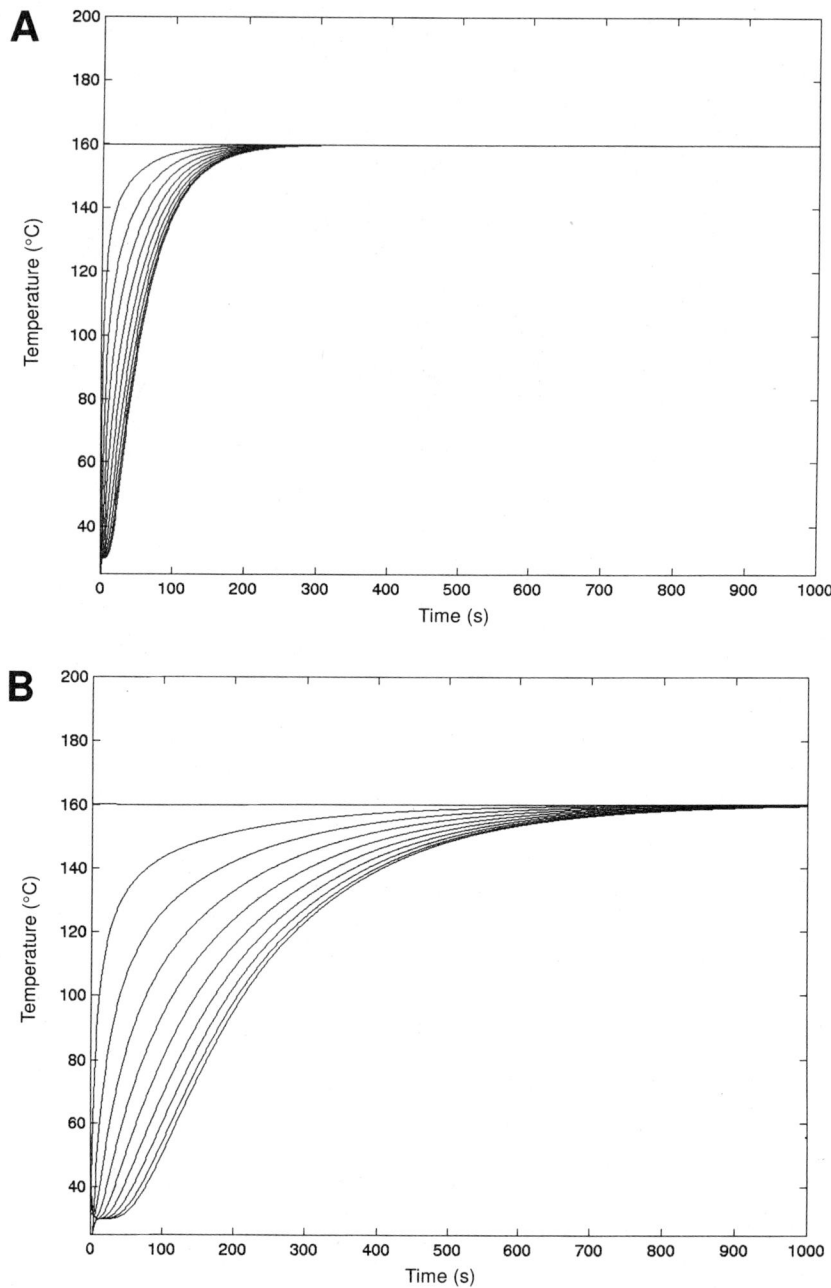

Fig. 1. Profiles showing the temperature vs time in seconds at each 0.1 radius for reactor with a 0.5-in. diameter ($r = 0.64$ cm) **(A)** and a 1.0-in. diameter ($r = 1.27$ cm) **(B)**.

Figure 2 shows the effect of temperature variations in Fig. 1 on xylan conversion for the 10 radial intervals. It is clear that hydrolysis in the inner rings (i.e. $r = 0.1R$ and $r = 0.2R$), begins significantly later than in the outer rings. However, because cross-sectional area and, therefore, relative xylan quantity is proportional to radius squared, the inner rings have relatively

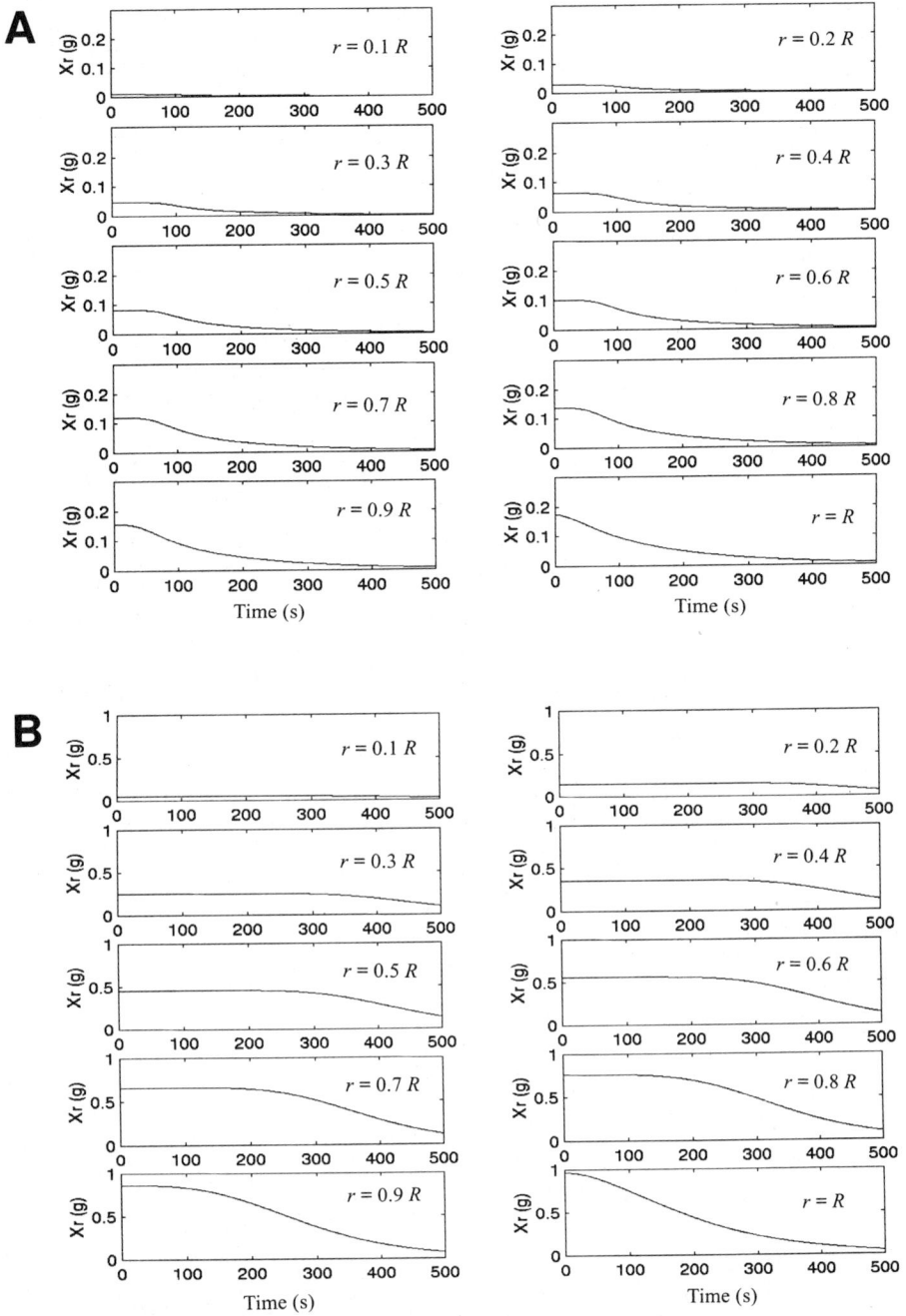

Fig. 2. Mass of xylan remaining (Xr) in grams in each concentric ring vs time in seconds for a reactor with a 0.5-in. diameter ($r = 0.64$ cm) **(A)** and a 1.0-in. diameter ($r = 1.27$ cm) **(B)**.

little influence on the total amount hydrolyzed. Thus, for many purposes, heating only 80–90% of the outside radius may be sufficient to achieve target yields because the outer rings have a much greater effect.

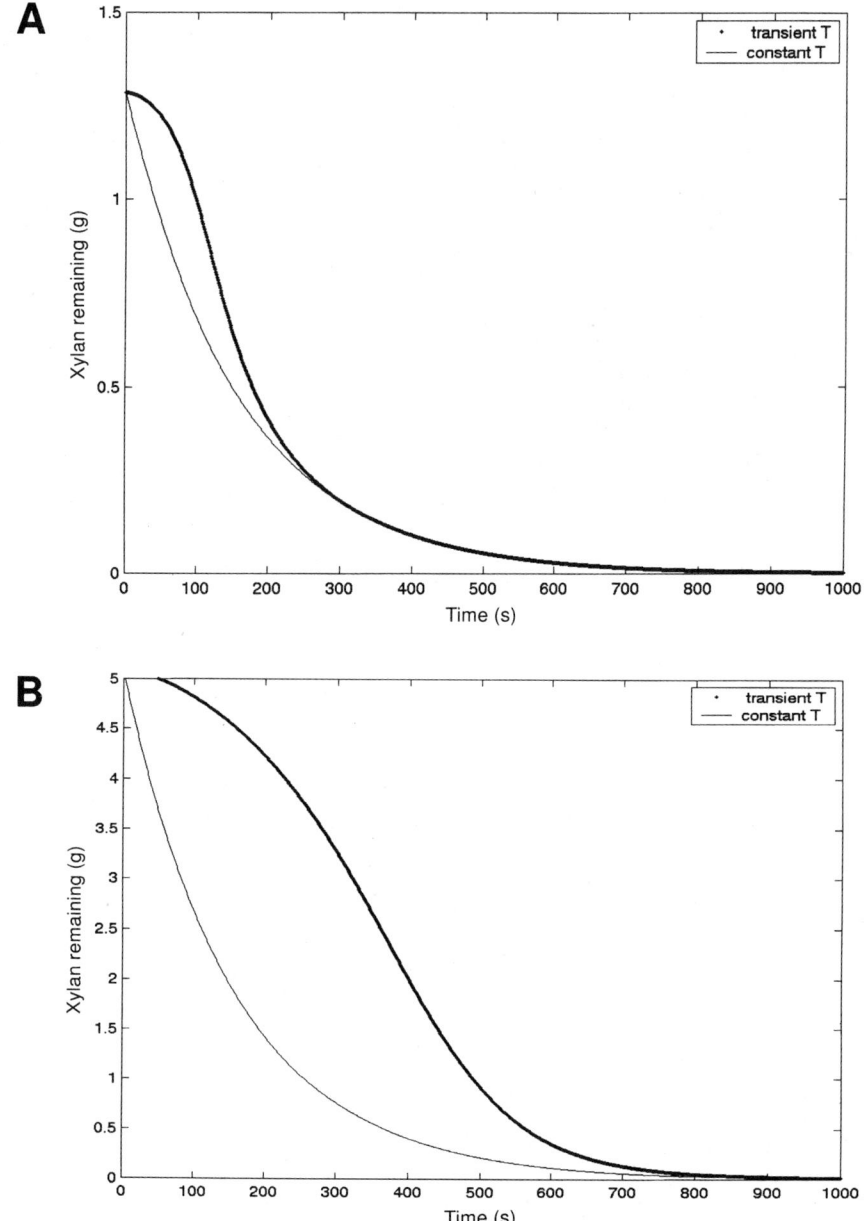

Fig. 3. Xylan remaining in grams vs time in seconds for a reactor with a 0.5-in. diameter (r = 0.64 cm) (**A**) and a 1.0-in. diameter (r = 1.27 cm) (**B**).

To allow comparison of the transient profile resulting from the delay in tube heat-up owing to heat transfer effects to a perfect system with virtually instant heat-up, the same kinetic equations were applied but at a constant temperature equal to the target value (e.g., 160°C in our previous example). Comparison of constant temperature profiles with transient temperature is shown in Fig. 3. Figure 3A shows that early in the reaction, even with a diameter of 0.5 in., the kinetic profile is somewhat affected by a heating lag. However, after about 50% conversion, the profiles for both cases are nearly identical. Figure 3B clearly indicates that the 1-in. diameter is too large to assume quick heating without effects on the xylan conversion profile.

Conclusion

The results from this model illustrate the value of analysis in designing experimental systems to ensure that unknown variations in key operational parameters that can affect results are avoided. In particular, a combined heat transfer and kinetic model has been developed to predict the effect of temperature transients owing to heat transfer on reaction kinetics and hemicellulose hydrolysis for a tubular batch reactor. Based on these results, we concluded that a reactor diameter of 0.5 in. or less is needed to ensure that the reaction performance is reasonably described by a constant temperature model. This result is important to keep in mind when reviewing reports of hydrolysis experiments reported in the literature. For larger tube diameters of 1.0 in. or more, the kinetics could be affected by heat transfer limitations unless measures are taken to specifically improve heating. One such option would be to employ two sand baths in series with the tubes submersed first in one heated to a higher temperature than targeted and then transferred into a second reactor held at the desired reaction temperature once the latter temperature is reached. Another option is to charge the reactor with preheated liquid to rapidly bring the system to the desired temperature. Furthermore, data obtained in even larger nontube batch reactors, usually with a volume of 1 to 2 L, could be even more significantly affected unless vigorous mixing is applied to ensure rapid heat transfer. However, even then, careful analysis is important to ensure that the temperature profile follows expectations.

This analysis also shows that the results are heavily weighted by the larger quantity of material contained in the outer portion of the tubes whereas the inner rings contain only a small fraction of the total biomass. This small amount does not significantly affect the overall reaction profile, and the performance will more closely match that for the outer wall temperature than might be expected based on a measurement of the temperature in the middle of a tube. Thus, measuring the temperature at a point other than at the center of a tube, as typical of most studies, may be a better indication of the effect of kinetic performance.

Note that because the heat transfer coefficient for a fluidized sand bath is very high, the model assumed that the wall of the reactor instantly reached the bath temperature. However, for other systems such as oil baths or steam, heat transfer to the outer wall could be important, and appropriate heat transfer coefficients would need to be applied for such a system.

Acknowledgment

This work was made possible through the generous support of the Thayer School of Engineering at Dartmouth College.

References

1. Lynd, L. R., Wyman, C. E., and Gerngross, T. U. (2000), *Biotechnol. Prog.* **15,** 777–793.
2. Grohmann, K., Himmel, M., Rivard, C., Tucker, M., and Baker, J. (1984), *Biotechnol. Bioeng. Symp.* **14,** 137–157.
3. Knappert, H., Grethlein, H., and Converse, A. O. (1980), *Biotechnol. Bioeng. Symp.* **11,** 67–77.
4. Lynd, L. R., Elander, R. T., and Wyman, C. E. (1996), *Appl. Biochem. Bioeng.* **57/58,** 741–761.
5. Torget, R., Hatzis, C., Hayward, T. K., Hsu, T.-A., and Philippidis, G. (1996), *Appl. Biochem. Bioeng.* **57/58,** 85–101.
6. Grohmann, K. and Torget, R. (1992), US Patent 5,125,977.
7. Torget, R. and Hsu, T.-A. (1994), *Appl. Biochem. Bioeng.* **45/46,** 5–21.
8. Jacobsen, S. E. and Wyman, C. E (2000), *Appl. Biochem. Bioeng.* **84–86,** 81–96.
9. Saeman, J. F. (1945), *Ind. Eng. Chem.* **37(1),** 43–52.
10. Perry, R. H. and Green, D. W., eds. (1997), *Perry's Chemical Engineer's Handbook*, 7th ed., McGraw-Hill, New York.
11. Carslaw, H. S. and Jaeger, J. C. (1959), *Conduction of Heat in Solids*, 2nd ed., Clarendon, Oxford.
12. Estaglalian, A., Hashimoto, A. G., Fenske, J. J., and Penner, M. H. (1997), *Bioresour. Technol.* **59,** 129–136.

Measurement of Bubble Size Distribution in Protein Foam Fractionation Column Using Capillary Probe with Photoelectric Sensors

LIPING DU, YUQING DING, ALEŠ PROKOP,
AND ROBERT D. TANNER*

*Chemical Engineering Department,
Vanderbilt University, Nashville, TN 37235,
E-mail: rtanner@vuse.vanderbilt.edu*

Abstract

Bubble size is a key variable for predicting the ability to separate and concentrate proteins in a foam fractionation process. It is used to characterize not only the bubble-specific interfacial area but also coalescence of bubbles in the foam phase. This article describes the development of a photoelectric method for measuring the bubble size distribution in both bubble and foam columns for concentrating proteins. The method uses a vacuum to withdraw a stream of gas-liquid dispersion from the bubble or foam column through a capillary tube with a funnel-shaped inlet. The resulting sample bubble cylinders are detected, and their lengths are calculated by using two pairs of infrared photoelectric sensors that are connected with a high-speed data acquisition system controlled by a microcomputer. The bubble size distributions in the bubble column 12 and 1 cm below the interface and in the foam phase 1 cm above the interface are obtained in a continuous foam fractionation process for concentrating ovalbumin. The effects of certain operating conditions such as the feed protein concentration, superficial gas velocity, liquid flow rate, and solution pH are investigated. The results may prove to be helpful in understanding the mechanisms controlling the foam fractionation of proteins.

Index Entries: Bubble size; capillary tube; photoelectric method; foam fractionation; ovalbumin.

Introduction

Information about bubble size distributions is important to mass transfer, heat transfer, and chemical reaction, which are very dependent on the

---

*Author to whom all correspondence and reprint requests should be addressed.

interfacial area in many types of chemical processing equipment, such as stirred-tank reactors and distillation columns *(1–7)*. Bubble size distribution is also of paramount importance in a bubble or a foam column during a foam fractionation process, since this type of adsorptive separation technique depends on the specific gas bubble interfacial area, *a* (square centimeters of area/cubic centimeters of gas) *(8–11)*. The specific interfacial area relates to the bubble diameter. The area-to-volume relation between a single bubble and its specific surface area, *a*, is given by Eq. 1 for an ideal spherical bubble:

$$a = \frac{4\pi R^2}{(4/3)\pi R^3} = \frac{3}{R} = \frac{6}{d} \tag{1}$$

in which *R* and *d* are the bubble radius and diameter, each measured in centimeters, respectively. In the foam phase, each bubble cell is generally close to a dodecahedron in shape *(8)*; thus, Eq. 1 is often modified to the more-descriptive Eq. 2 *(8)*:

$$a = \frac{6.59}{d} \tag{2}$$

Equations 1 and 2 show that the smaller the bubble size, the larger the specific interfacial area of each bubble. Smaller bubbles lead to larger interfacial areas and, thus, adsorb more solute than do larger bubbles in a foam column. It follows then that the enrichment (the ratio of protein concentration of the foamate to the protein concentration of the initial solution), a measure of concentration/purification performance, will be higher for a foam column with smaller bubbles than a column with larger bubbles assuming that the gas fraction and surface concentration remain the same.

The rheology and stability of the foam are strongly influenced by the foam's bubble size distribution and gas-liquid fraction *(11)*. Therefore, to understand the foaming process, it is necessary to know the bubble size distribution and how that distribution affects the flow properties and is itself affected by flow processes in a foam *(10)*. For example, Brown et al.'s *(12)* study showed that large bubbles cause high drainage (liquid flow in liquid films and in plateau borders formed between the bubbles) flow rates and thinner liquid films. Subsequently, large bubbles are more prone to rupture (owing to their thinner liquid films) in a foam column and, thus, destabilize the foam *(12)*. The increase in bubble size along the length of a foam column reflects the degree of bubble coalescence (two small bubbles become one large bubble because the film between them breaks) in the foam phase *(8,9,11,12)*.

Coalescence in a foam column cannot now be predicted from theory *(9)*. Therefore, the development of the proposed photoelectric capillary probe method may provide appropriate measurements of the bubble size distribution, which, in turn, can contribute empirical information on bubble coalescence needed for modeling a foam column *(9)*. Generally, bubble size is affected by solution and gas-liquid surface properties, such as solution pH, protein concentration, surface viscosity, surface tension, and some operating parameters, such as the superficial gas velocity and size of the sparger *(8,9,12,13)*.

Presently, the bubble size in a gas-liquid dispersion system can be determined by direct and indirect methods *(1)*. Indirect methods include the interfacial area method and the chemical method. These techniques give only the average bubble size in an entire system and cannot give either local or distribution information *(1)*. A direct photographic method is tedious and usually does not provide very high accuracy. A conductivity method only can give global information of liquid fraction *(14)*. The capillary probe (with a photoelectric sensor) method, however, can be developed to obtain the bubble size distribution directly in order to determine the bubble size distributions of the fermentation media *(2)*, a stirred-tank reactor *(1)*, and a stirred large-scale vessel *(3)*. Bae and Tavlarides *(15)* used a laser capillary spectrophotometer for the measurement of drop size distribution of reactive liquid-liquid dispersions. Research conducted in China *(4–7)* also independently included the use of the capillary photoelectric method for bubble size measurement in stirred gas-liquid dispersion systems. The capillary probe method, which uses a capillary probe and photoelectric sensors, combined with a high-speed microcomputer data acquisition system, can perform online measurements and give information on local bubble size distribution relatively quickly. To date, studies using this measurement method have focused on nonfoaming gas-liquid dispersions with large liquid holdups (>20%) *(1–7)* or liquid-liquid dispersions *(15)*.

Reports of bubble size distribution measurements in foam columns (liquid holdup <10%) are not common in the literature. The photography method has been used by Wong et al. *(13)* and Brown et al. *(12)* to obtain information on bubble size distribution in a foam column. However, a disadvantage of the photography method is that the measured two-dimensional picture reflects only the bubble size information on a transparent surface, which may be different from the actual bubble size information present inside a three-dimensional foam.

In this article, we describe our early work with the photoelectric capillary probe method in the foam column (liquid holdup <10%), in addition to the bubble column during foam fractionation of ovalbumin. It is apparently the first time that such measurements of bubble size in a foam column (near the interface with the bubble column) have been reported. Our measurement technique is similar to those previously reported in the literature *(1–7,15)* except that we use a pair of infrared phototransistors to obtain bubble signals within a capillary.

Principles of Photoelectric Suction Probe Method

The basic principle of the photoelectric suction probe method is that a bubble is reshaped into a cylindrical slug within a fine glass capillary probe when the bubble is sucked into the tube *(1)*. In the subsequent analysis, it is assumed that the bubble slug is uniform and occupies the whole

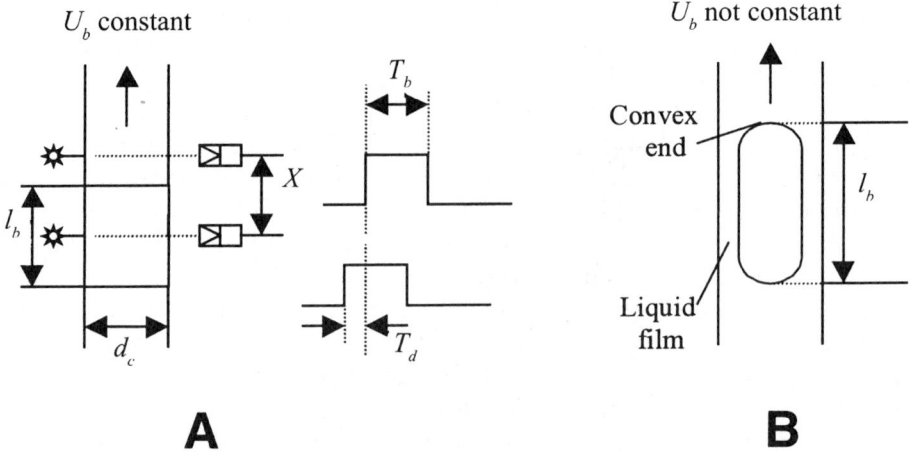

Fig. 1. Schematic of photoelectric method **(A)** and bubble flow inside a capillary **(B)** *(3)*.

cross section of the capillary and moves at constant velocity, U_b (Fig. 1A). Using these assumptions, the individual bubble length, l_b, and equivalent spherical diameter, d, can be determined by Eqs. 4 and 5. Equation 3 yields the bubble velocity, U_b, which is the quotient of distance and time:

$$U_b = X/T_d \tag{3}$$

$$l_b = U_b/T_b \tag{4}$$

$$d = \left[\frac{\pi l_b \left(\frac{d_c}{2}\right)^2}{\frac{1}{6}\pi} \right]^{1/3} = \left(\frac{3XT_b d_c^2}{2T_d} \right)^{1/3} \tag{5}$$

Here, X is the distance between two pairs of photoelectric sensors, d_c is the inside diameter of the capillary, T_d is the time it takes a bubble to travel the distance X, and T_b is the time it takes a whole bubble to pass through one sensor (as shown in Fig. 1A). Equation 5 results from equating the bubble volume to the cylinder volume in the capillary.

The Sauter mean diameter of the bubble size distribution, d_{32}, is then obtained (Eq. 6) *(1)*:

$$d_{32} = \frac{\sum_{i=1}^{N} d_i^3}{\sum_{i=1}^{N} d_i^2} \tag{6}$$

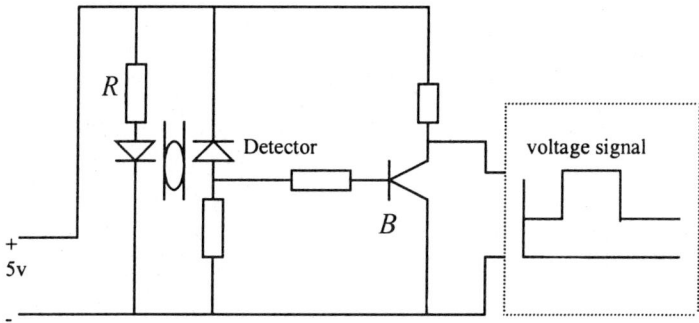

Fig. 2. Schematic of electric circuit.

in which N is the total number of bubbles sampled and d_i is the individual bubble diameter measured.

With the measured total time applied, the void fraction can now be calculated using Eq. 7:

$$\varepsilon_g = \frac{\sum_{i=1}^{N} T_{b,i}}{t_{\text{total}}} \tag{7}$$

Equation 7 assumes that the thickness of the bubble wall is negligible and bubble velocity, U_b, is constant.

On the right-hand sides of Eqs. 5 and 7, the only unknown values are T_b and T_d. In the photoelectric capillary probe method, T_b and T_d are measured and recorded automatically by the computer, with an installed high-speed data acquisition system. By applying a vacuum at the outlet of a capillary and immersing the other end into the gas-liquid dispersion system, a stream of the gas-liquid dispersion is sucked into the capillary and passed by two pairs of photoelectric sensors fixed in the apparatus surrounding the capillary. An electric circuit (Fig. 2) that connects the sensors will generate voltage signals whose magnitudes depend on which phase (liquid or gas) is detected. Typically, in a few seconds, several hundred bubbles pass by the sensors and the voltage signals are simultaneously recorded by the computer (Fig. 1). T_b and T_d are calculated from the time intervals between the gas and liquid phases (Fig. 1), thus obtaining the individual bubble diameters on line.

Theoretical Correction
to Individual Bubble Size Measured in Capillary

Actual bubble flow in a capillary is very complex *(1–7,16)* and deviates from the ideal description in Fig. 1A, because bubble slugs flowing inside a capillary tube are under the action of several forces: buoyancy, pressure gradient, inertial force, liquid viscosity force between the wall and the liquid film, and surface tension *(3)*. The actual shape of a bubble cylinder is not exactly an ideal cylinder because of its convex ends. Also, the cylinder does not occupy the whole cross section of the tube because of the thin liquid film surrounding the bubble (Fig. 1B). In addition, the volume of the cylinder may be different from the bubble volume in original dispersion owing to the expansion resulting from the pressure drop along the axis of the capillary. The velocity of the bubble is not constant along the capillary owing to the pressure gradient along its length. Even the tail velocity of a bubble slug can be different from its nose velocity *(3)*. Therefore, the volume of bubble measured in the capillary may be different from the true volume in the original dispersion, even though coalescence and breakup of slugs inside the tube are minimized when the suction pressure is in a suitable range *(4)*. To correct the volume error, either a direct calibration or a theoretical correction to the measured bubble size in the capillary is needed *(1–3)*.

Two-phase flow inside a capillary tube provides the theoretical basis for the correction of the measured bubble size in a capillary *(16)*. By neglecting the liquid film surrounding the bubble slug, Eq. 8 can be derived by equating the true bubble volume to the bubble slug volume (after pressure correction as ideal gas) in the capillary *(5)*:

$$d_b = \left(\frac{p}{p_0}\right)^{1/3} \left(\frac{3}{2}\frac{l_b}{d_c} - \frac{1}{2}\right)^{1/3} d_c \tag{8}$$

in which p is the pressure inside the capillary at the point where the bubble is detected. p is calculated from the pressure gradient equation based on two-phase flow inside a capillary tube. Yu et al. *(5)* found that when a capillary of 0.66-mm id was used with a vacuum of 40–53.3 kPa, the absolute error between the measured bubble diameter (after pressure correction) and the true value was <0.07 mm (a relative error <4.26%) for an individual surfactant bubble 1.0–3.0 mm in diameter. For the case in which the liquid film surrounding the bubble slug is not negligible, Zhang et al. *(7)* developed the bubble size correction formula as follows:

$$d_b = \left(\frac{3pl_b d_c^2}{2.38p_0}\right)^{1/3} = \left(\frac{p}{p_0}\right)^{1/3} \left(\frac{3l_b}{2.38d_c}\right)^{1/3} d_c \tag{9}$$

Equation 9 was obtained under the assumption of $S_k / S_b = 1.19$, in which S_k and S_b are cross-sectional areas of the capillary tube and bubble slug, respectively. This assumption is valid when Reynolds number is >3000 for plug flow (two phase) within a capillary *(16)*. Zhang et al.'s *(7)* simulation of the pressure drop along the length of the capillary tube showed that over a wide range of experimental conditions, the pressure drop is nearly linear with the length. Using a linear approximation to the pressure gradient and

Eq. 9, a 3% maximum deviation of bubble diameter (for the bubble range of 0.5–6 mm) from that obtained using a photographic method was observed with the photoelectric capillary probe method (1.8-mm diameter capillary).

Using a linear pressure drop along the axis of the capillary, considering the surrounding liquid film, Greaves and Kobbacy *(1)* developed a semiempirical formula for estimating the bubble volume (Eq. 10). Equation 10 showed a 30% on average overestimation of bubble volume (or 9% of diameter for 0.3- to 6-mm bubbles) over that calibrated by injecting a known volume bubble in the inlet of the capillary by a precision micrometer needle syringe *(1)*.

$$V_b = \frac{t_b^* U_b^* A}{P_A} \left[1 - \left(\frac{\mu U_b^*}{\sigma} \right)^{1/2} \right] \times \left[P_A - \frac{z}{l_c - t_b^* U_b^*} (P_A - P_V - 0.163 \rho U_b^{*2}) \right] \quad (10)$$

Equation 10 was applied in our work, along with the 9% overestimate correction for the spherical diameter (which was subtracted). We assume that the error of measured bubble diameter in our measurement will thus be <5%, because the diameter of the capillary tube we used, 0.5 mm, is very close to the 0.61 mm that Greaves and Kobbacy *(1)* used and the bubble dimensions in our bubble and foam column are close to those in his dispersion system (0.3–6 mm in diameter).

Improvement in Accuracy in Bubble Size Distribution

Even if the individual bubble size is measured accurately, the bubble size distribution results may be biased owing to the influence of sampling in a dynamic gas-liquid dispersion system. One problem is the coalescence or breakage that may occur at the inlet of the capillary. Another is the preferential attraction of bubbles of a certain size and liquid content into the capillary.

It has been established that the funnel-shaped capillary inlet tip could suppress the breakage of bubbles at the inlet of the capillary tube and that bubbles with radii smaller than the radius of curvature of the tip edge does not break on impact *(1–7)*. On the other hand, the funnel-shaped tip should not be too large; otherwise, significant disturbances of the two-phase flow pattern at the inlet would result. The tip size was therefore limited to suppress the breakage of the largest bubbles *(3)*. To minimize coalescence of bubbles at the inlet of the capillary, the residence time (on the order of milliseconds) of the bubbles at the inlet is shortened by increasing the suction velocity so that bubbles do not have enough time to coalesce (in milliseconds) before entering the capillary *(7)*. Greaves and Kobbacy *(1)* observed that a high sampling flow rate associated with a high suction velocity also minimizes the preferential attraction of specific size bubbles, thus making it possible for essentially all the bubbles around the sampling area to be sucked into the tube in a given time. When the suction speed rose too high, however, the bubble slug velocity inside the capillary increased so much that the slug expansion in the flow direction became significant and bubble size accuracy was likely to be less. Therefore, suction velocity needs to be controlled in a suitable range. The measures of funnel-shaped inlet capillary and control of suction velocity have been demonstrated to be very useful in measuring the bubble size distribution in a high-liquid holdup dispersion system *(1–7)*.

In a low-liquid holdup (<10%) foam column system, strong bubble-bubble interactions cause difficulties in accurately maintaining bubble size distributions using the photoelectric capillary probe method. In such cases, bubble sizes tend to be large relative to those in the bubble column of the same foam fractionation column owing to coalescence, and the liquid films are thinner owing to drainage in the foam column. If the same sized capillary tip as that used in the bubble column is also used for measuring the foam, the breakage rate may increase when bubbles enter the capillary tube. If the same suction velocity as that of a bubble column is used, bubble slugs inside the capillary may expand so much that breakage or extreme strain occurs. Therefore, accurate bubble size distributions of very dry foams with big bubbles may not be easily obtained using the same capillary tube as that used for a bubble column.

Materials and Methods

Figure 3 shows the foam fractionation and bubble size experimental setup. Continuous foam fractionation experiments were carried out in the glass column (14-cm id, 50-cm length). In our experiments, air is pumped into the column at a certain flow rate (superficial gas velocity of 0.05–0.2 cm/s), through a porous sparger (pore size 40–60 µm) mounted flush to the column bottom. To minimize loss of water in the air effluent stream, air is passed through a humidifier before it enters the column. The inlet end of the capillary tube is submerged vertically in the bubble or foam column. Then a vacuum is applied to the outlet end to suck the sample into the capillary for measurement of bubble size. The interface between the bubble column and foam phase is defined here as position 0. The bubble column level was kept constant at 28 cm by equating the continuous input feed solution rate with the continuous output residue solution rate in order to obtain a steady state. Sampling occurred at two positions below the interface (referred to as –12 cm and –1 cm in the bubble column) and two positions above the interface (referred to as +1 cm and +21 cm) (Fig. 4). The airflow rate was controlled by a flow rate regulator, and the liquid feed input was varied by changing the speed of the liquid pump

The internal diameter of the capillary was selected to enable the smallest bubble to be detected—it was 0.5 mm. Bubbles slightly smaller than the capillary can still be transformed into short cylinders because of the liquid film surrounding the slug. The diameter of the funnel-shaped tip was

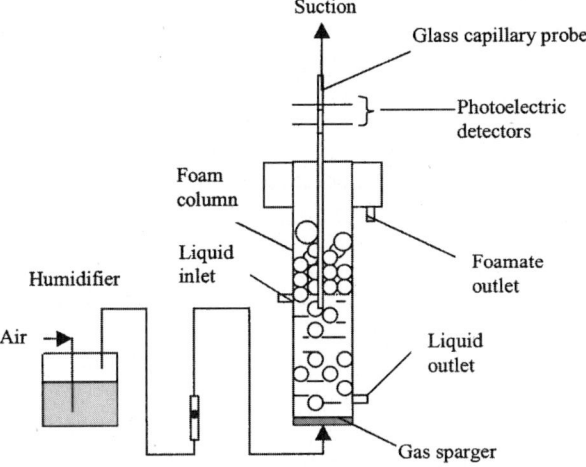

Fig. 3. Experimental setup of bubble size measurement and foam fractionation.

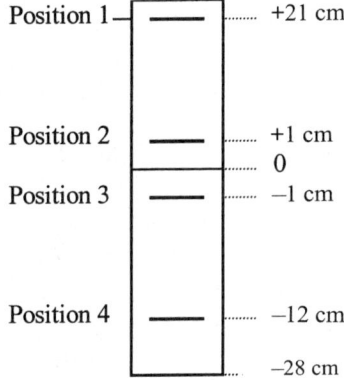

Fig. 4. Measurement points along the column. The 0 mark represents the interface between the bulk liquid phase (at the bottom) and the foam phase (at the top).

designed to be 6 mm based on our observation of bubble size in both the bubble and foam columns. The dimensions of the capillary tube are shown in Fig. 5. Two pairs of infrared photoelectric sensors, each of which comprises an infrared emitter and an infrared detector, were fixed (the distance between the two pairs of sensors is denoted as X) in the apparatus surrounding the capillary tube. At the position of each sensor pair, a very thin window was created by wrapping non-light-transparent material (here, TEFLON® tape) around the adjacent length along the capillary, in order to detect the very short slugs. The two-channel method was shown by Barigou and Greaves (3) to have low sensitivity to the velocity of bubbles inside the capillary. The vacuum pressure drop was controlled in a range of 1.5–1.75 psi (10.3–12.1 kPa) when the bubble size distribution in the bubble column was measured. This range was determined from our bubble size

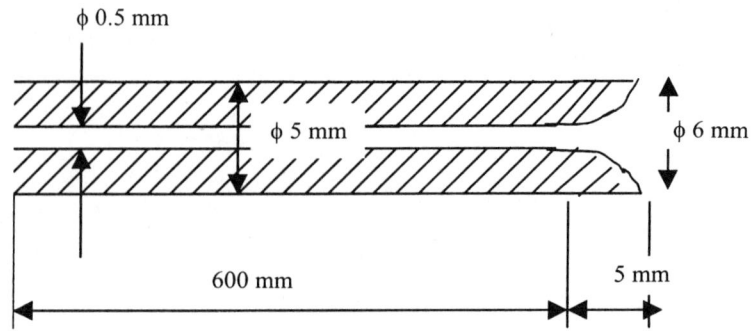

Fig. 5. Diagram of capillary probe (not to scale).

experiments at various vacuum pressure drops and to give the best relatively accurate results (unpublished data). Likewise, the vacuum pressure
drop was determined to be optimal at 0.5 psi (3.45 kPa) when the foam
column bubble size distribution was measured.

Protein (ovalbumin, grade II; Sigma, St. Louis, MO) aqueous solution
was used in the foam fractionation experiments. It was prepared by dissolving ovalbumin powder into deionized water. Ovalbumin concentration was determined by the Coomassie blue method (17). The pH of the
solution was adjusted by adding 1.0 M NaOH or HCl solution as needed.

The measurement hardware, in addition to the capillary probe and
photoelectric response circuit, included a PCI-6023E multifunction I/O
board from National Instrument. The board was mounted inside a Gateway 2000 computer. Software included the Data Acquisition Driver Software from National Instrument, and we developed an application interface
(Visual C) for measurement and data calculation and calibration (unpublished data).

Results and Discussion

Reproducibility

An important outcome of any experimental technique is the reproducibility of the results (1,3–5). Figure 6 compares two consecutive measurements under the same conditions in the foam column (position +1 cm).
Table 1 shows the average bubble size and void fraction results for these
two runs. It is seen from both Fig. 6 and Table 1 that the photoelectric probe
method used shows good reproducibility. Further investigation showed
that a log normal distribution was adequate to represent the bubble size
distribution data in Fig. 6 and all our other bubble size distribution data
(unpublished data). The log normal distribution of the bubble size in a
bubble or a foam column was also observed by Lage and Espósito (18),
Calvert and Nezhati (10), and Wong et al. (13) using the photography
method. This agreement showed at least that the sampling in our measure-

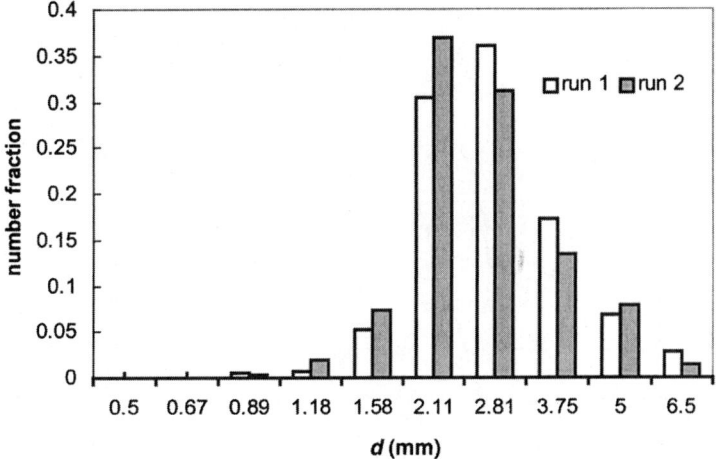

Fig. 6. Bubble size distribution at the bottom of the foam column (position +1 cm) under protein initial concentration of 56 mg/L, pH 9.7, superficial gas velocity of 0.1 cm/s, and liquid flow rate of 45 mL/min.

Table 1
Comparison of Two Measurement Results[a]

	Run 1	Run 2
Number of bubbles detected	418	453
Average bubble diameter (mm)		
By number	2.06	1.94
By surface area	2.16	2.04
By volume	2.28	2.15
Void fraction	0.89	0.90
Statistic average diameter (mm)	2.58	2.43
Deviation	1.02	0.87

[a]Bubble size distribution as Fig. 6.

ment is reliable and that the bubbles sampled are a good representation of those in bubble or a foam column.

Typical Bubble Size Distribution Results

Figure 7 shows the typical bubble size distribution results at two different positions: one in the bubble column just below (–1 cm) the interface, and the other in the foam column just above (+1 cm) the interface. We can see that below the surface, the bubble sizes are smaller and the bubble size distribution is narrower (deviation = 0.17), but above the surface, the bubbles become larger and the distribution is widened (deviation = 0.59). This change is attributed mainly to the coalescence of the bubbles at and above the interface (in the foam phase) *(19)*. The bubble size scale (0–5 mm) in Fig. 7 is in good agreement with the visual observation. Moreover, it is

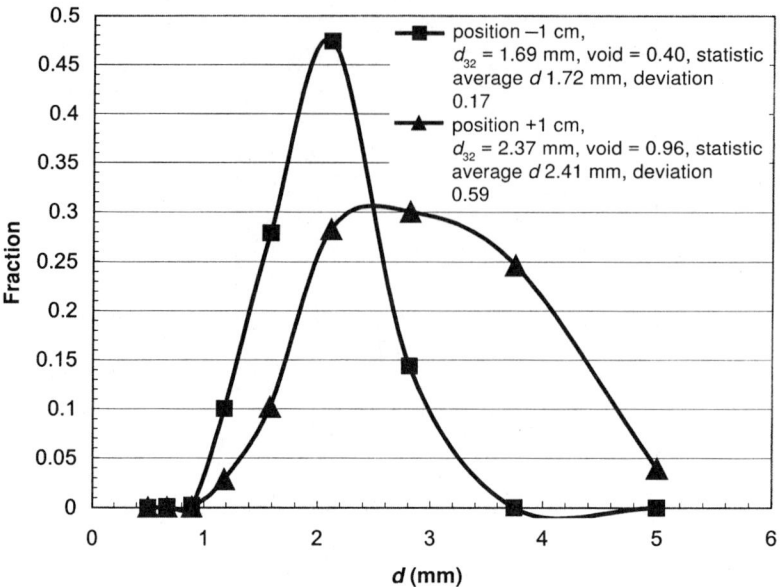

Fig. 7. Bubble size distribution at the top of the bulk liquid phase of bubble column and at the bottom of foam phase under protein initial concentration of 74.4 mg/L, pH 9.7, superficial gas velocity of 0.1 cm/s, and feed liquid flow rate of 45 mL/min.

close to the bubble size measured by Yu et al. *(5)*, Greaves and Kobbacy *(1)*, and Barigou and Greaves *(3)* in stirred gas-liquid dispersion systems using the photoelectric capillary probe method. It was also comparable to the bubble size measured by Lage and Espósito *(18)* in a similar bubble column, and Wong et al. *(13)* in a similar bovine serum albumin foam column using the photography method. In our previous work, the void fraction data obtained using the capillary probe method showed very good agreement with the weight method *(20)*, indicating that the capillary method used here is reasonably accurate and internally consistent.

Sauter Mean Diameter (d$_{32}$) and Void Fraction Variation with Position

Figure 8 shows the average bubble size, d_{32}, and the gas void fraction of gas change along the column. It can be seen that, in the bubble column, d_{32} and the void fraction grew gradually from position –12 cm to –1 cm. As the bubble crossed the interface, position 0, both d_{32} and the void fraction grew rapidly. In general, in a bubble column, the increase in the bubble size is mainly caused by the static pressure change at different liquid levels. For a very dilute solution, the density is approximately that of water. From position –12 cm to –1 cm, the static pressure decreases only slightly, so the bubble size change is negligible owing to this small static pressure change. When the bubbles rise up the bubble column to the interface, however, coalescence of some of these bubbles plays a significant role in the

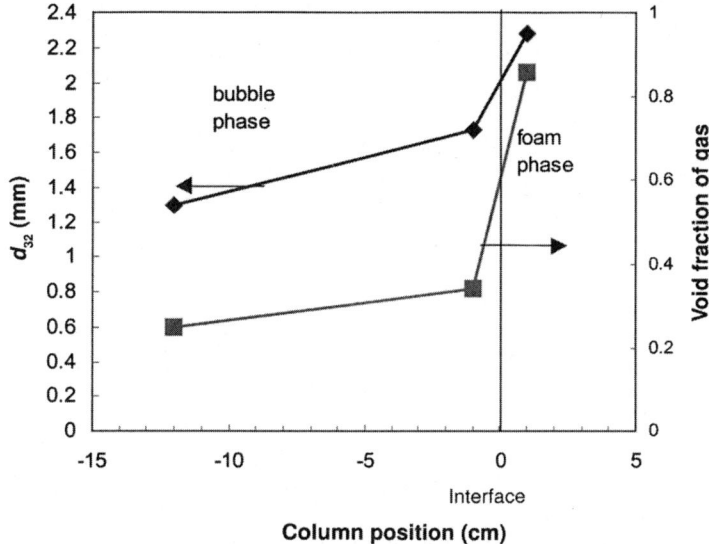

Fig. 8. Typical bubble size, d_{32}, and void fraction of the gas vs column position at an initial concentration of 35 mg/L, superficial gas velocity of 0.1 cm/s, and a liquid flow rate of 24 mL/min.

increase in bubble size. At the same time, because significant drainage occurs when the bubbles rise out of the bulk liquid phase, the void fraction increases rapidly. We were not able to accurately obtain the bubble size distribution at the top of the foam (+12 cm) using the same capillary tip owing to the very dry (liquid holdup <1%) foam and the large bubbles relative to the funnel-shaped inlet tip opening. Thus, the capillary tube and the method used here (without modification) were limited to measuring the bubble size only up to a low position in the foam column (such as 5 cm above the interface). Measuring the bubble sizes at higher positions needs to be investigated by enlarging the tip size and adjusting the vacuum pressure drop.

Sauter Mean Diameter (d $_{32}$) Varied with Protein Concentration

Figure 9 shows how d_{32} changes with the local ovalbumin concentration at the bubble column (position –1 cm). From the trend of the data, it is seen that as the ovalbumin increases in concentration, d_{32} decreases gradually and then levels off when the local protein concentration reaches 40 mg/L. This decrease in d_{32} can be explained by the surface concentration, Γ, which is a key variable in determining gas-liquid interfacial properties such as the surface viscosity and surface tension. The protein concentration (liquid) can be related to the surface concentration by an adsorption isotherm when the two variables are at equilibrium or by adsorption kinetics when they are not at equilibrium. If we assume that the surface concentration is at equilibrium with the protein liquid concentration in the form of a Langmuir isotherm, then with an increase in protein

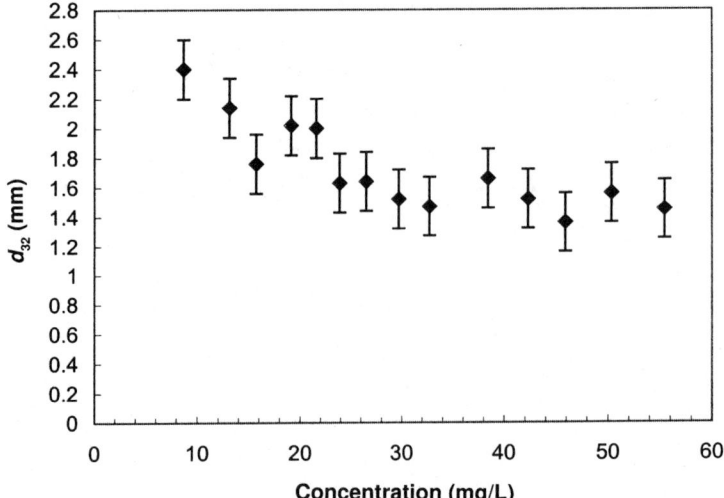

Fig. 9. Bubble diameter, d_{32}, vs local ovalbumin concentration at the higher position of the bubble column (position –1 cm) under superficial gas velocity of 0.1 cm/s, liquid flow rate of 24 mL/min, and pH 6.5 (original solution).

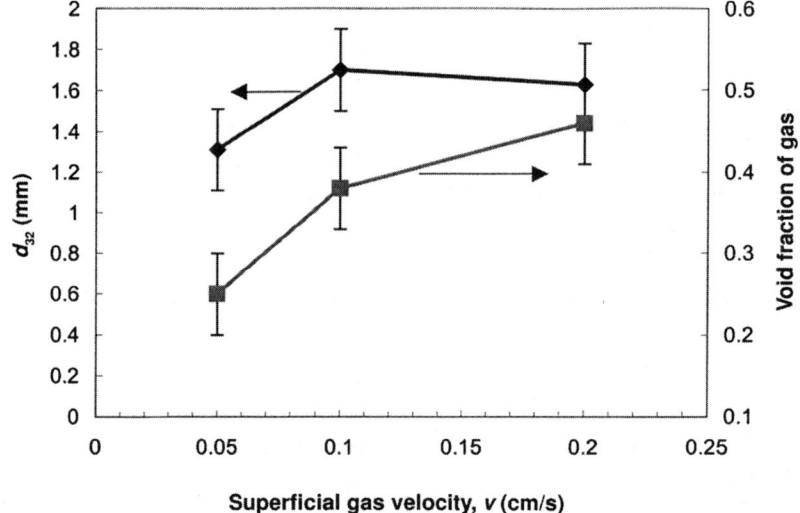

Fig. 10. Bubble size, d_{32}, and void fraction of gas vs superficial gas velocity at lower position of bubble column (position –12 cm) under initial ovalbumin concentration of 36.4 mg/L, liquid flow rate of 45 mL/min, and pH 6.5 (original solution).

concentration, the surface concentration increases. Once the protein liquid concentration reaches a given value, the surface concentration becomes saturated and will not increase further. Changes in surface tension will accompany changes in the surface concentration in an inverse relationship: the higher the surface concentration, the lower the surface tension. A lower surface tension may contribute to the formation of smaller bubbles. When

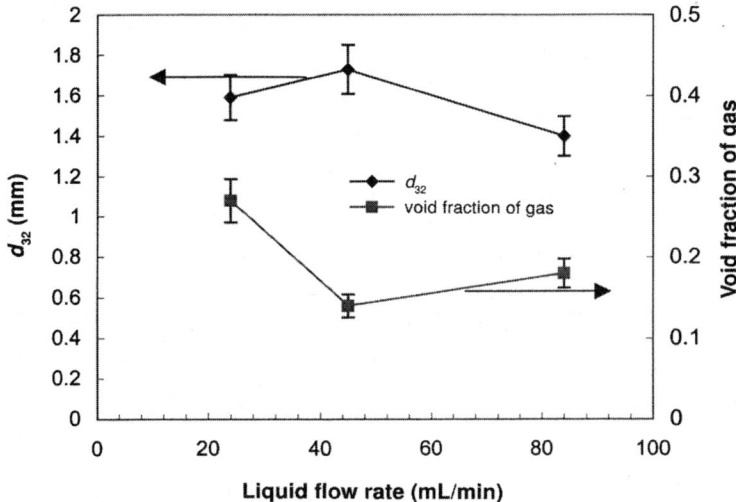

Fig. 11. Bubble size, d_{32}, and void fraction of gas vs liquid flow rate at higher bubble column position (position –1 cm) under initial ovalbumin concentration of 35 mg/L, superficial gas velocity of 0.1 cm/s, and pH 6.5 (original solution).

the surface concentration becomes saturated, the surface tension and the bubble size will no longer change.

Sauter Mean Diameter (d$_{32}$) and Void Fraction Varied with Superficial Gas Velocity

Figure 10 shows the results of the effect of superficial gas velocity on the bubble size and void fraction at position –12 cm. When the superficial gas velocity increases, the void fraction becomes larger. That follows because the gas flow rate increases as the gas velocity increases for the fixed diameter column and the increased sparging rate leads to a larger void fraction. The bubble size at this position, however, does not change as much with the increase in superficial gas velocity. Tentatively, it can be concluded that the superficial gas velocity is not a significant factor in affecting the bubble size in the lower bubble column (–12 cm). This is in agreement with the observation of Wong et al. *(13)*.

Effect of Feed Flow Rate

Figure 11 shows the effect of feed flow rate on the average bubble size (d_{32}) and void fraction in our experiments. It is seen that the feed flow rate influenced bubble size slightly. The void fraction decreased with an increase in feed flow rate. This is apparently the first time that bubble size and void fraction are investigated subject to the feed flow rate change.

Effect of pH on Bubble Size

Figure 12 shows the effect of pH on bubble size, d_{32}. The *p*I of ovalbumin is about 4.5 *(21)*. From Fig. 12, it is observed that at lower concentra-

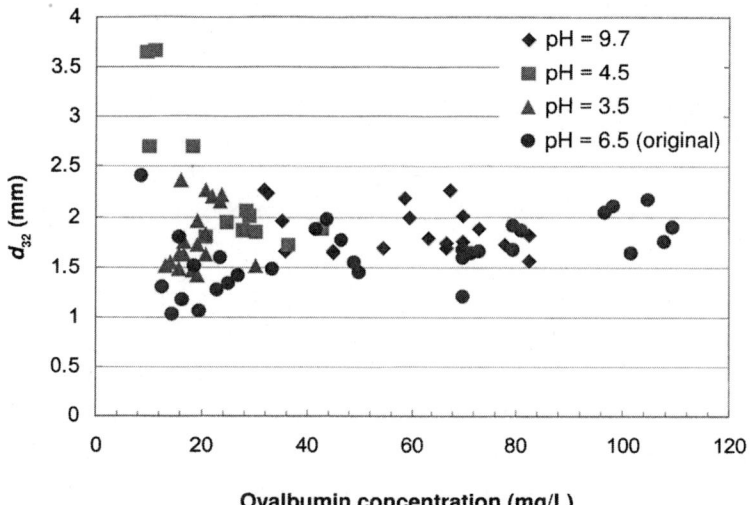

Fig. 12. Effect of pH on bubble size, d_{32}, at an upper position of the bubble column (position –1 cm) at different local ovalbumin concentrations. Other conditions were as follows: superficial gas velocity = 0.1 cm/s and liquid flow rate = 24 mL/min.

tions and pH 4.5, the bubble size is much larger than the average value at other conditions. The surface charge of proteins, which is determined by the solution pH, affects the rheology and stability of bubble films and, thus, the bubble size in the bubble and foam columns. At the pI, the net charge of protein molecules is 0 and the electric interaction between them is at a minimum. This may explain the large bubble size when the pH equals the pI. The detailed mechanism is not yet clear.

In summary, the bubble size distribution in the bubble and lower position of the foam column obtained is broad. A single average bubble diameter such as d_{32} may be inadequate for representing the bubble size distribution in the column. Because small bubbles control the interfacial area, while large bubbles control the drainage in the foam *(12)*, the bubble size distribution may better characterize the variation in bubble size in the column vs just one average bubble diameter.

Conclusion

The photoelectric capillary probe method was used to measure bubble size distribution, an important descriptive tool in elucidating mechanisms of the foam fractionation process, both in bubble and foam columns. The measurement is interpreted by applying a theoretical pressure calibration to the measured bubble size. For the foam fractionation column studied, a particular funnel-shaped inlet design in the capillary was selected, and a suitable vacuum pressure drop was chosen. The determined bubble size distributions in both the bubble and lower position of the foam columns were obtained with good reproducibility for the ovalbumin foam fractionation process. The measured bubble size distributions and void fractions are helpful in understanding the foam fractionation process of proteins. Further study is suggested to extend this method for measuring the bubble size at higher positions of the foam column.

Acknowledgment

This work was supported by the National Science Foundation grant no. CTS-9712486.

Nomenclature

A = internal cross-sectional area of capillary tube (m²)
d_b = bubble diameter (mm)
d_c = inside diameter of capillary (m)
l_b = length of bubble slug (m)
l_c = capillary tube length (m)
p = pressure at detection point of capillary (Pa)
P_A, p_0 = pressure at sampling point of dispersion (Pa)
P_v = pressure at capillary tube outlet, or vacuum pressure (Pa)
S_b = cross-sectional area of a bubble slug (m²)
S_k = cross-sectional area of a capillary tube (m²)
t_b^* = time for bubble slug to pass detection point on capillary tube (s)
U_b^* = bubble velocity at detection point (m/s)
V_b = bubble volume in dispersion (m³)
z = distance between point of observation and outlet of capillary tube (m)
μ = dynamic viscosity (Newton-seconds/m)
ρ = fluid density (kg/m)
σ = surface tension (N/m)

References

1. Greaves, M. and Kobbacy, K. A. H. (1984), *Chem. Eng. Res. Des.* **62**, 3–12.
2. Weiland, P., Brentrup, L., and Onken, U. (1980), *German Chem. Eng.* **3**, 269–302.
3. Barigou, M. and Greaves, M. (1991), *Meas. Sci. Technol.* **2**, 318–326.
4. Gao, D., Xu, Z., and Zhang, J. (1986), *J. East China Inst. Chem. Technol.* **12(Suppl.)**, 115–126 (in Chinese).
5. Yu, B., Deng, X., and Shi, Y. (1988), *J. East China Inst. Chem. Technol.* **14(5)**, 588–596 (in Chinese).
6. Jiang, H. and Gao, D. (1988), *J. East China Inst. Chem. Technol.* **14(5)**, 597–604 (in Chinese).
7. Zhang, Z., Dai, G., and Chen, M. (1989), *J. Chem. Eng. Chin. Univ.* **3(2)**, 42–49.
8. Uraizee, F. and Narsimhan, G. (1995), in *Bioseparation Processes in Foods*, Singh, R. K. and Rivzi, S. S. H., eds., Marcel Dekker, New York, pp. 175–225.
9. Uraizee, F. and Narsimhan, G. (1996), *Biotechnol. Bioeng.* **51**, 384–398.
10. Calvert, J. R. and Nezhati, K. (1987), *Int. J. Heat Fluid Flow* **8(2)**, 102–106.
11. Magrabi, S. A., Dlugogorski, B. Z., and Jameson, G. J. (1999), *Chem. Eng. Sci.* **54**, 4007–4022.
12. Brown, L., Narsimhan, G., and Wankat, P. C. (1990), *Biotechnol. Bioeng.* **36**, 947–959.

13. Wong, C. H., Hossain, M. D., Stanley, R. A., and Davies, C. E. (1996), in *CHEMECA'96, 24th Australian and New Zealand Chemical Engineering Conference Proceedings*, vol. 4, Sydney, Australia, pp. 105–110.
14. Wilde, P. J. (1996), *J. Colloid Interface* **178,** 733–739.
15. Bae, J. H. and Tavlarides, L. L. (1989), *AIChE J.* **35(7),** 1073–1084.
16. Wallis, G. B. (1969), in *One-Dimensional Two-Phase Flow*, McGraw-Hill, New York, pp. 212–314.
17. Bradford, M. M. (1976), *Analyt. Biochem.* **72,** 248–254.
18. Lage, P. L. C. and Espósito, R. O. (1999), *Powder Technol.* **101,** 142–150.
19. Brown, A. K., Kaul, A., and Varley, J. (1999), *Biotechnol. Bioeng.* **62(3),** 278–290.
20. Tanner, R. D., Parker, T., Ko, S., Ding, Y., Loha, V., Du, L., and Prokop, A. (2000), *Appl. Biochem. Biotechnol.* **84–86,** 835–842.
21. Hammershøj, M., Prins, A., and Qvist, K. B. (1999), *J. Sci. Food Agric.* **79,** 859–868.

Effect of a Natural Contaminant on Foam Fractionation of Bromelain

SAMUEL KO, JUSTIN CHERRY, ALEŠ PROKOP,
AND ROBERT D. TANNER*

*Chemical Engineering Department,
Vanderbilt University, Nashville, TN 37235
E-mail: rtanner@vuse.vanderbilt.edu*

Abstract

Foam fractionation is a simple, inexpensive method for separating and purifying proteins. Typically, a dilute bromelain solution with a pH ranging from 2.0 to 7.0 foams very well when bubbles are introduced into a foam fractionation column. It was observed, however, that the dilute enzyme solution only foamed between approximately pH 2.0 and 3.0 when the inner wall of the fractionation column was coated with a natural contaminant (okra residue). We studied the separation ratio and the protein mass recovery to explore the effect of a natural antifoaming agent on the foam fractionation of a dilute bromelain solution. The control variables used in this process were the initial bulk solution pH, which ranged from 2.0 to 7.0, and the superficial air velocity, which varied between 1.7 and 6.2 cm/s.

Index Entries: Bromelain; foam fractionation; protein; protein separation; protease recovery; antifoaming agent; natural antifoaming agent.

Introduction

Generally, naturally occurring foams in biologic processes (such as fermentation processes) are not desirable and are suppressed by antifoaming agents such as silicones and non-charged polymeric antifoaming agents. However, the addition of synthetic antifoaming agents to a process may raise the level of contamination and introduce new problems to the system. The substitution of a contamination problem in place of a foaming problem can perhaps be readily ameliorated if a bioreactor is operated such that the generated protein foam is continually removed from the process rather than suppressed with the antifoaming agents. Another approach is to use antifoaming additives that are compatible with the process and that, in turn, may be degraded by the process if they are natural materials.

---

*Author to whom all correspondence and reprint requests should be addressed.

In this study, we examined one such natural material, okra solution residue, as a possible antifoaming agent. This natural material, however, could create a problem if it were present in a foam fractionation process used to recover proteins.

Currently, foam fractionation can be used to remove dissolved organic wastes from water while increasing dissolved oxygen levels *(1)*. Foam fractionation is a simple and relatively inexpensive procedure that can also be used to separate and purify proteins *(2)*. This process has great potential for reducing protein recovery costs in the pharmaceutical and food industries. Nevertheless, when proteins are extracted from living organisms such as plants, existing natural contaminants in these organisms may act as antifoaming agents and suppress the desired foaming during foam fractionation.

Thus, it is important to investigate how natural contaminants can affect the foam fractionation of proteins when natural contaminants are inadvertently introduced into the fractionation column. This investigation may lead to a possible screening method for antifoaming agents. In this article, we report on the effect of the bulk solution pH and air superficial velocity on a dilute bromelain solution at two different conditions: (1) when the cylindrical fractionation column is coated with a natural contaminant (from okra), and (2) when the column is scrubbed clean prior to fractionation.

The enzyme we used is bromelain (a foaming protein), found in pineapples and other fruits. Bromelain is a group of proteolytic enzymes that are often used by people who suffer from malabsorption of food *(3,4)*. Recovery with minimal denaturation of this enzyme is an important first step in the industrial processing of bromelain from food wastes.

Materials and Methods

Materials

Bromelain (lot no. B-2252) and sodium hydroxide (lot no. 873487) were purchased from Sigma (St. Louis, MO). Coomassie brilliant blue G 250 (lot no. 23242) was purchased from Bio-Rad (Richmond, CA). Frozen okra (Kroger brand; Cincinnati, OH) was purchased from a local grocery store.

Experimental Procedure

A 100 mg/L bromelain solution was prepared by dissolving bromelain powder in deionized water. The dilute bromelain solution (originally at pH 4.5) was adjusted initially to the desired pH between 2.0 and 7.0 by adding HCl or NaOH. The initial volume of bromelain solution used for the batch experiments was 100 mL. The foam fractionation apparatus comprised an elongated glass column with a porous fitted glass sparger (pore size 40–60 µm) fitted flush to the bottom of the column, with a port at the top of the column for the foam to exit, as shown in Fig. 1. An air supply entering at the base of the column created bubbles in the bulk solution and

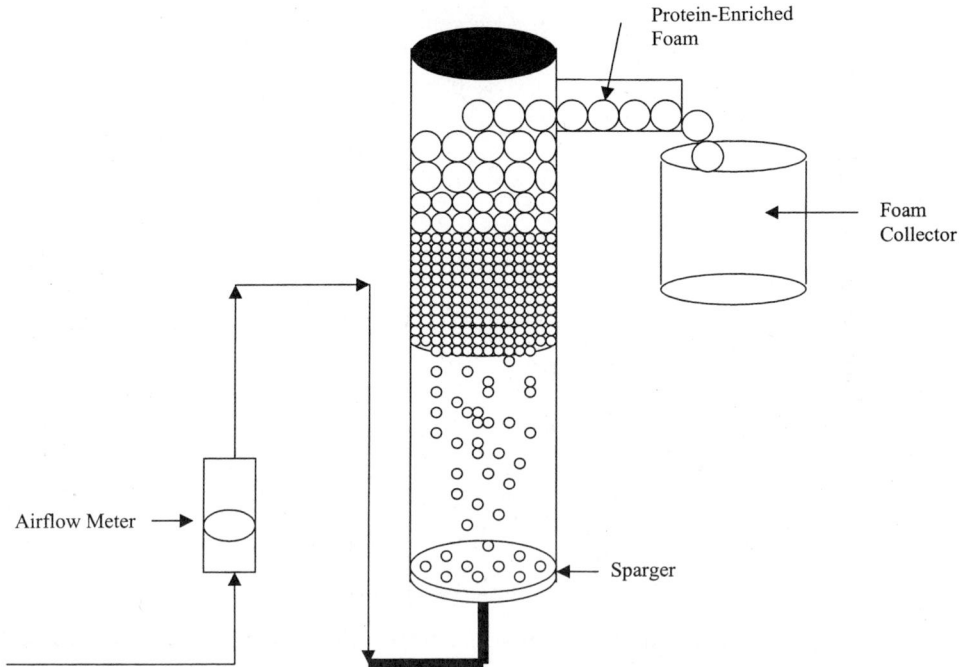

Fig. 1. Schematic of foam fractionation of a dilute bromelain solution.

foam at the air/liquid interface, which, in turn, rose up to the top of the column. The bubbles and foam were enriched with bromelain as the bubbles rose in the column. The air was humidified by bubbling it through water before feeding it into the column. The humidifier also served to trap undesired proteins found in the dust in the air. The air superficial velocity was measured with an inline rotameter. The foamate from the top of the column and the residue were collected and their volumes measured. The superficial air velocity was varied between 1.7 and 6.2 cm/s. Although this process was a rate-based collection, the recovery with varying time was not explored. The mass recovery (MR) of the total protein was analyzed with the collected foam until no more foam reached the top of the column. The okra coating on the column came from a previous experiment in which proteins from an okra solution were foamed. The column was cleaned with several rinsings of distilled water, but the residual protein coating the glass column was not scrubbed off with a laboratory brush. For the okra protein-free experiment, the glass column was scrubbed thoroughly.

Total Protein Assay

The total protein content in the bromelain solution was determined using the Coomassie blue (Bradford [5]) method with a Bausch and Lomb Spectronic 20 spectrophotometer set at 595 nm (5,6). In all the assays performed, 2 mL of Coomassie blue reagent was added to 3 mL of each sample solution. The optical absorbance was read 5 min after adding the reagent.

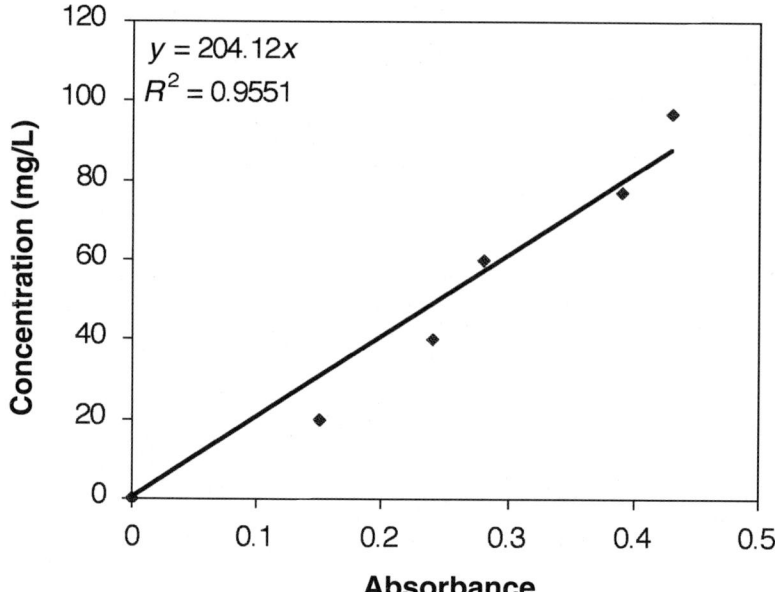

Fig. 2. A linear correlation between concentration of Coomassie blue (g/L) and absorbance at 595 nm.

The relationship between the optical absorbance and the known total protein content was correlated to determine the unknown total protein content during the experiment. The following calibration relationship was developed to determine the total protein content, as shown in Fig. 2 (drawn through the origin), and used for the bromelain assay:

$$\text{Bromelain (protein) concentration (mg/L)} = 204.12 \times (\text{absorbance @ 595 nm})$$

Results and Discussion

Without a contaminant on the glass wall of the foam fractionation column, bromelain foam was readily created in the column over the entire studied pH range of 2.0–7.0. The air superficial velocity was set at the intermediate value of 2.8 cm/s, as well as other values. The MR is one measure of the amount of foam produced; Figure 3 shows the okra-coated case. This foam fractionation can also be described by the partition coefficient, as shown in Fig. 4. A partition coefficient value of 1 indicates that there is no concentration of proteins in the foam, regardless of the amount of foam produced. The partition coefficient, K_p, is defined as the ratio of the protein (here, bromelain) concentration in the overhead collapsed foam (C_{foam}) to the remaining protein concentration in the remaining residual solution (C_{residue}):

$$K_p = [(C_{\text{foam}})/(C_{\text{residue}})]$$

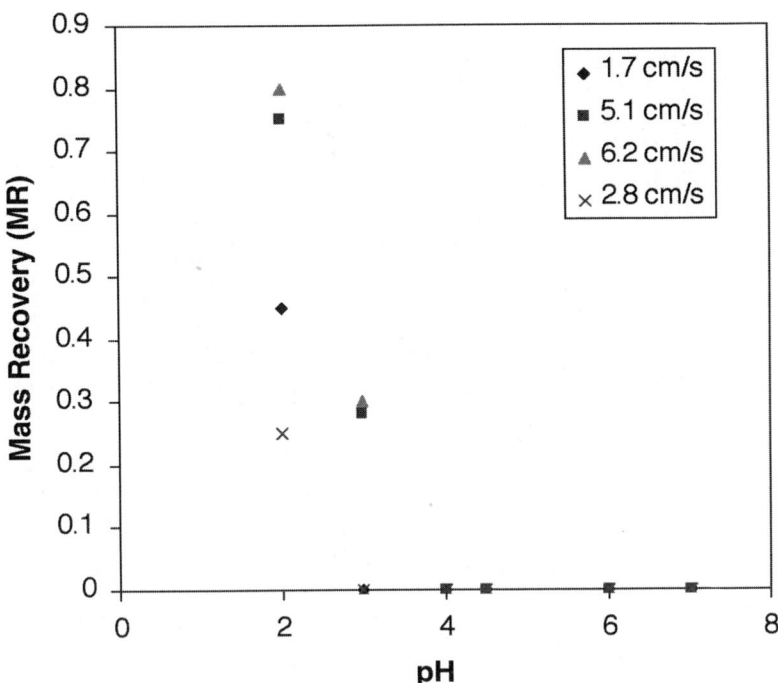

Fig. 3. Total partition coefficient, K_p, for the foam fraction at pH 2.0–7.0. The initial concentration of the bromelain-invertase mixed solution was 200 mg/L of total protein (100 mg/L of bromelain and 100 mg/L of invertase).

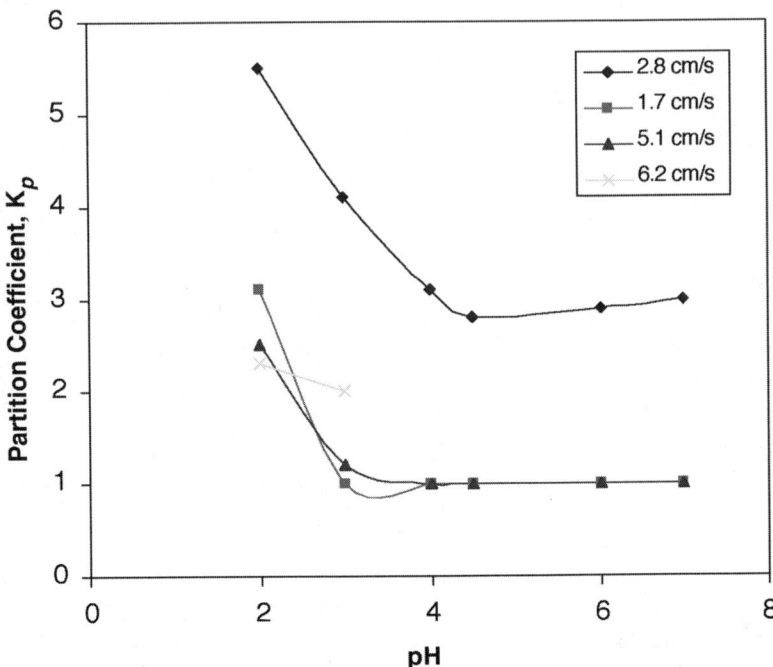

Fig. 4. Partition coefficient of proteins from the droplet fractionation experiment conducted at pH 2.0–7.0.

Here C is determined in milligrams of protein/liter. The maximum partition coefficient of 5.5 occurred at the low pH of 2.0, at which the bromelain may be denatured by adding an excessive amount of acid.

On the other hand, when okra residue (the natural contaminant/antifoaming agent) coated the foam fractionation column wall, foam was created only in a narrow pH range (between 2.0 and 3.0) at air superficial velocities ranging between 1.7 and 6.2 cm/s. With the okra residue, the partition coefficient reached a maximum value of 3.5 between pH 2.0 and 3.0, as shown in Fig. 4. In the pH 2.0–3.0 range, therefore, the partition coefficient with okra residue decreased by approx 50% relative to the partition coefficient without okra residue.

Figure 3 illustrates the effect of the bulk solution pH on the protein MR when the okra residue coats the foam fractionation glass column. MR is defined as the ratio between the recovered protein mass in the foamate and the initial protein mass in the foam fractionation column:

$$MR = (\text{protein mass in foamate})/$$
$$(\text{protein mass in initial bulk solution})$$

Here, the MR decreased with increasing bulk solution pH in the range between pH 2.0 and 3.0. It was observed that MR was negligible outside that range, at low airflow rates, since the bromelain solution did not foam significantly at pHs above 3.0 and because of the lower superficial air velocities. MR increased as the superficial air velocity increased. This appears to be owing to the enhanced foaming created by the additional aeration.

Conclusion

Based on our foam fractionation experiments, bromelain can be concentrated in the foamate phase from a dilute protein solution. When a natural contaminant coats the foam fractionation column, however, the foaming is significantly suppressed. The okra residue (natural contaminant) acts like an antifoaming agent and decreases the amount of foaming, leading to both a lower mass recovery and a lower partition coefficient in the pH 2.0–7.0 range. Negligible foaming occurred above pH 3.0 at air superficial velocities between 1.7 and 6.2 cm/s for the okra-inhibited system. A foam fractionation apparatus such as the one used here could be used to screen for natural antifoaming agents, which, in turn, could be compatible with commercial fermentation and other microbial processes.

Acknowledgment

This article is based on work supported by the National Science Foundation under Grant No. CTS-9712486.

References

1. <http://www.emperoraquatics.com/CommercialFiltration/foamcomm.htm>, Emperor Aquatics, Pottstown, PA.
2. Loha, V., Tanner, R. D., and Prokop, A. (1998), in *Advances in Biotechnology*, Pandey, A., ed., Educational Publishers and Distributors, New Delhi, pp. 245–356.
3. Izaka, K., Yamada, M., Kawano, T., and Suyama, T. (1972), *Jpn. J. Pharmacol.* **22,** 519–534.
4. <http://www.vitamins.com/encyclopedia/Supp/Bromelain.htm>, Healthnotes.
5. Bradford, M. M. (1976), *Anal. Biochem.* **72,** 248–264.
6. Ko, S., Loha, V., Du, L., Prokop, A., and Tanner, R. D. (1999), *Appl. Biochem. Biotechnol.* **77/79,** 501–510.

Application of Factorial Design to Study of Heavy Metals Biosorption by Waste Biomass from Beverage Distillery

MARISTELLA A. DIAS,[1] CARLOS A. ROSA,[1] VALTER R. LINARDI,[1] ROSA A. CONTE,[2] AND HEIZIR F. DE CASTRO*,[3]

[1]Department of Microbiology, Universidade Federal de Minas Gerais, PO Box 486, 30270-900, Belo Horizonte-MG, Brazil; and Departments of [2]Materials and [3]Chemical Engineering, Faculdade de Engenharia Química de Lorena, PO Box 116, 12600-000, Lorena-SP, Brazil, E-mail: decastro@easygold.com.br

Abstract

A full factorial design leading to 20 sets of sorption runs was conducted to study the influence of four variables (bleaching earth and biomass concentrations, pH, and sorption time) on the iron, nickel, and chromium removal from stainless steel effluent using waste biomass from a beverage industry. Similar factor effects and interactions were found for each metal involved in this biosorption study, and the main factors were pH (positive effect) and biomass concentration (negative effect). Response surface methodology was adopted and an empirical linear polynomial model constructed on the basis of the specific uptake (mg of metal/g of biomass as dry weight) for each metal species. Under optimized process conditions (pH 4.0, biomass concentration of 2.0 g/L, absence of Celite), uptake values of 155 mg of Fe/g, 38 mg of Cr/g, and 0.4 mg of Ni/g were achieved after 3 h. This corresponded to a reduction in heavy metals concentration of approx 94% for Cr, 57% for Fe, and 25% for Ni.

Index Entries: Biosorption; waste biomass; heavy metals; experimental design; effluent detoxification.

Introduction

Removal of heavy metals by either active or inactive microorganisms is a promising technique for detoxication of highly pollutant industrial wastes *(1,2)*. The use of active biomass can be a quite effective method by employing resistant microorganisms that are able to overcome the process

*Author to whom all correspondence and reprint requests should be addressed.

limitations including the metal toxicity to cell growth and the action of other toxic pollutants present in the effluent *(3,4)*. On the other hand, the use of inactive microorganisms can offer several advantages making this a method of choice from both economic and environmental points of view *(5–9)*. Inactive biomass is not ruled by physiologic restrictions and can be stored or used for extended periods at room temperature without the onset of putrefaction *(7,9)*. Besides having rapid and efficient metal uptake, the biomass behaves as an ion-exchange material and metals can be desorbed readily and recovered *(7)*. Moreover, if the biomass employed is a waste material (e.g., fermentation byproducts), biosorption represents a cheap alternative to conventional treatments, owing to the use of a low-cost sorbent material *(7,8)*.

Biomass samples obtained from a variety of waste streams have been reported to bind different metals, indicating that certain species might be better suited to particular metal pollutants *(1,2)*. For example, *Penicillium chrysogenum* and *Bacillus subtilis* have been shown to accumulate copper more effectively than many other microorganisms.

Interest in developing metal removal by biosorption using spent biomass has recently been indicated in the literature *(5–11)*. Suh et al. *(7)* compared the accumulation capacity of living and dead cells of *Saccharomyces cerevisiae* and *Aurebasidium pullulans* for lead. Stoll and Duncan *(8)* studied the metal sorptive properties of three types of immobilized nonviable *S. cerevisiae* biomass. Mattuschka and Straube *(9)* investigated the binding of several metals by pharmaceutical waste biomass (*Streptomyces noursei*), and Singleton and Simmons *(10)* examined the factors affecting silver biosorption by freeze-dried *S. cerevisiae* biomass produced in the brewing industry. Ferraz and Teixeira *(11)* used a flocculating brewer's yeast as biosorbent material for the removal of Cr and Pd from residual wastewater. They also determined the influence of physicochemical factors such as pH, biomass concentration, and the presence of a coion.

Our efforts in this area also concern the use of spent biomass as a biosorbent material. In a previous work, the biosorptive capacity of four types of waste biomass from beverage distilleries (*cachaça*) for different metal ions was investigated *(12)*. The metal affinities of these biomass types varied considerably, and best performance was achieved with spent biomass derived from Germana distillery, owing to its high capacity for retaining heavy metals from both synthetic and industrial effluent solutions.

Metal sorption performance also depends on external factors such as pH, other ions in solution (which may be in competition), organic materials in solution (such as complexing agents and cell metabolic products that can cause metal precipitation), and temperature *(1,2)*. In the present study we report the results, beyond our previous work *(12,13)*, of evaluating some of these parameters in terms of the capacity of the selected waste biomass to bind more than one metal simultaneously. The application of statistical design was used to determine the optimum operating conditions for the system. This technique is a powerful approach well suited for processes in which several variables must be considered simultaneously, a situation frequently found in industrial effluents that contain various metals ions that interfere with metal biosorption.

Materials and Methods

Sorbent Materials

Waste biomass was obtained from the stillage generated by a liquor distillery (Germana) situated in the state of Minas Gerais (Brazil). Stillage was collected and centrifuged for 20 min at 2000g at 25°C. The recovered biomass was washed three times with deionized water and dried at 60°C to a constant weight. Celite (341) was obtained from Aldrich (Milwaukee, WI).

Effluent and Biosorption Trials

Effluent was collected from a stainless steel company (Acesita) in the state of Minas Gerais (Brazil) and diluted with deionized water to reduce the iron (III) concentration to a level of 600 mg/L. Other metal species were also present: 2.7 mg/mL of nickel, 7.8 mg/mL of chromium (VI). Biosorption runs were carried out in 250-mL Erlenmeyer flasks containing 50 mL of industrial effluent solution. Before mixing with the waste biomass and/or Celite, the pH was adjusted to various values between 2.0 and 4.0. The flasks were shaken (150 rpm) at constant temperature (30°C) for a maximum of 6 h. Control runs were also performed without biomass to verify eventual metal precipitation. Flask contents were filtered using 0.45-µm Millipore filter membranes, and the filtrates were analyzed by flame atomic absorption spectrophotometry (Varian model AA-475) for residual metal content *(14)*.

Experimental Design

The influence of Celite concentration (X_1), biomass concentration (X_2), initial pH (X_3), and time course of biosorption (X_4) was studied using a 2^4 full factorial design. Table 1 gives the range and levels of the studied variables. The runs were performed at random. Four experiments were carried out at the center point level, for estimation of experimental error. Data processing and calculations were carried out using Statistica (version 5.0) software. The statistical significance of the regression coefficients was determined by student's t-test *(15)*, the model equation was determined by Fisher's test *(15)*, and the proportion of variance explained by the model obtained was given by the multiple coefficient of determination, R^2. The response of the process under investigation was the metal-specific uptake (q in milligrams/gram) calculated according to Eq. 1:

$$q = [(C_0 - C)/X] \tag{1}$$

in which q (mg of metal/g of biomass) is the metal-specific uptake, C_0 (mg/L) is the initial metal concentration, C is the residual metal concentration, and X (g/L) is the biomass concentration.

Table 1
Experimental Range and Levels
of Independent Process Variables According to 2^4 Full Factorial Design

Variable	Symbol	Level		
		−1	0	+1
Celite (g/L)	X_1	0	2.0	4.0
Biomass (g/L)	X_2	2.0	5.0	8.0
pH	X_3	2.0	3.0	4.0
Time (h)	X_4	3.0	4.5	6.0

Table 2
Experimental Design and Results According to 2^4 Full Factorial Design

	Variable				Response		
Run	X_1 (g/L)	X_2 (g/L)	X_3	X_4 (h)	Fe (mg/g)	Ni (mg/g)	Cr (mg/g)
1	0	2.0	2.0	3	0	0.15	1.20
2	4.0	2.0	2.0	3	0	0.05	0.31
3	0	8.0	2.0	3	10.4	0.05	1.25
4	4.0	8.0	2.0	3	5.2	0.02	0.60
5	0	2.0	4.0	3	155.5	0.40	38.05
6	4.0	2.0	4.0	3	40.5	0.11	12.40
7	0	8.0	4.0	3	43.2	0.08	9.10
8	4.0	8.0	4.0	3	25.7	0.06	6.13
9	0	2.0	2.0	6	3.2	0.10	2.80
10	4.0	2.0	2.0	6	8.3	0.05	1.85
11	0	8.0	2.0	6	0	0.02	0.45
12	4.0	8.0	2.0	6	0.2	0.02	0.25
13	0	2.0	4.0	6	153.0	0.40	37.85
14	4.0	2.0	4.0	6	38.1	0.05	12.10
15	0	8.0	4.0	6	41.7	0.08	9.27
16	4.0	8.0	4.0	6	29.0	0.05	6.16
17	2.0	5.0	3.0	4.5	36.3	0.08	7.58
18	2.0	5.0	3.0	4.5	30.4	0.08	7.64
19	2.0	5.0	3.0	4.5	19.1	0.05	4.00
20	2.0	5.0	3.0	4.5	20.4	0.05	4.00

Results

The effects of four variables (Celite and biomass concentrations, pH, and sorption time) on metal-binding affinity of the waste biomass were simultaneously investigated using a 2^4 full factorial design leading to 20 sets of experiments. The choice of variables was made based on their importance to this kind of process. The association of waste biomass (biosorbent) with inorganic sorbent (Celite) was used as a strategy to enhance the iron uptake as previously described by Dias et al. *(13)*. The design of this experiment is given in Table 2, together with the experimental results.

The results clearly showed that biomass adsorption capacity was strongly affected by initial pH, independently of the other variables. At pH 2.0 (runs 1–4 and 9–12) the biomass adsorption capacity was very low for Fe and Ni uptakes (<12 mg/g). This was likely owing to the nature of chemical interactions of each metal species and the waste biomass; at low pH, the overall surface charge on the biomass will become positive, which will inhibit the approach of positively charged metal cations. It is likely that protons then compete with metal ions for the binding sites and thereby decrease the interaction of biomass and metal ions *(7,8)*. By increasing the pH from 2.0 to 4.0 (runs 5–8 and 13–18), the biomass sorption capacity for all studied metal species reached their maximum values.

Another important factor in the biosorption performance was biomass concentration (Table 2). At a biomass concentration of 2 g/L, specific metal uptake values were maximized (runs 5 and 13). As the biomass concentration increased to the high level (8 g/L), a decrease in capacity was observed, suggesting an inverse relationship between biomass concentration and its biosorptive capacity. Similar behavior has been found for other biosorbent types, as reported by Singleton and Simmons *(10)*. Various reasons have been suggested to explain this effect, including limited availability of solute, electrostatic interactions, interference between binding sites, and reduced mixing at high biomass densities *(10)*. The time course of biosorption trials did not promote any positive effect on the parameters evaluated, showing that 3 h was sufficient to attain adsorption equilibrium. On the other hand, the addition of Celite to the biosorptive runs exerted a negative influence for all response variables. This suggests that, in the range investigated, association of sorbent materials is not necessary to attain high metal uptake.

The experimental results shown in Table 2 were used to estimate the main effects of variables and their interactions. The statistical analyses for each of the response variables evaluated—Fe, Ni, and Cr uptake—are summarized in Table 3. According to student's *t*-test results, the effects and interactions are similar for all metals involved in this biosorption. The most important variable was pH (X_3), since it presented a significant effect for Ni response (95% confidence level) and for both iron and chromium responses (99% confidence level). The effect of Celite (X_1) and biomass concentration (X_2) and their interactions with pH (X_1X_3, X_2X_3 and $X_1X_2X_3$) were also significant at the same confidence level ($p < 0.05$).

Based on the response evaluated, mathematical models were developed for each metal species. The main effects and their interactions for the responses were fitted by multiple regression analysis (Table 4) to a linear model since tests for curvature ($p > 0.05$) using center points showed that square terms were not important for the models.

By analyzing Table 4, it can be seen that polynomial models were adequate for describing the relationships among all the responses under study and the experimental factors. The regression models were highly significant ($p < 0.001$), at a 99% confidence level, and presented high deter-

418

Dias et al.

Table 3
Estimated Effects, Standard Errors, and Student's t-Test for Fe, Ni, and Cr Uptake (mg/g) Using 2^4 Full Factorial Design

Variable	Fe (mg/g)			Ni (mg/g)			Cr (mg/g)		
	Effect	Standard error	t-Value	Effect	Standard error	t-Value	Effect	Standard error	t-Value
Mean	33.00	±2.03	—	0.98	±0.11	—	8.15	±0.63	—
X_1	-32.50	±4.54	-7.15^a	-0.11	±0.25	-4.41^a	-7.52	±1.42	-5.28^a
X_2	-30.40	±4.54	-6.69^a	-0.12	±0.25	-4.72^a	-9.17	±1.42	-6.44^a
X_3	-62.42	±4.54	13.73^a	0.10	±0.25	3.91^b	15.29	±1.42	10.74^a
X_4	-0.87	±4.54	-0.19	-0.02	±0.25	-0.76	0.21	±1.42	0.15
X_1X_2	23.70	±4.54	-5.21^b	0.09	±0.25	3.60^b	5.78	±1.42	4.10^a
X_1X_3	-32.52	±4.54	-7.15^a	-0.06	±0.25	-2.59^b	-6.84	±1.42	-4.81^a
X_1X_4	1.92	±4.54	0.42	0.01	±0.25	0.05	0.01	±1.42	0.01
X_2X_3	-31.47	±4.54	-6.92^a	-0.05	±0.25	-2.28	-8.26	±1.42	-5.81
X_2X_4	-2.52	±4.54	-0.55	0.09	±0.25	0.35	-0.44	±1.42	-0.31
X_3X_4	0.10	±4.54	0.02	0.01	±0.25	0.05	-0.28	±1.42	-0.20
$X_1X_2X_3$	26.22	±4.54	5.77^b	0.06	±0.25	0.24^b	5.54	±1.42	3.89^b

$^a p < 0.01$.
$^b p < 0.05$.

Table 4
Analysis of Variance for Model Regression
That Represents Specific Metal Uptake

Source	Degrees of freedom	Sum of square			Mean square		
		Fe	Ni	Cr	Fe	Ni	Cr
Model	6	36,701.1	0.21	2215.9	5243.0	0.03	316.5
Residual	13	530.6	0.01	43.4	44.2	0.00	3.62
Total	19	37,231.7	0.22	2259.3			
F-ratio		118.6	37.5	87.43			
p-value		0.000	0.000	0.000			
R^2		0.98	0.95	0.98			

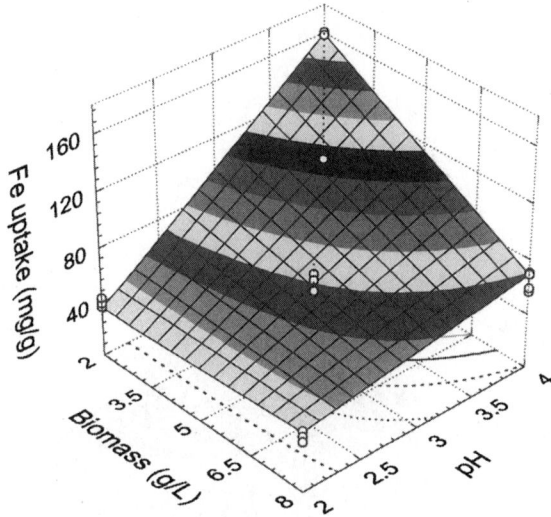

Fig. 1. Response surface described by the model $\hat{y}_1$ that represents specific Fe uptake by waste biomass.

mination coefficients ($R^2 > 0.95$). Thus, mathematical models representing the specific metal uptake in the range studied can be expressed by the following equations:

$$\hat{y}_1 = 33.0 - 16.2x_1 + 15.2x_2 + 31.2x_3 + 11.8x_1x_2 - 16.3x_1x_3 + 1.7x_2x_3 + 13.1x_1x_2x_3 \quad (2)$$

$$\hat{y}_2 = 0.09 - 0.5x_1 - 0.06x_2 + 0.05x_3 + 0.04x_1x_2 - 0.03x_1x_3 - 0.03x_2x_3 + 0.03x_1x_2x_3 \quad (3)$$

$$\hat{y}_3 = 8.1 - 3.8x_1 - 4.6x_2 + 7.6x_3 + 2.9x_1x_2 - 3.4x_1x_3 - 4.1x_2x_3 + 2.8x_1x_2x_3 \quad (4)$$

in which $\hat{y}_1, \hat{y}_2, \hat{y}_3$ are predicted values for Fe, Ni, and Cr uptake and $x_1, x_2,$ and x_3 are coded values for Celite concentration, biomass concentration, and pH, respectively.

The response surface described by the model equations are represented in Figs. 1–3, and the predicted values were found to be 155 mg/g for

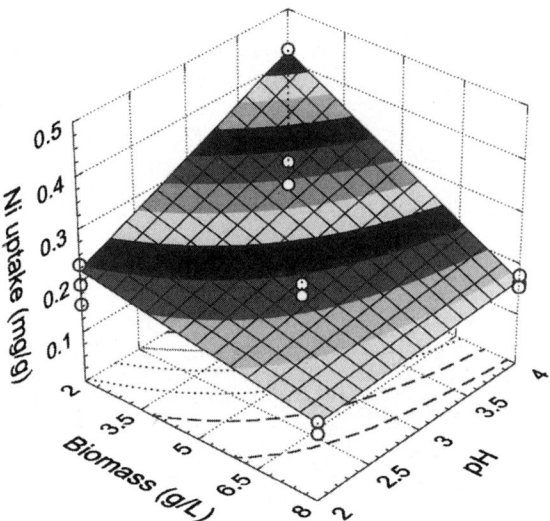

Fig. 2. Response surface described by the model $\hat{y}_2$ that represents specific Ni uptake by waste biomass.

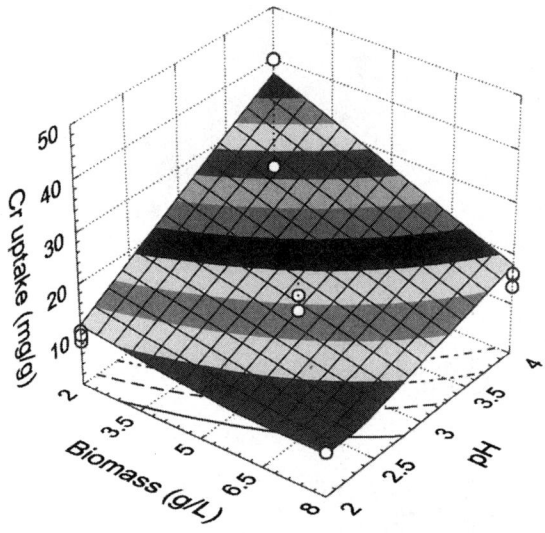

Fig. 3. Response surface described by the model $\hat{y}_3$ that represents specific Cr uptake by waste biomass.

Fe, 39 mg/g for Cr, and 0.4 mg/g for Ni working at a high level for pH, low level of biomass concentration, and absence of Celite.

Discussion

Biosorption of heavy metals to nonviable cells can be used as a potential technology for removal of toxic metals from industrial waste streams, in terms of efficiency and operational suitability. In the present study, waste biomass from the production of *cachaça* was used as a biosorbent to remove heavy metals from stainless steel production effluents. *Cachaça* is a typical Brazilian alcoholic beverage with an alcohol content in the range of 30–43%, and its manufacture by spontaneous fermentation of sugarcane juice is a traditional expertise in the state of Minas Gerais. Unlike fuel alcohol industries, the stillage generated by these distilleries has solid contents because the biomass is not recovered prior to the distillation step *(12,16)*. Therefore, the microbiota involved in the production of *aguardente* is very diversified including species from *Saccharomyces*, apiculate yeasts, *Candida*, and *Schizosaccharomyces* strains rendering to the waste biomass suitable characteristics to be applied to the removal of metals from industrial effluents, as shown in previous work *(12,13)*.

We extended the treatment of multimetal ion effluents by evaluating the effects of Celite and biomass concentrations, pH, and sorption time on the specific uptake (mg of metal/g of biomass as dry weight) by waste biomass employing a multivariate statistical approach. A 2^4 factorial design with center point was adopted for a full understanding of these effects and their interactions. The specific uptake was found to be dependent on both the operating conditions and metal species, although similar factor effects and interactions are found for each metal involved in this biosorption. Because the time course of biosorption runs was not involved in any significant value in terms of the models, the responses were well modeled by a linear function of the other three independent variables. The maximum metal uptake was obtained working at the highest level of pH (4.0), minimum level of biomass (2.0 g/L), and absence of Celite with a contact time of 3 h. Under such conditions, specific uptake values were found to be 155 mg/g for Fe, 38 mg/g for Cr, and 0.4 mg/g for Ni.

Sorption of multimetal ions by biomass is a complex function of the metal combination, levels of metal concentration, and specific metal uptake. In addition, there appears to be competition for adsorption sites on the biomass in such a way that higher specific metal uptake can be observed for single metal solutions.

In the particular case in which Fe was the dominant metal specie in the effluent solution, its removal was favored and lower adsorption of Cr was exhibited. This is typical antagonism behavior in which the effect of the mixture is less important than that of each individual effect of the constituents in the mixture *(1,2)*. On the other hand, no interaction between these metals with Ni was observed. According to the results obtained (reduction in the levels of about 57% for Fe, 25% for Ni, and 94% for Cr), two-stage treatment should be considered to remove completely the high level of Fe from this effluent stream.

References

1. Gomes, N. C. M., Mendonça-Hagler, I. C. S., and Savvaids, I. (1998), *Revista Microbiologia* **29**, 85–92.
2. Volesky, B. (1994), *FEMS Microbiol. Lett.* **14**, 291–392.

3. Gomes, N. C. M., Camargo, E. R. S., Dias, J. C. T., and Linadi, V. R. (1998), *World J. Microb. Biotechnol.* **14,** 149.
4. Sag, Y. and Kutsal, T. (1996), *Process Biochem.* **31,** 573–585.
5. Hu, M. Z.-C., Norman, J. M., Faison, B. D., and Reeves, M. E. (1996), *Biotechnol. Bioeng.* **51,** 237–247.
6. Hu, M. Z.-C. and Reeves, M. E. (1997), *Biotechnol. Prog.* **13,** 60–70.
7. Suh, J. A., Yun, J. W., and Jim, D. S. (1998), *Biotechnol. Lett.* **20,** 247–251.
8. Stoll, A. and Duncan, J. R. (1997), *Process Biochem.* **32,** 467–472.
9. Mattuschka, B. and Straube, G. (1993), *J. Chem. Technol. Biotechnol.* **58,** 57–63.
10. Singleton, I. and Simmons, P. (1996), *J. Chem. Technol. Biotechnol.* **65,** 21–28.
11. Ferraz, A. I. and Teixeira, J. A. (1999), *Bioprocess. Eng.* **21,** 431–437.
12. Dias M. A., Castro, H. F., Pimentel, P. F., Gomes, N. C. M., Rosa, C. A., and Linardi, V. R. (2000), *World J. Microb. Biotechnol.* **16,** 107, 108.
13. Dias, M. A., Castro, H. F., Rosa, C. A., and Linardi, V. R. (1999), in *XX Congresso Brasileiro de Microbiologia*, Book of Abstracts, p. 149.
14. American Public Health Association. (1992), in *Standard Methods for the Examination of Water and Wastewater*, 18th ed., American Public Health Association, Washington, DC, pp. 3–18.
15. Box, G. E. P., Hunter, W. G., and Hunter, J. S. (1978), in *Statistics for Experimenters: An Introduction to Design, Data Analysis and Model Building*, Wiley & Sons, New York, p. 653.
16. Morais, P. B., Rosa, C. A., Linardi, V. R., Pataro, C., and Maia, A. B. R. A. (1997), *World J. Microb. Biotechnol.* **13,** 241–243.

Model Compound Studies

Influence of Aeration and Hemicellulosic Sugars on Xylitol Production by Candida tropicalis

THOMAS WALTHERS,[1] PATCHAREE HENSIRISAK,[2] AND FOSTER A. AGBLEVOR*,[2]

[1]*Department of Mechanical Engineering, Technical University of Dresden, Dresden, Germany; and* [2]*Department of Biological Systems Engineering, Virginia Polytechnic Institute and State University, 212 Seitz Hall, Blacksburg, VA 24061-0303, E-mail: fagblevo@vt.edu*

Abstract

The influence of other hemicellulosic sugars (arabinose, galactose, mannose, and glucose), oxygen limitation, and initial xylose concentration on the fermentation of xylose to xylitol was investigated using experimental design methodology. Oxygen limitation and initial xylose concentration had strong influences on xylitol production by *Candida tropicalis* ATCC 96745. Under semiaerobic conditions, xylitol yield was highest (0.62 g/g), whereas under aerobic conditions volumetric productivity was highest (0.90 g/[L·h]). In the presence of glucose, xylose utilization was strongly repressed and sequential sugar utilization was observed. Ethanol produced from the glucose caused a 50% reduction in xylitol yield when the ethanol concentration exceeded 30 g/L. When complex synthetic hemicellulosic sugars were fermented, glucose was initially consumed followed by a simultaneous uptake of the other sugars. The highest xylitol yield (0.84 g/g) and volumetric productivity (0.49 g/[L·h]) were obtained for substrates containing high arabinose and low glucose and mannose contents.

Index Entries: Xylitol; fermentation; aeration; hemicellulose.

Introduction

Xylitol is a naturally occurring sugar alcohol. Its high sweetening power, anticariogenic properties, and possibilities for use in diabetic food products *(1)* makes xylitol an attractive sucrose substitute in a wide variety of foods and beverages. Xylitol can be produced by biologic reduction of xylose, a five-carbon sugar, using microorganisms such as *Candida guilliermondii (2)*, *Pichia stiptis*, *Pachysolen tannophilus (3)*, and *Candida tropicalis (4)*.

Xylose is a major component of hemicellulose, an abundant raw material. Hemicellulose hydrolyzes into a complex mixture of sugars that

*Author to whom all correspondence and reprint requests should be addressed.

Table 1
Composition of Hemicellulose Hydrolysates Expressed
as Percentage of Total Sugar

Hydrolysate	Xylose	Glucose	Arabinose	Galactose	Mannose	Ref.
Sugarcane	58	16	26	—	—	7
bagasse	75	14	11	—	—	2
Rice straw	67	21	12	—	—	2
Hardwood	70	14	5	5	5	3
	62	16	4	8	9	8
	27	11	5	14	43	9
Corn fiber	16	71	11	2	—	10
	31	41	25	4	—	5
Isolated corn fiber xylan	26–60	16–37	24–46	—	—	5

include arabinose, glucose, galactose, and mannose *(2,3,5)*. These sugars may influence xylitol yield and productivity during xylose fermentation.

Corncobs, hardwoods, sugarcane bagasse, the seed coats of rice, soybeans, and corn are sources of low-cost hemicellulose *(6)*. The composition of hemicellulose hydrolysates varies widely depending on the raw material used, hydrolysis procedures, and pretreatment methods employed *(5,7)*. Table 1 gives the compositions of several hemicellulosic substrates reported in the literature. For xylitol production using hemicellulose hydrolysate, the process is affected by the concentrations of the sugars in the fermentation medium, the ratios at which these sugars occur, as well as the toxic compounds released during the hydrolysis. For instance, high concentrations of monomeric sugars could cause osmotic stress, inhibit induction of xylose reductase enzymes, or lead to ethanol production that exceeds the tolerance level of the yeast. Additionally, the ratio of monomeric sugars may influence transport or enzyme kinetics, in cases in which both sugars compete for the same transport system or are metabolized simultaneously.

The goal of the present study was to use hemicellulose hydrolysate as the feedstock for the microbial production of xylitol. Because some hemicellulose hydrolysates contain high levels of glucose and arabinose in addition to other sugars, we conducted model sugar studies designed to simulate the influence of sugar composition and other fermentation parameters (in the absence of toxic hydrolysate components) on xylitol yield and productivity using *C. tropicalis* ATCC 96745 as a reference organism.

Materials and Methods

Microorganism

C. tropicalis ATCC 96745 was acquired from the American Type Culture Collection (Rockville, MD) and maintained on yeast extract, peptone, dextrose agar slants at 4°C. The microorganisms were subcultured every 2 wk.

Culture Media

The preculture medium contained 60 g/L of xylose, 10 g/L of yeast extract, 15 g/L of KH_2PO_4, 3 g/L of $(NH_4)_2HPO_4$, 1 g/L of $MgSO_4 \cdot 7H_2O$, and three drops of Sigma 289 antifoaming agent (Sigma, St. Louis, MO). The pH was adjusted to 5.0 using 1 *M* HCl *(4)*.

The production medium contained 20 g/L of yeast extract, 15 g/L of KH_2PO_4, 3 g/L of $(NH_4)_2HPO_4$, 1 g/L of $MgSO_4 \cdot 7H_2O$, and three drops of Sigma 289 antifoaming agent. The concentrations of xylose, glucose, arabinose, galactose, and mannose were adjusted according to the experimental design. Sugar and salt solutions were autoclaved separately for 20 min at 121°C. Weight loss, which occurred after sterilization, was made up with the addition of sterile water. The pH was adjusted to 4.0 using 1 *M* HCl. All chemicals were reagent grade, obtained from Sigma.

Fermentation Conditions

Precultures were grown at 30°C in 500-mL Erlenmeyer flasks containing 250 mL of medium that were agitated at 130 rpm on a rotary platform shaker (Innova 2050). The cells were harvested after 14–16 h during the midexponential growth phase. The cells were centrifuged (11,000g for 10 min), decanted, washed with sterile water, and recentrifuged. The inoculum was transferred to the production medium at an initial cell concentration of 0.5 g/L. Fermentation was carried out in 250-mL cotton-plugged Erlenmeyer flasks at 30°C and agitated at 130 rpm on a rotary platform shaker. Aeration levels were adjusted by varying the volume of medium *(11)* at a constant agitation speed of 130 rpm.

Analytical Methods

Dry cell mass was estimated from an optical density/dry cell weight calibration curve. Spectrophotometric measurements were carried out at 640 nm using a Spectronic 1001 instrument (Milton Roy, Rochester, NY). After determining the absorbance at 640 nm, the samples were dried to a constant weight at 105°C in a Thelco laboratory oven (Precision Scientific, Chicago, IL). The gravimetric and spectrophotometric data were used to develop a calibration curve. The fermentation samples for xylose and mixed sugars were diluted with deionized water 10- and 20-fold, respectively, before the spectrophotomeric analysis.

Sugar and sugar alcohols were analyzed using a high-performance liquid chromatograph (Shimadzu, Columbia, MD) equipped with a refractive index detector. A carbohydrate column (Supelcogel™ Ca, 30 cm × 7.8 mm; Supelco, Bellefonte, PA) was used for the analysis. Column temperature was 80°C and filtered deionized water was used as the mobile phase. The mobile phase flow rate was raised linearly over the course of

20 min, beginning at 0.5 mL/min with a final value of 2 mL/min. Peaks were detected by refractive index and were identified and quantified by comparison to retention times of authentic standards (xylose, glucose, mannose, galactose, arabinose, ethanol, and xylitol).

For high-performance liquid chromatography analysis, 400-μL aliquots were diluted with 1200 μL of deionized water in 2-mL plastic test tubes, centrifuged at 26,000g for 10 min, and decanted. The samples were filtered through a 0.2-μm syringe filter before injection of 20 μL into the column.

In cases in which the fermentation medium contained multiple sugar mixtures, the resolution of the mannose and galactose peaks on the Supelcogel Ca column was poor. Consequently, a complementary sugar analysis was carried out on a gas chromatograph (Shimadzu gas chromatograph, GC-14A) according to ASTM standard method E 1821-96. The following chromatographic conditions were used for the analysis: column, Supelco SP-2380 (30 m, 0.25 mm id, 0.2-μm film thickness); carrier gas, helium; column flow rate, 0.6 mL/min; total gas flow rate, 64 mL/min; split ratio, 101:1; detector, flame ionization at 220°C; injection temperature, 240°C; and sample size, 1 μL. Shimadzu CLASS-VP™ software was used for temperature programming and data retrieval.

Experimental Design

To quantify the influence of initial xylose concentration and aeration on the production of xylitol, a 2^2 factorial experimental design was applied with four star points ($a = 1.41$) and five replications at the center point *(12)*. Initial xylose concentration was varied from 23.5 to 156 g/L. The medium volume was varied from 50.5 to 150 mL to simulate microaerobic, semi-aerobic, and aerobic conditions (*see* Table 2).

In the fermentation of glucose/xylose mixtures, glucose concentrations were varied between 0 and 80 g/L, and initial xylose concentration was kept constant at 60 g/L. To simulate different aeration conditions, three levels of medium volume were used (65, 100, and 135 mL) in 250-mL Erlenmeyer flasks. The three medium levels were classified according to Nolleau et al. *(11)* as aerobic (65 mL), semiaerobic (100 mL), and microaerobic (135 mL).

For the complex sugar mixtures, a second-order experimental design was developed. This experimental design covered the possible variations in actual sugar concentrations as well as the variation in the ratios of these sugars found in most hemicellulose hydrolysates (Table 1). Table 3 gives the design parameters for glucose, arabinose, galactose, mannose, and medium volume.

With second-order polynomials, only one local extremum can be modeled. Hence, the amount of glucose was varied below a concentration of 60 g/L. Preliminary experiments using high arabinose concentrations showed that although it was not significantly fermented by the yeast, it had a stimulating effect on xylitol production (data not shown). Thus, the influence of arabinose was tested at concentrations as high as 80 g/L. Galactose

Table 2
Experimental Design and Results
for Xylitol Yield and Productivity from Xylose Fermentation

Coded value	Xylose A (g/L)[a]	Medium B (mL)[b]	Xylitol yield (g/g) observed	Xylitol productivity (g/[L·h]) observed
1	120	135	0.65	0.41
−1	40	135	0.37	0.12
−1	40	65	0.50	0.39
1	120	65	0.64	0.74
+a	156	100	0.69	0.70
−a	23.5	100	0.48	0.22
0	80	150	0.57	0.22
8	80	50.5	0.59	0.67
9	80	100	0.61	0.43
10	80	100	0.61	0.44
11	80	100	0.63	0.46
12	80	100	0.63	0.46
13	80	100	0.63	0.46

[a]A = initial xylose concentration.
[b]B = medium volume in shake flask.

Table 3
Experimental Design for Complex Sugar Mixtures
Showing Star (±a), Axis (±1), and Center (0) Point Values

Coded variable	Parameter	Unit	+a	1	0	−1	−a
A	Glucose	g/L	60	42.6	30	17.4	0
B	Arabinose	g/L	80	56.8	40	23.2	0
C	Galactose	g/L	10	7.1	5	2.9	0
D	Mannose	g/L	5	3.55	2.5	1.45	0
E	Medium	mL	130.7	110	95	80	59.3

and mannose are minor components of most hemicellulose hydrolysates (Table 1) except in some wood hydrolysates, in which they are significant. To limit osmotic stress, these two sugars were varied in proportion to the composition of most hemicellulose hydrolysates except those found in some wood hydrolysates. Table 3 lists limits of the design parameters and their corresponding coded values. Xylose concentration was kept constant at 60 g/L.

Results and Discussion

Effect of Initial Xylose Concentration and Volume of Medium

C. tropicalis utilized the accumulated xylitol and ethanol when xylose concentrations were very low (<1.0 g/L). Consequently, xylitol yields and

Table 4
Regression Equations
for Xylitol Yield and Productivity from Fermentation of Xylose

Parameter	Regression equation[a]	R^2
Xylitol yield (g/g)	$0.6220 + 0.0896A - 0.0185B + 0.0350A \times B - 0.0291A^2 - 0.0316B^2$	0.92
Xylitol productivity (g/[L·h])	$0.4500 + 0.1669A - 0.1546B$	0.99

[a]A = coded value for initial xylose concentration; B = coded value for medium volume.

productivities were determined at the maximum product concentration and not at the end of the run. The experimental results in Table 2 were used to estimate the main effects of the variables and their interactions. Polynomial models were used to establish the relationships between the dependent variables (xylitol yield and xylitol productivity) and the independent variables (initial xylose concentration and aeration). The data were fitted to the following model equations:

$$\text{Xylitol yield} = \beta_0 + \beta_1(A) + \beta_2(B) + \beta_3(A \times B) + \beta_4(A^2) + \beta_5(B^2)$$

$$\text{Xylitol productivity} = \alpha_0 + \alpha_1(A) + \alpha_2(B) + \alpha_3(A \times B) + \alpha_4(A^2) + \alpha_5(B^2)$$

The xylitol yield was fitted to the initial xylose concentration and volume of medium (aeration) by a second-order polynomial model. Analysis of variance (ANOVA) found that the linear, quadratic, and interaction terms were significant ($p < 0.05$). Table 4 gives the empirical model equations. Xylitol yield was correlated with high initial xylose concentration and low aeration ($R^2 = 0.92$). This trend indicates that xylitol is an overflow metabolite *(13)*. The highest xylitol yield calculated from the empirical model equation was 0.7 g/g, which was achieved at an initial xylose concentration of 156.5 g/L under semiaerobic conditions (117 mL of medium).

The positive interaction between initial xylose concentration and aeration can be explained in terms of cell density *(4)*. At high initial xylose concentrations and high aeration, the cells grew rapidly at the beginning of fermentation. This led to high cell densities and low oxygen levels in the later stages of the fermentation and resulted in high production rates. At lower initial xylose concentrations, cell densities were low and the level of dissolved oxygen remained high; therefore, less xylitol was accumulated.

The negative coefficient for the quadratic terms in the xylitol yield model also suggests that extremely high initial xylose concentrations will be detrimental to xylitol yields. This prediction could be attributed to osmotic stress, which could be induced in the microorganism by the excess amount of sugar in the medium. Thus, for this microorganism, there is an upper limit for the initial xylose concentration (156 g/L) that will not induce osmotic stress for low aeration rates. However, the osmotic stress can be counteracted by increased aeration. Thus, careful manipulation of

Table 5
Xylitol Yield and Productivity Data for Fermentation
of Mixtures Containing Xylose and Hemicellulosic Sugars[a]

Substrate	Fermentation time (h)	Ethanol concentration (g/L)	Y_{xyl} (g/g)	P_{xyl} (g/L·h)
Xylose	91.7	4.1	0.61	0.57
Xylose + 20 g/L of glucose	142	15.4	0.36	0.15
Xylose + 10 g/L of glucose	127	8.4	0.42	0.20
Xylose + 20 g/L of mannose	108.3	14.1	0.54	0.32
Xylose + 10 g/L of mannose	104.5	8.5	0.53	0.31
Xylose + 20 g/L of galactose	104.5	13.5	0.47	0.27
Xylose + 10 g/L of galactose	104.5	8.0	0.51	0.25
Xylose + 20 g/L of arabinose	129	4.2	0.54	0.37
Xylose + 20 g/L of arabinose	ND	ND	ND	ND

[a]All fermentations commenced with 60 g/L of xylose. ND, not determined.

both the aeration and initial xylose concentration probably can result in very high xylitol yields beyond those observed in these studies.

Xylitol productivity was fitted to initial xylose concentration and aeration by a linear regression model (Table 4). Xylitol productivity was correlated with high aeration rate and high initial xylose concentration ($R^2 = 0.99$). The highest productivity of 0.9 g/(L·h) was achieved at the highest aeration level (50.5 mL of medium volume).

In both the productivity and yield of xylitol, aeration appeared to have played a significant role. This is in agreement with the finding of Nolleau et al. *(11)*, who showed that by varying the volume of medium in a shaker flask, the aeration conditions could be varied sufficiently to enable investigations to be carried out. This approach was subsequently used to study the influence of aeration and hemicellulosic sugars on xylitol production.

Effect of Binary Sugar Mixtures

When *C. tropicalis* was cultivated in a medium containing single sugars as the carbon source, it fermented glucose, mannose, and galactose but was unable to utilize arabinose (data not shown). The yeast produced no xylitol from any of these sugars.

Similar to the xylose experiments, the xylitol yields and productivities for the xylose/glucose experiments were determined at the maximum product concentration instead of the end of the run. Table 5 gives the results of fermentation of binary sugar mixtures consisting of xylose (60 g/L) and other hemicellulosic sugars at two levels. Glucose strongly inhibited xylitol formation at all levels, and mannose and galactose caused moderate inhibition in xylitol production, whereas arabinose appeared to have a slightly positive effect on xylitol productivity.

The influence of initial glucose concentration and aeration regime on xylitol yield is shown in Fig. 1A. Xylitol productivity followed a trend

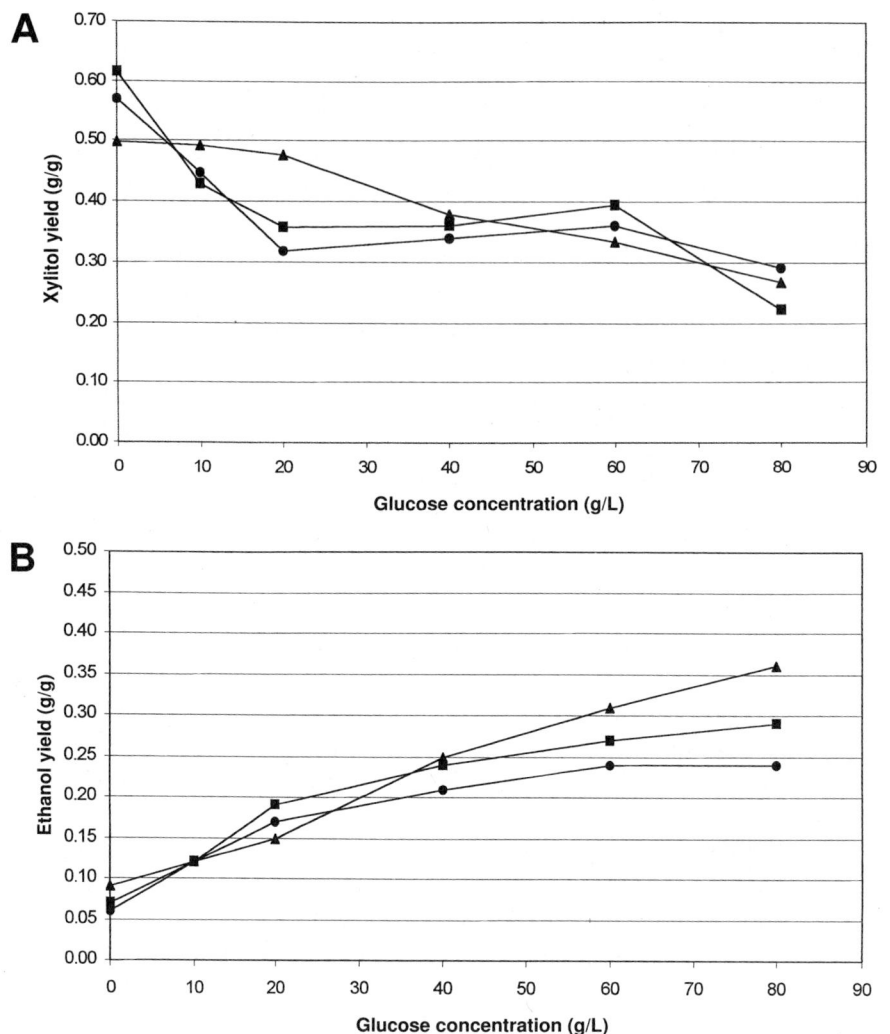

Fig. 1. **(A)** Observed xylitol yields in the fermentation of xylose/glucose mixtures. Aerobic, 65 mL of medium (●); semiaerobic, 100 mL of medium (■); and microaerobic, 135 mL of medium (▲). Initial xylose concentration was 60 g/L. **(B)** Observed ethanol yield from the fermentation of xylose/glucose mixtures. Aerobic, 65 mL of medium (●); semiaerobic, 100 mL of medium (■); and microaerobic, 135 mL of medium (▲). Initial xylose concentration was 60 g/L.

similar to that of the yield (data not shown). In the presence of glucose, xylitol yield was a function of the aeration regime. Xylitol yield was highest (0.62 g/g) under semiaerobic conditions in the absence of glucose. In the presence of glucose, xylitol yield was highest (0.50 g/g) under microaerobic conditions. Low initial glucose concentrations (10–20 g/L) did not have any appreciable effect on xylitol yield under microaerobic conditions, but above 20 g/L, xylitol yield decreased considerably. By contrast, for the aerobic and semiaerobic conditions, there were considerable decreases in xylitol yield at low initial glucose concentrations. The initial decrease in xylitol yields leveled off for both aeration regimes for glucose concentrations between 20 and 60 g/L. However, at 60 g/L or higher of glucose, xylitol yield decreased further.

Xylitol production was always accompanied by ethanol production. Ethanol yield increased with the concentration of glucose for all aeration regimes (Fig. 1B). However, ethanol yield was sensitive to the aeration regime. Ethanol yield was highest (0.36 g/g of substrate) under microaerobic conditions. In all cases, ethanol production was very rapid and occurred within 10 h after the start of the fermentation and remained almost constant throughout the run. This clearly showed that most of the ethanol was produced from the fermentation of glucose and probably sugars in the yeast extract.

Independent of either aeration regime or glucose concentration, the glucose in the medium was always consumed first before xylose. Thus, the growth curve showed a diauxic pattern when glucose was in the medium (Fig. 2A). The secondary lag period between glucose depletion and initiation of xylose consumption was more pronounced at higher glucose concentrations. Cell densities were about threefold that for xylose fermentation.

Note that during the early stages of glucose/xylose mixture fermentation, the cell densities were significantly higher than those for xylose fermentation (Fig. 2A). However, contrary to the predictions of Yahashi et al. *(14)*, the higher cell densities did not result in higher specific xylose uptake, higher specific xylitol yield, or productivity; instead, there was an overall decrease in xylitol yield relative to xylose fermentation.

These observations suggest the presence of an additional regulatory mechanism that affects the metabolism of glucose/xylose mixtures. Several plausible explanations could be adduced for the aforementioned phenomena. The most convincing evidence is the effect of ethanol on xylitol production. The influence of ethanol concentration on xylitol yield is shown in Fig. 2B. For the aerobic and semiaerobic fermentation, the presence of low levels of ethanol (10 g/L) resulted in a 45% decrease in xylitol yield. The microaerobic condition was less sensitive to low ethanol concentration. However, all three aeration regimes had more than a 50% decrease in xylitol yield when the ethanol concentration in the medium was >30 g/L.

The effect of ethanol was further investigated by the addition of similar concentrations of ethanol to xylose fermentation medium 24 h after the start of the experiment. There was a reduction in xylitol yield, but additionally, the yeast used ethanol as a cosubstrate to produce cell mass when the ethanol concentrations were <30 g/L. However, when the ethanol concentration was raised to 50 g/L, cell growth ceased and no xylitol was produced (data not shown).

The presence of ethanol in the medium could also account for the decrease in xylitol yield in the presence of the other hemicellulosic sugars (Table 5). The addition of glucose (20 g/L) caused the greatest reduction in xylitol yield (41%). Because glucose is a catabolite repressor, all the glucose

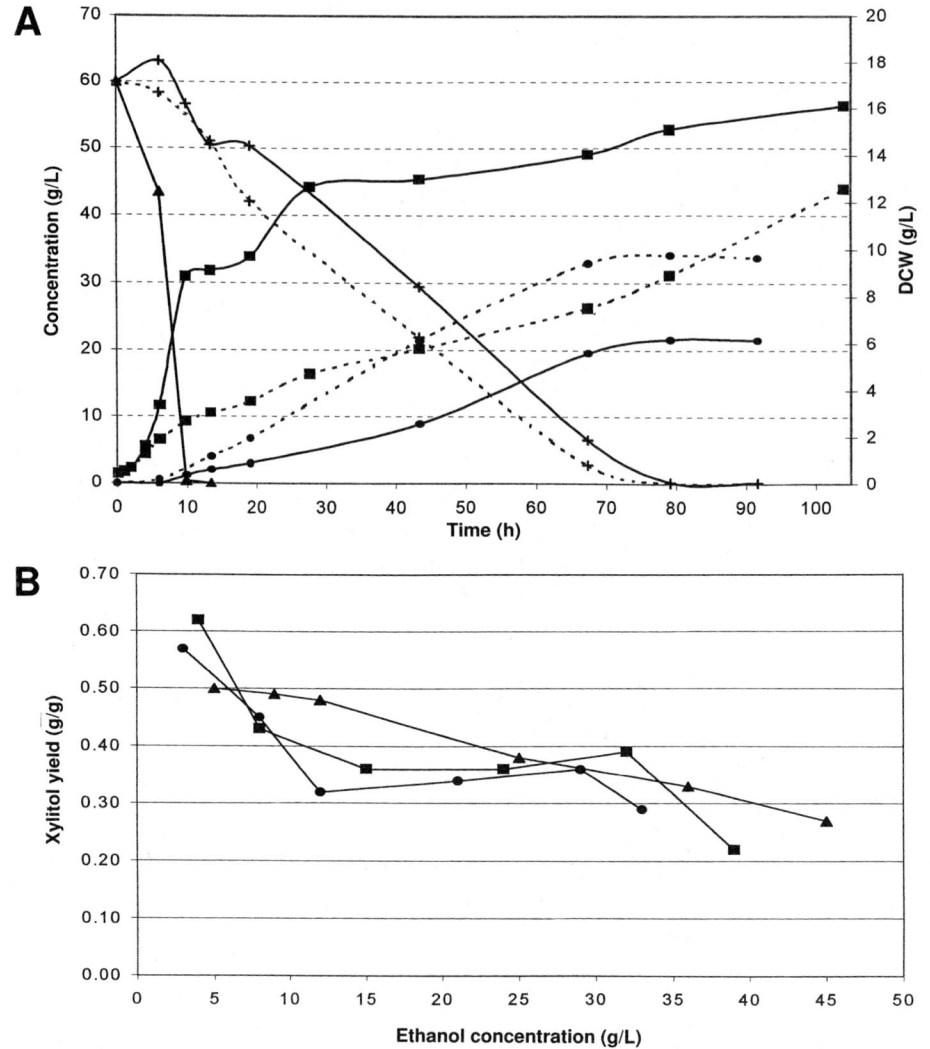

Fig. 2. **(A)** Influence of glucose on xylitol production. (– – –) Fermentation of xylose only; (——) fermentation of xylose/glucose mixture; (+) xylose consumption; (▲) glucose consumption; (■) cell concentration; and (●) xylitol formation. Initial xylose concentration was 60 g/L, glucose concentration was 60 g/L, and medium volume was 65 mL. **(B)** Influence of coproduct ethanol concentration on xylitol yield. Aerobic, 65 mL of medium (●); semiaerobic, 100 mL of medium (■); and microaerobic, 135 mL of medium (▲). Initial xylose concentration was 60 g/L. DCW, dry cell weight.

in the medium was converted to ethanol before xylose utilization started. The production of xylitol was therefore started in the presence of ethanol (approx 10 g/L). However, in the case of the fermentation of xylose with other sugars (mannose and galactose), there was a simultaneous uptake of xylose and the sugar, and, therefore, the concentration of ethanol in the medium was relatively low and rose gradually with time. Consequently, its

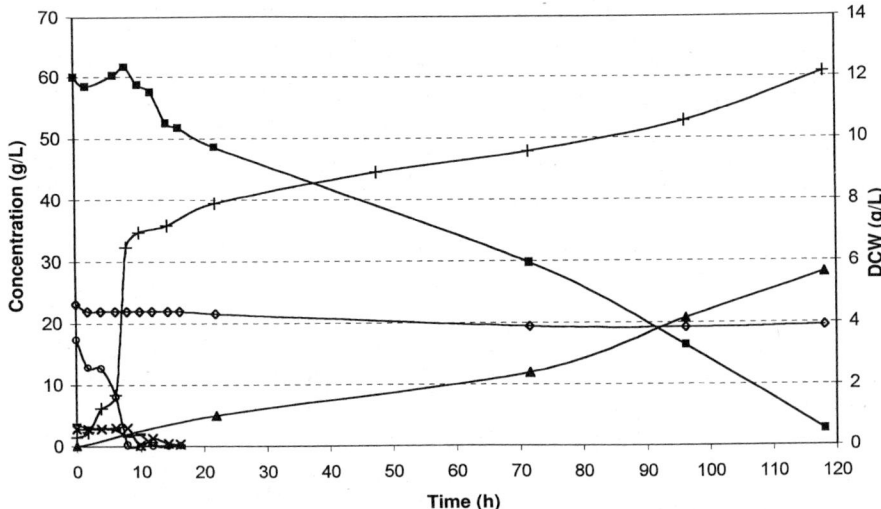

Fig. 3. Fermentation of complex sugar mixture with initial concentration of 60 g/L of xylose (■), 17.4 g/L of glucose (○), 23.2 g/L of arabinose (◇), 3.55 g/L of mannose (–), and 2.9 g/L of galactose (×). (+) Dry cell mass; (▲) xylitol. DCW, dry cell weight.

effect on xylitol yield was less drastic, and the overall xylitol yield decreased by 22% for 20 g/L of galactose and 11% for 20 g/L of mannose.

Complex Sugar Mixtures

In the fermentation of complex sugar mixtures, sugar utilization by *C. tropicalis* was sequential. Glucose was utilized first followed by the consumption of the other sugars. Mannose, galactose, and xylose were consumed simultaneously after glucose was depleted. Arabinose was not fermented significantly (Fig. 3). Interestingly, the pronounced diauxic growth pattern obtained for the binary sugars was almost absent from the complex sugar growth curve (Fig. 3). This growth pattern can be attributed to the simultaneous uptake of xylose, mannose, and galactose after the depletion of glucose in the medium and the ability of *C. tropicalis* to grow on mannose and galactose.

To predict and quantify the level of factors on xylitol yield and productivity, the experimental data were fitted to a second-order polynomial model. For xylitol yield, ANOVA found the influence of glucose, arabinose, mannose, and aeration to be significant ($p < 0.05$), whereas the contribution of galactose was statistically insignificant. Interactions between glucose and arabinose, mannose and arabinose, glucose and aeration, and mannose and aeration were also found to be statistically significant ($p < 0.05$). The regression model (Table 6) was significant ($p < 0.05$) with a good correlation ($R^2 = 0.90$). The model also indicates that except for arabinose/mannose, all the other interactions were negative. The antagonistic effect of glucose/arabinose, glucose/aeration, and mannose/aeration indicates that simultaneous increases in the levels of any of these factors

Table 6
Regression Equations for Fermentation of Complex Sugar Mixtures

Parameter	Regression equation[a]	R^2
Xylitol yield (g/g)	$0.5033 - 0.02362A + 0.0369B - 0.01094A \times B - 0.0178A \times E + 0.00781B \times D - 0.00969DE - 0.01252E^2$	0.90
Xylitol productivity (g/[L·h])	$0.27753 - 0.005A + 0.02226B + 0.01342C - 0.02324E - 0.00781A \times B - 0.01281A \times E - 0.00719B \times D - 0.01094D \times E + 0.00599A^2 - 0.01523E^2$	0.91

[a]A, B, C, D, and E are the coded values for glucose, arabinose, galactose, mannose, and medium volume, respectively, as given in Table 3.

will decrease xylitol yield. By contrast, a simultaneous increase in arabinose and mannose levels will improve xylitol yield.

Similar to xylitol yield, xylitol productivity was fitted to a second-order polynomial. ANOVA showed that the relationship between xylitol productivity and various factors was more complex than that for xylitol yield. The model was statistically significant ($p < 0.05$) with a satisfactory correlation ($R^2 = 0.91$). Glucose, arabinose, galactose, and aeration had a statistically significant influence on xylitol productivity ($p < 0.050$), whereas mannose had only an interactive effect. Interactions between glucose and arabinose, mannose and arabinose, glucose and aeration, and mannose and aeration were all statistically significant ($p < 0.05$). Unlike xylitol yield, interaction between various factors had a negative effect on xylitol productivity (Table 6).

The data show that for effective xylitol production, the glucose concentrations in the medium should be very low. When the fermentation medium contained glucose, higher yields and productivities were obtained under aerobic conditions, and in the absence of glucose, microaerobic conditions improved yields. This observation can be attributed to increased oxygen demand by the high cell densities achieved in the presence of glucose.

Arabinose appears to be a gratuitous inducer of xylose reductase enzymes; therefore, high arabinose concentrations stimulated both yield and productivity. Galactose consumption did not have any effect on xylitol yield but had a stimulating effect on xylitol productivity when its concentration was 10 g/L. Although the concentration of mannose was low, it caused a decrease in xylitol yield, because the yeast converted it to ethanol, which inhibits xylitol formation.

Conclusion

The empirical models developed can be used to estimate the achievable xylitol yields and productivities under different aeration conditions if the composition of the hemicellulose hydrolysate is known. For the hydrolysis of a raw material, the model can provide useful information for the design of the hydrolysis process, because varying the process conditions can influence the ratio and concentration of different sugars in the hydrolysate.

When glucose is present in the medium, high cell densities are obtained without the consumption of xylose, and xylitol formation is strongly inhibited by two mechanisms. Initially, xylose uptake is repressed and then the ethanol produced from the utilization of the glucose partially inhibits xylitol formation. Thus, for hemicellulosic feedstocks with high glucose content, it will be necessary to remove the initial ethanol formed from the fermentation of the glucose in order to attain high xylitol yield and productivity.

C. tropicalis utilized both accumulated xylitol and ethanol for cell growth when xylose concentration was very low (<1 g/L). It will, therefore, not be possible to ferment all the xylose in the medium without losing some of the xylitol accumulated.

References

1. Pepper, T. and Olinger, M. (1988), *Food Technol.* **10,** 98–106.
2. Roberto, I., Felipe, M. G. A., Mancilha, I. M., Vitolo, M., Sato, S., and da Silva, S. S. (1995), *Bioresour. Technol.* **51,** 255–257.
3. Perego, P., Converti, A., Palazzi, E., Del Borghi, M., and Ferraiolo, G. (1990), *J. Ind. Microbiol.* **6,** 157–164.
4. Horitsu, H., Yahashi, Y., Takamizawa, K., Kawai, K., Suzuki, T., and Watanabe, N. (1992), *Biotechnol. Bioeng.* **40,** 1085–1091.
5. Hespell, R. B., O'Bryan, P. J., Moniruzzaman, M., and Bothast, R. J. (1997), *Appl. Biochem. Biotechnol.* **62,** 87–96.
6. Whistler, R. L. (1993), in *Industrial Gums—Polysaccharides and Their Derivatives*, Whistler, R. D. and BeMiller, J. N., eds., Academic, San Diego, pp. 295–308.
7. Chen, L.-F. and Gong, C.-S. (1985), *J. Food Sci.* **50,** 226–228.
8. Jeffries, T. W. and Screenath, H. K. (1988), *Biotechnol. Bioeng.* **31,** 502–506.
9. Olsson, L. and Hahn-Hägerdahl, B. (1993), *Process Biochem.* **28,** 249–257.
10. Saha, B. C., Dien, B. S., and Bothast, R. J. (1998), *Appl. Biochem. Biotechnol.* **70–72,** 115–125.
11. Nolleau, V., Preziosi-Belloy, L., Delgenes, J. P., and Navarro, J. M. (1993), *Curr. Microbiol.* **27,** 191–197.
12. Box, G. E. P. and Draper, N. R. (1987), *Empirical Model Building and Response Surfaces*, John Wiley & Sons, New York.
13. Ojamo, H. (1994), *Yeast Xylose Metabolism and Xylitol Production*, Technical Research Centre of Finland, VTT Publications, Espoo, Finland.
14. Yahashi, Y., Horitsu, H., Kawai, K., Suzuki, T., and Takamizawa, K. (1996), *J. Ferment. Technol.* **81,** 148–152.

Nitrification and Denitrification Processes for Biologic Treatment of Industrial Effluents

CÉLIA REGINA GRANHEN TAVARES,*
RENATA RIBEIRO DE ARAÚJO ROCHA,
AND TEREZINHA APARECIDA GUEDES

*State University of Maringá, Paraná, Brazil, Av. Colombo 5790,
87020-900, Maringá, PR, Brazil, E-mail: celia@deq.uem.br*

Abstract

Nitrification process performance was evaluated using a three-phase fluidized-bed bioreactor. A synthetic effluent was used for this experiment containing 180–230 mg/L of chemical oxygen demand (COD), 25–30 mg/L of N-NH$_4^+$, 12 to 13 mg/L of total phosphorous, and micronutrients. The bioreactor used for denitrification behaved as completely mixed. The results indicate that the nitrification process was efficient, reaching efficiencies of about 98%. The best results related to the efficiency of the denitrification process were obtained when the processes were supplemented with the carbon source. The results indicated an efficiency of 86–98% COD removal.

Index Entries: Nitrification; denitrification; three-phase fluidized-bed bioreactor; *Nitrobacter*; *Nitrosomonas*.

Introduction

Biologic treatment is by far the most common method used to treat sewage, and two distinct process arrangements are used. Bacteria can be grown either in suspension (e.g., activated sludge) or attached to the surface of large solid medium (e.g., a biologic filter).

Biologic fluidized bed is a combination of these two processes because bacteria are grown on the surface of small solid particles that are held in fluidized suspension. The liquid to be treated is passed upward through a bed of solid medium at a velocity sufficiently high to fluidize the particles. A high concentration of bacteria leads to much greater rates of reaction per unit volume compared to either biologic filters or activated sludge. Consequently, a smaller reactor can be used, thereby reducing capital cost *(1)*.

Domestic sewage and many kinds of wastewater from pharmaceutical, agricultural, and food industries contain great amounts of carbonaceous

*Author to whom all correspondence and reprint requests should be addressed.

and nitrogenous substances as pollutants. Recently, much attention has been paid to the removal of ammonia from wastewater because ammonia promotes eutrophication in both terrestrial and littoral waters *(2)*. Ammonia is toxic to fish at very low concentrations. In general, it is recommended that no more than 0.02 mg/L of free ammonia be permitted in receiving waters *(3)*.

Biologic ammonium removal is a nitrification process, i.e., the conversion of ammonium into nitrate. Nitrification is initiated by two different functional groups of bacteria: the ammonium oxidizers, which convert ammonia into nitrite using an O_2-dependent ammonium monooxygenase; and the nitrite oxidizers, which oxidize nitrite into nitrate using a molybdenum-containing nitrite oxidoreductase.

Both physiologic groups contain Gram-negative bacteria with obligate chemolithoautotrophs that make use of the energy released from the two oxidation reactions. The nitrifying bacteria assimilate carbon dioxide into the cell material via the ribulosebisphosphate cycle. Both oxidations have relatively high redox potentials and these, coupled with the requirement for reverse electron flow for synthesis of reducing power, lead to low yields. These organisms also have low maximum specific growth rates with doubling times typically in the range of 10–24 h. These factors have limited studies on their physiology and growth kinetics, but a number of continuous-flow studies have been conducted *(4)*. The oxidation of ammonia into nitrite and its subsequent oxidation to nitrate are carried out by *Nitrosomonas* and *Nitrobacter*, respectively. The growth rate of nitrifying bacteria is controlled by substrate concentration, temperature, pH, and oxygen tension *(5)*.

The increase in nitrate concentration in public water supplies is becoming a serious problem in some parts of the world. Nitrate concentration in groundwater reached threateningly high levels 20 yr ago, and it has continued to increase ever since. Nitrate is a cause and an inorganic nutrient for the growth of algae, and it can represent a danger to public health if present in excessively high concentrations in drinking water. It was found that nitrate could cause methemoglobinemia in infants (blue baby syndrome) *(6)*.

Biologic denitrification of nitrates and nitrites present in wastewater is important and necessary. It is a process of nitrate and nitrite reduction in which nitrite serves as the terminal exogenous hydrogen acceptor when the oxygen tension in wastewater is sufficiently low. The normal end product of this nitrate and nitrite respiration is elementary nitrogen or nitrous oxide gas, which, being inert, can be allowed to escape into the atmosphere *(7)*.

The literature contains numerous data concerning the influence of different denitrification conditions on the rate of the process *(8)*. Environmental conditions that must be optimized for denitrification are temperature, pH, and type of carbon substrate. In the present work, the system of biologic treatment was based on the oxidation of organic and nitrogen matter of synthetic wastewater. Micronutrients were added for good per-

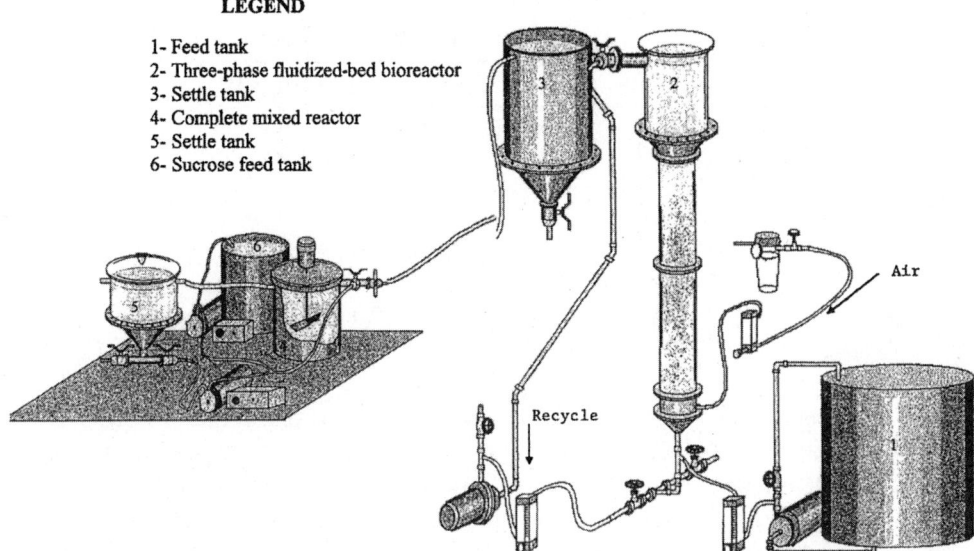

Fig. 1. Experimental unit.

formance of microorganisms. Dissolved oxygen, pH, and temperature were monitored for the nitrification process.

Materials and Methods

In the present work, the system of biologic treatment was based on the oxidation of organic and nitrogen matter ($N\text{-}NH_4^+$) of synthetic wastewater, and the pilot unit was operated for 50 d. For the removal of nitrogen matter, nitrification and denitrification processes were used, in series.

Experimental Setup

The experimental setup, as shown in Fig. 1, consisted of an acrylic column measuring 2 m high and 0.11 m in internal diameter (three-phase fluidized-bed nitrification reactor). In its upper part was an area of velocity reduction (0.5 m high and 0.24 m in internal diameter) that permitted separation of the three phases. There was a sedimentation column, made of polyvinyl chloride (PVC), that measured 0.8 m long and 0.24 m in internal diameter, coupled to the reduction area.

Concentrated substrate, stored in a reservoir, was diluted with tap water and continuously fed to the three-phase fluidized-bed reactor. The diluted synthetic effluent was distributed to the reactor by means of a gas-liquid distributor. This distributor consisted of a nylon cone through which the ascending liquid came in. On this cone there was a nylon tube whose diameter was the same as that of the column containing a gas distributor. This distributor was made of a 4-mm-diameter copper tube shaped like a spiral. Perforations of 0.5 mm in this spiraled tube allowed the gas to be dispersed in the surrounding liquid. Above the distributor was a

Table 1
Composition of Artificial Wastewater

Component	Concentration (mg/L)
Glucose	180–220 of COD
NH_4Cl	24–56 of $N-NH_4^+$
KH_2PO_4	50
$FeSO_4 \cdot 7H_2O$	13.7
$NaHCO_3$	750
Na_2CO_3	500
$CaCl_2 \cdot 2H_2O$	10

plate regularly perforated with 235 holes 3 mm in diameter, resulting in 16.10% free area. This plate had two main purposes: to improve fluid distribution in the fluidized-bed and to support the particles in the bed. The plate was flanged between the gas-liquid distributor and the fluidized bed.

Part of the flow that passed through the sedimentation column, coupled to the exit of the three-phase fluidized-bed, was recirculated to the base of the three-phase fluidized-bed reactor by means of a centrifugal pump, to promote the fluidization of the particles in suspension. Another part of the supernatant fed the denitrification reactor. Pumping was not necessary owing to the free fall of 2.5 m between the exit of the nitrification system and the denitrification process entrance.

The denitrification system consisted of a completely mixed reactor, which was made of PVC, with a volume of 35 L. A decanter was coupled to this reactor. The decanter was made of an acrylic column having at the base a stainless steel cone connected by a flange, with a total volume of 20 L. This system also had a storage reservoir for glucose, used as a supplementary source of carbon. A solution of glucose was then continuously fed to the denitrification reactor by a centrifugal pump. The sludge of the decanter from the denitrification system was recirculated to the denitrification reactor, to ensure a high retention time for the microbial cells.

Support

The support used in the three-phase fluidized-bed bioreactor for the development of the biofilm was cylindrical PVC particles, with a specific mass of 1.37 g/cm^3 and equivalent diameter of 2.94 mm.

Bacteria

The fluidized-bed support was inoculated in batch for 24 h with aerobic sludge from an effluent reservoir from Londrina, PR, Brazil, and nitrifying bacteria from activated sludge unit from the Refinary Getúlio Vargas (Araucária, PR, Brazil). Aerobic sludge and nitrifying bacteria were previously acclimatized to the synthetic substrate (Table 1) for 7 d.

The completely mixed bioreactor for denitrification was also inoculated in batch for 24 h. The inoculum used for the denitrification process was obtained from anaerobic sludge from an effluent treatment plant in Londrina, PR, Brazil. Anaerobic sludge was previously acclimatized for 7 d to a solution of KNO_3 (1.119 g/L), glucose (0.21 g/L), and KH_2PO_4 (0.02 g/L).

Wastewater

The composition of the synthetic wastewater used in this work is given in Table 1. In the experiments, sodium bicarbonate was used as a buffer for the medium during the nitrification process.

The formation of ion H^+, during ammonia oxidation, reduces the alkalinity of the medium. To sustain the optimal pH range, a well-buffered medium is necessary *(5)*. Bicarbonate was used as reported by others *(9–11)*. All the compounds that formed the synthetic effluent were dissolved daily with tap water in the feeding tank (500 L).

Operation

The experimental run was operated for 50 d by continuous flow.

Nitrification Process

The three-phase fluidized-bed reactor worked at a hydraulic retention time of 5 h and a gas flow of 6.72 L/h. The concentration of the dissolved oxygen in the liquid was from 3.5 to 4.5 mg/L, and the recycling flow was 313 L/h. The liquid temperature was controlled at 27–32°C by means of an electric heater. The pH value of the liquid was controlled at 7.0 to 8.0 by adding NaOH and $CaCO_3$ to the artificial substrate (feed tank) when needed. The dissolved oxygen of the wastewater was adjusted to 6 mg/L, which is optimal for the nitrifying bacteria *(5)*.

Denitrification Process

The complete mixed reactor was operated at a hydraulic retention time of 4 h and 30 min and a flow feed of 8 L/h. The liquid entering the denitrification process, which was the effluent of the nitrification process, was kept in an anoxic condition, and there was no need to manipulate oxygen flow inside the complete mixed reactor. The quantity of extra carbon source (glucose) was monitored, so as to give concentrations of 41–292 mg/L of COD for the process.

Analyses

COD was determined according to Micro-Methods from Tavares *(12)*. Values of ammonia were followed by Koroleff *(13)*. Nitrite and nitrate concentrations in the liquid were determined by the Test Kit (HACH) model NI-15 (cat. no. 21820-00) and model NI-11 (cat. no. 1468-03), respectively.

Attached biomass in the support was determined by protein and polysaccharide concentration according to the Lowry and Dubois methods in Tavares *(12)*, respectively. Each method made use of 60 particles taken from the interior of the three-phase fluidized-bed reactor, and the extrac-

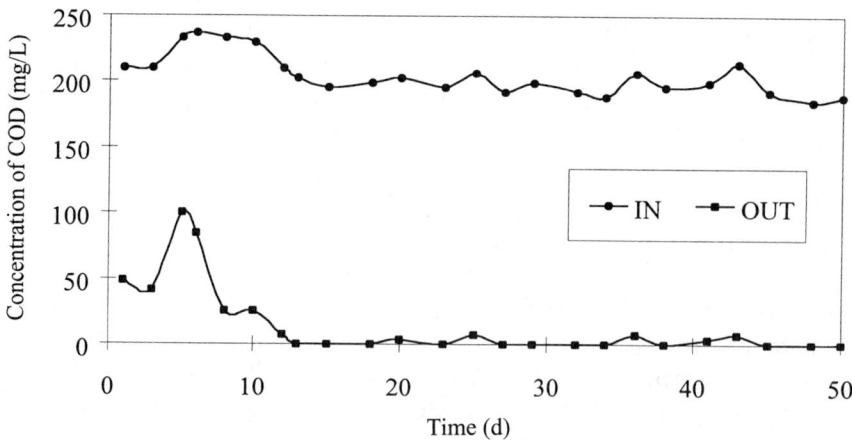

Fig. 2. COD reduction concentration vs time: nitrification process.

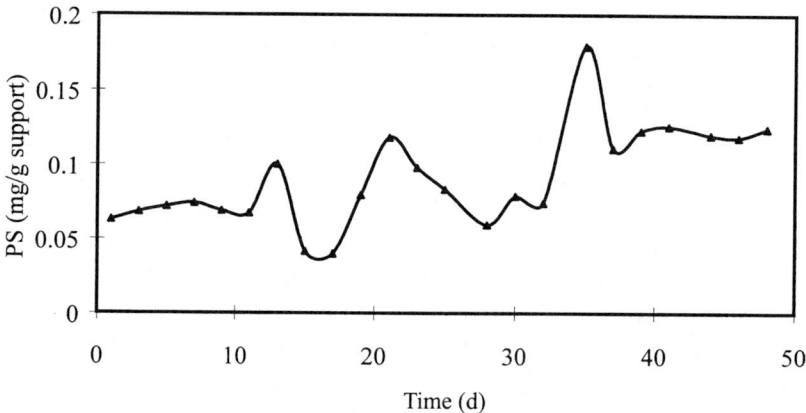

Fig. 3. Variation in polysaccharide (PS) content in the attached biofilm with reactor operation time.

tion was carried out in water bath at 80°C with specific reactants for each analysis (*see* ref. *12*).

Results

Figure 2 shows COD concentration in the influent and effluent wastewater vs time (days) during the nitrification process. Mean values of COD concentration fell from 204 to 15 mg/L. Biofilm was estimated by the values of the polysaccharides and protein attached to the support vs time (Figs. 3 and 4, respectively). Removal efficiency was calculated as (CODin – CODout)/CODout.

Figure 5 shows the evolution of the nitrification process. These data indicate that the ammonia was converted into nitrate and nitrite. The results of COD reduction concentration vs time for the denitrification process are presented in Fig. 6; mean values fell from 173 to 95 mg/L. Figure 7 shows

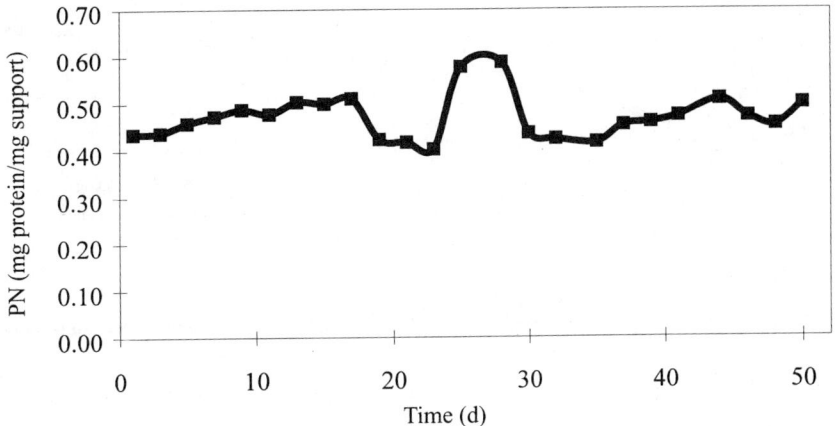

Fig. 4. Variation in protein (PN) content in the attached biofilm with reactor operation time.

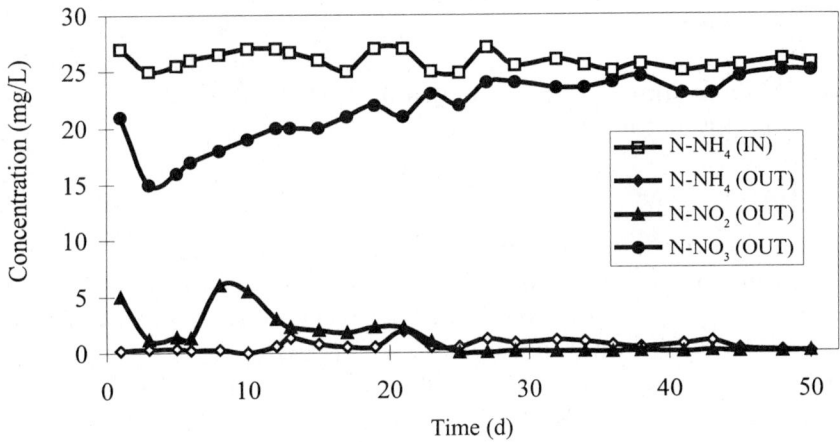

Fig. 5. Evolution of nitrification process.

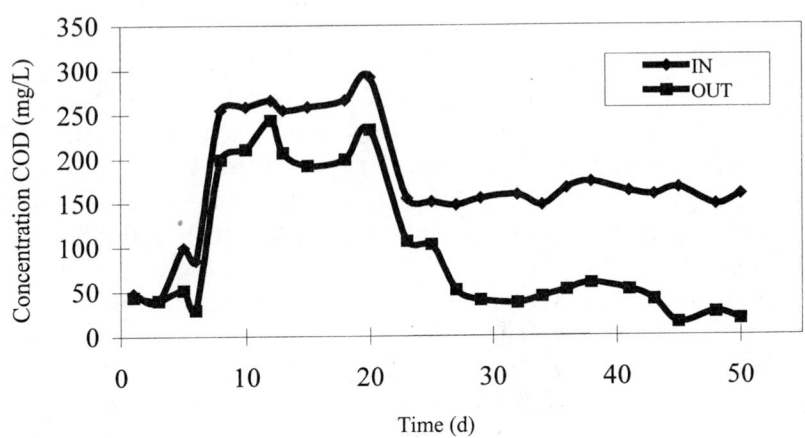

Fig. 6. COD reduction concentration vs time: denitrification process.

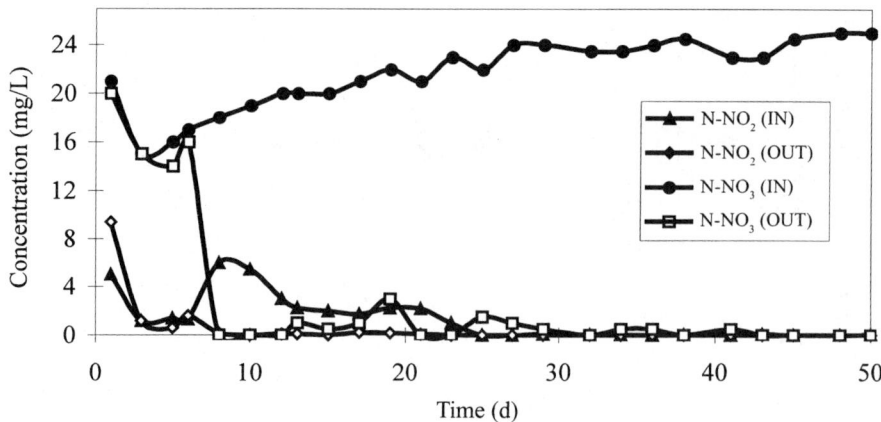

Fig. 7. Evolution of denitrification process.

the evolution of the denitrification process, in which inhibition of the process was verified between 1d and 6d. After 6d, additional carbon source (glucose) was added, and mean values fell from 148 to 292 mg/L. With a sufficiently high concentration of carbon, the nitrate and nitrite were converted to nitrogen gas.

Discussion

Throughout nitrification no organic substance was needed; however, some organic compounds may be assimilated to a limited extent. Perhaps the microbiologic oxidation of organic carbon, contained in the substrate that fed the nitrification process, was also produced by nitrifying organisms (autotrophic), even though most of it was produced by heterotrophic organisms from the fluidized-bed reactor. This hypothesis, though, would need further experiments to be confirmed.

It is a known fact that nitrification takes place at a considerably slower reaction rate than microbial oxidation of organic carbon *(3)*. To induce and increase nitrification, it was necessary to operate the fluidized bed with a high hydraulic retention time (4 h).

The dissociation balance for ammonium-ammonia is dependent on the temperature and pH. Thus, it is necessary to control the pH and temperature in the nitrification process. According to Abeling and Seyfried *(14)*, the nonionized forms of ammonium and nitrate have—as ammonia (NH_3) and as nitrous acid (HNO_2)—an inhibition effect on the *Nitrosomonas* and on *Nitrobacter*.

In the present study, we observed that all the ammonia consumed was converted into nitrate or nitrite and a small amount used for assimilation into nitrifiers. In terms of the performance of a denitrification system, optimal pH and anoxic conditions are important environmental conditions. A nonoptimal pH condition will prolong the lag phase of nitrite reductase in the denitrification process. Limiting of the carbon source would

Table 2
t-Test for Dependent Sample of Fluidized-Bed Inflow and Outflow

	COD (mg/L)	N-ammonium (mg/L)
Average values of the reactor inflow	204.88	25.91
Average values of the reactor outflow	15.06	0.59
p^a	0	0

[a]Marked differences are significant at $p < 0.05$.

Table 3
t-Test for Dependent Sample of Denitrification Bioreactor Inflow and Outflow

	COD (mg/L)	N-nitrite (mg/L)	N-nitrate (mg/L)
Average values of the reactor inflow	173.81	1.40	21.56
Average values of the reactor outflow	95.66	0.54	3.00
p^a	0.00	0.03	0.00

[a]Marked differences are significant at $p < 0.05$.

cause the accumulation of nitrate, which is dangerous to health. Therefore, we must be able to supply enough carbon source to reach a high efficiency of nitrite and nitrate removal of about 98%. A small amount of nitrite and nitrate incorporated into the sludge was considered, following the assumption that no stripping out occurs in the denitrification bioreactor.

The biologic fluidized bed has been shown to be technically viable for an effluent that needs upgrading to complete nitrification.

A *t*-test was carried out to evaluate the efficiency of the process of COD, nitrification, and denitrification removal. The *t*-tests for matched data were carried out to verify whether the inlet media values (μ_{in}) were equal to the outlet media values (μ_{out}):

$$H_0: \mu_{in} = \mu_{out}$$
$$H_1: \mu_{in} \neq \mu_{out} \ (\mu_{in} > \mu_{out})$$

in which H_0: is the null hypothesis and H_1: is the no null hypothesis. Tables 2 and 3 show the inlet and outlet media values of the fluidized-bed reactor and of the complete mixed bioreactor. They also show the significance levels (p) of the values.

All p values are <0.05. This indicates that for all the variables, the treatments were efficient, once the inlet media values of the process were statistically greater than the outlet ones.

Conclusion

The results indicate that the nitrification and denitrification processes in series were efficient for reducing ammonia in the treatment of effluents.

The nitrification process reached efficiencies of about 98%. The best results related to the efficiency of the denitrification process were obtained when a carbon source (glucose) was supplemented to the process. COD removal was satisfactory, showing the efficiency of the three-phase fluidized-bed bioreactor in the biologic treatment of effluents. The results indicated an efficiency of 86–98% COD removal.

References

1. Cooper, P. F. and Williams, S. C. (1990), *Water Sci. Technol.* **22,** 431–442.
2. Boongorsrang, A., Kenichi, S., and Yoshimichi, M. (1982), *J. Ferment. Technol.* **60,** 357–362.
3. Fang, H., Chou, M., and Huang, C. (1993), *Water Res.* **27,** 1761–1765.
4. Prosser, J. I. (1989), *Adv. Microb. Physiol.* **30,** 125–182.
5. Rosa, M. F. (1997), PhD thesis, DEQ/DEB-EQ, UFRJ, Rio de Janeiro-RJ, Brazil.
6. Yatong, X. (1995), *Water Treat.* **10,** 81–88.
7. Narjari, N. K., Khilar, K. C., and Mahajan, S. P. (1984), *Biotechnol. Bioeng.* **26,** 1445–1448.
8. Mazierski, J. (1994), *Water Res.* **28,** 1981–1985.
9. Cheng, S. and Chen, W. (1994), *Water Sci. Technol.* **30,** 131–142.
10. Siegrist, H. and Gujer, W. (1987), *Water Res.* **21,** 1481–1487.
11. Szwerinski, H., Arvin, E., and Harremoes, P. (1986), *Water Res.* **20,** 971–976.
12. Tavares, C. R. G. (1992). PhD thesis, COPPE/UFRJ, Rio de Janeiro-RJ, Brazil.
13. Koroleff, K. (1983), in *Methods of Seawater Analysis*, Grasshoff, E. and Kremling, S., eds., Verlag Chemie, Weinhein, Germany, pp. 126–127.
14. Abeling, U. and Seyfried, C. F. (1992), *Water Sci. Technol.* **26,** 1007–1015.

Removal and Recovery
of Copper (II) Ions by Bacterial Biosorption

Mui F. Wong,[1] Hong Chua,[1]
Waihung Lo,*[,2] Chu K. Leung,[2] and Peter H. F. Yu[2]

*Departments of [1]Civil and Structural Engineering
and [2]Applied Biology and Chemical Technology and Open Laboratory
of Chiral Technology, The Hong Kong Polytechnic University,
Hung Hom, Hong Kong SAR, China, E-mail: bctlo@polyu.edu.hk*

Abstract

Studies were conducted to investigate the removal and recovery of copper (II) ions from aqueous solutions by *Micrococcus* sp., which was isolated from a local activated sludge process. The equilibrium of copper biosorption followed the Langmuir isotherm model very well with a maximum biosorption capacity (q_{max}) of 36.5 mg of Cu^{2+}/g of dry cell at pH 5.0 and 52.1 mg of Cu^{2+}/g of dry cell at pH 6.0. Cells harvested at exponential growth phase and stationary phase showed similar biosorption characteristics for copper. Copper uptake by cells was negligible at pH 2.0 and then increased rapidly with increasing pH until 6.0. In multimetal systems, *Micrococcus* sp. exhibited a preferential biosorption order: Cu ~ Pb > Ni ~ Zn. There is virtually no interference with copper uptake by *Micrococcus* sp. from solutions bearing high concentrations of Cl^-, SO_4^{2-}, and NO_3^- (0–500 mg/L). Sulfuric acid (0.05 M) was the most efficient desorption medium, recovering >90% of the initial copper sorbed. The copper capacity of *Micrococcus* sp. remained unchanged after five successive sorption and desorption cycles. Immobilization of *Micrococcus* sp. in 2% calcium alginate and 10% polyacrylamide gel beads increased copper uptake by 61%. Biomass of *Micrococcus* sp. may be applicable to the development of potentially cost-effective biosorbent for removing and recovering copper from effluents.

Index Entries: Biosorption; copper removal; bioremediation; immobilization; metal adsorption.

Introduction

The presence of copper ions in water poses serious environmental and human health hazards because of their toxicity, tendency to bioaccumulate,

*Author to whom all correspondence and reprint requests should be addressed.

and abundance and persistence in the environment *(1)*. In Hong Kong and South China, copper pollution arises mainly from the effluents discharged from the electroplating and printed circuit-board factories. It has been reported that for a typical printed circuit-board factory, about 55 kg of copper was discharged per month into Hong Kong waters before enactment of the Water Control Ordinance. A high level of copper has also been found in the sediment of Victoria Harbor.

Metal-processing industries must pretreat or detoxify metal-rich effluents before discharging them into the aqueous environment. Current technologies for copper removal such as chemical precipitation, electrochemical treatment, and ion exchange provide only partially effective treatment and are costly to implement and use, especially when the metal concentration is low. The use of biological materials for the removal or recovery of heavy metals has gained importance in the last decade owing to their good performance and low cost. Microorganisms including bacteria *(2,3)*, microalgae *(4)*, and fungi *(5)* could accumulate large amounts of heavy metals with high selectivity even when they are dead or inactive. Accumulation of metals without active uptake is known as biosorption, which can be considered a collective term for a number of passive accumulation processes including ion exchange, coordination, complexation, adsorption, and microprecipitation *(4)*. These biomasses are capable of removing even trace levels of metal ions. Moreover, they can also be used to recover rare, precious, or strategic metals from waste solutions. The abundant bacterial biomass can be obtained inexpensively, because it is a waste byproduct of large-scale industrial processes such as fermentation and activated sludge wastewater treatment. Hence, biosorption coupled with desorption may provide an economic and effective alternative for the removal and recovery of heavy metals.

It is important to identify more microbial strains that can uptake metals with high efficiency and specificity as well as to design better bioprocess that effectively remove or recover heavy metals from aquatic systems. To optimize design and operation of biosorbent systems for removal and recovery of metals, a thorough understanding of biosorption behavior and desorption kinetic characteristics of microbial cells is needed. This motivated us to evaluate the feasibility and ability of microorganisms indigenous to biological treatment systems to remove metals in wastewater *(6)*.

The present study was conducted to characterize copper biosorption and desorption behavior of *Micrococcus* sp., which was isolated from a local activated sludge system. *Micrococcus* sp. was found to have high copper biosorption capacity *(6)*. The effects of initial copper concentrations, cell age, pH, competing cations, anions, desorption, and immobilization on the copper biosorption capacities by *Micrococcus* sp. have been studied systematically and extensively. The ultimate goals of this study were to develop novel and economical processes for removing and recovering copper (II) ions from aqueous wastes using bacterial-based biosorbents.

Materials and Methods

Isolation Procedures and Identification

Micrococcus sp. was isolated from fresh activated sludge collected from the return sludge channel at the Shatin Sewage Treatment Works in Hong Kong. The sludge was serially diluted in distilled deionized water. Aliquots (0.1 mL) were spread on nutrient agar (Difco) and cultivated in an incubator at 30°C for 3 d. Colonies were picked up and maintained on the same medium for the subsequent metal biosorption test. The isolates were identified using the MIDI Sherlock Microbial Identification System.

Preparation of Biosorbents

Colonies from freshly prepared agar plates were used to inoculate several sterilized 1-L conical flasks, each containing 400 mL of nutrient broth. The cultures were grown at pH 7.0 and 37°C on an orbital shaker at 250 rpm for 48 h. Cells were harvested and washed twice with distilled deionized water by centrifuging at $9000g$ force and 4°C for 20 min. The biomass was then resuspended in 200 mL of distilled deionized water for preparing a biomass stock solution. The concentration of biomass in stock was estimated by oven drying a designated portion of cells.

Biosorption Studies

Batch experiments for determining metal biosorption isotherms were carried out with the initial copper concentrations ranging from 1 to 150 mg/L and biomass concentrations ranging from 1 to 2 g of dry cell/L. The total reaction volume was 50 mL in 500-mL polyethylene bottles. The solution pH was adjusted to 5.0 by adding 0.1 M NaOH and 0.1 M HNO$_3$. Reaction bottles were agitated on an orbital shaker at 25°C for 12 h, which is more than ample time to reach metal biosorption equilibrium (7). Bacteria were always the last component to be added to the reaction mixtures, and all experiments were performed in duplicate. At designated intervals, samples were taken and the biomass was separated by centrifuging. Copper concentrations in the supernatants were determined using a model 100 Perkin-Elmer atomic absorption spectrophotometer.

Effects on Competing Cations

To ascertain whether there was any competition between different metal ions for uptake by *Micrococcus* sp., biosorption of copper, zinc, nickel, and lead by *Micrococcus* sp. from single and multimetal ion solutions was studied. The initial concentration of each single metal ion for biosorption was 2 mM. Our previous study (6) has shown that a metal concentration of 2 mM would be high enough to saturate all the binding sites of *Micrococcus* sp. under the experimental conditions. For the binary metal systems, total metal concentration was 4 mM, whereas for the four-metal systems, the total metal concentration was 8 mM. The experiment was conducted according to the biosorption methodology described earlier at pH 5.0. All metal ion solutions were prepared by using nitrate salts to prevent possible salt precipitation.

Effects of Anions

Experiments examining the effect of various anions on copper removal by *Micrococcus* sp. were carried out by adding sodium salts of chloride, sulfate, or nitrate to obtain the additional anion concentrations of 0, 50, 100, and 500 mg/L, to solutions containing 100 mg/L of copper nitrate. Experiments were conducted according to the biosorption methodology described earlier.

Biosorption and Desorption Cycles

The bacterium *Micrococcus* sp. with a final concentration of 1.5 g of cell/L was suspended in solutions containing 50 mg of Cu^{2+}/L in centrifuge tubes. The final volumes were made up to 25 mL. The pH of the resulting mixtures was adjusted to 6.0 by adding HNO_3 and NaOH. The tubes were shaken at 250 rpm and 25°C for 6 h. Then, the copper-loaded biomass was centrifuged, rinsed with distilled deionized water, and resuspended in 10 mL 0.05 *M* separately for 45 min to recover the copper ions from cells. The regenerated biomass was again suspended in copper solutions for the next biosorption run. The biosorption and desorption steps just described were repeated five times. The copper concentrations in the supernatants were determined after the biosorption and desorption steps.

Immobilization Studies

In the exponential phase of growth, *Micrococcus* sp. cells were centrifuged, washed, and resuspended in solution A containing 4% sodium alginate, 18.2% acrylamide, and 1.8% *N,N'*-methylene-*bis*-acrylamide. The volume ratio of cell suspension to solution A was 1:1. distilled deionized water was added instead of cell suspension for preparing the gel beads for controls. The resulting mixtures were dropped into solution B, which consisted of 4% calcium chloride, 0.1% *N,N,N',N'*-tetramethylene-diamine, and 0.1% ammonium persulfate. The drops gelled into 2.3 ± 0.3 mm diameter spheres on contact with solution B. The immobilized cell particles were kept in solution B at room temperature with constant and gentle stirring for at least 2 h to complete gel formation. The immobilized cell particles were prepared under aseptic conditions.

Results and Discussion

Biosorption Studies

Copper biosorption isotherms for *Micrococcus* sp. at pH 5.0 and 6.0 are presented in Fig. 1. The calculated model parameter values and the correlation coefficients for Langmuir and Freundlich isotherms are given in Table 1.

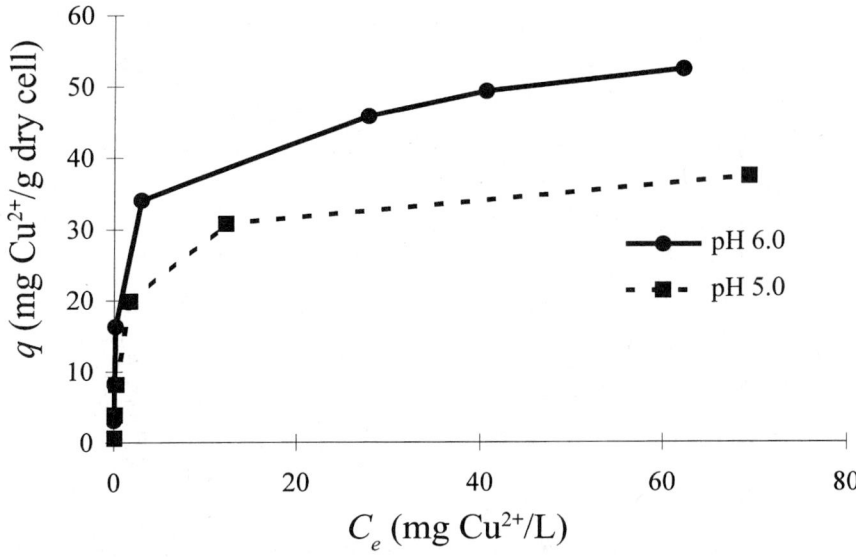

Fig. 1. Copper removal isotherm by *Micrococcus* sp. at pH 5.0 and 6.0.

Table 1
Calculated Parameters of Langmuir and Freundlich Isotherm Models
for Copper Biosorption by *Micrococcus* sp. at pH 5.0 and 6.0

	Langmuir equation			Freundlich equation		
	Q_{max} (mg/g)	b (L/mg)	r^2	K	n	r^2
pH 5.0	36.5	0.61	0.9993	8.24	0.49	0.8008
pH 6.0	52.1	0.66	0.9893	13.6	0.61	0.7827

The isotherms were found to follow the typical Langmuir adsorption pattern with r^2 of the linearized Langmuir isotherm equal to 0.999. The linearized Langmuir isotherm can be represented as follows:

$$C_e/q = 1/(q_{max} \cdot b) + C_e/q_{max}$$

in which q_{max} and b are the maximum biosorption capacity and affinity, respectively. The Freundlich isotherms did not fit the biosorption data very well, as shown by the r^2 of the linearized Freundlich plot (0.801). The results demonstrated that copper ions were well adsorbed by the mobilized *Micrococcus* sp. The maximum biosorption capacity (q_{max}), estimated by the Langmuir model, reached 36.5 mg of Cu^{2+}/g of dry cell at pH 5.0 and 52.1 mg of Cu^{2+}/g of dry cell at pH 6.0. The Freundlich coefficients K at pH 5.0 and 6.0 were 8.24 and 13.56, respectively. The parameters q_{max} and K reflect different characteristics: the Langmuir parameter, q_{max}, represents the saturation level of sorbed copper at high solution concentrations, whereas the Freundlich K represents the amount of copper sorbed when the solution concentration in the equilibrium is unity. Both the Freundlich

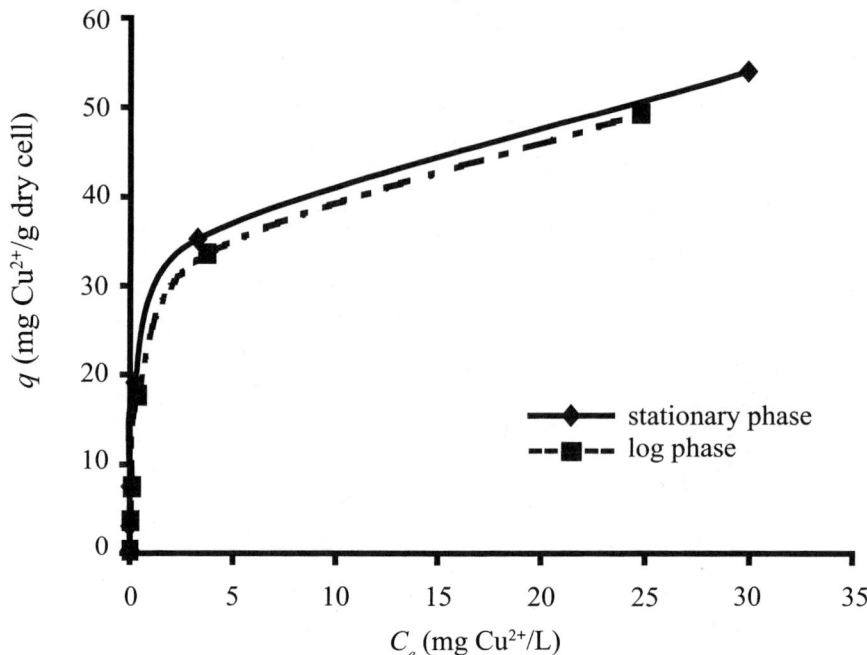

Fig. 2. Effect of culture age on copper biosorption.

parameter K and the Langmuir parameter b measure the effectiveness of copper biosorption at low copper concentration in solution. Higher values of b imply a higher biosorption level at low solution concentrations. The Freundlich parameter n measures the extent of impact on biosorption of a change in residual solution concentration from unity. High values of n imply a relatively large change in sorbed copper when the residual copper concentration deviates either above or below unity.

Effect of Culture Age on Copper Biosorption

Cells of two different culture ages were used to remove copper from metal-contaminated solutions (Fig. 2). Results showed that there was virtually no difference in copper uptake between cells harvested during log phase and those during stationary phase. Volesky and Holan *(4)* reported that metal ions bound mainly to the anionic functional groups present on the cell wall, such as carboxyl, hydroxyl, sulfate, phosphate, and amino groups. This may suggest that the surface composition and properties of the cell wall did not change much during the stationary phase.

Effect of pH

Because biosorption is a predominantly physio-chemical process taking place between positively charged metal ions and anionic groups of the cell surface *(8)*, metal removal is strongly influenced by the experimental conditions of solution pH, specific surface properties of adsorbent, the

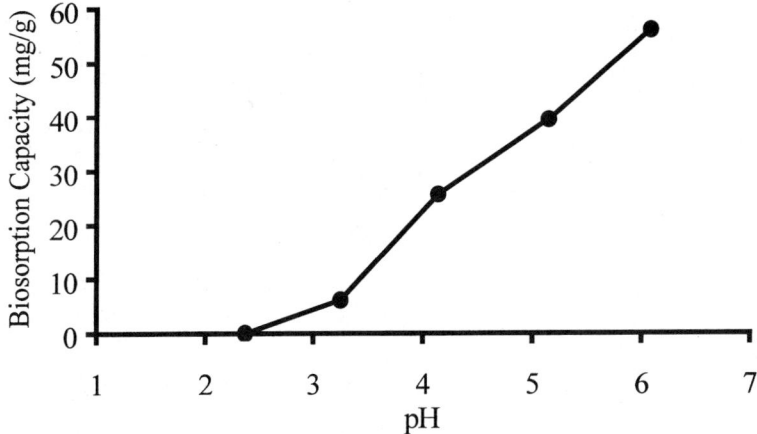

Fig. 3. Effect of pH on 100 mg/L copper biosorption by *Micrococcus* sp.

concentration of adsorbate, and the presence of coions in solutions. It has been consistently reported that pH is the dominant solution parameter controlling biosorption and that cation biosorption increases as solution pH increases *(6,9,10)*.

Copper biosorption by *Micrococcus* sp. was strongly affected by solution pH, as indicated in Fig. 3. Copper uptake was negligible at pH 2.0 and then increased rapidly with increasing pH. It is very likely that hydrogen ions compete with copper ions for the sorption sites of cells *(11)*. At lower pH, biosorption of hydrogen ions was preferred over that of copper ions, but at higher pH, more copper ions were taken up. The pH biosorption studies were not conducted at pH values above 6.0, because insoluble copper hydroxide precipitates from the solution at higher pH values, making true biosorption studies impossible. The pH profile suggested that copper binding to cell walls and external surfaces was most likely one of the dominant removal processes.

Effects on Competing Cations

As with hydrogen ion on pH effect, the cocations existing in the solution have the same effect. Table 2 gives the percentage decrease in the metal removal by *Micrococcus* sp. in the presence of one or three other metals. The presence of additional metal ions does not significantly impair the biosorption of copper. Copper biosorption was more sensitive to the presence of lead than zinc and nickel. On the other hand, removal of zinc and nickel by *Micrococcus* sp. was affected conspicuously by the presence of copper. When all four metals were present, the inhibition influence on copper biosorption by *Micrococcus* sp. did not accumulate proportionally. On the contrary, the removal capacities for copper and lead were increased when compared with their two metal systems. *Micrococcus* sp. bound the heavy metals in the following order: $Cu \sim Pb > Ni \sim Zn$. These results showed high similarity to the study carried out by Shuttleworth and Unz

Table 2
Percentage Decreases in Metal Removal by *Micrococcus* sp.
in Presence of Two-, Three-, or Four-Metal Systems

	Cu + Zn	Cu + Pb	Cu + Ni	Cu + Zn + Pb + Ni
Cu	20.0	39.1	22.8	29.9
Zn	88.2	—	—	88.9
Pb	—	30.9	—	18.8
Ni	—	—	80.3	90.0

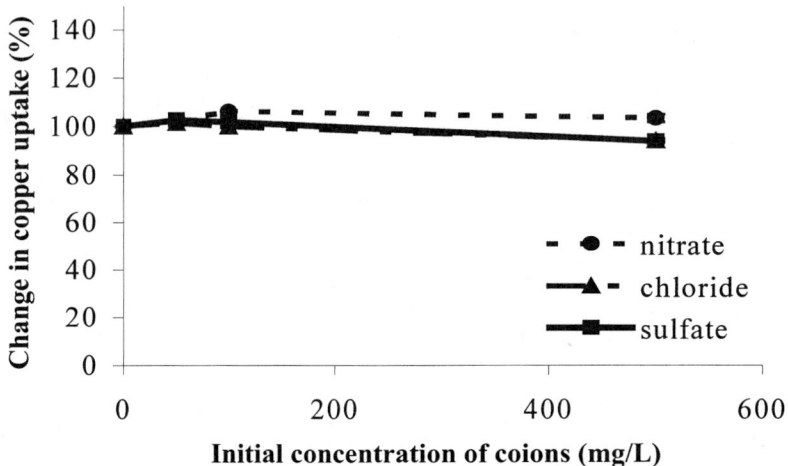

Fig. 4. Effects of three anions (chloride, nitrate, and sulfate) on the biosorption of 100 mg/L of copper by *Micrococcus* sp.

(10). In their investigation, the affinity of the filamentous bacterium *Thiothrix* strain A1 for metals was Cu > Zn ~ Ni.

The concentration of each metal ion used in this experiment was 2 mM. This value was known from our previous study *(6)* to be high enough for saturating all the binding sites of *Micrococcus* sp. under the experimental conditions. If the metal ion-binding sites on biosorbents have not been attained for saturation, no or an insignificant effect on the competition of metals may be observed. Hence, misleading results would be obtained.

Effects of Anions

Anions such as chloride, sulfate, and nitrate are often found in industrial wastewater together with heavy metal ions. The presence of these ligands is usually assumed to reduce the sorption of metals to bacteria because only free metal ions are bioavailable. Figure 4 illustrates the effect of anions (Cl$^-$, SO$_4^{2-}$, and NO$_3^-$) on the biosorption of copper by *Micrococcus* sp. There was virtually no interference with copper uptake by *Micrococcus* sp. from the three anions (Cl$^-$, SO$_4^{2-}$, and NO$_3^-$), although a wide range of

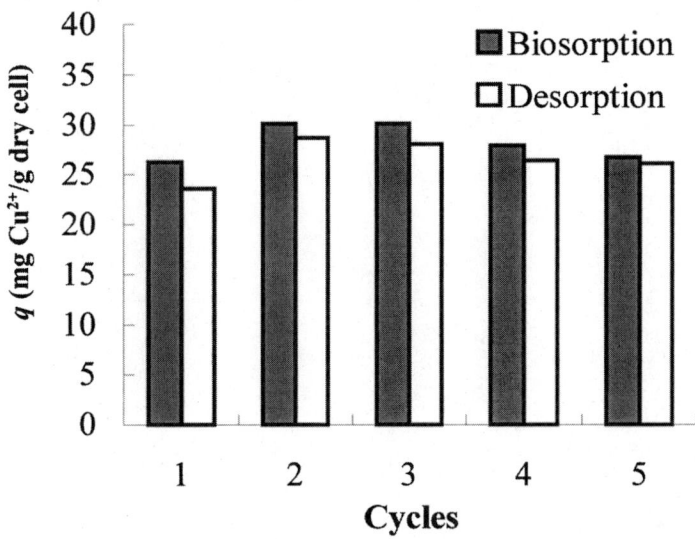

Fig. 5. Five regeneration cycles by 0.05 *M* sulfuric acid.

anion concentrations (0–500 mg/L) was tested. The result obtained was promising for applying *Micrococcus* sp. to detoxification of heavy metals bearing industrial wastewaters.

Biosorption and Desorption Cycles

Repeated biosorption and desorption operations were performed to examine the reusability and metal recovery efficiency of the biomass. Five consecutive regeneration cycles by *Micrococcus* sp. using 0.05 *M* sulfuric acid as a desorption medium are illustrated in Fig. 5. The copper biosorption capacity of the biomass had no significant difference from cycle 1 to cycle 5. For all runs, more than 90% of sorbed copper could be recovered and concentrated in a small volume by the desorption eluent. The results indicated that *Micrococcus* sp. possessed high reuse potential for removal and recovery of copper (II) ions from wastewater. A slight enhancement in copper uptake was observed in the second operation in both experiments. It is likely that the cell particles became much finer after the first treatment of desorption medium. Thus, the increase in surface area of the biomass may be a cause of the increase in removal capacity. Chang et al. *(9)* and Wong et al. *(14)* reported similar observations. They attributed the behavior to HCl-induced structural changes, which increased surface binding of the metals. Because suspended cells were used in our study, about 35% loss in cell concentration from the working solution was observed after five regeneration cycles. This loss of biomass is most likely owing to repeated centrifugation and rinse operations. Using immobilized cell systems can minimize the problems associated with the loss of biomass owing to solid-liquid phase separations.

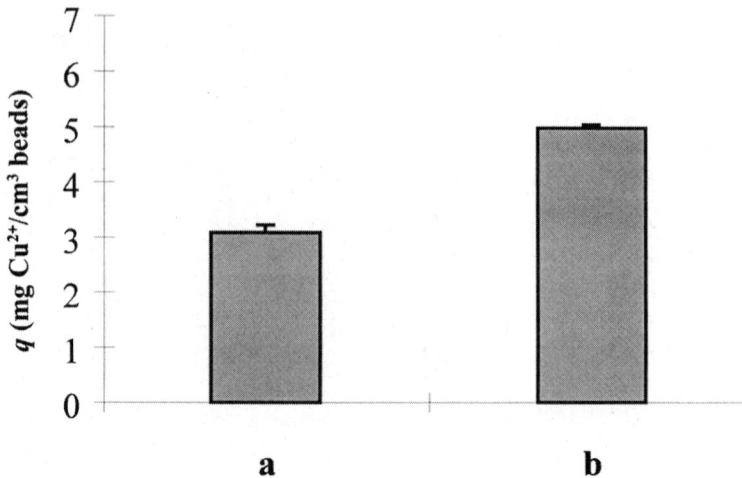

Fig. 6. Copper removal capacity by 2% calcium alginate and 10% polyacrylamide:
(a) 10% polyacrylamide and 2% calcium alginate (without cells); (b) 10% polyacrylamide and 2% calcium alginate mixed with 42.3 g dry cell/L.

Immobilization Studies

The high surface-to-volume ratio of the microbial cells allows better
contact between the biosorbent surface and soluble ions and molecules
(1,12). However, because of the difficulty of separating cells from the solution phase, cell immobilization procedures may be required for the development of large-scale recovery processes. The most common strategies
used to attain this goal include the colonization of cells on a solid substrate
(i.e., biofilm), or their immobilization into a polymeric or porous matrix
(2,13,14).

Micrococcus sp. was immobilized in 2% calcium alginate and 10% polyacrylamide gel beads to remove copper (II) ions from wastewater. As shown
in Fig. 6, the copper removal capacity of the biomass increased by 61%
when it was immobilized into gel beads. Polyacrylamide-entrapped cells
have good mechanical properties and are inert to microbial degradation
(15). Calcium alginate is also a well-known biopolymer. It is biocompatible,
chemically resistant, inexpensive, and easy to regenerate. In addition, it has
a loose structure for overcoming diffusion limitations and provides a good
model system for adsorption. Hence, the combined calcium alginate–
polyacrylamide-based immobilization matrix offers promising potential for
whole-cell immobilization in order to detoxify the metals-bearing solutions.

Conclusion

Copper biosorption by *Micrococcus* sp. was strongly dependent on the
pH and copper concentration of the solution. Cells harvested at exponential growth phase and stationary phase showed similar biosorption characteristics for copper. The biosorption of copper decreased with decreasing

pH from 6.0 to 2.0, which suggested that metal cations and protons compete for the same binding sites on the cell wall as pH decreased. In multimetal systems, *Micrococcus* sp. exhibited a preferential biosorption order: Cu ~ Pb > Ni ~ Zn. There is virtually no interference with copper uptake by *Micrococcus* sp. from solutions bearing a high concentration of Cl^-, SO_4^{2-}, and NO_3^- (0–500 mg/L). Sulfuric acid (0.05 M) was an efficient desorption medium. The copper capacity of *Micrococcus* sp. remained after five successive sorption and desorption cycles. Immobilization of *Micrococcus* sp. in 2% calcium alginate and 10% polyacrylamide gel beads increased copper uptake by 61%. These results show that biomass of *Micrococcus* sp. may be applied to the development of potentially cost-effective biosorbents for the removal and recovery of copper from effluents.

Acknowledgments

We gratefully acknowledge financial support from the Hong Kong Polytechnic University Central Research Grant and the Hong Kong Research Grants Council (grant no. PolyU 5001/00E).

References

1. Eccles, H. (1995), *Int. Biodeter. Biodegrad.* 35, 5–16.
2. Wang, L., Chua, H., Wong, P. K., Lo, W., Yu, P. H. F., and Zhao, Y. G. (2000), *Water Sci. Technol.* **41(12),** 241–248.
3. Hu, M. Z. C., Norman, J. M., Faison, B. D., and Reeves, M. E. (1996), *Biotechnol. Bioeng.* **51,** 237–247.
4. Volesky, B. and Holan, Z. R. (1995), *Biotechnol. Prog.* **11,** 235–250.
5. Lo, W., Chua, H., Lam, K. H., and Bi, S. P. (1999), *Chemosphere* **39(15),** 2723–2736.
6. Leung, W. C., Wong, M. F., Chua, H., Lo, W., Yu, P. H. F., and Leung, C. K. (2000), *Water Sci. Technol.* **41(12),** 233–240.
7. Wong, M. F. (2001), M Phil thesis, The Hong Kong Polytechnic University, Hong Kong SAR, PRC.
8. Muraleedharan, T. R., Iyengar, L., and Venkobachar, C. (1991), *Curr. Sci.* **61(6),** 379–385.
9. Chang, J. S., Law, R., and Chang, C. C. (1997), *Water Res.* **31(7),** 1651–1658.
10. Shuttleworth, K. L. and Unz, R. F. (1993), *Appl. Environ. Microbiol.* **59(5),** 1274–1282.
11. Zhang, L., Zhao, L., Yu, Y., and Chen, C. (1998), *Water Res.* **32(5),** 1437–1444.
12. Butter, T. J., Evison, L. M., Hancock, I. C., Holland, F. S., and Matis, K. A. (1996), *Med. Fac. Landbouww. Univ. Gent.* **61/4b,** 1863–1870.
13. Hu, M. Z. C. and Reeves, M. E. (1997), *Biotechnol. Prog.* **13,** 60–70.
14. Wong, P. K., Lam, K. C., and So, C. M. (1993), *Appl. Microbiol. Biotechnol.* **39,** 127–131.
15. Nakajima, A., Horikoshi, T., and Sakaguchi, T. (1982), *Appl. Microbiol. Biotechnol.* **16,** 88–91.

Production of Biosurfactant from a New and Promising Strain of *Pseudomonas aeruginosa* PA1

L. M. Santa Anna,*,[1] G. V. Sebastian,[2] N. Pereira, Jr.,[3] T. L. M. Alves,[4] E. P. Menezes,[5] and D. M. G. Freire[6]

[1,2]*Centro de Pesquisas da Petrobras (Petrobras Research Center-CENPES),* [3]*Escola de Química, UFRJ,* [4]*PEQ/COPPE/UFRJ, Rio de Janeiro, Brazil;* [5]*Fundação Tropical André Tosello, Campinas, São Paulo, Brazil; and* [6]*Faculdade de Farmácia, UFRJ, Rio de Janeiro, Brazil*

Abstract

The *Pseudomonas aeruginosa* PA1 strain, isolated from the water of oil production in Sergipe, Northeast Brazil, was evaluated as a potential rhamnolipid type of biosurfactant producer. The production of biosurfactants was investigated using different carbon sources (*n*-hexadecane, paraffin oil, glycerol, and babassu oil) and inoculum concentrations (0.0016–0.008 g/L). The best results were obtained with glycerol as the substrate and an initial cell concentration of 0.004 g/L. A C:N ratio of 22.8 led to the greatest production of rhamnolipids (1700 mg/L) and efficiency (1.18 g of rhamnolipid/g of dry wt).

Index Entries: Production of biosurfactants; glycolipids; rhamnolipids; *Pseudomonas aeruginosa*; surface tension.

Introduction

The potential for the application of biosurfactant's has increased considerably in the last few years owing to their utilization in several areas. The most important advantage of a microbial surfactant in relation to a chemical one lies in its ecologic acceptance, because it is biodegradable and nontoxic in natural environments. Furthermore, biosurfactants can be applied in extreme temperature, pH, and salinity conditions, making them more versatile than industrial surfactants found in the market *(1)*.

---

*Author to whom all correspondence and reprint requests should be addressed.

Current address: Centro de Pesquisas e Desenvolvimento, Leopoldo A. Miguez de Mello, CENPES—PETROBRAS, Av. Hum, Quadra 7, Cidade Universitária, 1 Ilha do Fundão, Rio de Janeiro, Brazil, CEP 21949-900.

In situ investigations have shown the capacity of biosurfactants in removing pollutants from the marine environment. *Pseudomonas aeruginosa* SB30 biosurfactants were used to remove oil from gravel in the *Exxon Valdez* tanker oil spill in Alaska. A 1% solution of biosurfactants was enough to remove two times more oil than water in temperatures ranging from 10 to 80°C (*2*). However, for biosurfactants to conquer a significant share in the market, they should be produced at low cost. Therefore, there is a need to better understand the metabolism, physiology, and industrial process parameters, in addition to using a cheaper substrate (*3*).

Bacteria of the genus *Pseudomonas* can use different substrates such as glycerol, mannitol, fructose, glucose, *n*-paraffins, and vegetable oils to produce rhamnolipid-type biosurfactants (*4*). Many studies have been conducted to determine the best carbon, nitrogen, phosphorus, and iron concentrations, to improve the bioprocess performance. The optimization of the ratio of carbon to nitrogen (C:N) has been researched in continuous cultures of *P. aeruginosa*, and ratios between 15 and 23 have been indicated as optimum values to yield a high rhamnolipid-specific productivity, using glucose and vegetable oil, respectively (*5*). It has been determined that nitrogen depletion in the culture medium, cellular metabolism is directed toward rhamnolipid production, because several studies have detected their production after the exponential growth phase (*6*).

The objective of the present study was to examine the production and molecular characterization of a rhamnolipid-type biosurfactant, produced by a bacteria isolated from petroleum environments.

Materials and Methods

Isolation, Identification, and Preservation of Microorganism

Isolation of the microorganism was carried out using samples from petroleum wells of the Northeast Brazilian region. Isolation was performed using the successive dilution method of the water sample and placed on Cetrimide agar plates (Merck, Darmstadt). Plates were incubated at 30°C for 48 h and the bacterial culture was isolated. Identification was done using the BIOLOG™ (Biolog, Hayward, CA) automated identification system for Gram-negative bacteria. Results were compared with the Microlog software database for determining the similarity coefficient for the type established in the identification system (*7*). The isolated and identified bacteria was preserved by freezing with liquid nitrogen in a glycerol solution (10% [v/v]) at –196°C.

Preinoculum

The bacterium was reactivated in a trypticase soy agar (Merck) medium, cultivated at 30°C for 48 h, and then transferred to 250-mL conical flasks containing 50 mL of mineral medium with the following composition: 1.0 g/L of $(NH_4)_2SO_4$, 3.0 g/L of KH_2PO_4, 7.0 g/L of K_2HPO_4,

0.2 g/L of MgSO$_4$·7H$_2$O, and 1% (v/v) glycerol at pH 7.0. The flasks were incubated at 30°C and 250 rpm for 20 h *(6)*.

Fermentation

Rhamnolipid production was studied during 7- to 10-d fermentation periods in 500-mL conical flasks containing 100 (medium volume:flask volume [MV:FV] of 0.2) or 200 mL (MV:FV of 0.4) of mineral medium as previously mentioned and 0.004 g/L of initial inoculum. The flasks were maintained at 30°C and agitated at 120 rpm in an incubator shaker (New Brunswick model G24; Edison, NJ). Carbon sources used, keeping a C:N ratio of 22.8, were *n*-hexadecane (Merck), paraffin oil (from production wells in the Buracica, BA, Brazil, containing 32% paraffins, 23% aromatics, 36% resins and 9.1% asphaltenes), glycerol (PA, Merck), and babassu oil (Du Reino).

Besides the effect of several carbon sources, the effect of inoculum concentration (0.0016, 0.004, and 0.008 g/L), the C:N ratio from the best carbon source, and the MV:FV ratio on rhamnolipid production were also investigated.

Analyses

Biomass Concentration

Bacterial growth was monitored by measuring the optical density (OD) of the prepared sample using a spectrophotometer (model B442; Micronal, Brazil) at 500 nm. A 50-mL sample was taken from the fermentation flask at the end of cultivation and OD was achieved by measuring after dilution. The sample was centrifuged at 7000 rpm for 20 min (model J2-21; Beckman). Centrifuged cells were suspended in 5 mL of distilled water. The dry weight of the biomass was found from a constructed calibration curve. The biomass was expressed as grams/liter.

Rhamnose and Glycerol Assay

Rhamnolipid quantification expressed as rhamnose (milligrams/liter) was evaluated in the cell-free spent medium using the phenol sulfuric acid method *(8)*. Glycerol was evaluated by an enzymatic-colorimetric method for the determination of triglycerides (CELM, São Paulo, Brazil) *(9)*.

Measurement of Surface Tension

Surface tension was measured in the cell-free spent medium using the SIGMA 70 digital tensiometer (KSV Instruments, Helsinki, Finland), at a temperature of 25°C, by the Du Nouy method *(10)*.

Extraction and Purification of Biosurfactant

After 7 d of fermentation, the culture medium was centrifuged at 4400 rpm for 15 min, and to the floating matter was added a solution of H$_2$SO$_4$ (5 N) until a final pH of 2.0 was obtained for the rhamnolipid precipitation. The precipitate was recovered in a 2:1 chloroform:ethanol solution.

After evaporation of the solvent, part of the residue was mixed with a KBr tablet to obtain its infrared absorption spectrum with the Fourier transform (Perkin-Elmer–2000). The remaining residue was used for the molecular characterization of rhamnolipid using the nuclear magnetic resonance (NMR) technique (Varian Inova 300 spectrophotometer).

Results and Discussion

Identification of Bacteria

The bacteria were identified as *P. aeruginosa* in the BIOLOG automated system with a 99% similarity to the standard strain *P. aeruginosa* and received the PA1 code. The bacteria used carbon sources such as fructose, D-glucose, mannitol, mannose, glycerol, and lactic acid, substances well known as good carbon sources for rhamnolipid production *(7)*.

Effect of Carbon Source

Experimental results for rhamnolipid production from *P. aeruginosa* PA1 using substrates such as hexadecane, paraffin oil, babassu oil, and glycerol are given in Table 1. *P. aeruginosa* PA1 was capable of using *n*-hexadecane as the sole carbon and energy source, producing 130 mg/L of rhamnolipid and a variation in the surface tension of 47.4% by the end of 7 d of fermentation. Suk et al. *(11)* used 3% hexadecane as the carbon and energy source, and after 3 d of fermentation with *P. aeruginosa* the surface tension of the medium was reduced from 72 to 30 D/cm with a reduction of 58%.

The utilization of paraffin oil, which has a complex and heterogeneous nature, resulted in reasonable rhamnolipid production expressed as rhamnose (260 mg/L); however, there was almost no change in surface tension by the end of the cultivation (4.4%). This fact is probably related to the formation of an emulsion during fermentation, which interfered with the determination of surface tension and made impossible the quantification of surface tension in a suitable manner. It was also shown from research by Hisatsuka et al. *(12)* that rhamnolipids in hydrocarbon fermentation contribute to stabilizing oil-in-water emulsions.

The utilization of vegetable oil and glycerol as carbon sources for rhamnolipid production seems to be a rather interesting and probably a low-cost alternative *(4)*. The bacteria produced 200 mg/L of rhamnolipid when babassu oil was used as the carbon source, and there was a decrease of 31% in surface tension by the end of cultivation.

However, Table 1 shows that the greatest rhamnolipid production (690 mg/L) with the best tensoactive characteristics (48% of decrease in surface tension) was achieved with glycerol, a carbon source easily assimilated. In addition to the quantity of rhamnolipid obtained in the medium (three to four times greater than those obtained for other sources), there was a quite large formation of foam in the cultivation floating material.

Table 1
Rhamnolipid Quantification and Surface Tension Measurement
After 7 d of Fermentation of *P. aeruginosa* PA1
Using *n*-Hexadecane, Paraffin Oil, Babassu Oil, and Glycerol
as Carbon Sources, for a C:N Ratio of 22.8

Carbon source	Rhamnose (mg/L)	Initial surface tension (D/cm)	Surface tension (D/cm)	Surface tension reduction (%)
n-Hexadecane	130	53.90	28.35	47.4
Paraffin oil	260	54.00	51.60	4.4
Babassu oil	200	40.00	27.60	31.0
Glycerol	690	53.00	27.46	48.2

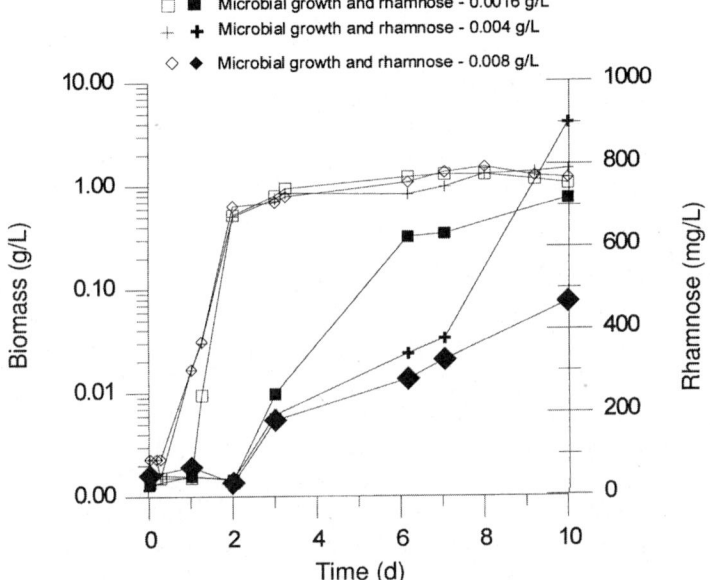

Fig. 1. Microbial growth curve and rhamnolipid production from the fermentation of *P. aeruginosa* PA1, over 10 d, using glycerol (C:N ratio of 22.8) with initial inocula of 0.0016, 0.004, and 0.008 g/L.

In agreement with our findings, studies conducted with a *P. aeruginosa* CFTR-6 have demonstrated a great capacity of the strain for glycolipid production (620 mg/L) when glycerol (2% [w/v]) is used as a source of carbon and energy *(6)*.

Effect of Inoculum Concentration

The effect of the initial inoculum concentration was studied using glycerol as the substrate. Figure 1 shows the microbial growth kinetics and rhamnolipid production in a 10-d fermentation with inoculum ratios of 0.0016, 0.004, and 0.008 g/L.

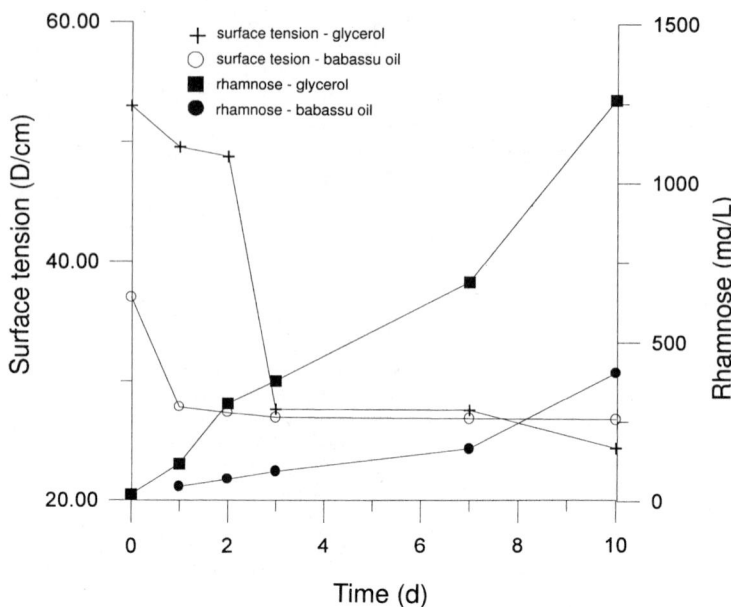

Fig. 2. Rhamnolipids production and surface tension measurements along *P. aeruginosa* PA1, fermentation using babassu oil and glycerol with a C:N ratio of 22.8.

The microbial growth profiles for the three different inocula were very similar with practically identical specific growth rates of $0.165\,h^{-1}$ and the stationary phase beginning between 48 and 72 h. In the beginning of the stationary phase, we observed a substantial increase in the rhamnolipid concentration at the end of 10 d of fermentation, 720 and 900 mg/L, with 0.004 and 0.008 g/L of inoculum concentration, respectively. Similar to our study, other researchers have observed that the increase in rhamnolipid production occurs when microbial growth reaches the stationary phase, indicating its characteristic as a secondary metabolite *(5)*.

Effect of Ratio MV:FV

A ratio of MV:FV of 0.2, twice as great as in the previous experiments, was used, employing glycerol and babassu oil as substrates at a C:N ratio of 22.8. The fermentations were inoculated with a cell concentration corresponding to a inoculation of 0.004 g/L. Figure 2 shows the kinetic profiles of rhamnolipid production and surface tension over 10 d of fermentation.

Glycerol was still the best carbon source for rhamnolipid production; however, the use of a higher MV:FV ratio led to a twofold greater production of rhamnolipid (1400 mg/L), reaching a decrease in the surface tension of 24.6 D/cm at the end of 72 h of fermentation. When the MV:FV increased twofold, probability had increased oxygen supply in the medium just greater than the surface area in the flask.

Table 2
Rhamnolipid and Biomass Concentrations
After 7 d of Fermentation at Different Concentrations of Glycerol

Glycerol concentration (%)	C:N	Biomass (mg/L)	Rhamnose (mg/L)
0.5	11.4	1500	410
0.71	16.2	1950	710
1	22.8	2500	760

In the fermentation using babassu oil as the carbon source, production did not increase beyond 500 mg/L of rhamnolipid in 10 d of cultivation. Although surface tensions using either glycerol or babassu oil had reached similar values in 72 h of fermentation (24.6 D/cm), the rhamnolipid concentration was three times greater when glycerol was used as the substrate. Other researchers have obtained a similar reduction in the surface tension (25 D/cm) when glycerol was used in higher concentrations of approx 2% *(13)*.

At the beginning of fermentation with babassu oil, the culture medium already presented a low surface tension of 38 D/cm (Fig. 2). This fact can be ascribed to the presence of free constituent fatty acids in the oil with surfactant activity.

Effect of C:N Ratio

Table 2 gives the results for biomass production and rhamnolipid quantification at the end of 7 d of fermentation for various C:N ratios. The smallest C:N ratio used (0.5% [v/v] of glycerol) was limited in terms of carbon, since cell growth was about 60% lower compared with that using the highest C:N ratio. While the fermentation with the intermediate glycerol concentration (0.71% [v/v]) obtained good rhamnolipid production (710 mg/L), the fermentation using 1% (v/v) of glycerol yielded the greatest production of biomass (2500 mg/L) and rhamnolipid (760 mg/L).

The microbial growth kinetics and rhamnolipid production in the fermentation with a 1% (v/v) concentration of glycerol (C:N ratio = 22.8), MV:FV of 0.2, and inoculum of 0.004 g/L are represented in Fig. 3. The specific growth rate of the *P. aeruginosa* PA1 strain was 0.09 h^{-1}. The stationary phase was reached after 72 h of fermentation at the same time rhamnolipid production was increased. The rhamnolipid and biomass concentrations after 10 d were 1700 and 1470 mg/L, respectively. Glycerol was entirely consumed within 7 d of fermentation, and the rhamnolipid concentration peaked after another 3 d. The production of this rhamnolipid is typically of a secondary metabolite and increased considerably along the stationary phase.

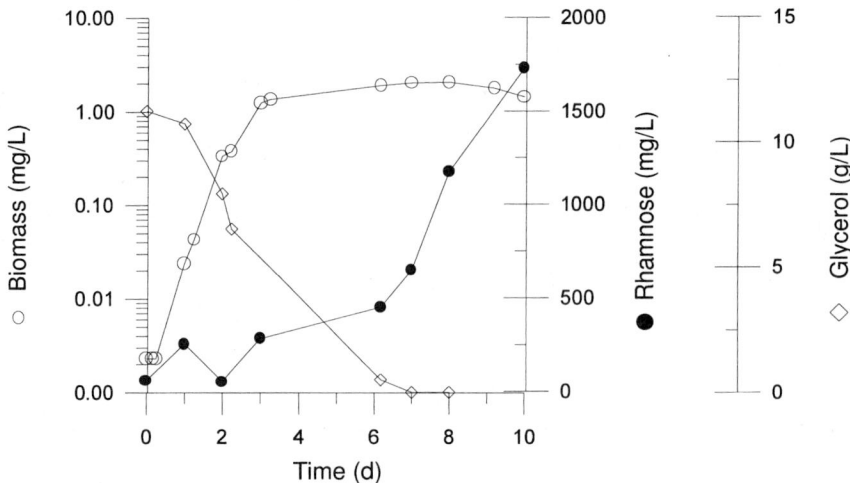

Fig. 3. Microbial growth curve, rhamnolipid production, and consumption of glyc-erol from the fermentation of *P. aeruginosa* PA1, over 10 d, using a 1% (v/v) glycerol concentration.

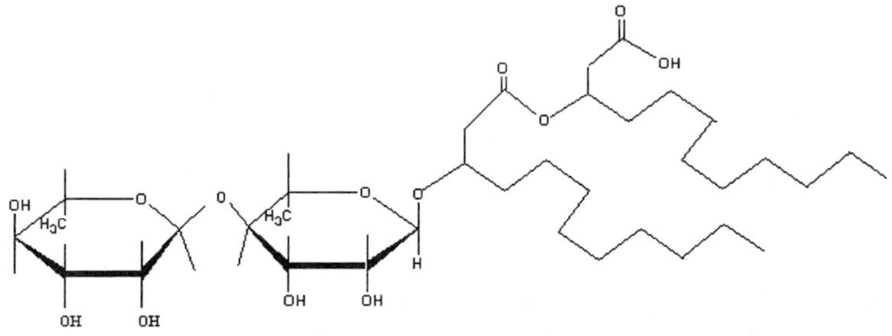

Fig. 4. Rhamnolipid structure of *P. aeruginosa* PA1 obtained from NMR analysis.

Molecular Characterization of Rhamnolipid

The structure of the rhamnolipid produced in the fermentation by the *P. aeruginosa* PA1 strain with 1% (v/v) glycerol (C:N ratio = 22.8), MV:FV of 0.2, and inoculum of 0.004 g/L is represented in Fig. 4.

The characterization of the molecular structure of the rhamnolipid was investigated by the combination of ^{13}C NMR and infrared spectroscopy *(14)*. Preliminarily compatible structures have been proposed. Two pos-sible average structures were reached. One was a diester, and the other was an acid-ester molecule, which is shown in Fig. 4. Spectra of standard mol-ecules in the literature could not allow a clear distinction among the carbo-nyls in the possible structures. The C=O deformation, which could be a definite difference in infrared spectra, was broad in the analysis of the samples and could not yield the necessary resolution to distinguish the possibilities. Therefore, the molecule has this undetermined structure. Another important aspect was the bond between the two rhamnose rings.

This could be compatible with a β 1.4 bond as in lactose and cellobiose, since the NMR spectrum showed resonance in the chemical shift of 103 ppm. However, that was not enough to ensure the bond position. Future studies will focus on the chromatographic purification of the rhamnolipids and on the definite proposition of the molecular conformation. Syldatk et al. *(15)* have proposed a very similar structure, with an α 1.2 bond between the cycles and with fewer carbons in the aliphatic chain of the esters of the rhamnolipid molecule.

Conclusion

The strain isolated from oil environments was identified as *P. aeruginosa*, showing a capacity of using carbon sources such as fructose, lactic acid, D-glucose, mannitol, mannose, and glycerol. This strain is capable of producing rhamnolipid-type biosurfactants from substrates such as *n*-hexadecane, paraffin oil, babassu oil, and glycerol. The rhamnolipid production kinetics by *P. aeruginosa* PA1 is typical of a secondary metabolite. The greatest rhamnolipid production (1700 mg/L) was obtained after 10 d of fermentation with 1% (v/v) glycerol (C:N ratio = 22.8) as the sole carbon source, MV:FV of 0.2, and initial inoculum of 0.004 g/L. The molecular characterization shows that the rhamnolipid is a disaccharide esterified in C-1, with acid having a chain size of 12 carbon atoms.

Acknowledgment

The molecular structure was determined by the group in the chemical sector at the Petrobras Research Center in Brazil for a future publication.

References

1. García, M. A. (1992), *Revista Instituto Mexicano Petroleo* **24**, 68.
2. Harvey, S., Elashvili, L., Valdes, J. J., Kamely, D., and Chakrabarty, M. (1990), *Biotechnology* **8**, 228–338.
3. Fiechter, A. (1992), *Tibtech* **1**, 208.
4. Boulton, C. and Ratledge, C. (1987), *Biosurfactants Biotechonol.* **25**, 47.
5. Ochener, U. A., Hembach, T., and Fiechter, A. (1995), *Adv. Biochem. Eng. Biotechnol.* **53**, 89.
6. Venkata Ramana, K. and Karanth, N. G. (1989), *J. Chem. Technol. Biotechnol.* **45**, 249.
7. Biolog (1993), MicroStation™ System Release, version 3.50.
8. Dubois, M., Gilles, K. A., Hamilton, J. K., Rebers, P. A., and Smith, F. (1956), *Anal. Chem.* **28**, 350–356.
9. Trinder, P. (1969), *Ann. Clin. Biochem.* **6**, 24–27.
10. American Society for Testing Materials. (1999), ASTM D971–99[a] Standard Test Method for Interfacial Tension of Oil Against Water by the Ring Method, American Society for Testing Materials.
11. Suk, W.-S., Son, H.-J., Lee, G., and Lee, S.-J. (1999), *J. Microbiol. Biotechonol.* **9(1)**, 56–61.
12. Hisatsuka, K., Nakahara, T., Sano, N., and Yamanda, K. (1971), *Agric. Biol. Chem.* **35**, 686–692.
13. Banat, I. M. (1995), *Acta Biotechnol.* **15(3)**, 251–267.
14. Ribeiro, A., Zhou, A., and Raetz, C. R. H. (1999), *Magnet. Reson. Chem.* **37**, 620–630.
15. Syldatk, C., Lang, S., and Matulovic, U. (1985), *Z. Naturforsch.* **40(1)**, 61–67.

Cassava Starch Maltodextrinization/ Monomerization Through Thermopressurized Aqueous Phosphoric Acid Hydrolysis[†]

José D. Fontana,*[,1] Mauricio Passos,[1]
Madalena Baron,[2] Sabrina V. Mendes,[1] and Luiz P. Ramos[3]

[1]LQBB—Biomass Chemo/Biotechnology Laboratory,
Department of Biochemistry, Biological Sciences Sector,
UFPR—Federal University of Parana, PO Box 19046, Curitiba, PR, Brazil,
81531l-990, E-mail: jfontana@bio.ufpr.br; [2]Pharmaceutical Sciences/
UNIANDRADE, and [3]CEPESQ—Research Center in Applied Chemistry,
Department of Chemistry, UFPR, PO Box 19081,
Curitiba, PR, Brazil, 81531l-990

Abstract

Kinetic conditions were established for the depolymerization of cassava starch for the production of maltodextrins and glucose syrups. Thin-layer chromatography and high-performance liquid chromatography analyses corroborated that the proper H_3PO_4 strength and thermopressurization range (e.g., 142–170°C; 2.8–6.8 atm) can be successfully explored for such hydrolytic purposes of native starch granules. Because phosphoric acid can be advantageously maintained in the hydrolysate and generates, after controlled neutralization with ammonia, the strategic nutrient triplet for industrial fermentations (C, P, N), this pretreatment strategy can be easily recognized as a recommended technology for hydrolysis and upgrading of starch and other plant polysaccharides. Compared to the classic catalysts, the mandatory desalting step (chloride removal by expensive anion-exchange resin or sulfate precipitation as the calcium-insoluble salt) can be avoided. Furthermore, properly diluted phosphoric acid is well known as an allowable additive in several popular soft drinks such as colas since its acidic feeling in the mouth is compatible and synergistic with both natural and artificial sweeteners. Glycosyrups from phosphorolyzed cassava starch have also been upgraded to high-value single-cell protein such as the pigmented yeast biomass of Xanthophyllomyces dendrorhous (Phaffia rhodozyma), whose astaxanthin (diketo-dihydroxy-β-carotene) content may reach 0.5–1.0 mg/g

*Author to whom all correspondence and reprint requests should be addressed.
[†]A patent request concerning this article's subject and its unfolding was addressed to Instituto Nacional da Propriedade Intelectual, Brazil, in May 2000 as protocol PI 0002001-0.

of dry yeast cell. This can be used as an ideal complement for animal feeding as well as a natural staining for both fish farming (meat) and poultry (eggs).

Index Entries: Starch; hydrolysis; phosphoric acid; maltodextrinization; astaxanthin; byproducts.

Introduction

Following cellulose, starch is the most widespread natural source of the monosaccharide glucose in a polymeric form. Chitin, the second most abundant glycopolymer in nature, bears a modified monomeric unit (2-deoxy-2-acetylamino-D-glucopyranose) *(1)*. Although fully linear as a homo-β-D-glucopyranan, cellulose is exceptionally resistant to strong acid hydrolysis with mineral acids owing to the maximized inter- and intramolecular hydrogen bonding that holds the chains together in bundles. Even enzymatic depolymerization through fungal and bacterial cellulolytic complexes seldomly leads to complete monomerization. These optimized cellulase complexes, at the most, accumulate cellobiose since β-glucosidases are not a dominant or balanced component within naturally occurring cellulolytic complexes, mainly composed of isoforms of endo- and exoglucanases including cellobiohydrolases *(2)*. By contrast, starch is amenable to both mineral acid and enzymatic hydrolyses since the same kind of glycosidic linkage is built on the α-anomer in both the linear homopolymeric 1,4-linked amylose fraction and the 1,6-ramified fraction amylopectin, thus favoring helicoidal conformations rather than the particularly rigid crystalline structure found in cellulose.

A number of enzymatic preparations are commercially available for starch bioprocessing and upgrading. The most intensively employed enzymes are the thermoresistant α-amylases (maltogenic for amylose, malto- and dextrinogenic for amylopectin) and amyloglucosidases or glucoamylases (glucogenic for both starch fractions). The combination of both types of enzyme, taking advantage of the thermostability of *Bacillus* spp. amylases, allows for the simultaneous gelatinization and saccharification of starch granules. At lower temperatures, amyloglucosidases complete the splitting of any remaining 1,4- and 1,6-linkages (branching points) of short dextrins, branched oligosaccharides, and maltose *(3)*.

Alternatively, the acid-catalyzed depolymerization of starch, most often using HCl (and less often H_2SO_4), is in fact still applied extensively in starch-processing factories around the world, despite the full commercial availability, at higher costs, of several enzymatic preparations for starch hydrolysis. HCl-mediated hydrolysis, depending on the acid strength and temperature (pressurization), may convert the slurry from native starch granules to either maltooligosaccharides (maltodextrins) or glucose *(3)*. In the latter case, owing to the partial overlapping of the kinetic constants for the breakdown of glycosidic linkages *(4)* and for the subsequent free hexose dehydration/degradation, glucose syrups are undesirably accompanied by some nonsugar byproducts. In addition, if the starch preparations (commercial feculas) have some degree of contamination with (hemi)cellulosic fibers (e.g., arabinoxylans), the level of degradation products increases with the accumulation of another dehydration byproduct from pentoses—furfural. These parallel color- and taste-generating reactions require additional industrial steps for clarification of glucose syrups (e.g., activated charcoal), a situation different from the kitchen caramels *(5)* resulting from light or deep browning of sucrose on excessive heating in which color and a particular taste are intended to replace or combine the basic sweetening power of table sugar.

This article deals exactly with the replacement of strong mineral acids—HCl or H_2SO_4—by a milder catalyst, H3PO4. The pursued advantages for this new technology are broached in the Results and Discussion.

Materials and Methods

Starch Sources

Fresh tubers of cassava (variety Fibra Branca) were obtained in a local market, peeled, and quickly submerged in cold water to avoid the browning phenolization. Each root segment was coarsely sliced with a sharpened knife. The slices were combined with 2 to 3 vol of cold water and triturated in a Waring blendor for 30 s. The thick and milky suspension of starch granules, containing (hemi)cellulosic fibers and released protein, was filtered through a double layer of cheesecloth in which most of the foam material was retained along with coarser fragments. Cassava starch granules settled very quickly as a deep white and firm bed, leaving most of the protein in the yellowish supernatant, which was then discarded. To complete the removal of residual free protein and also the smallest and less dense starch granules, the referred bed was resuspended in an excess of water, agitated to disperse the starch granules, and filtrated through a fourfold bent layer of cheesecloth repeatedly until there was no noticeable turbidity in the supernatant. The final stock solution of cassava starch was shown to have at least 35% solids concentration on dry basis, as determined by freeze-drying of the starch slurry.

In some cases, cassava and potato feculas from local producers were used, and for comparative purposes, soluble starch (Reagen) and wet-milled corn kernels (no SO_2 addition) served as representative samples of purified tuber (potato) and crude grass starch, respectively.

Acid and Enzymatic Hydrolyses

The aforementioned starch slurries were acidified with 1:10 and 1:100 dilutions of commercial phosphoric acid (85% [w/w]) (Merck, Brazil) under strong agitation (magnetic bar) to avoid settlement of starch granules. This addition of acid corresponded to a range of 12.5–425 mg of acid/g of starch (dry basis). The minor buffering interference of the residual protein from the intact starch granules was taken into account, and this motivated the extensive granule washing procedure described in the previous section. Reference incubations of soluble starch (Reagen) with amylolytic enzymes from Novo Industri (*Fungamyl®* and amyloglucosidase) followed the experimental procedure recommended by the supplier.

Thermopressurized Acid Hydrolysis

A homemade steel reactor vessel (V_t = 15 L), equipped with both temperature and pressure control systems and a purge valve, was used for the hydrolysis operations. After hydrolysis, starch suspensions were contained in Pyrex glassware with loosely fitted caps. The heating period to reach the desired peak temperature (or corresponding pressure) corresponded to about 20 min. The employed thermopressurization ranges varied from 2.8 to 6.8 atm, and the residence time at the peak temperature was from 5 to 10 min.

Analytical Procedures

Partially or completely phosphorolyzed starch slurries were diluted from 1:10 to 1:100 for the chromatographic analyses. Thin-layer chromatography (TLC) was carried out on silica gel 60 chromatoplates (Merck-Darmstadt, Germany) irrigated with isopropanol:ethyl acetate:nitroethane:acetic acid:water (30:5:5:0.5:7 [v/v]) as the mobile phase for double developments. Free glucose and maltooligosaccharides were revealed by first spraying the plates with orcinol (0.5 g% [w/v]) in a sulfuric acid:methanol solution (5:95 [v/v]) and then heating them until full color development (blue violet) was achieved. Densitometry of developed TLC plates was carried out in a CS-9301PC from Shimadzu (Tokyo, Japan) under the flying spot operational mode.

High-pressure liquid chromatography (HPLC) was performed in a Shimadzu HPLC system, model LC10AD, provided with an SIL10A autosampler and an RID10A refractive index detector. HPLC analysis was performed at 65°C in an Aminex HPX-87H column (Bio-Rad) eluted with 8 mM H$_2$SO$_4$ at a flow rate of 0.6 mL/min, a condition that does not lead to any on-line hydrolysis of maltooligosaccharides. Soluble sugars were quantified by calibration using maltose (retention time, R_T, of 6.9 min), glucose (8.7 min), and acetic acid (14.7 min) as external standards.

Upgrading of Phosphoric Hydrolysates
to Carotenoid-Enriched Biomasses

Following pH adjustment of starch digests with aqueous ammonia to 5.0 and dilution to a theoretical total sugar content of 4 g%, the substrate was sterilized and supplemented with 75 mg/L of yeast extract. All individual assays were normalized with respect to the final content of ammonium phosphate. Inoculation proceeded with the basidiomycetous yeast *Xanthophyllomyces dendrorhous* (ATCC *Phaffia rhodozyma* strain 24202) or with an amylolytic yellowish bacterium isolate from rot dahlia tubers

(*AMRL*). Cultures were grown in 50-mL Erlenmeyer flasks (medium:total volume = 1:5) for 5 d at 25 to 26°C at moderate oxygen transfer rate (100 rpm on an orbital shaker).

The yeast or bacterial cell mass was collected by centrifuging at 3000g, lyophilized, and then treated with dimethyl sulfoxide (2 vol for good swelling) for carotenoid extraction with an excess of acetone (two times). Comparative yields of astaxanthin (yeast) or β-carotene plus xanthophylls (bacterium) were evaluated in each 10-mL normalized organosolvent extract by spectrophotometry at 470 nm, assuming that 0.25 U of absorbance corresponds to 1 μg of carotenoid/mL *(5)*.

Results

The cassava business currently is an important world concern. Although Nigeria (Africa) is the biggest producer (31 million t/yr), Thailand occupies the leading position for exports of native and modified cassava starches. In Brazil (12% of the world production), cassava is mainly cultivated in the southern state of Paraná and in the Amazon region (Pará State), each affording crops of about 3.5 million t/yr. In the former state, an industrial plant (INDEMIL, Paranavaí, PR) is reported to be the only one producing glucose syrups from cassava in the Occident (about 100 t/mo). Catalysis for this is carried out with HCl, the same acid procedure also utilized by larger plants operating with corn starch.

Figure 1 demonstrates the effectiveness of bringing cassava starch slurries to dextrinization (lanes 1 and 2), oligomerization (lanes 3 and 4), or monomerization (lane 5) using phosphoric acid as the acid catalyst. It is readily observed that the progressive increase in acid strength favored monomerization for any prefixed thermopressurization condition (160°C and 5.1 atm, in the present case, but with a residence time at the peak temperature enlarged to 10 min). Minor amounts of degradation products were detected in the most severe hydrolytic condition, together with the progressive accumulation of glucose (R_f = 0.6). The amount of released glucose increased accordingly (Fig. 2; analyzed samples from TLC of Fig. 1 were those corresponding to lanes 10 and 12), as measured by densitometry, and the amount of degradation products (main peak at R_f > 0.8) accounted for <5% of the recovered free monosaccharide in the case of the strongest hydrolysis condition. In one case, even a trace amount of maltononaose was visible, but by increasing the catalyst concentration, the detection or recording limit shifted to maltopentaose. The distribution ratio of glucose:maltose:maltotriose for four phosphoric strengths, except for the mildest condition, are reported in Table 1; it increased (proportionally with acid strength) from 35:31:34 to 61:17:22. Also, the hydrolysis products profile obtained with phosphoric acid superseded that obtained with *Fungamyl*® (an *Aspergilllus* sp. amylolytic preparation from Novo Nordisk, Denmark) and approached that provided by amyloglucosidases (last two lanes in Fig. 1).

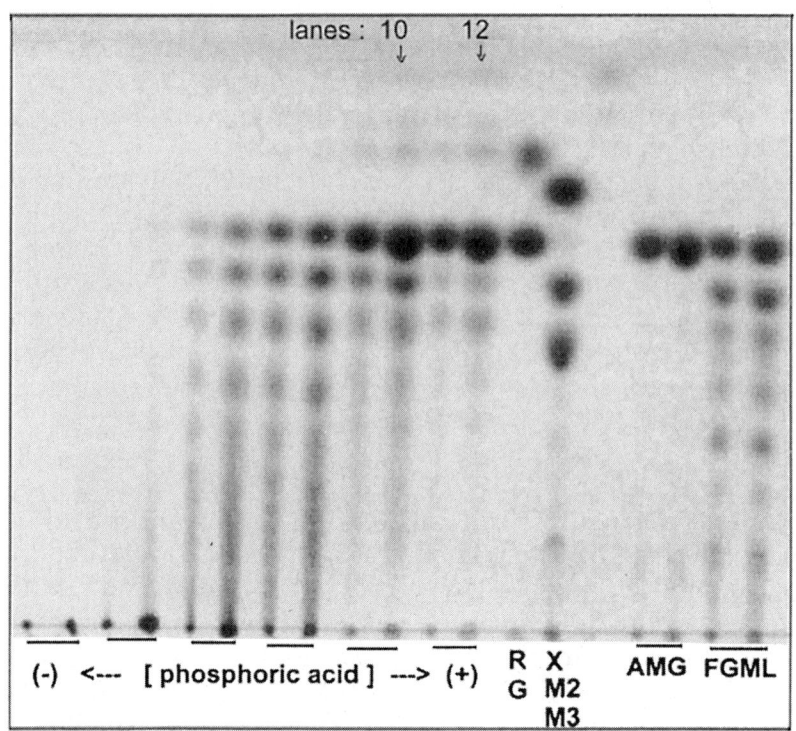

Fig. 1. Thin-layer chromatogram of cassava starch hydrolysis with thermo-pressurized aqueous phosphoric acid. (–) <— [phosphoric acid] —> (+) = hydrolysis with increasing phosphoric acid strength from 10.8 to 425 mg of acid/g of dry cassava native starch; R, G = rhamnose and glucose standard, respectively; X, M2, M3 = xylose, maltose, and maltotriose standard (in order of decreasing mobility, respectively); AMG = hydrolysis with amyloglycosidase; FGML = hydrolysis with *Fungamyl*, the latter two incubations using soluble starch as substrate.

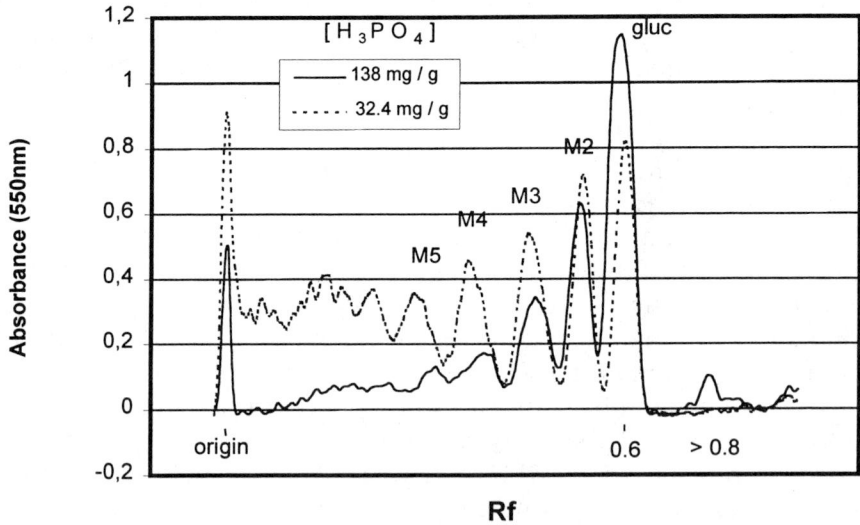

Fig. 2. Densitometric analyses at 540 nm of the developed thin-layer chromatogram (see Fig. 1, lanes 10 and 12) from the thermopressurized aqueous phosphoric digest of cassava starch. Apparatus: Shimadzu CS-9301PC densitometer; multipeak dashed and solid lines: starch (dry basis) processed at acid catalyst ratios of 32.4 and 138 mg/g, respectively.

Table 1
Monomeric to Trimeric Percentage Distribution of Reducing Sugars
After Thermopressurized Diluted Phosphoric Acid Hydrolysis
of Cassava Starch

Acid addition (mg/g starch)	Glucose	Maltose	Maltotriose
32.4	35.0	31.2	33.8
138	54.4	24.4	21.2
425	60.7	17.2	22.1

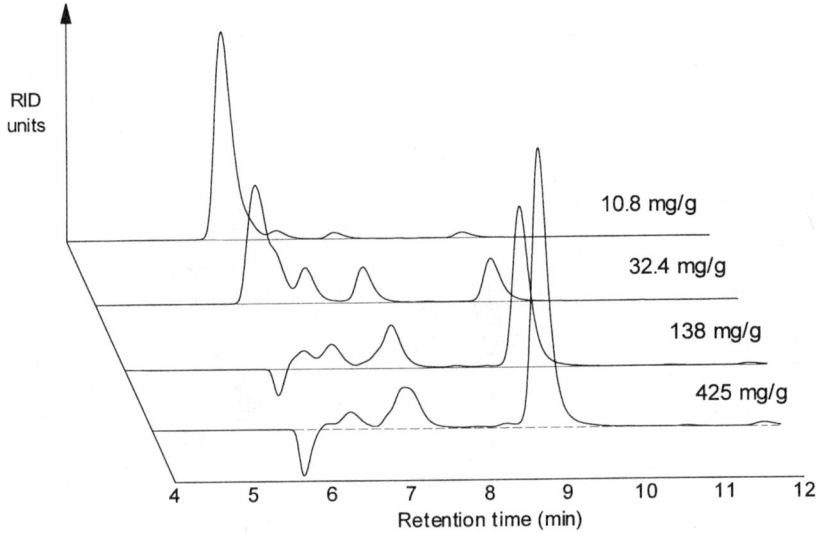

Fig. 3. HPLC of cassava starch hydrolysis with thermopressurized aqueous phosphoric acid (from top to bottom: increasing [phosphoric acid] from 10.8 to 425 mg of acid/g of dry cassava native starch). Chromatographic conditions are as described in Materials and Methods. RID, refractometer index detector.

HPLC analysis (Fig. 3) of starch hydrolysates corroborated the aforementioned interpretation, despite the lower resolution for maltooligomers with degree of polymerization (DP) higher than 3 (maltotriose). Similar results were obtained when commercial feculas, soluble starch, or even corn starch were used in substitution for our laboratory-made native starch granules preparation. The purity and granule integrity of the prepared starch were assessed by optical microscopy of the fresh material or following staining of the intact granules with iodine/iodide or Coomassie blue. Transmission electron microscopy has also been used for this purpose (6).

The suitability of phosphoric acid–hydrolyzed starch for fermentation procedures was successfully verified with two different microorganisms: the astaxanthinogenic GRAS yeast *X. dendrorhous* (formerly *P. rhodozyma*) and a bacterial isolate. In both cases, the usual level of micro-

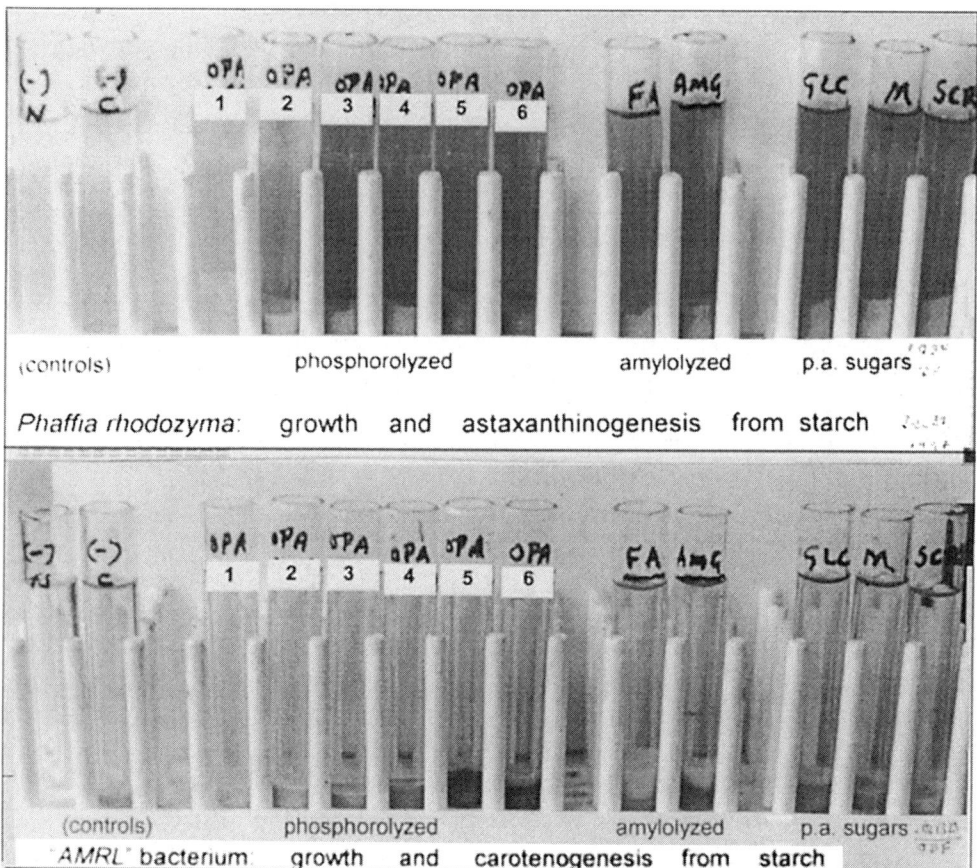

Fig. 4. Carotenogenesis from phosphoric acid–hydrolyzed starch with *X. dendrorhous* (**top**) or with the bacterial isolate *AMRL* (**bottom**). (–)N and (–)C denote the controls where nitrogen or carbohydrate were omitted in the fermentation. Assays 6, 5, 4, and 3 refer to starch thermopressurized phosphoric pretreatments with 425, 138, 32.4, and 10.8 mg of acid/g of dry starch at 160°C/5.1 atm, respectively. For assay 2, starch was mildly acidulated to pH 3.85 before digestion, and assay 1 is the control with no acid addition. FA and AMG correspond to enzymatic digests with *Fungamyl* or amyloglucosidase. GLC, M, and SCR are the fermentative runs carried out with glucose, maltose, or sucrose standard sugars, respectively. Supernatants contain the carotenoid material after organosolvent extraction of each formed yeast or bacterial biomass. Precipitates contain either only depigmented microbial cells (**right**) or the same with residual starch (**left**). See Materials and Methods for other details.

bial biomass production was obtained, and, more important, carotenoid pigmentation (Fig. 4) was indeed attained just after the onset of the stationary phase of growth.

Discussion

Starch hydrolysis with HCl or H_2SO_4 requires, for purposes such as food application of glucose- or maltose-enriched syrups, catalyst removal by anion exchange or precipitation of insoluble salts (e.g., calcium sulfate), respectively. Also, in the former case, the chloride anion is seldom beneficial (if not inhibitory) as compared to the sulfate anion when syrups are used for microbial upgrading to solvents, organic acids, antibiotics, and related fermentation goods. Here resides the main advantage of using phosphoric acid as an alternative catalyst because it may be directly incorporated, regardless of the particular concentration, to both food and fermentative applications. A partial neutralization with ammonia is recommended both to increase the pH to a more appropriate physiologic condition for fermentation and to provide an input of N source. In the case of lower sweetening syrups enriched in maltooligosaccharides or short dextrins (obtained with more dilute catalyst for food uses), no neutralization step would be required as long as the final acid concentration is <1.5 mM. Ammonium phosphate, together with the released reducing sugars themselves, constitutes the basic triplet of mandatory macronutrients in any fermentation procedure (C, N, and P). Also differentiating it from the stronger mineral acids, phosphoric acid (and its salts) may be considered by far more (if not exclusively) compatible with human physiology, acting as a "physiologic catalyst." Although HCl plays a role in stomach digestion, no consumer would accept any food acidified with HCl because both the mouth and pylorus lack the natural protection available in the stomach walls for this particular acid. Furthermore, the presence of small amounts of phosphoric acid (final pH of about or above 3.0) in food materials is widely accepted in the market, such as in the case of established soft drinks or colas.

Fermentation results matched those obtained in parallel runs with pure glucose, maltose, or sucrose. No attempt was made for the exact gravimetric expression of biomass yield (Y) because, in those runs in which starch was barely hydrolyzed (e.g., in the case of 11.25 mg of catalyst/g of starch), the removal of residual starch from the centrifuged cells proved very tedious. Conversely, the biosynthesized carotenoid (mostly if not completely remaining associated to the microorganism cells) was estimated, after organosolvent extraction, to be between 0.5 and 1.0 mg/g of dry cells (results not shown) based on previously reported literature procedures (7,8), and, as expected, astaxanthin was the major highly oxygenated carotenoid (>70%) produced by *X. dendrorhous*.

The strategy described herein for the phosphoric acid–mediated hydrolysis of starch confirmed our pioneering work carried out on sugarcane and sorghum bagasses (9) (xylan as the selectively hydrolyzed target) and dahlia inulin (10). Xylose and fructose were the respective dominant and sole hydrolytic products in those preliminary studies.

If no quantitative yield of glucose is pursued (which would demand much larger amounts of phosphoric acid as compared to HCl, since the pK_{a1} of the former is 2.15), phosphoric acid catalysis looks rather advantageous. And in dealing with acid catalysis, HCl being the most often employed at a final concentration of 20 mM, the higher price for phosphoric acid would not account for significant differences for two reasons: (1) the commercial provision refers to 37% HCl and to 85% H_3PO_4, both expressed as weight/weight; and (2) the acid catalyst is itself a secondary component of the cost breakdown. In fact, a much higher cost component is the removal of degradation products, which also brings additional advantages to phosphoric acid, since it leads to less colored starch hydrolysates. Taking into account the local market prices for bulk purchases for 32% HCl (w/w) and 85% H_3PO_4 (w/w) and computing the effective and respective dry acid concentrations, the cost is very similar for both catalysts. Work is in progress to refine the kinetic aspects of this chemical technology, whereas the improved enzymatic technology employing unusual isolates (e.g., extremophiles) or novel enzymes (e.g., maltogenic amylases from lactobacilli acting on crude nongelatinized starch granules [6]) goes to maturity.

Conclusion

Phosphoric acid was successfully tested as an alternative acid catalyst for starch depolymerization to glucose and maltooligosaccharides. Fermentation culture media based on these hydrolysates, with no need for catalyst removal, led to good biomass generation and normal physiology of the tested microorganisms. *X. dendrorhous* experienced the usual pigmentation expressed mainly as astaxanthin (diketo-dihydroxy-β-carotene), which increased in proportion to starch DP reduction.

Acknowledgments

This work was supported by CNPq/PADCT-World Bank program (subprograms SBIO and QEQ for Biotechnology and Chemical Engineering, respectively) and from UFPR.

References

1. Krassig, H., Steadman, R. G., Schliefer, K., and Albrecht, W. (1986), *Ullmann's Encyclopedia of Organic Compounds*, 5th ed., A5, VCH.
2. Ramos, L. P. and Saddler, J. N. (1994), in *Enzymatic Conversion of Biomass for Fuels Production*, Himmel, M. E., Baker, J. D., and Overend, R. P., eds., American Chemical Society/ACS Symposium Series 566, Washington, DC, pp. 325–341.
3. Blanchard, P. H. and Katz, F. R. (1995), in *Food Polysaccharides*, Stephen, A. M., ed., Marcel Dekker, New York, pp. 99–122.
4. BeMiller, J. N. (1965), in *Starch Chemistry and Technology vol. 1—Fundamental Aspects*, Whistler, R. L. and Paschall, E. F., eds., Academic, New York, pp. 495–520.
5. Berk, Z. (1976), *Braverman's Introduction to the Biochemistry of Foods*, Elsevier SPC, Amsterdam, pp. 149–167.
6. Florencio, J. A., Raimbault, M., Guyot, J. P., Stofella, D. E. E., Soccol, C. R., and Fontana, J. D. (1999), *Appl. Biochem. Biotechnol.* **84/86,** 1–13.
7. Fontana, J. D. (2000), in *Food Microbiology Protocols*, Spencer, J. F. T., Whelan, W. L., Brown, J., and Ragout de Spencer, A., eds., Humana, Totowa, NJ.
8. Fontana, J. D., Guimarães, M. F., Martins, N. T., Fontana, C. A., and Baron, M. (1996), *Appl. Biochem. Biotechnol.* **57/58,** 413–422.
9. Fontana, J. D., Correa, J. B. C., Duarte, J. H., Barbosa, A. M., and Blumel, M. (1984), *Biotechnol. Bioeng. Symp.* **14,** 175–186.
10. Hauly, M. C. O., Bracht, A., Beck, R., and Fontana, J. D. (1992), *Appl. Biochem. Biotechnol.* **34/35,** 292–308.

SESSION 4

Bioenergy and Bioproducts:
Forum on Recent Government Initiatives

Bioenergy and Bioproducts

Forum on Recent Government Initiatives

Robert A. Harris[1] and Bruce E. Dale[2]

[1]U. S. Department of Energy and [2]Michigan State University

This session was a roundtable forum on the President's Executive Order, the Bioenergy Initiative, the Technology Roadmap for Renewables Vision 2020 and other initiatives. The forum consisted of remarks by the panelists followed by a roundtable discussion with audience participation on the recent Bioenergy Initiative, the Executive order, and other recent thrusts.

Mr. Harris began with the political impacts of the Lugar bill and Executive Order 1314 challenging of a threefold increase in biomass utilization by 2010. He also reported on DOE-led efforts with industry in roadmapping for a Bioenergy and Bioproducts industry.

Dr. Dale discussed the National Research Council's "Bio-Based Industrial Products" study. Don Johnson of Grain Processing Corporation gave his perspectives on the challenge to make ethanol at a low enough cost (about $0.50/gal) to push the market. He also brought up challenges in supply assurance for the potential biomass feedstocks.

Dr. Robert Bloksberg-Fireovid, National Institute of Standards and Technology, presented the ATP and NIST programs to support deployment. The evening concluded with a lively discussion with the audience.

SESSION 5
Industrial Chemicals

Industrial Chemicals

Michael C. M. Cockrem[1] and Manoj Kumar[2]

[1]KiwiChem International, Madison, WI
and [2]Genencor International, Palo Alto, CA

The session on industrial chemicals was a reflection of the change and challenges facing the global chemical industry today. Presentations offered future solutions to these urgent needs. The U. S. chemical industry has identified key biotechnology needs for the future. These include: a) biocatalysts that perform equal to or better than current chemical catalysts, b) application to the generation of high performance and value-added products, c) process integration with chemical processing, d) continuous processing, and e) better capital efficiency. Advances in genetic engineering, DNA technology, fermentation, and cell physiology play an important role in improved bioconversion of sugars and other substrates. Improved understanding of chemistry during separation processing allows a wider variety of products to be potentially produced at low cost. Improved separations support process development. Advances such as those presented at this year's session will no doubt continue the upward path toward commercialization.

Simple molecules such as ethanol and lactic acid are of almost perennial interest. The former conserves energy present in the glucose, while the latter preserves mass and has a potential mass yield approaching 100%. While ethanol and hydrogen are primarily of interest for fuel, lactic acid is of great interest as a chemical intermediate and polymer feedstock. Lactic acid can either be produced largely as a salt and then separated by various means, or with the aid of biotechnology produced largely as a free acid. Water splitting electrodialysis is one strategy to separate the acid from the salt (see paper by Toräng et al.). Use of engineered yeast to ferment to low pH may be able to reduce or eliminate the need for a neutralizing salt in the fermentation and thus reduce the energy demand for techniques such as electrodialysis (an unpublished industrial presentation by Lievense of Tate & Lyle).

A more complex and expensive organic acid is 2-keto-L-gulonic acid, a valuable intermediate in the synthesis of vitamin C. Engineering a whole-cell biosynthetic system to produce such an intermediate requires significant metabolic pathway engineering to direct the carbon flow to the desired product (see paper by Dodge et al.). Other possible platform chemicals from renewables may include levulinic acid being developed by Bozell at NREL. Aromatics are structural building blocks for a variety of polymers and higher value products. Two possible bio-routes to aromatics are either via fermentation of sugars or via hydrolysis of lignins. Currently in the paper industry, lignins are often burnt to recover their energy value alone. Therefore, yield constraints in a chemical conversion and separation sequence are not as severe as in the case of sugar conversion: unrecovered material can be burnt in current equipment. High-pressure alkaline hydrolysis of lignins followed by distillation can give an overall yield up to 20% in total of phenol, guaiacol and 2,6-dimethoxyphenol (see paper by Gonçalves and Schuchardt).

One approach to producing polymers involves producing chemical intermediates such as aromatics or lactic acid. Another approach is to form them directly via fermentation of sugars or waste streams containing sugars. For example, bacterial cellulose has high purity, an amorphous structure, high surface area and tensile strength (see paper by Thompson and Hamilton).

Use of new developing technologies in the industry itself was evident from the work on application of solid/gas biocatalysis in esterification using mixed solvents and nonaqueous enzyme-based catalysis. This approach may yield the development of a new, cleaner continuous process for the synthesis of esters. John Barton of ORNL presented work in this area to substitute for Legoy.

There were very interesting poster presentations encompassing wide areas of industrial biotechnology indicating that this field has begun to mature and yield the fruits of work done in the last few decades. Some of the examples of note are lipopeptide production work by Eric Akpa et al., production of succinate using immobilization microorganism technology by Jung and Won, and genetic engineering work to improve poly-hydroxyalkanoates production by Chua and Fu.

The Effect of Pretreatments on Surfactin Production from Potato Process Effluent by *Bacillus subtilis*

David N. Thompson,* Sandra L. Fox, and Gregory A. Bala

*Biotechnology Department,
Idaho National Engineering and Environmental Laboratory, PO Box 1625,
Idaho Falls, ID 83415-2203, E-mail: thomdn@inel.gov*

Abstract

Pretreatments of low-solids potato process effluent were tested for their potential to increase surfactin yield. Pretreatments included heat, removal of starch particulates, and acid hydrolysis. Elimination of contaminating vegetative cells was necessary for surfactin production. After autoclaving, 0.40 g/L of surfactin was produced from the effluent in 72 h, vs 0.24 g/L in the purified potato starch control. However, surfactin yields per carbon consumed were 76% lower from process effluent. Removal of starch particulates had little effect on the culture. Acid hydrolysis decreased growth and surfactant production, except 0.5 wt% acid, which increased the yield by 25% over untreated effluent.

Index Entries: *Bacillus subtilis*; biosurfactant; surfactin; alternate feedstock; enhanced oil recovery.

Introduction

Numerous investigators have examined chemical surfactants to enhance *in situ* removal of hydrocarbons, pesticides, and polychlorinated biphenyls *(1–3)*. Chemically synthesized surfactants, however, are not always environmentally benign *(4,5)* and are frequently expensive. Biosurfactants have been suggested as replacements for synthetic surfactants in environmental remediation, as well as for food emulsifiers, detergents, and use in tertiary oil recovery *(6)*. Biosurfactants are potentially desirable vs synthetic surfactants on the basis of biodegradability, lowered toxicity, and the potential use of renewable substrates for their production *(6)*.

*Author to whom all correspondence and reprint requests should be addressed.

Surfactin is a powerful cyclic lipopeptide antibiotic biosurfactant produced by *Bacillus subtilis (7)*. Purified surfactin has an aqueous critical micelle concentration of 25 mg/L and lowers the surface tension to 27 mN/m *(7)*. Surfactin has been produced from glucose and other monosaccharides in amounts ranging from 0.1 to 0.7 g/L *(8–11)*. Foam fractionation techniques and the addition of iron or manganese can improve yields to 0.8 g/L *(7)*. High medium and separation costs limit surfactin's use in lower-value applications such as *in situ* bioremediation and enhanced oil recovery *(7,12)*. However, high-purity surfactants are not needed for environmental or enhanced oil recovery applications, and thus substrate costs and product yields become overriding constraints *(12)*.

In a previous study, we showed that *B. subtilis* ATCC 21332 produces surfactin from a low-solids (LS) potato-processing effluent *(13)*. In the present study, we examined the effects of several pretreatments on surfactin production from LS potato process effluent. Tested were the effects of autoclaving, removal of particulates, and dilute-acid hydrolysis at several levels of severity. The data show that although it is necessary to initially kill contaminating vegetative cells in the effluent, no significant benefit in surfactin production is gained over the unamended effluent by further pretreatments. However, surfactin yield was increased slightly by dilute-acid hydrolysis of the substrate before use.

Materials and Methods

Potato Substrates

LS potato process effluent was obtained from a southeast Idaho potato processor. An average composition of the effluent is presented in Table 1. Effluent slurry was diluted 1:10 by volume with nanopure water before use. Diluted effluent was pretreated and tested for surfactin production. Control experiments utilized both the diluted effluent and a purified potato starch obtained from Sigma (St. Louis, MO).

Pretreatments

Pretreatments included autoclaving, filtration, and dilute-acid hydrolysis. Autoclaving was done at 121°C for 20 min. To prepare the filtered effluent, diluted LS was centrifuged for 10 min at 1180*g*. The slurry in the lower half of the tube was discarded, and the supernatant slurry was filtered through P8 filter paper (average pore size of 20 μm; Fisher). The filtrate (FLS) was used as the final substrate after autoclaving. Dilute-acid hydrolysis of diluted LS was done by autoclaving after adding 1.42, 2.85, or 5.69 mL of concentrated H_2SO_4 to the undiluted effluent and adding sufficient nanopure water to give 500 mL of 1:10 diluted LS containing 0.5, 1.0, or 2.0 wt% H_2SO_4. The acid-hydrolyzed substrates were neutralized with 10 *N* NaOH before use. Acid-hydrolyzed substrates were designated ALS½, ALS1, and ALS2, respectively. Untreated and pretreated media are summarized in Table 2 with their initial data.

Table 1
Average Composition
of Low-Solids Potato Process Effluent
as Received from the Processor[a]

Effluent component	Concentration (g/L)
Soluble starch	128.
Insoluble starch/fiber	14.
Glucose	5.2
Fructose	3.0
Galactose	<0.30
Sucrose	9.7
Maltose	9.1
Lactose	<0.30
Protein	72.
Ca	0.98
Cu	0.0021
Fe	0.012
Mg	1.1
Mn	0.0044
P	2.0
K	23.
Na	1.3
Zn	0.011
Total NH_3-nitrogen	12.
Total ash	52.

[a]Effluent media for the experiments were prepared using 1:10 (v/v) dilutions of this effluent.

Controls

Controls for surfactin production included abiotic and biotic controls using purified starch in an optimized medium. In each case, the pH 7.0 medium (PS) contained 5.0 g/L of potato starch (Sigma), rendered soluble by boiling in distilled water for 30 min, and trace minerals as previously described *(13)*. All media were autoclaved at 121°C for 20 min. Abiotic controls (A-PS) were not inoculated, and biotic controls (B-PS) were inoculated to 1 vol% with *B. subtilis* seed inoculum.

Because indigenous spore formers were found to survive autoclaving of diluted LS effluent *(13)*, controls employing diluted LS medium, pH 7.0, were included. The first of these controls was an "abiotic" control (A-LS) that was autoclaved but not inoculated with *B. subtilis*; it was called abiotic to indicate that it was not inoculated with *B. subtilis*. The second of these controls was a "biotic" control (B-LS) that was not autoclaved but was inoculated to 1 vol% with *B. subtilis* seed inoculum; it was called biotic to indicate that *B. subtilis* was added to the culture. The abiotic control was used to show the growth of germinated bacteria in autoclaved samples and

Table 2

Substrate Characterization Data for Purified Potato Starch Control and Pretreated Potato Media[a]

Medium[b]	Pretreatment	Inoculated with *B. subtilis*?	Glucose (g/L)	Soluble starch (g/L)	Insolubles (g/L)	Specific growth rate (h^{-1})
Uninoculated controls						
A-PS	Autoclaved	No	0.009	4.88	0.00	0
A-LS	Autoclaved	No	1.29	18.9	12.2	0.084 ± 0.029
Inoculated controls						
B-PS	Autoclaved	Yes	0.009	4.88	0.00	0.177 ± 0.014
B-LS	None	Yes	1.41	14.0	5.50	0.419 ± 0.007
Pretreated substrates						
LS	Autoclaved	Yes	1.13	16.5	10.8	0.348 ± 0.012
FLS	Filtered and autoclaved	Yes	1.15	20.3	11.2	0.369 ± 0.006
ALS½	0.5 wt% H_2SO_4 and autoclaved	Yes	2.37	14.3	10.2	0.402 ± 0.048
ALS1	1.0 wt% H_2SO_4 and autoclaved	Yes	2.52	12.6	10.2	0.362 ± 0.011
ALS2	2.0 wt% H_2SO_4 and autoclaved	Yes	5.37	11.3	10.3	0.333 ± 0.005

[a]The B-LS control, which was not autoclaved, is the untreated 1/10-diluted process effluent. Some variation among batches was seen in this stream. The final column is the observed specific bacterial growth rates at 12 h seen in experiments using each substrate (A-LS and B-LS are at 24 h). The growth rates are for total bacteria in the cultures, and include the 95% confidence intervals (±2.5σ) assuming a normally distributed population.

[b]A-PS, abiotic purified starch control; B-PS, biotic purified starch control; A-LS, abiotic low-solids effluent control; B-LS, Biotic low-solids effluent control; LS, low-solids effluent medium; FLS, filtered low-solids effluent medium; ALS½, ½ wt% acid-hydrolyzed low-solids effluent medium; ALS1, 1 wt% acid-hydrolyzed low-solids effluent medium; ALS2, 2 wt% acid-hydrolyzed low-solids effluent medium. Unless otherwise indicated, all media were autoclaved.

the lack of surfactin production in the absence of *B. subtilis*. The biotic control was used to determine the competitiveness of *B. subtilis* and the production of surfactin in the presence of high numbers of vegetative contaminating bacteria. Initial substrate data for diluted LS and its controls and for the purified potato starch (PS) controls are also presented in Table 2.

Cultures and Maintenance

Bacterial Strains

B. subtilis 21332 was obtained from the American Type Culture Collection and cultured as previously described *(13)*. Freezer stocks were prepared from cells grown in maintenance broth *(14)*. Seed cultures containing $4.0 \pm 0.6 \times 10^8$ cells/mL were prepared from the freezer stocks and were used to inoculate surfactin production tests. The seed inocula were grown on Difco nutrient broth *(15)* as previously described *(13)*.

Experimental Procedure

The surfactin production tests were performed in 250-mL Erlenmeyer flasks on a gyratory shaker at 30°C, 150 rpm, for 72 h, as previously described *(13)*. Media used in the tests are listed in Table 2. All media were adjusted to pH 7.0 before autoclaving except for the acid-pretreated substrates, which were adjusted just before use.

Analytical Methods

Cell Numbers

Cell numbers were determined immediately after sampling using direct visual microscopic count techniques, as previously described *(13)*.

Glucose and pH

Glucose was measured, after removing cells and particulates by centrifugation for 3 min at 13,214*g*, using a YSI Model 2700 Glucose Analyzer (Yellow Springs Instrument, Yellow Springs, OH). Culture pH was measured using a standard pH probe.

Soluble Starch

Soluble starch was estimated as previously described *(13)*, after removing cells and particulates by centrifugation for 3 min at 13,214*g*, and using the phenol-sulfuric acid assay for total carbohydrates *(16)*. The estimates of soluble starch included all reducing sugars (see Table 1) except for the measured amounts of free glucose, which were subtracted from the total.

Insolubles

Frozen samples saved for surface tension analyses were thawed and centrifuged for 8 min at 4811*g*. Insolubles were estimated as previously described *(13)*, using the lyophilized pellet weight and the estimated weight of cells in each sample assuming an average per-cell mass of 10^{-12} g *(17)*.

As defined here, "Insolubles" includes not only insoluble starch, but also cellulose, insoluble proteins, ash, etc. (but not cells). The supernatant was used for surface tension measurement.

Surface Tension

Surface tensions were measured by video image analysis of inverted pendant drops as previously described *(18)*. All measurements were made on cell-free supernatants obtained by centrifugation.

Isolation of Surfactin and Critical Micelle Concentration

Crude surfactin was isolated by precipitation *(9)*, as previously described *(13)*. The crude lyophilized powder was then used to estimate the critical micelle concentration, in nanopure water, as previously described *(19,20)*.

Results

Pretreatments

The effects of the pretreatments on initial substrate composition are summarized in Table 3. The B-LS control (unautoclaved) represents the LS medium without pretreatment. There were some small variations in the initial glucose and starch levels of undiluted LS effluent as obtained from the processor. Initial glucose in the B-LS medium was 1.41 g/L, with soluble starch and insolubles at 14 and 5.5 g/L, respectively. Autoclaving the effluent lowered the initial glucose by about 20%, increased the soluble starch by 18%, and nearly doubled the insolubles content. As expected, filtering the diluted LS before autoclaving had little effect on glucose content. However, the soluble starch content increased by 45%, and the total insolubles content was essentially unchanged. Acid hydrolysis increased the glucose content from hydrolysis of the starch, thereby decreasing the soluble starch content. Again, the insolubles content increased relative to the unautoclaved effluent.

Cell Growth

In all cases except the B-PS and A-LS controls, log phase growth was complete by 12 h. For the B-PS control, 24 h were required. The specific growth rates for the log phase are presented along with the initial substrate data in Table 2. For LS-based media, growth cannot be specifically attributed to *B. subtilis*, since contaminating bacteria were present (germination and growth of these bacterial spores was observable in the A-LS control). Thus, the growth rates in Table 2 represent the sum of contaminant bacteria and *B. subtilis*.

Removal of particulates by filtration had little effect on growth. The specific growth rate averaged over 12 h for the FLS culture was slightly higher than that for the LS culture. The addition of acid to the LS medium and autoclaving increased the specific growth rate when 0.5 wt% H_2SO_4

Table 3
Effect of Pretreatments
on Relative Substrate Concentrations in Pretreated Diluted LS Effluent[a]

Pretreatment	Medium	Fraction of initial concentration (after pretreatment)		
		Glucose	Soluble starch	Insolubles
None	B-LS	1.00	1.00	1.00
Heat	LS	0.801	1.18	1.96
Filtered + heat	FLS	0.816	1.45	2.04
0.5 wt% H_2SO_4 + heat	ALS½	1.68	1.02	1.86
1.0 wt% H_2SO_4 + heat	ALS1	1.79	0.900	1.86
2.0 wt% H_2SO_4 + heat	ALS2	3.81	0.807	1.87

[a]All pretreatments included autoclaving the substrate at 121°C for 20 min. The B-LS substrate was not autoclaved. Note that "Soluble Starch" includes all reducing sugars other than free glucose.

was added. To 95% confidence, there was a very slight overlap of ALS½ and ALS1 growth rates. However, doubling the acid concentration decreased the specific growth rate relative to 0.5 wt%, although this was about equal to the 0% acid medium (LS). The addition of acid to 2 wt% also had a detrimental effect on cell growth, decreasing the specific growth rate to below that for the LS medium.

Culture pH

In all autoclaved media inoculated with *B. subtilis*, culture pH remained essentially constant at about 7.0 over the first 48 h of the experiment (data not shown). After 48 h, all but the B-PS control culture showed an increase in pH to near 8.0; the B-PS control culture pH remained constant at 7.0 for the entire experiment. With autoclaving but without inoculation of *B. subtilis* (A-LS), the pH stayed essentially constant at about 7.0 for 8 h and then decreased to 6.5. Finally, the inoculated, unautoclaved culture pH decreased to 4.5–5.0 over the first 8 h of culture and remained low over the rest of the experiment.

Glucose Consumption

Glucose consumption in the controls and pretreated media is presented in Fig. 1. The B-PS control culture initially used all glucose released from the added starch but began to accumulate glucose to a small degree after 8 h. The A-LS culture showed an 8 h lag before use of the 1.3 g/L of free glucose but eventually began utilizing much of the free glucose. Glucose in the B-LS culture quickly dropped to about 0.25 g/L over 8 h and then remained relatively constant. The LS and FLS media essentially mirrored one another, first accumulating glucose over 12 h and then slowly utilizing the glucose over the remainder of the culture. Finally, the acid-pretreated LS media showed 8- to 12-h lag times before utilization of the free glucose,

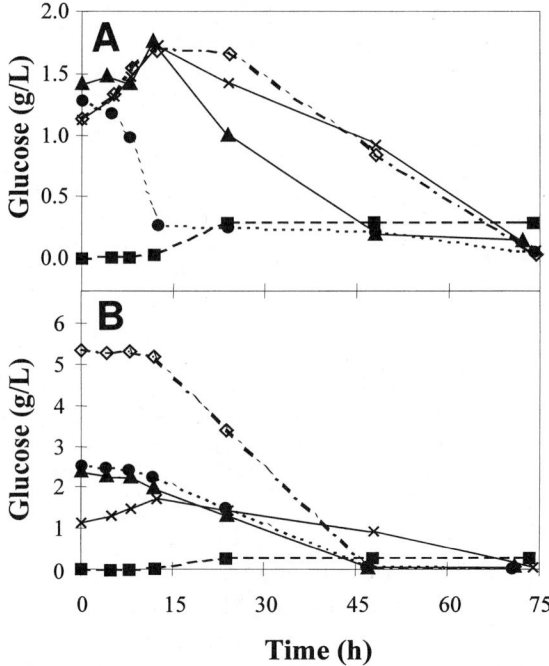

Fig. 1. Glucose consumption vs time for growth on control and pretreated potato effluent. **(A)** Controls, autoclaving, and filtration pretreatments: (■) B-PS control; (▲) A-LS; (●) B-LS; (◇) FLS; (×) LS. **(B)** Potato starch control and dilute-acid pretreatments: (■) B-PS control; (×) LS (0% acid); (▲) ALS½; (●) ALS1; (◇) ALS2.

but after the lag, each culture utilized the glucose at an essentially linear rate until it was gone.

Soluble Starch

Soluble starch consumption for all media is presented in Fig. 2. In the B-PS control medium, there was a short lag of 4–8 h in degradation of soluble starch, with apparently linear degradation thereafter. The A-LS medium showed a slight increase in soluble starch over the first 4 h of culture and again showed linear degradation afterward. The B-LS control displayed an initial increase in soluble starch, and then nonlinear degradation. The LS and FLS medium behaved similarly, peaking at 8 h. However, acid-pretreated LS media showed essentially a linear degradation of the starch over the entire culture.

Insolubles

The time courses of insolubles concentration in the cultures are presented in Fig. 3. The B-PS control did not contain initial insolubles and therefore is not included. In all LS-based media, a fraction of the insolubles was quickly solubilized, leveling off after about 8 h. In all but the A-LS and FLS cultures, the final insolubles concentration was 5 to 6 g/L. The final

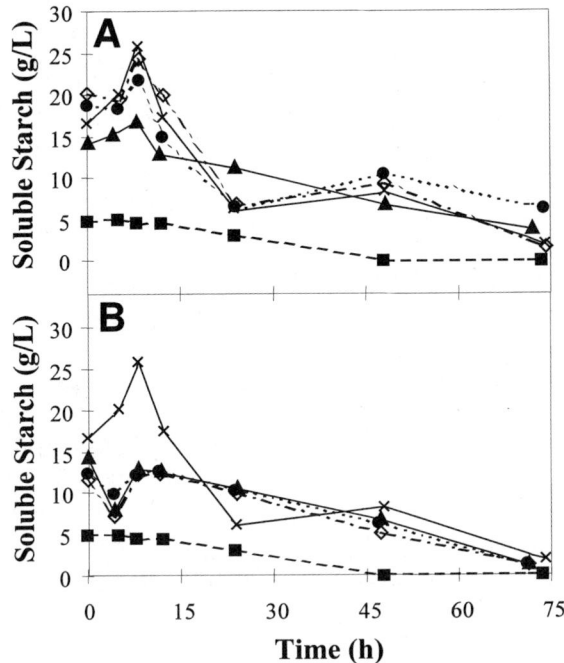

Fig. 2. Soluble starch vs time for growth on control and pretreated potato effluent. **(A)** Controls, autoclaving, and filtration pretreatments: (■) B-PS control; (▲) A-LS; (●) B-LS; (◇) FLS; (×) LS. **(B)** Potato starch control and dilute-acid pretreatments: (■) B-PS control; (×) LS (0% acid); (▲) ALS½; (●) ALS1; (◇) ALS2.

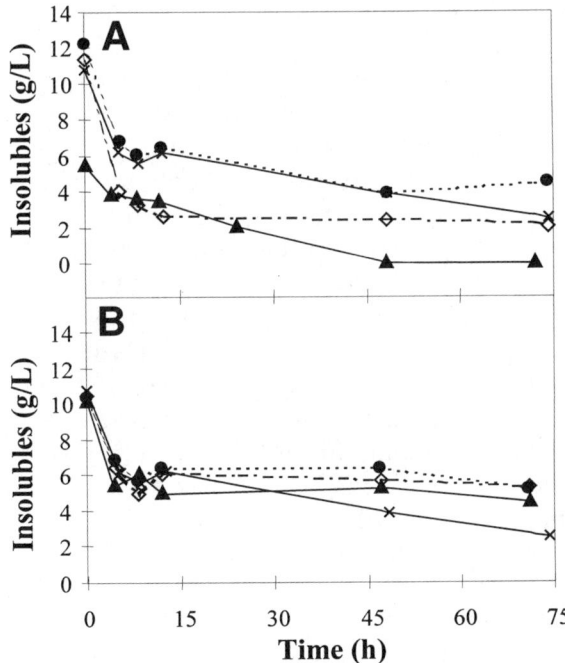

Fig. 3. Insolubles vs time for growth on pretreated potato effluent. **(A)** Controls, autoclaving, and filtration pretreatments: (▲) A-LS; (●) B-LS; (◇) FLS; (×) LS. **(B)** Potato starch control and dilute-acid pretreatments: (×) LS (0% acid); (▲) ALS½; (●) ALS1; (◇) ALS2.

A-LS insolubles concentration was <1 g/L, and that in the FLS culture was 2 to 3 g/L. Most of the decrease in insolubles took place while soluble starch was increasing, indicating solubilization of the starch fraction of the insolubles.

Surface Tension

The surface tensions of the effluent-based control media were initially in the range of 60–65 mN/m. The surface tensions of the LS and acid-hydrolyzed effluent media were in the range of 50–55 mN/m. The B-PS control medium began with a surface tension of 72 mN/m. The B-PS control culture reached 31 mN/m in 24 h but did not change much thereafter. The surface tension in the A-LS culture remained essentially constant, with a small drop after 24 h to about 55 mN/m. In the B-LS control, the surface tension increased slightly from 60 to 65 mN/m. The LS and FLS cultures again behaved identically, reaching 29 mN/m in 24 h and 25–27 mN/m after 72 h of culture. The behavior of the acid-pretreated media was slightly different from that of the LS culture, slowly increasing from the initial surface tension of 55 to 62 mN/m, and then decreasing to 29 mN/m at 48 h; no change in surface tension was seen after 48 h of culture.

Surfactin Isolation and Critical Micelle Concentration

Surfactin recovery data and critical micelle concentrations are given in Table 4. The B-PS control produced 2200 mg/L of crude surfactant, and the B-LS control produced 870 mg/L. Since the surface tension did not change in the B-LS control, the 870 mg/L is an acid-insoluble fraction present in the LS effluent. This cannot be verified with the A-LS medium because no recovery was done. The FLS medium produced 4000 mg/L of crude surfactant vs 3600 mg/L from the LS medium. Somewhat less solid was recovered from the acid-pretreated media, at about 2700 mg/L for each.

The critical micelle concentrations indicate that there was substantial carryover of nonsurfactant acid-precipitable solids into the crude surfactant. The lowest measured critical micelle concentration was 141 mg/L for ALS$\frac{1}{2}$, which indicates that about 17 wt% of the crude surfactant was surfactin. The ALS$\frac{1}{2}$ was the best pretreatment, producing nearly 0.50 g/L of surfactin at 72 h. The nearest result was for autoclaved LS, at about 0.40 g/L of surfactin. The B-PS control produced 0.24 g of surfactin/L by 72 h. The B-PS control produced 0.154 g of surfactin/g of carbon consumed, compared with 0.037 g of surfactin for LS and 0.051 g from ALS$\frac{1}{2}$. Removing the large particulates from the LS decreased the yield from the effluent to 0.025 g of surfactin/g of carbon, and stronger acid pretreatments also substantially decreased surfactin yields from glucose.

Discussion

PS Medium: Purified Starch Control

Growth of *B. subtilis* on PS medium was poorer than cell growth on LS effluents. The lag observed in cultures with low initial glucose occurs while

Table 4
Crude Surfactant Recovery, Critical Micelle Concentration, and Yield for Each Pretreatment[a]

	Recovered solid (g)	Solid concentration at 72 h (mg/L)	Measured critical micelle concentration (mg/L)	Surfactin yield at 72 h (g/L)	Surfactin yield at 72 h (g/g carbon)
Nanopure H_2O	0	0	∞	—	0
A-PS	0	0	ND	0	0
B-PS	0.46	2200	231	0.238	0.154
A-LS	ND	ND	ND	—	—
B-LS	0.21	870	∞	—	0
LS	0.76	3600	228	0.395	0.037
FLS	0.85	4000	340	0.294	0.025
ALS½	0.58	2700	141	0.479	0.051
ALS1	0.56	2600	265	0.245	0.030
ALS2	0.57	2700	458	0.147	0.017

[a]Crude surfactant was recovered from the combined 72-h culture fluid from the three replicates. ND, not determined.

the amylase system is induced, as verified by glucose and soluble starch data. The surface tension in the B-PS control culture reached 31 mN/m in 24 h but did not change much thereafter. Crude surfactant recovery from the B-PS control was 2200 mg/L, with a critical micelle concentration of 231 mg/L. The concentration of surfactin at 72 h, estimated as previously (13) from the critical micelle concentrations of pure surfactin and of the crude precipitate, was 0.238 g/L, which is in the range of 0.1–0.7 g/L previously reported from monosaccharides (8–11). This suggests that much of the glucose consumed went to production of the amylase system. The yield per gram of carbon consumed was the highest from any of the media tested in this study, at 154 mg of surfactin.

B-LS, A-LS, and LS Media: Effect of Autoclaving

The A-LS and B-LS controls both supported growth, indicating significant contaminating microbial activity in the effluent. However, the purpose of autoclaving was not to sterilize the effluent but to minimize cost by simply allowing *B. subtilis* to compete successfully in the culture. After autoclaving, inoculated *B. subtilis* grew well and produced surfactant and thus could apparently compete for resources. The contaminant bacteria were likely fermentative, since pH in the A-LS and B-LS controls quickly dropped to 4.5, well below that required by *B. subtilis* 21332.

Glucose in the B-LS control dropped over 8 h and remained constant thereafter, which correlates with the bottoming out of culture pH. Soluble starch was degraded in all cultures; thus, contaminant cells expressed amylase activity. Soluble starch degradation was slower in the A-LS culture than in the others but was to a greater extent than in either the B-LS or the LS cultures. Both B-LS and LS cultures initially showed increasing soluble starch, corresponding with high rates of degradation of insolubles. As in the other media, insolubles leveled off at 5 to 6 g/L, indicating a recalcitrant or nonstarch fraction.

Surface tensions in the A-LS control did not change appreciably, verifying that surface tension changes in *B. subtilis*–inoculated cultures were attributable to *B. subtilis*. The surface tension also did not change in the B-LS control nor was there a measurable critical micelle concentration. Since the pH of this culture quickly dropped to 4.5, the *B. subtilis* was not able to produce surfactin and may not have been able to grow significantly.

The LS medium produced 0.395 g/L of surfactin at 72 h, which was within the reported range for monosaccharides (8–11), and above that observed for the B-PS control. However, the estimated yield of surfactin from carbon consumed was only 24% of that from B-PS medium, at 0.037 g of surfactin. It is likely that additions of medium or complete sterilization of the effluent could make up this yield loss. However, because the aim is to keep costs low, it is unclear whether the economics of the process through the final separation step would favor feedstock additions or treatments.

FLS Medium: Effect of Filtration

The principle reasons for the filtration pretreatment were to remove large particulates that may serve as carriers for spores that survive autoclaving, and that could plug oil reservoirs if carried over to the surfactin product. Filtration had little effect on cell growth, as expected. The great majority of the particulates was clearly able to pass through the 20-μm filter paper, evidenced by the essentially unchanged insolubles concentration.

Culture pH, glucose consumption, and soluble starch consumption in the FLS medium all paralleled those of the LS medium. Insolubles consumption in the FLS medium was somewhat higher than that in the LS medium, ending at 2 to 3 g/L of insolubles. Thus, it is likely that some of the recalcitrant insolubles were removed during the filtration. The lack of differences in LS and FLS cultures suggests that removal of particulates from the diluted LS effluent before autoclaving had no effect. This was again seen in the surface tensions, which mirrored one another over the course of the runs. However, the critical micelle concentration of the crude surfactant from the FLS culture was substantially higher than that from the LS medium. The estimated surfactin concentration at 72 h was 25% lower than that observed from the unfiltered LS medium, and the yield of surfactin per gram of carbon consumed was 32% lower than from LS and 84% lower than from B-PS. Thus, it is preferable to leave the particles in the LS medium during surfactin production.

ALS Media: Effect of Dilute-Acid Pretreatments

The addition of a small amount of acid to the medium before autoclaving (0.5 wt%) slightly increased cell growth rates, although higher acid concentrations adversely affected cell growth (Table 2). It is likely that the higher acid concentrations caused some decomposition of the glucose released, forming 5-hydroxymethyl-2-furfuraldehyde, levulinic acid, and formic acid *(21)*. These decomposition reactions are common in acid hydrolysis of cellulosic biomass *(22)*, and acid hydrolysis products of lignocellulose have been shown to be somewhat toxic to yeasts used for ethanol fermentations *(23)*. It is probable that similar decomposition products are formed during starch hydrolysis and that these products could be toxic to *B. subtilis* 21332.

Culture pH in the acid-pretreated LS media again mirrored that in the other LS-based media. The acid-pretreated LS media all showed 8- to 12-h lag times in glucose consumption, but after the lag each culture utilized the glucose at a linear rate. Soluble starch consumption was also linear, indicating balanced glucose consumption and soluble starch degradation. There was no difference in the rates of soluble starch degradation with increasing severity of pretreatment, indicating that initial glucose concentrations from 2.3 to 5.3 g/L had little effect on amylase induction and production. Insolubles consumption in the acid-pretreated cultures again bottomed out near 5 to 6 g/L.

The critical micelle concentration of the crude surfactant increased with pretreatment severity, indicating that more acid-precipitable compounds were present after hydrolysis with higher amounts of acid. The 0.5 wt% acid treatment had the highest estimated surfactin concentration of any of the media tested, at 0.479 g/L, which is 68% of the highest reported value without foam removal (60% of that with foam removal) *(7)*. The 0.5 wt% acid treatment also had the highest surfactin yield from glucose, at 0.051 g of surfactin/g of carbon consumed, as compared with 0.030 and 0.017 g/g for the 1 and 2 wt% pretreatments, respectively. This again suggests that an inhibitory product of the acid hydrolysis limits surfactin production and also suggests that the lowered surfactin production seen in a previous work in which corn steep liquor (CSL) was added to the process effluent *(13)* was owing to an inhibitory compound present in the CSL.

Conclusion

Autoclaving of the process effluent before use as a substrate for surfactant production is absolutely required. If removal of particulates is necessary, this step would be better placed after surfactant production. Dilute-acid hydrolysis of the diluted LS effluent with 1 wt% acid or higher has a detrimental effect on growth, rate of production, and total amount of surfactant produced. Pretreatment with 0.5 wt% acid modestly increased surfactin yield over untreated LS. All media performed poorly on a yield-per-carbon consumed basis when compared with the optimized control culture. While it is likely that additions of medium or complete sterilization of the effluent could make up this yield loss, the question remains: Do the economics of the process through the final separation step favor feedstock additions or complete sterilization? Further studies that include separation of the surfactin will be necessary to answer this question.

Acknowledgments

We thank Rick Scott at the Idaho National Engineering and Environmental Laboratory for performing the surface tension measurements. This work was supported by the US Department of Energy, Assistant Secretary for Fossil Energy, Office of Fossil Energy, under contract number DE-AC07-99ID13727.

References

1. Abdul, A. S., Gibson, T. L., and Rai, D. N. (1990), *Ground Water* **28,** 920–926.
2. Ang, C. C. and Abdul, A. S. (1991), *Ground Water Monitor. Rev.* **11,** 121–127.
3. Ellis, W. D., Payne, J. R., and McNab, G. D. (1985), EPA/600/S2-85/129, United States Environmental Protection Agency, Office of Research and Development, Washington, DC.
4. Rudolph, P. (1989), in *Aquatic Toxicity Data Base*, Federal Environmental Agency, Berlin.
5. Schröder, H. F. (1993), *J. Chromatogr.* **647,** 219–234.

6. Zajic, J. E. and Seffens, W. (1984), *CRC Crit. Rev. Biotechnol.* **1(2),** 87–107.
7. Rosenberg, E. (1986), *CRC Crit. Rev. Biotechnol.* **3(3),** 109–132.
8. Arima, K., Kakinuma, A., and Tamura, G. (1968), *Biochem. Biophys. Res. Commun.* **31,** 488–494.
9. Cooper, D. G., McDonald, C. R., Duff, S. J. B., and Kosaric, N. (1981), *Appl. Environ. Microbiol.* **42,** 408–412.
10. Besson, F. and Michel, G. (1992), *Biotechnol. Lett.* **14(11),** 1013–1018.
11. Georgiou, G., Lin, S.-Y., and Sharma, M. M. (1992), *Bio/Technology* **10,** 60–65.
12. Lin, S.-Y. (1996), *J. Chem. Technol. Biotechnol.* **66,** 109–120.
13. Thompson, D. N., Fox, S. L., and Bala, G. A. (2000), *Appl. Biochem. Biotechnol.* **84–86,** 917–930.
14. Gherna, P. and Pienta, P., eds. (1989), in *American Type Culture Collection Catalogue of Bacteria and Phages,* 17th ed., American Type Culture Collection, Rockville, MD, p. 403.
15. Atlas, R. M. (1993), in *Handbook of Microbiological Media,* Parks, L. C., ed., CRC Press, Boca Raton, FL, p. 672.
16. Gerhardt, P., Murray, R. G. E., Wood, W. A., and Krieg, N. R., eds. (1994), in *Methods for General and Molecular Bacteriology,* American Society for Microbiology, Washington, DC, pp. 518, 519.
17. Bailey, J. E. and Ollis, D. F. (1986), in *Biochemical Engineering Fundamentals,* 2nd ed., Verina, K. and Martin, C. C., eds., McGraw-Hill, New York, p. 5.
18. Herd, M. D., Lassahn, G. D., Thomas, C. P., Bala, G. A., and Eastman, S. L., (1992), in *Proceedings of the DOE Eighth Symposium on Enhanced Oil Recovery,* SPE/DOE 24206, Tulsa, OK.
19. Gerson, D. F. and Zajic, J. E. (1979), *Process Biochem.* **14,** 20–29.
20. Sheppard, J. D. and Mulligan, C. N. (1987), *Appl. Microbiol. Biotechnol.* **27,** 110–116.
21. Grethlein, H. E. (1978), *Biotechnol. Bioeng.* **20,** 503–525.
22. Marsden, L. M. and Gray, P. P. (1986), *CRC Crit. Rev. Biotechnol.* **3(3),** 235–276.
23. Lee, J. (1997), *J. Biotechnol.* **56,** 1–24.

Production of Bacterial Cellulose from Alternate Feedstocks

DAVID N. THOMPSON* AND MELINDA A. HAMILTON

*Biotechnology Department,
Idaho National Engineering and Environmental Laboratory, PO Box 1625,
Idaho Falls, ID 83415-2203, E-mail: thomdn@inel.gov*

Abstract

Production of bacterial cellulose by *Acetobacter xylinum* ATCC 10821 and 23770 in static cultures was tested from unamended food process effluents. Effluents included low-solids (LS) and high-solids (HS) potato effluents, cheese whey permeate (CW), or sugar beet raffinate (CSB). Strain 23770 produced 10% less cellulose from glucose than did strain 10821 and diverted more glucose to gluconate. Unamended HS, CW, and CSB were unsuitable for cellulose production by either strain, and LS was unsuitable for production by strain 10821. However, strain 23770 produced 17% more cellulose from LS than from glucose, indicating that unamended LS could serve as a feedstock for bacterial cellulose.

Index Entries: Bacterial cellulose; *Acetobacter xylinum*; potato effluent; beet raffinate; whey permeate.

Introduction

Bacterial cellulose has significant advantages over plant cellulose. Bacterial cellulose fibrils are randomly oriented and the product is highly amorphous *(1,2)*. It is generated as a never-dried membrane in a nearly pure form *(3)* that contains 99.1 wt% water, of which 0.3 wt% is bound and 98.8 wt% is free water *(4)*. It has more than 200 times greater surface area than isolated softwood cellulose *(1)* and has a tensile strength similar to that of steel *(5)*. Many potential high-value markets exist for thin film bacterial cellulose, including acoustic diaphragms *(6)*, artificial skin *(7,8)*, artificial blood vessels *(9)*, liquid-loaded medical pads *(10)*, supersorbers *(11)*, and specialty membranes *(12)*. Potential markets for bacterial cellulose produced as pellets in agitated culture include foods and the mining, oil, and pulp and paper industries *(1,2)*.

*Author to whom all correspondence and reprint requests should be addressed.

The most studied producer of bacterial cellulose is *Acetobacter xylinum* *(13)*, a Gram-negative, obligately aerobic bacterium *(10,14,15)*. The optimum pH for cellulose production is 5.0 *(13,16)*, and production is associated with and proportional to growth *(17)*. Cellulose production has been demonstrated from glucose, sucrose, fructose, glycerol, mannitol, arabitol, and many other substrates *(5,13,16,18)*. Excess glucose is oxidized to gluconate in wild-type *A. xylinum*, which lowers the pH and inhibits cellulose production *(13)*. Genetically altered strains with substantially reduced ability to form gluconate have been developed *(5)*. Methionine and lactate stimulate cell growth in the early stages, allowing higher rates of cellulose production *(19)*.

Low production rates and high medium costs limit commercial use of bacterial cellulose *(3)*. In this article, we report the testing of unamended food process effluents, which typically represent economical and environmental liabilities to the producers, as substrates for bacterial cellulose production in static culture.

Materials and Methods

Food Process Effluents

Potato effluents *(20,21)*, cheese whey permeate (CW), and concentrated sugar beet raffinate (CSB) were obtained from Idaho processing plants. Two potato process effluents were tested: high-solids (HS) and low-solids (LS) effluents. To lower carbohydrate concentrations to levels similar to that of the control medium (described subsequently), the effluents were diluted with distilled water. HS effluent was diluted 1:10 by weight, whereas LS effluent, CW, and CSB were diluted 1:10 by volume. Each diluted effluent was autoclaved at 121°C for 20 min, and the pH was adjusted to 5.0 with HCl prior to use. Control experiments were conducted using an optimized pH 5.0 glucose medium containing 20 g/L of glucose; 5 g/L of yeast extract; 5 g/L of peptone; 2.7 g/L of Na_2HPO_4; and 1.15 g/L of citrate (Schramm and Hestrin's medium) *(15,22)*. Initial carbohydrate data for the control and the diluted effluents are presented in Table 1.

Cultures and Maintenance

A. xylinum 10821 and 23770 were obtained from the American Type Culture Collection (ATCC) (Manassas, VA). Several generations were grown in 25 g/L of mannitol; 5 g/L of yeast extract; and 3 g/L of peptone, pH 5.0 *(23)*. Reference cultures were maintained at 4°C on malt agar slants containing Schramm and Hestrin's medium *(15,22)* and 15 g/L of agar (Difco) and were subcultured monthly. Frozen seed stocks were prepared as follows. The cellulose pellicle was removed from the reference slant with sterile distilled water, vortexed for 5 min, and sonicated for 20 min to remove the cells from the pellicle. The pellicle was removed, and the liquid was added to 500 mL of Schramm and Hestrin's medium and cultured with

Table 1
Typical Initial Carbohydrate Data for Control Medium
and Diluted HS and LS Potato Effluents, Diluted CSB, and Diluted CW[a]

Carbohydrate	Control medium	Composition (wt%)			
		Effluents diluted 1:10			
		HS	LS	CSB[b]	CW
Glucose	2.0	0.002	0.04	0.04	0.19
Sucrose	—[d]	—	—	2.2	—
Lactose	—	—	—	—	1.3[e]
Soluble starch	—	0.47	1.5	—	—
Insolubles[c]	—	1.6	0.66	—	—

[a]Effluents varied slightly among batches obtained from the food-processing plants.

[b]Although not measured, sugar beet raffinate also contains minute amounts of raffinose and fructose.

[c]Insolubles include starch and nonstarch components. Insoluble starch accounted for up to 85% of the insolubles in both HS and LS, depending on the particular batch from the processor.

[d]None.

[e]This value includes small amounts of galactose.

occasional gentle mixing for 7–14 d at 30°C, until the cell number exceeded 2×10^8 cells/mL. Cells were then removed from the resulting pellicle by vortexing and sonication, and 10-mL aliquots of the culture liquid were frozen at –80°C in 15 vol% glycerol as previously described (23).

Experimental Procedures

Experiments were conducted in static 100-mm storage dishes (cat. no. 08-782; Fisher) containing 100 mL of medium. Abiotic controls contained Schramm and Hestrin's medium and were not inoculated, and biotic controls were prepared as follows. Frozen seed stocks were thawed and added at 10 mL/250 mL of Schramm and Hestrin's medium (15,22). The cultures were grown as described for preparation of seed cultures. When the cell number exceeded 2×10^8 cells/mL, cells were removed from the pellicle and inoculated into a large excess of Schramm and Hestrin's medium to give 1×10^7 cells/mL. The medium was mixed well and transferred in 100-mL aliquots to sterile 100-mm dishes. All controls were incubated at 30°C for 14 d. For each medium, this amounted to 16 separate cultures, all starting at the same conditions and inoculum size. Two of these cultures were sacrificed every second or third day and analyzed for substrate, cell numbers, pH, cellulose weight, and pellicle thickness. This was done because measurement of cellulose weight and cell counts required disruption of the static culture and removal of the cells from the pellicle. Therefore, data at each time point are averages from separate cultures. Experiments utilizing unamended effluents were conducted as described earlier, except the final transfer was completed using the autoclaved, diluted pH 5.0 effluent medium.

Analytical Methods

Cell Numbers and pH

Cells were removed from the pellicle using vortexing and sonication. Liquid samples were then taken for substrate analyses and cell counts. Cell numbers were then determined by direct count of 5 µL of culture fluid on a Petroff-Hauser hemocytometer slide (average of 10 separate counts). pH was measured using a standard pH probe and meter.

Substrate Concentrations

Samples were measured for substrate levels after filtering through a 0.22-µm polysulfone membrane filter to remove cells and cellulose fibers. Glucose, sucrose, and lactose were measured using a YSI Model 6200 Glucose Analyzer (Yellow Springs Instrument, Yellow Springs, OH) fitted with the appropriate membrane(s) (glucose, sucrose, and lactose; cat. no. 2365, 2703, and 2702, respectively), buffers, and standard solutions. Soluble starch was estimated after removal of cells and particulates as previously described *(20)*, using the phenol–sulfuric acid assay for total carbohydrates *(24)*. For potato effluent cultures, an aliquot of well-mixed culture fluid was filtered through a tared cellulose filter paper (Fisher P8, average pore size of 20 µm), and insolubles were estimated by weight difference after drying to constant weight at 105°C.

Pellicle Weight and Thickness

Cellulose pellicle thickness was measured without removal of the absorbed water present in the pellicle at harvest. The pellicle was placed onto a smooth, flat board and spread out. The thickness was measured at three different positions around the periphery to the nearest 0.5 mm, and the values were averaged. The pellicle was then placed into 300 mL of 2 wt% NaOH and autoclaved for 20 min at 121°C. After cooling, the pellicle was neutralized with dilute H_2SO_4 and washed by filtration (Fisher P8 filter paper) until the filtrate pH was 7.0–7.5. The pellicle was then collected by filtration on a tared filter paper and dried to constant weight at 105°C.

Results

Comparison of Strain

Data from 14-d control cultures of *A. xylinum* 10821 and 23770 are shown in Figs. 1 and 2, respectively. Glucose consumption by strain 23770 was much more rapid than that observed for strain 10821, with essentially all the glucose gone in 9 d. Only 37% of the glucose was consumed by strain 10821 in the same period, and only 64% was consumed by 14 d. Growth rates for both strains were essentially the same over the course of the experiment. The pH of the strain 23770 cultures dropped to 3.2 after 4 d of culture, and strain 10821 cultures slowly decreased to pH 4.5. Cellulose production was similar for the two strains over the initial 4 d of culture, at which time

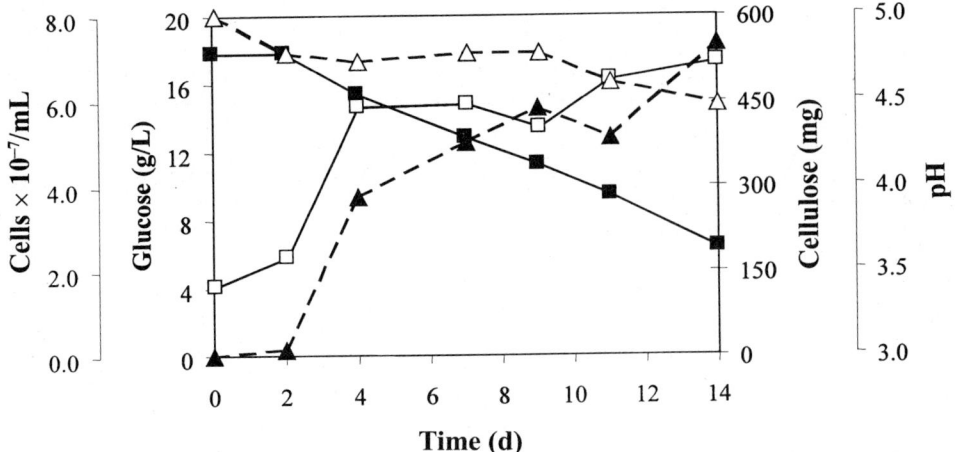

Fig. 1. Culture data for 14-d stationary control cultures of *A. xylinum* 10821. Each data point represents the average of duplicate sacrificed cultures. (□) Cell number; (■) glucose; (▲) dry weight of the cellulose pellicle; (△) pH.

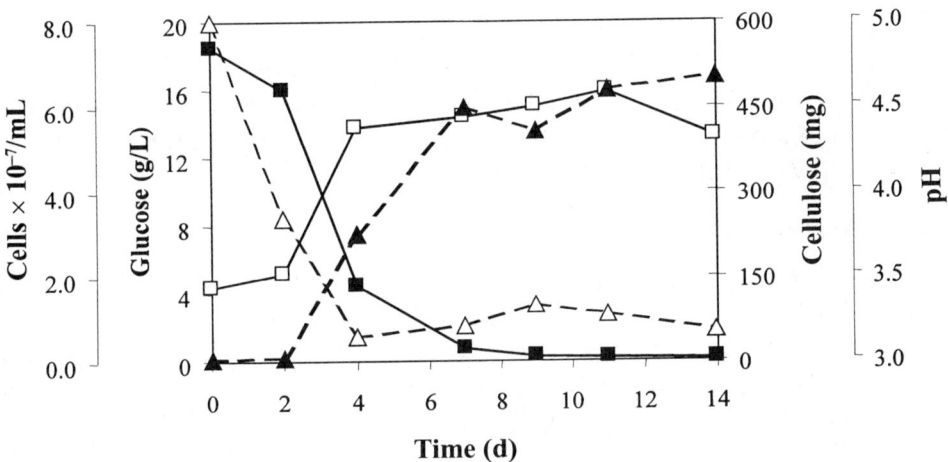

Fig. 2. Culture data for 14-d stationary control cultures of *A. xylinum* 23770. Each data point represents the average of duplicate sacrificed cultures. (□) Cell number; (■) glucose; (▲) dry weight of the cellulose pellicle; (△) pH.

both strains had produced 220–280 mg of cellulose. Linear production was observed after d 4 for strain 10821, to 550 mg on d 14. Strain 23770 reached 450 mg on d 7, after which only 50 mg more of cellulose was produced.

Net cellulose production rates and yields from glucose (g cellulose produced per g glucose consumed) are presented in Table 2. Cellulose production rates were initially high for strain 10821 but decreased slowly over time to a production rate of 5.0 g of cellulose/(d·m^2) of surface and a yield of 48% after 14 d. For strain 23770, the glucose was essentially depleted after 7 d, at which point the net production rate was 8.1 g/(d·m^2) and the

Table 2
Cellulose Production Rates and Yields
for Glucose Control Cultures of *A. xylinum* 10821 and 23770

	A. xylinum 10821		*A. xylinum* 23770	
Day	Net production rate (g/[d·m²])[a]	Yield (%)[b]	Net production rate (g/[d·m²])[a]	Yield (%)[b]
0	0.00	0.0	0.00	0.0
2	0.69	0.0	0.09	0.5
4	8.93[c]	116[c]	7.00	15.8
7	6.84	76.0	8.06	25.1
9	6.19	66.8	5.68	22.1
11	4.45	45.9	5.49	26.0
14	5.00	48.1	4.51	27.1

[a]Net production rates (g of cellulose/[d·m² of surface]) were each calculated from time zero.
[b]Yield is defined as (g of cellulose produced)/(g of glucose consumed) × 100%.
[c]These values are indicative of measurement error in dry weights for samples containing very small amounts of cellulose.

yield was 25%. After 14 d, the net production rate was 4.5 g/(d·m²), with an average production rate of 5.2 g/(d·m²) after d 7. Overall cellulose production in 14 d of culture (volumetric basis) was slightly higher from strain 10821, with 11% more cellulose produced.

Cellulose Production from Effluents

Cell growth data, culture pH, and cellulose production data for the unamended effluents are given in Table 3. Control data are included for comparison. Growth was observed in all media. Cell numbers in the potato effluent cultures were an order of magnitude higher than those observed in the other effluents or in the control, which was found to be from spores in the effluent that survived the autoclaving (20,21). Some contaminating cells that survived the autoclaving step were also seen in the other effluent media, but these cells did not seem to grow as well at pH 5.0 as those in the potato effluents.

There were also differences in the trends of culture pH between the two strains. For strain 10821, there was a general slight decrease in pH to the 4.0–4.5 range for HS and CW, whereas the pH of the LS culture was essentially unchanged. In the CSB culture, the pH increased to about 7.5 over the 14-d period. For strain 23770, the pH increased in all media except CW, for which it did not change significantly. The largest increase was for the LS effluent, to pH 7.3. This is in stark contrast to the glucose control, which decreased to pH 3.2 over the 14-d period.

Comparison of the cellulose production data in Table 3 indicates that neither strain was capable of producing significant bacterial cellulose from unamended HS, CSB, or CW. Values from replicate sacrificed cultures were

Table 3
Culture Data for Production of Bacterial Cellulose
from Diluted Unamended HS and LS Potato Effluents, Diluted Unamended CSB, and Diluted Unamended CW

| | A. xylinum 10821 | | | | | A. xylinum 23770 | | | | |
Day	Control	HS	LS	CSB	CW	Control	HS	LS	CSB	CW
	Cell counts ($\times 10^{-7}$ cells/mL)									
0	1.66	2.19	1.31	1.30	1.78	1.78	2.07	1.26	2.07	1.83
2	2.33	17.8	6.25	3.84	3.88	2.11	22.3	6.48	3.06	4.28
5	5.84	44.6	16.7	4.34	5.09	5.55	32.8	23.3	8.80	7.00
7	5.92	38.7	13.8	5.13	6.70	5.78	15.5	20.7	4.38	6.70
9	5.36	45.4	47.3	8.72	5.08	6.02	43.1	48.9	9.05	8.56
12	6.48	67.8	29.6	5.00	14.4	6.39	58.8	25.1	5.27	17.1
14	6.91	58.9	32.8	15.6	4.02	5.31	70.8	32.3	8.86	6.73
	Culture pH									
0	5.00	5.00	5.00	5.00	5.00	5.00	5.00	5.00	5.00	5.00
2	4.78	3.87	4.87	5.29	4.15	3.85	3.62	4.59	5.21	3.70
5	4.73	4.02	4.96	5.32	4.18	3.16	4.63	4.85	5.27	4.37
7	4.78	4.13	4.91	5.26	3.96	3.22	5.56	5.10	5.29	4.62
9	4.78	4.64	6.05	5.79	4.03	3.34	5.39	5.62	5.39	4.92
12	4.61	4.76	5.31	7.24	4.01	3.29	5.84	7.14	5.32	4.95
14	4.48	4.65	5.09	7.45	3.93	3.19	5.65	7.27	5.28	4.87
	Mass of cellulose produced (mg)									
0	0.00	0.00	0.00	0.00	0.00	0.00	0.00	0.00	0.00	0.00
2	10.9	19.1	0.00	2.40	0.00	1.40	1.76	17.4	0.00	7.05
5	281	21.5	36.4	1.15	2.15	220	84.4	335	6.00	108
7	376	0.00	28.8	0.00	1.65	443	0.00	553	0.400	97.7
9	437	0.00	94.6	9.55	0.45	401	0.00	498	18.6	17.7
12	385	13.5	35.1	30.3	4.15	474	26.9	483	18.1	17.6
14	550	0.00	48.4	33.0	1.30	496	25.9	581	0.700	113

somewhat scattered, but in all cases, very little cellulose was produced from these unamended effluents. The maximum amounts of cellulose produced in any of the sacrificed cultures utilizing these diluted effluents by strain 10821 were 22, 33, and 4 mg of cellulose from HS, CSB, and CW, respectively. By comparison, strain 23770 produced maximum amounts of 84, 19, and 113 mg of cellulose from the same media. Results with diluted LS effluent were markedly different for the two strains. A maximum 95 mg of cellulose was produced by strain 10821, whereas strain 23770 produced up to 580 mg of cellulose from LS in 14 d. This amounts to an estimated cellulose yield (g cellulose produced per g glucose consumed) from diluted LS effluent of about 27%, which is nearly the same yield as that obtained by strain 23770 from optimized glucose medium.

Discussion

Comparison of Strains

Cell growth on Schramm and Hestrin's medium was similar for the two strains, but strain 23770 consumed the glucose much more quickly. The rapid utilization of glucose and the drop in pH observed for strain 23770 suggest that the excess glucose was converted to gluconate. Excess glucose not used for cellulose synthesis is oxidized by *A. xylinum* to gluconate *(13)*. This allows some accumulation of gluconate, which can lower the pH and inhibit cellulose production at high glucose concentrations *(13)*.

Forng et al. *(25)* demonstrated that glucose concentrations above about 1% give relatively smaller increases in cellulose produced, and Masaoka et al. *(13)* observed no difference in the rate of cellulose production from 0 to 2% glucose, with inhibition observed at 4% glucose. Strain 23770 produced cellulose at a markedly higher initial rate than strain 10821, but at only about half the yield. When initial carbon source concentrations below 2 wt% are used, a production efficiency of 45% (weight of cellulose per carbon source consumed) can be obtained *(25)*. This magnitude was observed for strain 10821 (48% on the same basis), but not for strain 23770 (27%). After the glucose was consumed in strain 23770 cultures, a low rate of cellulose production continued. This production was likely from the citrate in the medium, since it has been shown that citrate is eventually used as a carbon source for growth and cellulose production, but only after the glucose is consumed *(27)*. In the present study, overall cellulose production in 14 d of culture (volumetric basis) by strain 10821 was slightly higher (11%) than that produced by strain 23770.

Cellulose Production from Effluents

All of the unamended food process effluents were able to support growth of both strains of *A. xylinum*, and at least limited cellulose production. Strain 10821 produced 5%, 21%, 7%, and 0.6% normalized yields (weight basis) of bacterial cellulose from HS (insoluble starch), LS (starch), CSB (sucrose), and CW (lactose), respectively, relative to glucose (100%).

Strain 23770 produced 15%, 96%, 3%, and 30% normalized yields (weight basis) of bacterial cellulose from the same effluent media, relative to glucose (100%). For comparison, Masaoka et al. *(13)* showed 18, 33, and 16% normalized yields of bacterial cellulose from starch, sucrose, and lactose, respectively, relative to glucose (100%). The Masaoka et al. experiments *(13)* consisted of 3-d cultures of *A. xylinum* IFO 13693 on a basic medium (pH 6.0) consisting of 2% peptone, 0.5% yeast extract, 0.5% glucose, 0.1% $MgSO_4 \cdot 7H_2O$, and 0.2% ethanol. Glucose was simply replaced by the other carbon sources for the tests.

We cannot make a direct comparison of the HS medium results with those of Masaoka et al. *(13)*, since although it is not specifically stated, we have assumed that their starch substrate was likely to have been soluble. From the LS medium, our results show poorer than expected cellulose yields with strain 10821, and much better than expected cellulose yields with strain 23770. Our results with the CSB effluent are much lower than those reported from sucrose by Masaoka et al. *(13)*, which may be due to the presence of toxic heavy metals in the CSB *(28)*. Finally, cellulose yields from CW effluent were very low from strain 10821 and about twice that expected from strain 23770. Of the unamended, diluted effluents tested, only the LS effluent medium supported substantial cellulose production, and only by strain 23770.

Given the similarities between cell numbers and overall cellulose production by the two strains on Schramm and Hestrin's medium, it is not immediately clear to the authors why strain 10821 was unable to produce substantial cellulose from the LS medium. The most quantifiable differences between the two cultures on LS medium, other than cellulose production, are the different trends in pH. The LS effluent is known to contain indigenous bacteria as received from the processor *(20,21)*. Some of these bacteria survived the autoclaving step and were present in the LS cultures, as evidenced by the much higher cell numbers observed, and by uninoculated controls (data not shown). These bacteria are known to lower medium pH to 4.0 if left unchecked *(20,21)*, suggesting that they are fermentative bacteria. Because they are indigenous to the potato effluents, they would be expected to possess an amylase system and to convert starch to glucose. This would help to increase cellulose yields from the starch in the effluent through supplementation of the ability of *A. xylinum* to convert starch to glucose.

In addition, pH in the strain 23770 culture on LS medium did not drop as it did in pure culture on glucose medium. This suggests that indigenous bacteria consumed the gluconate as it was being produced. Removal of the gluconate would minimize its inhibitory effects on cellulose production. Gluconate, which is not a substrate for cellulose production, may have been degraded by the indigenous bacteria to smaller organic acids, which both stimulate *(29)* and serve as substrates *(13)* for cellulose production. The strain 10821 culture, which did not produce significant amounts of gluconate (as evidenced by the lack of a drop in pH on glucose medium), would not experience this benefit, because the gluconate would not be present for indigenous bacteria to consume. The observed increase in pH could then be due to consumption of the organic acids in the unbuffered LS medium. Although not a part of the scope of the present study, these possibilities should be examined because they suggest a synergistic mixed culture strategy for production of bacterial cellulose from LS effluent.

Because it has been shown that *A. xylinum* can produce cellulose from a myriad of carbon sources (13), it is likely that additions could be made to the media that would allow higher cellulose production from each of the effluents. Pretreatments of the effluents could be done in order to increase free carbohydrate concentrations, or to remove potentially toxic components such as heavy metals in sugar beet molasses (28). In addition, it is probable that contaminating bacteria already present in the effluents also serve to lower yields, because all the media contained indigenous bacteria that were resistant to heat sterilization by autoclaving.

Because medium costs, among other things such as production rate and process footprint, limit commercial production of bacterial cellulose in static cultures (3), it is important to minimize carbon, nutrient, and energy additions wherever possible. Thus, it is generally undesirable to make medium additions, to use pretreatments, or to use extensive or energy-intensive sterilization procedures beyond the absolute minimum required. The estimated yield of cellulose from unamended, diluted LS effluent by strain 23770 was not significantly different from that on Schramm and Hestrin's medium. This was true even though substantial glucose was diverted to gluconate and the expected yield from starch is only 18% of that from glucose. This suggests that *A. xylinum* 23770 is more suitable than 10821 for the production of bacterial cellulose from LS.

Conclusions

A diluted, unamended LS potato process effluent was found to support substantial production of bacterial cellulose by *A. xylinum* ATCC 23770 in static culture. Diluted, unamended HS potato process effluent, CW, and CSB did not support significant cellulose production by either *A. xylinum* 10821 or 23770. *A. xylinum* 10821, while diverting less glucose to gluconate than 23770, was unable to produce a significant amount of cellulose from the LS effluent medium. Because the yield of cellulose by *A. xylinum* 23770 was equivalent from diluted, unamended LS potato process effluent and from optimized Schramm and Hestrin's medium, this strain is more appropriate to use for cellulose production from this effluent.

Acknowledgment

This work was supported through the Laboratory Directed Research & Development Program at the Idaho National Engineering and Environmental Laboratory under DOE Idaho Operations Office Contract DE-AC07-99ID13727.

References

1. Krieger, J. (1990), *Chem. Eng. News* **68,** 35–37.
2. Johnson, D. C. and Winslow, A. R. (1990), *Pulp Paper* **64(6),** 105–107.
3. Brown, R. M. Jr. (1989), in *Cellulose: Structural and Functional Aspects*, Kennedy, J. F., Phillips, G. O., and Williams, P. A., eds., Ellis Horwood, Chichester, pp. 145–151.
4. Okiyama, A., Motoki, M., and Yamanaka, S. (1992), *Food Hydrocoll.* **6,** 479–487.
5. Ross, P., Mayer, R., and Benziman, M. (1991), *Microbiol. Rev.* **55,** 35–58.
6. Nishi, Y., Uryu, M., Yamanaka, S., Watanabe, K., Kitamura, N., Iguchi, M., and Mitsuhashi, S. (1990), *J. Mater. Sci.* **25,** 2997–3001.
7. Fontana, J. D., de Souza, A. M., Fontana, C. K., Torriani, I. L., Moreschi, J. C., Gallotti, B. J., de Souza, S. J., Narcisco, G. P., Bichara, J. H., and Farah, L. F. X. (1990), *Appl. Biochem. Biotechnol.* **24/25,** 253–264.
8. Farah, L. F. X. (1990), US patent 4,912,049.
9. Yamanaka, S., Ono, E., Watanabe, K., Kusakabe, M., and Suzuki, Y. (1990), European patent application EP 0 396 344.
10. Ring, D. F., Nashed, W., and Dow, T. (1986), US patent 4,588,400.
11. Chatterjee, P. K. (1989), in *Proceedings of the Nisshinbo International Conference on Cellulosics Utilization in the Near Future*, Inagaki, H. and Phillips, G. O., eds., Elsevier Science, New York, pp. 12–17.
12. Shibazaki, H., Kuga, S., Onabe, F., and Usuda, M. (1993), *J. Appl. Polym. Sci.* **50,** 965–969.
13. Masaoka, S., Ohe, T., and Sakota, N. (1993), *J. Ferment. Bioeng.* **75,** 18–22.
14. Zaar, K. (1979), *J. Cell Biol.* **80,** 773–777.
15. Watanabe, K. and Yamanaka, S. (1995), *Biosci. Biotechnol. Biochem.* **59,** 65–68.
16. Oikawa, T., Morino, T., and Ameyama, M. (1995), *Biosci. Biotechnol. Biochem.* **59,** 1564, 1565.
17. Ishikawa, A., Matsuoka, M., Tsuchida, T., and Yoshinaga, F. (1995), *Biosci. Biotechnol. Biochem.* **59,** 2259–2262.
18. Oikawa, T., Ohtori, T., and Ameyama, M. (1995), *Biosci. Biotechnol. Biochem.* **59,** 331, 332.
19. Matsuoka, M., Tsuchida, T., Matsushita, K., Adachi, O., and Yoshinaga, F. (1996), *Biosci. Biotechnol. Biochem.* **60,** 575–579.
20. Thompson, D. N., Fox, S. L., and Bala, G. A. (2000), *Appl. Biochem. Biotechnol.* **84–86,** 917–930.
21. Thompson, D. N., Fox, S. L., and Bala, G. A. (2001), *Appl. Biochem. Biotechnol.* **91–93,** 487–501.
22. Schramm, M. and Hestrin, S. (1954), *J. Gen. Microbiol.* **11,** 123–129.
23. Gherna, P. (1989), in *American Type Culture Collection Catalogue of Bacteria and Phages*, 17th ed., American Type Culture Collection, Rockville, MD, p. 403.
24. Gerhardt, P., Murray, R. G. E., Wood, W. A., and Krieg, N. R., eds. (1994), in *Methods for General and Molecular Bacteriology*, American Society for Microbiology, Washington, DC, pp. 518, 519.
25. Forng, E. R., Anderson, S. M., and Cannon, R. E. (1989), *Appl. Environ. Microbiol.* **55,** 1317–1319.
26. De Wulf, P., Joris, K., and Vandamme, E. J. (1996), *J. Chem. Technol. Biotechnol.* **67,** 376–380.
27. Geyer, U., Klemm, D., and Schmauder, H.-P. (1994), *Acta Biotechnol.* **14,** 261–266.
28. Roukas, T. (1998), *Process Biochem.* **33,** 805–810.
29. Dudman, W. F. (1959), *J. Gen. Microbiol.* **21,** 327–337.

Production of Polyhydroxybutyrate by *Bacillus* Species Isolated from Municipal Activated Sludge

KIN-HO LAW,[1] YUN-CHUNG LEUNG,[1]
HUGH LAWFORD,[2] HONG CHUA,[3] WAI-HUNG LO,[1]
AND PETER HOIFU YU*,[1]

*Department of [1]Applied Biology and Chemical Technology,
Hong Kong Polytechnic University, Hung Hom, Hong Kong, China;
[2]C Department of Biochemistry, University of Toronto, Toronto, Canada;
and [3]Civil and Structural Engineering, Hong Kong Polytechnic University,
Hung Hom, Hong Kong, China*

Abstract

Plastic wastes are considered to be severe environmental contaminants causing waste disposal problems. Widespread use of biodegradable plastics is one of the solutions, but it is limited by high production cost. Biologic wastewater treatment generates large quantities of biomass as activated sludge. Only a few reports focus on the potential of utilizing resident *Bacillus* species from activated sludge in polyhydroxbutyrate (PHB) production as well as the production of PHB from food wastes. They have attractive properties such as short generation time, absence of endotoxins, and secretion of both amylases and proteinases that can well utilize food wastes for nutrients, which can further reduce the cost of production of polyhydroxyalkanoates (PHAs). Two PHA-producing strains, HF-1 and HF-2, were isolated from activated sludge. HF-1 outperformed HF-2 in terms of growth and PHB production in hydrolyzed soy and malt wastes. The isolated bacteria was characterized by DNA sequence alignment. Cell extracts of HF-1 were also compared to *Bacillus megaterium* cell extracts on sodium dodecyl sulfate polyacrylamide gel electrophoresis. The biopolymers accumulated were analyzed by gas chromatography, nuclear magnetic resonance, and Fourier transform infrared methods.

Index Entries: Polyhydroxybutyrates; malt waste; soy waste; *Bacillus*; activated sludge; Fourier transform infrared; inclusion body; nuclear magnetic resonance.

*Author to whom all correspondence and reprint requests should be addressed.

Introduction

Plastics have become an integral part of our lives, but the generation of plastic wastes has increased dramatically. In 1986–1998, about 15% of total domestic wastes or commercial and industrial wastes in Hong Kong were plastics *(1)*. The petroleum-derived plastics are not easily degraded by microorganisms. Plastic wastes are therefore considered to be severe environmental contaminants causing waste disposal problems.

Polyhydroxyalkanoates (PHAs) are polyesters of hydroxyalkanoates synthesized by numerous bacteria as intracellular carbon and energy storage compounds and accumulated as granules in the cytoplasm *(2)*, usually when essential nutrients such as N or P are limited in the presence of an excess carbon source *(3)*. The first PHA discovered was polyhydroxybutyrate (PHB), which is the most abundant form of PHA in nature.

By comparing the cost of BIOPOL™ (P[3HB-co-3HV] produced from fermentation by *Alcaligenes eutrophus*) (US$16/kg) with that of polypropylene (less than US$1/kg), the conventional plastics *(3)*, one can see that the cost of PHAs is much higher. The carbon source should be inexpensive because it is the major contributor to the total substrate cost (up to 50% of the total operating cost) *(4)*, and, thus, an expensive carbon source is not practical in large-scale industrial production. Several studies have investigated the use of low-cost substrate for PHA production, such as xylose *(4)*, molasses *(5)*, and malt waste and soy waste *(6)*.

Biologic wastewater treatment is the largest application of microorganisms in the service sector and generates large quantities of biomass as activated sludge. Our investigations showed the presence of different types of biopolymers in the activated sludge and the yield obtained was 6.5 g/L *(7)*. Only a few *Bacillus* strains were examined for their ability to accumulate PHAs, which naturally occur in biodegradable, biocompatible, and microbial thermoplastic. *Bacillus* strains have the attractive properties of short generation time, absence of endotoxins, and presence of both amylase and proteinase secretion that can well utilize food wastes. In this article, the potential of resident *Bacillus* species from activated sludge in PHB production as well as the production of PHB from food wastes are reported.

Materials and Methods

Activated Sludge

Activated sludge was collected from the Sha Tin Wastewater Sewage Treatment Plant in Hong Kong.

Microorganisms

Bacillus megaterium ATCC strain 11561 was kindly provided by Prof. Maura C. Cannon of University of Massachusetts. PHAs producing bacteria screened from activated sludge were named HF-1 and HF-2. The isolated strains were identified by Microbial ID (Newark, DE).

Media

Preparation of Growth Medium

Medium A consisted of a nutrient-rich medium LB broth (10 g/L of tryptone, 5 g/L of yeast extract, 10 g/L of NaCl) and 10 g/L of glucose.

Preparation of PHA Accumulation Medium

Medium B consisted of a medium for bioplastics accumulation containing 3.57 g/L of Na_2HPO_4, 0.25 g/L of $(NH_4)_2SO_2$, 1.50 g/L of KH_2PO_4, 0.20 g/L of $MgSO_4 \cdot 7H_2O$, and 20 g/L of glucose.

Preparation of Food Wastes Medium

Malt waste, mostly semisolids of spent barley and millet refuse, was obtained from Carlsberg, a beer brewery in Hong Kong. Soy waste, chiefly semisolid cellular residues of soy beans, was collected from Vitasoy International Holdings, a soy milk company in Hong Kong. The ratio of the C and N contents of the malt and soy wastes were 7:1 and 8:1, respectively, as determined by total organic carbon (TOC) *(8)* and total Kjeldahl nitrogen (TKN) *(8)* methods.

Two hundred grams of the waste was hydrolyzed with 1 L of 0.5 *M* HCl. The mixture was incubated at 90°C for 8 h. The resultant mixture was centrifuged at 14,333*g* for 20 min. The supernatant was filtered to remove debris and adjusted to pH 7.0 by adding NaOH. The medium was autoclaved for 30 min at 121°C and used as the substrate for the growth of bacteria.

Preparation of Trace Elements Solution

One hundred milliliters of water consisted of 0.60 g of $FeCl_3 \cdot H_2O$, 0.1 g of $CaCl_2 \cdot 2H_2O$, 0.03 g of H_3BO_3, 0.002 g of $CoCl_2 \cdot 6H_2O$, 0.010 g of $ZnSO_4 \cdot 7H_2O$, 0.003 g of $MnCl_2 \cdot 4H_2O$, 0.003 g of $Na_2MoO_4 \cdot 2H_2O$, 0.0024 g of $NiCl_2 \cdot 6H_2O$, and 0.001 g of $CuSO_4 \cdot 5H_2O$.

Screening of PHA-Producing Bacteria from Activated Sludge

Activated sludge (50 mL) was transferred into conical flasks. The flasks were placed in a water bath at 87°C for 8 min to kill all nonspore-forming cells *(9)*, leaving the surviving *Bacillus* spores that are heat resistant. The sludge sample was centrifuged at 2610*g* for 10 min. The supernatant was discarded, and the pellet was resuspended with 100 mL of sterile medium A and incubated for 18 h at 37°C with 250 rpm shaking to enrich the cell population. The culture from medium A (25 mL) was centrifuged at 2610*g* for 10 min. The supernatant was discarded. The pellet was resuspended in 100 mL of medium B with 0.1 mL of trace elements solution. The culture was incubated at 37°C for 16 h with 250 rpm shaking for PHB accumulation.

Serial dilutions of $10^{-1} - 10^{-7}$ of cell culture from medium B were prepared (the culture was diluted with 0.9% NaCl solution). The culture from each dilution was streaked on production agar (medium B with 20 g/L of glucose, 1 mL/L of trace elements solution, and 1.5% agar) plates. The plates were incubated for 16 h at 37°C. The cells from each single colony were subjected to Fourier transform infrared (FTIR) analysis.

The selected PHA-producing bacteria were cultivated and subjected to Gram staining, endospore staining, and microscopic morphology examination to select the potential PHA-producing *Bacillus* strains.

Fermentation

Glycerol stocks of all bacterial cells were used as inoculum and were prepared by cell culture with a final glycerol concentration of 15% (v/v) and stored at −80°C. It was first inoculated into 5 mL of 2 XYT medium (1.6 g of tryptone, 1.0 g of yeast extract, and 0.5 g of NaCl in 100 mL of distilled water) in universal bottles. A 1% inoculum was used in flask fermentations. Cultures were grown on a rotary shaker at 250 rpm at 37°C.

Batch fermentation was carried out in the computer-controlled Bioengineering 3.7-L fermentor (Bioengineering, Switzerland) with growth conditions set at 37°C and pH 7.0. The pH was adjusted by adding 2 M HCl and 2 M NaOH. One hundred milliliters of seed culture of the target strain (4% of fermentation medium) was inoculated into 2.5 L of malt waste medium.

Extraction of Biopolymers

After fermentation, the culture was centrifuged at 14,333g for 25 min at 4°C, washed with distilled water, and freeze-dried. One gram of the freeze-dried cell powder was treated with a dispersion containing 15 mL each of chloroform and 30% NaOCl solution.

The mixture was incubated at 37°C with 250 rpm agitation for 1 h, and then centrifuged at 2610g for 15 min, which resulted in three phases. The upper phase was a hypochlorite solution, the middle phase contained the non-PHB cell material and undisrupted cells, and the bottom phase was chloroform-containing PHB.

The bottom chloroform layer was filtered and allowed to concentrate by evaporation to a final volume of 5 mL. Pure PHB was obtained by nonsolvent precipitation (chloroform:methanol at a ratio of 1:9). Finally, the white precipitate was dried and weighed.

Isolation of PHB Inclusion Bodies

The culture of *B. megaterium* 11561 and the HF-1 after fermentation were pelleted at 6000g in 4°C for 20 min and resuspended in 5 mL of 10 mM Tris-Cl, pH 8.0; 1 mM EDTA; 20 mM MgSO$_4$; and 0.25 M sucrose at 4°C. Lysozyme was added to a final concentration of 1.5 mg/mL and the solution was incubated at 37°C for 15 min and room temperature for 10 min. The cells were broken by sonication for 10 min.

Aliquots of 1 mL of lysate were loaded on sucrose step gradients in 5-mL ultracentrifuge tubes and consisted of 0.9 mL of each of the following sucrose concentrations: 2.0 M, 1.66 M, 1.33 M, 1.0 M, and 0.4 mL of 0.66 M

in TE (10 m*M*, Tris-Cl, pH 8.0; 1 m*M* EDTA). The tubes were centrifuged at 150,000*g* for 1 h at 10°C. The inclusion bodies, which banded about midtube, were collected, washed in 20 vol of TE, and pelleted at 20,000*g*. The sucrose gradient steps were repeated for further purification, and the purified inclusion bodies were stored in TE buffer at 4°C.

Sodium Dodecyl Sulfate Polyacrylamide Gel Electrophoresis

Resuspended the pelleted inclusion bodies by TE with 2% sodium dodecyl sulfate (SDS). An equal volume of 2X sample buffer was added prior to boiling for 5 min, samples were centrifuged for 3 min to pellet the PHA, and the supernatant was loaded on an SDS 12% polyacrylamide gel. Coomassie Brilliant Blue R-250 staining and silver staining (Bio-Rad) were used after the gel electrophoresis for analysis.

DNA Sequencing

The DNA fragment from polymerase chain reaction (PCR) of HF-1 was sequenced from both ends, using designed primers based on the sequence of *pha* gene cluster from *B. megaterium*, following the protocol of an ABI Prism DNA Sequencing Kit (Perkin-Elmer). By using the dye-terminator chemistry, cycle sequencing, and an ABI Prism 310 sequencer (Perkin-Elmer). Sequence assembly was performed by using the software Advanced BLAST (National Center for Biotechnology Information).

Analytical Methods

TOC Analysis

Fermentation medium was analyzed with an Astro 2000 TOC Analyzer. The method was according to APHA (4500-Norg) *(8)*.

TKN Analysis

Fermentation medium was analyzed with a Kjeltec Auto 1030 Analyzer. The method was according to APHA (5310C) *(8)*.

Gas Chromatography Analysis

One milliliter of esterification solution (3 mL of 95–98% H_2SO_4, 0.29 g of benzoate, and 97 mL of methanol), freeze-dried cells, and 1 mL of chloroform were heated at 100°C for 4 h; and distilled water was added and vortexed to enhance phase separation. One milliliter of distilled water was added to the cooled mixture, which was vortexed for phase separation. A 1-µL portion of the lower organic phase was subjected to gas chromatography (GC) analysis. GC analysis was performed on a 5890 Series II Gas Chromatograph (Hewlett Packard), using an Ultra 2 (crosslinked 5% Ph Me silicone) Capillary Column 0.2 mm in diameter and 25 m long (Hewlett Packard).

Nitrogen was chosen as the carrier gas. Analysis was started at 70°C for 3 min and was increased to 120°C at a rate of 10°C/min. After reaching 120°C, the temperature was kept stable for 15 min to remove all nonvolatile component.

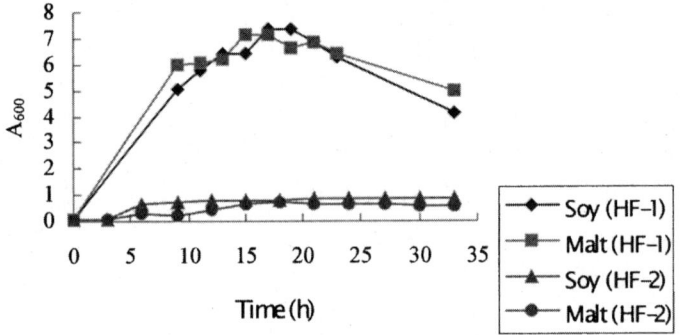

Fig. 1. Cell growth of HF-1 and HF-2 during flask fermentation in malt and soy wastes.

Analysis by FTIR Spectroscopy

Two to five milliliters of the cell culture was centrifuged at 2610g for 15 min. An appropriate amount of cells was transferred on an IR window (ZnSe Disc; Spectratech) and dried on it. A mirror was used to give the reflected IR to the horizontal laid window. With a scan of 32, resolution of 16, and autogain, spectra were recorded at wave numbers (cm^{-1}) from 400 to 4000 using a Mangna-IR spectrometer 750 (Nicolet).

Analysis by ^{1}H Nuclear Magnetic Resonance

^{1}H nuclear magnetic resonance (NMR) analysis was carried out on a DPX-400 Spectrometer (Bruker). ^{1}H NMR spectra was recorded at room temperature from a deuteriated chloroform (CDCl$_3$) solution of the extracted biopolymers. The 400 MHz ^{1}H NMR spectra were recorded.

Results and Discussion

Screening PHA-Producing Strains from Activated Sludge

In the first part of the experiment, two microbial strains (HF-1 and HF-2) were successfully screened from the activated sludge. The potential of the strains for the conversion of food wastes into bioplastic was investigated. The data of cell growth during the fermentation in malt and soy wastes are shown in Fig. 1. The OD$_{600}$ of HF-2 was maintained at a low level throughout the fermentation in both malt waste and soy waste as media when compared with strain HF-1. This might be because the composition of hydrolyzed food wastes was not suitable for the growth of strain HF-2.

Comparison of the yield of PHB produced by the two strains from the results of the flask experiment shows that HF-1 accumulated the higher amount of PHB recovery (19.22% of its cell dry wt) than HF-2 (10.18% of its cell dry wt). Thus, further investigation was mainly concentrated on strain HF-1.

During fermentation, the PHB content of the HF-1 cells was monitored by FTIR analysis, and the yield of PHB was estimated by the height ratio of

Table 1
Height Ratio of PHB Peak
to Protein Peak on FTIR Absorbance Spectra

Time (h)	Malt waste	Soy waste
9	0.676	0.641
11	0.720	0.693
13	0.738	0.606
15	0.740	0.488
17	0.720	0.420
19	0.700	0.286
21	0.556	0.321
23	0.556	0.216
33	0.520	0.047

The highest PHB accumulation in flask fermentation occurred at 15 h when malt waste was used as the medium, but at 11 h when soy waste was used. The spectra had a characteristic sharp PHA absorption band at a narrow range around 1726–1740 cm^{-1}, and the pattern is similar to that of previous studies *(10,11)*. This sharp absorption band is assigned to the stretching vibration for the ester carbonyl of the PHB.

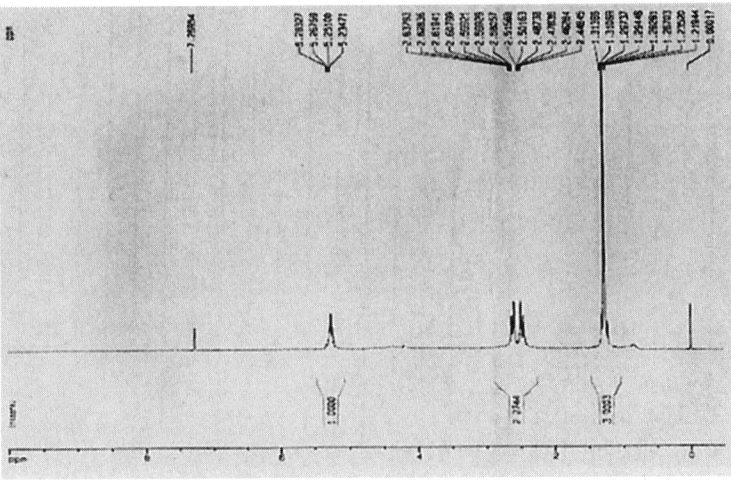

Fig. 2. NMR spectra of extracted biopolymer of HF-1.

PHB to protein, as shown in the absorbance spectra of FTIR and as listed in Table 1.

Data from ^{1}H NMR analysis of the extracted biopolymer produced by fermentation of HF-1 is displayed in Fig. 2.

The ^{1}H-NMR spectrum showed the presence of three groups of characteristic signals of the homopolymer PHB. The signal at about 1.26 ppm

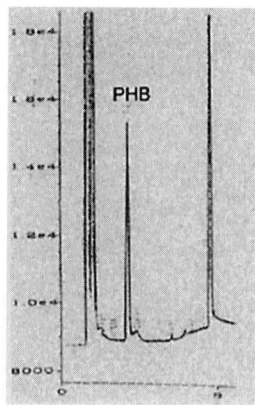

Fig. 3. GC spectrum of the HF-1 cells after fermentation.

was attributed to the methyl group coupled to one proton, the signal at about 2.52 ppm was attributed to a methylene group, and a multiplet at 5.25 ppm was attributed to a methyne group. The signal at about 7.27 ppm was owing to the chloroform. The spectra showed the CH–, CH$_2$– and CH$_3$– groups in the molecule, but no –COOH and –OH groups; thus, so the compound cannot be β-hydroxybutyrate and should be a polymer.

Identification of Screened Strain

The isolated strains were identified by Microbial ID. Identification was based on the fatty acids composition profiles of the bacteria and comparison with the profiles of the other bacteria in their library and expression of similarity by Similarity Index of how closely they matched the other known bacteria. Two strains that are the most similar to HF-1 are *Brevibacillus laterosporus* (0.745) and *Bacillus megaterium* (0.736).

Data from the GC analysis from freeze-dried cells of HF-1 after fermentation are shown in Fig. 3. The PHB peak was present at a retention time of 3.363 min but there was no PHV peak when compared to the spectra of the standard.

There were two choices of strains provided by Microbial ID; therefore further investigation was made. The DNA fragment after the PCR reaction was subjected to DNA sequencing in both forward and backward primers. The percentage identity was 92 and 94%, respectively, to those sequences of *B. megaterium*. This suggests that HF-1 was quite similar to *B. megaterium* 11561. The slightly unequal DNA sequencing might be owing to the sequencing error or the difference in subspecies strain.

PHA inclusion body–associated proteins were also analyzed by SDS-polyacrylamide gel electrophoresis to identify similarities between the strain HF-1 and *B. megaterium* (Fig. 4). Proteins that associated with PHA inclusion bodies were separated. There were at least 20 such proteins present in various quantities. The two most abundant proteins had

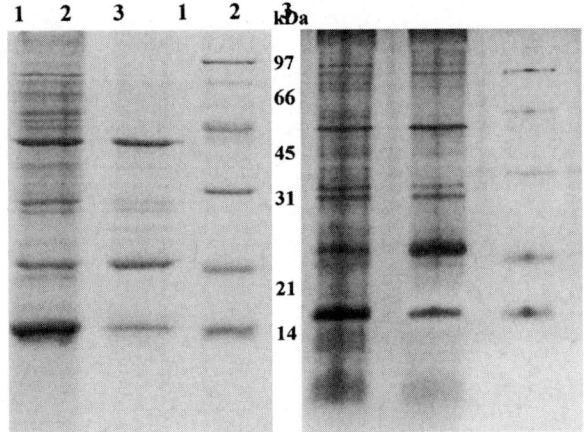

Fig. 4. PHA inclusion body–associated proteins. **(Left)** Coomassie blue staining; **(right)** silver staining. Lane 1, proteins from inclusion bodies of *B. megaterium* 11561; lane 2, proteins from inclusion bodies of HF-1; lane 3, molecular markers (14,400, 21,500, 31,000, 45,000, 66,200, 97,400 Daltons).

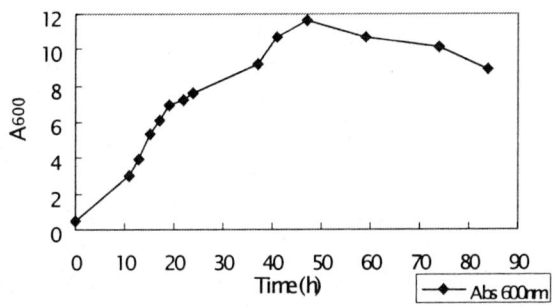

Fig. 5. Cell growth of HF-1 during fermentor fermentation in malt wastes.

molecular weights of approx 20 and 41 kDa and were found in both strains. The 14-kDa protein was lysozyme. The pattern of the visualized proteins bands after Coomassie blue and silver staining was similar in both strains excepts two bands. Thus, it could be suggested that HF-1 is closely related to *B. megaterium*.

Production of PHB from Malt Waste Using HF-1 in a 3-L Fermentor

The growth of HF-1 during fermentation is presented graphically in Fig. 5. According to the height ratio of PHB band to protein band on FTIR absorbance spectra (Table 2), the optimum PHB production time of HF-1 in fermentor fermentation with malt waste as medium was 15 h. The time was the same as for the flask experiment (Fig. 1), but the A_{600} was much higher. At 24 h, the PHB content had dropped dramatically owing to the consumption of the PHB by HF-1.

Table 2
Height Ratio of PHB Peak
to Protein Peak
on FTIR Absorbance Spectra

Time (h)	Malt waste
11	0.611
13	0.732
15	0.879
17	0.813
19	0.750
22	0.712
24	0.579

Conclusion

Two PHA-producing bacteria were successfully screened from municipal activated sludge. The results of the fermentation showed that the bacteria HF-1 well utilized food wastes for nutrients to produce biopolymer. The use of inexpensive carbon substrates to produce bioplastics would be beneficial in lowering the cost of PHA production. The potential of the screened bacteria on microbial biopolymer production was subjected to fermentor fermentation. The results showed that only PHB homopolymer could be produced. The biopolymer produced was characterized by FTIR, GC, NMR, and genetic analysis.

Acknowledgment

We wish to express our sincere thanks to the Hong Kong Polytechnic University and the University Grant Council of Hong Kong for the support of a grant for this research and the research assistance of Dr. K. Hong, H. Liu, C. Y. Chung, and C. Y. Heung.

References

1. Hong Kong Environmental Protection Department. (1999), *Environment Hong Kong 1999*, Hong Kong Government Press, HKSAC.
2. Lee, S. Y. (1996), *Biotechnol. Bioeng.* **49,** 1–14.
3. Lee, S. Y. (1996), *Trends Biotechnol.* **14,** 431–438.
4. Lee, S. Y. (1998), *Bioprocess Eng.* **18,** 397–399.
5. Liu, F., Li, W., Ridgway, D., and Gu, T. (1998), *Biotechnol. Lett.* **20(4),** 345–348.
6. Yu, P. H., Chua, H., Huang, A. L., Lo, W., and Chen, G. Q. (1998), *Appl. Biochem. Biotechnol.* **70–72,** 603–614.
7. Yu, P. H., Chua, H., and A. L. Huang (1999), *Macromol. Symp.* **148,** 415–424.
8. Greenberg, A. E., Clesceri, L. S., and Eton, A. D. (1992), *Standard Methods for the Examination of Water and Wastewater*, 18th ed., American Public Health Association, Washington, DC.
9. Brock, D., Madigan, T., Martinko, M., and Parker, J. (1994), *Biology of Microorganisms*, Prentice-Hall, Englewood Cliffs, NJ.
10. Szewcyk, E. and Mikucki, J. (1989), *FEMS Microbiol. Lett.* **61,** 279–284.
11. Blackwood, C. and Agene, E. (1957), *J. Bacteriol.* **74,** 266, 267.

Characterization of Bioconversion of Fumarate to Succinate by Alginate Immobilized *Enterococcus faecalis* RKY1

Hwa-Won Ryu[*,1] and Young-Jung Wee[2]

*Departments of [1]Biochemical Engineering and [2]Biomedical Engineering,
Chonnam National University, Kwangju 500-757, Korea,
E-mail: hwryu@chonnam.ac.kr*

Abstract

In this study, the immobilization characteristics of *Enterococcus faecalis* RKY1 for succinate production were examined. At first, three natural polymers—agar, κ-carrageenan, and sodium alginate—were tried as immobilizing matrices. Among these, sodium alginate was selected as the best gel for immobilization of *E. faecalis* RKY1. Efficient conditions for immobilization were established to be with a 2% (w/v) sodium alginate solution and 2-mm-diameter bead. The bioconversion characteristics of the immobilized cells at various pH values and temperatures were examined and compared with those of free cells. The optimum pH and temperature of the immobilized cells were the same as for free cells, 7.0 and 38°C respectively, but the conversion ratio was higher by immobilization for all the other pH and temperature conditions tested. When the seed volume of the immobilized cells was adjusted to 10% (v/v), 30 g/L of fumarate was completely converted to succinate (0.973 g/g conversion ratio) after 12 h. In addition, the immobilized cells maintained a conversion ratio of >0.95 g/g during 4 wk of storage at 4°C in a 2% (w/v) $CaCl_2$ solution. In repetitive bioconversion experiments, the activity of the immobilized cells decreased linearly according to the number of times of reuse.

Index Entries: *Enterococcus faecalis* RKY1; succinate; fumarate; immobilized cells; bioconversion.

Introduction

Succinic acid is a C_4-dicarboxylic acid produced as an intermediate of the tricarboxylic acid cycle and also as one of the fermentation products of

*Author to whom all correspondence and reprint requests should be addressed.

anaerobic metabolism *(1,2)*. It has many industrial applications as a raw
material for food, medicine, plastics, cosmetics, textiles, plating, and waste-
gas scrubbing *(3,4)*. Currently, it is produced commercially by petrochemical processes. Recently, however, great interest has been focused on the
enhanced production of succinic acid through an industrial fermentation
process *(5)*. Production of succinic acid using a fermentation process represents an alternative synthesis route via the utilization of renewable feedstocks such as corn *(6)*. Microorganisms have the ability to modify
chemically a wide variety of organic compounds, referred to as bioconversion. One such example is the previously reported bioconversion of fumarate to succinate by *Enterococcus faecalis* RKY1 *(4,7)*. This strain is able to
produce succinic acid at a high yield if cultured anaerobically with glycerol
as a hydrogen donor and fumaric acid as a hydrogen acceptor.

Immobilized microorganism technology is increasingly used for producing biochemicals and useful products, such as ethanol and organic
acids *(8)*. However, the industrial application of immobilized cells for
succinic acid production has rarely been reported. Immobilization is the
restriction of cell mobility within a defined space *(9)*. Immobilized cell
cultures have the following potential advantages over suspension culture: high cell concentration, high productivity, cell reuse, and reduced
cost for cell recovery and recycling *(8,10)*. Despite these advantages, the
development of an immobilization technique for succinic acid production has rarely been studied.

Calcium alginate gels are now one of the most widely used supports
for the immobilization of whole microbial cells *(11,12)*. Entrapment of cells
in alginate is one of the simplest methods of immobilization *(13)*. Algi-
nates are available commercially as water-soluble sodium alginate and
have been used for more than 65 yr in the food and pharmaceutical industries as thickening, emulsifying, film forming, and gelling agents *(9)*. Entrapment in insoluble calcium alginate gels is recognized as a rapid, nontoxic,
inexpensive, and versatile method for immobilization of cells *(14)*.

In the present study, *E. faecalis* RKY1 was used for bioconversion of
fumarate to succinate by cell immobilization. Sodium alginate was used as
a supporting material because it forms gels with divalent ions like calcium.
We investigated the optimum conditions for whole-cell immobilization of
E. faecalis RKY1 cells and examined the effect of various culture conditions
on succinate production.

Materials and Methods

Microorganism and Medium

E. faecalis RKY1 was isolated from our laboratory culture with respect
to its ability to convert fumarate to succinate at a high yield *(15)*. The
medium for cell growth contained the following: 10 g of glycerol (Yakuri,
Osaka, Japan), 22 g of fumaric acid (Yakuri), 15 g of yeast extract (Difco,
Detroit, MI), 10 g of K_2HPO_4 (Yakuri), 1 g of NaCl (Junsei, Tokyo, Japan),

20 g of Na_2CO_3 (Yakuri), 0.05 g of $MgCl_2 \cdot 6H_2O$, 0.01 g of $FeSO_4 \cdot 7H_2O$, and 1 L of distilled water. The medium for bioconversion contained the following: 20 g of glycerol, 22 g of fumaric acid, 15 g of yeast extract, 0.5 g of K_2HPO_4, 1 g of NaCl, 20 g of Na_2CO_3, 0.05 g of $MgCl_2 \cdot 6H_2O$, 0.01 g of $FeSO_4 \cdot 7H_2O$, and 1 L of distilled water. The seed culture broth used in this experiment was transferred to the new media every 6 h or 12 h for 2 d. Cell stocks were made by mixing the culture with sterile glycerol and then stored at $-20°C$.

Cell Immobilization

Cells were grown at 38°C for 5 to 6 h and then harvested by centrifugation (Vision Scientific, Taejon, Korea) at 18,600g for 10 min. After discarding the supernatant, the harvested cells were washed twice with 0.1 M Tris-HCl buffer (pH 7.0) and then resuspended to a cell concentration of 2.5 g of dry cell weight/100 mL of buffer solution in 0.1 M Tris-HCl buffer. One hundred milliliters of the cell suspension was mixed with 100 mL of a 1–4% (w/v) sodium alginate (Junsei) solution. This mixture of alginate and cells was dropped into a sterile cold 2% (w/v) $CaCl_2$ (Kanto, Tokyo, Japan) solution using a blunt-ended needles connected to a peristaltic pump. Calcium alginate gel beads of approx 2–4 mm in diameter were obtained by using needles of various diameters. After immobilization, the beads were washed with sterilized distilled water, placed into a 2% (w/v) $CaCl_2$ solution, and gently agitated in order to increase the strength of the beads. They were stored in a 2% (w/v) $CaCl_2$ solution at 4°C until use. The agar (Junsei) and κ-carrageenan (Sigma, St. Louis, MO) gels were prepared according to the methods of Han and Chung (16).

Culture Conditions

Batch cultures for cell growth were conducted in 100-mL vials containing 80 mL of growth medium. These cultures were incubated in a shaking incubator (Vision Scientific) at 200 rpm and 38°C. Bioconversion by free cells was prepared by inoculating 0.6 mL of resting cells into the culture medium (15 mL) in 20-mL vials, followed by incubation at 38°C and 200 rpm for 6 h or 12 h in a shaking incubator. For the bioconversion experiments using the immobilized cells, 10% (v/v) of the immobilized cells was inoculated into 20-mL vials containing 15 mL of medium and 100-mL vials containing 80 mL of medium according to experimental condition. These bioconversions were conducted at 38°C and 150 rpm for 12 h in a shaking incubator. Before the cultivations of free and immobilized cells, the inoculated vials were vacuum degassed for 2 min, and high-purity CO_2 gas was charged in these vials for 2 min to incubate the strain anaerobically.

Analytical Methods

Succinic acid and fumaric acid were quantitatively analyzed by high-pressure liquid chromatography with a pump (Waters 510; Millipore), a

Bio-Rad Aminex HPX-87H ion-exclusion column (7.8×300 mm; Bio-Rad, Hercules, CA), and an ultraviolet (UV) detector (Waters 486; Millipore). The column was eluted with 5 mM sulfuric acid at a flow rate of 0.6 mL/min and column temperature of 35°C. In this study, the amount of succinic acid produced was expressed as sodium succinate (MW 162.14) and residual fumaric acid as sodium fumarate (MW 160.04).

Cell growth of the free cells was measured as the optical density at 660 nm (OD_{660}) with a UV spectrophotometer (UV-160A; Shimadzu, Japan). One optical unit for *E. faecalis* RKY1 is equivalent to 1.118 g of dry cell weight/L. To estimate the effluent cell concentration from immobilized cells, 1 mL of sample was centrifuged at 16,000g for 10 min, the supernatant discarded, and the pellet washed twice with deionized water. The pellet was then resuspended in 1 mL of deionized water. This resuspension was used to determine the OD_{660}.

Results and Discussion

Determination of Gel Matrix

Immobilized cells entrapped in three natural polymers typically used for immobilization of enzymes and cells—agar, κ-carrageenan, and sodium alginate *(9)*—were compared for their conversion ratio and effluent cells. Table 1 gives the effect of each immobilizing carrier on succinate production and effluent cell concentration. The conversion ratio (grams/gram) term is expressed as the value of succinate produced per initial fumarate. It was shown that with an initial fumarate concentration of 30 g/L, about 0.96 g/g (28.9 g/L) was converted to succinate after a 12-h incubation with the alginate-entrapped cells, whereas when the cells were entrapped in the agar or κ-carrageenan gels, only about 0.60 g/g (18.7 and 18.0 g/L, respectively) of fumarate was converted to succinate after the 12-h incubation. Effluent cells from agar, Ca-alginate, and κ-carrageenan gels were 2.3, 0.29, and 0.31 g/L, respectively. Since the agar and κ-carrageenan gels were formed by lowering the temperature, the strength of these gels was somewhat lower and, therefore, the effluent cell concentrations from these gels were higher. Furthermore, these gels have disadvantages in that the preparation of beads with these types of gels is quite difficult and that many of the microbial cells that were immobilized were killed or damaged owing to the high temperature necessary to keep the gel molten. By contrast, Ca-alginate gels have a higher rigidity and the preparation of their beads is quite easy. Thus, sodium alginate was selected for the immobilization of *E. faecalis* RKY1.

Concentration of Sodium Alginate

The matrix for Ca-alginate bead was prepared using a solution of sodium alginate between 1 and 4% (w/v). The beads of 1% (w/v) sodium alginate were not rigid enough and, therefore, were not spherical. The 4% (w/v) sodium alginate solution was difficult to work with because of

Table 1

Effect of Gel Matrix Used for Cell Immobilization
on Bioconversion of Fumarate to Succinate and Effluent Cell Concentration[a]

Type of matrix	Succinate concentration (g/L)	Residual fumarate (g/L)	Conversion ratio (g/g)	Effluent cell concentration (g/L)
Free cell	29.5	0.5	0.983	4.5
Ca-alginate[b]	28.9	1.1	0.965	0.29
Agar[c]	18.7	11.3	0.622	2.3
κ-Carrageenan[c]	18.0	12.0	0.599	0.31

[a]Bioconversion was conducted in a 20-mL vial with 15 mL of medium at 38°C for 12 h. The seed volume of immobilized cells was 10% (v/v). The initial fumarate concentration was 30 g/L. The conversion ratio (g/g) is expressed as the value of succinate produced per initial fumarate.
[b]Bead-type gel of 3 mm in diameter.
[c]Cubic-type gel $3 \times 3 \times 3$ mm.

Table 2

Effect of Sodium Alginate Concentration
on Bioconversion of Fumarate to Succinate and Effluent Cell Concentration[a]

Sodium alginate concentration (% [w/v])	Succinate concentration (g/L)	Residual fumarate (g/L)	Conversion ratio (g/g)	Effluent cell concentration (g/L)
1.0	29.5	0.5	0.983	0.30
2.0	29.2	0.8	0.973	0.27
3.0	28.7	1.3	0.958	0.17
4.0	28.6	1.4	0.954	0.17

[a]Bioconversion was conducted in a 20-mL vial with 15 mL of medium at 38°C for 12 h. The seed volume of immobilized cells was 10% (v/v). The initial fumarate concentration was 30 g/L. The conversion ratio (g/g) is expressed as the value of succinate produced per initial fumarate.

its high viscosity. As shown in Table 2, the 2% (w/v) sodium alginate concentration was determined to be the optimum concentration for efficient bioconversion. When the concentration of the sodium alginate solution was 2% (w/v), about 0.97 g/g of the initial fumarate, 30 g/L, was converted to succinate and the effluent cell concentration was 0.27 g/L. Since an increase in the sodium alginate concentration would reduce the pore size of the Ca-alginate beads, both the conversion ratio and effluent cell concentration would also decrease. Klein et al. (11) also reported that as the pore size of Ca-alginate beads decreased, the mass transfer rate within the beads was decreased accordingly.

CaCl$_2$ Concentration

CaCl$_2$ must be added to the medium in order to harden the Ca-alginate beads during cultivation (12). Therefore, the effect of the amount of CaCl$_2$

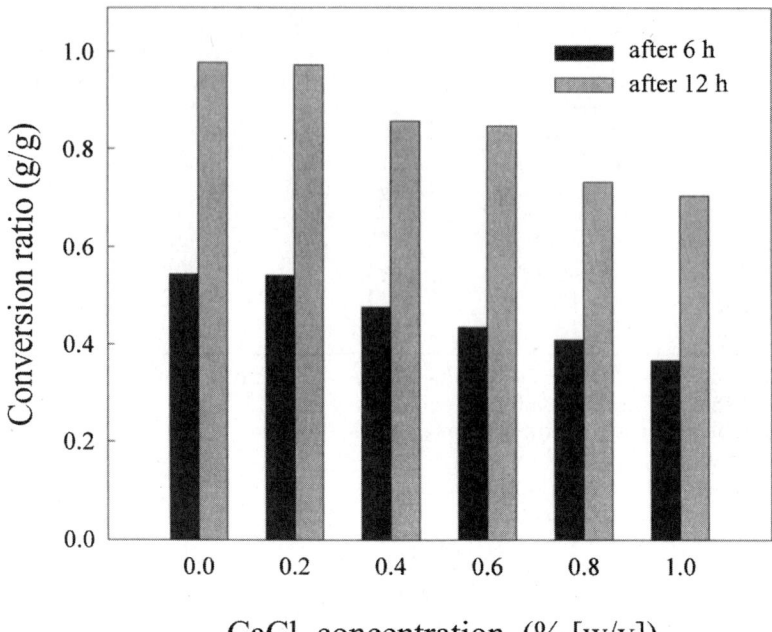

Fig. 1. Effect of CaCl$_2$ concentration added to the medium on bioconversion of fumarate to succinate by alginate-entrapped *E. faecalis* RKY1. Bioconversion was conducted in a 20-mL vial with 15 mL of medium at 38°C for 6 h and 12 h. The seed volume of immobilized cells was 10% (v/v), and the initial fumarate concentration was 30 g/L.

added to the medium on the conversion ratio was investigated. The CaCl$_2$ concentration used was between 0.2 and 1.0% (w/v). As shown in Fig. 1, the conversion ratio decreased slightly as the CaCl$_2$ concentration added in the medium increased after both 6 and 12 h of incubation with the conversion ratios for 0.2% CaCl$_2$ being 0.54 and 0.97 g/g, respectively. As the CaCl$_2$ concentration increased above 0.25% (w/v), however, the conversion ratio gradually decreased further.

Seed Volume of Immobilized Cells

To investigate the effect of the seed volume of the immobilized cells, 5, 10, 20, 30, 40, and 50% (v/v) samples were examined at 38°C for 12 h in 20-mL vials with 15 mL of medium; Figure 2 presents the results. The optimum seed volume of immobilized cells was found to be 10% (v/v) with a conversion ratio of 0.97 g/g. As the seed volume of the immobilized cells was increased further, above 10% (v/v), the conversion ratio did not change much during the first 6 h of incubation but significantly decreased after 12 h. These results suggest that the immobilized cells at the higher concentrations consumed more substrate for cellular maintenance and that adjustment of the seed volume is essential for efficient bioconversion. According to these results, the most appropriate seed volume of the immobilized cells was 10% (v/v).

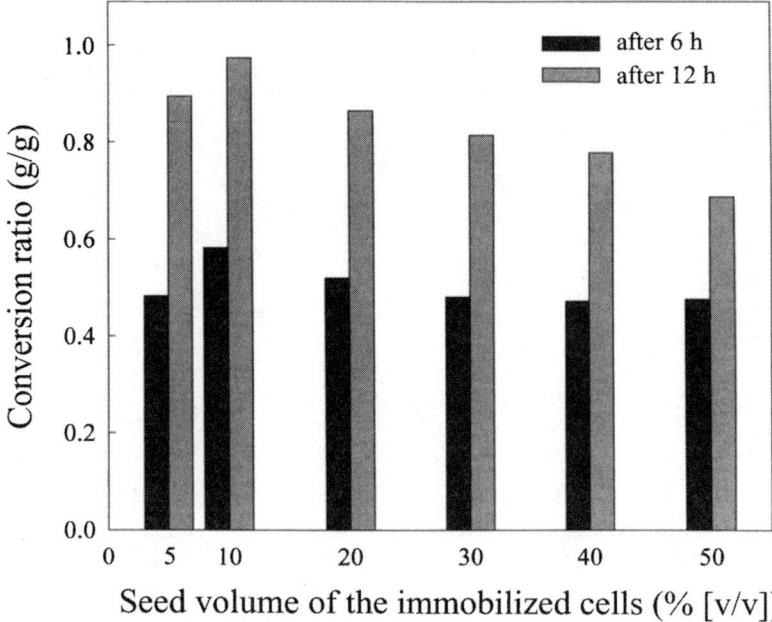

Fig. 2. Effect of seed volume of immobilized cells on bioconversion of fumarate to succinate by alginate-entrapped *E. faecalis* RKY1. Bioconversion was conducted in a 100-mL vial with 80 mL of medium at 38°C for 6 and 12 h. The initial fumarate concentration was 30 g/L.

Immobilized Bead Size

The effect of the immobilized bead diameter on succinate production and fumarate consumption was examined. The bead diameter used in this experiment was 2, 3, and 4 mm; Figure 3 presents the results. The succinate produced using beads with a 2 and 3 mm diameter was 29.2 and 28.9 g/L (a 0.973 and 0.965 g/g conversion ratio), respectively, after a 12-h incubation. When the immobilized bead diameter was 4 mm, the succinate concentration and conversion ratio after a 12-h incubation were 26.7 g/L and 0.89 g/g, respectively, showing that the conversion ratio of the 4-mm bead was lower than that of the 2- and 3-mm beads. Aksu and Bülbül (8) reported that when *Pseudomonas putida* cells were entrapped in Ca-alginate gels for the degradation of phenol, as the diameter of the immobilized bead was increased the degradation rate of phenol decreased owing to the diffusion limitation of substrates and products. According to our study, the optimum diameter for the Ca-alginate bead was found to be 2 mm.

Initial pH and Temperature

The effect of the initial pH on the conversion ratio with free and immobilized cells was examined. Bioconversion was conducted at 38°C for 12 h under anaerobic conditions and with a seed volume of 10% (v/v). The conversion ratios of the free-cell culture in the pH range of 5.0–8.5 were

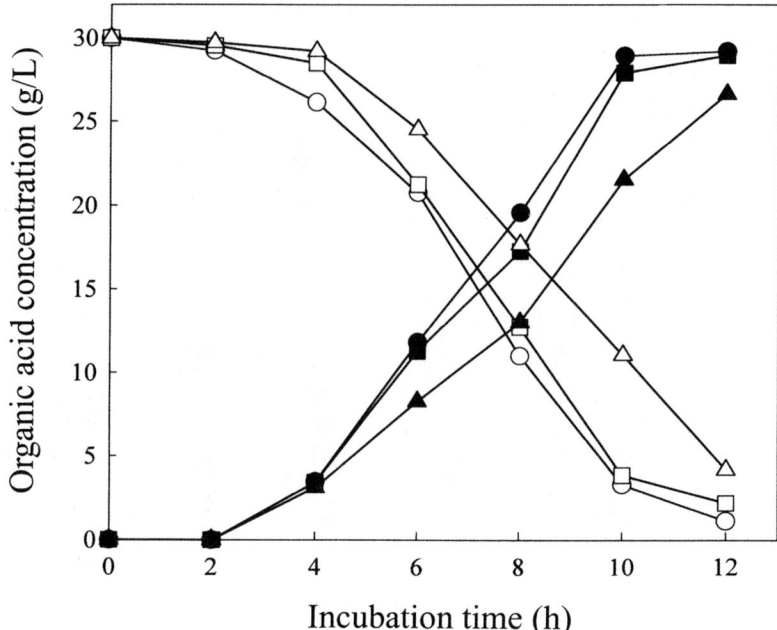

Fig. 3. Effect of immobilized bead size on succinate production and fumarate consumption. Bioconversion was conducted in a 100-mL vial with 80 mL of medium at 38°C for 12 h. The seed volume of immobilized cells was 10% (v/v), and the initial fumarate concentration was 30 g/L. Solid symbols indicate the succinate produced and open symbols the residual fumarate (—●—, —○—, 2 mm; —■—, —□—, 3 mm; —▲—, —△—, 4 mm).

similar to those of immobilized cell culture with the optimum pH for both the free and immobilized bioconversions being 7.0. However, the conversion ratio for the immobilized cells was higher than that of the free cells for all pH values tested.

To determine the effect of temperature on succinate production, experiments were conducted at various temperatures for 12 h under anaerobic conditions with a 10% seed volume. The best succinate conversion after a 12-h incubation was seen at 38°C for both the free and immobilized cells (0.983 g/g for the free cells and 0.973 g/g for the immobilized cells). Above 40°C, however, the immobilized cells were able to produce succinate more efficiently than the free cells. When the cells were incubated at 50°C, the conversion ratio of the free and immobilized cells were 0.121 and 0.38 g/g, respectively. These results suggest that the Ca-alginate gel provides a protective barrier against heat transfer, as was reported elsewhere *(12)*. Krisch and Szajani *(17)* reported that when *Acetobacter aceti* cells were immobilized in Ca-alginate gels, the immobilized cells produced more acetic acid than free cells at higher temperatures.

Storage Stability of Immobilized Cells

The immobilized cells were stored in a 2% (w/v) $CaCl_2$ solution at 4°C for 10 wk to investigate their storage stability. As shown in Fig. 4, the

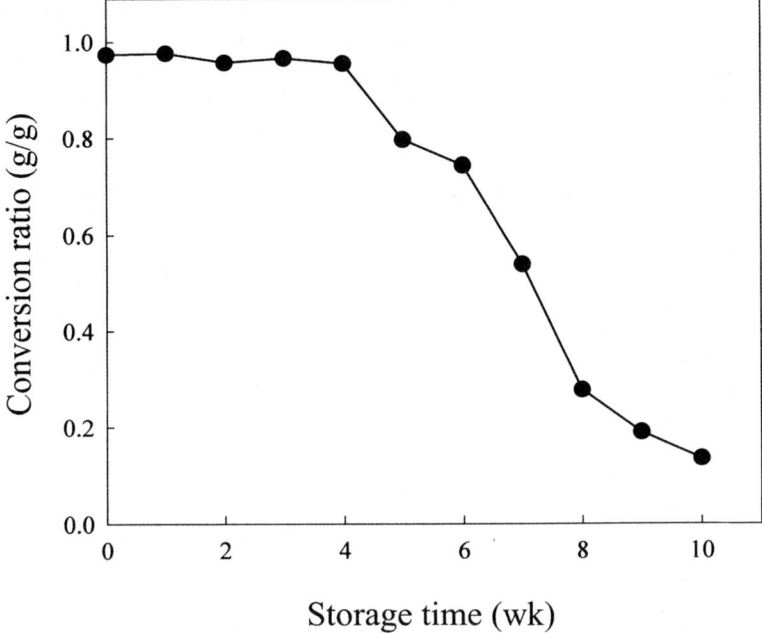

Fig. 4. Storage stability of immobilized cultures of *E. faecalis* RKY1. The immobilized cells were stored in a 2% (w/v) CaCl$_2$ solution until used. Bioconversion was conducted in a 20-mL vial with 15 mL of medium at 38°C for 12 h per each batch. The seed volume of immobilized cells was 10% (v/v), and the initial fumarate concentration was 30 g/L.

conversion ratio for a 12-h incubation after storage for 4 wk was >0.95 g/g. When the immobilized cells were stored for 5 or 6 wk, the conversion ratio decreased by about 0.20 g/g of its initial value. When stored for 10 wk, the conversion ratio of immobilized cells decreased to only 0.14 g/g. Therefore, *E. faecalis* RKY1 cells entrapped in Ca-alginate gels can be stored for up to 4 wk without noticeable loss of cellular activity.

Repetitive Batch Operation

The operational stability of the immobilized cells was examined by repetitive batch bioconversions. Each batch reaction was conducted at 38°C in a 100-mL vial containing 80 mL of medium and under anaerobic conditions. After one batch was finished, the Ca-alginate beads were removed and washed using the sterilized mesh and water and then inoculated into fresh medium. The initial concentration of fumarate used was 30 g/L, and the time for one batch was set at 12 h; Figure 5 presents the results. Immobilized cells were repeatedly used for five batches, or about 60 h. Until the third batch bioconversion, the conversion ratio was above 0.80 g/g. However, it decreased to 0.77 and 0.57 g/g by the fourth and fifth batches, respectively. These results suggest that as the batch bioconversion was

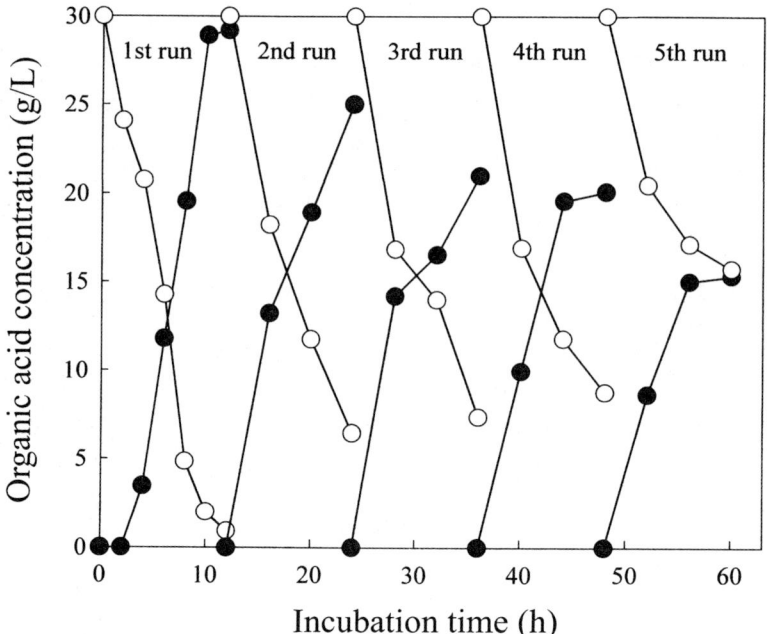

Fig. 5. Repetitive batch bioconversion profiles for cultures of immobilized *E. faecalis* RKY1. Repetitive bioconversion was carried out in a 100-mL vial with 80 mL of medium at 38°C for 12 h per each batch. The seed volume of immobilized cells was 10% (v/v), and the initial fumarate concentration was 30 g/L. (—●—), Succinate produced; (—○—), residual fumarate.

repeated, the *E. faecalis* RKY1 cells entrapped in the Ca-alginate gels seemed to be damaged by the low pH values in the matrix at the end of each batch.

Conclusion

E. faecalis RKY1 cells, a newly isolated strain, were immobilized in various gelling media in order to characterize the bioconversion of fumarate to succinate through the use of immobilized cells. Among the various natural polymers tested—agar, κ-carrageenan, and sodium alginate— sodium alginate was the best material for whole-cell immobilization of *E. faecalis* RKY1. An enhanced production of succinate was achieved when the sodium alginate concentration and the diameter of the calcium alginate beads were adjusted to 2% (w/v) and 2 mm, respectively. The results with *E. faecalis* RKY1 showed that immobilization can also enhance the stability of cells and protect them from the effects of alterations in pH and temperature. The efficient bioconversion of fumarate to succinate was demonstrated through immobilization of *E. faecalis* RKY1 in alginate.

Acknowledgment

We wish to acknowledge the financial support of the Korea Research Foundation made in the program year of 1998 (KRF, 1998-020-E00016).

References

1. Gottschalk, G. (1986), *Bacterial Metabolism*, 2nd ed., Springer-Verlag, New York.
2. Lee, P. C., Lee, W. G., Kwon, S., Lee, S. Y., and Chang, H. N. (1999), *Enzyme Microb. Technol.* **24,** 549–554.
3. Datta, R. (1992), US patent 5,143,833.
4. Ryu, H. W., Kang, K. H., and Yun, J. S. (1999), *Appl. Biochem. Biotechnol.* **77/79,** 511–520.
5. Zeikus, J. G., Jain, M. K., and Elankovan, P. (1999), *Appl. Microbiol. Biotechnol.* **51,** 545–552.
6. Wang, X., Gong, C. S., and Tsao, G. T. (1998), *Appl. Biochem. Biotechnol.* **70/72,** 919–928.
7. Kang, K. H. and Ryu, H. W. (1999), *J. Microbiol. Biotechnol.* **9,** 191–195.
8. Aksu, Z. and Bülbül, G. (1999), *Enzyme Microb. Technol.* **25,** 344–348.
9. Bickerstaff, G. F. (1997), in *Immobilization of Enzymes and Cells*, Bickerstaff, G. F., ed., Humana, Totowa, NJ, pp. 1–11.
10. Büyükgüngor, H. (1992), *J. Chem. Technol. Biotechnol.* **53,** 173–175.
11. Klein, J., Stock, J., and Vorlop, K.-D. (1983), *Appl. Microbiol. Biotechnol.* **18,** 86–91.
12. Smidsrød, O. and Skjak-Bræk, G. (1990), *Trends Biotechnol.* **8,** 71–78.
13. Kierstan, M. and Bucke, C. (1977), *Biotechnol. Bioeng.* **11,** 387–397.
14. Nilsson, K. (1987), *Trends Biotechnol.* **5,** 73–78.
15. Ryu, H. W., Yun, J. S., and Kang, K. H. (1998), *Kor. J. Appl. Microbiol. Biotechnol.* **26,** 545–550.
16. Han, M. S. and Chung, D. H. (1992), *J. Microbiol. Biotechnol.* **20,** 459–469.
17. Krish, J. and Szajani, B. (1996), *Biotechnol. Lett.* **18,** 393–396.

Modeling and Simulation of Cephalosporin C Production in a Fed-Batch Tower-Type Bioreactor

RENATA M. R. G. ALMEIDA,[1] ANTONIO J. G. CRUZ,*,[1]
MARIA LUCIA G. C. ARAUJO,[2] ROBERTO C. GIORDANO,[1]
AND CARLOS O. HOKKA[1]

[1]Departamento de Engenharia Química da Universidade Federal
de São Carlos, Cx. Postal 676 CEP 13565-905, São Carlos, SP, Brazil,
E-mail: ajgcruz@deq.ufscar.br; and [2]Departamento de Bioquímica
e Tecnologia Química, Instituto de Química–UNESP,
Campus de Araraquara, Cx. Postal 355, CEP 14801-970,
Araraquara, SP, Brazil

Abstract

Immobilized cell utilization in tower-type bioreactor is one of the main alternatives being studied to improve the industrial bioprocess. Other alternatives for the production of β-lactam antibiotics, such as a cephalosporin C fed-batch process in an aerated stirred-tank bioreactor with free cells of *Cephalosporium acremonium*, or a tower-type bioreactor with immobilized cells of this fungus, have proven to be more efficient than the batch process. In the fed-batch process, it is possible to minimize the catabolite repression exerted by the rapidly utilization of carbon sources (such as glucose) in the synthesis of antibiotics by utilizing a suitable flow rate of supplementary medium. In this study, several runs for cephalosporin C production, each lasting 200 h, were conducted in a fed-batch tower-type bioreactor using different hydrolyzed sucrose concentrations. For this study's model, modifications were introduced to take into account the influence of supplementary medium flow rate. The balance equations considered the effect of oxygen limitation inside the bioparticles. In the Monod-type rate equations, cell concentrations, substrate concentrations, and dissolved oxygen were included as reactants affecting the bioreaction rate. The set of differential equations was solved by the numerical method, and the values of the parameters were estimated by the classic nonlinear regression method following Marquardt's procedure with a 95% confidence interval. The simulation results showed that the proposed model fit well with the experimental data, and based on the

*Author to whom all correspondence and reprint requests should be addressed.

experimental data and the mathematical model, an optimal mass flow rate to maximize the bioprocess productivity could be proposed.

Index Entries: Cephalosporin C; tower-type bioreactor; fed-batch; modeling; simulation.

Introduction

The β-lactam family of antibiotics is the most important group among pharmaceutical products and makes up the largest part of the world's multibillion-dollars antibiotic market. Approximately 60% of the total worldwide production of antibiotics belongs to the β-lactam type *(1)*. Cephalosporin C is one of the most important antibiotics in this group. Its molecule is synthesized and produced as a secondary metabolite by strains of a strictly aerobic filamentous fungus, *Cephalosporium acremonium*. The cephalosporin C in its natural form has a relatively low antibiotic activity. However, its molecule can be modified through chemical or enzymatic methods to produce different semisynthetic cephalosporins, which have clinical use and a high market worth and are important to pharmaceutical industries *(2)*.

Industrial production of cephalosporin C is still carried out via conventional batch fermentation in aerated stirred-tank bioreactors utilizing submerged cultures of *C. acremonium*. The unfortunate feature of this fungus growth is the high viscosity of the culture growth, which results in an increase in energy costs in order to keep the dissolved oxygen concentration above the critical level. Tower-type bioreactor without mechanical agitation have appeared as an alternative process and are improving the production of cephalosporin C and reducing these costs *(3)*. Furthermore, there is a trend to use immobilized cells in beads of inert and biocompatible materials such as natural pellets in this type of bioreactor. Despite the mass transfer limitation, the whole-cell immobilization has the advantages of minimizing the broth viscosity and allowing greater cell longevity in continuous and semicontinuous processes such as the fed-batch process *(4)*.

The fed-batch process with a continuous or intermittent nutrient feed rate has been used to avoid high concentrations of some substrates that are inhibitors or cause undesirable precipitation. It is also used to regulate the metabolism of the microorganism, thus improving product formation. Regulatory mechanisms for cephalosporin C production in this process include catabolite repression of the β-synthetases exerted by rapidly utilized carbon sources such as glucose. These carbon sources are essential for cell growth but are prejudicial for antibiotic production *(5)*. The fed-batch process to produce cephalosporin C has been shown to be more efficient than the batch process that minimizes the repressive effects exerted by glucose and utilizes a suitable flow rate of supplementary medium.

Regarding modeling and simulating cephalosporin C process, some important studies should be mentioned. Matsumura et al. *(6)* developed a kinetic model of cephalosporin C production. The proposed model was based on morphologic differentiation of the fungus *C. acremonium*, on the fact that methionine strongly stimulated cephalosporin C production, and on the catabolite repression exerted by glucose. The model proposed by Chu and Constantinides *(7)* for cephalosporin C production considered that antibiotic formation should be directly associated with the production of enzymes that affect the synthesis of this compound. Araujo et al. *(8)* proposed a kinetic model for cephalosporin C production with free and immobilized cells based on stoichiometric equations that represent the main phases of cell variation, substrate, and product concentrations. Araujo et al. *(9)* studied the effect of oxygen transfer on the effectiveness factor of the process rates of cephalosporin C production with immobilized cells. The study concluded that although oxygen limits the production rate, it is only slightly lower than the free-cell system. Cruz et al. *(10)* elaborated a mathematical model in fed-batch production of cephalosporin C that uses the main features of all the aforementioned models.

In general, the mathematical models used in bioprocesses are formed by a set of nonlinear differential equations with several unknown parameters. These parameters should be estimated using experimental results. Marquardt *(11)* developed the classic algorithm to estimate parameters by the nonlinear least-squares method. Nihtilä and Virkkunen *(12)*, using Marquardt's method, developed parameters for models that described the growth of and glucose consumption by *Trichoderma viride*. They observed difficulties in obtaining acceptable parametric values owing to the large number of parameters to be estimated in relation to the amount of experimental data. Matsumura et al. *(6)* estimated the kinetic parameters of a cephalosporin C production model. First, they estimated growth phase parameters and then the production phase parameters. Zangirolami et al. *(13)* formulated a structural model for fed-batch penicillin production. In each parametric estimate, the experimental data of the cell, penicillin, and substrate concentration were compared with the set of data obtained in the simulation.

In the present study, the model proposed by Araujo et al. *(8)* was modified to simulate cephalosporin C production in a fed-batch tower-type bioreactor with immobilized cells of *C. acremonium* and took into account the influence of supplementary medium flow rate. The balance equations described the reactions involved in this process, taking into consideration the effect of oxygen limitation inside the bioparticles. In the Monod-type rate equations, in addition to cell and substrate concentrations, dissolved oxygen was included as a reactant affecting the bioreaction rate. The set of differential equations was solved by the numerical method and the values of the parameters were estimated by the classic nonlinear regression method following Marquardt's procedure *(11)* with a 95% confidence interval. The simulation results show that the proposed model fit the experimental data, and based on the experimental data and the mathematical model, an optimal mass flow rate maximizing the bioprocess productivity could be proposed.

Table 1
Hydrolyzed Sucrose Concentrations

Run	Hydrolyzed sucrose (g/L)
4	115.2
5	86.4
6	144.0
7	107.5

Materials and Methods

Microorganism

C. acremonium ATCC 48272 (C-10) was used throughout this study.

Culture Media

Inoculum Preparation

A synthetic medium containing the following components was used: glucose (30.0 g/L); ammonium acetate (8.8 g/L); DL-methionine (5.0 g/L); oleic acid (1.5 g/L); $CaCO_3$ (2.0 g/L); KH_2PO_4 (2.0 g/L); and traces of inorganic salts $Fe(NH_4)_2(SO_4)_2 \cdot 6H_2O$ (0.16 g/L), Na_2SO_4 (0.81 g/L), $MgSO_4 \cdot 7H_2O$ (0.384 g/L), $CaCl_2 \cdot 2H_2O$ (0.08 g/L), $MnSO_4 \cdot H_2O$ (0.032 g/L), $ZnSO_4 \cdot 7H_2O$ (0.032 g/L), $CuSO_4 \cdot 5H_2O$ (0.002 g/L). The pH was adjusted to 7.0 ± 0.1.

Main Fermentation

A synthetic medium defined by Demain et al. *(14)* and modified by Araujo et al. *(8)* containing the following components was used: glucose (27.0 g/L), DL-methionine (3.0 g/L), KH_2PO_4 (1.5 g/L), $CaCl_2$ (1.0 g/L), and other components used in the inoculum preparation. The pH was adjusted to 7.0 ± 0.1.

Supplementary Medium

Following glucose exhaustion, a supplementary medium containing hydrolyzed sucrose (glucose + fructose) and the same nutrients as the initial batch, except glucose, was added continuously. The sucrose was previously hydrolyzed at 40°C in 10^{-2} M sodium acetate buffer solution at a pH of 4.5 *(15)* and utilized invertase enzyme (Novo Ferment). The hydrolyzed sucrose concentration was different for each run. Table 1 gives the concentrations used.

Immobilization

The gel used for immobilization was composed of sodium alginate (20.0 g/L) and alumina (15.0 g/L); these components were mixed with the cells from the inoculum. Details of the preparation are found in Araujo et al. *(9)*.

Analysis

Glucose Concentration

Glucose concentration was measured using the enzymatic glucose oxidase method.

Cell Concentration

Free and immobilized cell-mass concentration was evaluated as dry weight (grams/liter) at 105°C for 24 h and as volatile suspended solids (grams/liter) at 600°C for 1.5 h.

Cephalosporin C Concentration

Cephalosporin C titers were determined by high-performance liquid chromatography.

Procedure

Fed-batch experiments were carried out in a 1.6-L tower bioreactor and lasted approx 200 h. Temperature was maintained at 26°C and air flow rate at 170 L/h (21.1°C and 1 atm). All runs started with 1.4 L of main fermentation culture medium. After glucose depletion, supplementary medium was added using a peristaltic pump at established flow rates. The same volumetric flow rate of 1.8 mL/h was maintained for all runs, and the invert sugar concentration varied for each run (Table 1). Samples were withdrawn periodically to determine pH, biomass, glucose, and antibiotic concentrations. Figure 1 shows a bioreactor during the fermentative process.

Mathematical Model

A kinetic model based on the model proposed by Araujo et al. *(8)* was used to describe the behavior of the fed-batch process. In this model, modifications were introduced to take into account the influence of supplementary medium flow rate. The stoichiometric equations developed in the present study are based on those proposed by Cruz et al. *(10)*. These equations were adapted to represent the process with immobilized cells, and they are composed of kinetic expressions of *C. acremonium* in this culture medium and consider the mass transfer through the gel particle. The balance equations also considered the effect of oxygen limitation inside the bioparticles. In the Monod-type rate equations, in addition to cell and substrate concentrations, dissolved oxygen was included as a reactant affecting the bioreaction rate.

In this model, it was assumed that during the growth phase and glucose consumption the cells under catabolite repression (cells X_1) were able to produce enzymes responsible for biomass formation. When glucose concentrations fall below a certain critical value (C_{s1c}), cells X_1 are transformed into derepressed cells (X_2). As pointed out by Chu and Constantinides *(7)*, these cells are able to produce large amounts of the enzymes. (*E*) responsible for cephalosporin C synthesis. Antibiotic formation is regulated by

Fig. 1. **(A)** Bioreactor photograph during the fermentative process; **(B)** bioreactor detail.

reactions catalyzed by these enzymes. It was assumed that cephalosporin C and these enzymes suffer degradation. The stoichiometric equations and mass balance equations were described by Almeida et al. *(16)*. The differential equations were solved by numerical method. Regarding estimation of parameters, the classic nonlinear regression method following Marquardt's procedure *(11)* with 95% confidence interval was applied.

Results and Discussion

The model is composed of 6 differential equations, 6 state variables, and 16 parameters. Three of the five state variables were measured: total immobilized cell (C_x), glucose (C_{s1}) and cephalosporin C (C_p) concentrations. The growth phase and glucose consumption parameters were optimized for all fed-batch runs. The estimated parameters were μ_{max1}, k_x, Y_{xs}, k_{d2}, and m. Table 2 gives the parameters optimized in this process. Some parameters, such as k_{d3}, k_{d4}, and β, were determined by fitting to the experimental data. The parameters k_T and k_1 are related to morphologic differentiation present in the microorganism, and the adopted value was that used by Cruz et al. *(17)*. The parameters related to respiration and oxygen concentration were based on those estimated by Araujo et al. *(8)*. The coefficient α must be estimated empirically for each run because the relationship between the productivity and mass flow rate of glucose varied for each fed-batch run (Eq. 1). Table 3 gives estimated values of α.

Table 2
Estimated Parameters for Cephalosporin C Production Process
in Fed-Batch Tower-Type Bioreactor with Immobilized Cells of *C. acremonium*

Cell, glucose, and product parameters	Symbols	Values
Maximum specific growth rate[a]	μ_{max1} (h^{-1})	0.04875 ± 0.0018
Contois constant[a]	k_{x1} (g S_1/g X_1)	0.04381 ± 0.0078
Yield coefficient[a]	Y_{xs1} (g X_1/g S_1)	0.5554 ± 0.0609
Death rate constant of cells X_1	k_{d1} (h^{-1})	0.0
Death rate constant of cells X_2[a]	k_{d2} (h^{-1})	0.0064 ± 0.0009
Kinetic constants	k_T (g X_1/L·h)	6.01
(morphologic differentiation	K_1 (g X_1/L)	0.01
rate constant)		
Decomposition rate of enzyme	k_{d3} (h^{-1})	0.002
Decomposition rate of cephalosporin C	k_{d4} (h^{-1})	0.002
Maintenance coefficient[a]	m (g S_1/g X_2·h)	0.0150 ± 0.0041
Empirical coefficient β	β (–)	0.010
Maximum specific respiration rate[b]	R_{max} (mmol O_2/g X·h)	1.055 ± 0.219
Kinetic constant (respiration rate)[b]	K_{O2} (mmol O_2/L)	0.00277 ± 0.00073
Volumetric mass transfer coefficient	$K_L a$ (h^{-1})	100
Oxygen concentration of saturation	C_L^a (mmol O_2/L)	0.22
Critical glucose concentration	C_{s1c} (g S_1/L)	0.8

[a]This parameter was estimated using nonlinear regression with 95% confidence limits.
[b]This parameter was estimated by Araujo et al. *(1)*.

Table 3
Values of Empirical Coefficient α

Run	Coefficient α
4	16.0
5	19.0
6	10.0
7	15.0

$$\varepsilon_{gel} \cdot \frac{\partial C_P}{\partial t} = \frac{1}{r} \frac{\partial}{\partial r}\left(r^2 \cdot De_P \frac{\partial C_P}{\partial r}\right) + \beta \cdot C_{X2} \cdot \frac{R_{O_2}}{R_{max}} \cdot (\alpha \cdot \mu \cdot C_{X2} - k_{d3} \cdot C_E) - k_{d4} \cdot C_P \quad (1)$$

Several runs with different sugar mass feed rates were carried out according to the proposed model, and parameters estimate that it was possible to simulate the glucose consumed, cell growth, and product formation for all runs. The experimental data were compared with the simulated curves. Figure 2 shows the experimental results and the simulated curve of run 4 (115.2 g/L of hydrolyzed sucrose).

Figure 3 illustrates the experimental results and the simulated curve of run 5 (86.4 g/L of hydrolyzed sucrose). Figures 4 and 5 present the

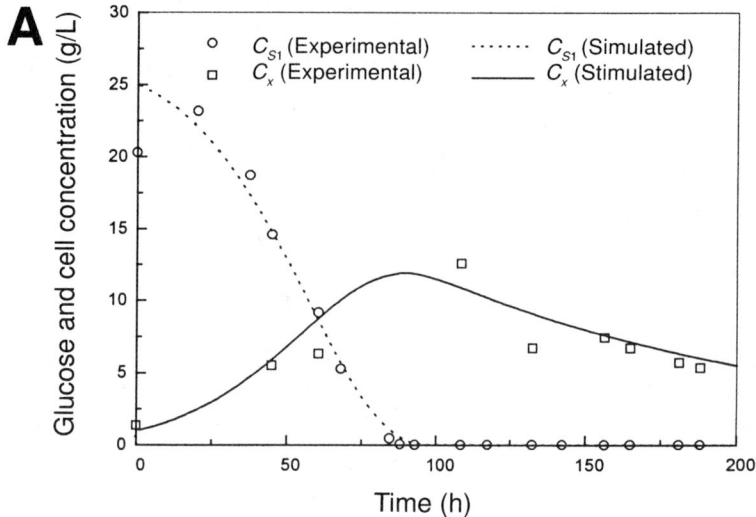

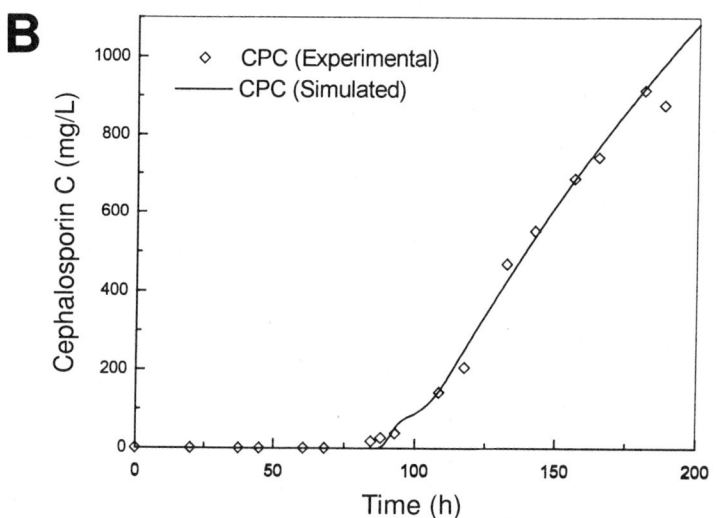

Fig. 2. Experimental data and simulation results of cephalosporin C fed-batch production: **(A)** glucose consumed and biomass formation; **(B)** product formation during run 4 (115.2 g/L of hydrolyzed sucrose). CPC, cephalosporin C.

experimental results and the simulated curve of run 6 (144.0 g/L of hydrolyzed sucrose) and run 7 (107.5 g/L of hydrolyzed sucrose), respectively.

The simulated results in Figs. 2–5 show that the proposed model fit well to the experimental data for glucose concentration, cell mass, and cephalosporin C concentration in the broth. It was observed that run 4, with 115.2 g/L of hydrolyzed sucrose and 1.8 mL/h of volumetric flow rate, led to the higher cephalosporin C production. In runs 5 and 7, when the hydrolyzed sucrose concentrations were 25 and 7% smaller than in run 4, respectively, the microorganism was able to produce antibiotic for 65 h after the

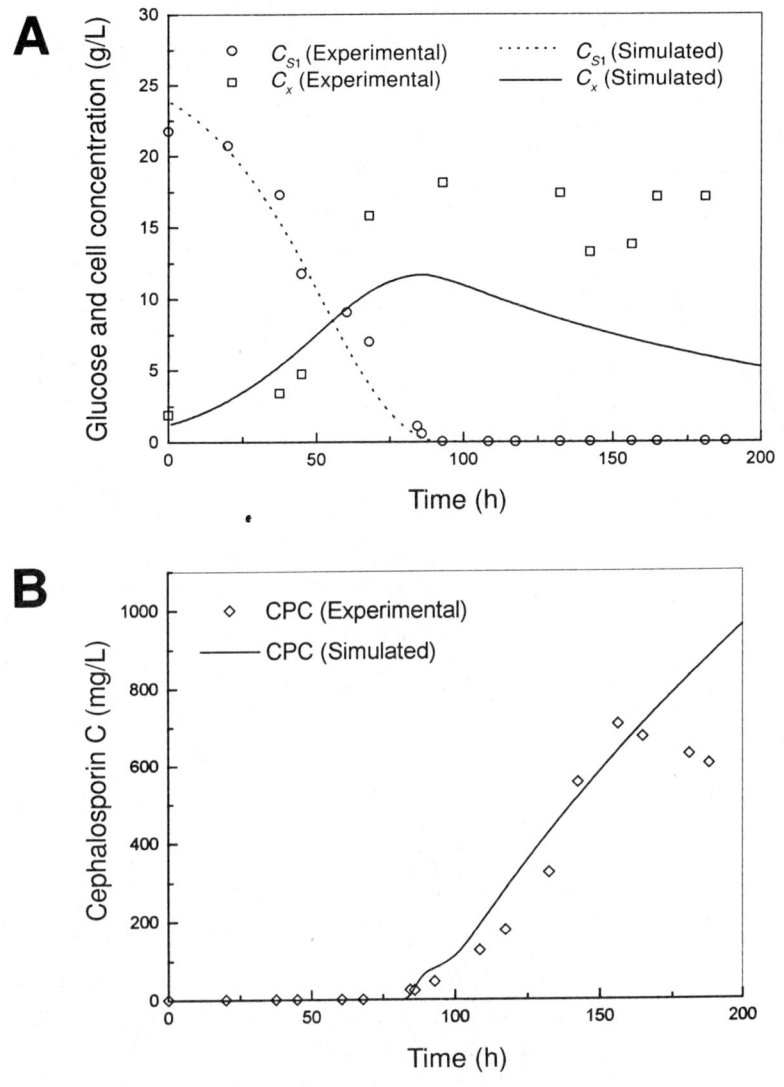

Fig. 3. Experimental data and simulation results of cephalosporin C fed-batch production: **(A)** glucose consumed and biomass formation; **(B)** product formation during run 5 (86.4 g/L of hydrolyzed sucrose). CPC, cephalosporin C.

beginning of the feed. Consequently, glucose concentration had little increase in these runs, as observed in Fig. 6.

Note the inflection that occurs in the product concentration curve at the beginning of the idiophase (*see* Fig. 2B at approx 90 h, Fig. 3B at approx 90 h, Fig. 4B at approx 100 h, and Fig. 5B at approx 75 h). This behavior is owing to the mass transfer diffusion intraparticle resistance. Glucose concentrations inside the gel particles are lower than in the broth, and this triggers the beginning of the production of antibiotic. When the supplementary feed is turned on, there is a dilution effect, which is over-

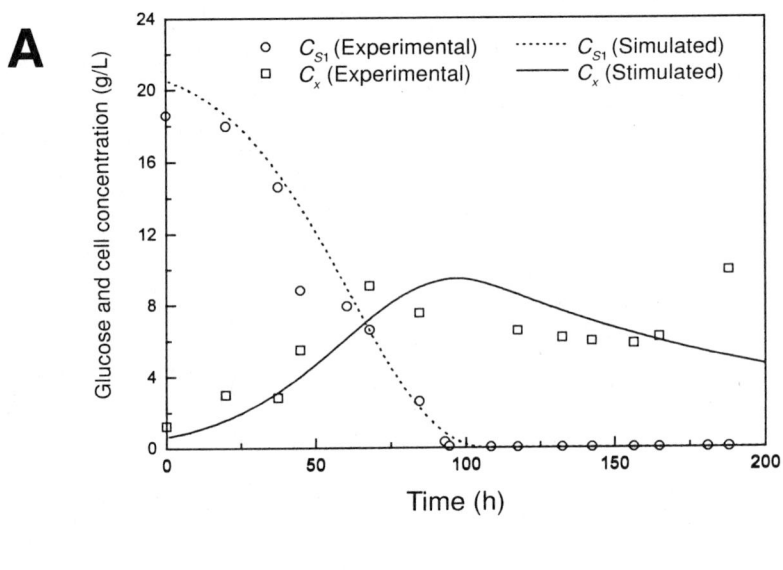

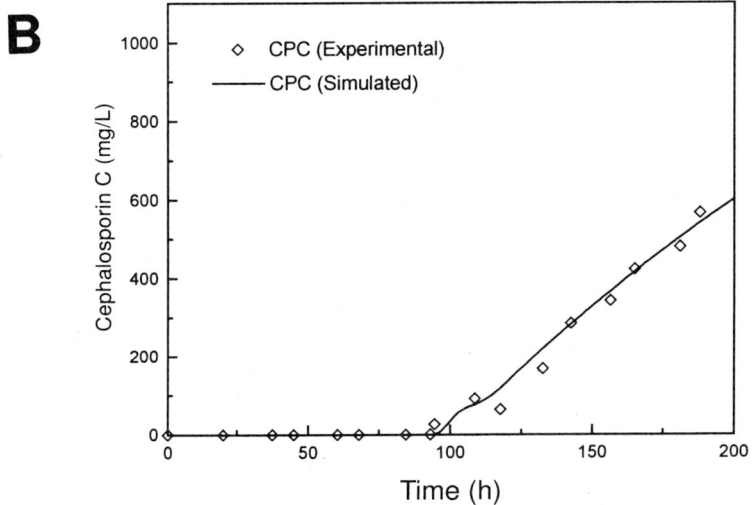

Fig. 4. Experimental data and simulation results of cephalosporin C fed-batch production: **(A)** glucose consumed and biomass formation; **(B)** product formation during run 6 (144.0 g/L of hydrolyzed sucrose). CPC, cephalosporin C.

come by the increasing rate of cephalosporin C production. The model has a response consistent with this phenomenon.

The model did not explain the decline in the production observed in runs 5 and 7. Probably, in these runs a shortage of carbon source occurred, causing metabolic damage in cells. Adding supplementary medium in a crescent flow rate would minimize these damages.

In run 6, when the hydrolyzed sucrose concentration used was 25% higher than for run 4, product degradation was not observed. However, a slightly lower production rate of cephalosporin C compared with run 4 was

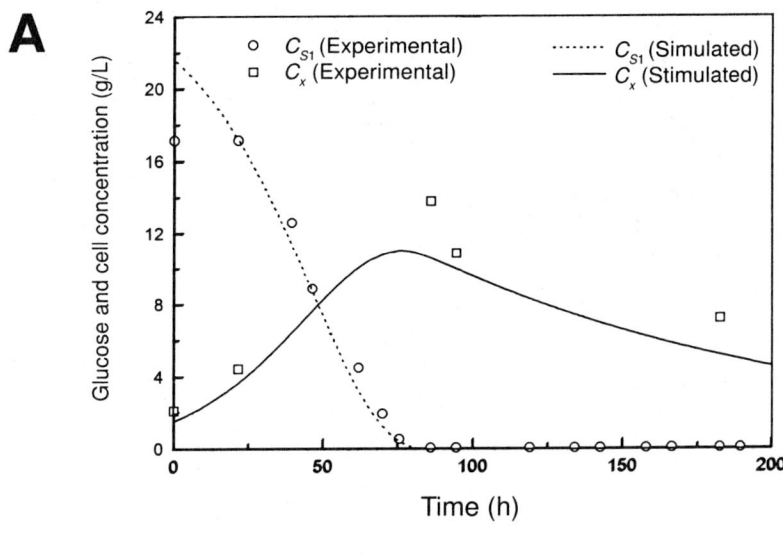

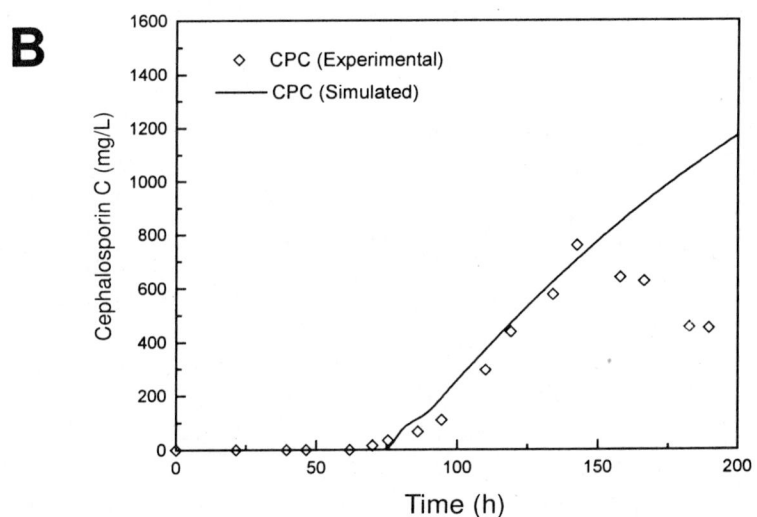

Fig. 5. Experimental data and simulation results of cephalosporin C fed-batch production: **(A)** glucose consumed and biomass formation; **(B)** product formation during run 7 (107.5 g/L of hydrolyzed sucrose). CPC, cephalosporin C.

observed, indicating that catabolite repression begins to exert its influence on the process.

Conclusion

It was evident from the results that the model proposed to simulate the fed-batch process of cephalosporin C production in a tower-type bioreactor fit the experimental data satisfactorily and providing valuable information for further process optimization.

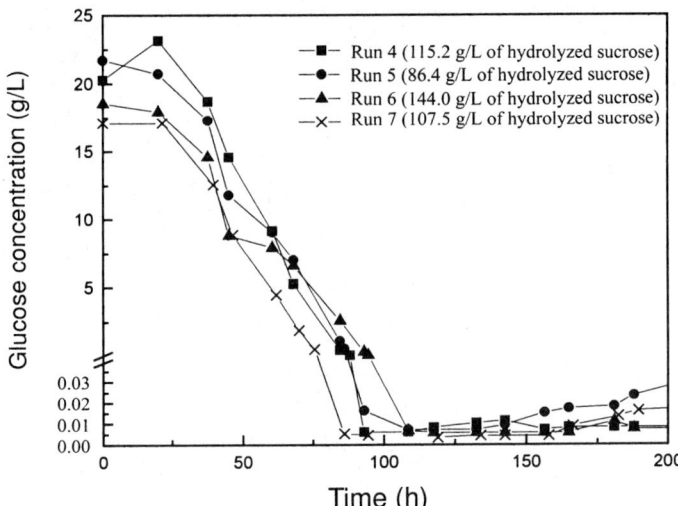

Fig. 6. Experimental results of glucose concentration in runs 4, 5, 6, and 7 with 115.2, 86.4, 144.0, and 107.5 g/L of hydrolyzed sucrose, respectively.

In runs 5 and 7 it was observed that the low sugar concentration in the supplementary medium minimized the repressive effect exerted by glucose, resulting in a high production of antibiotic in the beginning of the feed. After a certain interval of time, however, there was a shortage of carbon source, and the microorganism was no longer able to synthesize the antibiotic. To overcome this problem, the use of a crescent flow rate of supplementary medium or a small increase in sugar concentration is suggested.

The model was a good fit for the concentration profiles of cephalosporin C for runs 4 and 6 but could not predict the decline observed at the end of runs 5 and 7. We are presently working on a hybrid neural network algorithm to infer on-line the concentration of the antibiotic, with the purpose of optimizing fermentation time and interrupting the process when the maximum yield is achieved.

Acknowledgments

We gratefully acknowledge the financial support of CNPq and FAPESP (Process 96/5918-9 and 96/03170-7), respectively.

References

1. Ghosh, A. C., Bora, M. M., and Dutta, N. N. (1996), *Bioseparation* **6**, 91–105.
2. Bayer, T., Zhou, W., Holzhauer, K., and Schügerl, K. (1989), *Appl. Microbiol. Biotechnol.* **30**, 26–33.
3. König, B., Schügerl, K., and Seewald, C. (1982), *Biotechnol. Bioeng.* **24**, 259–280.
4. Kundu, S., Mahapatra, A. C., Srivastava, P., and Kundu, K. (1992), *Process Biochem.* **27**, 347–350.
5. Zanca, D. M. and Martín, J. F. (1983), *J. Antibiot.* **36(6)**, 700–708.

6. Matsumura, M., Imanaka, T., Yoshida, T., and Taguchi, H. (1981), *J. Ferment. Technol.* **59(2)**, 115–123.

7. Chu, W. Z. and Constantinides, A. (1988), *Biotechnol. Bioeng.* **32**, 277–288.

8. Araujo, M. L. G. C., Oliveira, R. P., Giordano, R. C., and Hokka, C. O. (1996), *Chem. Eng. Sci.* **51(11)**, 2835–2840.

9. Araujo, M. L. G. C., Giordano, R. C., and Hokka, C. O. (1999), *Biotechnol. Bioeng.* **63(5)**, 593–600.

10. Cruz, A. J. G., Silva, A. S., Araujo, M. L. G. C., Giordano, R. C., and Hokka, C. O. (1999), *Chem. Eng. Sci.* **54**, 3137–3142.

11. Marquardt, D. W. (1963), *J. Soc. Indust. Appl. Math.* **2(2)**, 431–441.

12. Nihtilä, M. and Virkkunen, J. (1977), *Biotechnol. Bioeng.* **19(12)**, 1831–1850.

13. Zangirolami, T. C., Johansen, C. L., Nielsen, J., and Jorgensen, S. B. (1997), *Biotechnol. Bioeng.* **56(6)**, 593–604.

14. Demain, A. L., Newkirk, J. F., and Hendlin, D. (1963), *J. Bacteriol.* **85**, 339–344.

15. Ribeiro, E. J. (1989), PhD thesis, FEA-UNICAMP, Campinas, SP, Brazil.

16. Almeida, R. M. R. G., Cruz, A. J. G., Araujo, M. L. G. C., Giordano, R. C., and Hokka, C. O. (1999), *Proceedings of the XXVII Congresso Brasileiro de Sistemas Particulados*, ENEMP-99, Campos do Jordão, SP, Brazil, in press.

17. Cruz, A. J. G., Araujo, M. L. G. C., Giordano, R. C., and Hokka, C. O. (1998), *Appl. Biochem. Biotechnol.* **70–72**, 579–592.

Influence of Culture Conditions on Lipopeptide Production by *Bacillus subtilis*

ERIC AKPA,*,[1] PHILIPPE JACQUES,[1] BERNARD WATHELET,[2]
MICHEL PAQUOT,[2] REGINE FUCHS,[3] HERBERT BUDZIKIEWICZ,[3]
AND PHILIPPE THONART[1]

[1]*CWBI (Unité de Bio-Industries) and* [2]*Unité de Chimie Biologique,
Faculté Universitaire des Sciences Agronomiques,
5030 Gembloux, Belgium, E-mail: akpa.e@fsagx.ac.be;
and* [3]*Institut für Organische Chemie, 50939 Köln, Germany*

Abstract

Bacillus subtilis produces various families of lipopeptides with different homologous compounds. To produce "new molecules" with improved activities and to select strains that produced a reduced number of homologs or isomers, we studied the effects of different media on the nature of the synthesis of fatty acid chains for each lipopeptide family. This study focused on two *B. subtilis* strains cultivated in flasks. Optimized medium for lipopeptide production and Landy medium modified by replacing glutamic acid with other α-amino acids were used. We found that the intensity of production of homologous compounds depends on the strain and the culture medium. Analysis of these lipopeptides by high-performance liquid chromatography showed that the strain *B. subtilis* NT02 yielded various homologous compounds when cultivated in Landy medium (L-Glu), but primarily one homologous product in high relative amounts when cultivated in the optimized medium. Mass spectrometric analysis and determination of the amino acid composition of this molecule enabled us to identify it as Bacillomycine L c15.

Index Entries: *Bacillus subtilis*; lipopeptide; biosurfactant; bacillomycin; high-performance liquid chromatography; culture medium.

Introduction

The *Bacillus subtilis* lipopeptides are members of a particular antibiotic class formed by the iturin, surfactin, and fengycin families. The general structure of these lipopeptides is a peptide cycle of 7 (iturin and surfactin)

*Author to whom all correspondence and reprint requests should be addressed.

or 10 amino acids (fengycin) linked to a fatty acid chain. The length of the fatty acid chains can vary from C-13 to C-16 for surfactins, from C-14 to C-17 for iturins, and from C-14 to C-18 for fengycins, giving different homologous compounds and isomers (*n, iso, anteiso*) for each lipopeptide *(1–5)*. Such molecules are of great interest because of their biologic and physicochemical properties, which can be exploited in food, oil, and pharmaceutical industries *(6–9)*. Iturin A and fengycin have a wide antifungal activity *(5,10,11)*. In addition, iturin A possesses an antibacterial spectrum, though it is restricted to a few bacteria such as *Micrococcus luteus (12,13)*. Surfactin exhibits a larger antibacterial spectrum *(14)*, but it is primarily one of the most powerful biosurfactants known *(2,15)*. At their critical micellar concentration, surfactin S1 and iturin A diminish surface tension of water from 72 to 31 and 54.5 mN/m, respectively *(16)*. Furthermore, surfactin exhibits antitumoral, antiviral, antimycoplasma, and fibrin clot-inhibiting activities *(2,17,18)*. No interfacial studies have been available until recently for fengycin.

Several studies have shown the existence of a relationship between the structure of these molecules and their properties. Hbid *(19)* and Bland et al. *(20)* demonstrated that the antifungal and hemolytic activities of these agents are enhanced with increasing number of carbon atoms of their fatty acid side chains, presumably owing to stronger interactions with biomembranes. Concerning the effect of peptide molecular attributes, the substitution of L-asparagine 1 in iturin A by acid L-aspartic in iturin C and the esterification of tyrosine 2 residue in iturin A diminish drastically their biologic activities *(10,21)*. Thimon et al. *(22)* demonstrated that the Glu-γ-methylester of surfactin has a much lower critical micellar concentration of 30 μM than natural isomer (240 μM) and showed 100% hemolysis at an appreciable lower concentration (12 μM) compared with the nonesterified compound (200 μM).

Unfortunately, applications of these biomolecules are limited by the cost of their production and purification. In addition, some *B. subtilis* strains coproduced various families and homologs *(23)*, causing additional purification problems.

To produce "new molecules" with improved activities and to select strains that produce a reduced number of homologous compound or isomers, we studied the effect of α-amino acids on the nature of the fatty acid chains for each lipopeptide family. This provided information about specificity of the biosynthetic enzymatic system for each lipopeptide family. Two *B. subtilis* strains were chosen among a collection of *Bacilli* conserved in our laboratory. *B. subtilis* S499 was chosen for its ability to produce the three lipopeptide families in order to determine the effect of α-amino acid pool on the production of each lipopeptide family of homologous compounds. Also chosen was *B. subtilis* NT02, which produces only iturins and is sensitive to the culture medium. The results obtained are presented herein.

Table 1
Optimized Medium Composition
per Liter of Distilled Water

Medium components	Quantity
Sucrose	20 g
Peptone	30 g
Yeast extract	7 g
KH_2PO_4	1.9 g
$MgSO_4$	0.450 g
Trace elements solution	9 mL[a]

[a]Composition of trace elements per liter of distilled water is 0.001 g of $CuSO_4$, 0.005 g of $FeCl_3$, 0.004 g of $NaMnO_4$, 0.002 g of KI, 0.014 g of $ZnSO_4$, 0.01 g of H_3BO_3, 0.0036 g of $MnSO_4$, 10 g of citric acid.

Materials and Methods

Strains

B. subtilis S499 was a gift from L. Delcambe (CNPEM, Liège, Belgium). This strain was collected in Ituri, Congo (formally Zaïre) *(1)*, and *B. subtilis* NT02 was isolated in CWBI (Wallon Center of Industrial Biology, Belgium) from *netetu*, a Senegalese fermented food.

Culture Media

Strains were cultivated on Jacques et al. *(23)* medium called optimized medium (Table 1), Landy medium *(24)*, and Landy media modified by replacing the L-glutamic acid with various L-α-amino acids at the same concentration (5 g/L). Landy medium contained the following per liter of distilled water: 20 g of glucose, 5 g of L-glutamic acid, 0.5 g of $MgSO_4$, 0.5 g of KCl, 1 g of KH_2PO_4, 0.0012 g of Fe_2SO_3, 0.0014 g of $MnSO_4$, 0.0016 g of $CuSO_4$. The solutions were brought to pH 7.0 with 5 N KOH or 5 N H_3PO_4, according to whether the amino acid used was acid or basic, before sterilizing.

Bacterial Culture and Extraction of Lipopeptide

The bacteria were grown in 100 mL of medium contained in 500-mL Erlenmeyer flasks stirred at 130 rpm at 30°C. For culture seeding, 10 mL of a preculture grown under the same conditions for 16 h was used. The culture was stopped after 96 h and the cell material was removed by centrifugation at 15,300*g*. Lipopeptides were extracted from the culture supernatant by solid-phase extraction on bond elut C18 (5 g) (Varian, CA) as described by Razafindralambo et al. *(25)*. The extract was brought to dryness, the residue was redissolved in $CHCl_3/CH_3OH$ (2:1 [v/v]) and separated by chromatography on silicagel 60 with the solvent $CHCl_3/CH_3OH/H_2O$ (65:25:4 [v/v/v]). Subsequently, the fengycins were eluted with $CHCl_3/$

$C_2H_5OH/CH_3OH/H_2O$ (7:3.5:3:1.5 [v/v/v/v/v]). The flow rate was about 2 mL/min, and the glass column was 30 cm long and had a 2-cm id.

Purification and Measurement of Lipopeptide

The lipopeptide extract was dissolved in methanol and analyzed by high-performance liquid chromatography (HPLC) on a C18 column (5 μm, 1 × 25 cm) (Chrompack, Middelburg, The Netherlands). Each family of lipopeptides was separately analyzed. Iturins were analyzed with the solvent acetonitrile/water/trifluoroacetic acid (TFA) (40:60:0.05 [v/v/v]), surfactins with the solvent acetonitrile/water/TFA (80:20:0.05 [v/v/v]), and fengycins with a gradient of 0.05% TFA in water and acetonitrile (60:40 for 25 min, 65:35 for 15 min). Solvents were HPLC grade. TFA was from Sigma (St. Louis, MO). The flow rate was 1 mL/min, detection wavelength was 214 nm (HP 1100; Waldbronn, Germany), and sample injection volume was 20 μL. Mass spectra analysis was performed on a VG platform (Fison, Manchester, UK) with an electrospray source. Samples were dissolved in 1 mL of a mixture of $CH_3OH/H_2O/CH_3COOH$ (50:50:0.1 [v/v/v]). Amino acid determinations were carried out according to the Stein and Moore method (Pharmacia Alpha Plus, Uppsala, Sweden).

Results

Production of Lipopeptide by B. subtilis S499

The production of different homologous compounds from the three families of lipopeptides produced by *B. subtilis* S499 in the optimized medium, Landy medium, and Landy media modified with L-leucine, L-valine, L-isoleucine, and L-threonine was analyzed. To avoid complex purification procedures, a proportion of the different homologous compounds was evaluated by ESI-MS. The behavioral similarity of these homologous congeners during the ionization process of mass spectrometry (MS) was confirmed by MS analysis of known samples of iturin and surfactin. This test related to *B. subtilis* S499 cultivated in the optimized medium. Figure 1A,B shows the proportions of iturin and surfactin of various homologous compounds analyzed by HPLC and electrospray ionization (ESI)-MS. The general tendency (height of the homologous peaks) for the two types of analysis is similar for surfactin (c15 > c14 > c13) and iturin homologous compounds (c14 > c15 > c16). This comparison could not be made for fengycins because their homologous compounds are not formally identified yet by hplc analysis.

Surfactin Homologous Compounds

Figure 2A–C shows the proportions of lipopeptide homologous compounds analyzed by ESI-MS. The use of Val, Leu, Ile, Thr, and optimized medium modified the percentages of the homologous compounds of surfactin (Fig. 2A). The rich medium allowed the synthesis of homologous

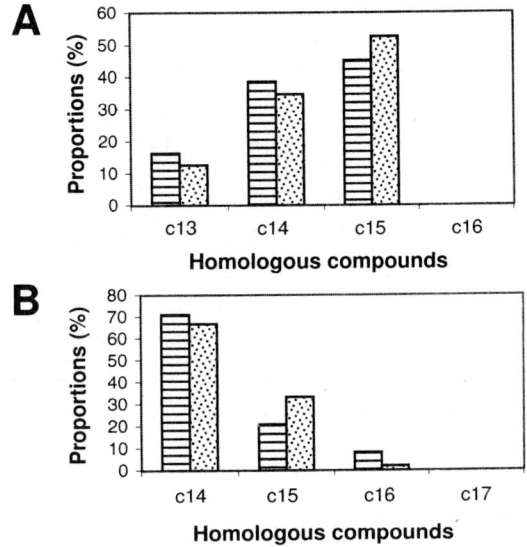

Fig. 1. Proportions of iturin and surfactin determined by HPLC (▦) and MS (▤) analysis. (**A**, surfactin homologs; **B**, iturin homologs.)

compounds c13, c14, and c15 with a higher proportion of c15 β-hydroxy fatty acid followed by the c14 chain. We did not observe production of surfactin homologous compound with the 16 carbon atom chain. The addition of Val gave equal proportions of c13 and c15 β-hydroxy fatty acid, which were less than for the c14 carbon chain. Unlike the optimized medium, we observed production of c16 homologous compound. These results agree with the literature, since Val is known to be a precursor of even fatty acid carbon chains (26). The homologous S2 c13 produced with this medium is an isoform of the standard surfactin S1 c13. In this homologous S2, one of the Leu residues of surfactin S1 is replaced by a Val residue (27). The addition of Leu and Ile, respectively, increases the proportion of c13 and c15 β-hydroxy fatty acid chain. This can be explained by the fact that these two α-amino acids are known to be precursors of odd β-hydroxy fatty acid chains (28). Nevertheless, we observed the production of homologous congener with 16 carbon atoms. The more prominent increase was observed with c15 β-hydroxy fatty acid when we used Thr. We observed also the best proportion of c16 and the production of a small quantity of c17. Thus, this is the first time that the presence of such a long carbonaceous chain has been described according to surfactin. This positive effect of Thr on the long chains of surfactin cannot be easily explained.

Iturin Homologous Compounds

On the other hand, with the use of the optimized medium, Ile and Thr gave a strong percentage of the c14 β-amino fatty acid (>60%) of the iturin family (Fig. 2B). The proportion of c15 fatty acid chain was slightly higher

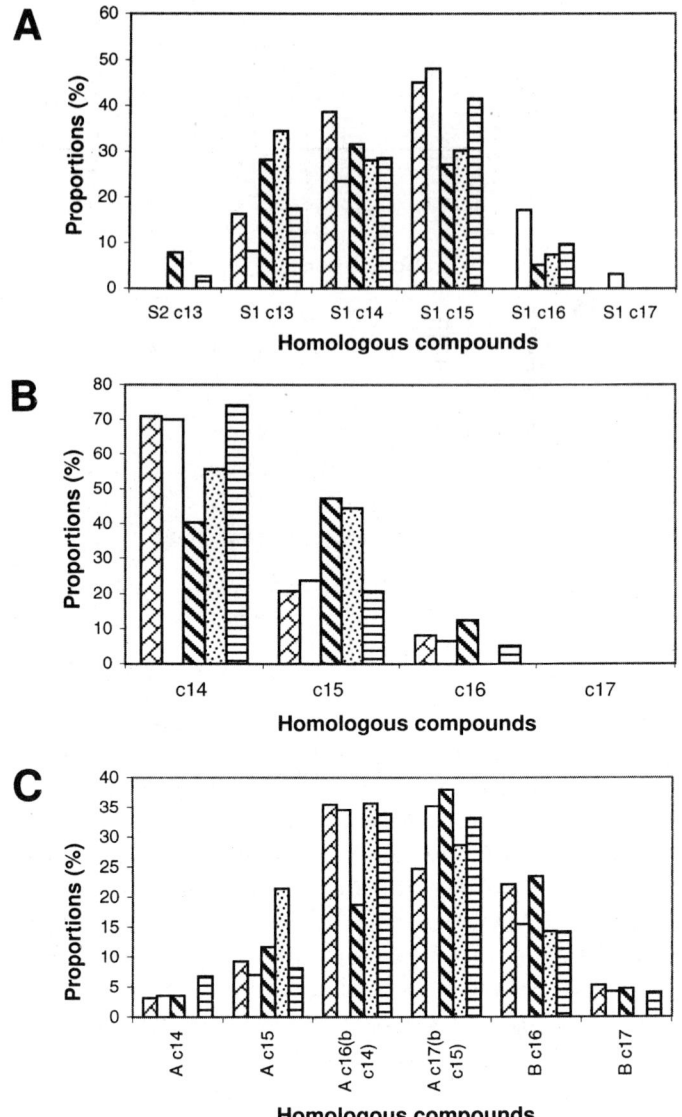

Fig. 2. MS analysis of homologous compounds produced by *B. subtilis* s499 grown in various culture media. **(A)** Surfactin homologous compounds; **(B)** iturin homologous compounds; **(C)** fengycin homologous compounds. ▨, optimized medium; □, Landy L-threonine; ◩, Landy L-valine; ▨, Landy L-leucine; ▤, Landy L-isoleucine.

than 20% and the c16 was <10%. The use of Leu increased the rate of c15 β-amino fatty acid (44%), but this rate remained lower than its c14 congener (56%). This result is in agreement with the effect previously reported for cellular fatty acid biosynthesis of *B. subtilis (26)*. In addition, leucine has been demonstrated to be the precursor of *iso*-c15 β-amino fatty acids *(29)*. The addition of Val gave the best production of iturin c16. This is understandable because Val is known to be a precursor of even fatty acid, but we had an unexpected result with a decrease in the c14 rate (40%) and an increase in the c15 rate (47%).

Fengycin Homologous Compounds

Concerning the fengycin homologous compounds (Fig. 2C), the analysis showed a tendency toward a "bell shape" according to the intensity of production of the various homologous. Fengycin A c16 (B c14) and A c17 (B c15) had the most intense peaks. They were flanked on both sides by the homologous A c14, A c15, B c16, and B c17. The most intense part of the peaks can be explained by the simultaneous presence of the homologous compounds fengycin A c16 and B c14, on the one hand, and fengycin A c17 and B c16, on the other hand. These two different types of fengycin are owing to the presence of an alanine residue (molecular weight of 71 kDa) in fengycin A and a valine residue (molecular weight of 99 kDa) in fengycin B, giving a difference of 28 kDa. The use of optimized medium and Leu medium resulted in fengycin A c16 (B c14) having the most intense production (35%), followed by the homologous A c17 (B c16), with 24–28%. In addition, the Leu medium allowed production in a strong proportion of fengycin A c15. The addition of Val gave an unexpected result, because fengycin A c17 (B c15) production was the most intense (38%). The proportions of fengycin homologous A c16 (B c14) and A c17 (B c15) were simultaneously most intense with Thr (34%) and Ile (34%). Then, the proportion decreased gradually on both sides. Note that results in the case of fengycin homologous compounds are difficult to explain because there is an overlapping of molecular mass of the homologous congeners of fengycin A and B.

Characterization of Iturin Homologous Compounds a, b, and c Produced by B. subtilis NT02

B. subtilis NT02 produces only antibiotic molecules belonging to the iturin family, not the surfactin or fengycin families. It was grown for 96 h in Landy medium. HPLC of the homologous compounds produced is shown in Fig. 3. To identify these homologous iturins whose retention times do not correspond to those of iturin A, they were purified with semipreparative HPLC. These purified peaks were analyzed by ESI-MS and amino acid composition was determined. The molecular ions $(M+K)^+$ of the iturin homologous compounds a, b, and c produced were, respectively, at m/z 1059.3, 1073.3, and 1073.3. Table 2 gives the amino acid composition of these various peaks. Six different amino acids were given with a constant ratio. Regarding these results, we concluded that these homologous compounds are, respectively, Bacillomycin L c14 and Bacillomycin L c15. The two m/z 1073.3 were owing to the presence of two isomers (*iso*, *anteiso*, or normal) of the c15 homolog.

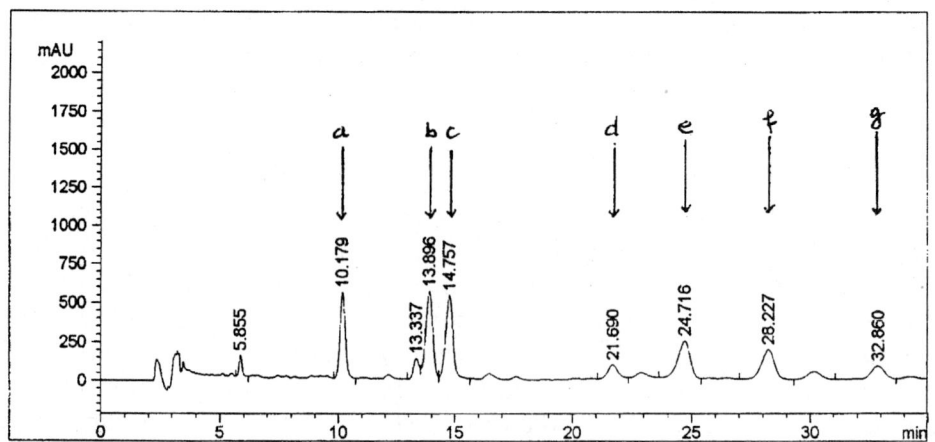

Fig. 3. HPLC analysis of iturin family lipopeptide produced by *B. subtilis* NT02 cultivated in Landy medium.

Table 2
Amino Acid Composition, Molecular Weight, and Identification
of Main Different Molecules of Bacillomycin Purified by HPLC

Peak	Amino acid composition	Molecular weight $(M+K)^+$	Identification
a	2Asx, 1Tyr, 2Ser, 1Glx, 1Thr	1059	Bacillomycin L c14
b	2Asx, 1Tyr, 2Ser, 1Glx, 1Thr	1073	Bacillomycin L c15
c	2Asx, 1Tyr, 2Ser, 1Glx, 1Thr	1073	Bacillomycin L c15

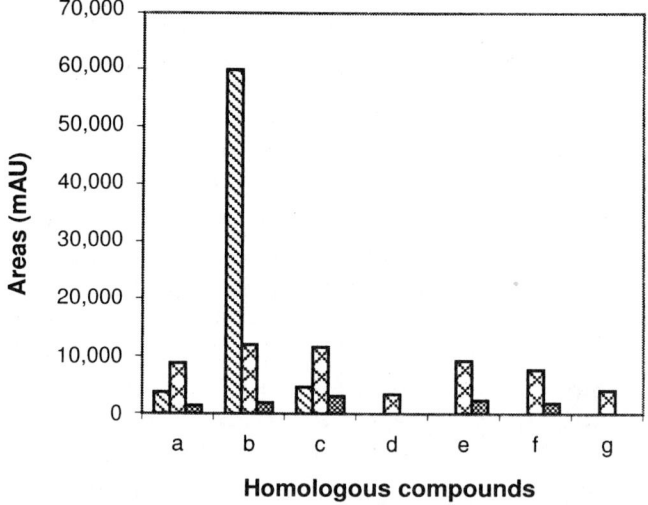

Fig. 4. Iturin family homologous compounds produced by *B. subtilis* NT02 cultivated in various media. ◪, optimized medium; ⊠, Landy ʟ-glutamic acid; ▦, Landy ʟ-serine.

Production of Iturin Homologous Compounds by Strain NT02 According to Culture Medium

To study the influence of the culture medium on the production of iturin homologous compounds, strain NT02 was grown for 96 h in the optimized medium, Landy L-serine, and Landy L-proline media. The intensities of the principal peaks analyzed by HPLC are represented in Fig. 4. Several iturin homologs were produced by NT02 when cultivated in Landy medium and in Landy media modified with L-serine and L-proline (data not shown). However, NT02 produced mostly one homologous compound when it was cultivated in the optimized medium.

Discussion

This work consisted of studying the influence of the culture medium on lipopeptide production by *B. subtilis*. The aim was to produce "new molecules" with improved activities and to select strains that produce a reduced number of homologous compounds or isomers according to the culture medium. It was shown that the modification of the culture media influences the proportion of surfactin homologous compounds produced by *B. subtilis* S499. This influence, as opposed to what Hourdou (30) revealed, can relate to the length of the carbonaceous chain, as observed with the medium of Landy L-threonine. Nevertheless, this variation in proportion does not mean the total suppression of one of the principal surfactin homologous compounds (c13, c14, c15). Nor does it produce an unusual number of carbon, except for the homologous c17 observed in a very small quantity in the production resulting from the medium Landy L-Thr, which constitutes a first.

Concerning the iturins, *B. subtilis* S499 invariably produced more homologous compound with 14 carbon atoms than 15 or 16 carbons, excluding production in Landy L-valine medium. However, this handicap was quickly overcome by the intensity of production.

Overall, the influence of the culture medium in the production of iturin homologous molecules was less significant than that of the surfactins. This was owing to the fact that the enzymatic complex responsible for surfactin biosynthesis would be less sensitive to the intracellular fatty acid chain pool than that of iturin A. Indeed, it has been shown that the enzymatic system of the biosynthesis of the surfactins is not very specific (27) but little is known about biosynthesis of the iturin family. Some amino acids such as threonine probably would be favorable to the lengthening of the fatty acid chain and others (ramified α-amino acid) rather than the formation of isomers *i, ai,* and *n* (31). However, these α-amino acids would not have a great influence in producing high carbonaceous chains. It has been suggested that the responsible enzyme system for β-amino acid synthesis selects among the cellular fatty acids those that correspond to the carbon chain of β-amino acid (29).

Of the two studied strains, *B. subtilis* NT02 was shown to be the most efficient for obtaining less homologous compounds. This strain yielded various compounds when cultivated in Landy media but primarily one homologous product in very strong intensity when cultivated in optimized medium. MS analysis and determination of amino acid composition of this compound showed that it is Bacillomycin L c15. This molecule belongs to the iturin family and possesses biologic properties similar to those of iturin A *(11)*. The length of carbon chain of this principal molecule is a great advantage. This will enable us to reduce purification problems and improve biologic properties, as suggested by Hbid *(19)*.

Conclusion

Modification of the culture medium influences the proportion of surfactin homologous compounds produced by *B. subtilis* S499. Threonine would be favorable to the lengthening of the fatty acid chain. The influence of the culture medium in the production of iturin homologous molecules was less significant than that of the surfactins. This would be owing to the fact that the enzymatic complex responsible for surfactin biosynthesis would be less sensitive to intracellular fatty acid chain pool than that of iturin A. *B. subtilis* NT02 yielded various compounds when cultivated in Landy media but primarly one homologous product (Bacillomycin L c15) in very strong intensity when cultivated in optimized medium. This will enable us to reduce purification problems. The length of carbon chain would be a great advantage for biologic properties.

Acknowledgments

We thank C. Cornélius and B. N'Dir for providing *B. subtilis* NT02. This work received financial support from FNRS (Belgium; FRFC project no. 2.4558.98 and Credit aux Chercheurs no. 1.5.193.99) and CGRI (Senegal). E. A. gratefully acknowledges the government of Ivory Coast for his study bill.

References

1. Delcambe, L. (1965), *Bull. Soc. Chim. Belges* **74,** 329–340.
2. Arima, K., Kakinuma, A., and Tamura, G. (1968), *Biochem. Biophys. Res. Commun.* **31(3),** 488–494.
3. Kakinuma, A., Sugino, H., Isono, M., Tamura, G., and Arima, K. (1969), *Agric. Biol. Chem.* **33(6),** 973–976.
4. Peypoux, F., Michel, G., and Delcambe, L. (1976), *Eur. J. Biochem.* **63,** 391–398.
5. Vanittanakom, N., Loffler, W., Koch, U., and Jung, G. (1986), *J. Antibiot.* **39(7),** 888–901.
6. Saotoshi, K. (1988), in *Proceedings of the World Conference on Biotechnology for the Fats and Oils Industry*, Appelwhite, T. H., ed., American Oil Chemist's Society, pp. 195–201.
7. Shigeo, I. (1988), in *Proceedings of the World Conference on Biotechnology for the Fats and Oils Industry*, Appelwhite, T. H., ed., American Oil Chemist's Society, pp. 206–210.

8. Brown, J. M. and Moses, V. (1988), in *Proceedings of the World Conference on Biotechnology for the Fats and Oils Industry*, Appelwhite, T. H., ed., American Oil Chemist's Society, pp. 202–205.
9. Jacques, P., Hbid, C., Vanhentenryck, F., Destain, J., Bare, G., Razafindralambo, H., Paquot, M., and Thonart, P. (1993), *Prog. Biotechnol.* **3**, 1067–1070.
10. Maget-Dana, R. and Peypoux, F. (1994), *Toxicology* **87**, 151–174.
11. Besson, F., Peypoux, F., Michel, G., and Delcambe, L. (1979), *J. Antibiot.* **32(8)**, 828–833.
12. Besson, F, Peypoux, F., and Michel, G. (1979), *Biochimica Biophysica Acta* **552**, 558–562.
13. Peypoux, F., Besson, F., Michel, G., and Delcambe, L. (1979), *J. Antibiot.* **32**, 136–140.
14. Bernheimer, A. W. and Avigad, L. S. (1970), *J. Gen. Microbiol.* **61**, 361–369.
15. Razafindralambo, H., Paquot, M., Baniel, A., Popineau, Y., Hbid, C., Jacques, P., and Thonart, P. (1997), *Food Hydrocolloids* **11(1)**, 59–62.
16. Thimon, L., Peypoux, F., Maget-Dana, G., and Michel, G. (1992), *JAOCS* **69**, 92, 93.
17. Nissen, E., Pauli, G., Vater, J., and Vollenbroich, D. (1997), *In Vitro Cell. Dev. Biol.-Anim.* **33**, 414–415.
18. Vollenbroich, D., Pauli, G., Vater, J., Ozel, M., and Kamp, R. M. (1997), *Biologicals* **3**, 289–297.
19. Hbid, C. (1996), PhD thesis, Université de Liège.
20. Bland, J., Lax, A., and Klich, M. (1993), *Proceedings of the Twenty-Second European Peptide Symposium*, Schneider, C. H. and Eberle, A. N., eds., ESCOM, Leiden, The Netherlands, pp. 332–333.
21. Harnois, I., Maget-Dana, R., and Ptak, M. (1989), *Biochimie* **71**, 111–116.
22. Thimon, L., Peypoux, F., Das, B. C., Wallach, J., and Michel, G. (1994), *Biotechnol. Appl. Biochem.* **20**, 415–423.
23. Jacques, P., Hbid, C., Destain, J., Razafindralambo, H., Paquot, M., De Pauw, E., and Thonart, P. (1999), *Appl. Biochem. Biotechnol.* **77–79**, 223–233.
24. Landy, M., Warren, G. H., Rosenman, S. B., and Colio, L. G. (1948), *Proc. Soc. Exp. Biol. Med.* **67**, 530–541.
25. Razafindralambo, H., Paquot, M., Hbid, C., Jacques, P., Destain, J., and Thonart, P. (1993), *J. Chromatogr.* **639**, 81–85.
26. Kaneda, T. (1977), *Bacteriol. Rev.* **41**, 391–418.
27. Peypoux, F. and Michel, G. (1992), *Appl. Microbiol. Biotechnol.* **36**, 515–517.
28. Besson, F., Tenoux, I., Hourdou, M.-L., and Michel, G. (1992), *Biochimica Biophysica Acta* **1123**, 51–58.
29. Besson, F. and Hourdou, M.-L. (1987), *J. Antibiot.* **60(2)**, 221–223.
30. Hourdou, M.-L. (1989), PhD thesis, Université Claude Bernard-Lyon 1, France.
31. Hourdou, M.-L., Besson, F., and Michel, G. (1988), *J. Antibiot.* **61(2)**, 207–211.

Mathematical Modeling of Controlled-Release Kinetics of Herbicides in a Dynamic-Water-Bath System

Félix M. Pereira, Adilson R. Gonçalves,
André Ferraz, Flávio T. Silva,
and Samuel C. Oliveira*

*Departamento de Biotecnologia, Faculdade de Engenharia Química de Lorena, CP 116, 12.600-000, Lorena, SP, Brazil,
E-mail: scoliveira@debiq.faenquil.br*

Abstract

Release of herbicides from lignin-based formulations follows a diffusion-controlled mechanism. For mathematical modeling of diffusive transport, the conventional approach is to assume sink conditions at both surfaces of polymeric matrix. This boundary condition proved to be inadequate to describe experimental data obtained in a water dynamic bath system. However, satisfactory descriptions for this system were obtained when a stagnant unstirred layer of herbicide solution was used as the boundary condition. The adequacy of the model incorporating this new boundary condition was statistically tested using the Fisher test at a confidence level of 95% and plotting the residual distribution.

Index Entries: Mathematical modeling; controlled release; herbicides; lignin; anetryn; diuron; diffusion.

Introduction

Lignin is a byproduct of several industrial processes, mainly pulping and production of fermentable sugars from biomass. The use of lignin as matrix for controlled-release systems with a large number of pesticides has been extensively studied (1) and can be synergistic with chemical production from cellulose.

The release kinetics of herbicides (active ingredient [AI]) from lignin matrices is a very important aspect to be investigated in order to predict

*Author to whom all correspondence and reprint requests should be addressed.

both dosage of AI used at the time and the AI destination when applied in acceptor systems containing an aqueous phase such as the soil.

Release kinetics studies usually are carried out by monitoring the amount of herbicide released into a static or dynamic water bath *(2–4)*. In the static water bath system (SWBS), the controlled-release formulations (CRFs) are immersed in deionized water contained in flasks. In this system, the sampling is performed by changing all the water of the flasks in a defined time interval. On the other hand, in the dynamic water bath system (DWBS), the CRFs are placed in cylindrical tubes of glass, with sintered discs in the top and bottom of the columns for retention of the particles. Water is continuously pumped into the column. Samples of the effluent of the column are periodically collected in order to determine herbicide concentration *(4)*.

In previous studies, a mathematical model was developed for the release kinetics based on a diffusive transport mechanism described by the second Fick's law *(5,6)*. In this model, sink conditions were adopted: the concentrations of AI in both slab surfaces are assumed to be zero during the entire time range of herbicide release. In a preliminary analysis, the model incorporating this boundary condition proved to be valid to describe the release kinetic data obtained in the SWBS. However for the DWBS, this approach was not able to describe the kinetic data satisfactorily. This result is dubious because the concentration in the slab surface in the DWBS is much lower than in the SWBS. A hypothesis to explain this fact might be that the static system is all unstirred and thus more susceptible to being modeled as a single process. To avoid the lack of fit of the model to the experimental data in the DWBS, another boundary condition at the slab surface is used. This new boundary condition incorporates the existence of a stagnant unstirred layer at the interface between the slab surface and the release medium.

Materials and Methods

The CRFs employed in release experiments in the DWBS were manufactured using lignin extracted from sugarcane bagasse pretreated by the steam explosion process in a pilot plant in our laboratory. Lignin was recovered by precipitation with HCl and characterized as described by Ferraz et al. *(3)*. The lignin and herbicide were melted in equal amounts in stainless-steel concave recipients and immersed in a silicon bath with temperatures between 170 and 200°C, and the mixture was homogenized. After complete homogenization, the recipients were cooled at room temperature. The formulations were ground and sieved to select a range of granule size from 0.71 to 1.00 mm. Two types of CRFs were prepared: CRFD (with diuron: 3-[3,4-dichlorophenyl]-1,1-dimethylurea; 94.8% purity) and CRFA (with ametryn: 2-ethylamino-4-isopropylamino-6-methylthio-1,3,5-triazine; 97.8% purity) *(4)*.

Release Experiments in a DWBS

In the DWBS the CRF samples were weighed and allocated in glass columns 40 mm long and 18 mm in diameter that contained sintered disks in the top and bottom for retention of the particles. The system was maintained at 30°C and fed on the bottom with deionized water at a flow rate of 2 mL/min. The effluent was collected on the top and stored in glass flasks that were changed periodically. The herbicide concentration in the flasks was determined by high-performance liquid chromatography using a 1.8 × 200 mm C-18 column (HP-RP18). Diuron was eluted with a methanol/water solution (7:3 [v/v]) at 0.7 mL/min and detected at 280 nm. Ametryn was eluted with a methanol/sodium phosphate buffer (pH 5.0) solution (7:3 [v/v]) at 0.8 mL/min and detected at 250 nm *(4)*.

Modeling of Release Kinetics

The release rate of AI is controlled by a diffusion mechanism, mathematically described by the second Fick's law *(5–7)*. According to this law, the spatial and temporal profiles of AI concentration are governed by Eq. 1:

$$\frac{\partial c_p}{\partial t} = D_{eff} \frac{\partial^2 c_p}{\partial x^2}$$ (1)

in which c_p is the AI concentration in the pores (g/cm), D_{eff} is the effective diffusion coefficient (cm^2/d), x is the spatial coordinate (cm), and t is the time (d) *(7)*.

To solve Eq. 1, it is necessary to formulate the initial and boundary conditions. The initial condition is given by

$$c_p(x, t = 0) = \frac{M_0}{\varepsilon A L}$$ (2)

in which M_0 is the initial amount of the content of AI in the slab (g), ε is the matrix porosity, A is the slab area (cm^2), and L is the matrix thickness (cm) *(4)*.

In the present study, two types of boundary conditions were used at the slab surface: sink conditions and stagnant unstirred layer.

Model Incorporating Sink Conditions at Slab Surface

The sink conditions at both slab surfaces are described by Eq. 3:

$$c_p(0, t) = c_p(L, t) = 0$$ (3)

By solving Eq. 1 using initial and boundary conditions (Eqs. 2 and 3), Eq. 4 is obtained:

$$c_p(x,t) = \frac{4M_0}{\varepsilon A L \pi} \sum_{n=0}^{\infty} \left\{ \frac{1}{2n+1} \exp\left[-\frac{(2n+1)^2 \pi^2 D_{eff}}{L^2} \right] \sin\left[\frac{(2n+1)\pi x}{L} \right] \right\}$$ (4)

From Eq. 4, Eq. 5, describing the cumulative amount of AI released at time t (M_t), can be developed (5,7):

$$\frac{M_t}{M_0} = 1 - \frac{8}{\pi^2} \sum_{n=0}^{\infty} \left\{ \frac{\exp\left[-\frac{(2n+1)^2 \pi^2 D_{eff} t}{L^2} \right]}{(2n+1)^2} \right\} \tag{5}$$

Since granule size was the same for all CRFs used in the release experiments, the value of slab thickness L can be assumed constant and the D_{eff} parameter can be replaced by D^* (d^{-1}), which incorporates the following constants (5):

$$D^* = \frac{\pi^2 D_{eff}}{L^2} \tag{6}$$

Model Incorporating a Stagnant Unstirred Layer as Boundary Condition at Slab Surface

The mass transfer of herbicide through a stagnant unstirred layer at both slab surfaces is described by Eq. 7:

$$-D_{eff} \frac{\partial c_p}{\partial x} = k(c_p^* - c_b) \tag{7}$$

in which c_p^* is the concentration in the bulk fluid (g/cm), c_b is the concentration in the surrounding medium (g/cm), and k is the parameter of mass transfer by convection (cm/d).

If the slab $0 < x < L$ is initially at uniform concentration c_0 and if the law of mass transfer at both surfaces follows Eq. 7, the solution of Eq. 1 is:

$$\frac{c_p - c_0}{c_p^* - c_0} = 1 - \sum_{n=0}^{\infty} \frac{2Bi \cos(\beta_n x/2L) \exp(-\beta_n^2 D_{eff} t/4L^2)}{(\beta_n^2 + Bi^2 + Bi) \cos \beta_n} \tag{8}$$

in which β_n are the positive roots of the equation:

$$\beta \tan \beta = Bi \tag{9}$$

in which

$$Bi = Lk/D_{eff} \text{ (Biot number)} \tag{10}$$

The six first roots of Eq. 9 are listed in Table 1 (8).

Table 1
Six First Roots of $\beta \tan \beta = Bi$

Bi	β_0	β_1	β_2	β_3	β_4	β_5
0	0	3.1416	6.2832	9.4248	12.5664	15.7080
0.01	0.0998	3.1448	6.2848	9.4258	12.5672	15.7086
0.1	0.3111	3.1731	6.2991	9.4354	12.5743	15.7143
0.2	0.4328	3.2039	6.3148	9.4459	12.5823	15.7207
0.5	0.6533	3.2923	6.3616	9.4775	12.6060	15.7397
1.0	0.8603	3.4256	6.4373	9.5293	12.6453	15.7713
2.0	1.0769	3.6436	6.5783	9.6296	12.7223	15.8336
5.0	1.3138	4.0336	6.9096	9.8928	12.9352	16.0107
10.0	1.4289	4.3058	7.2281	10.2003	13.2142	16.2594
100.0	1.5552	4.6658	7.7764	10.8871	13.9981	17.1093
∞	1.5708	4.7124	7.8540	10.9956	14.1372	17.2788

For this approach, the cumulative amount of AI released at time t (M_t) is given by (7):

$$\frac{M_t}{M_0} = 1 - \sum_{n=0}^{\infty} \frac{2Bi^2 \exp(-4\beta_n^2 D^* t/\pi^2)}{\beta_n^2 (\beta_n^2 + Bi^2 + Bi)} \qquad (11)$$

An asymptotic analysis of the meaning of the boundary condition represented by Eq. 7 indicates that for high values of k (convective parameter), the thickness of stagnant unstirred layer is decreased and c_0 tends to c_p. In a limit case, when k tends to an infinite value, Bi also tends to an infinite value. By analyzing Eq. 11 when $Bi \rightarrow \infty$, one obtains

$$\lim_{Bi \rightarrow \infty} \frac{M_t}{M_0} = \lim_{Bi \rightarrow \infty} \left[1 - \sum_{n=1}^{\infty} \frac{2Bi^2 \exp(-4\beta_n^2 D^* t/\pi^2)}{\beta_n^2 (\beta_n^2 + Bi^2 + Bi)} \right] = 1 - \frac{\infty}{\infty} \qquad (12)$$

The ∞/∞ indetermination in Eq. 12 can be avoided by using L'Hôpital's rule. The final result of this mathematical procedure is

$$\lim_{Bi \rightarrow \infty} \frac{M_t}{M_0} = 1 - \sum_{n=1}^{\infty} \frac{2 \exp(-4\beta_n^2 D^* t/\pi^2)}{\beta_n^2} \qquad (13)$$

When $Bi \rightarrow \infty$, it can be demonstrated that

$$\beta_n = \frac{\pi}{2}(2n + 1) \qquad (14)$$

By substituting Eq. 14 in Eq. 13, Eq. 5 is obtained again. This result indicates that for high Bi values, the external resistance to mass transfer can be neglected and Eq. 11 is equivalent to Eq. 5.

A

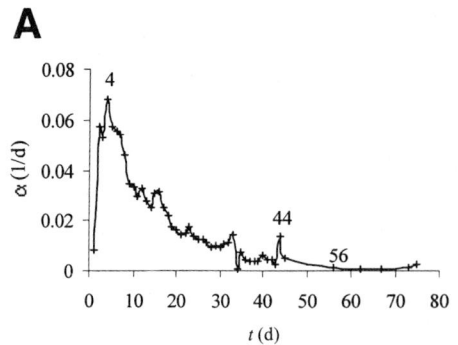

B

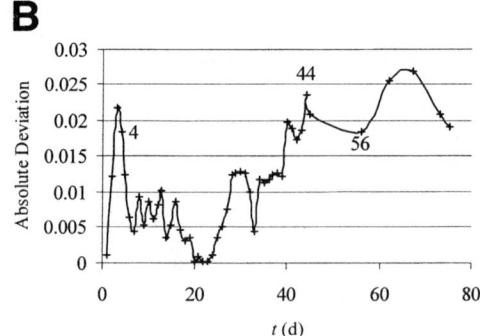

Fig. 1. Graphics used during the elimination procedure of experimental data: **(A)** average rate of herbicide cumulative amount released and **(B)** absolute deviation.

Results and Discussion

Prior to the mathematical modeling, an analysis of the experimental data was carried out aiming to remove the inadequate experimental data as well as to delimit the time range for which Eqs. 5 and 11 are valid.

Exclusion of Inadequate Experimental Data

The utilization of the original experimental data set resulted in a significant lack of fit of the model for both boundary conditions used at the slab surface. Consequently, the experimental data without an expected behavior were removed from the original experimental data set, according to the following graphic procedure: Figure 1A shows the behavior of the ratio $\alpha = [(M_t/M_0)_i - (M_t/M_0)_{i-1}]/(t_i - t_{i-1})$ as a function of the time t_i (in which $i = 1, 2, 3, \ldots, n$ and n is the number of experimental data). The ratio α can be interpreted as an average rate of the cumulative amount of herbicide released in the time interval between t_i and $t_i - 1$. Figure 1B shows the absolute value of the difference between two data of $(M_t/M_0)_i$ obtained in replicated experiments as a function of the time t_i. The absolute value of difference provides a rough estimate of the experimental error.

Figure 1A,B clearly indicates that some experimental data must be removed from the original set. Figure 1A shows the existence of an initial period (0–4 d) during which the average rate increases and decreases oscillating at the time. Taking into account that during the diffusive process the release rate decays asymptotically at the time, the oscillations observed in the initial period may be attributed to the occurrence of two phenomena: slab entumesciment and dissolution of herbicide initially existing on the slab surface. Thus, the experimental data concerning the initial period were discharged. Experimental datum obtained on d 44 is seen as a peak in Fig. 1A and is inaccurate according to Fig. 1B (high value of deviation). For these reasons, this experimental datum was also excluded. After 56 d, the release rate is very close to zero, resulting in inaccurate determinations of herbicide concentration in the medium. This fact explains the high values

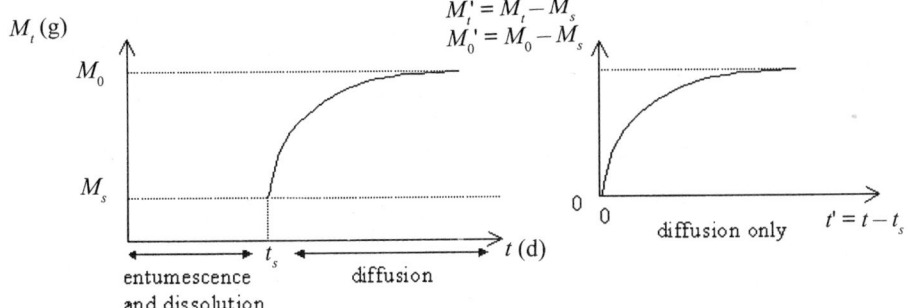

Fig. 2. Schematic representation of change in variables.

of deviation in Fig. 1B for times above 56 d. The data obtained in this experimental region were not used in the mathematical modeling.

Delimitation of Time Range
During Which Diffusive Models Are Valid

Equations 5 and 11 describe only the diffusion of herbicide from slab to acceptor medium. However, other phenomena occur during the first days of the release experiment: the slab entumescence and dissolution of herbicide initially existing on the slab surface. Because the diffusive models take into account that in the start of release ($t = 0$), $M_t/M_0 = 0$, it is necessary to write Eqs. 5 and 11 in terms of the corrected variables t' and M_t', in order to describe exclusively the diffusive phenomenon as shown schematically in Fig. 2.

To correct the time variable, it is necessary to realize the following subtraction:

$$t' = t - t_s \tag{15}$$

in which t' is time data corrected and t_s is the initial time during which the entumescence and dissolution phenomena occur.

Correction of the cumulative mass released may be realized according to the following relationship:

$$\frac{M_t'}{M_0'} = \frac{M_t - M_s}{M_0 - M_s} \tag{16}$$

in terms of cumulative amount

$$\frac{M_t'}{M_0'} = \frac{M_t/M_0 - M_s/M_0}{1 - M_s/M_0} \tag{17}$$

in which M_t'/M_0' is the corrected cumulative amount of herbicide released and M_s/M_0 is the cumulative mass released by dissolution.

Table 2
Values of Parameters M_s and t_s
Used for Correction of Release Curves

CRFs	M_s/M_0	t_s (d)
CRFD	0.074	6
CRFA	0.195	4

Table 3
Diffusion Coefficients Estimated
for FLCs of Diuron and Ametryn
Considering Sink Conditions
on Slab Surface (Eq. 5)

CRFs	D^* (d^{-1})
CRFD	0.0067 ± 0.0003
CRFA	$0.055 \ \pm 0.001$

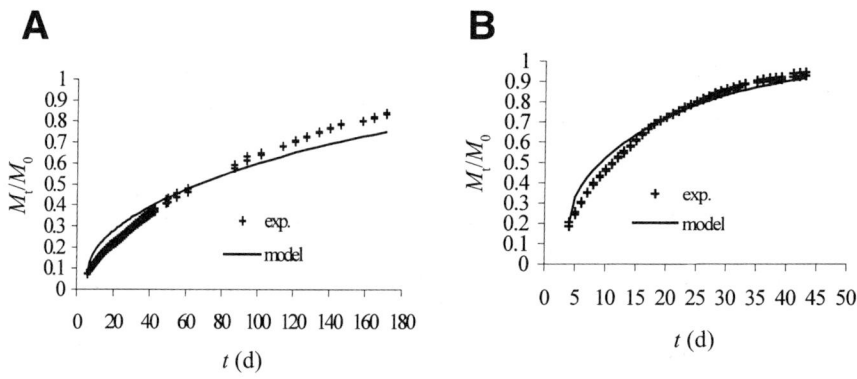

Fig. 3. Plot of experimental and calculated values for the model incorporating sink conditions at slab surface: **(A)** CRFD; **(B)** CRFA.

The M_s/M_0 and t_s values for the replicated experiments were estimated by inspection of similar plots shown in Fig. 1A. These values are presented in Table 2.

Mathematical Modeling Incorporating Sink Conditions at Slab Surface

The values of the diffusion coefficient (D^*) estimated by nonlinear regression *(9)* are presented in Table 3 for the two CRFs used, and a plotting of experimental data and model predictions is shown in Fig. 3. Figure 3 shows that there was a clear lack of fit of the model to the experimental data for both CRFs used. To quantify statistically this lack of fit, the tables of analysis of variance (ANOVA) were built for the two herbicide formula-

Table 4
ANOVA for Fit of Eq. 5 to Corrected Experimental Data Obtained with CRFD

Source	Sum of squares	Degrees of freedom	Mean square	F ratio	$F_{95\%}$ (tabulated)
Regression	3.554	1	3.554	1204	3.932
Residues	0.307	104	0.0030		
Lack of fit	0.304	51	0.0060	132	1.583
Pure error	0.0024	53	0.000045		
Total	3.861	105			
R^2	0.9205				
R^2 maximum	0.9995				

Table 5
ANOVA for Fit of Eq. 5 to Corrected Experimental Data Obtained with CRFA

Source	Sum of squares	Degrees of freedom	Mean square	F ratio	$F_{95\%}$ (tabulated)
Regression	3.024	1	3.024	2422	3.974
Residues	0.090	72	0.0012		
Lack of fit	0.087	35	0.0025	28	1.739
Pure error	0.0032	37	0.000088		
Total	3.114	73			
R^2	0.9711				
R^2 maximum	0.9992				

Table 6
Parameter Estimates for Model Incorporating
a Stagnant Unstirred Layer
as Boundary Condition at Slab Surface (Eq. 11)

CRF	D^* (d^{-1})	Bi
CRFD	0.0209 ± 0.0001	2.0 ± 0.1
CRFA	$0.118 \ \pm 0.005$	2.8 ± 0.2

tions (Tables 4 and 5). For both CRFs, the F ratio value calculated was greater than the corresponding tabulated value. This analysis, however, does not show the nature of this inadequacy, which is attributed here to the use of an inappropriate boundary condition at the slab surface.

Mathematical Modeling Incorporating
a Stagnant Unstirred Layer
as Boundary Condition at Slab Surface

The parameter estimation results for the model incorporating a stagnant layer as boundary condition on the slab surface are presented in Table 6

A

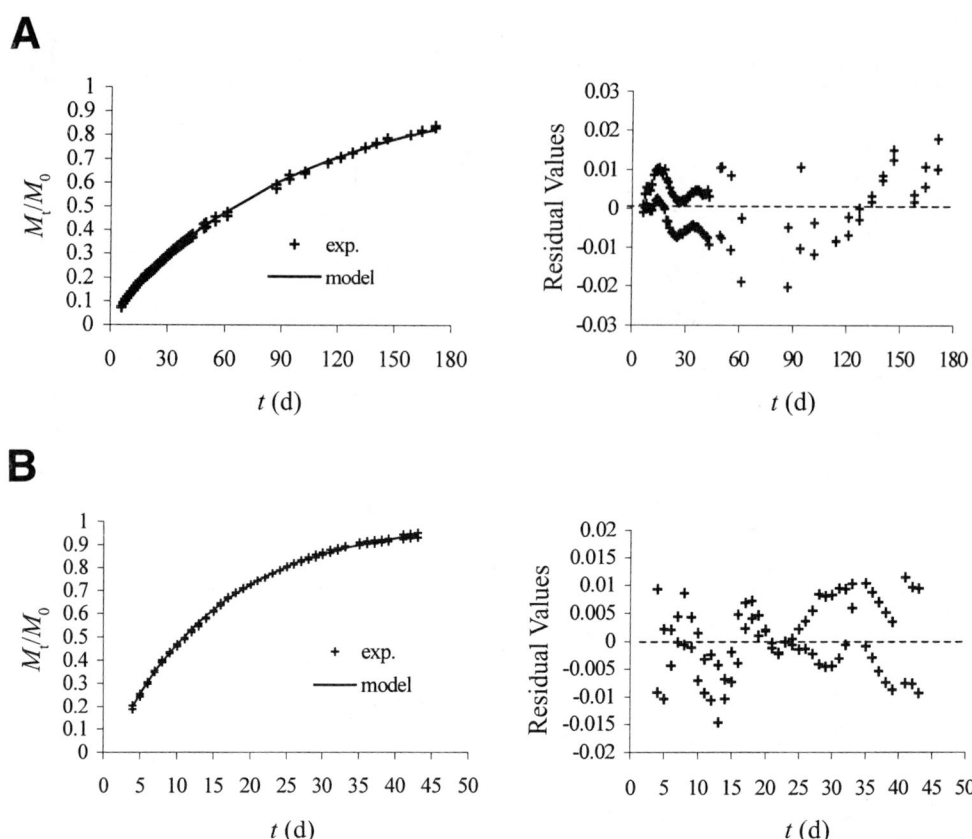

Fig. 4. Plot of experimental and calculated values for the model incorporating a stagnant unstirred layer as boundary condition: **(A)** CRFD; **(B)** CRFA.

for both CRFs used. A plot of the experimental data and model predictions and the distribution of the residual values are shown in Fig. 4. A good fit of the model to the experimental data may be observed for both CRFs. Moreover, the residuals distribution is randomized in both cases. To confirm the adequacy of the model, ANOVA is presented in Tables 7 and 8 for each herbicide formulation employed. For both CRFs investigated, the high values of F ratio for regression and the small values of F ratio for lack of fit indicate that the model incorporating a stagnant unstirred layer as the boundary condition at both slab surfaces is adequate to describe the experimental data obtained in the DWBS.

Relevant information obtained from the analysis of data presented in Tables 3 and 6 is that the value of D^* estimated using a stagnant unstirred layer at the slab surface is higher than the ones estimated using the sink conditions. This can explain some of our experimental data on release in soil, which show that the release rate is faster in comparison with that observed in the SWBS and DWBS. This result suggests that the stagnant unstirred layer in the slab surface is reduced in the soil system, probably

Table 7
ANOVA for Fit of Eq. 11 to Experimental Data Obtained with CRFD

Source	Sum of squares	Degrees of freedom	Mean square	F ratio	$F_{95\%}$ (tabulated)
Regression	5.623	1	5.523	109435	3.932
Residues	0.0053	104	0.000051		
Lack of fit	0.0029	51	0.000058	1.317	1.583
Pure error	0.0024	53	0.000045		
Total	5.629	105			
R^2	0.9991				
R^2 maximum	0.9996				

Table 8
ANOVA for Fit of Eq. 5 to Experimental Data Obtained with CRFA

Source	Sum of squares	Degrees of freedom	Mean square	F ratio	$F_{95\%}$ (tabulated)
Regression	4.0922	1	4.0923	5.5396	3.974
Residues	0.0053	72	0.000074		
Lack of fit	0.0021	35	0.000059	0.678	1.739
Pure error	0.0032	37	0.000087		
Total	4.0976	73			
R^2	0.9987				
R^2 maximum	0.9992				

owing to an adsorption mechanism of herbicide in the soil. To obtain secure conclusions, new experiments must be conducted to understand this phenomenon.

Conclusion

The utilization of a boundary condition incorporating a stagnant unstirred layer of herbicide solution at both slab surfaces provided a good description of release kinetics in the DWBS. The model considering sink conditions at the slab surface is a simplification that is not valid for describing release kinetics data obtained in the DWBS. The results obtained in this study are an advancement in the search of a relationship coupling the controlled release kinetic data obtained in a simple system (water bath) with those obtained in a complex system (soil). However, the practical contribution of the model incorporating a stagnant unstirred layer as the boundary condition at both slab surfaces depends on the extent to which the model allows one to obtain, from laboratory studies, fundamental values of diffusion coefficients that can be extrapolated from the laboratory to the field.

Acknowledgments

This work was supported by Fundação de Amparo à Pesquisa do Estado de São Paulo and Programa de Incentivo à Capacitação Docente e Técnica, Fundação Coordenação de Aperfeiçoamento de Pessoal de Nível Superior.

References

1. Wilkins, R. M. (1990), in *Controlled Delivery of Crop-Protection Agents*, Wilkins, R. M., ed., Taylor & Francis, London, pp. 149–165.
2. Cotteril, J. V., Wilkins, R. M., and Silva, F. T. (1995), *J. Controlled Release* **40,** 135–142.
3. Ferraz, A., Souza, J. A., and Silva, F. T. (1992), in *Proceedings of the 2nd Brazilian Symposium on the Chemistry of Lignins and Other Wood Components*, vol. 3, UNICAMP, Campinas, Brazil, pp. 173–177.
4. Reis, M. F. (1999), MSc thesis, Faculdade de Engenharia Química de Lorena, Departamento de Biotecnologia, Lorena, Brazil.
5. Pereira, F. M., Gonçalves, A. R., Silva, F. T., Ferraz, A., and Oliveira, S. C. (2000), in *Proceedings of the 6th Brazilian Symposium on the Chemistry of Lignins and Other Wood Components*, Silva, F. T. and Ferraz, A., eds., in press.
6. Oliveira, S. C., Pereira, F. M., Ferraz, A., Silva, F. T., and Gonçalves, A. R. (2000), *Appl. Biochem. Biotechnol.* **84–86,** 595–615.
7. Siegel, R. A. (1989), *Controlled Release of Drugs: Polymers and Aggregate Systems*, Morton Rosoff ed., VCH, New York, pp. 1–9.
8. Crank, J. (1976), *The Mathematics of Diffusion*, 2nd ed., Oxford University Press, Bristol, England.
9. Marquardt, D. W. (1963), *J. Soc. Ind. Appl. Math.* **11(2),** 431–441.

Do Cellulose Binding Domains Increase Substrate Accessibility?

**ALI R. ESTEGHLALIAN,[1] VINIT SRIVASTAVA,[2]
NEIL R. GILKES,[2] DOUGLAS G. KILBURN,[2]
R. ANTONY J. WARREN,[2] AND JOHN N. SADDLER*,[1,2]**

[1]*Chair of Forest Products Biotechnology,
Department of Wood Science, 4th Floor, Forest Sciences Center;
and* [2]*Department of Microbiology and Immunology,
The University of British Columbia, Vancouver, British Columbia
V6T 1Z4, Canada, E-mail: saddler@interchange.ubc.ca*

Abstract

This article provides an overview of various theories proposed during the past five decades to describe the enzymatic hydrolysis of cellulose highlighting the major shifts that these theories have undergone. It also describes the effect of the cellulose-binding domain (CBD) of an exoglucanase/xylanase from bacterium *Cellulomonas fimi* on the enzymatic hydrolysis of Avicel. Pretreatment of Avicel with CBD_{Cex} at 4 and 37°C as well as simultaneous addition of CBD_{Cex} to the hydrolytic enzyme (Celluclast, Novo, Nordisk) reduced the initial rate of hydrolysis owing to irreversible binding of CBD proteins to the substrate's binding sites. Nonetheless, near complete hydrolysis was achieved even in the presence of CBD_{Cex}. Protease treatment of both pure and CBD_{Cex}-treated Avicel reduced the substrates' hydrolyzability, perhaps owing to proteolysis of the hydrolyzing enzyme (Celluclast) by the residual Proteinase K remaining in the substrate. Better protocols for complete removal of CBD proteins from the substrate need to be developed to investigate the effect of CBD adsorption on cellulose digestibility.

Index Entries: Cellulose; enzymatic hydrolysis; cellulases; cellulose-binding domain; *Cellulomonas fimi*.

Introduction

Cellulose hydrolysis has been studied for many decades, while recently new and specific applications have been refueling this interest. The production of low-cost, high-quality animal feed from wood or agricultural residues has been one of the first applications of cellulose digestion using

*Author to whom all correspondence and reprint requests should be addressed.

simple treatments. Today, however, enzymatic digestion of cellulose has applications in the textile industry; detergent manufacturing; and the production of energy, chemicals, and fuels from lignocellulosics. The wide range of applications for cellulose modification by biologically derived enzymes (cellulases) is a great example of how biotechnology can either replace or improve the traditional industrial processes while offering additional benefits, e.g., improving product quality and reducing environmental impacts.

The agricultural sector has always been interested in studying the mechanism of hydrolysis of cellulose. An understanding of the digestion of cellulosic materials (e.g., forage) in the digestive tract of feedlot cattle and other farm animals has enabled agricultural scientists to produce high-quality, low-cost feed from agricultural products and residues (e.g., feed crops, plant residues, or even wood). Cellulases can be added to grain-based fodder to reduce the feed viscosity and water uptake in animal digestive systems, thereby enhancing the amount of nutrients obtained from the feed *(1)*.

In the textile industry, hydrolysis of cellulose by fungal and bacterial enzymes has created new market opportunities and many economic benefits. Biopolishing of denim and other natural fabrics using cellulases has replaced, for the most part, the stonewashing operation for producing fabrics with softer texture and aged appearance, both in demand by the fashion market. Cellulases are also added to laundry detergents to remove "pills" formed on fabric surfaces during washing, hence improving the appearance of garments *(1)*. More technical studies have shown that changing the ratio of different components in an enzyme mixture, i.e., cellobiohydrolases to endoglucanases, can produce different results. Endoglucanase-rich mixtures are more effective in biopolishing (aging of denims) while complete enzyme mixtures enhance the depilling property in detergents without causing excessive abrasion *(2)*. Commercial cellulases have also been shown to enhance the whiteness, brightness, and color characteristics of cotton fabrics *(3)*.

The pulp and paper industry also takes advantage of the potential of cellulases in improving paper and fiber quality, reducing papermaking energy requirements, and removing ink from recycled paper. Cellulases have been shown to preferentially attack and hydrolyze shorter wood fibers, i.e., fines, produced during the refining operation. Refining, a mechanical action necessary for improving the physical properties of primary or secondary fibers, can generate small particles (fines) that can reduce the pulp's water drainage rate during papermaking operations *(4)*. By preferential hydrolysis of fines, enzymatic treatment can improve the pulp's drainage property and reduce the energy requirements of papermaking operations. A combination of cellulase treatment and refining has also been shown to provide a means for using coarse wood fibers (e.g., Douglas fir) to produce finer paper products *(5)*.

Enzymatic hydrolysis of cellulose is also a key step in the production of fuels and chemicals from lignocellulosic materials such as wood, grass, and waste paper. In a biomass-to-ethanol bioconversion process, chemical and mechanical treatments remove the majority of hemicelluloses from the feedstock and produce a highly digestible, cellulose-rich substrate that needs to be further hydrolyzed to obtain monomeric glucose for fermentation to ethanol.

Understanding and improving the interactions between cellulose and cellulase enzymes can help reduce the use of costly enzymes and improve the economy of the overall process. A more economic process will reduce the price of ethanol fuel, hence enabling it to compete with gasoline as a transportation fuel for road vehicles. Many social, economic, and environmental benefits can result from using renewable resources for energy production *(6)*.

Hydrolysis of cellulose by cellulase enzyme continues to be a research priority for scientists from a variety of disciplines owing to its diverse applications and its multifaceted nature (e.g., the enzyme components and properties, substrate characteristics, reactor configurations and process design).

In this article, we provide an overview of the theories that have been proposed to describe the enzymatic hydrolysis of cellulose and highlight the major shifts that these theories have undergone during the past five decades. We also report our observations of the interactions between an isolated cellulose-binding domain (CBD) from *Cellulomonas fimi* exoglucanase/xylanase, a hypothetically hydrolysis-inducing agent, and a model cellulosic substrate, Avicel.

Enzymatic Hydrolysis of Cellulose

Depolymerization of cellulose into its monomeric glucose units by cellulase enzymes is a complex phenomenon mediated by inter- and intramolecular synergism among the main components of the cellulase system—endoglucanases, cellobiohydrolases, and β-glucosidase. Both the enzyme mixture and the solid substrate have attributes that can strongly influence the dynamics of the reaction *(7)*. In addition, the reaction heterogeneity arising from the interaction between a solid substrate and the liquid enzyme further adds to the complexity of an enzymatic hydrolysis system. Adsorption of enzyme onto the solid surface, penetration into the crystalline structure, occurrence of the chemical reaction, desorption/ readsorption or movement of proteins on the cellulose structure, and, finally, diffusion of the hydrolysis products (glucose, cellobiose, or short oligomer chains) out of the porous structure seems to be the logical order of events during hydrolysis.

Simultaneous or consecutive occurrence of these events on various reaction sites on the substrate creates an intense nonlinear dynamic within the first few hours of the reaction. At any moment, different fractions of the enzyme are engaged in different actions. It has been shown that almost all the enzyme activities, except β-glucosidase, are adsorbed onto the substrate very rapidly. Swelling of the substrate accommodates the penetration of enzyme proteins into the structure. Through concerted action of endo- and exo-acting enzymes, cellulose chains are cleaved and cellobiose units are released into the solution. Once in the liquid phase, cellobiose is further cleaved by the action of β-glucosidase to produce soluble glucose monomers.

During the first few hours of the reaction, the rate of glucose production is linear with time. The intensity of reaction, however, subsides after a few hours and the reaction gradually slows down, hence exhibiting the typical biphasic pattern observed in any cellulose hydrolysis profile, i.e., of glucose production (or percentage of hydrolysis) vs time. The cause of this gradual drop in the reaction rate is not fully understood, but it has been postulated that both enzyme- and substrate-related properties contribute to this effect *(7)*. Examples of substrate-related factors include the degree of crystallinity and polymerization of cellulose chains, extent of accessible surface area, substrate porosity, and presence of extraneous materials such as lignin and hemicellulose in cellulosic substrates. Enzyme-related factors include thermal or mechanically induced deactivation of enzymes due to mixing or exposure to high temperatures, sieving (separation) of enzyme components by the substrate and the loss of synergism following this separation, as well as product inhibition due to accumulation of cellobiose and glucose in the reaction medium. Conflicting observations have prevented scientists from identifying a single factor as the sole cause of gradual loss of efficiency during the hydrolysis of cellulose. This is further confounded by the fact that research has not been able to provide a detailed and all-encompassing mechanism for hydrolysis of cellulose by cellulolytic proteins.

The first attempt at establishing a mechanistic model for enzymatic hydrolysis of cellulose was made by Elwyn Reese in 1950 *(10)*. His theory, known as the C_1-C_x model, suggested that a mechanical action (by the C_1 factor) precedes the actual hydrolytic action (by the C_x factor). Although groundbreaking in its own time, this theory did not describe the nature of C_1 and C_x factors. Building on this model, numerous attempts were made to attribute the C_1 and C_x activities to different components of an enzyme mixture. For example, Wood and McCrae *(11)* equated the C_1 and C_x factors with exo- and endo-acting enzymes, respectively, suggesting that exoglucanases that have a higher affinity for the cellulose chains start the reaction and that the action of endoglucanases would follow. While maintaining the same nomenclature (C_1 = exo and C_x = endo), in a later revision of their original proposition, Wood and McCrae *(12)* proposed that it is the endoglucanases (C_x) that initiate the attack on the cellulose chains as they are more random in their action and can create susceptible chain ends suitable for exoglucanases (C_1) to act on. This was a clear deviation from the original assumption of the C_1-C_x model that C_1 acts first and C_x follows. While it can be argued that the differences are solely in the interpretations and have no bearing on the actual theory of hydrolysis, we find it worth

Table 1
Highlights of Various Concepts Developed During the Past Five Decades
to Describe Mechanism by which Crystalline Cellulose Is Hydrolyzed

Year	Reference	Proposed mechanism
1950	10	The C_1-C_x model: A nonhydrolytic component (C_1) splits the structure of native cellulose and creates short linear chains and a hydrolytic component (C_x) depolymerizes the short chains into glucose.
1972	11	C_1 and C_x factors are, respectively, the exo- and endo-acting enzymes; that is, the exoglucanases initiate the attack.
1979	12	Endoglucanases (C_x) act more randomly and therefore are more likely to initiate the attack (C_x followed by C_1— the reverse of original Reese's model).
1985	13	The amorphogenesis action that swells, segments, and/or destratifies the substrate precedes any hydrolytic action by either exo or endo enzymes. Amorphogenesis may be mediated by some nonhydrolytic factors (*see* Table 2) or by H_2O_2 produced by other enzymes.
1986	21	C_1 is not a distinct component but a specific property of different enzymes, i.e., binding ability. The extent of catalysis is directly related to how tightly the enzyme is bound to the substrate.
1989	23	Synergism among CBHs and endoglucanases with varying stereospecificities cause the disaggregation of crystalline cellulose, which is then followed by the catalytic action.
1990	22	Mechanochemical action: Tightly bound enzymes attack the disturbed regions and disperse the crystalline cellulose by penetrating into the structure and opening new sites for the action of weakly absorbed enzymes.
1991	20	C_1 activity resides not in a system distinct from C_x but in a discrete domain of each enzyme. Cellulose binding modules initially defibrillate the substrate and render it more susceptible to the action of the catalytic core.

highlighting as it appears to be an interesting turning point in the evolution of cellulose hydrolysis theory, as summarized in Table 1.

In 1985, Coughlan (13) introduced the concept of amorphogenesis as a prerequisite for the hydrolytic action of the enzyme. His suggestion was that amorphogenesis (i.e., swelling, segmentation, or destratification of the substrate) renders the crystalline cellulose more accessible to the enzymes for hydrolytic dissolution. Thus, amorphogenesis was the equivalent of C_1-induced mechanical dispersion proposed by Reese (10).

Amorphogenesis has been attributed to the H_2O_2 produced by some enzymes (14), some iron-containing proteins in fungal filtrates (15), or even to CBH I produced by *Trichoderma reesei* (16). Table 2 summarizes some of

Table 2

Evidence for Existence of Nonhydrolytic Component in Cellulase Mixtures[a]

Reference	Nonhydrolytic factor	Substrate	Method of monitoring	Observation
17	MGF from *T. reesei* (~5 kDa)	Filter paper and corn leaf	SEM Glucose:phenol–sulfuric acid	MGF increased microfibril content of filter paper (SEM). Incubation of filter paper with 4 mL of MGF solution (23 µg/mL) for 24 h released 5.9 mg of soluble carbohydrates but did not enhance filter paper's enzymatic digestibility.
20	CBD_{CenA} (~11 kDa)	Ramie fiber (~72% crystalline)	SEM to examine fibers Carbohydrates: orcinol–sulfuric acid UV absorbance at 600 nm to measure particles	Treatment with CBD_{CenA} alone disrupted fiber structure and roughened the surface (SEM), and released some small particles (A_{600nm}) but no sugar.
18	CBH II$_{cp}$ from *T. reesei* (~36 kDa)	Avicel and cotton linters	TEM Reducing sugar by DNS	Incubation of 2.5 mg of cotton fibers with 194 µg of CBH II$_{cp}$ at 23°C for 18 h dispersed the fibers (TEM). Pretreatment with CBH II$_{cp}$ enhanced Avicel's enzymatic digestibility but did not produce any reducing sugars from Avicel or cotton.
19	FFP from *T. reesei* (11 to 12 kDa)	Filter paper (Whatman no. 1)	PCM UV absorbance at 600 nm Sugars by phenol–sulfuric acid	Incubation of 20 mg of filter paper with 0.3 mL of FFP (50–100 µg) at 10°C for 24 h caused partial disruption of filter paper (PCM) and produced free fibrils (A_{600nm}) but no sugars.

[a]MGF, microfibril-generating factor; CBD_{CenA}, CBD from *C. fimi* endoglucanase; CBH II$_{cp}$, catalytic domain of CBH II; FFP, fibril-forming protein; SEM, scanning electron microscopy; TEM, transmission electron microscopy; DNS, dinitrosalicylic acid; PCM, phase contrast microscopy.

the more recent studies that have tried to elucidate the nature of a nonhydrolytic factor responsible for mechanical dispersion of a substrate prior to catalytic reaction. Several groups *(17–19)* have been able to isolate a component or a portion of a component from *T. reesei* filtrate that had the ability to cause dispersion in the substrate. The microfibril-generating factor (MGF; 5 kDa) isolated by Krull et al. *(17)* increased the microfibril content of filter paper as observed by scanning electron microscopy. It also released soluble carbohydrates from filter paper but did not improve the digestibility of filter paper after a 24-h pretreatment at 50°C.

In a similar study, Banka et al. *(19)* isolated a larger nonhydrolytic component, referred to as fibril-forming protein (FFP; 11 to 12 kDa), from *T. reesei* that was able to disrupt filter paper as detected by phase contrast microscopy and produce free fibrils. Thus, the MGF and FFP seemed to play the role of C_1 factor in Reese's model *(10)*. Another interesting observation by Woodward et al. *(18)* showed that the catalytic domain (36 kDa) of CBH II from *T. reesei* (CBH II$_{CP}$) had no catalytic activity toward Avicel and cotton linters; nevertheless, transmission electron microscopy revealed that the cotton fibers were dispersed upon treatment with this protein. This nonhydrolytic effect makes the CBH II$_{CP}$ a good candidate for the C_1 activity and also implies that the hydrolytic activity of this particular catalytic domain is contingent on the presence of its related binding domain. Din et al. *(20)* observed that the catalytic domain of *C. fimi* endoglucanase could hydrolyze cotton to a limited extent, even in the absence of its binding domain. This is perhaps owing to the differences in the origin of the two enzymes and how their domains were isolated (e.g., intensity of papain digestion).

While the search for an individual component that can fit the role of Reese's C_1 factor continued, an interesting study by Klyosov et al. *(21)* shifted the concept of C_1 factor as a "component" to a "property." Klyosov suggested that "C_1 factor is not an individual substance or enzyme with a particular specificity but rather a property of already known enzymes, namely the capacity of cellulases for binding onto the surface of insoluble cellulose." He further elaborated his theory *(22)* and proposed that the tightly bound enzymes initiate the attack at the disturbed regions of the crystalline cellulose and disperse the structure through a mechanochemical action, creating more-accessible areas of attack for the weakly bound, more-mobile enzymes that will carry out the catalytic reaction. At the amorphous regions, however, the quality of binding is not an issue, and the quantity of enzyme available will determine how fast and how much of the substrate will be hydrolyzed.

Klyosov's *(22)* theory also had interesting implications in terms of synergism. It is known that during the hydrolysis of cellulose multiple components of an enzyme mixture act in harmony to produce a synergistic effect that is greater than the sum of individual effects. The complementary role of different components in an enzyme mixture has been attributed to their different stereospecificities *(23)*; however, Klyosov *(22)* proposes that the synergism occurs because of the existence of multiple enzyme components with various adsorption capacities. The presence of a tightly bound endoglucanase ensures effective dispersion of the substrate and entails an efficient hydrolysis, even if other components (e.g., CBHs) are only weakly bound. Thus, the tightness of binding is equated with C_1 factor as the per-quisite for an efficient catalytic degradation.

The pioneering work of Din et al. *(20)* in 1991 with the cellulose-binding domain (CBD) of endoglucanase A (CenA) from the bacterium *C. fimi* brought yet another shift to the way the nonhydrolytic component was perceived. This work suggested that the C_1 activity resides not in a system distinct from C_x but in a discrete domain of each enzyme. It was shown that the isolated CBD_{CenA} causes a roughening of the surface of cellulosic fibers (cotton and ramie) and releases small particles into the solution that can be detected spectrophotometrically at 600 nm. It was also proposed that this protein could penetrate into structure and create free chain ends; however, it could not be verified whether the protein is indeed capable of destabilizing the hydrogen bonds within the crystal. This nonhydrolytic effect was observed only with the binding domain, and not the isolated catalytic core of the same enzyme.

Recently, a 50-kDa protein called swollenin was reported in *T. reesei* *(24)*. Incubation of cotton fibers with a crude swollenin preparation appears to result in an opening and swelling of fiber structure without hydrolysis. The protein comprises an N-terminal CBD joined by a linker to a region with sequence similarity to expansins (molecules implicated in cell wall extension in higher plants).

The primary role of carbohydrate-binding domains is to increase the "local" concentration of enzyme, thereby enhancing the hydrolysis reaction rate. However, it has been postulated, and in some cases shown, that the binding domains of cellulases perform functions other than physical or chemical binding *(20,25)*. These additional functions also seem important for initiating and maintaining an efficient hydrolysis. For example, it has been proposed that CBDs of exoglucanases can act as a plough, delaminating cellulose layers and releasing free chain ends. The enzymes' catalytic domains will then act on these free ends and depolymerize the cellulose structure *(25)*.

The study by Din et al. *(20)* also showed that "small particles" from cotton were released into the medium as a result of treatment with CBD_{CenA}. These particles are assumed to be small cellulose fragments that are noncovalently bound to the fiber surface and sloughed off during CBD binding. The isolated catalytic domain of this enzyme, on the other hand, was shown to have a polishing effect on the fiber surface. It has been suggested that this effect is caused by the hydrolysis of glycosidic bonds at fibers' mechanically damaged regions, which, in turn, removes the short fibers attached to these areas and smooths the substrate's surface.

Jervis et al. *(26)* did not observe such disruptive effects in treatment of highly crystalline cellulose from *Valonia ventricosa* with CBD_{CenA}. However, they showed that the majority (70%) of bound CBDs are mobile on the crystalline surface, hence undermining the "binding-site exclusion" theory suggested by McGhee and von Hippel *(27)*. Thus, this study showed that binding of a CBD molecule does not permanently exclude the corresponding binding sites from the pool of substrate-binding sites. Instead, the mobility of CBD makes those sites available for new binding upon its displacement. In this study, surface diffusion, however, was not found to be the rate-limiting factor during hydrolysis of cellulose, although it might have important implications in terms of processivity of cellulase action on cellulose.

Creagh et al. *(28)* studied the structural changes of bacterial microcrystalline cellulose (BMCC) after treatment with increasing amounts of CBD_{Cex} using confocal microscopy. The degree of dispersion of the BMCC fibrils was found to be dependent on the concentration of bound CBD_{Cex}.

The work presented herein is our first step in testing the hypothesis that the dispersing effect of CBDs can increase the interfibrillar space, thereby increasing the available surface area and enhancing the rate and extent of hydrolysis of a cellulosic substrate.

Materials and Methods

CBD_{Cex} Protein

The CBD of *C. fimi* exoglucanase/xylanase (Cex) was produced at the University of British Columbia Biotechnology Laboratory according to protocols described previously *(29)*. The protein solution was concentrated to 1–1.5 mg/mL by ultrafiltration using a YM1 (1 kDa) membrane (Millipore, Bedford, MA). Avicel was used as the solid substrate in all experiments.

Adsorption Assays

For quantitative analysis, CBD_{Cex} (1–30 µ*M*) was mixed end-over-end with 5 mg of Avicel (1% slurry) in a final volume of 1 mL of buffer (50 m*M* potassium phosphate, pH 7.0) at 4 or 37°C. After 90 min, at 4 or 37°C, cellulose was removed by centrifugation (13,000*g*, at respective temperature), and the free protein left in the supernatant was measured (280 nm) and used to calculate the amount of CBD_{Cex} bound to Avicel. All experiments were performed in duplicate. The results were fitted to a Langmuirian equation in which the affinity binding constants (K_a) and the saturation capacity (N_o) were obtained.

Determination of Protein Concentration

The concentration of CBD_{Cex} was determined by ultraviolet (UV) absorbance at 280 nm, in which the molar extinction coefficient of 0.027625 µM^{-1}cm^{-1} was used.

Proteolysis by Proteinase K

Saturating amounts of CBD_{Cex} protein (20 µ*M*) were mixed (4°C) end-over-end with 5 mg of Avicel in a final reaction volume of 1 mL of buffer

(50 mM potassium phosphate, pH 7.0). After 90 min, cellulose was spun down by centrifugation (13,000g, 4°C), and the concentration of free protein was measured using a fluorimeter (Perkin-Elmer LS50 Luminescence Spectrometer), which, in turn, was used to measure the amount of CBD$_{Cex}$ bound to Avicel. This entailed constructing a standard curve of the fluorescence emission at 537 nm vs CBD$_{Cex}$ concentration. The cellulose pellet was rinsed with Tris buffer five times with centrifugation (13,000g, 4°C) following each rinse. The pellet was then treated with Proteinase K (50 µg/mL, 37°C, 16 h) to remove bound protein. After overnight incubation, the pellet was centrifuged (13,000g, 4°C, 15 min), and the free CBD$_{Cex}$ was measured using the fluorimeter. No fluorescence was detected from Proteinase K alone. The cellulose pellet was further rinsed with potassium phosphate buffer (pH 7.0) five times and centrifuged (13,000g, 4°C, 5 min) after each rinse. The pellet was then treated with guanidinium hydrochloride (6 M, room temperature, 5 h) to remove any bound protein remaining after the protease treatment; this method was used to estimate the amount of CBD$_{Cex}$ removed after Proteinase K treatment. The pellet was centrifuged (13,000g, 4°C, 15 min), and the free protein was measured using the fluorimeter.

Hydrolysis of CBD$_{Cex}$-Treated Avicel

Avicel slurry (1%) was prepared in sodium acetate buffer, pH 4.8, to which CBD$_{Cex}$ was added (final concentration of 2 or 20 µM) in a reaction volume of 10 mL of buffer (0.05 M sodium acetate, pH 4.8). After overnight incubation at either 4 or 37°C, cellulose was removed by centrifugation (14,000g, room temperature, 15 min), and the pellet was rinsed with buffer (sodium acetate, pH 4.8) and transferred volumetrically into 10-mL flasks. Celluclast (40 filter paper units [FPU]/g cellulose) was added to the mixture, which was then supplemented with Novozyme 188 (80 cellobiase units [CBU]/g cellulose). The final volume was brought to 10 mL by the addition of 0.05 M sodium acetate buffer, pH 4.8. The samples were transferred to 50-mL Erlenmeyer flasks and incubated in a 45°C shaker bath. Samples (300 µL) were taken at 1, 5, 24, 50, 75, and 100 h and boiled for 5 min prior to storage at –20°C. Controls included samples with no CBD$_{Cex}$ pretreatment (0.05 M sodium acetate buffer, pH 4.8 only) followed by Celluclast, and samples that received CBD$_{Cex}$ treatment but no Celluclast (0.05 M sodium buffer, pH 4.8 only). Also, there were samples that did not receive any protein treatment but consisted only of Avicel and buffer (0.05 M sodium acetate buffer, pH 4.8). All samples were run in duplicate. Enzymes used in this study, i.e., Celluclast (complete cellulase system, 98.06 FPU/mL) and Novo 188 (β-glucosidase, 462.6 IU/mL) were provided by Novo Nordisk (Denmark).

Sugar Analysis

Glucose concentration was measured by high-performance liquid chromatography (HPLC). The HPLC system (Dionex DX-300; Dionex, CA) was equipped with an ion-exchange PA1 (Dionex) column, a pulsed

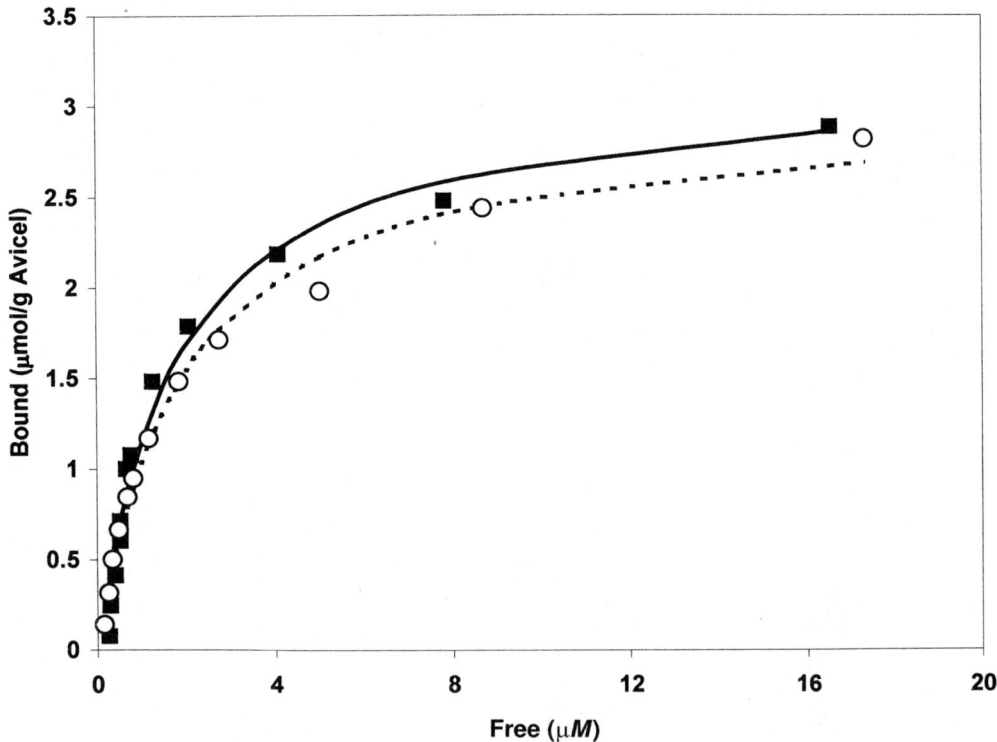

Fig. 1. Adsorption isotherms of CBD_{Cex} on Avicel. Lines represent Langmuirian fit to the experimental data. Five milligrams of Avicel (1% slurry) was incubated with 500 μL of CBD_{Cex} solution (1–30 μM) at 4 (■) or 37°C (○) overnight. Free protein concentration was measured by UV absorption at 280 nm ($\varepsilon = 0.027625$ L/[μM · cm]), and bound protein concentration was calculated by subtracting the free from total protein added.

amperometric detector with a gold electrode, and a Spectra AS3500 autoinjector (Spectra-Physics, CA). Prior to injection, samples were filtered through 0.45-μm HV filters (Millipore, Bedford, MA) and a volume of 20 μL was loaded. The column was equilibrated with 250 mM NaOH and eluted with deionized water at a flow rate of 0.8 mL/min.

Defibrillation Experiments

Twenty-milligram pieces of filter paper (Whatman no. 1) were treated with 300 μL of CBD_{Cex} solution (20 μM) at room temperature. Samples were taken at 24 and 48 h and were analyzed for glucose content, small particle release (adsorption at 600 nm), and filter paper weight loss after being freeze-dried.

Results and Discussion

We generated two isotherms for adsorption of CBD_{Cex} on Avicel at 4 and 37°C. These isotherms (Fig. 1) revealed that temperature did not have

Table 3
Kinetic Parameters of Langmuir Equation
for Binding of CBD_{Cex} to Avicel at 4 and 37°C[a]

Substrate	K_a (M^{-1})	N_o (mol/g Avicel)
Avicel		
4°C	0.4×10^6	3.3×10^{-6}
37°C	0.5×10^6	3.0×10^{-6}
Solka floc (4°C)	0.5×10^6	2.9×10^{-6}

[a]The experimental data were fitted to a Langmuir-type adsorption model: $[B] = [N_o]K_a[F]/1 + K_a[F]$, in which $[B]$ and $[F]$ are the concentrations of bound (mol/g) and free (mol/L) protein in the supernatant, respectively; N_o is the initial concentration of binding sites (mol/g); and K_a is the affinity constant (L/mol). At either 4 or 37°C, 5 mg of Avicel or Solka Floc (1% slurry) was mixed with 1–30 µM CBD_{Cex} (for 90 min) in 50 mM potassium phosphate (pH 7.0) to a final volume of 1 mL. The concentration of free protein $[F]$ was measured spectrophotometrically $(A_{280\,nm})$, and the amount of bound protein was determined by subtracting $[F]$ from the protein initially added.

an appreciable effect on the adsorption capacity of Avicel for CBD_{Cex}. At both temperatures (4 and 37°C), saturation was achieved with about 2.7 µmol of protein/g of Avicel. The experimental data were fitted to a Langmuir-type adsorption model:

$$[B] = [N_o]K_a[F]/1 + K_a[F]$$

in which $[B]$ and $[F]$ are the concentrations of bound (mol/gram) and free (mol/liter) protein in the supernatant, respectively. N_o is the initial concentration of binding sites (mol/gram) and K_a is the affinity constant (liters/mol). Table 3 presents the kinetic parameters for these two isotherms. The kinetic values also reveal that Solka Floc had a lower adsorption capacity (12% at 4°C) for the CBD_{Cex} while exhibiting the same affinity (K_a) for the protein as Avicel.

We pretreated Avicel with saturating amounts of CBD_{Cex} hoping that adsorption of this protein onto Avicel would defibrillate the substrate as observed by Din et al. *(20)* and enhance the hydrolyzability of the substrate. However, within the constraints of this study, this hypothesis could not be proven, and, in fact, the opposite was observed. As shown in Figs. 2 and 3, presaturation of Avicel with CBD_{Cex} at both 4 and 37°C initially hindered the hydrolysis reaction as manifested by lower reaction rates within the first 24 h. After the first day, however, the reaction rates were restored and hydrolysis yields similar to those of controls were achieved. It seems reasonable to assume that this phenomenon is due to the irreversible binding of CBD_{Cex} onto the substrate. Tomme et al. *(30)* have shown that the adsorption of CBD_{Cex} to BMCC is irreversible. We also tried desorbing this protein from Avicel (results not shown) by either diluting or replacing the supernatant (at equilibrium) with buffer, but no desorption could be observed.

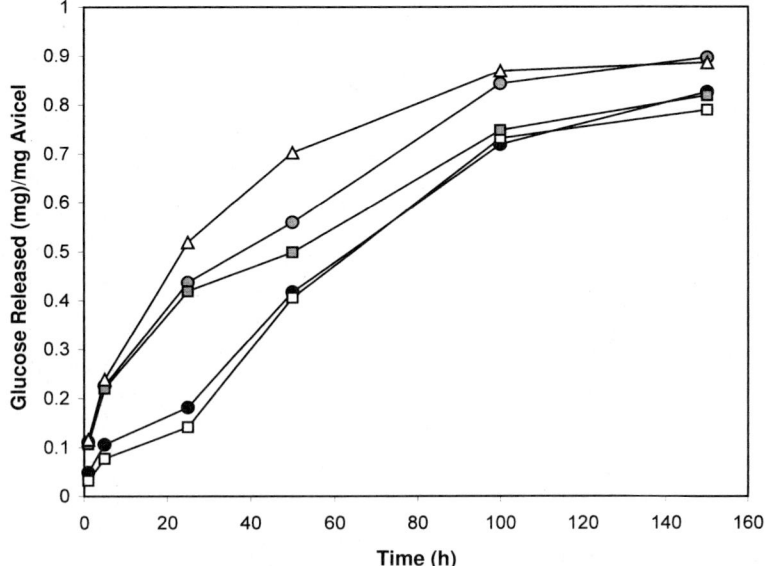

Fig. 2. Rate and extent of glucose production from enzymatic hydrolysis of untreated and CBD_{Cex}-treated Avicel (4°C). In simultaneous runs, CBD_{Cex} and hydrolyzing enzymes were added at the same time. Prior to hydrolysis (40 FPU/g of Avicel; CBU:FPU = 2:1) with Celluclast and Novo 188 (Novo) at 45°C, Avicel was incubated at 4°C with enough CBD_{Cex} solution to achieve either full saturation or 10% saturation. Glucose released was measured by HPLC.

We postulate that the presence of CBD_{Cex} on the surface and within the pores limits the accessibility of the substrate by the components of the hydrolyzing enzyme (Celluclast). The gradual movement of CBD_{Cex} proteins, as observed by Gilkes et al. *(29)*, or the competition between Celluclast components and the preadsorbed proteins for the same adsorption sites initiates the hydrolysis reaction although with a slower rate. This initial hydrolysis disperses the substrate and exposes new surfaces that have not been precoated with CBD_{Cex}. Once the available surface area is sufficiently large, the proportion of surfaces covered by CBD_{Cex} becomes insignificant and the hydrolysis reaction proceeds with the same rate as with pure Avicel.

We removed the CBD_{Cex} from the substrate by proteolysis so that it would not block the substrate pores and limit enzyme accessibility. Our hypothesis was that if CBD_{Cex} adsorption is able to alter the pores and enhance the accessibility of cellulose structure, upon its removal using a noninvasive method such as proteolysis, the substrate should exhibit better hydrolyzability because of the increase in its pore size. However, the

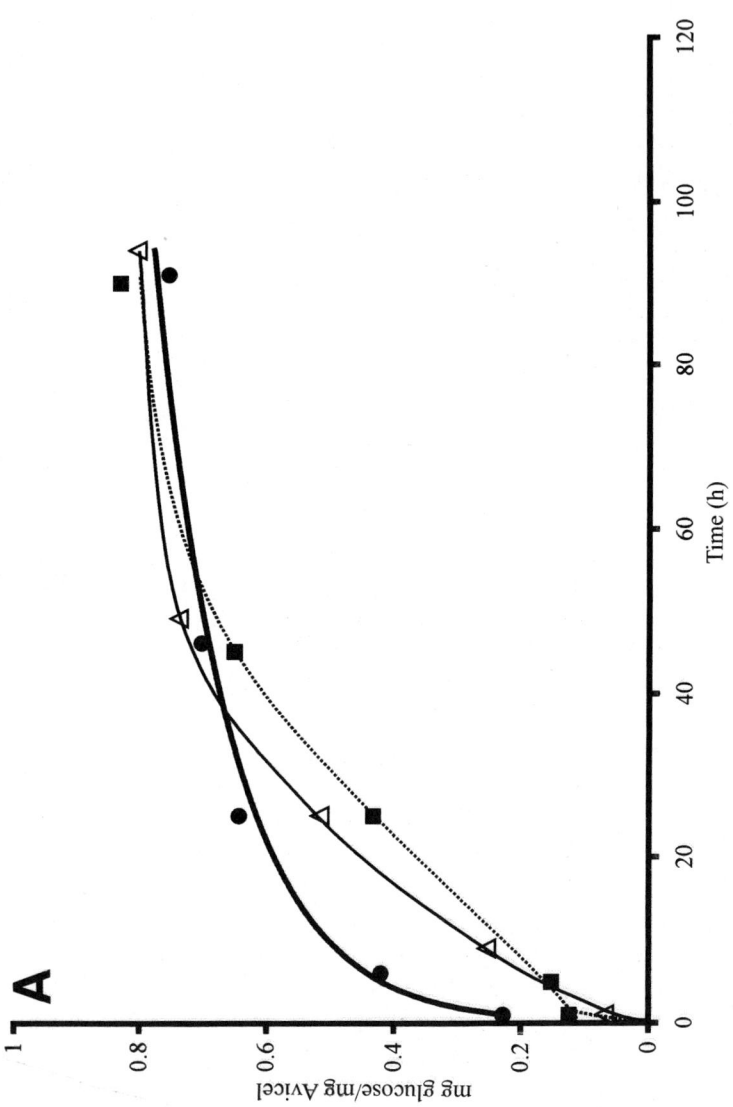

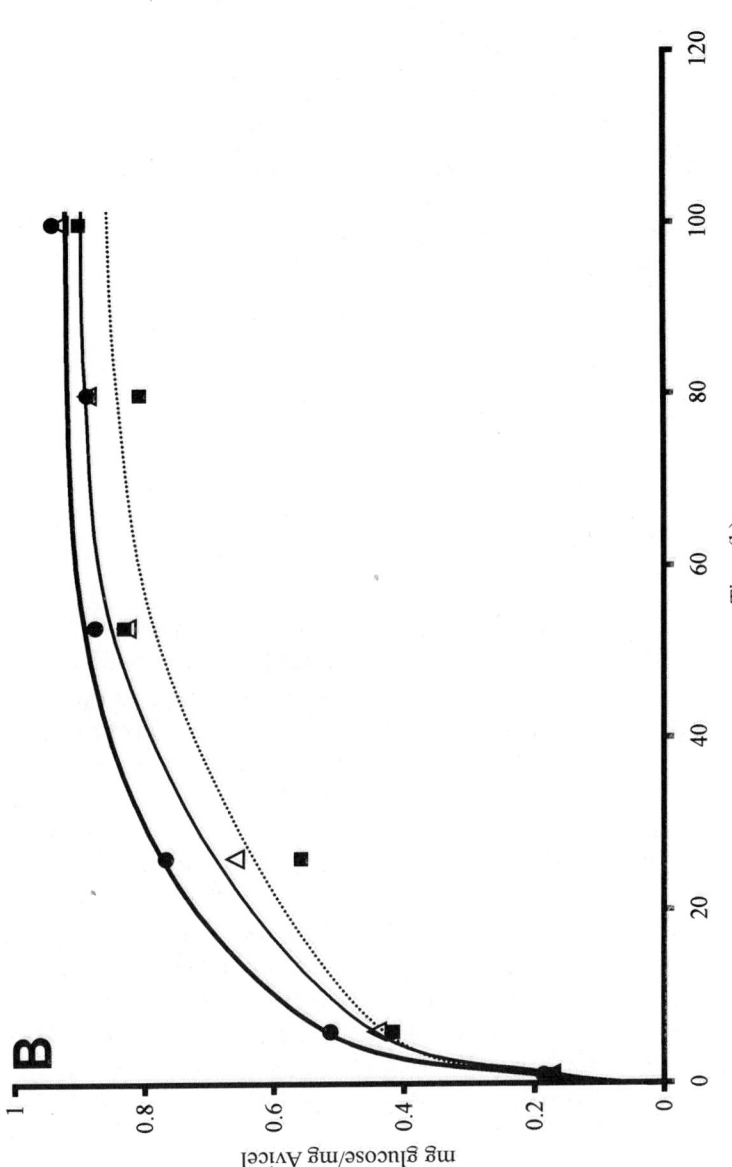

Fig. 3. Rate and extent of glucose production from enzymatic hydrolysis of untreated (——) and CBD$_{Cex}$-treated (- - ■ - -) Avicel at 37°C. In simultaneous runs (—Δ—), CBD$_{Cex}$ and hydrolyzing enzymes were added at the same time. Prior to hydrolysis (40 FPU/g of Avicel; CBU:FPU = 2:1) with Celluclast and Novo 188 (Novo) at 45°C, Avicel was incubated at 37°C with enough CBD$_{Cex}$ solution to achieve either full saturation (**A**) or 10% saturation (**B**). Glucose released was measured by HPLC.

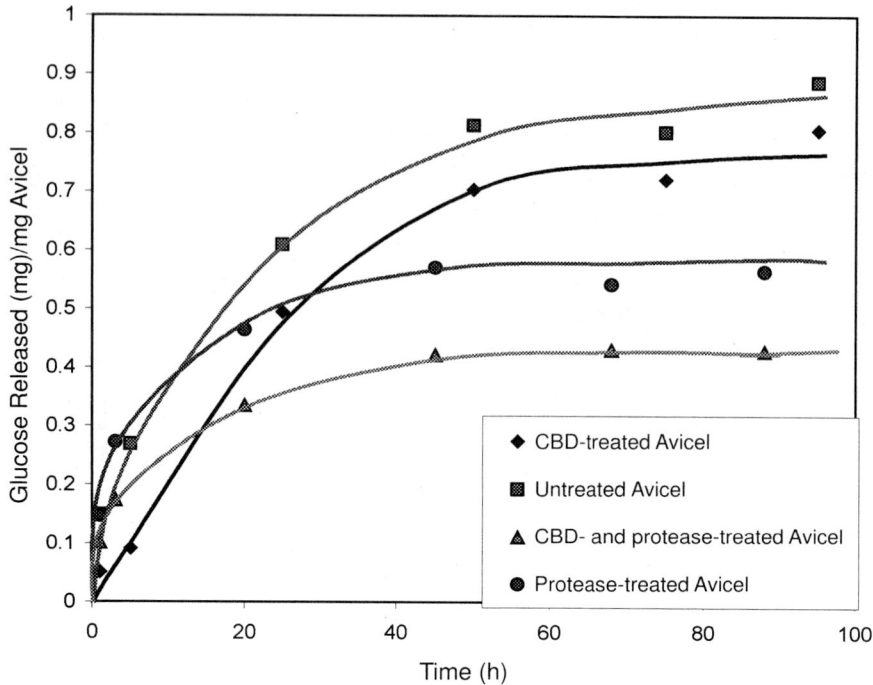

Fig. 4. Hydrolysis of Avicel treated with CBD_{Cex} (20 μM, 37°C, 24 h) and protease.

contrary was observed (Fig. 4). Treatment of both untreated (control) and CBD_{Cex}-treated Avicel with Proteinase K reduced the extent of hydrolysis significantly. This seems to be due to the entrapment of protease in the substrate and its subsequent impact on the hydrolyzing enzyme (Celluclast). Multiple washing could not remove the residual protease, and apparently its release during hydrolysis rendered Celluclast less effective. We are currently testing other chemical methods for removing the protein from Avicel to elucidate the impact of CBD_{Cex} adsorption on substrate hydrolyzability.

Extended incubation of small pieces of filter paper with CBD_{Cex} solution according to the protocol suggested by Banka et al. *(19)* did not release any glucose or small particles (as measured by A_{600nm}) into the solution. However, we observed a 2.8% weight loss in all filter paper samples treated with CBD_{Cex}. It is likely that the amount of fragments separated from the substrate was not significant enough to be detected by spectrophotometery.

Conclusion

The irreversible nature of CBD_{Cex} adsorption onto Avicel initially reduced the effective interaction between the substrate and Celluclast enzyme. The presence of CBD proteins reduced the rate of hydrolysis reaction within the first 24 h. After 24 h, however, these rates were restored to the control level, perhaps due to the dispersion of Avicel and creation of new surface areas. Ultimately, the CBD-treated samples achieved near-complete hydrolysis. Removal of CBD_{Cex} by proteolysis could not clearly elucidate the impact of CBD adsorption, because the presence of residual protease, which could not be removed by repeated washing, degraded the hydrolyzing enzymes and produced lower hydrolysis yields even in the control sample (Avicel treated with protease). The question that whether CBD_{Cex} adsorption can enhance substrate porosity and enzymatic hydrolyzability remains unanswered. Complete removal of adsorbed protein from the substrate prior to any further investigation will be necessary. It must also be ensured that the protein removal agent will not affect the hydrolytic activity of the hydrolyzing enzyme.

Acknowledgments

We would like to thank Dr. Bill Cruickshank of Natural Resources Canada for his support of this work. We also wish to thank Josephine Chow and Mahina Bilodeau from Simon Fraser University (Vancouver, British Columbia, Canada) for their technical assistance.

References

1. Tolan, J. S. and Foody, B. (1999), *Adv. Biochem. Eng./Biotechnol.* **65**, 41–67.
2. Cavaco-Paulo, A. (1998), in *Enzyme Applications in Fiber Processing*, Eriksson, K. E. and Cavaco-Paulo, A., eds., ACS Symposium Series 687, American Chemical Society, Washington, DC, pp. 180–189.
3. Csiszar, E., Szakacs, G., and Rusznak, I. (1998), in *Enzyme Applications in Fiber Processing*, Eriksson, K. E. and Cavaco-Paulo, A., eds., ACS Symposium Series 687, American Chemical Society, Washington, DC, pp. 204–211.
4. Eriksson, L., Heitmann, J., and Venditti, R. (1998), in *Enzyme Applications in Fiber Processing*, Eriksson, K. E. and Cavaco-Paulo, A., eds., ACS Symposium Series 687, American Chemical Society, Washington, DC, pp. 41–54.
5. Mansfield, S., Swanson, D. J., Roberts, N., Olson, J., and Saddler, J. (1999), *Tappi* **5**, 152–158.
6. Sheehan, J. and Himmel, M. (1999), *Biotechnol. Prog.* **15**, 817–827.
7. Mansfield, S., Mooney, C., and Saddler, J. N. (1999), *Biotechnol. Prog.* **15**, 804–816.
8. Boussaid, A. and Saddler, J. (1999), *Enzyme Microb. Technol.* **24**, 138–143.
9. Yu, A., Lee, D., and Saddler, J. (1995), *Biotechnol. Appl. Biochem.* **21**, 203–216.
10. Reese, E. T., Siu, R. G. H., and Levinson, H. S. (1950), *J. Bacteriol.* **59**, 485–497.
11. Wood, T. and McCrae, S. (1972), *Biochem. J.* **128**, 1183.
12. Wood, T. and McCrae, S. (1979), in *Hydrolysis of Cellulose: Mechanism of Enzymatic and Acid Catalysis*, Brown, R. D. and Jurasek, L., eds., ACS Series 181, American Chemical Society, Washington, DC, pp. 181–209.
13. Coughlan, M. (1985), *Biotech. Genet. Eng. Rev.* **3**, 39–109.
14. Koenigs, J. W. (1975), *Biotech. Bioeng. Symp.* **5**, 151–159.
15. Griffin, H., Dintzis, F. R., Krull, L., and Baker, F. L. (1984), *Biotech. Bioeng.* **26**, 296–300.
16. Chanzy, H., Henrissat, B., Vuong, R., and Schulein, M. (1983), *FEBS Lett.* **153**, 113–117.
17. Krull, L., Dintzis, F., Griffin, H., and Baker, F. (1988), *Biotech. Bioeng.* **31**, 321–327.
18. Woodward, J., Affholter, K., Noles, K., Troy, N., and Gaslightwala, S. (1992), *Enzyme Microb. Technol.* **14**, 625–630.
19. Banka, R., Mishra, S., and Ghose, T. (1998), *World J. Microbiol. Biotechnol.* **14**, 551–558.
20. Din, N., Gilkes, N., Tekant, B., Miller, R., Warren, R., and Kilburn, D. (1991), *Bio/Technology* **9**, 1096–1099.

21. Klyosov, A., Mitkevich, O., and Sinitsyn, A. (1986), *Biochemistry* **25,** 540–542.
22. Klyosov, A. (1990), *Biochem.* **29(47),** 10,577–10,585.
23. Wood, T. (1989), in *Enzyme Systems for Lignocellulose Degradation*, Coughlan, M. P., ed., Elsevier, London, pp. 17–36.
24. Swanson, B. A., Ward, M., Pentilla, M., and Saloheimo, M. (1999), International Patent, WO 99/02693.
25. Teeri, T., Reinikainen, T., and Ruohonen, L. (1992), *J. Biotechnol.* **24,** 169–176.
26. Jervis, E. J., Haynes, C. A., and Kilburn, D. G. (1997), *J. Biol. Chem.* **272,** 24,016–24,023.
27. McGhee, J. D. and von Hippel, P. H. (1974), *J. Mol. Biol.* **86,** 460–489.
28. Creagh, L., Ong, E., Jervis, E., Kilburn, D., and Haynes, C. (1996), *Proc. Natl. Acad. Sci. USA* **93,** 12,229–12,234.
29. Gilkes, N. R., Warren, R., Miller, R. C., and Kilburn, D. G. (1988), *J. Biol. Chem.* **263,** 10,401–10,407.
30. Tomme, P., Boraston, A., McLean, B., Kormos, J., Creagh, A., Sturch, K., Gilkes, N., Haynes, C., Warren, R., and Kilbrun, D. (1998), *J. Chromatogr. B* **715,** 283–296.

SESSION 6

Enzymatic Processes and Enzyme Production

Enzymatic Processes and Enzyme Production

DAVID SHORT[1] AND JEFFREY TOLAN[2]

[1]DuPont Central Research and Development, Newark, DE
and [2]Iogen Corporation, Ottawa, Canada

The use of enzymes in industrial processes is growing rapidly. For example, the 1990s brought the introduction of xylanase enzymes for bleaching pulp and cellulase enzymes for stonewashing denim. Other chemical industry examples include the use of nitrile hydratase to convert acryloniltrile to acrylamide, currently carried out in Japan at the 10,000 tons per year scale, and of course the large-scale commercial use of glucose isomerase for the production of high-fructose corn syrup. All of these processes have seen the use of enzymes displace toxic or corrosive chemicals or aggressive chemical process conditions, thereby improving the plant operating conditions and/or the environment around the plant. These processes are also cost effective if they are to be used for an extended period of time, and in some cases have displaced older, classical chemical processes.

There are several distinct stages in the development of enzymatic processes. First, the enzyme effect must be discovered and the appropriate enzyme isolated and identified. In some cases, the natural producing microbe makes adequate quantities for a commercial process, but often the key enzymes must be cloned and overexpressed. This is the second step, identifying the genes encoding the enzyme and overexpressing the enzyme in a production strain. These commercial-scale expression systems employing suitable hosts must be developed and shown to be cost-effective by economic analysis. Third, the enzymes themselves must often be improved. This involves modification of the protein by chemical addition, site-directed mutagenesis, or random mutagenesis. Fourth, the actual enzymatic process must be defined and proven on the pilot scale. Enzymatic processes do not occur in isolation, but take place after physical and/or chemical preparation of the substrate, and are followed by complex separation, purification, and recovery processes. The pilot effort can identify further developments possible or required within the enzyme conversion step or in other parts of the process. Fifth, if piloting is successful, the process must be demonstrated at the semi-commercial or commercial scale.

This session reported examples of most of these important stages of enzyme-based processes. The broad scope of this research is shown by the far-reaching scope of the authors, who represented five different countries in seven papers. The breadth of disciplines represented, ranging from molecular and microbiology through chemistry and process engineering, also demonstrates this broad scope. And so does the breadth of scales involved, ranging from laboratory-scale fundamental research to a large-scale, fully integrated, commercial demonstration plant.

The first symposium paper, entitled "Expression of Thermostable Nitrile Hydratase in *Escherichia coli* for Acrylamide Production," by Patrick Oriel of Michigan State University, described the early stages of the enzyme development process in which the genes are cloned to improve the production of a key enzyme in the synthesis of acrylamide. The second paper, "Control of the Growth Rate in Cellulase Hyperproducers Derived from *Trichoderma reesei* by Haploidization and Autopolyploidization Techniques," by Hideo Toyama of Minamikyushu University in Japan, also described the overexpression of genes, in this case for cellulase production.

The third paper, "Preparation and Characterization of Metallized Cellulose Binding Domains," by Barbara Evans of Oak Ridge National Laboratory, described the properties of cellulase enzymes used to hydrolyze cellulose. In this case, the authors are improving the activity of the enzymes by attaching a metal ion complex. The fourth paper, "Interaction of *Trichoderma reesei* Cellulases and Lignocellulosic Materials," by Hetti Palonen of VTT Biotechnology in Finland, follows the track of characterizing cellulase kinetics and performance as a means of improving the enzymes.

The fifth paper, "Two-Step Steam Pretreatment of Softwood Impregnated with Dilute Sulfuric Acid for Ethanol Production," by Johanna Soderstrom of Lund University in Sweden, moved to the next stage of development by focusing on the need for a better cellulose pretreatment prior to enzymatic hydrolysis. The sixth paper, entitled "Bioethanol Production Using Simultaneous Saccharification and Cofermentation: Impact of Cellulase Enzyme Quality," by Jim McMillan of the National Renewable Energy Laboratory, described the kinetics of cellulase hydrolysis of cellulose and presented preliminary process scale-up data based on use of an improved cellulase.

The seventh and final paper was "Progress Report: Iogen's Demonstration Plant for Conversion of Cellulosic Biomass to Ethanol by Enzymatic Hydrolysis," by Brian Foody of Iogen Corporation. This paper described the ethanol demonstration plant that Iogen has built in Ottawa, Canada, which uses enzymatic hydrolysis as a key step in the process. There were also 25 poster presentations, many of which are presented in these Proceedings.

To conclude, the session chairs thank the authors for their contributions to a timely and well-received session. With this degree of attention and sharing of technical insights, enzymatic processes will continue to develop in the years ahead.

Cobalt Activation of *Bacillus* BR449 Thermostable Nitrile Hydratase Expressed in *Escherichia coli*

SANG-HOON KIM, RUGMINI PADMAKUMAR, AND PATRICK ORIEL*

Department of Microbiology, Michigan State University, 40A Giltner Hall, East Lansing, MI 48824, E-mail: oriel@pilot.msu.edu

Abstract

Expression of nitrile hydratase enzymes utilized in a new "green" process for acrylamide production has proven difficult in *Escherichia coli* owing to lack of a cobalt transport system to introduce the required cobalt ion into this host. We describe the expression of a thermostable nitrile hydratase from a moderate thermophile *Bacillus* sp. BR449 in *E. coli* in which the cobalt required for enzyme activation is introduced by incubation of the apoenzyme in the presence of Co^{++} ion at 50°C, yielding active and thermostable enzyme.

Index Entries: Nitrile hydratase; cobalt; thermophile; acrylamide; enzyme activation.

Introduction

Hydration of acrylonitrile catalyzed by acid or mixed metals is a key step in the production of acrylamide, a commodity chemical utilized as polyacrylamide in waste treatment, secondary oil recovery, and textile applications. More recently, the utilization of nitrile hydratase enzyme catalysts to reduce catalyst costs and environmental impact has been initiated by Nitto Chemical in Japan *(1)*. Initial production by Nitto catalyzed by an Fe^{+++}-containing nitrile hydratase from *Rhodococcus* sp. N-774 has been replaced by a more effective Co^{+++}-containing nitrile hydratase from *Rhodococcus rhodochrous* J1 *(2)*. Investigations of the genes encoding the α- and β-subunits for both iron- and cobalt-containing nitrile hydratase enzyme families have revealed extensive sequence similarities and common participation of three cysteine or oxidized cysteine residues in the α-subunit, which are believed to position and stabilize the metal ion at the active site (reviewed in ref. *3*).

*Author to whom all correspondence and reprint requests should be addressed.

Cloning of genes for nitrile hydratase into *Escherichia coli* desirable for obtaining increased enzyme content, eliminating contaminating amidases, and genetically improving enzyme characteristics has proven difficult owing to formation of insoluble inclusion bodies and difficulties in the provision and insertion of cobalt into nitrile hydratase *(4,5)*. Wu et al. *(6)* have described cloning of a cobalt-containing nitrile hydratase from *Pseudomonas putida* 5B in which expression of active enzyme in *E. coli* is aided by coexpression of a downstream gene encoding an "activator" protein. Nojiri et al. *(7)* achieved increased expression of the iron-containing nitrile hydratase of *Rhodococcus* sp. N-771 using a similar strategy. The function of the activator protein is not yet known but may be involved in the introduction of cobalt ion into the cell *(3)* or into the enzyme *(6,7)*. We previously described the cloning of genes encoding a thermostable nitrile hydratase from a moderately thermophilic *Bacillus* strain BR449 into *E. coli (8)*. Although growth at 32°C allowed high expression levels for both nitrile hydratase α- and β-subunits, activity levels were limited even when cobalt ion was provided in the growth medium and were not increased by coexpression of an open reading frame associated with the amidase/nitrile hydratase gene cluster. In this article, we describe effective activation of the overexpressed nitrile hydratase from BR449 in *E. coli* using enzyme incubation at an elevated temperature in the presence of Co^{++} ion. We also describe purification and initial characterization of the enzyme.

Materials and Methods

Bacterial Strain and Cultivation

Construction of the *E. coli* recombinant EC463, which contains the nitrile hydratase α- and β-subunit genes from the acrylonitrile-resistant moderate thermophile *Bacillus* sp. BR449 on a 2.4-kb insert has been described *(8)*. EC463 was grown in Luria Bertani broth containing 50 μg/mL of ampicillin at 32°C in a gyrorotatory water bath.

Enzyme Assay

Nitrile hydratase was assayed using formation of acrylamide from 0.5 M acrylonitrile in 0.05 M phosphate buffer, pH 7.5, at 50°C using high-performance liquid chromatography as described previously *(9)*. One unit is defined by 1 μmol/min formation of acrylamide under these conditions.

Enzyme Activation

Activated cell extracts were produced from EC463 cells grown to late exponential phase at 32°C. Following washing with 0.05 M potassium phosphate buffer and resuspension to a concentration of 5 mg/mL (dry weight), cell extracts were produced by sonication with a Cole-Parmer 4710 sonifier (Cole-Parmer, Chicago, IL) for 10 min (five 2-min bursts with 2-min cooling intervals on ice). The sonicate was centrifuged at 12,000g for 20 min and the supernatant adjusted to 5 µM CoCl$_2$ and incubated for 30 min at 50°C. For experiments measuring cobalt activation under anoxic conditions, extracts were sparged with helium for 15 min following cobalt addition and incubated at 50°C in a screw-cap vial with minimal head space. A similarly incubated tube containing unsparged enzyme extract served as the control.

Purification of Enzyme

Nitrile hydratase enzyme from EC463 cell extracts was purified to electrophoretic homogeneity utilizing cobalt-activated EC463 cell extract. Forty milliliters of cell extract was adjusted to 1 M ammonium sulfate in 50 mM potassium phosphate buffer, pH 7.5, and applied to a column containing 70 mL of Phenyl-Sepharose (Pharmacia, Uppsala, Sweden) preequilibrated with 1 M ammonium sulfate in 50 mM potassium phosphate buffer, pH 7.5. Following removal of unbound protein using elution with this buffer, nitrile hydratase was eluted using a 200-mL gradient of 1 to 0 M ammonium sulfate in 50 mM potassium phosphate buffer, pH 7.5, at a flow rate of 2 mL/min. Fractions containing active enzyme determined by assay were concentrated by ultrafiltration and dialyzed against 50 mM potassium phosphate buffer, pH 7.5.

Protein Electrophoresis

Protein purification was monitored and subunits were examined using a Bio-Rad mini-Protean gel II electrophoresis (Bio-Rad, Hercules, CA). Sodium dodecyl sulfate (SDS) polyacrylamide gel electrophoresis was performed using a 4% stacking gel and 12% separating gel according to the manufacturer's directions. Native protein electrophoresis utilized a 4% stacking gel and 12% separating gel. SDS and native electrophoresis protein standards were obtained from Sigma (St. Louis, MO).

Visible and Ultraviolet Spectrum of Nitrile Hydratase

Spectral data for holoenzyme and apoenzyme forms of BR449 nitrile hydratase were measured using a Beckman DU7500 spectrophotometer (Beckman, Fullerton, CA) at a concentration of 20 mg/mL for the visible range and 1 mg/mL for the ultraviolet (UV) range.

Results

Activation of Nitrile Hydratase with Co^{++}

After initial discovery of cobalt stimulation of intact cells and cell extracts, conditions for optimization of cobalt activation were determined. As seen in Table 1, stimulation of enzyme activity in the presence of 5 µM Co^{++} was optimal with incubation at 50°C. An incubation time of 30 min was optimal with longer times or increased cobalt concentrations not improving activation. Attempts to utilize other ions including Fe^{++},

Table 1
Cobalt Activation of BR449 Nitrile
Hydratase at Varied Temperatures[a]

Temperature (°C)	NHase activity (U/mg protein)
20	7
30	6
40	58
50	327
60	2

[a]Activation time was 30 min at the temperature shown followed by standard assay.

Table 2
Heat Activation of Nitrile Hydratase with Co^{++}

	Unheated	Heated separately	Heated combined	Heated combined (anoxic)
NHase activity (U/mg protein)	6	15	307	310

Table 3
Purification of BR449 Nitrile Hydratase

Purification step	Volume (mL)	Protein concentration (mg/mL)	Protein (mg)	Specific activity (U/mg)	Purification (fold)
Cell extract	50	2.3	115	271	1.0
Phenyl-Sepharose	1	21.0	21	852	3.1

Ni^{++}, Mn^{++}, and Cu^{++} instead of cobalt ion for enzyme activation resulted in inactive extracts (data not shown). Separate incubation of cobalt and apoenzyme at 50°C followed by cooling and mixing at 18°C prior to assay indicated that activation required incubation of combined reactants at elevated temperature, and that incubation under anoxic conditions did not affect the activation at this elevated temperature (Table 2).

Purification of Nitrile Hydratase

Purification of BR449 nitrile hydratase from extracts of EC465 was accomplished in a single-step, Phenyl-Sepharose affinity purification as shown in Table 3, allowing a threefold purification. The specific activity of the purified BR449 enzyme compares favorably with those reported for other nitrile hydratase enzymes (3,10), although differences in temperature and other conditions of assay preclude exact comparison. The high specific

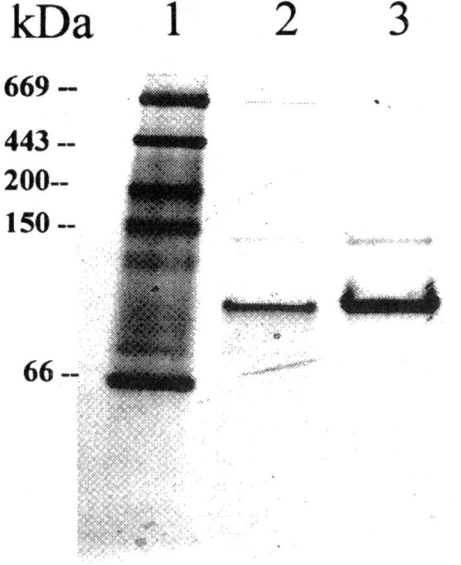

Fig. 1. Comparison of BR449 nitrile hydratase holoenzyme (lane 2) and apoenzyme (lane 3) molecular weights. Electrophoresis molecular weight markers are shown in lane 1.

activity obtained for purified enzyme suggests extensive, if not complete, activation of the enzyme. SDS slab gel electrophoresis of the purified nitrile hydratase showed the expected equal amounts of 25-kDa α- and 27.5-kDa β-subunits described previously using the *Bacillus* parent *(8)*.

Molecular Weight of Enzyme

The molecular weight of the BR449 holoenzyme was estimated using gel electrophoresis at 100 kDa (Fig. 1). This suggests that the holoenzyme is composed of two α- and two β-subunits as found for the nitrile hydratase of *R. rhodochrous* J1. Molecular weight determination of the apoenzyme utilizing the same purification but without cobalt activation yielded an inactive apoenzyme of 100-kDa molecular weight (Fig. 1). This indicates that cobalt ion is not required for subunit assembly.

Enzyme Absorption Spectrum

The visible and UV absorption spectrum of the purified nitrile hydratase apoenzyme and holoenzyme were compared (Fig. 2). In addition to the similar UV absorption bands below 300 nm expected from aromatic amino acids, the holoenzyme shows a weak absorption band at 340 nm, which is not observed for the apoenzyme. Similar weak absorption bands have been observed for Co^{+++}-containing nitrile hydratases from *R. rhodochrous* J1 *(11)* and *Pseudomonas putida (10)* and attributed to ligand:metal charge transfer.

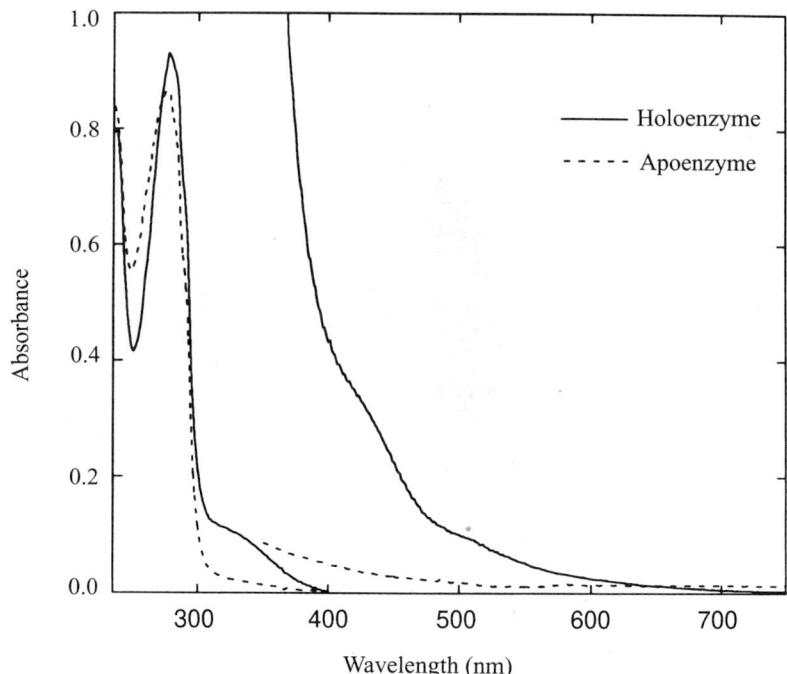

Fig. 2. Comparison of BR449 nitrile hydratase apoenzyme (dashed line) and holoenzyme (solid line) visible and UV absorption spectra. Visible spectra utilized an
enzyme concentration of 20 mg/mL and UV spectra utilized a concentration of 1 mg/mL.

Discussion

The ability to activate BR449 nitrile hydratase to high activity levels
following overexpression using Co^{++} ion is unexpected, because the active-
site cobalt in nitrile hydratases has been reported to be tightly bound in the
Co^{+++} state (11,12). Although the nature of the activation is not yet understood, it requires incubation of combined Co^{++} and apoenzyme at elevated
temperature. If the cobalt ion oxidizes to Co^{+++} during insertion into the
active site as suggested by the spectral data, the undiminished activation
under anoxic conditions would indicate participation of an oxidant other
than oxygen. Consistent with this interpretation, Nojiri et al. (13) recently
reported partial activation of a cobalt-substituted Fe-type nitrile hydratase
with the chemical-oxidizing agent potassium hexacyanoferrate. Although
details of the chemistry remain to be elucidated and are under study, the
simple and effective nitrile hydratase heat activation procedure described
in this article is of value in circumventing the requirement for cobalt introduction into E. coli during growth and expression, and the complexities
associated with coexpression of activation genes. It remains to be seen
whether the procedure can be utilized with nitrile hydratase enzymes of
mesophilic origin.

As described previously, the BR449 nitrile hydratase demonstrates
good thermostability in buffer but is rapidly inactivated by acrylonitrile,

probably reflecting the proclivity of acrylonitrile for protein alkylation *(14)*. The ability to express high levels of active enzyme in *E. coli* offers the opportunity for utilization of new genetic techniques such as directed evolution or site-directed mutagenesis to improve alkylation resistance of this nitrile hydratase in industrial applications.

Acknowledgment

This research was supported by SNF Floerger (SHK), and the National Science Foundation under Grant NSF BES 9817621 (RP).

References

1. Ashina, Y. and Suto, M. (1993), in *Industrial Applications of Immobilized Biocatalysts*, Tanaka, A., Tosa, T., and Kobayashi, T., eds., Marcel Dekker, New York, pp. 91–107.
2. Yamada, H. and Kobayashi, M. (1996), *Biosci. Biotech. Biochem.* **60**, 1391–1400.
3. Kobayashi, M. and Shimizu, S. (1998), *Nat. Biotechnol.* **16**, 733–736.
4. Ikehata, O., Nishiyama, M., Horinoushi, S., and Beppu, T. (1989), *J. Biochem.* **181**, 563–570.
5. Kobayashi, M., Snishiyama, M., Nagasawa, T., Horinoushi, S., Beppu, T., and Yamaday, H. (1991), *Biochim. Biophys. Acta* **1129**, 23–33.
6. Wu, S., Fallon, R. D., and Payne, M. S. (1997), *Appl. Microbiol. Biotechnol.* **48**, 704–708.
7. Nojiri, M., Yohda, M., Masafumi, O., et al. (1999), *J. Biochem.* **125**, 696–704.
8. Kim, S. H. and Oriel, P. (2000), *Enz. Microb. Technol.* **27**, 492–501.
9. Padmakumar, R. and Oriel, P. (1999), *Appl. Biochem. Biotechnol.* **77**, 671–680.
10. Payne, M. S., Wu, S., Fallon, R. D., Tudor, G., Stieglitz, B., Turner, I. M., Jr., and Nelson, M. J. (1997), *Biochemistry* **36**, 5447–5454.
11. Nagasawa, T., Takeuchi, K., and Yamada, H. (1991), *Eur. J. Biochem.* **196**, 581–589.
12. Brennan, B. A., Alms, G., Nelson, M. J., Durney, L. T., and Scarrow, R. C. (1996), *J. Am. Chem. Soc.* **118**, 9194–9195.
13. Nojiri, M., Nakayama, H., Odaka, M., Yohda, M., Takio, K., and Endo, I. (2000), *FEBS Lett.* **465**, 173–177.
14. Friedman, M., Calvins, J. F., and Wall, J. S. (1965), *J. Am. Chem. Soc.* **87**, 3672–3682.

Effect of Agitation and Aeration on Production of Hexokinase by *Saccharomyces cerevisiae*

Daniel Pereira Silva,[1] Adalberto Pessoa, Jr.,[1]
Inês-Conceição Roberto,[2] and Michele Vitolo*,[1]

[1]*Department of Biochemical and Pharmaceutical Technology,
Faculty of Pharmacy, University of São Paulo, Av. Prof. Lineu Prestes,
580, B.16. 05508-900, São Paulo/SP, Brazil, E-mail: michenzi@usp.br;
and* [2]*DEBIQ-FAENQUIL, Lorena/SP, Brazil*

Abstract

A batch culture of *Saccharomyces cerevisiae* for the production of hexokinase was carried out in a 5-L fermentor containing 3 L of culture medium, which was inoculated with cell suspension (about 0.7 g/L), and left fermenting at 35°C and pH 4.0. The aeration and agitation were adjusted to attain $k_L a$ values of 15, 60, 135, and 230 h[1]. The highest hexokinase productivity (754.6 U/[L·h]) and substrate-cell conversion yield (0.21 g/g) occurred for a $k_L a$ of 60 h. Moreover, the formation of hexokinase and cell growth are coupled events, which is in accordance with the constitutive character of this enzyme. Hexokinase formation for $k_L a > 60$ h[1] was not enhanced probably owing to saturation of the respiratory pathway by oxygen.

Index Entries: Hexokinase; *Saccharomyces cerevisiae*; agitation and aeration; fermentation; volumetric coefficient of oxygen transfer.

Introduction

Hexokinase (EC 2.7.1.1) is the first enzyme of glycolysis that catalyzes the phosphorylation of glucose into glucose 6-phosphate (G6P). The G6P, in turn, is a key intermediate for several pathways such as gluconeogenesis, shunt of pentoses, and glycogen metabolism. This enzyme is sensitive to the catabolic repression and has a role in the glucose uptake mechanism through the plasma membrane *(1)*.

Hexokinase, besides its role in the cell metabolism and importance to biochemical studies, is used in analytical methods to measure glucose, fructose, mannose, adenosine triphosphate (ATP), and creatin-kinase activity

*Author to whom all correspondence and reprint requests should be addressed.

(2); in the phosphorylation of pyranose and furanose analogs of glucose *(3)*; and in the wine and fruit juice industries for the detection of illegal sugar addition to the final products *(4)*.

Because hexokinase is a constitutive enzyme in all living cells, its production from a microorganism is linked to the amount of biomass obtained through a fermentative process. However, attaining a high amount of cells depends on the pH, temperature, components of the culture medium, and the availability of oxygen to the cells *(5)*. For a facultative microorganism such as *Saccharomyces cerevisiae*, oxygen has a crucial role in its overall metabolism, because it participates in the generation of energy, through the respiratory chain inside mitochondria, which is fundamental for attaining a significant specific growth rate. According to Gregory et al.*(6)*, there is a strong correlation among specific growth rate, oxygen transfer rate, and hexokinase production rate in yeasts. Although hexokinase could be attained from several microbial species, in Brazil, the use of *S. cerevisiae* as a source of this enzyme and other products *(7)* is practical owing to the large experience in handling this strain in industrial plants. In addition, coupling the yeast processing with the ethanol production probably should have a positive effect on distillery profits.

The present study deals with the influence of oxygen transfer rate, expressed as $k_L a$ (volumetric coefficient of oxygen transfer), on the production of hexokinase by *S. cerevisiae*

Materials and Methods

Chemicals

Glucose 6-phosphate dehydrogenase, ATP, nicotinamide adenine dinucleotide phosphate (NADP), phenylmethylsulfonyl fluoride (PMSF), β-mercaptoethanol, G6P, glucose, and sucrose were obtained from Sigma (St. Louis, MO). Yeast extract and peptone were obtained from Difco (Detroit, MI). All other chemicals were of analytical grade.

Preparation of Inoculum

S. cerevisiae (isolated from pressed yeast cake) was maintained at 4°C in agar slant tubes containing 23.0 g/L of nutrient agar (Difco) and 1.0 g/L of glucose. The cells were transferred to 250-mL Erlenmeyer flasks containing 50 mL of autoclaved (121°C for 15 min) growth medium, and the culture was carried out in a shaker (New Brunswick, Edison, NJ) for 11 h at 35°C, agitation of 175 rpm, and initial pH of 4.5 (adjusted with 0.5 M H_2SO_4 or 0.1 M NaOH). The growth medium composition was as follows: 3.8 g/L of glucose, 28.5 g/L of sucrose, 3.0 g/L of yeast extract, 5.0 g/L of peptone, 2.4 g/L of $Na_2HPO_4 \cdot 12H_2O$, 0.075 g/L of $MgSO_4 \cdot 7H_2O$, and 5.1 g/L of $(NH_4)_2SO_4$.

Batch Fermentation

A volume of 0.45 L of inoculum (about 4.7 g of dry mass/L) was poured into a 5-L fermentor (NBS-MF 105, coupled with NBS dissolved oxygen controller, DO-81; New Brunswick) containing 2.55 L of the culture medium: 3.0 g/L of yeast extract, 5.0 g/L of peptone, 20.0 g/L of glucose, 2.4 g/L of $Na_2HPO_4 \cdot 12H_2O$, 0.075 g/L of $MgSO_4 \cdot 7H_2O$, and 5.1 g/L of $(NH_4)_2SO_4$. The fermentation was carried out at 35°C and pH 4.0 (controlled automatically by adding 0.5 M H_2SO_4 or 0.1 M NaOH), and agitation and aeration were automatically adjusted to attain an initial k_La of 15 h^1 (0.7 vvm and 200 rpm), 60 h (1.7 vvm and 400 rpm), 135 h (2.3 vvm and 600 rpm), and 230 h^1 (2.3 vvm and 800 rpm). The pair agitation/aeration related to a desired initial k_La was fixed against distilled water and evaluated by using a dissolved oxygen controller (NBS-DO81; New Brunswick) associated with a galvanic dissolved oxygen probe (Mettler-Toledo, São Paulo, SP, Brazil). The initial k_La was calculated through the conventional Pirt's mathematical model (8). To control the foam, a 10% (w/w) aqueous silicone emulsion (Dow Corning, FG-10; New York) was added dropwise. Samples were collected periodically to follow the variation in cell, glucose, and ethanol concentrations as well as hexokinase activity (after cell disruption).

Measurement of Cell, Glucose, and Ethanol Concentrations

One milliliter of fermenting medium was centrifuged (8.720 g for 10 min) and the supernatant discharged. Then, the cell cake was suspended in 50 mL of distilled water and the turbidity was measured with a spectrophotometer at 600 nm (Perkin-Elmer 552; Bethesda, MD). The cell concentration was obtained by comparing the optical density (OD) of cell suspension against a standard OD × dry cell mass curve.

Ten milliliters of fermenting medium was centrifuged (2.880 g for 10 min). The cell cake was stored at 4°C until submitted to disruption. In the supernatant, the glucose and ethanol concentrations were measured using a GOD/POD enzymatic kit (kit no. 02200; Laborlab, Guarulhos, São Paulo, SP, Brazil) and the reducing dichromate back-titration method (9), respectively.

Measurement of Hexokinase Activity

The cell cake was suspended in 50 mM Tris-HCl buffer (pH 7.5) containing 5 mM $MgCl_2$, 10 mM β-mercaptoethanol, 2 mM aminocaproic acid, 1 mM phenylmethylsulfonyl fluoride, and 0.2 mM EDTA. Then, the cells were disrupted by submission to a vortex (Phoenix AP 56; São Paulo, SP, Brazil) in the presence of glass beads (diameter = 0.5 mm), and the mixture was maintained below 10°C all the time. The wet cell cake, glass beads, and Tris-HCl buffer were always mixed in a volumetric proportion of 1:1:1. Cell debris and glass beads were removed by centrifugation (2.880 g for 10 min) and the supernatant was collected. The hexokinase activity in the supernatant was measured through the continuous reduction of NADP at 30°C in a spectrophotometer (λ = 340 nm) as described by Bergmeyer (2). One hexokinase unit was defined as the amount of enzyme catalyzing the reduction of 1 μmol of NADP/min under the assay conditions. Each determination was made in triplicate and the standard deviation was 2%. No significant decrease in hexokinase activity was detected after storing the supernatant for 5 h on ice at room temperature.

Hexokinase production was calculated using the following equation:

$$P = (v/X_e) \times X \tag{1}$$

in which P = concentration of hexokinase (U/L$_{medium}$); v = hexokinase activity in the supernatant (U/mL); X_e = cell concentration in the suspension submitted to disruption (g$_{cell}$/mL); and X = cell concentration in the fermenting medium (g$_{cell}$/L$_{medium}$).

Calculation of Kinetic Parameters

The specific cell growth rate (μ_x), specific substrate consumption rate (μ_S), and specific hexokinase production rate (q_p) were defined as follows:

$$\mu_x = (1/X) \times dX/dt \tag{2}$$

$$\mu_S = (1/X) \times dS/dt \tag{3}$$

$$q_p = (1/X) \times dP/dt \tag{4}$$

The derivates dX/dt, dS/dt, and dP/dt were calculated according to the method proposed by Le Duy and Zajic(10). The substrate-to-cell conversion factor ($Y_{X/S}$) and the substrate-to-hexokinase conversion factor ($Y_{HK/S}$) were calculated as the inclination obtained between the variation in cell concentration (ΔX) or hexokinase production (ΔP) to the glucose consumed (ΔS), respectively. The maximum enzyme (Pr_{HK}) and cell (Pr_X) productivities were calculated, respectively, as $\Delta P/\Delta t$ and $\Delta X/\Delta t$, in which Δt = interval of cultivation time.

Results and Discussion

From Fig. 1 we clearly observe that the k_La affected the hexokinase activity produced by the yeast. The highest hexokinase concentration, which was about 4.200 U/L$_{medium}$, occurred at a k_La of 60 h^1 after 7 h of fermentation. At this point, all glucose was consumed and the cell concentration was equal to 4.61 g/L (Figs. 2 and 3).

Nevertheless, at the extreme values of k_La employed (15 and 230 h^1), the glucose concentration in the medium became negligible after 11 and 7 h, respectively (Fig. 2). Furthermore, the hexokinase concentration and cell concentration were markedly low for a k_La of 15 h^1 compared with any other k_La employed (Figs. 1 and 3). This result indicates that glucose consumption and hexokinase and cell production are coupled events and strongly depend on the aeration/agitation of the medium. In the present case, we can assume that the k_La of 60 h^1 guaranteed an adequate amount of dissolved oxygen in the medium to fine-tune the metabolic pathways related to growth and glucose conversion. This fact is clearly noted through

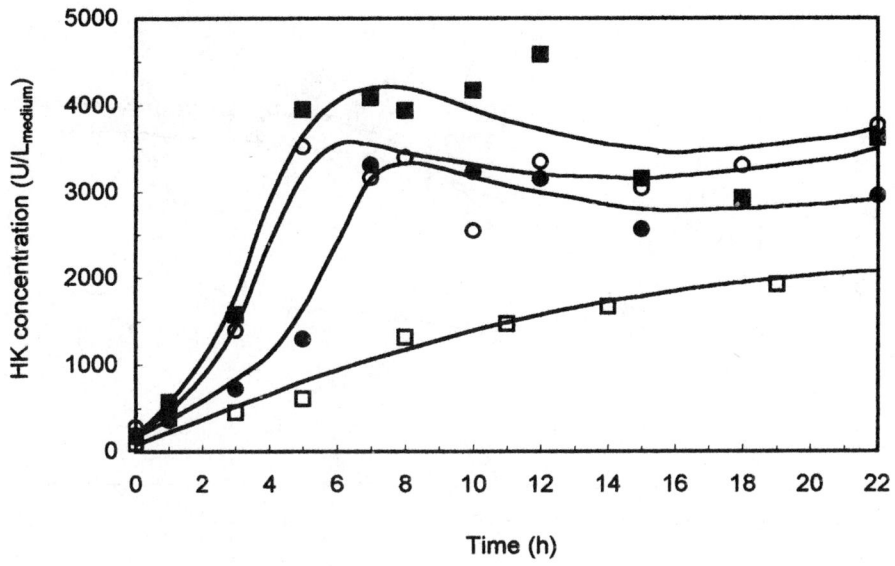

Fig. 1. HK production against time for *S. cerevisiae* grown under different $k_L a$ (h¹) values: 15 (□), 60 (■), 135 (○), and 230 (●).

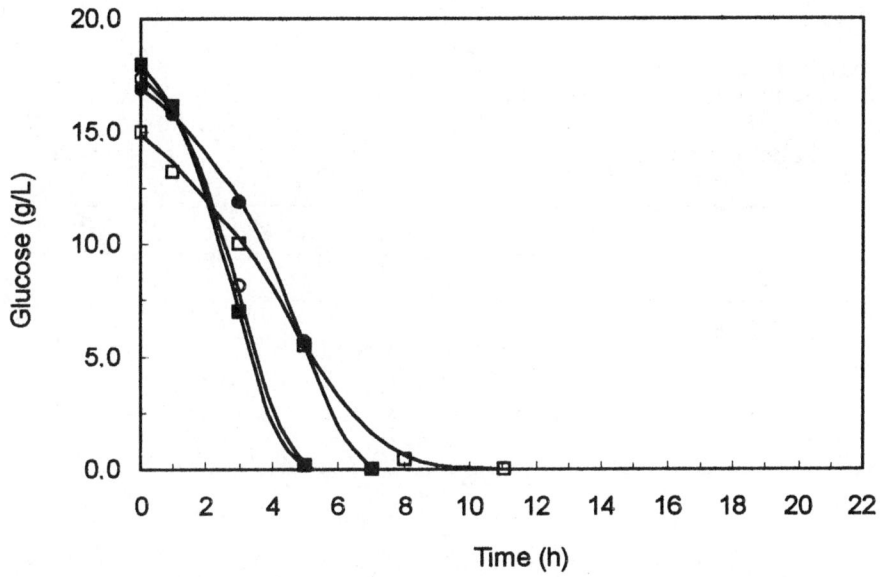

Fig. 2. Variation in glucose concentration against time for *S. cerevisiae* grown under different $k_L a$ (h⁻¹) values: 15 (□), 60 (■), 135 (○), and 230 (●).

the production parameters shown in Table 1, which at a $k_L a$ of 60 h¹, $Y_{X/S}$ (0.21), $Y_{HK/S}$ (221.5 U/g$_{glu}$), Pr_X (0.69 g$_{cell}$/[L·h]), and Pr_{HK} (754.6 U/[L·h]) were the highest values attained. In addition, for this $k_L a$ the generation time (*tg*), calculated as proposed by Borzani *(11)*, was the lowest and equal to 1.8 h (Table 1).

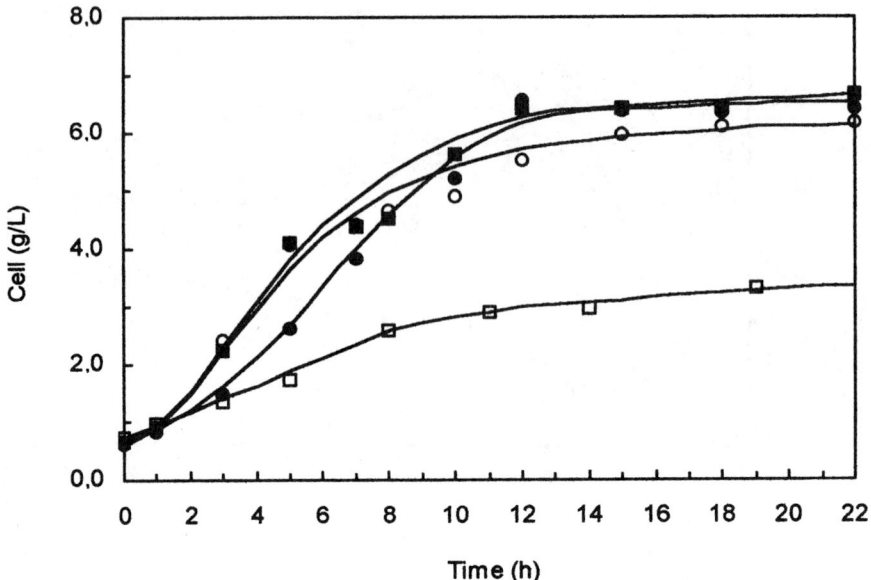

Fig. 3. Variation in cell concentration against time for *S. cerevisiae* grown under different k_La (h⁻¹) values: 15 (□), 60 (■), 135 (○), and 230 (●).

Table 1
Generation Time (*tg*), Glucose-to-Cell Conversion Factor ($Y_{X/S}$),
Glucose-to-Hexokinase Conversion Factor ($Y_{HK/S}$), and Maximum Cell (Pr_X)
and Hexokinase (Pr_{HK}) Productivities Related to Batch Fermentation
of *S. cerevisiae* Conducted at Different k_La Values

Parameter	k_La (h⁻¹)			
	15	60	135	230
tg (h)	4.3	1.8	1.8	2.5
$Y_{X/S}$ (g_{cell}/g_{glu})	0.14	0.21	0.21	0.19
$Y_{HK/S}$ (U/g_{glu})	82.8	221.5	192.9	175.8
Pr_X (g_{cell}/[L·h])	0.24	0.69	0.69	0.49
Pr_{HK} (U/[L·h])	300.5	754.6	647.0	445.7

From Table 1 it can also be seen that the parameters related to growth (*tg*, $Y_{X/S}$, and Pr_X) had the same values for either a k_La of 60 or 135 h, whereas the hexokinase related parameters ($Y_{HK/S}$ and Pr_{HK}) differed about 15% at those k_La values. Most likely, for high k_La the intracellular ATP concentration is so high that it becomes an inhibitor of hexokinase biosynthesis (*12*), although such a condition should favor cell growth. Values of k_La higher than 135 h⁻¹ did not enhance cell growth owing probably to saturation of the respiratory pathway by oxygen.

It must be borne out that in all the experiments cell growth continued even when all the glucose initially present in the mash was consumed (Fig. 3).

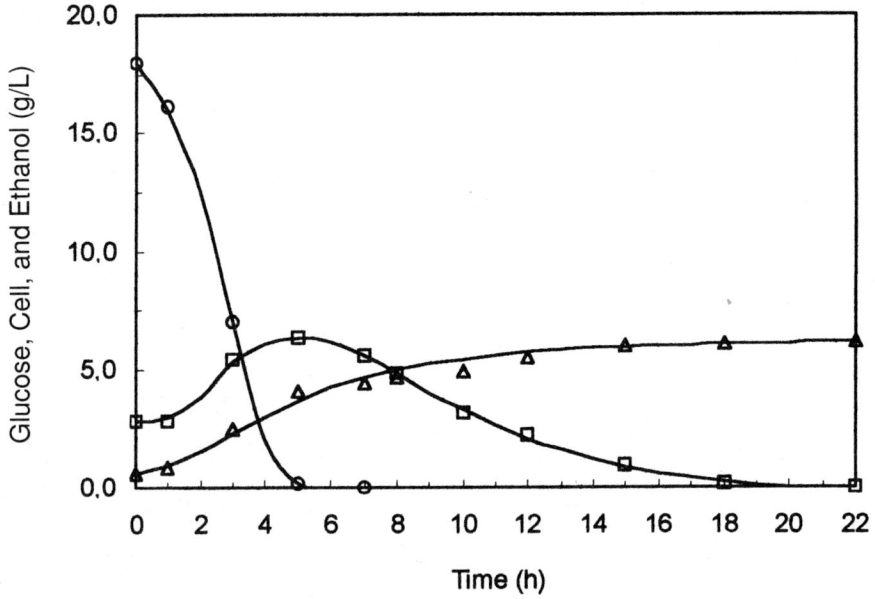

Fig. 4. Variation against time of glucose (○), cell (△), and ethanol (□) concentrations for *S. cerevisiae* grown at a $k_L a$ of 60 h^{-1}.

This can easily be understood if we consider that after glucose consumption another substance becomes a substrate to the yeast. In this case, a probable one should be ethanol, which accumulates into the medium because glucose has been taken up by the yeast (Fig. 4). According to Rose and Harrison *(13)* and Rehm and Reed *(14)*, the production of ethanol by *S. cerevisiae*, even under high aeration, is a characteristic of this specie of yeast. In other words, the Pasteur effect does not occur plentifully in *S. cerevisiae*.

Because hexokinase is a constitutive and growth-related enzyme, a viable process for its production would be a steady-state continuous culture, as already stated for other enzymes with the same characteristics, such as invertase *(15)*. In this case, according to Doin *(16)*, an estimation of the more suitable dilution rate (*D*) for the continuous process must be made from the data attained in batch. This can be accomplished by plotting the variation in growth rate (dX/dt) vs cell concentration (*X*) (Fig. 5). By setting 7 h of batch fermentation, as the period in which the highest hexokinase production was attained (Fig. 1), the correspondent cell concentration was 4.61 g/L ($k_L a$ = 60 h). According to Doin *(16)*, the angular coefficient of the line drawn between the points (0, 0) and (4.61, 0.38) would be a reasonable estimation of *D*, which in this case should be about 0.08 h^{-1}. Of course, to produce hexokinase by direct extraction from residual distillery yeast is also an open opportunity for ethanol-producing countries (e.g., Brazil).

In short, to understand the data presented, we must take into account that the dependence of hexokinase on the availability of oxygen and cell growth derives from the fact that it is a constitutive-type enzyme as well as

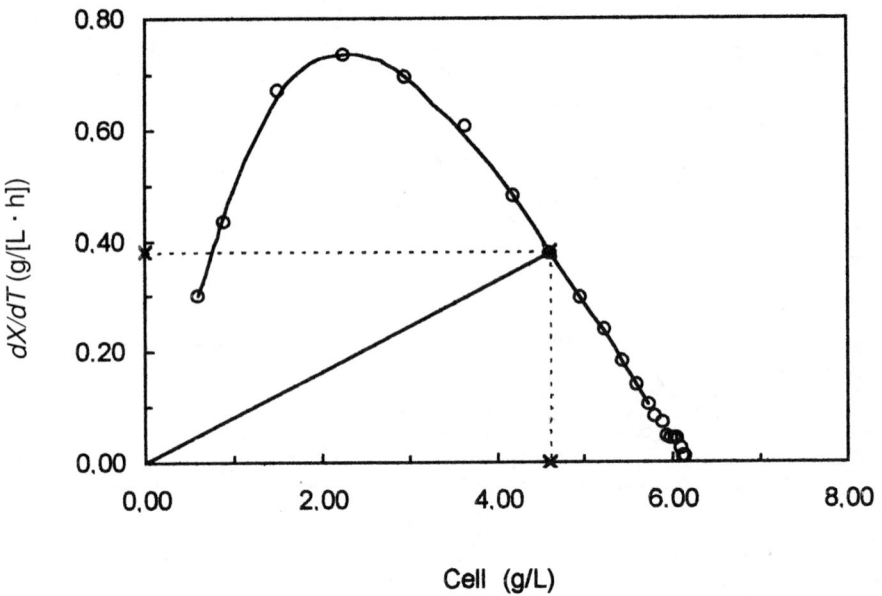

Fig. 5. Growth rate (dX/dt) vs cell concentration (X) for estimating the dilution rate (D) for a steady-state continuous culture based on the data obtained from a discontinuous culture carried out at an initial k_La of 60 h^{-1}.

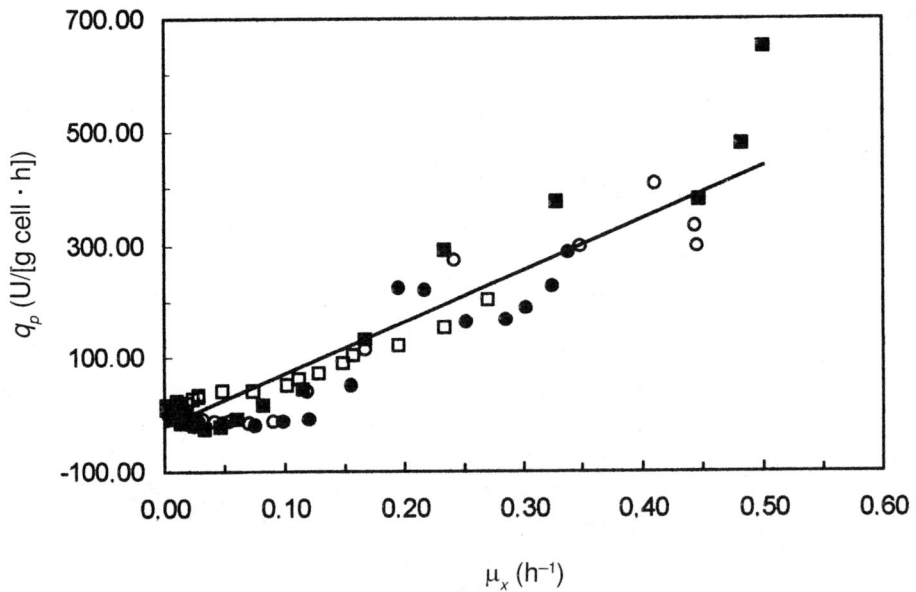

Fig. 6. Variation in HK specific production rate (q_p) against specific growth rate (μ_x) for *S. cerevisiae* grown under different k_La (h^{-1}) values: 15 ($\square$), 60 ($\blacksquare$), 135 ($\bigcirc$), and 230 ($\bullet$).

the G6P, the product of the hexokinase-catalyzed reaction, which is converted either in pyruvate (glycolysis pathway) or in 6-phosphoglucono lactone (6PL). The pyruvate is mainly directed to ATP generation (through

Krebs cycle and phosphorylative oxidation, both depending on oxygen), and 6PL is converted through the pentose pathway in sugars intimately related to growth and intracellular metabolism. However, a fraction of pyruvate is diverted to ethanol, which avoids the quantification of the correlation between the formation of hexokinase and cell growth by methods such as those of Luedeking and Piret *(17)* and Moser *(18)*. Figure 6, in which are plotted the data of all tests realized, shows a large points dispersion, an indication of the formation of at least one byproduct other than cells.

Conclusion

The data presented lead to two main conclusions. First, the high formation of hexokinase by *S. cerevisiae* occurred at a $k_{L}a$ of 60 h^{-1} after 7 h of batch culture. Second, because the formation of hexokinase is related to cell growth, a perspective should be to produce the enzyme through a continuous culture at a dilution rate of 0.08 h^{-1}, as estimated from the batch process.

Acknowledgments

We wish to thank Prof. Dr. José Abrahão Neto for valuable suggestions. D.P.S. acknowledges the financial support of FAPESP/São Paulo, Brazil, in the form of a Master of Science fellowship.

References

1. Kovak, L., Nelson, B. D., and Ernster, L. (1986), *Biochem. Biophys. Res. Comm.* **134(1),** 285–291.
2. Bergmeyer, H. U. (1984), *Methods of Enzymatic Analysis,* 3rd ed., Verlag Chimie, Weinheim, Germany.
3. Chenault, H. K., Mandes, R. F., and Hornberger, K. R. (1997), *J. Org. Chem.* **62,** 331–336.
4. Whitaker, J. R. (1991), *Food Enzymology,* Elsevier Applied Science, New York.
5. White, J. (1954), *Yeast Technology,* Chapman & Hall, London.
6. Gregory, M. E., Bulmer, M., and Bogle, I. D. L. (1996), *Bioproc. Eng.* **15(5),** 237–245.
7. Godfrey, T. and West, S. (1996), *Industrial Enzymology,* 2nd ed., MacMillan, London.
8. Wise, W. S. (1951), *J. Gen. Microbiol.* **5,** 167–177.
9. Joslyn, M. A. (1970), *Methods in Food Analysis,* 2nd ed., Academic, New York.
10. Le Duy, A. and Zajic, E. J. (1973), *Biotechnol. Bioeng.* **15,** 805–810.
11. Borzani, W. (1975), in *Engenharia Bioquímica,* vol. 3, Borzani, W., Lima, U. A., and Aquarone, E., eds., Universidade de São Paulo, São Paulo, pp. 168–184.
12. Garrett, R. H. and Grisham, C. M. (1995), *Biochemistry,* Saunders College Publishing, San Diego.
13. Rose, A. H. and Harrison, J. S. (1989), *The Yeasts,* 2nd ed., Academic, New York.
14. Rehm, H. J. and Reed, G. (1993), *Biotechnology,* 2nd ed., VCH, Weinheim, Germany.
15. Vitolo, M., Vairo, M. L. R., and Borzani, W. (1985), *Biotechnol. Bioeng.* **27,** 1229–1235.
16. Doin, P. A. (1975), in *Engenharia Bioquímica,* vol. 3, Borzani, W., Lima, U. A., and Aquarone, E., eds., Universidade de São Paulo, São Paulo, pp. 112–134.
17. Luedeking, R. and Piret, E. L. (1959), *J. Biochem. Microbiol. Technol. Eng.* **1,** 393–401.
18. Moser, A. (1985), in *Biotechnology: Fundamentals of Biochemical Engineering,* vol. 2, Rehm, H. J. and Reed, G., eds., VCH, Weinheim, Germany.

Comparison of Catalytic Properties of Free and Immobilized Cellobiase Novozym 188

Luiza P. V. Calsavara, Flávio F. De Moraes,
and Gisella M. Zanin*

*Chemical Engineering Department, State University of Maringá,
Av. Colombo, 5790, Bloco E46-09, Maringá-PR, CEP 87020-900, Brazil,
E-mail: gisellazanin@maringa.com.br*

Abstract

The enzyme cellobiase from Novo was immobilized in controlled pore silica particles by covalent binding with the silane-glutaraldehyde method with protein and activity yields of 67 and 13.7%, respectively. The activity of the free enzyme (FE) and immobilized enzyme (IE) was determined with 2 g/L of cellobiose, from 40 to 75°C at pH 3.0–7.0 for FE and from 40 to 70°C at pH 2.2–7.0 for IE. At pH 4.8 the maximum specific activity for the FE and IE occurred at 65°C: 17.8 and 2.2 micromol of glucose/(min · mg of protein), respectively. For all temperatures the optimum pH observed for FE was 4.5 whereas for IE it was shifted to 3.5. The energy of activation was 11 kcal/mol for FE and 5 kcal/mol for IE at pH 4.5–5, showing apparent diffusional limitation for the latter. Thermal stability of the FE and IE was determined with 2 g/L of cellobiose (pH 4.8) at temperatures from 40 to 70°C for FE and 40 to 75°C for IE. Free cellobiase maintained its activity practically constant for 240 min at temperatures up to 55°C. The IE has shown higher stability, retaining its activity in the same test up to 60°C. Half-life experimental results for FE were 14.1, 2.1, and 0.17 h at 60, 65, and 70°C, respectively, whereas IE at the same temperatures had half-lives of 245, 21.3, and 2.9 h. The energy of thermal deactivation was 80.6 kcal/mol for the free enzyme and 85.2 kcal/mol for the IE, suggesting stabilization by immobilization.

Index Entries: Immobilized enzyme; cellobiase; cellobiose; thermal stability; energy of activation; energy of deactivation.

Introduction

The search for new, renewable raw materials as energy sources useful in biotechnology has led to increasing interest in the study of enzymatic

*Author to whom all correspondence and reprint requests should be addressed.

hydrolysis of cellulosic biomass. However, lignocellulose is difficult to hydrolyze because cellulose is closely associated with hemicellulose and lignin, as well as other biomass constituents *(1)*.

Cellulose is a linear polymer of D-glucose units linked by 1,4-β-D-glucosidic bonds. The enzyme system for the conversion of cellulose to glucose comprises endo-1,4-β-glucanase, cellobiohydrolase, and β-glucosidase (cellobiase). Cellulolytic enzymes in conjunction with β-glucosidase act sequentially and cooperatively to degrade crystalline cellulose to glucose. The cellobiase is generally responsible for the regulation of the entire cellulolytic process and is a rate-limiting factor during enzymatic hydrolysis of cellulose, since both endoglucanase and cellobiohydrolase activities are often inhibited by cellobiose. Thus, the cellobiase not only produces glucose from cellobiose but also reduces cellobiose inhibition, allowing the cellulolytic enzymes to function more efficiently *(2)*.

Cellobiose hydrolysis rates depend on both reaction conditions and catalyst activity; therefore, the knowledge of the cellobiase thermal deactivation through time is a prerequisite to obtain useful design equations *(3)*.

Because the substrate of cellobiase, i.e., cellobiose, is water soluble, an immobilized (water-insoluble) cellobiase preparation could be used to supplement commercial cellulase/cellulose mixtures *(4)*. Advantages are the possibility of using continuous processes with insoluble enzymes, facilitation of enzyme recovery, and in some cases modification of the properties of the enzyme. It has been claimed that immobilization of an enzyme on an insoluble support will result in an enhanced stability against the denaturing effect of heat. However, there seems to be no clear-cut correlation between the observed stability and the way in which the enzyme was fixed to the support. Immobilization may thus result in stabilization, destabilization, or no effect at all *(5)*.

The technology of immobilized enzyme (IE) offers technical and economical advantages, such as the following *(6,7)*:

1. Enzyme consumption is reduced, since once immobilized the enzyme can be used for a much longer period than in the soluble form.
2. IE can lead to preferred continuous processes that may use either fixed or fluidized-bed reactors.
3. It is possible to use higher enzyme dosage per volume of reactor than in the soluble enzyme process, and this contributes to high reaction rates and, consequently, small reactor sizes.
4. These technical advantages allow a reduction in the process operational and capital costs, if the IE half-life is sufficiently long.

This article covers experimental determination of the activity, energy of activation, thermal stability, and energy of deactivation of free and immobilized Novo Nordisk cellobiase.

Materials and Methods

Substrate

The substrate was cellobiose from Sigma (St. Louis, MO), and it contained a very low level contamination of glucose (0.133% [w/w]).

Enzyme

Novozym 188, a β-glucosidase produced by the microorganism *Aspergillus niger*, containing 170 mg/mL of protein and a specific activity of 9.5 micromol of glucose/(min·mg of protein) at 50°C and pH 4.8 was kindly supplied by Novo Nordisk (Copenhagen, Denmark).

Carrier

The carrier used for immobilization was controlled pore silica (CPS), a gift from Corning Glass Works, having a mean pore size of 37.5 nm and average particle diameter of 0.351 mm.

Enzyme Immobilization

Cellobiase was immobilized in CPS by the covalent method of Weetall *(8)* with the following steps:

1. Silanization of the carrier, with a 0.5% (v/v) solution of γ-aminopropyltrietoxisilane, for 3 h at 75°C.
2. Washing with distilled water and drying for 15 h at 105°C.
3. Activation with a 2.5% (v/v) solution of glutaraldehyde (pH 7.0) for 45 min at 20°C.
4. Washing with water.
5. Contact of the activated carrier (50 g) with a solution of the enzyme (250 mL, 20 mg of protein/mL) for 15 h at 20°C.
6. Washings of the IE with distilled water and stocking under sodium acetate buffer (0.2 *M*), pH 4.5, at 4°C.

Procedure for Assaying Enzymatic Activity

Activity tests were conducted using the method of initial rate *(9)*, using a jacketed glass batch thermo-controlled reactor, equipped with magnetic stirring. A volume of 20 mL of cellobiose solution containing 1 mg/mL of sodium benzoate was incubated at the current pH and temperature with the enzyme β-glucosidase at a concentration of 95 µL of enzyme/L of solution, whereas for the immobilized cellobiase, 0.06 g of dry wt IE was used inside a stainless steel screen basket. Half-milliliter samples were collected at 3-min intervals for a period of 18 min and were boiled and stocked at 4°C for later glucose assay.

Analytical Methods

Glucose was assayed with the enzymatic method GOD-PAP *(10)*, and protein was measured according to the method of Lowry et al. *(11)* using bovine serum albumin as protein standard.

Specific Activity as a Function of pH and Temperature

The specific activity of the free enzyme (FE) and IE was determined using a 2 g/L solution of cellobiose (5.85 mM) at 40–70°C and pH 3.0–7.0 for FE and pH 2.2–7.0 for IE. This span of pH values was covered with the McIlvaine buffer *(12)* at a final concentration of 10 mM.

Enzyme Activity as Function of Temperature

The increase in activity as temperature is raised in an enzymatic reaction is modeled with the Arrhenius equation:

$$V = V_0 \exp(-E_a/RT) \tag{1}$$

in which E_a = energy of activation (cal/mol); R = universal law gas constant (1.987 cal/mol K); T = absolute temperature (K); V = enzymatic activity measured by the initial rate of reaction (µmol of glucose/[min·mg of protein]); and V_0 = Arrhenius preexponential constant.

The Arrhenius equation applies only for temperatures below that in which thermal denaturation of the enzyme becomes severe. From Eq. 1, it can be seen that the Arrhenius plot of log of activity (V) as a function of the inverse of the absolute temperature (T) results in a straight line in which the angular coefficient multiplied by R gives the energy of activation (E_a). For the higher temperatures where thermal deactivation becomes too fast, the experimental points in the Arrhenius plot deviate from the straight line, giving lower activities and shaping the curve with a maximum, associated with the optimum temperature for maximum activity. At this point, the gains in activity given by higher temperatures are offset by decreased concentration in active enzyme, caused by thermal denaturation *(6,9,13)*.

Enzyme Thermal Stability

The thermal stability of the FE and IE was determined at pH 4.8 and temperatures from 40 to 70°C for FE and 40 to 75°C for IE. The enzyme was incubated at the specified temperature and pH in a 2 g/L solution of cellobiose, and samples were collected every 40 min to measure the residual enzymatic activity at 50°C and pH 4.8. For the FE, at 70°C sampling time was shortened to 5-min intervals because of the rapid thermal denaturation observed at this temperature.

Energy of Thermal Deactivation

With the data obtained for the thermal deactivation of cellobiase, the energy of thermal deactivation of the enzyme was calculated. This follows the assumption that enzyme thermal denaturation is a reaction in which the rate of enzyme deactivation (r_d) is first order in relation to the concentration of the active enzyme (E):

$$r_d = -K_d E \tag{2}$$

and the deactivation constant (K_d) is a function of temperature as given by the following Arrhenius equation *(14)*:

$$K_d = K_d^0 \exp(-E_d/RT) \tag{3}$$

in which E_d is the energy of thermal deactivation (cal/mol), and R and T are defined as for Eq. 1.

For a batch reactor of constant liquid density, the rate of reaction equals the time derivative of the concentration, and therefore it follows from Eq. 2 that

$$\frac{dE}{dt} = -K_d E \tag{4}$$

which integrated gives:

$$E = E_0 \exp(-K_d t) \tag{5}$$

in which E_0 is the initial active enzyme concentration, and t is the time elapsed during the reaction.

When the enzyme is present in catalytic quantities, i.e., in low concentrations, the residual enzyme activity (A_r) is directly proportional to the concentration of the active enzyme (E):

$$\frac{A_r}{A_0} = \frac{E}{E_0} \tag{6}$$

in which A_0 is the initial enzyme activity observed with the initial enzyme concentration (E_0).

By combining Eqs. 5 and 6, the residual enzyme activity results as follows:

$$A_r = A_0 \exp(-K_d t) \tag{7}$$

This result is the exponential decay model. Therefore, by plotting residual activity data in the form of log of A_r/A_0 against time, the deactivation constant (K_d) is obtained as the angular coefficient of the adjusted straight line.

From Eq. 3, and as observed by experiment, it can be seen that the deactivation constant increases with temperature. Values obtained for K_d for various test temperatures are plotted in the form of an Arrhenius plot, i.e., ln of K_d against the inverse of absolute temperature (T), yielding the energy of deactivation (E_d), as the angular coefficient of the adjusted straight line times R, the universal gas constant.

It is of interest to calculate also the enzyme half-life ($t_{1/2}$), i.e., the time period necessary for the residual enzymatic activity to decrease to half of its initial value. If the enzyme thermal denaturation follows Eq. 7, then there is an inverse relation between the half-life of the enzyme and the deactivation constant *(15)*:

$$t_{1/2} = \ln 0.5/(-K_d) = 0.693/K_d \tag{8}$$

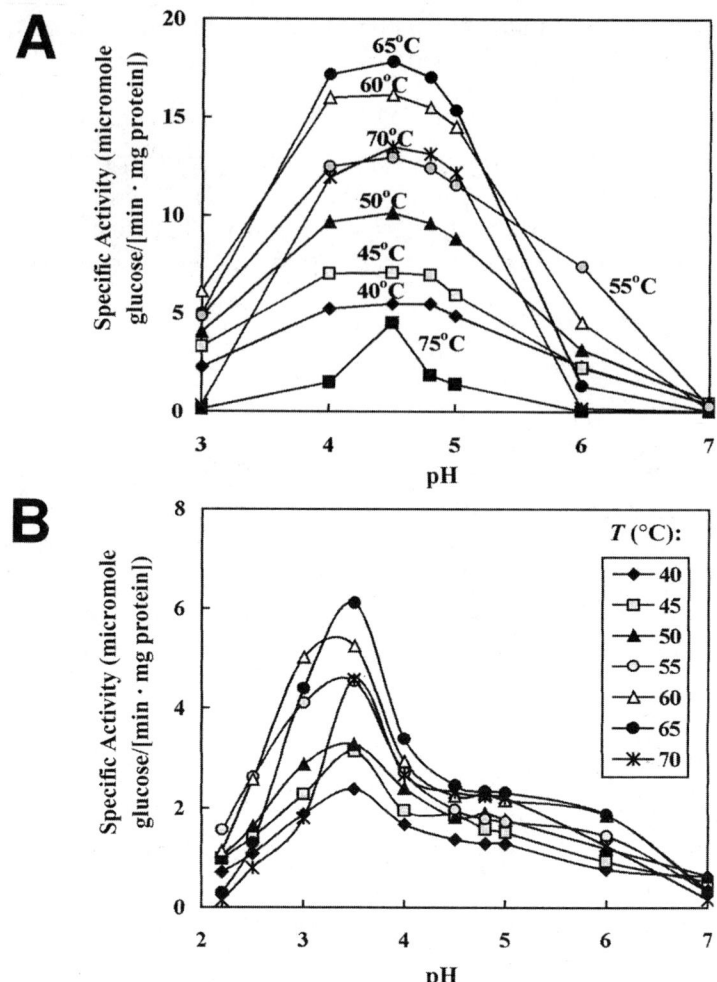

Fig. 1. Specific activity as a function of T (°C) and pH (2 g/L of cellobiose as substrate) for **(A)** free Novozym cellobiase at a concentration of 95 µL/L, pH 3.0–7.0; and **(B)** immobilized Novozym cellobiase at a concentration of 0.06 g of dry wt IE/20 mL of substrate, pH 2.2–7.0.

Results and Discussion

Immobilization

The quantity of protein fixed on the CPS support on immobilization was 74.9 mg of protein/g of dry support, representing a protein yield of 67% (i.e., 67% of the theoretical maximum that corresponds to the total enzyme offered for immobilization).

Activity as a Function of Temperature and pH

Figure 1A shows the specific activity of the Novo free cellobiase as a function of pH and temperature. Figure 1B gives the same kind of results but for the IE.

Table 1

Comparison of Energy of Activation for Enzymatic Hydrolysis of Cellobiose
with Novozym 188 Cellobiase for Different pH Values for FE and IE[a]

pH	$V = V_0 \exp(-E_a/RT)$	Energy of activation (kcal/mol)
	Free enzyme	
3.0	$V = 1.627 \times 10^7 \exp(-9782/RT),\ r = 0.9917$	9.78
4.0	$V = 6.961 \times 10^8 \exp(-11,633/RT),\ r = 0.9993$	11.64
4.5	$V = 5.266 \times 10^8 \exp(-11,435/RT),\ r = 0.9972$	11.44
4.8	$V = 2.610 \times 10^8 \exp(-11,005/RT),\ r = 0.9985$	11.01
5.0	$V = 7.747 \times 10^8 \exp(-11,766/RT),\ r = 0.9948$	11.77
6.0	$V = 1.225 \times 10^8 \exp(-15,495/RT),\ r = 0.8830$	15.50
7.0	$V = 3.163 \times 10^8 \exp(-7037/RT),\ r = 0.9806$	7.04
	Immobilized enzyme	
2.2	$V = 4.752 \times 10^6 \exp(-9773/RT),\ r = 0.9600$	9.78
2.5	$V = 7.422 \times 10^6 \exp(-9790/RT),\ r = 0.9662$	9.79
3.0	$V = 4.419 \times 10^7 \exp(-10,588/RT),\ r = 0.9937$	10.59
3.5	$V = 1.058 \times 10^6 \exp(-8086/RT),\ r = 0.9820$	8.09
4.0	$V = 2.200 \times 10^4 \exp(-5889/RT),\ r = 0.9927$	5.89
4.5	$V = 3.824 \times 10^3 \exp(-4931/RT),\ r = 0.9973$	4.93
4.8	$V = 3.455 \times 10^3 \exp(-4881/RT),\ r = 0.9526$	4.88
5.0	$V = 3.220 \times 10^3 \exp(-4861/RT),\ r = 0.9748$	4.86
6.0	$V = 1.864 \times 10^6 \exp(-9192/RT),\ r = 0.9978$	9.19

[a]Calculated by Fitting Eq. 1 to the activity data of Figs. 1A,B (V = specific activity of the enzyme, V_0 = Arrhenius constant, E_a = energy of activation of the enzyme).

In Fig. 1A maximum activities were observed in the pH range 4.0–4.8. The maximum activity, 17.8 micromol of glucose/(min·mg of protein), was found at 65°C and pH 4.5. These optimum conditions are in exact agreement with those of Woodward et al. *(4)* and Dekker *(16)*. For the IE, Fig. 1B shows that the optimum pH is observed at pH 3.5 for all temperatures studied in this experiment, and the optimum temperature was 65°C in the pH range 3.5–6.0. The activity at 65°C was 6.1 micromol of glucose/(min·mg of protein) at pH 3.5. Disregarding the shift in pH, the activity yield on immobilization is low: 34.0%. For the same pH (pH 4.5), the activity yield is lower: 13.7%. FE data can also be found in an earlier publication *(17)*; the objective here is the comparison of the free and immobilized cellobiase.

Comparison of Fig. 1A,B shows that the immobilization of cellobiase in CPS by the covalent method shifts the optimum pH of the enzyme about 1.0 points, from 4.5 to 3.5. Similar results were obtained by Bergamasco et al. *(18)* for immobilized invertase in CPS: the pH optimum was shifted from 5.0 to 4.5.

The energy of activation was calculated by fitting Eq. 1 to the activity data of Fig. 1A,B, and the hydrolysis rate can be written in the Arrhenius form as presented in Table 1 for FE and IE.

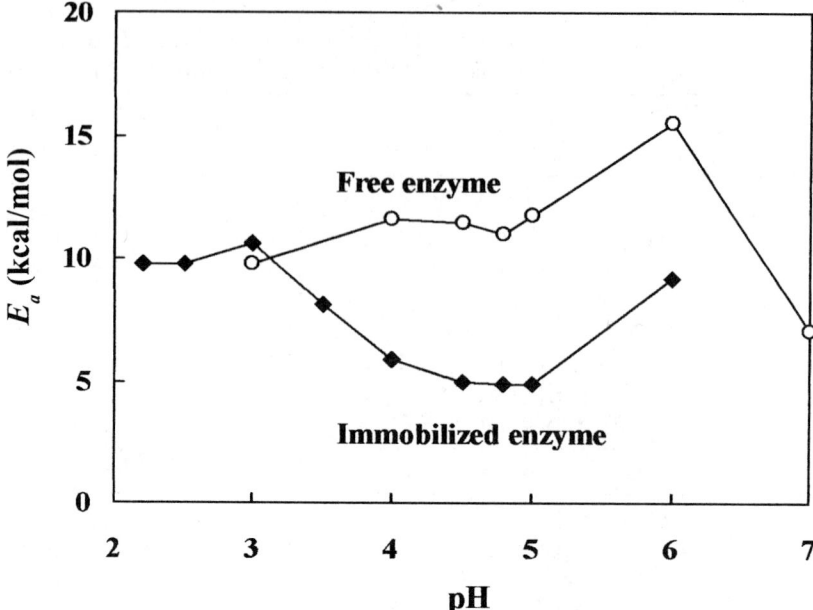

Fig. 2. Comparison of the energy of activation as a function of pH, for free and immobilized Novozym cellobiase, 2 g/L of cellobiose, 95 μL of enzyme/L for FE, and 0.06 g of dry wt IE/20 mL of substrate for IE.

The plot of E_a as a function of pH in Fig. 2 reveals a dependence of the energy of activation on pH and shows that E_a for the IE is generally lower. Figure 2 shows that the energy of activation for the hydrolysis of cellobiose with FE is about 11 kcal/mol in the pH range 4.5–5.0, in accordance with published data for the same enzyme *(16,19,20)*, which show values for the energy of activation in the range of 10–12.5 kcal/mol. In the same range of values, other researchers have found equivalent results for cellobiases produced by different microorganisms *(21–23)*. Figure 2 also shows that in the same pH range, the energy of activation for the IE is about 5 kcal/mol. Therefore, the IE energy of activation observed is half that of the FE, demonstrating an apparent diffusional limitation with the IE *(22,24,25)*. Bisset and Sternberg *(22)* immobilized β-glucosidase of *Aspergillus phoenicis* on chitosan and obtained 11.95 and 7.93 kcal/mol for FE and IE, respectively. They also interpreted their results as a diffusional limitation in the case of the IE.

The experimental results obtained with the thermal stability test for FE are shown in Fig. 3A, and for IE in Fig. 3B. It can be observed in Fig. 3A that up to 55°C, and a thermal denaturation period of 240 min, the FE maintained its activity practically constant, whereas at 70°C the enzyme was almost totally deactivated in 40 min. The IE (Fig. 3B) has shown higher stability, maintaining its activity in a similar test up to 60°C; at 70°C, 40% of its activity was retained after 240 min.

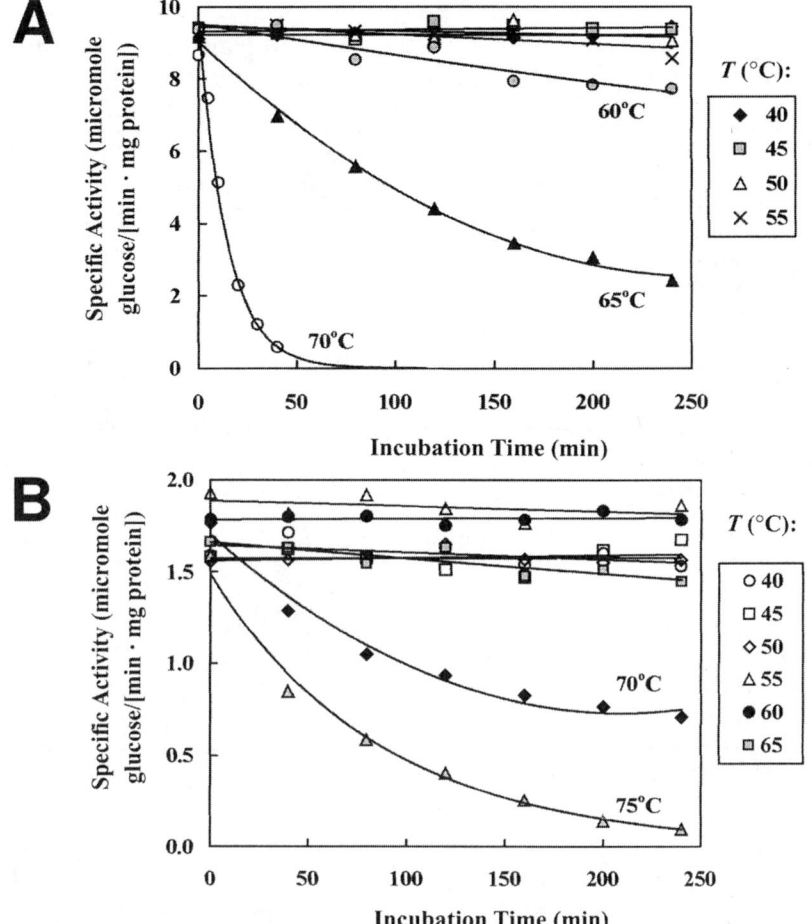

Fig. 3. Residual enzymatic activity incubated in 2 g/L of cellobiose at pH 4.8 for **(A)** free Novozym cellobiase at a concentration of 95 µL/L and **(B)** immobilized Novozym cellobiase at a concentration of 0.06 g of dry wt IE/20 mL of substrate.

Residual activity data for temperatures from 55 to 70°C for FE and from 60 to 75°C for IE were applied to Eq. 7, giving the experimental results for the deactivation constant (K_d) and half-lives ($t_{1/2}$) shown in Table 2 (first K_d and $t_{1/2}$ column). Figure 4 compares the Arrhenius plot of the deactivation constant (K_d) for free and immobilized cellobiase. Equation 3 adjusted for these data gives:

Free cellobiase

$$K_d = 5.728 \times 10^{51} \exp(-80,573/RT) \qquad r = 0.9789 \qquad (9)$$

Immobilized cellobiase

$$K_d = 3.175 \times 10^{53} \exp(-85,238/RT) \qquad r = 0.9885 \qquad (10)$$

Table 2
Comparison of Deactivation Constant (K_d) and Half-Life ($t_{1/2}$) Values
for Free and Immobilized Cellobiase Novozym 188, Adjusted
for a Single Temperature and for All Temperatures[a]

T (°C)	Fitting equation $y = \ln(A_r/A_0) = -K_d t + C_1$ (t = time [h], C_1 = constant)	Adjusted for a single temperature		Adjusted for all temperatures	
		K_d (h^{-1})	$t_{1/2}$ (h)	K_d (h^{-1})	$t_{1/2}$ (h)
	Free enzyme				
		Eq. 7	Eq. 8	Eq. 9	Eq. 8
55	$y = -0.0192t + 0.0238$, $r = 0.8348$	0.0192	36.09	0.0124	80.65
60	$y = -0.0492t + 0.0158$, $r = 0.9773$	0.0492	14.09	0.0789	8.78
65	$y = -0.3288t + 0.0452$, $r = 0.9969$	0.3288	2.11	0.4771	1.45
70	$y = -4.1678t + 0.0910$, $r = 0.9969$	4.1678	0.166	2.7382	0.253
	Immobilized enzyme				
		Eq. 7	Eq. 8	Eq. 10	Eq. 8
60	$y = -0.00283t + 0.0104$, $r = 0.8273$	0.00283	244.9	0.0038	182.4
65	$y = -0.0326t + 0.0086$, $r = 0.9404$	0.0326	21.26	0.0255	27.18
70	$y = -0.2412t + 0.1080$, $r = 0.9691$	0.2412	2.87	0.1620	4.28
75	$y = -0.6991t + 0.013$, $r = 0.9986$	0.6991	0.991	0.9756	0.710

[a]The enzyme was incubated in 2 g/L of cellobiose at pH 4.8.

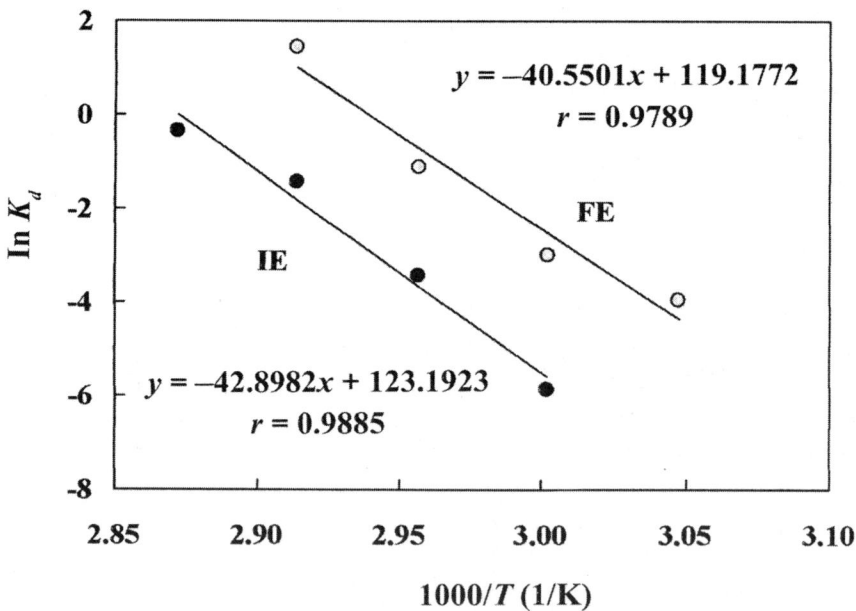

Fig. 4. Arrhenius plot for the energy of thermal deactivation of FE (95 µL/L) and IE (0.06 g of dry wt IE/20 mL of substrate), incubated in 2 g/L of cellobiose at pH 4.8.

Thus, the experimentally observed energy of deactivation (E_d) is approx 80.6 kcal/mol for FE and 85.2 kcal/mol for IE. There is a 5.8% increase in the energy of deactivation for IE, demonstrating that immobilization increases enzyme stability.

From Eqs. 9 and 10 the values of the deactivation constant, adjusted for all temperatures, were obtained and, then, with Eq. 8, the adjusted half-life. The results are shown in Table 2 (second K_d and $t_{1/2}$ column). Comparison of the experimental values obtained for $t_{1/2}$ shows that the IE half-lives are 18.8 times greater on average than half-lives shown by FE, confirming that immobilization confers more stability to the enzyme. The ratio of the half-lives decreases with temperature from 20.8 at 60°C to 16.9 at 70°C. Note that thermal stabilization may, on occasion, be an apparent result. Under mild conditions only the external layer of IE may be working. As deactivation sets in, deep layers of the IE become active, replacing the deactivated ones. However, the same methodology of immobilization as used here was employed with very low enzyme loading (4.12 mg of protein/g of support) with amyloglucosidase *(15)* and stabilization by immobilization was observed. We believe the same phenomenon is occurring here.

Bisset and Sternberg *(22)* obtained the following half-lives for the cellobiase derived from *A. phoenicis* QM 329 using 7.5 m*M* cellobiose as substrate at pH 4.8: 216 h (55°C), 8.6 h (60°C), 0.5 h (65°C), and 0.04 h (70°C). By comparing these results with the experimental half-lives for free enzyme presented in Table 2, it can be observed that cellobiase Novozym 188 is more stable than the enzyme derived from *A. phoenicis* QM 329 at temperatures between 60 and 70°C.

Conclusion

During the immobilization of the enzyme in CPS with the silane-glutaraldehyde method, 67% of the offered enzyme was fixed on the support. The method gives a relatively low activity yield: 13.7% at 65°C and pH 4.5. For all temperatures the enzyme showed optimum activities in the pH range of 4.0–4.8 for FE and about pH 3.5 for IE. Maximum activity was 17.8 micromol of glucose/(min·mg of protein) at 65°C and pH 4.5 for FE and 6.1 micromol of glucose/(min·mg of protein) at 65°C and pH 3.5 for IE. The energy of activation of the enzyme was 11.0 kcal/mol for FE and 5 kcal/mol for IE, at pH values from 4.5 to 5.0, demonstrating an apparent diffusional limitation for the IE.

Cellobiase Novozym 188 when thermally denatured free in solution in a 2 g/L cellobiose solution (pH 4.8) was stable up to 55°C for a period of 4 h. Immobilized in CPS by the silane-glutaraldehyde covalent method, the same enzyme was stable up to 60°C, under the same conditions.

The enzyme thermal deactivation followed reasonably the exponential decay model, giving $K_d = 5.728 \times 10^{51} \exp(-80,573/RT)$ for FE and $K_d = 3.175 \times 10^{53} \exp(-85,238/RT)$ for IE. Finally, the immobilized cellobiase showed half-lives almost 20 times greater on average than the half-lives observed with the FE, suggesting that immobilization confers more stability to this enzyme.

Acknowledgments

We wish to thank Novo Nordisk (Denmark) for kindly providing the enzyme samples. We are also thankful to CAPES, CNPq, PADCT, and the State University of Maringá for financial support.

References

1. Busto, M. D., Ortega, N., and Perez-Mateos, M. (1995), *Process Biochem.* **30(5),** 421–426.
2. Saha, B. C., Freer, S. N., and Bothast, R. J. (1994), *Appl. Environ. Microbiol.* **60(10),** 3774–3780.
3. Aguado, J., Romero, M. D., Rodríguez, L., and Calles, J. A. (1995), *Biotechnol. Prog.* **11,** 104–106.
4. Woodward, J., Koran, L. J., Jr., Hernandez, L. J., and Stephan, L. M. (1993), *Am. Chem. Soc.* **533,** 240–250.
5. Lenders, J.-P., Germain, P., and Crichton, R. R. (1985), *Biotechnol. Bioeng.* **27,** 572–578.
6. Hartmeier, W. (1988), *Immobilized Biocatalysts—An Introduction*, Wieser, J., trans., Springer–Verlag, Berlin.
7. Zanin, G. M. and de Moraes, F. F. (1994), *Appl. Biochem. Biotechnol.* **45/46,** 627–639.
8. Weetall, H. H. (1993), *Appl. Biochem. Biotechnol.* **41,** 157–188.
9. Dixon, M. and Webb, E. C. (1979), *Enzyme*, 3rd ed., Longman Group Limited, London.
10. Trinder, P. (1969), *Ann. Clin. Biochem.* **6,** 24–27.
11. Lowry, O. H., Rosebrough, N. J., Farr, A. L., and Randall, R. J. (1951), *J. Biol. Chem.* **193,** 265–275.
12. Morita, T. and Assumpção, R. M. V. (1972), *Manual de Soluções Reagentes e Solventes—Padronização—Preparação—Purificação*, 2nd ed., Ed. Edgard Blücher Ltda., São Paulo, Brazil.
13. Ballesteros, A., Boross, L., Buchholz, K., Cabral, J. M. S., and Kasche, V. (1994), in *Applied Biocatalysis*, Cabral, J. M. S., Best, D., Boross, L., and Tramper, J., eds., Hardwood Academic, Switzerland, pp. 237–278.
14. Chaplin, M. F. and Bucke, C. (1992), *Enzyme Technology*, Cambridge University Press, Cambridge, UK.
15. Zanin, G. M. and de Moraes, F. F. (1998), *Appl. Biochem. Biotechnol.* **70–72,** 383–394.
16. Dekker, R. F. H. (1986), *Biotechnol. Bioeng.* **28,** 1438–1442.
17. Calsavara, L. P. V., de Moraes, F. F., and Zanin, G. M. (1999), *Appl. Biochem. Biotechnol.* **77–79,** 789–806.
18. Bergamasco, R., Bassetti, F. J., de Moraes, F. F., and Zanin, G. M. (1997), *Symposium on Biotechnology for Fuels and Chemicals*, Colorado Springs.
19. Beltrame, P. L., Carniti, P., Focher, B., Marzetti, A., and Sarto, V. (1983), *Chimica L'Industria* **65(6),** 398–401.
20. Alfani, F., Cantarella, L., Gallifuoco, A., Pezzullo, L., Scardi, V., and Cantarella, M. (1987), *Ann. NY Acad. Sci.* **501,** 503–507.
21. Maguire, R. J. (1977), *Can. J. Biochem.* **55,** 19–26.
22. Bisset, F. and Sternberg, D. (1978), *Appl. Environ. Microbiol.* **35(4),** 750–755.
23. Sundstrom, D. W., Klei, H. E., Coughlin, R. W., Biederman, G. J., and Brouwer, C. A. (1981), *Biotechnol. Bioeng.* **23,** 473–485.
24. Engasser, J. M. and Horvath, C. (1976), in *Applied Biochemistry and Bioengineering*, vol. 1, Wingard, L. B., Jr., Katchalski-Katzir, E., and Goldstein, L., eds., Academic, New York, p. 182.
25. Pitcher, W. H., Jr. (1975), in *Immobilized Enzyme for Industrial Reactors*, Messing, R. A., ed., Academic, New York, p. 170.

Influence of Operating Conditions and Vessel Size on Oxygen Transfer During Cellulase Production

DANIEL J. SCHELL,* JODY FARMER, JENNY HAMILTON,
BOB LYONS, JAMES D. MCMILLAN, JUAN C. SÁEZ,
AND ARUN THOLUDUR

*National Renewable Energy Laboratory, 1617 Cole Boulevard,
Golden, CO 80401, E-mail: dan_schell@nrel.gov*

Abstract

The production of low-cost cellulase enzyme is a key step in the development of an enzymatic-based process for conversion of lignocellulosic biomass to ethanol. Although abundant information is available on cellulase production, little of this work has examined oxygen transfer. We investigated oxygen transfer during the growth of *Trichoderma reesei*, a cellulase-producing microorganism, on soluble and insoluble substrates in vessel sizes from 7 to 9000 L. Oxygen uptake rates and volumetric mass transfer coefficients (k_La) were determined using mass spectroscopy to measure off gas composition. Experimentally measured k_La values were found to compare favorably with a k_La correlation available in the literature for a non-Newtonian fermentation broth during the period of heavy cell growth.

Index Entries: Oxygen transfer; mass transfer coefficient; cellulase; cellulose; ethanol.

Introduction

A key step in the production of ethanol from lignocellulosic biomass is converting cellulose to sugars prior to fermentation of these sugars to ethanol. One promising technology being considered for this process is based on using cellulase. Cellulase is a multicomponent enzyme system that can effectively hydrolyze cellulose to glucose. Cellulase is typically produced by submerged cultivation using the fungus *Trichoderma reesei*. This aerobic microorganism requires good oxygen transfer to achieve good growth and subsequent production of cellulase. Although some information is available on the influence of operating conditions (e.g., agitation and

*Author to whom all correspondence and reprint requests should be addressed.

aeration rates) on cellulase production *(1–3)*, little of this work has specifically examined oxygen transfer rates (OTRs). The usual approach is to maintain the dissolved oxygen (DO) concentration above 20% air saturation by manipulating agitation and aeration rates or by increasing the oxygen concentration in the sparge air.

Cellulase cultivations are non-Newtonian because of the hyphal-fungal morphology and the high concentrations of solid substrates sometimes used in these cultivations. Researchers have investigated oxygen transfer in non-Newtonian fungal cultivations *(4–6)*, during production of highly viscous polysaccharides (e.g., xanthan gum) *(7,8)* or in the presence of solids *(9)*. Schugerl *(10)* reviewed the many mechanisms by which high viscosity unfavorably influences mixing and oxygen transfer. With increasing viscosity higher power inputs are required to achieve similar levels of mixing, gas/liquid interfacial area is reduced because of increasing bubble size, gas/liquid interfacial area is reduced by bubble coalescence, stirring efficiency is reduced because the impeller is more easily flooded, the residence time of large bubbles decreases, and the mass transfer coefficient k_L decreases.

OTRs or oxygen uptake rates (OURs) have been measured and reported by many researchers. OTRs are determined by vessel-operating conditions (e.g., agitation and aeration rates, viscosity), and OURs are a measure of the oxygen utilization rate of the microorganisms. OTR equals OUR at steady state. Atkinson and Mavituna *(11)* summarized OTR data obtained in both small and large vessels using a variety of different methods (e.g., sulfite oxidation, oxygen balance, or biologic outgassing). In general, values reported using the sulfite method are higher than values reported by other methods *(11)*; that is, the sulfite method overpredicts OTR for biologic systems. In small bioreactors (<100 L), the reported sulfite values range from 50 to 500 mol/(L·h). For measurements performed in small bioreactors on active biologic systems (i.e., using oxygen balance technique or biologic outgassing), OURs as high as 140 mmol/(L·h) are reported. OURs up to 22 mmol/(L·h) are reported for 20,000- to 42,000-L bioreactors, also determined using biologic systems.

In more recent studies, an OUR of 30–35 mmol/(L·h) was measured in a 2-L vessel (1.3-L working volume, 730 rpm, 0.31 vessel volumes/min [vvm]) for a bacterial fermentation *(12)*. In a vessel configuration very similar to the 1500-L fermentor used in the present study (see Materials and Methods), Junker et al. *(13)* measured OURs of 28–50 mmol/(L·h) depending on impeller type in an 800-L vessel (600-L working volume, 200 and 275 rpm, 0.37–0.83 vvm) during a *Streptomyces* cultivation. The best performance (50 mmol/[L·h]) was obtained with downward-pumping, Lightnin A315 impellers (tank diameter to impeller diameter ratio [T/D] of 2.0), and the worst performance (28 mmol/[L·h]) was obtained with the Prochem Maxflo T impellers (T/D = 2.3). Rushton impellers (T/D = 2.7) achieved 41 mmol/(L·h) and Prochem Maxflo T impellers (T/D = 2.0) achieved 40 mmol/(L·h). Amanullah et al. *(14)* studied oxygen transfer during a xanthan gum fermentation in a 6-L bioreactor equipped with three Rushton impellers (T/D = 2.0). Agitation and aeration rates were held at 1000 rpm and 0.5 vvm, respectively, and DO concentration was maintained at 20% of air saturation by supplementation with oxygen. Typical OURs ranged from 5 to 14 mmol/(L·h). In another study, Amanullah et al. *(15)* used a 150-L bioreactor to investigate the effect of impeller type on oxygen transfer during a xanthan gum fermentation. Aeration rate was fixed at 0.5 vvm and agitation speed was varied to maintain DO above 15% of air saturation. OUR values up to approx 10 mmol/(L·h) were measured and the best performance was obtained using a Prochem Maxflo T impeller.

In a fungal cultivation to produce xylanase from 2% birch wood xylan (similar to cellulase production using a solid substrate), Hog et al. *(16)* measured OURs up to 25 mmol/(L·h) in a 15-L vessel (10-L working volume, 200 rpm, 1.0 vvm). Lejeune and Baron *(2)* performed cellulase production using *T. reesei* QM 9414 in a 20-L vessel (15-L working volume, 130–400 rpm, 0.2 vvm) on a 1% (w/v) Avicel medium. Although OURs were not directly measured, carbon dioxide evolution rates (CERs) were measured and should be nearly equal to OURs during aerobic metabolism. A maximum CER of 8 mmol/(L·h) was measured at an agitation rate of 300 rpm. Marten et al. *(3)* performed cellulase production using *T. reesei* RUT-C30 grown on 5% (w/v) Solka-Floc in a 16-L vessel (10-L working volume, 250 rpm). They measured volumetric mass transfer coefficients ($k_L a$) of 500–600 h^{-1} at the beginning of the cultivation, and then $k_L a$ decreased to near 300 h^{-1} by 24 h and remained at this value for the rest of the cultivation.

While conducting numerous studies to investigate cellulase production using cellulosic substrates, we had the opportunity to measure OURs at a variety of vessel sizes. In particular, we conducted runs in 7-L bench-scale vessels and in pilot-scale vessels ranging from 160 to 9000 L using both soluble (glucose) and insoluble (Solka-Floc) substrates. Our objectives were to determine whether existing $k_L a$ correlations could be used to predict OURs during cellulase production and to examine how oxygen transfer is affected by vessel size. This information can ultimately be used to estimate capital and operating costs for the cellulase production section of a biomass-to-ethanol conversion facility.

Materials and Methods

Microorganism

The microorganism used was *T. reesei* L27 *(17)* grown on potato dextrose agar plates until sporulation occurred. Spores were suspended in 15% (w/v) glycerol and stored in vials at –70°C.

Vessels

Bench-scale cultivations were performed using New Brunswick Bioflo 3000 fermentation systems (Edison, NJ) in 7-L vessels. Airflow to the vessels was controlled using external MKS Instruments (Andover, MA) type

Table 1
Vessel and Impeller Dimensions

	Vessel size (L)			
	7	160	1500	9000
T (m)[a]	0.165	0.51	1.07	1.83
H (m)[a]	0.038	0.17	0.38	0.89
$H1$ (m)[a]	0.076	0.32	0.61	0.97
Baffle width (m)[b]	0.016	0.044	0.089	0.20
Bottom impeller				
Type	Rushton	Rushton	Rushton	CBI[c]
Diameter (m)	0.076	0.20	0.46	0.97
Width (m)	0.016	0.05	0.10	0.20
Top impeller				
Type	Rushton	CBI[c]	CBI[c]	CBI[c]
Diameter (m)	0.076	0.30	0.56	0.91
Width (m)	0.016	0.05	0.13	0.20

[a]See Fig. 1 for definitions.
[b]All vessels contain four equally spaced baffles.
[c]Curved Blade Impeller (Prochem Maxflo T).

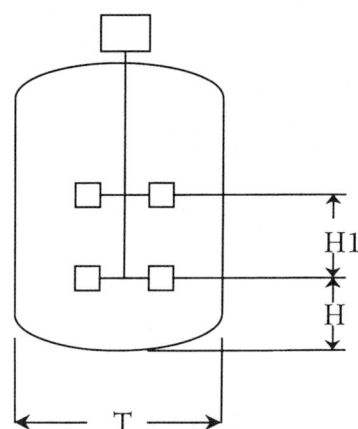

Fig. 1. Dimensions of vessels used in this study.

1159B mass flow controllers. Data for all cultivations in the bench- or pilot-scale vessels were collected and logged by computers at 5-min intervals.

    Pilot-scale cultivations were performed in vessels ranging from 160 to 9000 L using fully instrumented, packaged fermentation systems manufactured by Associated Bio-Engineers and Consultants (Allentown, PA). Table 1 and Fig. 1 give dimensions for vessels used in this work. All vessels were equipped with Rushton turbines, Prochem (Robbins and Myers, Dayton, OH) Maxflo T hydrofoils, or a combination of the two impellers. The Prochem hydrofoil is designed to enhance bulk mixing but has also been shown to be more efficient for oxygen transfer in viscous mycelial cultivations *(18)*. Seed growth for pilot-scale cultivations was performed in a New Brunwick Bioflo IV 20-L fermentation system.

Bench-Scale Cultivations

The medium for inoculum growth and cellulase production was a modified Mandel medium as previously reported *(19)*, but sodium citrate was only used for shake-flask cultivation and not in the larger vessels. A first-stage seed culture was grown by inoculating a 250-mL shake flask (50-mL working volume) containing 2% (w/v) glucose (only the concentration of the carbon source is reported from now on because the other medium components remained constant) with a single vial of frozen stock culture. The flask was incubated at 28°C in a temperature-controlled shaking incubator operating at 200 rpm for 36 h.

A second stage of seed growth is required to condition the microorganism to enzyme production. Fifteen milliliters of first-stage culture (5% [v/v] inoculum) was transferred to a 1-L shake flask (300-mL working volume) containing 1% (w/v) Solka-Floc (a purified cellulose from Fiber Sales and Development, Urbana, OH) and incubated at 28°C and 200 rpm for 48 h. The 7-L vessel (4-L working volume) containing 5% (w/v) Solka-Floc was inoculated with 200 mL of second-stage culture (5% [v/v] inoculum). When more than one vessel was used during a run (up to four vessels could be operated simultaneously), second-stage culture from separate shake flasks was combined before inoculation of the multiple vessels. The temperature in the 7-L vessels was controlled at 28°C and pH was maintained at 4.8 by the addition of either $4\,N\,NH_4OH$ or $2\,N\,H_3PO_4$. The airflow rate was set at 5 L/min (1.25 vvm), and agitation was allowed to vary between 450 and 600 rpm to maintain the DO concentration above 30% of air saturation. Sterile diluted (1:10) antifoam (Ucon Lubricant LB-625; Union Carbide, Danbury, CT) was added manually as needed to control foaming. These runs lasted 7 d.

There was a slight modification to the aforementioned procedures during one experiment (consisting of two vessels) when glucose was used as the carbon source instead of Solka-Floc. In this case, the second-stage culture was inoculated with a 10% (v/v) first-stage culture, the medium for second-stage seed growth contained 3% (w/v) glucose, and the second stage was incubated for only 24 h. The medium for the 7-L vessels contained 5% (w/v) glucose. All other conditions were as reported except that the run was terminated at 70 h after all the glucose was consumed.

Pilot-Scale Cultivations

Two experiments were conducted in the pilot-scale equipment using conditions listed in Table 2. The pilot-scale cultivations used most of the procedures described for the bench-scale cultivations, except that more stages for seed production were required. In the first experiment, *T. reesei* L27 was grown in the 1500-L vessel on 5% (w/w) Solka-Floc. Four stages

Table 2
Operating Conditions for Pilot-Scale Cultivations

	Vessel size (L)			
	20	160	1500	9000
Common operating conditions				
Temperature (°C)	28	28	28	28
pH	Uncontrolled	4.8	4.8	4.8
Absolute pressure (kPa)	80	180	180	180
Dissolved oxygen (%)[a]	Uncontrolled	>20	>20	>20
Acid-H_3PO_4	—	—	70% (w/w)	70% (w/w)
Base-NH_4OH	—	4 N	30% (w/w)	30% (w/w)
First experiment				
Working volume (L)	10	110	1100	
Solka-Floc concentration (% [w/w])	1.0	1.0	5.0	
Culture time (h)	28	124	168	
Inoculum (% [v/v])	6.0	9.0	10.0	
Agitation (rpm)	150–300	150–200	70–130	
Aeration (L/min)[b]	5	40–50	300–1000	
Second experiment				
Working volume (L)	15		600	6000
Glucose concentration (% [w/w])	2.0		2.0	5.0
Culture time (h)	24		28	50
Inoculum (% [v/v])	4.0		2.5	10.0
Agitation (rpm)	300		100	45–105
Aeration (L/min)[b]	7.5		300	1900–2100

[a]Agitation and aeration were adjusted to maintain target dissolved oxygen level.
[b]Flow at standard conditions of 21°C and 1.0 atm pressure.

of seed production were necessary to produce the required volume of inoculum. Two stages were done in shake flasks using procedures discussed above for bench-scale cultivations, followed by growth in the 20 and 160-L vessels at conditions summarized in Table 2.

In the second experiment, *T. reesei* L27 was grown on 5% (w/w) glucose in the 9000-L vessel. Again, four stages of seed production were used, but in this case only the first stage was done in shake flasks. Two 1000-mL shake flasks (300-mL working volume) were inoculated with spores from a frozen vial. This culture was grown for 48 h and then used to inoculate the 20-L vessel that was subsequently used to inoculate the 1500-L vessel. The conditions listed in Table 2 were used.

Measurement of OUR

OURs (mmol/[L·h]) were calculated from measurements of inlet and outlet gas flow rate and gas composition using a VG Prima 600 mass spectrometer (Fisons, Middlewich, UK), according to Eq. 1:

$$\text{OUR} = \frac{1}{V}(q_i Y_i^{02} - q_o Y_o^{02}) \tag{1}$$

in which V is the fermentor working volume (L), q_i is the inlet airflow rate (mmol/h), q_o is the outlet flow rate (mmol/h), Y_i^{O2} is the oxygen concentration in the inlet air (mol%), and Y_o^{O2} is the oxygen concentration in the outlet gas (mol%). The outlet flow rate is not measured but can be readily calculated from a nitrogen mass balance. Since nitrogen is not consumed, a mass balance reduces to the following simple expression:

$$q_o = q_i \left(\frac{Y_i^{N2}}{Y_o^{N2}} \right) \tag{2}$$

in which Y_i^{N2} and Y_o^{N2} are the nitrogen concentration in the inlet air and outlet gas (mol%), respectively. Substituting Eq. 2 into Eq. 1 yields the following final expression for OUR:

$$\text{OUR} = \frac{q_i}{V}\left[Y_i^{02} - Y_o^{02}\left(\frac{Y_i^{N2}}{Y_o^{N2}} \right) \right] \tag{3}$$

The airflow rate q_i is calculated from the volumetric airflow rate F_i (L/min) measured at standard conditions using the ideal gas law:

$$q_i = (P_s F_i / RT_s) \tag{4}$$

in which P_s (1.0 atm) and T (21°C for the pilot-scale fermentor systems and 0°C for the MKS mass flow controllers) are standard pressure and temperature, respectively, for the particular mass flowmeter used; and R is the ideal gas constant (0.08203 [atm·L]/[mol·K]).

Estimation of OTR for Pilot-Scale Vessels

The $k_L a$ (h^{-1}) and the driving force for oxygen transfer ($c^* - c$) govern the rate of oxygen transfer according to Eq. 5:

$$OTR = k_L a\, (c^* - c) \tag{5}$$

in which c^* is the saturated oxygen concentration or solubility (mmol/L) determined from Henry's Law:

$$p_i = Hc^* \tag{6}$$

and c is the actual DO concentration (mmol/L), p_i is the partial pressure of oxygen in the sparge gas (atm), and H is Henry's coefficient (1.0 atm/[mmol·L]) (20). The $k_L a$ is usually given by a correlation of the following form:

$$k_L a \propto \left(\frac{P_g}{V} \right)^n u_s^m \tag{7}$$

in which (P_g/V) is gassed power input per unit volume and (u_s) is the gas superficial velocity. The exponents n and m and a proportionality constant are determined by fitting experimental data.

Power Input

McCabe and Smith *(21)* give an ungassed power correlation for a non-Newtonian (pseudoplastic) liquid in a vessel equipped with a single six-bladed turbine (Rushton) and four baffles. The correlation relates the power number (N_p) to the modified Reynolds number (Re_m). These terms are defined as follows:

$$Re_m = \frac{ND^2\rho}{\mu_a} \tag{8}$$

in which N is the agitation rate (s^{-1}), D is the impeller diameter (m), ρ is the fluid density (kg/m^3, assumed to be 1.02 g/cm^3 in all calculations), and μ_a is the apparent viscosity (kg/[m·s]). The power number is

$$N_p = \frac{P_o g_c}{N^3 D^5 \rho} \tag{9}$$

in which P_o is the ungassed power for a single impeller (W); and g_c is the Newton's Law proportionality factor, which is 32.174 lb-ft/s^2-lb$_f$ in English units and is unity and dimensionless in SI units. Since power correlations depend on impeller type, tank and impeller dimensions, and impeller placement, any correlation will only give approximate results. The total ungassed power input (P_T) is approximated as the single impeller power times the number of impellers.

On aeration, the power drawn by an impeller decreases because of an effective decrease in fluid density caused by the holdup of air under highly aerated conditions. This effect can reduce the gassed power input to as little as 35% of the ungassed power input. Gassed power input was estimated from published experimental data for a non-Newtonian fermentation broth *(22)*.

$k_L a$ Correlations

Correlations for $k_L a$ are available for both Newtonian liquids and a non-Newtonian fermentation broth.

Bailey and Ollis *(23)* give the following correlation for coalescing water, a highly Newtonian liquid:

$$k_L a \ (s^{-1}) = 0.0026 \left(\frac{P_g}{V}\right)^{0.4} u_s^{0.5} \tag{10}$$

in which the power input per unit volume is expressed in W/m^3 and u_s is the superficial velocity (m/s). This equation is valid for $V < 2600$ L and $500 < P_g/V < 10{,}000$ W/m^3 (~5–10 hp/1000 gal).

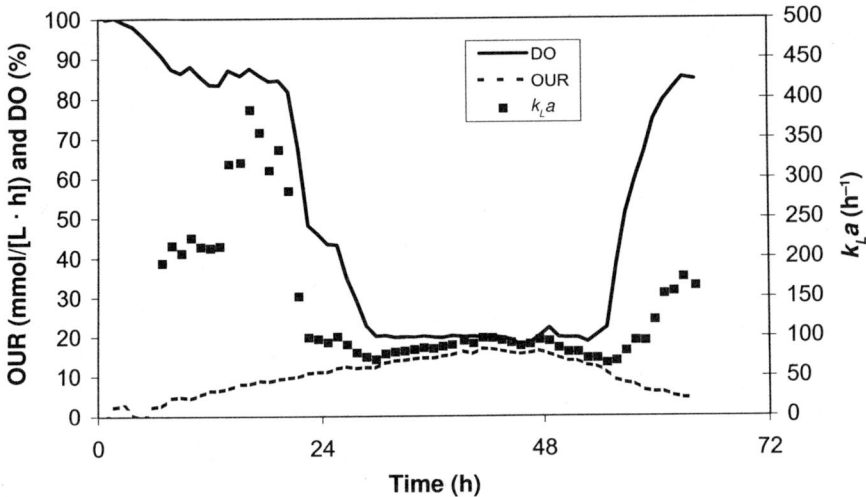

Fig. 2. DO, OUR, and measured $k_L a$ during growth on glucose in a 7-L vessel.

Wang et al. *(22)* present the following $k_L a$ correlation for a non-Newtonian fermentation broth for vessel sizes from 20 to 30,000 L during a fungal cultivation:

$$k_L a \text{ (h}^{-1}) = 8.42 \left(\frac{P_g}{V} \right)^{0.33} u_s^{0.56} \tag{11}$$

in which power input per unit volume is expressed in hp/1000 L and the superficial gas velocity in cm/min.

Results

Bench-Scale Results

Figure 2 shows profiles for DO, OUR, and measured $k_L a$ values for one of the glucose cultivations (one of two vessels). During the first 12 h of the cultivation, $k_L a$ was about 200 h^{-1} but increased to 300–400 h^{-1} for the next 12 h even though both aeration and agitation rates remained constant. After the start of rapid cell growth, $k_L a$ dropped to much lower values between 60 and 90 h^{-1}. During the period of high oxygen uptake from 30 and 54 h, DO concentration was maintained at 20% by increasing the agitation rate. This is reflected by the slight increase in measured $k_L a$; we suspect that a larger increase was not seen because the rapidly increasing cell concentration (data not shown) was also causing broth viscosity to increase. The DO increases only after glucose is nearly consumed when the culture begins to sporulate and lyse, producing a noticeable drop in viscosity. This is also reflected by an increase in $k_L a$ after 54 h.

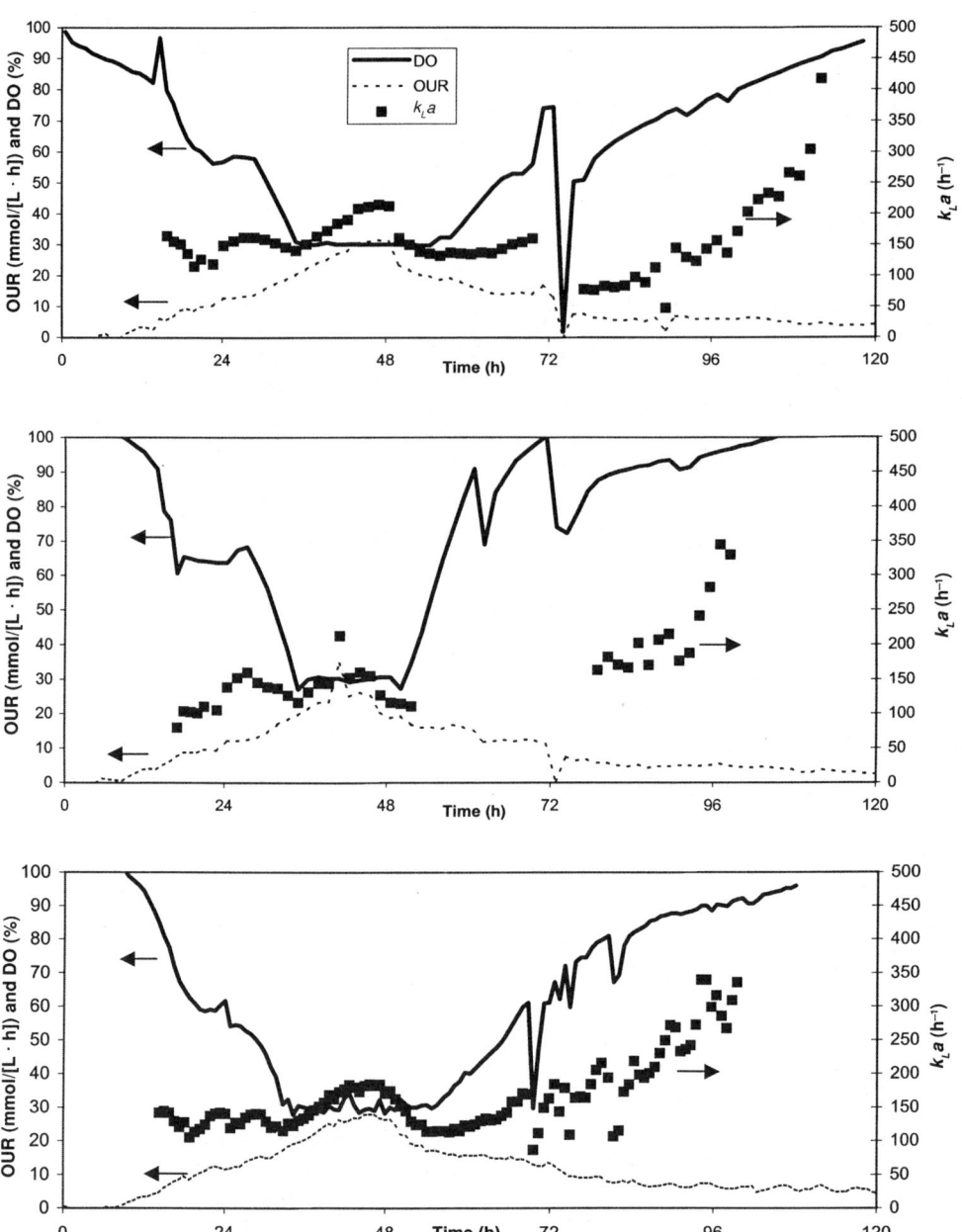

Fig. 3. DO, OUR, and measured $k_L a$ during growth on Solka-Floc for three runs in 7-L vessels.

Figure 3 presents typical results for the bench-scale Solka-Floc runs (three of six cultivations). Some $k_L a$ data are missing because operating problems (foaming that caused loss of airflow and erratic pressure fluctuations) did not allow calculation of good values. The DO, OUR, and $k_L a$ profiles for all of the runs are remarkably similar and generally display the

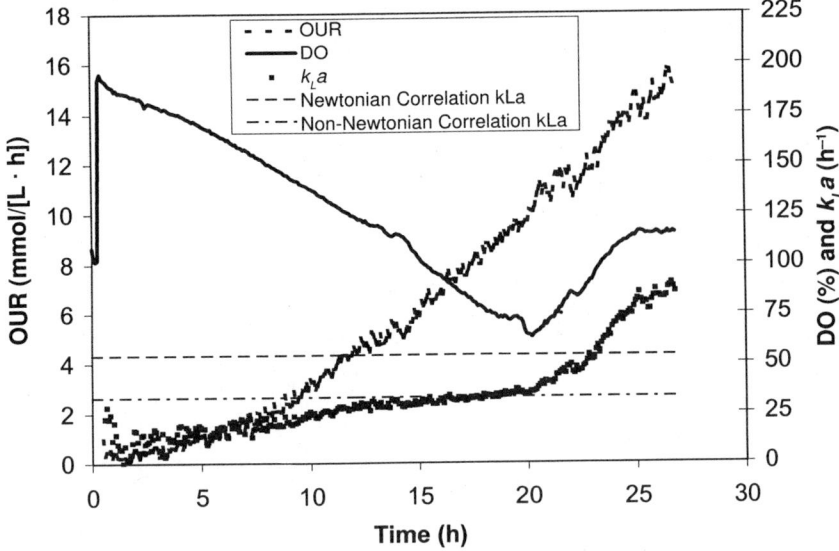

Fig. 4. DO, OUR, and measured k_La during a seed growth on glucose in the 1500-L vessel and values of k_La calculated from Eq. 10 (Newtonian) and Eq. 11 (non-Newtonian).

same trends seen during glucose cultivations. However, during the period when agitation rate was increased to control DO concentration at 30% (typically between 36 and 54 h), there was a noticeable increase in k_La values. As expected, because the thick Solka-Floc slurry hinders mass transfer, the k_La values during the initial stages of these cultivations were lower than values measured for the glucose cultivation. The k_La values were between 100 and 200 h^{-1} during the first 3 d of cultivation and, similar to what was observed in the glucose cultivations, began to rise after the carbon source was consumed and cell lysis began to thin the broth. The maximum OUR for these cultivations ranged from 30 to 35 mmol/(L·h), which is quite similar to OURs reported by other researchers for these vessel sizes.

Pilot-Scale Results

Data from pilot-scale cultivations were obtained during inoculum growth and in the final production vessels (1500- or 9000-L vessels). Figure 4 shows DO, OUR, and measured k_La profiles during inoculum growth on 2% (w/w) glucose in the 1500-L vessel. The inoculum was harvested after 27 h with the broth still containing 10 g/L of glucose. Figure 4 also shows the k_La values calculated from Eqs. 10 and 11 for Newtonian and non-Newtonian broth rheology assuming an apparent viscosity of 1500 cP. Junker et al. *(13)* measured apparent viscosities ranging from 500 to 2200 cP for a *Streptomyces* cultivation depending on impeller type. Viscosity values from 600 to 1000 cP were obtained for Maxflo T impellers and values of 2200 for Rushton impellers, so an intermediate value of 1500 cP was used

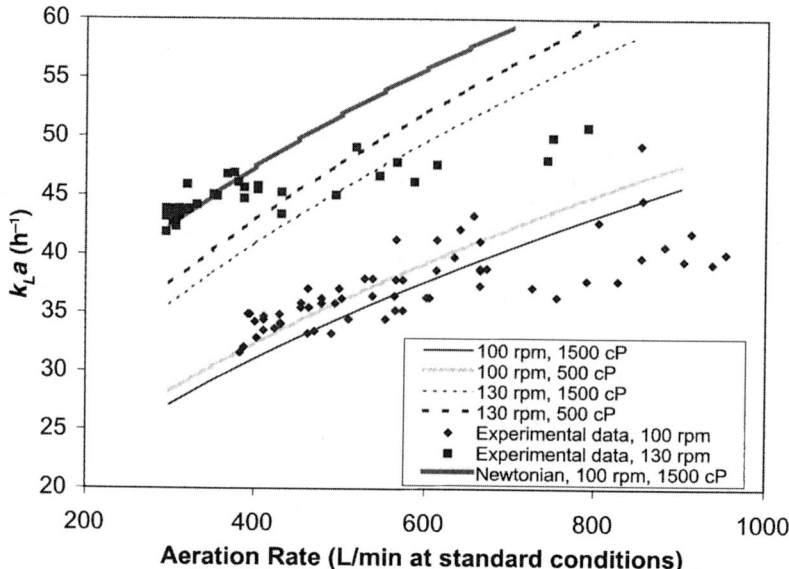

Fig. 5. Measured k_La values as a function of aeration rate during growth on Solka-Floc in the 1500-L vessels. k_La values are calculated from Eq. 11 and shown for the two agitation rates utilized during this run at two different viscosities. k_La values are also calculated from Eq. 10 at one agitation rate and viscosity.

in our calculations; but as shown later (see Fig. 5), calculated k_La values are not extremely sensitive to viscosity.

Since both agitation rate (100 rpm) and aeration rate (300 L/min at standard conditions) were held constant for the entire 27-h seed cultivation, Eqs. 10 and 11 predict constant k_La. The trend of increasing k_La values during this run suggests that during the early stages of this cultivation, the broth rheology is changing to enhance mass transfer. Although unexplained, there are indications of this same type of behavior during the early stages of the bench-scale glucose cultivation (Fig. 2). The trend appears to reverse after approx 25 h, but the cultivation was not run long enough to confirm this trend. Although the non-Newtonian correlation gives lower values for k_La, neither correlation does well at predicting the experimental results during the early part of the cultivation. The k_La correlations were not used to model bench-scale data because the agitation power correlations are based on data obtained from larger vessels.

Figure 5 presents measured and predicted k_La values for the 5% (w/w) Solka-Floc cultivation in the 1500-L vessel. Predicted k_La values are shown for apparent viscosities of 500 and 1500 cP using the non-Newtonian correlation and for an apparent viscosity of 1500 cP using the Newtonian correlation. The experimental data are for the period from 48 to 66 h, during which reliable data acquisition and control occurred. DO was controlled at 30% by adjusting the aeration rate using one of two different agitation rates (130 rpm was lowered to 100 rpm at 54 h). As seen by comparing Fig. 5 to 3, these results were collected during a period of heavy cell growth and

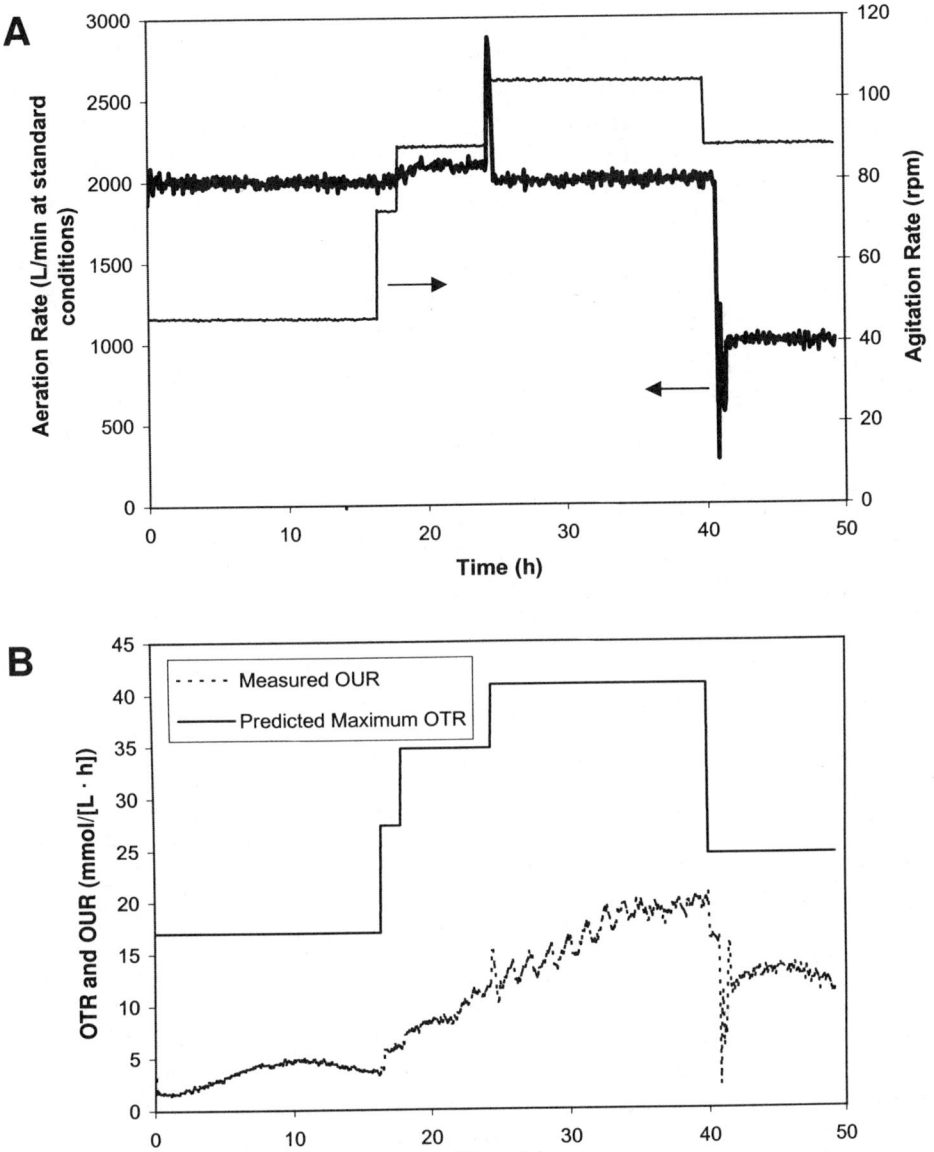

Fig. 6. **(A)** Aeration and agitation rates during growth on glucose in the 9000-L vessel and **(B)** measured OUR compared to the predicted maximum OTR calculated from Eq. 11 for this cultivation.

substrate utilization, before substrate depletion and cell lysis significantly decrease broth viscosity.

Although the trend of measured k_La values is not well predicted by the correlations, the measured values, even at the extremes are within ±25% of the predicted values from the non-Newtonian correlation at 1500 cP over a broad range of aeration rates. This is certainly within the expected accuracy of these types of correlations. However, as seen from previous results, this correlation only appears to work reasonably well during the period of heaviest cell growth and oxygen utilization and not as well during the early and late stages of cultivation. As expected, the Newtonian correlation did not fit the data. Significant changes in apparent viscosity have little effect on the predicted k_La values because changes in viscosity have only a minor effect in the power calculation. However, the equations do not account for the other ways in which viscosity can significantly influence oxygen transfer, as previously discussed, e.g., by influencing both the mass transfer coefficient (k_L) and interfacial area (a).

Because of mechanical failure of a DO probe during the 5% (w/w) glucose cultivation in the 9000-L vessel, the k_La values were not measured during this run. However, Fig. 6A presents the agitation and aeration rate profiles during this run. Using this information, the maximum OTRs were calculated using the non-Newtonian correlation assuming a DO concentration of 0. The maximum OTRs are compared to measured OURs in Fig. 6B. We believe that the DO was actually never near 0. Thus, we expected that measured OUR would be lower than calculated OTR if the correlation predicted reasonable OTR. Since the measured OURs are less than the predicted maximum OTR, we have some confidence that the correlation is effective and will work for even larger vessels.

Discussion

This study shows that a non-Newtonian correlation for k_La can be used to estimate OTR in cellulase production runs during the period of heavy cell growth and oxygen utilization for the vessel sizes used (160–9000 L). For an ethanol production facility, the cellulase production tanks are likely to be at least 500,000 L. It is unlikely that the correlation will work at this large scale, since it has been suggested that the exponents in k_La correlations are functions of scale *(24)*. Nevertheless, the non-Newtonian k_La correlation was utilized to explore the effect of vessel size on oxygen transfer using two commonly used scale-up criteria: constant shear or impeller tip speed (4.04 m/s) and constant power (gassed) input per unit volume (500 W/m³). Figure 7 presents the results of these calculations for 9000-, 100,000-, and 500,000-L vessels. These calculations assume a DO concentration of 0, an average pressure of 2.0 atm, and a working volume of 75% of the total vessel volume. These calculations also assume that the vessels have similar geometries (i.e., vessel height/vessel diameter = 2.0, vessel diameter/impeller diameter = 1.9).

The results show that regardless of the scale-up criteria used, it is possible to achieve better oxygen transfer in the larger vessels. Oxygen transfer is enhanced at larger scales because higher superficial velocities are achieved when the dimensional ratios of the vessels are held constant. As size increases, however, it may be impractical to construct a vessel with a 2:1 height-to-diameter ratio. In addition, the agitation rates required to

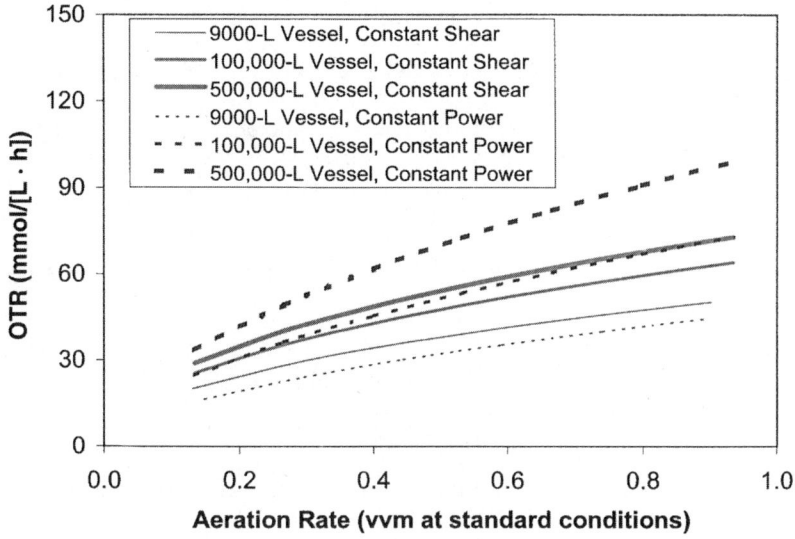

Fig. 7. OUR as a function of vessel size using either constant shear or constant power as the scale-up criterion.

achieved these power inputs may not be practical at very large scale or may produce shear rates that are too high. Oxygen transfer is lower in larger vessels when using shear rate as the scale-up criteria because less power input per unit volume is needed in larger vessels to achieve the same impeller tip speed. Scale-up based on shear rate may be more appropriate for cellulase production because of the reported shear sensitivity of cellulase *(25)*.

Unfortunately, it is probably not possible to simplify scale-up of cellulase production using the correlation presented herein. Accurate predictions of oxygen transfer at large scales are difficult because of the large changes that occur in broth rheology as well as the mounting problems with bulk mixing, efficient bubble breakup, and gas distribution. One researcher has suggested a modification to $k_L a$ correlations to include a viscosity term *(26)*. Additionally, impeller flooding is known to occur at much lower aeration rates in larger vessels, particularly in viscous cultivations *(24)*. In practice, it is probably not possible to achieve higher OTRs in larger vessels than can be achieved in smaller vessels. Our study nevertheless provides some useful oxygen transfer rate data and insight into some of the scale-up issues for cellulase production. This information supports efforts to improve economic analysis of enzyme-based ethanol production processes.

Acknowledgments

This work was funded by the Biochemical Conversion Element of the Department of Energy's Office of Fuels Development.

References

1. Mukataka, S., Kobayashi, N., Sato S., and Takahashi, J. (1988), *Biotechnol. Bioeng.* **32,** 760–763.
2. Lejeune, R. and Baron, G. (1995), *Appl. Microbiol. Biotechnol.* **43,** 249–258.
3. Marten, M., Velkovska, S., Khan, S., and Ollis, D. (1996), *Biotechnol. Prog.* **12,** 602–611.
4. Steel, R. and Maxon, W. (1966), *Biotechnol. Bioeng.* **8,** 97–108.
5. Manfredini, R., Cavallera, V., Marini, L., and Donite, G. (1983), *Biotechnol. Bioeng.* **25,** 3115–3131.
6. Konig, B., Schugerl, K., and Seewald, C. (1982), *Biotechnol. Bioeng.* **23,** 259–280.
7. Li, G., Qiu, H., Zheng, Z., Cai, Z., and Yang, S. (1995), *J. Chem. Tech. Biotechnol.* **62,** 385–391.
8. Amanullah, A., Tuttiett, B., and Nienow, A. (1998), *Biotechnol. Bioeng.* **57,** 198–210.
9. Roman, R. and Tudose, R. (1997), *Bioproc. Eng.* **17,** 361–365.
10. Schugerl, K. (1981), *Adv. Biochem. Eng.* **19,** 71–174.
11. Atkinson, B. and Mavituna, F. (1991), *Biochemical Engineering and Biotechnology Handbook,* 2nd ed., Stockton Press, New York.
12. Ferreira, B., van Keulen, F., and da Fonseca, M. (1998), *Bioproc. Eng.* **19,** 289–296.
13. Junker, B., Stanik, M., Barna, C., Salmon, P., and Buckland, B. (1998), *Bioproc. Eng.* **19,** 403–413.
14. Amanullah, A., Tuttiet, B., and Nienow, A. (1998), *Biotechnol. Bioeng.* **57,** 198–210.
15. Amanullah, A., Serrano-Carreon, L., Castro, B., Galindo, E., and Nienow, A. (1998), *Biotechnol. Bioeng.* **57,** 95–108.
16. Hoq, M., Hempel, C., and Deckwer, W. (1994), *J. Biotechnol.* **37,** 49–58.
17. Shoemaker, S., Watt, K., Tsitovsky, G., and Cox, R. (1983), *Bio/Technology* **1,** 687–690.
18. Buckland, B., Gbewonyo, K., DiMasi, D., Hunt, G., Westerfield, G., and Nienow, A. (1988), *Biotechnol. Bioeng.* **31,** 737–742.
19. Hayward, T., Hamilton, J., Templeton, D., Jennings, E., Ruth, M., Tholudur, A., McMillan, J., Tucker, M., and Mohagheghi, A. (1999), *Appl. Biochem. Biotechnol.* **77–79,** 293–309.
20. Wooley, B., Ruth, M., Sheehan, J., Ibsen, K., Majdeski, H., and Galvez, A. (1999), NREL/TP-580-26157, National Renewable Energy Laboratory, Golden, CO.
21. McCabe, W. and Smith, J. (1976), *Unit Operations of Chemical Engineering,* 3rd ed., McGraw-Hill, New York.
22. Wang, D., Cooney, C., Demain, A., Dunnill, P., Humphrey, A., and Lilly, M. (1979), *Fermentation and Enzyme Technology,* John Wiley & Sons, New York.
23. Bailey, J. and Ollis, D. (1986), *Biochemical Engineering Fundamentals,* 2nd ed., McGraw-Hill, New York.
24. Humphrey, A. (1998), *Biotechnol. Prog.* **14,** 3–7.
25. Ganesh, K., Joshi, J., and Sawant, S. (2000), *Biochem. Eng. J.* **4,** 137–141.
26. Ryu, D. and Humphrey, A. (1972), *J. Ferment. Technol.* **50,** 424.

Characterization
of Cyclodextrin Glycosyltransferase
from *Bacillus firmus* Strain No. 37

Graciette Matioli,[1] **Gisella M. Zanin,**[2]
and Flávio F. De Moraes[*,2]

*[1]Pharmacy and Pharmacology Department
and [2]Chemical Engineering Department,
State University of Maringá, Av. Colombo, 5790, BL D-90,
87020-900 Maringá, PR, Brazil, E-mail: flavio@maringa.com.br*

Abstract

The enzyme cyclodextrin glycosyltransferase (CGTase), EC 2.4.1.19, which produces cyclodextrins (CDs) from starch, was obtained from *Bacillus firmus* strain no. 37 isolated from Brazilian soil and characterized in the soluble form using as substrate 100 g/L of maltodextrin in 0.05 M Tris-HCl buffer, 5 mM CaCl$_2$, and appropriate buffers. Enzymatic activity and its activation energy were determined as a function of temperature and pH. The activation energy for the production of β- and γ-CD was 7.5 and 9.9 kcal/mol, respectively. The energy of deactivation was 39 kcal/mol. The enzyme showed little thermal deactivation in the temperature range of 35–60°C, and Arrhenius-type equations were obtained for calculating the activity, deactivation, and half-life as a function of temperature. The molecular weight of the enzyme was determined by sodium dodecyl sulfate polyacrylamide gel electrophoresis, giving 77.6 kDa. Results for CGTase activity as a function of temperature gave maximal activity for the production of β-CD at 65°C, pH 6.0, and 71.5 mmol of β-CD/(min · mg of protein), whereas for γ-CD it was 9.1 mmol of γ-CD/(min · mg of protein) at 70°C and pH 8.0. For long contact times, the best use of the enzymatic activity occurs at 60°C or at a lower temperature, and the reaction pH may be selected to increase the yield of a desired CD.

Index Entries: Cyclodextrin glycosyltransferase; cyclodextrins; activation energy; deactivation energy.

Introduction

The cyclodextrins (CDs) are cyclic oligosaccharides formed by residues of glucopyranose linked by α-1,4 bonds. The most common are the

---

*Author to whom all correspondence and reprint requests should be addressed.

α-, β-, and γ-CDs that present 6, 7, and 8 U of glucopyranose, respectively. CDs are usually produced from starch by the reaction of cyclization of linear chains of glucopyranose by the enzyme cyclodextrin glycosyltransferase (CGTase). A mixture of CDs is usually produced, and the ratio of these CDs formed depends on the origin of the enzyme and on the reaction time *(1,2)*.

CDs have the form of a truncated cone with an interior relatively nonpolar compared to water that allows them to form inclusion complexes with organic substances *(3)*. The formed complex can improve the properties of the complexed molecule such as its solubility and chemical and thermal resistance *(1,4)*. Because of these characteristics, CDs can be used in a range of industrial applications *(5,6)*.

CGTase is a monomeric enzyme, with a molecular weight on the order of 74.5 kDa, that presents a sequence of amino acids that reveals a structural similarity to the enzyme α-amylase *(7)*. In addition to the cyclization reaction, CGTase catalyzes the coupling reaction and the disproportionation of linear maltodextrins *(8)*.

The first microorganism described in the literature as a producer of CGTase was *Bacillus amylobacter*. In 1891, using this microorganism, A. Villiers produced the first Franz Schardinger dextrins *(9)*. In 1903, Schardinger defined the *Bacillus macerans* as a good producer of CGTase *(1)*.

Since the discovery of CGTase in the middle of a culture of *B. macerans* in 1903, the production of that enzyme has been studied in several lineages of bacteria, such as *B. megaterium*, *B. macerans*, *Klebsiella pneumoniae*, and *B. stearothermophillus*. The enzymes obtained from those microorganisms present different properties, such as thermal stability, optimal pH, molecular weight, and capacity for the formation of CDs *(10)*. The great majority of CGTases isolated to date produce preferably α- or β-CD, with traces of γ-CD *(11)*.

In the literature are mentioned more than 15 species of bacteria that produce CGTase, and most of them can be classified into two great groups: (1) α-CGTase, which produces mainly α-CD in the initial instants of the reaction and among the extracellular CGTase isolated, that from *B. macerans* is prominent; and (2) β-CGTase, which produces β-CD initially at a higher ratio *(4)*. CGTases that are γ-CGTase (i.e., produce a higher initial ratio of γ-CD) are rare and avidly sought because of the great interest shown by the pharmaceutical industry *(4,12,13)*. When 10% maltodextrin was used as substrate, the CGTase from *Bacillus firmus*, isolated from Brazilian soil and to be fully characterized in this study, produced γ- to β-CD in the ratio of 0.156 and a small amount of α-CD and therefore, is a β-CGTase *(13)*.

The study of CDs, their production, and application is increasing worldwide, because their applications are numerous. Technologic progress in the production of CDs has resulted in significant reductions in their costs. However, several of their potential applications will become a reality for large-scale use only if their production costs are further reduced *(14)*.

In spite of the common use of CGTases to produce a mixture of CDs, the selection of a strain that would produce a specific CGTase with high selectivity for the β- or γ-CD would greatly facilitate downstream processing and reduce production costs. This is one of the directions for which research on CGTase and selection of strains have been aimed *(10)*.

The present study presents results of the characterization of a CGTase, obtained from an alkalophylic microorganism isolated from soil, regarding its activity as a function of pH and temperature, activation energy, thermal stability, and deactivation energy.

Materials and Methods

Enzyme

CGTase enzyme was obtained from an alkalophylic microorganism isolated from Brazilian soil (*B. firmus*). The microorganism was cultivated in 250 mL of a liquid medium, pH 10.0, with the following composition ([w/v]): 2.0% soluble starch, 0.5% polypeptone, 0.5% yeast extract, 0.1% K_2PO_4, 0.02% $MgSO_4 \cdot 7H_2O$, 1.0% Na_2CO_3. The cultivation was accomplished at 37°C for 5 d, with agitation at 150 rpm. The cells were removed by centrifugation at 8800g for 10 min. The cell-free supernatant received ammonium sulfate (80% of saturation) and the mixture was maintained at 4°C for 48 h. The precipitate obtained was separated by centrifugation at 8800g for 20 min under refrigeration. Next, it was dissolved in a 50 mM Tris-HCl buffer solution, pH 8.0, purified by biospecific affinity chromatography using β-CD as ligand *(15–17)* and concentrated by ultrafiltration with a 30-kDa cutoff *(17)*. The protein contents were determined by the method of Bradford *(18)*, using bovine serum albumin as standard, giving 0.171 mg of protein/mL of stock solution.

CGTase Activity Assay

One unit of activity corresponds to the amount of CGTase that liberates 1 μmol of β-CD/min in the reaction conditions. The activity assay conditions consisted of a substrate solution containing 10% (w/v) maltodextrin in 50 mM Tris-HCl buffer and 5 mM $CaCl_2$, pH 8.0, at 50°C. The diluted buffer solutions of enzyme and substrate were separately heated to 50°C. In six test tubes 1 mL of substrate solution was placed and soon after 1 mL of enzyme solution was added. The tubes were agitated and incubated for 30 min at 50°C, and one tube was removed every 5 min. CGTase was inactivated by heating the tubes in boiling water for 10 min *(13,17)*. The reaction time and enzyme dilution were selected in a manner allowing a linear relationship between the formed CD and the time, seeking to reduce the effect of the inhibition of the reaction products, according to criteria established by the method of the initial velocities *(19)*. For assay of the CGTase activity for producing β-CD, a 1:400 dilution of the stock enzyme was used (4.27×10^{-4} mg of protein/mL), whereas for γ-CD a 1:50 dilution

$(3.42 \times 10^{-3}$ mg of protein/mL) was used because of the lower activity of the enzyme for producing γ-CD.

The β-CD produced in the assay was determined by the method of dye extinction, i.e., color reduction that occurs after complexation with β-CD using phenolphthalein at 550 nm *(20)*. The γ-CD concentration was determined by the bromocresol green colorimetric method, and in this case there is an increase in the color of the solution after complexation with γ-CD, which is measured at 620 nm *(20)*.

Enzymatic Activity as a Function of Temperature and pH

The influence of pH and temperature was separately determined for the formation of β-CD and γ-CD, since the amount of γ-CD produced is quite small compared with the β-CD produced. In the β-CD formation test, the substrate solution was prepared in the following pH values: 4.0, 5.0, 6.0, 7.0, 8.0, 9.0, and 9.5; for γ-CD formation, the pH values were 5.0, 6.0, 7.0, 8.0, 8.5, 9.0, 9.5, and 10.0. Disodium phosphate–citric acid buffer was used for pH 4.0–7.0, and boric acid–potassium chloride was used for pH 8.0–10.0. Final buffer concentration was 50 mM. The tests were performed, as in the case of the enzymatic activity assay, at 50°C. The influence of temperature on the activity of the enzyme was determined in the temperature range of 35–70°C with an interval of 5°C. The tests were performed, also as in the case of the enzymatic activity assay, at pH 8.0. The CDs produced were determined as given in CGTase Activity Assay.

Thermal Stability of CGTase

The residual activity of the CGTase, during incubation for 240 min at temperatures from 35 to 80°C with an interval of 5°C, was determined by the method of initial velocities *(19)*. A solution of diluted enzyme (1:40) was prepared in a solution of 10% (w/v) maltodextrin, 50 mM Tris-HCl buffer, and 5 mM CaCl$_2$, pH 8.0. This diluted enzyme solution was maintained in a selected temperature, and every 40 min a 0.8-mL aliquot was taken and added to 7.2 mL of distilled water. One milliliter of this enzyme solution (now diluted 1:400) was added to the six test tubes followed by the addition of 1 mL of a 20% substrate solution, and the residual activity was determined at 50°C. The substrate solution was prepared at a concentration of 20% (w/v) in 50 mM Tris-HCl buffer and 5 mM CaCl$_2$, pH 8.0. This procedure was repeated until 240 min of incubation of the enzyme in each selected temperature was completed. The tubes resulting from the residual activity test were maintained at 4°C for later determination of the produced CDs by the colorimetric methods indicated in CGTase Activity Assay *(20)*.

Determination of Molecular Weight

The molecular weight of the CGTase was determined according to Weber and Osborn *(21)*, by sodium dodecyl sulfate polyacrylamide gel electrophoresis (SDS-PAGE), using the molecular weight reference kit

(Pharmacia, Uppsala, Sweden) of six standard proteins with molecular weights ranging from 14.4 to 94 kDa. The relation between log molecular weight and relative mobility was established, and the molecular weight of the CGTase was determined through this relation.

Results and Discussion

Activity of CGTase Enzyme of B. firmus as Function of pH

In the determination of the β-CD produced by the CGTase from strain no. 37, as a function of pH, and maintaining the temperature of 50°C, the maximal specific activity occurred at pH 6.0 (Fig. 1A), giving 104.1 mmol of β-CD/(min · mg of protein). This value is in accordance with the range of pH reported in the literature, mainly for CGTases from Bacilli. The same value of optimal pH was not obtained for γ-CD production by the CGTase from strain no. 37. In this case, the maximal specific activity occurred at pH 8.0 (Fig. 1B), giving 5.0 mmol of γ-CD/(min · mg of protein).

The ratio of β-CD and γ-CD formed is presented in Fig. 1C. The relationship of γ-CD/β-CD produced reached a maximum at pH 8.0, within the pH range of 5.0–9.5, and the largest initial relative rate of production of γ-CD at this pH was 0.071.

Kato and Horikoshi *(22)* worked with a strain of Bacillus that produced mainly γ-CD, and whose pH value is close to the value obtained herein. Sato and Yagi *(23)* studied a CGTase of *B. macerans* that displayed a behavior similar to that of the CGTase in the present study, i.e., with different pH values for the maximal production of each type of CD. Existing different pH values for the maximal production of β- and γ-CD imply that the degree of ionization of the different groups of the catalytic site of the enzyme demands different states, and possibly different conformations, for the greatest production of each product. In practice, this verification is quite advantageous, because it propitiates a new method of addressing the production of CDs for increasing the selectivity toward one of the CDs that is accomplished through the choice of appropriate pH.

The optimal pH of other CGTases reported in the literature varies according to the microorganism species that produces the CGTase. For CGTases from Bacilli, the range of optimal pH is quite wide, varying from 4.0, as in the case of the CGTase studied by Techaiyakul et al. *(24)*, to 12, as in the case of the CGTase studied by Horikoshi *(25)*.

Activity of CGTase Enzyme of B. firmus as Function of Temperature

Figure 2 presents the activity of the CGTase of strain no. 37, as a function of temperature, for the production of β-CD (Fig. 2A), the production of γ-CD (Fig. 2B), and the ratio of γ-CD/β-CD produced at different temperatures (Fig. 2C). As the temperature increased the ratio of γ-CD/β-CD produced increased (i.e., an increase in the temperature favors the production of γ-CD in relation to β-CD). At 70°C the ratio of γ-CD/β-CD production (0.15) was nearly double that observed at 50°C (0.077).

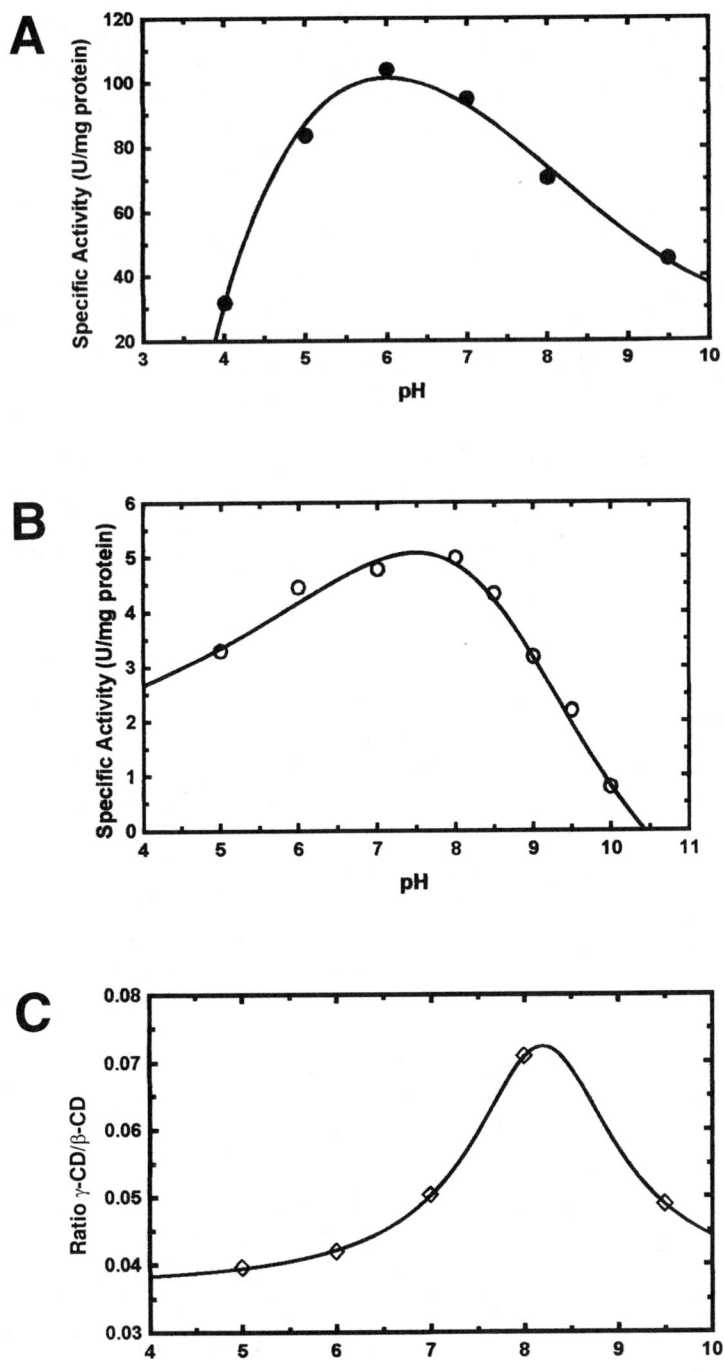

Fig. 1. Specific activity for the production of **(A)** β-CD and **(B)** γ-CD and **(C)** their ratio, γ-CD/β-CD, as a function of pH for the CGTase from *B. firmus* strain no. 37. Conditions: substrate is 10% (w/v) maltodextrin, in 50 m*M* buffer and 5 m*M* CaCl$_2$, 50°C. Buffers: Disodium phosphate–citric acid buffer for pH 4.0–7.0 and boric acid–potassium chloride buffer for pH 8.0–10.0.

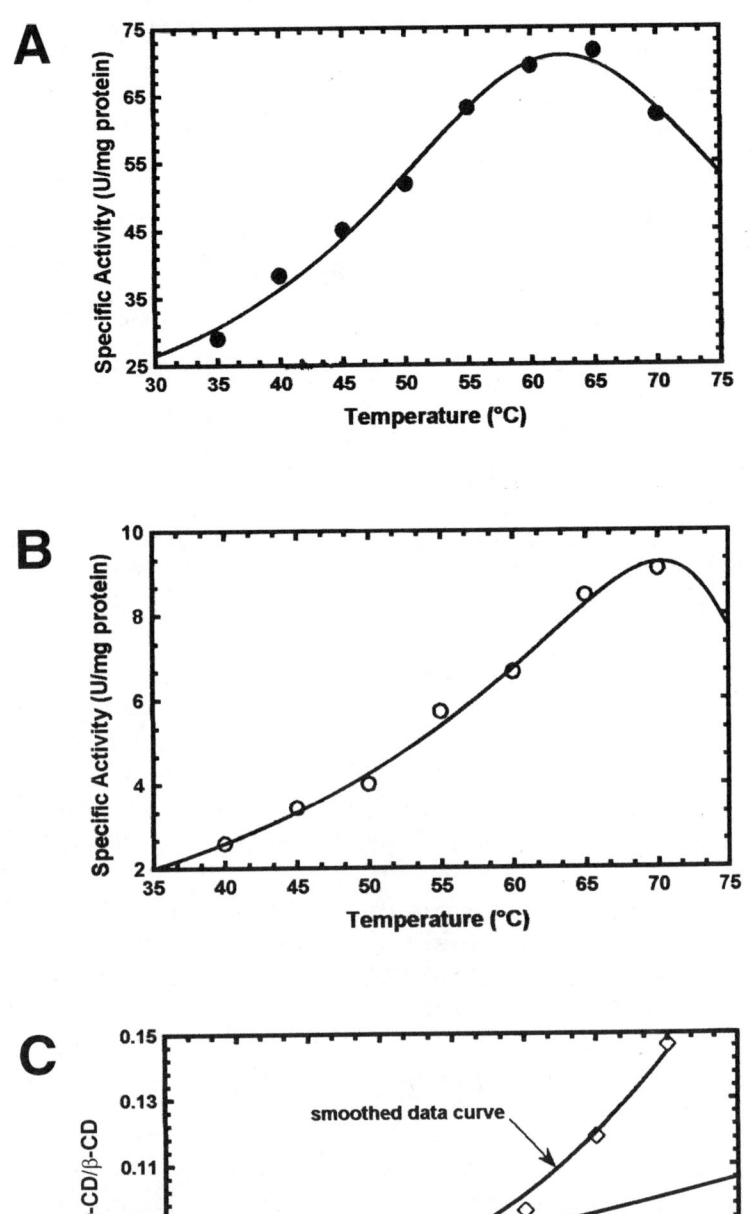

Fig. 2. Specific activity for the production of **(A)** β-CD and **(B)** γ-CD and **(C)** their ratio, γ-CD/β-CD, as a function of temperature for the CGTase from *B. firmus* strain no. 37. Conditions: substrate is 10% (w/v) maltodextrin, in 50 m*M* Tris-HCl buffer and 5 m*M* CaCl$_2$, pH 8.0.

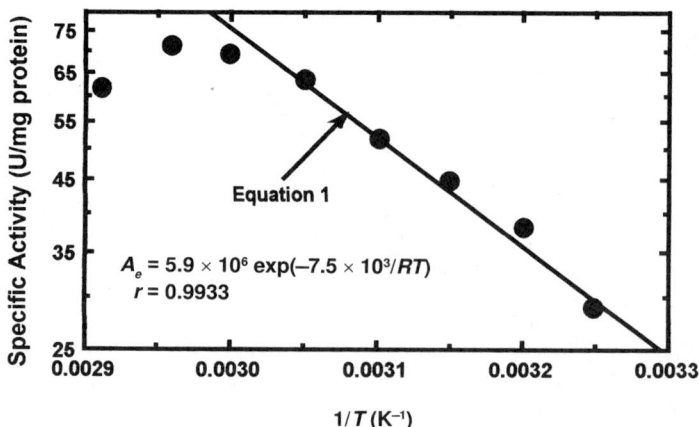

Fig. 3. Arrhenius plot of the specific activity for the production of β-CD as a function of the inverse of the absolute temperature for the CGTase from *B. firmus* strain no. 37. Conditions: substrate is 10% (w/v) maltodextrin, in 50 mM Tris-HCl buffer and 5 mM CaCl$_2$, pH 8.0.

In the determination of the β-CD produced by the CGTase of strain no. 37, as a function of temperature, the maximal specific activity was found at 65°C (71.5 mmol of β-CD/[min · mg of protein]). For the γ-CD, the maximal specific activity was observed at 70°C (9.1 mmol of γ-CD/[min · mg of protein]). Therefore, both the optimal pH and optimal temperature values, of maximal specific activity of the enzyme, were different for the products β- and γ-CD.

The values of optimal temperature obtained herein are in agreement with those found in the literature, and the optimal temperature for known Bacilli is found in the range of 45–70°C. CGTase of *B. firmus* no. 324, studied by Yim et al. *(26)*, has also shown an optimal temperature of 65°C.

Energy of Activation of CGTase Enzyme of B. firmus

The effect of temperature on an enzymatic reaction can be analyzed through the Arrhenius equation. Therefore, this equation was adjusted to the experimental points of Fig. 2, and it allowed the determination of the activation energy for the reaction of β- (Fig. 3) and γ-CD production (Fig. 4), giving 7.5 and 9.9 kcal/mol, respectively. The larger activation energy found for the reaction of γ-CD production, together with the results presented in Fig. 2C, which show that larger temperatures favor the production of γ-CD, confirms the following general rule: "larger temperatures favor the reactions of larger activation energy" *(27)*.

Within the range of validity, the enzymatic activity for the production of β- and γ-CD and their ratio can be calculated with the adjusted Arrhenius-type Eqs. 1–3 plotted in Figs. 3 and 4:

$$\text{β-CD: } A_e = 5.9 \times 10^6 \exp(-7.5 \times 10^3/RT); \quad T \le 55°C \tag{1}$$

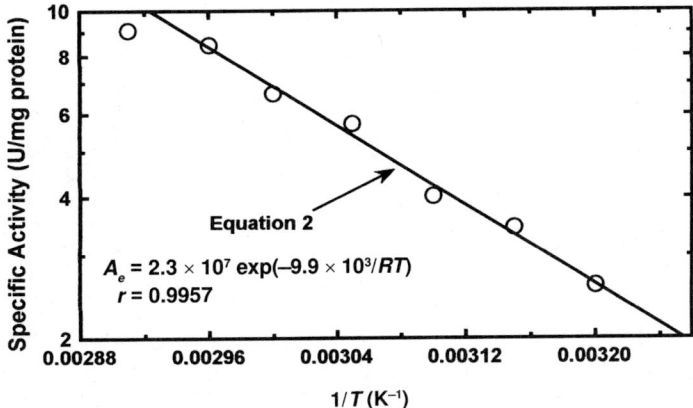

Fig. 4. Arrhenius plot of the specific activity for the production of γ-CD as a function of the inverse of the absolute temperature for the CGTase from *B. firmus* strain no. 37. Conditions: substrate is 10% (w/v) maltodextrin, in 50 m*M* Tris-HCl buffer and 5 m*M* CaCl$_2$, pH 8.0.

$$\gamma\text{-CD: } A_e = 2.3 \times 10^7 \exp(-9.9 \times 10^3/RT); \quad T \le 65°C \quad (2)$$

$$\gamma\text{-CD}/\beta\text{-CD: production ratio} = 3.9 \exp(-2.5 \times 10^3/RT); \quad T \le 55°C \quad (3)$$

in which R is the ideal gas constant (1.987 cal/[mol·K]) and T is the absolute temperature in degrees kelvin.

Thermal Stability of CGTase from B. firmus

Figure 5 presents the residual specific activity for β-CD production as a function of time for the CGTase of *B. firmus* strain no. 37. Although the data presented in Fig. 2 show that the enzyme presents a high specific activity for β-CD production at 65°C, and at 70°C for γ-CD, its thermal stability in these conditions is relatively low, as observed in Fig. 5. This result makes the use of this CGTase unfeasible for periods of time longer than 4 h at 70°C, which was shown as the best temperature for maximizing the ratio of γ- to β-CD production (Fig. 2C). Whereas at 65°C and pH 8.0, the enzyme showed a value of 71.5 U/mg of protein, the specific activity at 60°C and the same pH was equal to 56.4 U/mg of protein, which is 1.2 times less than at 65°C. In spite of the lesser activity at 60°C, it is advisable to use this temperature instead of 65°C because at 60°C, the enzyme presents an activity still relatively high that is associated with a higher thermal stability, which is necessary if the enzyme is to be reacted for long periods such as 24–48 h used in industrial processes. These results are compatible with those obtained by other CGTases *(22)*.

Energy of Thermal Deactivation of the CGTase of B. firmus

Figure 5 shows that the CGTase of *B. firmus* strain no. 37 practically did not show thermal deactivation in the temperature range of 35–60°C for a reaction period of 4 h. At 65°C and above, the enzyme presented increas-

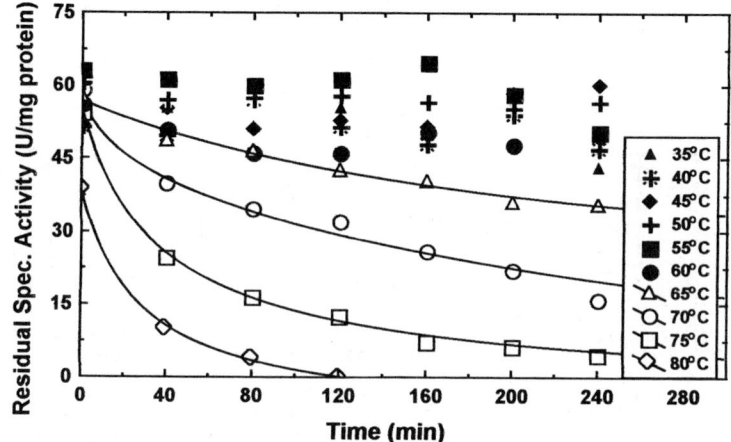

Fig. 5. Residual specific activity for the production of β-CD as a function of time for the CGTase from *B. firmus* strain no. 37. Conditions: substrate is 10% (w/v) malto-dextrin, in 50 mM Tris-HCl buffer and 5 mM CaCl$_2$, pH 8.0; enzyme concentration is 4.3×10^{-4} mg of protein/mL.

ingly higher thermal deactivation, particularly at 80°C, where after 120 min it did not show any residual activity.

The half-life of the CGTase enzyme was calculated by the exponential model *(17,28)*. For the enzyme incubated in a solution of 10% (w/v) maltodextrin, pH 8.0, the half-life was higher than 11 h for temperatures lower than 60°C and dropped to just 40 min at 80°C. For long reaction periods, these results delimit the best use of the enzymatic activity to temperatures lower than 60°C.

It is usually assumed that the kinetics of enzyme thermal deactivation is first order in relation to the concentration of active enzyme, and that the coefficient of thermal inactivation (K_d) is a function of the temperature as given by the Arrhenius law. The slope of the adjusted straight line that correlates to the natural logarithm of the coefficient of thermal inactivation (K_d) with the inverse of the absolute temperature (T) is the energy of thermal deactivation *(17,28)*. This graph is shown in Fig. 6 and the energy of deactivation obtained was 39 kcal/mol. The Arrhenius-type equation adjusted to the data allow the determination of the coefficient of thermal inactivation (K_d) and the half-life ($t_{1/2}$) for all temperatures *(17,28)*:

$$K_d = 2.1 \times 10^{24} \exp(-39 \times 10^3 / R \cdot T) \qquad (4)$$

$$t_{1/2} = \ln(2) / K_d \qquad (5)$$

For example, at 65°C the half-life is 5.2 h whereas at 50°C it is 76.6 h. Additionally, the deactivation of the CGTase enzyme by heating in boiling water (100°C) can now be calculated, and it takes 8.9 min to lower the enzymatic activity to 1% of its initial activity. Therefore, the deactivation of the enzyme by boiling for 10 min was more than satisfactory. This enzyme is less stable

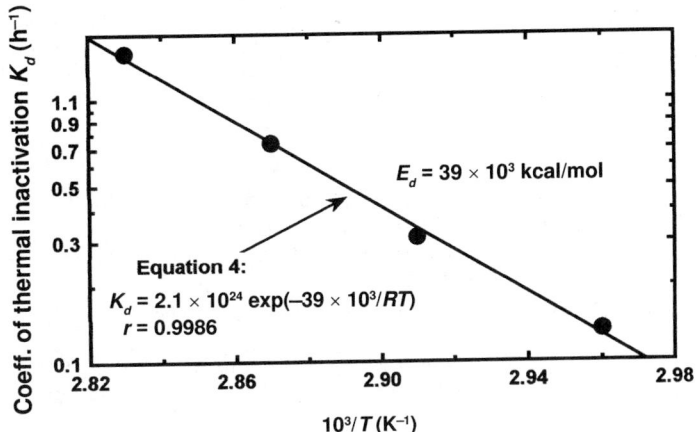

Fig. 6. Arrhenius plot of the residual activity for the production of β-CD as a function of the inverse of the absolute temperature for the CGTase from *B. firmus* strain no. 37. Conditions: substrate is 10% (w/v) maltodextrin, in 50 m*M* Tris-HCl buffer and 5 m*M* CaCl$_2$, pH 8.0; enzyme concentration is 4.3×10^{-4} mg of protein/mL.

than Novo Nordisk (Copenhagen, Denmark) amyloglucosidase, which has an energy of deactivation of 50.6 kcal/mol *(28)*.

Molecular Weight of Enzyme

The molecular weight of the CGTase enzyme was determined by SDS-PAGE and found to be 77.6 kDa. This value is within the usual range of molecular weight obtained for CGTases from different microorganisms (66–80 kDa).

Conclusion

The characterization of the CGTase from *B. firmus* strain no. 37 led to a mol wt of 77.6 kDa. At 50°C the maximum specific activities of this CGTase for β-CD production occurred at pH 6.0, whereas for γ-CD they occurred at pH 8.0. When the pH was fixed at 8.0, the temperatures for maximum specific activities were 65°C for β-CD, giving 71.5 U/mg, and 70°C for γ-CD, giving 9.1 U/mg, respectively. Therefore, both optimum values for pH and temperature for the production of β- and γ-CD were different. This is advantageous because greater production of one of the products can be achieved by appropriate choice of the operating conditions. Although the enzyme was more active at 65°C for the production of β-CD, it was more stable at 60°C. With the CGTase of *B. firmus* strain no. 37, the reactions that produce CDs from the substrate 10% (w/v) maltodextrin at pH 8.0 had an energy of activation of 7.5 and 9.9 kcal/mol for the production of β-CD and γ-CD, respectively. This enzyme when incubated in the same substrate solution had a half-life >11 h for temperatures below 60°C. The energy of thermal deactivation of this enzyme was 39 kcal/mol.

Acknowledgments

We thank CNPq, CAPES, PADCT-II, and the State University of Maringá for financial support.

References

1. Szejtli, J. (1988), in *Cyclodextrin Technology*, Szejtli, J., ed., Kluwer Academic, Dordrecht, The Netherlands, pp. 1–78 and 79–185.
2. Bekers, O., Uijtendaal, E. V., Beijnen, J. H., Bult, A., and Underberg, W. J. M. (1991), *Drug Dev. Ind. Pharm.* **17,** 1503–1549.
3. Duchêne, D., Debruères, B., and Brétillon, A. (1984), *Labo-Pharma—Probl. Tech.* **32,** 843–850.
4. Englbrecht, A., Harrer, G., Lebert, M., and Schmid, G. (1990), in *Minutes of the 5th International Symposium on Cyclodextrins*, Duchêne, D., ed., Editions de Santé, Paris, pp. 25–31.
5. Lee, J. H., Choi, K. H., Lee, Y. S., Kwon, I. B., and Yu, J. H. (1992), *Enzyme Microb. Technol.* **14,** 1017–1020.
6. Duchêne, D. (1987), *Cyclodextrins and Their Industrial Uses*, Edition de Santé, Paris.
7. Nakamura, A., Haga, K., and Yamane, K. (1994), *FEBS Lett.* **337,** 6–70.
8. Mattsson, P., Mäkelä, M., and Korpela, T. (1988), in *Proceedings of the Fourth International Symposium on Cyclodextrin*, Huber, O. and Szejtli, J., eds., Kluwer Academic, Dordrecht, The Netherlands, pp. 65–70.
9. French, D. (1957), *Adv. Carbohydr. Chem.* **12,** 189–260.
10. Horikoshi, K. (1988), in *Proceedings of the Fourth International Symposium on Cyclodextrin*, Huber, O. and Szejtli, J., eds., Kluwer Academic, Dordrecht, The Netherlands, pp. 7–17.
11. Jamuna, R., Saswathi, N., Sheelar, R., and Ramakrishina, V. (1993), *Appl. Biochem. Biotecthnol.* **43,** 163–176.
12. Mori, S., Hirose, S., Oya, T., and Kitahata, S. (1994), *Biosci. Biotechnol. Biochem.* **58,** 1968–1972.
13. Matioli, G., Zanin, G. M., Guimarães, M. F., and de Moraes, F. F. (1998), *Appl. Biochem. Biotechnol.* **70–72,** 267–275.
14. Bender, H. (1986), *Adv. Biotechnol. Processes* **6,** 31–71.
15. László, E., Banky, B., Seres, G., and Szejtli, J. (1981), *Starch* **33,** 281–283.
16. Berna, P., de Moraes, F. F., Barbotin, J. N., Thomas, D., and Vijayalaksmi, M. A. (1996), in *Advances in Molecular and Cell Biology*, vol. 15B, Bittar, E. E., series ed., Danielson, B. and Bülow, L., guest eds., JAI Press, pp. 521–535.
17. Matioli, G. (1997), PhD thesis, UFPR, Curitiba, PR, Brazil.
18. Bradford, M. (1976), *Anal. Biochem.* **72,** 248.
19. Dixon, M. and Webb, E. C. (1979), *Enzymes*, 3rd ed., Longman Group Limited, London.
20. Hamon, V. and de Moraes, F. F. (1990), in *Etude Preliminaire a L'immobilisation de L'enzyme CGTase WACKER*, Laboratoire de Tecnologie Enzymatique, Université de Tecnologie de Compiègne.
21. Weber, K. and Osborn, M. (1969), *J. Biol. Chem.* **244,** 4406–4412.
22. Kato, T. and Horikoshi, D. (1986), *Agric. Biol. Chem.* **50,** 2161, 2162.
23. Sato, M. and Yagi, Y. (1991), in *Biotechnology of Amylodextrin Oligosaccharides*, Friedman, R. B., ed., American Chemical Society, Washington, DC, pp. 125–137.
24. Techaiyakul, W., Pongssawasdi, P., and Mongkolkl, P. (1990), in *Minutes of the 5th International Symposium on Cyclodextrins*, Duchêne, D., ed., Editions de Santé, Paris, pp. 50–54.
25. Horikoshi, K. (1971), *Agric. Biol. Chem.* **35,** 1407–1414.
26. Yim, D. G., Sato, H. H., Park, Y. H., and Park, Y. K. (1997), *J. Ind. Microbiol. Biotechnol.* **18,** 402–405.
27. Levenspiel, O. (1972), *Chemical Reaction Engineering*, 2nd ed., John Wiley & Sons, New York, p. 239.
28. Zanin, G. M. and de Moraes, F. F. (1998), *Appl. Biochem. Biotechnol.* **70–72,** 383–394.

Physiological Aspects Involved in Production of Xylanolytic Enzymes by Deep-Sea Hyperthermophilic Archaeon *Pyrodictium abyssi*

CAROLINA M. M. CARVALHO ANDRADE,[*,1]
WILSON BUCKER AGUIAR,[1] AND GARO ANTRANIKIAN[2]

[1]*Universidade Federal do Rio de Janeiro, Escola de Química, Departamento de Engenharia Bioquímica, 21.949-900 Rio de Janeiro, R.J., Brazil, E-mail: carolinaandrade@uol.com.br; and [2]Technical University Hamburg-Harburg, Institute of Technical Microbiology, Denickestr. 15, D-21071 Hamburg, Germany*

Abstract

Xylanases (EC 3.2.1.8) catalyze the hydrolysis of xylan, the major constituent of hemicellulose. The use of these enzymes could greatly improve the overall economics of processing lignocellulosic materials for the generation of liquid fuels and chemicals. The hyperthermophilic archaeon *Pyrodictium abyssi*, which was originally isolated from marine hot abyssal sites, grows optimally at 97°C and is a prospective source of highly thermostable xylanase. Its endoxylanase was shown to be highly thermostable (over 100 min at 105°C) and active even at 110°C. The growth of the deep-sea archaeon *P. abyssi* was investigated using different culture techniques. Among the carbohydrates used, beech wood xylan, birch wood glucuronoxylan and the arabinoxylan from oats pelt appeared to be good inducers for endoxylanase and β-xylosidase production. The highest production of arabinofuranosidase, however, was detected in the cell extracts after growth on xylose and pyruvate, indicating that the intermediate of the tricarboxylic acid cycle acted as a nonrepressing carbon source for the production of this enzyme. Electron microscopic studies did not show a significant difference in the cell surface (e.g., xylanosomes) when *P. abyssi* cells were grown on different carbohydrates. The main kinetic parameters of the organism have been determined. The cell yield was shown to be very low owing to incomplete substrate utilization, but a very high maximal specific growth rate was determined ($\mu_{max} = 0.0195$) at 90°C and pH 6.0. We also give information on

*Author to whom all correspondence and reprint requests should be addressed.

the problems that arise during the fermentation of this hyperthermophilic archaeon at elevated temperatures.

Index Entries: Xylanases; *Pyrodictium abyssi*; hyperthermophilic; archaea.

Introduction

Xylan, a 1,4-β-glycoside-linked polymer of D-xylose, is one of the most widespread carbohydrates in nature. The polymer can be catabolized by the synergistic action of several hydrolytic enzymes including endo-xylanases, β-xylosidases, and debranching enzymes. Examples of debranching enzymes are α-L-arabinofuranosidase, α-glucuronidase, and acetyl-xylan-esterase, which liberate the side chains α-L-arabinose, glucuronic acid, and acetate respectively, producing pentoses, which can be further metabolized.

The hyperthermophilic archaeon *Pyrodictium abyssi*, which was originally isolated from marine hot abyssal sites, grows optimally at 97°C *(1)* and is a prospective source of highly thermostable xylanases *(2,3)*. Its endoxylanase was shown to be highly thermostable (over 100 min at 105°C) and active even at 110°C.

Hyperthermophilic microorganisms (those that grow above 90°C and optimum temperature of at least 80°C), which are mostly archaea *(4,5)*, have been investigated for clues to evolutionary processes as well as to uncover biologic strategies underlying life at elevated temperatures. Most of the focus has been on characterization of new isolates and on the mechanisms responsible for thermostability. Although just as important, the physiology of this novel group has been less studied. This is not surprising because hyperthermophiles are generally difficult to culture, in addition to the fact that no generic systems are available for directed analysis of cellular phenomena *(6)*.

The development of cultivation protocols for hyperthermophilic microorganisms presents some interesting problems that are not encountered when working with more conventional organisms growing at mesophilic temperatures. Probably the most significant problem is the relatively little information generated to date on the growth and metabolism of hyperthermophiles. Biochemical and enzymologic research, however, is often limited by the low biomass yields that can be reached for many hyperthermophiles. Consequently, research on the sugar metabolism of hyperthermophiles has been carried out particularly with a few well-culturable species, such as *Sulfolobus* sp., *Thermotoga maritima*, and *Pyrococcus furiosus (7)*.

Along these lines, production of large amounts of biomass presents the unfavorable prospect of very poor volumetric efficiency of fermentors with the additional problem of dealing with the hazards and corrosivity associated with high levels of biologically generated hydrogen sulfide. Thus, difficulties with cultivation of hyperthermophiles represent the key technologic roadblock *(8)*.

The studies reported herein were undertaken to better characterize some parameters involved in the cultivation of the hyperthermophilic crenarcheote *P. abyssi*. The regulation of xylan assimilation of *P. abyssi* and xylanase production were also investigated.

Materials and Methods

Chemicals

All chemicals were of analytical grade and obtained from Merck (Darmstadt, Germany) unless otherwise stated. *p*-Nitrophenylglycosides, oat-spelt xylan, and arabinose were purchased from Fluka (Switzerland); birch wood xylan was from Carl Roth GmbH (Karlsruhe, Germany); beech wood xylan was from Lenzing (Lenzing, Austria); yeast extract was from Gibco (Eggenstein, Germany); Pefabloc and pyruvate were from Boehringer (Mannheim, Germany), and cationized ferritin from was Sigma (München, Germany).

Organism and Growth Conditions

The anaerobic extremely thermophilic archaeon *P. abyssi* (DSM 508) was obtained from the German Collection of Microorganisms and Cell Cultures (DSMZ, Braunschweig, Germany). The cells were cultivated anaerobically in modified SME medium *(9)* and the pH was adjusted to 5.5 with H_2SO_4. Cultures were grown in 100- and 1000-mL serum bottles at 97°C without shaking (gas phase: H_2/CO_2, 80:20; 0.1–0.2 MPa of over-pressure).

Induction Experiments

The induction experiments were carried out in 1000- and 2000-mL serum bottles containing 500 and 1000 mL of basal medium (initial pH 5.5) supplemented with 0.2% (w/v) mono- or polysaccharides. Initial cell concentration in each culture was about 1.0×10^6 cells/mL. Growth in the presence of different substrates was conducted using H_2/CO_2 atmosphere (80:20, 0.1–0.2 MPa of overpressure). The cultures were incubated at 97°C for 48 h.

To localize xylanase activity (cell-associated or cell-free xylanase), sonicated cell pellets and the corresponding concentrated culture supernatant were prepared and examined for relative levels of xylanolytic activity.

Batch Production of Endoxylanases

The basic synthetic medium SME was used and oat-spelt xylan (0.5% [w/v]) or arabinose (0.5% [w/v]) was used as the carbon source. The cultures were grown in 2000-mL serum bottles at 97°C for 48 h and used as inoculum for a 16-L bioreactor (Bioengeneering, Wald, Switzerland) with a working volume of 13 L. At different time intervals, 20-mL samples were taken and enzyme activity, amount of reducing sugars, and pH were monitored. The

initial pH of the medium was 5.5 and was not adjusted during the fermentation. Fermentation was carried out at 90°C, without stirring, with a gas
phase of H_2/CO_2 (80:20) and 0.1–0.12 MPa of overpressure.

Enzyme Assays

Xylanase activity was determined by the method of Somogyi *(10)* using
birch glucuronoxylan (Carl Roth GmbH) as substrate in universal buffer
($0.12 M$, pH 6.0). The reaction mixture was incubated at 95°C for 15–30 min,
the reaction was stopped on ice, and the liberated reducing sugars were
assayed. One unit of enzyme activity was defined as the amount of enzyme
that released 1 µmol of reducing sugar (xylose as standard)/min under
the assay conditions specified. Xylosidase and arabinofuranosidase were
assayed using *p*-nitrophenyl-β-D-xylopyranoside and *p*-nitrophenyl-α-L-
arabinofuranoside, respectively. One unit of enzyme activity was defined
as the amount of enzyme that released 1 µmol of *p*-nitrophenol/min under
the assay conditions described. The total amount of reducing sugars was
determined colorimetrically at 540 nm by the dinitrosalicylic acid method
(11), using xylose as standard. Protein concentration was determined using
the method of Bradford *(12)*, with bovine serum albumin as standard.

Scanning Electron Microscopy

To evaluate the presence of xylanosomes, *P. abyssi* cells grown on
glucose and xylan as the sole carbon source were examined. Cell samples
(10 mL) were centrifuged for 20 min at 12,000*g*, washed, recentrifuged, and
resuspended with 1-mL aliquots of $0.15 M$ saline solution. Cells were
labeled with cationized ferritin suspension (1 mg/mL), the specimens were
dried with a series of graded (20–100%) ethanol solutions, and then they
were critical point dried with liquid CO_2. Cells were coated with gold and
viewed with a Leitz electron microscope. Controls were also prepared as
just described, except that they were not incubated with cationized ferritin.

Prediction of Growth Model Parameters and Statistical Analysis

The kinetic parameters of the Monod, Contois, and Tessier models
(Eqs. 1–3) were estimated using a nonlinear method. The calculated
responses (cell mass and substrate concentration) were obtained from the
numerical solution of the equations set composed by each of the growth
models plus the relation between substrate uptake and growth rates (Eq. 4).

$$\frac{dX}{dt} = \frac{\mu_{max} \cdot X \cdot S}{K_s + S} \tag{1}$$

$$\frac{dX}{dt} = \frac{\mu_{max} \cdot X \cdot S}{B \cdot X + S} \tag{2}$$

$$\frac{dX}{dt} = \mu_{max} \cdot \left[1 - e^{(-S/K_S')} \right]$$ (3)

$$\frac{dS}{dt} = -\frac{1}{Y_{X/S}} \cdot \frac{dX}{dt}$$ (4)

For statistical analysis, response variables were assumed in a linearized context, in such a way that the confidence limits for correct responses are given by (Eq. 5):

$$y - t_{v,1-\alpha/2} \left\{ S_R^2 diag \left[J(J^tJ)^{-1}J^t \right] \right\}^{1/2} < \eta < y + t_{v,1-\alpha/2} \left\{ S_R^2 diag \left[J(J^tJ)^{-1}J^t \right] \right\}^{1/2}$$ (5)

The standard deviations of the parameters were the square roots of the variances drawn from the main diagonal of the covariance matrix of estimated parameters (Eq. 6):

$$\mathbf{Cov}(\widehat{\beta}) = \sigma_\varepsilon^2 \cdot (J^t \cdot J)^{-1}$$ (6)

in which the coherent and unbiased estimator used for the error variance (σ_ε^2) was

$$S_R^2 = \frac{\sum_i^n (y_i - \widehat{y}_i)^2}{n - p}$$ (7)

All these calculations were coded in MATLAB 4.2 c.1 (Mathworks).

Results and Discussion

Effect of S^0 on Cell Growth

P. abyssi grew well in the medium containing S^0 as a terminal electron acceptor with proteinaceous material (yeast extract) as the sole carbon source, as previously reported *(1)*. When the concentration of yeast extract was increased to 0.25%, the final cell density was at least one order of magnitude higher. As shown in Table 1, growth was observed also in the absence of S^0. When 0.5% xylan was added to the medium containing 0.05% yeast extract, cell concentration increased from 4.0×10^7 to 1.18×10^8. Growth on insoluble beech xylan was not affected by the absence of sulfur in the medium (Table 1).

P. abyssi showed the best growth rate, with a doubling time of 2.5 h, on medium containing 0.5% yeast extract in the absence of S^0. As shown in Table 2, xylanase activity was not present when growth was performed with a medium without xylan. The addition of xylan to the minimal medium (0.05% yeast extract) was accompanied by the formation of endoxylanase

Table 1
Effect of S^0 and Carbon Source on Growth of *P. abyssi*[a]

Medium[b]	Substrate concentration	Incubation time (h)	Cells/mL[c]
SME[d]	0.05% YE, 0.05% S^0	40	2.12×10^7
Minimal[e]	0.05% YE	40	4.00×10^7
YE	0.25% YE	24	3.20×10^8
Tryptone	0.25% Try	24	3.10×10^8
Beech xylan	0.05% YE, 0.5% Xyl	40	1.18×10^8

[a]YE, yeast extract; S^0, sulfur; Try, tryptone; Xyl, insoluble xylan (Lenzing A.G.).

[b]Cultures were grown in 100- and 500-mL serum bottles at 97°C without shaking (gas phase H_2/CO_2, 80:20; 0.1-MPa overpressure).

[c]Cell number was estimated in a Neubauer counting chamber; a starting inoculum of 1.5×10^6 was used.

[d]SME medium *(9)*.

[e]SME medium without sulfur.

Table 2
Effect of Carbon Source
on Xylanase Production, Growth Rate, and Doubling Time of *P. abyssi*

Fermentation[a]	Initial cell number (cell/mL)	Endoxylanase[b] (U/L)	Growth rate (h^{-1})	Doubling time[c] (h)
0.5% YE	1.12×10^7	0 (14 h)	0.41	2.5
0.05% YE + 0.2% beech xylan	1.53×10^7	0.27 (16 h)	0.28	3.5
0.05% YE + 0.5% xylose	1.37×10^7	0.29 (19 h)	0.21	4.7
0.05% YE + 0.5% arabinose	0.12×10^7	0.14 (93 h)	0.12	8.1

[a]YE, yeast extract. The 16-L fermentor was operated without stirring with a gas phase of H_2/CO_2 (80:20) and 0.1–0.12 MPa of overpressure at 93°C.

[b]Fermentation was stopped at the stationary phase and xylanase activity was measured in the cell extract and concentrated supernatant. Enzyme activity is expressed as described in Materials and Methods.

[c]Doubling time was calculated from the slopes of the growth curve (not shown).

(Table 2). Poor growth was observed when xylan was replaced with arabinose.

The obligate chemoorganotrophic *P. abyssi* grows organotrophically with complex substrates such as xylan, while many extremely thermophilic archaea grow with sulfur and H_2 as the sole energy substrates *(13)*, indicating that *P. abyssi* does not necessarily have to reduce sulfur. Fermentation of organic substrates in the absence of elemental sulfur has been reported for several extremely thermophilic sulfur reducers *(5)*. Usually, the final cell densities of these bacteria in batch cultures are low *(14)*, from 10^6 to

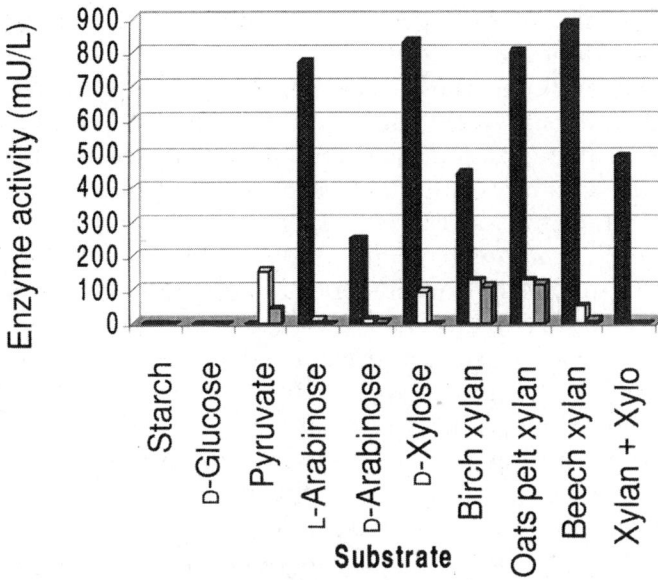

Fig. 1. Production of endoxylanase (■), arabinofuranosidase (□) and betaxylosidase (▨) by *P. abyssi*. Growth on the different substrates was determined in the presence of H_2/CO_2 atmosphere (80:20, 0.1–0.2 MPa of overpressure). The initial cell concentration was about 1.0×10^6 cells/mL and the cultures were incubated at 97°C for 48 h. Sonicated cell pellets and the corresponding clarified culture supernatant were prepared and examined for activity. Enzyme activity is expressed as described in Materials and Methods.

10^7 cells/mL, which was also observed for *P. abyssi*. Higher cell densities were observed for *P. abyssi*, when the organism was cultivated on high concentrations of yeast extract or starch (>10^8 cells/mL).

Influence of Various Carbon Sources on Xylanase Activity

Figure 1 shows total (extracellular and cell-bound) endoxylanase, arabinofuranosidase, and β-xylosidase production in the extreme thermophilic archeon *P. abyssi* after 48 h of growth with the various carbon sources tested. Enzyme production in *P. abyssi* seems to be inducible in the presence of various xylans and xylose. The xylanolytic enzymes endoxylanase, arabinofuranosidase, and β-xylosidase were not synthesized in the presence of starch or glucose (Fig. 1).

P. abyssi degrades xylan by the coordinate action of a complex of hydrolyzing enzymes, which act together in a concerted manner to effectively degrade the substrate. The findings in this study suggest that xylandegrading enzymes, including endoxylanase, arabinofuranosidase, and β-xylosidase, were predominantly associated with the cell during growth on xylan, except under conditions of xylan limitation in which endoxylanase became predominantly extracellular. Immunologic and electron microscopic studies with monoclonal antibodies raised against endoxylanase will be necessary to provide conclusive evidence for the cellular location of the xylanolytic enzymes.

Although starch was the best carbon source for growth of the organism, it was a poor substrate for xylanase production. The xylanolytic enzymes seem to be regulated by induction and repression. Similar results were observed for the mesophilic actinomycete *Streptomyces avermitilis (15)*, thermophilic actinomycete *Thermonospora curvata (16)*, and thermophilic fungus *Thermomyces lanuginosus (17)*. Synthesis of the xylanolytic enzymes in *P. abyssi* is regulated by the presence of substrates containing pentose. The levels of xylanolytic enzymes in crude culture supernatants varied greatly in response to the carbon source used for growth. Since the secreted enzyme was highly stable, showing activity even after 100 min of incubation at 105°C, enzyme synthesis but not enzyme turnover is likely to play a major role in mediating the level of extracellular activity.

The regulatory systems for biosynthesis of xylanase studied so far are mainly from mesophilic organisms, including the filamentous fungi of the genera *Aspergillus* and *Trichoderma*. Several thermophilic organisms have been reported to produce thermostable xylanases, but little information is available concerning the biosynthetic regulation of xylanases in thermophilic organisms. The regulatory mechanisms involved in xylan degradation by thermophilic anaerobic bacteria are still not well understood. In previous studies on the regulation of xylanase synthesis, two mechanisms have been proposed. First, in many xylan-degrading microorganisms, the xylanolytic enzymes appear to be inducible *(18,19)*. The synthesis of xylanolytic enzymes can be inhibited by the presence of glucose or some other sugars in the growth medium. The synthesis is then regulated by an induction-repression mechanism. Because polysaccharides cannot enter the cells, a soluble oligosaccharide such as xylobiose or xylotriose is considered to act as a direct inducer of xylanase synthesis. The oligosaccharides are formed by the hydrolysis of xylan in the medium by low amounts of enzyme that are produced constitutively. Induction is also possible with some synthetic alkyl-, aryl-, and methyl-β-D-xylosides, as well as positional isomers of xylobiose such as 1,2-β-xylobiose *(18,19)*. Second, in other xylan-degrading bacteria, the xylanase synthesis is considered to be constitutive but repressed by conditions favoring optimal growth. The xylanase synthesis is subject to control by growth rate–dependent repression by readily metabolized carbon sources *(18)*.

For the anaerobic bacterium *Bacteroides xylanolyticus*, pyruvate appeared to be a stronger inducer *(18)*. Extracellular xylanase production in the anaerobic thermophilic *Dictyoglomus* sp. B1 was achieved at high levels using insoluble beech wood xylan *(20)*. On the other hand, the β-xylosidase from *Trichoderma reesei* was induced by xylose and beech wood xylan, while L-arabinose induced enzyme production at a very low level *(19)*. By contrast, the thermophilic *T. lanuginosus* has shown low constitutive levels of enzymes, endoxylanase, and β-xylosidase, using a variety of substrates, including birch xylan, maltose, xylose, and bagasse *(17)*. L-Sorbose induced xylanase activity as well as cellulase in *T. reesei*. It induced a higher level of xylanase activity than sophorose and xylose did *(21)*. The arabinofuranosidase from the mesophilic anaerobic bacterium *B. xylanolyticus* X5-1 *(18)* was induced by L-arabinose and xylose but was poorly induced by oat-spelt xylan.

In the case of *P. abyssi*, the highest levels of endoxylanase were obtained with beech wood and oat-spelt xylan. The highest levels of β-xylosidase production were observed with the naturally occurring xylan, oat spelt, and birchwood xylan. The highest production of arabino-furanosidase, however, was detected in the cell extracts after growth on xylose and pyruvate, indicating that the intermediate of the tricarboxylic acid cycle acted as a nonrepressing carbon source for the production of this enzyme.

Although other bacteria are able to utilize L-arabinose anaerobically *(18)* or even aerobically *(22)*, bacteria do not commonly metabolize D-ara-binose. Interestingly, the hyperthermophilic archaeon *P. abyssi* is able to induce xylanase production with both enantiomeric forms. Although xylan is the dominant pentosan and glucomannan the dominant hexosan, the level of arabinan is significant in some biomass materials *(23)*. The ability of microorganisms to ferment L-arabinose in the context of conversion of hemicellulosic sugars is extremely important. In the fermentation of pentoses two pathways are possible: direct breakdown of xylulose 5-phosphate by the enzyme phosphoketolase or the conversion of xylulose 5-phosphate to hexose phosphate prior to metabolism in the pentose phosphate pathway. The pentose phosphate pathway is the most common mechanism of pentose catabolism in anaerobic bacteria *(24,25)*.

The availability of xylan and other plant-derived polysaccharides in the marine hot abyssal environments is likely and may transiently provide resident cells with a growth advantage. Regulation of xylanase syntheses in the hyperthermophile *P. abyssi* provides an energy-efficient means for the utilization of such polysaccharides.

The specific activity of xylanases produced by *P. abyssi* in batch culture was lower than 0.1 mU/L for β-xylosidase and arabinofuranosidase. The specific activity of endoxylanase on xylan was 0.24–0.41 mU/mg and on arabinose 0.64 mU/mg (Table 3). Usually, the specific activities determined for crude xylanase preparations from different xylan-degrading organisms grown in batch cultures are relatively low, usually ranging from 0.06 to 50 U/mg *(26,27)*. However, higher specific activities up to 200 U/mg were already described for a thermostable xylanase from *Clostridium stercorarium (28)* and for the xylanase from *Dictyoglomus* sp. *(20)*. The specific activity measured for the xylanase of *P. abyssi* was very low, but comparable with those obtained by Canganella et al. *(13)* for amylases and pullulanases from archaea. The low specific activities could be owing to the yeast extract concentration in the medium that was required to obtain acceptable growth yield in the absence of elemental sulfur.

Table 3
Effect of Growth Substrates on Specific Activities of Xylanolytic Enzymes of *P. abyssi*[a]

Substrate (0.2%)	Protein (mg/L)[b]	Endoxylanase[c]		Arabinofuranosidase[c]		β-Xylosidase[c]	
		(mU/L)	Specific activity (mU/mg)	(mU/L)	Specific activity (mU/mg)	(mU/L)	Specific activity (mU/mg)
L-Arabinose	1254	810	0.64	10	0.008	0	0
D-Arabinose	1068	251	0.23	18	0.016	>10	0
D-Xylose	1710	308	0.18	177	0.10	0	0
Birch xylan	1822	442	0.24	46	0.025	92	0.050
Beech xylan	2364	984	0.41	67	0.028	>10	0
Oat-spelt xylan	2730	886	0.32	68	0.025	12	0.004

[a]Growth on the different substrates was determined in the presence of H_2/CO_2 atmosphere (80:20, 0.1–0.2 MPa of overpressure). The cultures were incubated at 97°C for 48 h.
[b]Protein was measured by the method of Bradford (12) and represents the total protein (supernatant and cell extract).
[c]Enzyme activity is expressed as described in Materials and Methods.

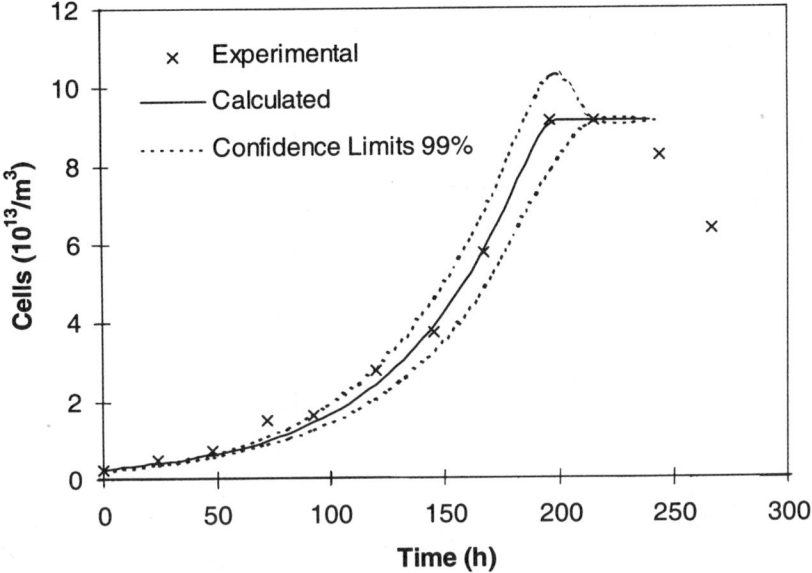

Fig. 2. Growth model parameter prediction of *P. abyssi* during growth on 0.5% xylan. The medium contained 0.5% (w/v) oat-spelt xylan. Fermentation was carried out in the 16 L fermentor at 90°C, without stirring, with a gas phase of H_2/CO_2 (80:20) and 0.1–0.12 MPa of overpressure. Total extracts were evaluated for xylanase activity and reducing sugars as described in Materials and Methods.

Cultivation of P. abyssi *in Batch Culture*

To investigate xylanase production in batch experiments, L-arabinose and oat-spelt xylan were used as substrates. During growth on oat-spelt xylan, the maximum xylanase activity reached 390 mU/L after 120 h. After this point the endoxylanase activity decreased and was 120 mU/L after 264 h. From the experimental data (Fig. 2), the kinetic parameters of Monod (Eq. 5), Contois (Eq. 6), and Tessier (Eq. 7) models were estimated (Table 4). In Fig. 2 the solid lines show estimated responses for the Contois model and the dotted lines delimit the confidence region for the correct responses (Eq. 14) at the 99% level. The calculated responses of these models with their estimated parameters were only slightly different and were represented by the same determination coefficient (R^2) values (Table 4).

During cultivation on arabinose, endoxylanase activity was lower, reaching 142 mU/L after 120 h and decreasing gradually to 23 mU/L after an additional 100 h. The total reducing sugar concentration at the end of the fermentation was 22 mM, indicating that sugars were consumed very slowly (data not shown). The turbidity caused by the turbid xylan-containing medium prevented the measurement of cell growth by absorbance. In experiments with arabinose, the formation of Maillard products did not allow the determination of growth using turbidimetric methods.

The adequacy of the Monod equation form (Eq. 5) to describe substrate-limited growth is normally related to low cell population (29). On the

Table 4
Growth Parameter Prediction by Different Models

Parameter	Monod			Contois		Tessier		
	μ_{max} (h^{-1})	K_s (kg/m^3)	μ_{max} (h^{-1})	B (kg S/10^{13} cells)	μ_{max} (h^{-1})	K'_s (kg/m^3)	μ_x (h^{-1})
Estimate	0.0195	0.1532	0.0189	0.0163	0.0188	0.0803	0.0174
Standard deviation	0.012	0.29	9.3×10^{-3}	0.029	7.7×10^{-3}	23	—
R^2	0.9898			0.9898		0.9898		

[a]K_s, Monod saturation constant; K'_s, Tessier apparent saturation constant; μ_{max}, maximum specific uptake rate.

other hand, the Tessier form (Eq. 7) seems to present similar behavior since its apparent Michaelis constant (K_s') can be directly related to the true one by means of $K_s = \ln(2) \cdot K_s'$. According to this model, it can be shown that the specific growth rate is half of the maximum when $S = \ln(2) \cdot K_s'$.

The Contois model (Eq. 6) usually appears more appropriate for high-density cell cultures *(29)* since it has a pseudo–Michaelis constant proportional to biomass concentration. The standard deviations of estimated parameters (Table 4), obtained from square roots of the main diagonal of Eq. 13, showed smaller values for parameters of the Contois model than the other ones. This implies smaller variability for Contois estimates than for the Monod and Tessier models. The physical significance of its parameters and its acceptability for high cell density cultures pointed to the choice of the Contois model as the most indicated to predict the growth of *P. abyssi* on xylan.

The cultivation of *P. abyssi* in a fermentor without stirring was successful, although agitation is usually necessary to maintain the medium homogeneity in order to avoid the formation of large metabolic inactive pellets. Most hyperthermophiles investigated so far grow optimally at an agitation speed of 2000 rpm *(8)*. However, shearing forces may also disrupt fragile microbial cell networks, such as the network of *P. abyssi*, and can have a marked influence on xylanase production. *P. abyssi* cells are reminiscent of fungal mycelia *(1)*, and disruption of the network causes cessation of growth with consequent hindrance of xylanase production. This is the first report on fermentation experiments to improve production and secretion of xylanases in archaea.

Presence of Xylanosomes

The presence of xylanosomes on the cell surface of *P. abyssi* was investigated after growth on xylan and glucose. Cationized ferritin-labeled cells grown on glucose were compared with those grown on xylan by scanning electron microscopy. No xylanosomes were observed under the conditions tested (data not shown).

Many anaerobic cellulolytic bacteria posses high molecular weight, multisubunit cellulases that are often associated with the cell surface or sedimentable membranous fragments *(30,31)*. In *Thermoanaerobacter* B6A, the finding of cell-bound endoxylanase activity, a strong affinity of the cells for the substrate, and the presence of cell-surface protuberances suggests the presence of xylanosomes, a structure analogous to the cellulosome *(32)*. However, such structures were not detected in *P. abyssi*. The absence of xylanosomes in this archeon suggests that different strategies could be employed to bind the enzyme to the cell surface, and it is possible that the enzymes from *P. abyssi* are associated with the network of tubules.

Although there are many reports focusing on the microbial degradation of xylan by xylanolytic enzymes, xylanase activity recently has been found also in another archaeon, *Thermococcus zilligii*, which grows optimally at 75°C *(33)*. We were able to show that the hyperthermophilic

P. abyssi can produce xylanolytic enzymes consisting of endoxylanase, arabinofuranosidase, and β-xylosidase, and we have demonstrated that this system is inducible by different carbon sources. The regulation of endoxylanase production appears to be independent of cell growth rate. The synthetic production medium selected for this study with oat-spelt xylan as the carbon source seems to be a good medium for xylanase production. The optimization of the medium composition for a desired enzyme activity profile may provide a way for customized enzyme cocktail production for special purposes such as that required in pulp and paper applications.

Nomenclature

B = Contois apparent saturation constant

$diag(\mathbf{M})$ = vector extracted from main diagonal of generic matrix $\mathbf{M}$

$\mathbf{J}$ = Jacobian matrix of estimated parameters

K_s = Monod saturation constant

K_s' = tessier apparent saturation constant

n = number of responses

p = number of parameters

S = substrate concentration

S_R^2 = estimate of error variance

t = time

$t_{\nu,1-\alpha/2}$ = t-Student abscissa for ν degrees of freedom at the significance level $\alpha/2$.

X = cell mass concentration

y = experimental response variable (cell mass or substrate concentration)

$\hat{y}$ = estimated response variable

$\mathbf{y}$ = estimated responses vector

$Y_{X/S}$ = cell mass yield factor

$\hat{\beta}$ = estimated parameters vector

η = correct responses vector

μ_{max} = maximum specific uptake rate

σ_ε^2 = error variance

References

1. Pley, U., Schipka, J., Gambacorta, A., Jannasch, H., Fricke, H., Rachel, R., and Stetter, K. O. (1991), *System. Appl. Microbiol.* **14,** 245–253.
2. Antranikian, G. and Sjøholm, C. (1997), US Patent no. 05688668.
3. Antranikian, G. and Sjøholm, C. (1999), US Patent no. 05912150.
4. Andrade, C. M. M. C., Pereira, N. Jr., and Antranikian, G. (1999), *Rev. Microbiol.* **30,** 325–336.
5. Stetter, K. O. (1996), *FEMS Microbiol. Rev.* **18,** 149–158.
6. Halio, S. B., Blumentals, I. I., Short, S. A., Merrill, B. M., and Kelly, R. M. (1996), *J. Bacteriol.* **178(9),** 2605–2612.

7. Kengen, S. W. M., Stams, A. J. M., and deVos, W. M. (1996), *FEMS Microbiol. Rev.* **18,** 119–137.

8. Krahe, M., Antranikian, G., and Märkl, H. (1996), *FEMS Microbiol. Rev.* **18,** 271–285.

9. Stetter, K. O., König, H., and Stackebrandt, E. (1983), *System. Appl. Microbiol.* **4,** 535–551.

10. Somogyi, M. (1952), *J. Biol. Chem.* **195,** 19–23.

11. Miller, G. L. (1969), *Anal. Chem.* **31,** 426–428.

12. Bradford, M. (1976), *Anal. Biochem.* **72,** 248–254.

13. Canganella, F., Andrade, C. M., and Antranikian, G. (1994), *Appl. Microbiol. Biotechnol.* **42,** 239–245.

14. Schauder, R. and Kröger, A. (1993), *Arch. Microbiol.* **159,** 491–497.

15. Garcia, B. L., Ball, A. S., Rodriguez, J., Perez-Leblic, M. I., Arias, M. E., and Copa-Patino, J. L. (1998), *FEMS Microbiol. Lett.* **158(1),** 95–99.

16. Hostalka, F., Moultrie, A., and Stutzenberger, F. (1992), *J. Bacteriol.* **174(21),** 7048–7052.

17. Damaso, M. C. T., Andrade, C. M., and Pereira, N. Jr. (2000), *Appl. Biochem. Biotechnol.* **84–86,** 1–14.

18. Schyns, P. J. Y. M. J. and Stams, A. J. M. (1992), in *Xylan and Xylanases*, Visser, J., Beldman, G., Kusters-van Someren, M. A., and Voragen, A. G. J. eds., Elsevier Science, Amsterdam, pp. 295–300.

19. Kristufek, D., Zeilinger, S., and Kubicek, C. P. (1995), *Appl. Microbiol. Biotechnol.* **42,** 713–717.

20. Adamsen, A. K., Lindhagen, J., and Ahring, B. K. (1995), *Appl. Microbiol. Biotechnol.* **44,** 327–332.

21. Xu, J. P., Nogawa, M., Okada, H., and Morikawa, Y. (1998), *Biosci. Biotechnol. Biochem.* **62(8),** 1555–1559.

22. Gobbetti, M., Lavermicocca, P., Minervini, F., De Angelis, M., and Corsetti, A. (2000), *J. Appl. Microbiol.* **88(2),** 317–324.

23. McMillan, J. D. and Boynton, B. L. (1994), *Appl. Biochem. Biotechnol.* **45/46,** 569–582.

24. Cook, G. M., Janssen, P. H., Russel, J. B., and Morgan, H. W. (1994), *FEMS Microbiol. Lett.* **116,** 257–262.

25. Biesterveld, S., Kok, M. D., Dijkema, C., Zehnder, J. B., and Stams, A. J. M. (1994), *Arch. Microbiol.* **161,** 521–527.

26. Nakanishi, K., Marui, M., and Yasui, T. (1992), *J. Ferment. Bioeng.* **74,** 392–394.

27. Khasin, A., Alchanati, I., and Shoham, Y. (1993), *Appl. Environ. Microbiol.* **59(6),** 1725–1730.

28. Bérenger, J. F., Frixon, C., Bigliardi, J., and Creuzet, N. (1985), *Can. J. Microbiol.* **31,** 635–643.

29. Asenjo, J. Á. (1995), *Bioreactor System Design*, Marcel Dekker, New York.

30. Ponpium, P., Ratanakhanokchai, K., and Kyu, K. L. (2000), *Enzyme Microbiol. Technol.* **26(5–6),** 459–465.

31. Ohara, H., Karita, S., Kimura, T., Sakka, K., and Ohmiya, K. (2000), *Biosci. Biotechnol. Biochem.* **64(2),** 254–260.

32. Lee, Y., Lowe, S. E., and Zeikus, J. G. (1993), *Appl. Environ. Microbiol.* **59,** 3134–3137.

33. Uhl, A. M. and Daniel, R. M. (1999), *Extremophiles* **3(4),** 263–267.

Preliminary Kinetic Characterization of Xylose Reductase and Xylitol Dehydrogenase Extracted from *Candida guilliermondii* FTI 20037 Cultivated in Sugarcane Bagasse Hydrolysate for Xylitol Production

LUCIANE SENE,[1] MARIA G. A. FELIPE,[2]
SILVIO S. SILVA,[2] AND MICHELE VITOLO*,[1]

[1]*Department of Biochemical and Pharmaceutical Technology,
Faculty of Pharmacy, University of São Paulo, Av. Prof. Lineu Prestes,
580, B-16, 05508-900, São Paulo, SP, Brazil, E-mail: michenzi@usp.br;
and [2]FAENQUIL—DEBIQ-125600-000, PO Box 116, Lorena, SP, Brazil*

Abstract

Candida guilliermondii FTI 20037 was cultured in sugarcane bagasse hydrolysate supplemented with 2.0 g/L of $(NH_4)_2SO_4$, 0.1 g/L of $CaCl_2 \cdot 2H_2O$, and 20.0 g/L of rice bran at 35°C; pH 4.0; agitation of 300 rpm; and aeration of 0.4, 0.6, or 0.8 vvm. The high xylitol production (20.0 g/L) and xylose reductase (XR) activity (658.8 U/mg of protein) occurred at an aeration of 0.4 vvm. Under this condition, the xylitol dehydrogenase (XD) activity was low. The apparent K_M for XR and XD against substrates and cofactors were as follows: for XR, $6.4 \times 10^{-2}M$ (xylose) and 9.5×10^{-3} mM (NADPH); for XD, $1.6 \times 10^{-1}M$ (xylitol) and 9.9×10^{-2} mM (NAD+). Because XR requires about 10-fold less xylose and cofactor than XD for the condition in which the reaction rate is half of the V_{max}, some interference on the overall xylitol production by the yeast could be expected.

Index Entries: Xylitol; xylose reductase; xylitol dehydrogenase; *Candida guilliermondii*.

Introduction

Xylose reductase (XR) (EC 1.1.1.21) and xylitol dehydrogenase (XD) (EC 1.1.1.9) catalyze the first steps of the pentose phosphate pathway in

*Author to whom all correspondence and reprint requests should be addressed.

eukaryotes *(1)*. Microorganisms having an active pentose phosphate pathway are capable of converting D-xylose-containing materials (e.g., lignocellulosic hydrolysates) to xylitol, a polyalcohol with commercial value owing to its sweetening and anticarious properties *(2)*.

Currently, xylitol is produced by the catalytic hydrogenation of pure xylose extracted from wood hydrolysates *(3)*. However, there is growing interest in producing xylitol through a microbial process, because it can be carried out directly in the lignocellulosic hydrolysate, without prior purification of xylose as required in the chemical process *(2)*. However, the fermentative process also presents handicaps. Low xylitol yield is caused by factors such as O_2 limitation, pH, temperature, and inhibitors (e.g., furfural, acetic acid), which perturb cell metabolism *(2)*. Alterations in metabolism, in the case of xylitol production, reflect the biosynthesis and catalytic performances of XR and XD. When aiming for a high yield in xylitol, the culture conditions (pH, temperature, dissolved oxygen [DO], agitation) should be managed so that the XR activity is enhanced over XD. Another possibility would be the direct conversion of xylose in xylitol by using isolated XR, provided that the enzyme was readily available and inexpensive.

Although the enzymes XR and XD are well characterized for several species of yeasts *(4–6)*, little information specifically related to *Candida guilliermondii* FTI 20037 grown in sugarcane bagasse hydrolysate is available for either enzyme. The present study was undertaken to verify the kinetic behavior of XR and XD present in crude extract from *C. guilliermondii* against pH, temperature, and substrate concentrations. We also evaluated the effect of aeration on xylitol production and XR and XD activities of *C. guilliermondii*.

Materials and Methods

Microorganism

The experiments were carried out with *C. guilliermondii* FTI 20037 described by Barbosa et al. *(7)*. The yeast was maintained on malt-extract agar slants at 4°C.

Preparation of Sugarcane Bagasse Hydrolysate

Sugarcane bagasse was hydrolyzed in a 250-L reactor at 121°C for 10 min with H_2SO_4 (solid:liquid ratio of 1:10). A portion of the hydrolysate was further concentrated under vacuum at 70°C to increase xylose concentration threefold. The hydrolysates were then treated in order to reduce the concentrations of toxic substances. The initial pH was raised to 7.0 with CaO (commercial powder) and acidified to pH 5.5 with H_3PO_4. Subsequently, 2.4% (w/v) activated charcoal (refined powder) was added to the hydrolysates, which were then left under agitation (200 rpm) at 30°C for 1 h *(2)*. The precipitates resulting from all the stages of the treatment were removed by vacuum filtration.

Preparation of Inoculum and Medium and Fermentation Conditions

Inoculum was grown in nonconcentrated bagasse hydrolysate containing 18.8 g/L of xylose, 1.7 g/L of glucose, and 3.2 g/L of acetic acid, at an initial pH of 5.5. The tests were conducted with concentrated bagasse hydrolysate containing 51.7 g/L of xylose, 4.1 g/L of glucose, and 5.0 g/L of acetic acid. Both hydrolysates were supplemented with the following nutrients: 2.0 g/L of $(NH_4)_2SO_4$, 0.1 g/L of $CaCl_2 \cdot 2H_2O$, and 20.0 g/L of rice bran extract.

For the preparation of inoculum, cells were grown in 125-mL Erlenmeyer flasks containing 50 mL of medium and incubated in a rotary shaker (200 rpm) at 30°C for 24 h. The initial cell concentration in all the experiments was 1 g/L. Fermentations were carried out for 48 h in a 1.5-L fermentor (NBS-BIOFLO III; equipped with electrodes for monitoring pH and dissolved oxygen) containing 1.25 L of supplemented concentrated bagasse hydrolysate at 35°C; pH 4.0; agitation of 300 rpm; and aeration of 0.4, 0.6, or 0.8 vvm. At the end of fermentation, the cells were separated by centrifugation (5000g for 10 min), rinsed twice with distilled water, and resuspended in 0.1 M potassium-phosphate buffer (pH 7.2). The final suspension (15.0 g of cells/L) was stored in a freezer.

Preparation of Cell-Free Extracts

The cell suspension was thawed and submitted to sonication (equipped with a 20-kHz probe, Vibra Cell 100W; Sonics & Materials) at pulsing/resting cycles of 1 s for 35 min. During all disruption procedures, the suspension was left in a bath at –10°C. Cell homogenates were then centrifuged at 6700g for 10 min at 4°C, and the supernatant solution was used for enzymatic assays. The cell disruption was considered complete, since intact cells were not detected through microscopic observation in a Neubauer chamber (area = 1/400 mm²; height = 0.100 mm). The samples were taken from three disruption procedures conducted independently under the same conditions.

Analytical Methods

Glucose, xylose, arabinose, xylitol, and acetic acid concentrations were determined by liquid chromatography, and the cell concentration was measured by turbidimetry as described elsewhere (*2*).

XR and XD activities were determined (always in duplicate) using a Beckman DU-640 spectrophotometer at 340 nm at room temperature (about 25°C) as described by Alexander (*8*). The reaction media used were as follows: for XR, 570 µL of water, 200 µL of 0.5 M xylose, 80 µL of 1 M potassium-phosphate buffer (pH 7.2), 100 µL of 1.2 mM NADPH, and 150 µL of 0.1 M β-mercaptoethanol; for XD, 450 µL of water, 150 µL of 0.5 M xylitol, 200 µL of 0.5 M Tris buffer (pH 8.6), 150 µL of 1.26 mM NAD, and 150 µL of 0.1 M β-mercaptoethanol. The reactions were initiated by the addition of 200 µL of crude extract into the cuvet (final volume of 1.3 mL). One XR or

XD unit was defined as the amount of enzyme catalyzing, respectively, the formation of 1 μmol of NADP/min or 1 μmol of NADPH/min. Specific activities were expressed as units/milligram of protein, based on protein determinations (always in triplicate) according to the method of Bradford *(9)* using bovine serum albumin as the standard. The mean protein concentration in the crude extracts was 0.400 ± 0.02 mg/mL. Because the enzyme activities were measured in crude extracts, controls, in which no substrate was added (xylose or xylitol), were performed to ensure that the cofactor was not significantly consumed during 1 min of enzymatic reaction by other competing oxi-red reactants eventually present in the extract.

The apparent kinetic constants for XR and XD were determined through the Lineweaver-Burke method by changing the substrate and cofactor concentrations at intervals of 0.01–0.25 M (xylose), 0.025–0.25 M (xylitol), 0.03–0.18 mM (NADPH), and 0.05–0.18 mM (NAD+). The effect of pH and temperature on the enzyme activities and stabilities were evaluated by changing individually these parameters as follows: for XR, pH 5.5–8.0 (0.5 M KH$_2$PO$_4$-K$_2$HPO$_4$ buffer) and 30–70°C; for XD, pH 7.0–9.0 (0.5 M Tris buffer) and 30–60°C. While investigating the stability, the cell-free crude extract containing XR and XD was incubated at 30°C for 45 min at the desired pH, or for 10 min at the desired temperature.

Results and Discussion

Figure 1 shows that glucose was consumed up to 12 h in all fermentations, whereas xylose consumption and xylitol production depended on the aeration of the culture. After 36 h at 0.4 vvm, nearly 90% of initial xylose was consumed and the xylitol concentration in the medium was 20 g/L. However, Felipe et al. *(10)* observed that the high xylose consumption occurred at 0.8 vvm for the same strain grown in sugarcane bagasse hydrolysate but at pH 5.3. These results indicate some correlation between aeration and pH of the culture medium, as already described by Silva et al. *(11)*. From fermentations carried out without pH control, we observed that the mash became more acidic at 0.4 than at 0.6 or 0.8 vvm (Fig. 2). Moreover, the decrease in pH at 0.4 vvm resulted from the accumulation of acetic acid in the mash *(12)*, which under high aeration (e.g., 0.6 or 0.8 vvm) would be metabolized by the yeast, leading to an increase in pH. An analogous behavior was observed for *Pichia stipitis (13)*, *Candida blankii*, *Pachisolen utilis (14)*, and *Peacicomyces variotti (15)*.

The uptake of arabinose by the yeast, which started after 36 h of culture when xylose in the mash was low, was apparently more favored at 0.4 than at 0.6 or 0.8 vvm (Fig. 1). This is in accordance with Roberto et al. *(16)*, who verified a significant arabinose consumption by the same strain grown in rice bran, whose arabinose/xylose ratio is higher than that of bagasse hydrolysate. This result is clearly related to the preferential uptake of sugars by microorganisms, a well-known concept.

The high xylitol productivity (0.55 g/[L·h]) occurred at 0.4 vvm after 36 h of fermentation (Table 1). The increase in xylitol production under

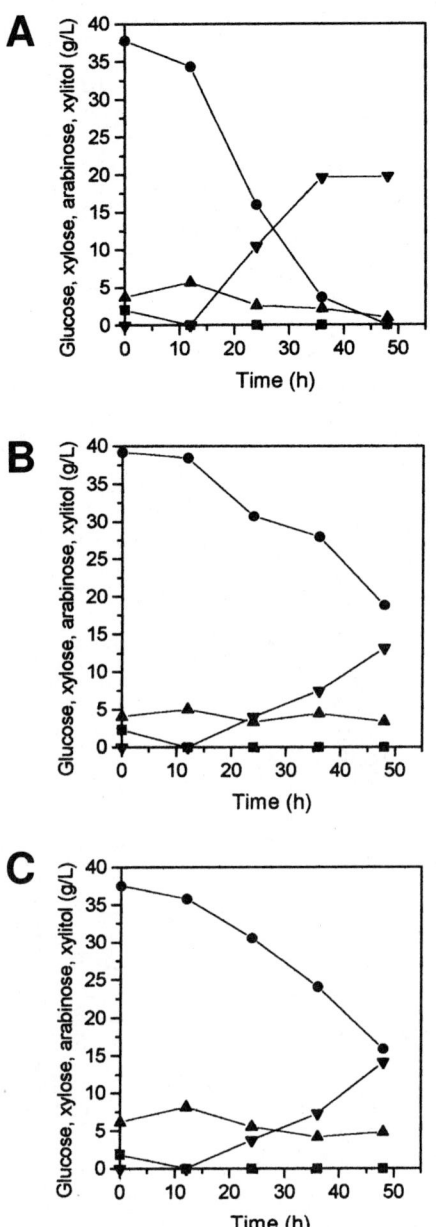

Fig. 1. Variation in glucose (■), xylose (●), arabinose (▲), and xylitol (▼) concentration during the culture of *C. guilliermondii* FTI 20037 in sugarcane bagasse hydrolysate at aeration of **(A)** 0.4, **(B)** 0.6, and **(C)** 0.8 vvm.

limited DO in the mash has also been verified by several researchers *(11,17–19)*, which suggests that under low DO the ratio NADPH:NAD+ is increased, so that the NADPH-dependent XR activity is favored against NAD+-dependent XD. Thus, the cofactor balance pending to XR would lead to the overproduction and excretion of xylitol by the yeast.

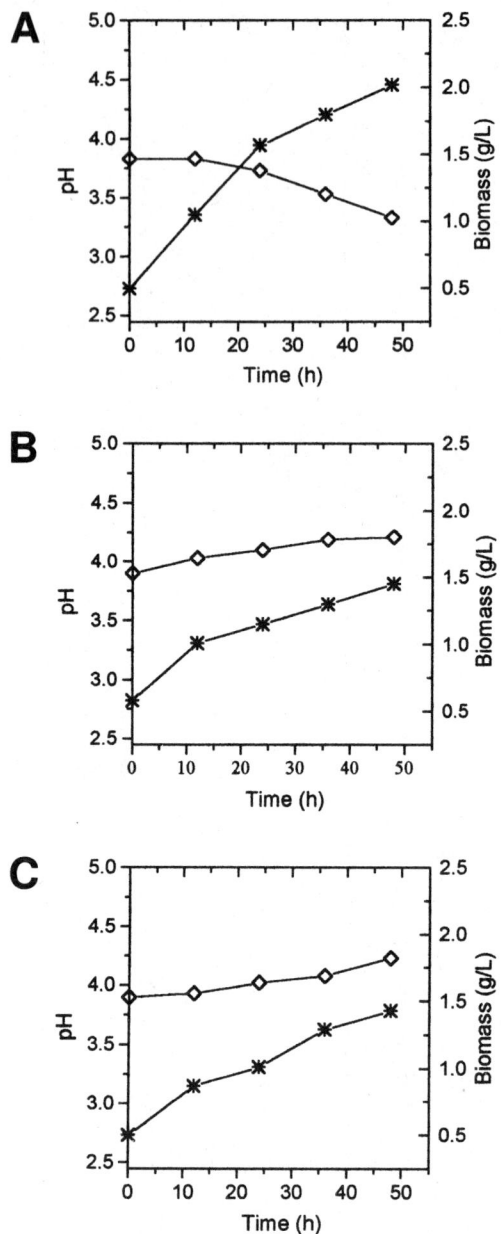

Fig. 2. Variation in pH (◇) and cell concentration (✳) during the culture of *C. guilliermondii* FTI 20037 in sugarcane bagasse hydrolysate at aeration of **(A)** 0.4, **(B)** 0.6, and **(C)** 0.8 vvm.

The final cell concentration was favored under aeration of 0.4 vvm (Fig. 2). This result does not match with those described by Vandeska et al. *(17)* and Girio et al. *(20)*, who found that high DO led to an increase in biomass. The disagreement would probably arise from the different culture media employed. As in the present study, sugarcane bagasse hydrolysate

Table 1

Effect of Aeration on Productivity (Q_p) of Xylose-Xylitol Conversion
by *C. guilliermondii* FTI 20037 Cultivated in Sugarcane Bagasse Hydrolysate

Aeration (vvm)	Q_p (g/[L·h]), time (h)				
	t_0	t_{12}	t_{24}	t_{36}	t_{48}
0.4	0	0	0.44	0.55	0.41
0.6	0	0	0.17	0.21	0.28
0.8	0	0	0.16	0.21	0.27

Table 2

Effect of Aeration on XR and XD Specific Activities (U/mg of protein)
from *C. guilliermondii* FTI 20037 Cultivated in Sugarcane Bagasse Hydrolysate

Time (h)	Aeration (vvm)								
	0.4			0.6			0.8		
	XR	XD	XR/XD	XR	XD	XR/XD	XR	XD	XR/XD
0	518	350	1.48	495	118	4.19	369	104	3.55
12	252	118	2.14	183	78.1	2.34	111	19.0	5.84
24	583	239	2.44	203	41.5	4.88	145	74.6	1.94
36	659	336	1.96	143	76.4	1.87	210	160	1.31
48	459	321	1.43	204	161	1.27	302	198	1.53

was used, and, therefore, the presence of an inhibitor or the insufficient amount of some ingredient would lead to a limitation on the biomass formation.

The specific activities of XR and XD present in the crude cell-free extract were affected by the aeration of the culture medium under which the yeast was grown (Table 2). The highest XR (659 U/mg of protein) and XD (336 U/mg of protein) occurred after 36 h of fermentation at 0.4 vvm. However, taking into account the XR:XD ratio, we observe that XR was about six times higher than XD after 12 h at 0.8 vvm, a condition that should lead to xylitol production more significant than at 0.4 vvm (XR:XD = 2.44), at least in thesis. Yet, according to the data attained, the best condition for xylitol production was 0.4 vvm (Fig. 1); hence, some other factor is influencing the overall xylose/xylitol conversion such as substrate limitation or the presence of an inhibitor (a real possibility when the hydrolysate of a residue is the component of the culture medium). A clue on this aspect can be found if we consider that the cell growth, which is expected to increase as aeration rate increases, was higher at 0.4 than at 0.6 or 0.8 vvm (Fig. 2).

After 12 h of fermentation, the activities of both enzymes decreased sharply (at least 25%) under all aeration conditions. Because glucose was present in the original mash at a concentration of 4.2 g/L and it is the sugar preferentially taken up by the yeast, a catabolic repression of both enzymes

Table 3

Effect of pH on Activity and Stability
of XR and XD Present in Crude Cell-Free Extract Obtained
from Disrupted Cells of *C. guilliermondii* FTI 20037

pH	Activity[a]		Stability[b]	
	XR	XD	XR	XD
4.0	85	—	—	—
5.0	329	—	276	—
5.5	338	—	277	—
6.0	330	17	98	—
6.5	307	26	87	50
7.0	230	33	86	57
7.5	197	70	81	74
8.0	144	78	54	77
8.5	—	85	—	66
9.0	—	56	—	53

[a]Values represent the activities of XR and XD (expressed as
U/mg of protein) at indicated pH of the reaction medium.
[b]Values represent the residual activities of XR and XD (expressed
as U/mg of protein) after an aliquot of the crude extract was left for
45 min in a buffer solution at the desired pH. For the stability
evaluation, the residual activities of XR and XD were measured
at pH 7.2 and 8.6, respectively.

could occur. This point is borne out by the fact that all initial glucose was
consumed up to 12 h, followed by intensification of xylose consumption
(Fig. 1).

From Table 3 we can see that the pH intervals at which XR and XD had
high specific activities were 5.0–6.0 and 8.0–8.5, respectively. This behavior
observed for both enzymes is common among other yeast species *(21,22)*.
Regarding stability, Table 3 also shows that the residual specific activity of
XR decayed sharply (about 70%) at a pH higher than 5.5 and remained
constant up to pH 7.5, whereas the residual specific activity of XD presented a clear maximum at pH 8.0. Because we employed a crude extract,
a full comparison of the data described in the literature is not possible for
two main reasons. First, there is little information on the kinetic performance of XR and XD related to the strain *C. guilliermondii* FTI 20037,
although there are many data regarding other yeast species and strains.
Second, it is possible that by changing the pH of the buffer solution, other
factors, such as the activation of proteases present in the crude extract,
could cause structural damage to the molecular structures of both enzymes.
Thus, the real effect of pH on the stability of both enzymes should be
masked. This observation can also be extended to explain the high thermal
sensitivity presented by these enzymes, since after 10 min at 30°C activities
of XR and XD decreased 31 and 40%, respectively (Table 4). However, the
optimum temperatures for specific activities of XR and XD were 65 and

Table 4
Effect of Temperature on Activity and Stability
of XR and XD Present in Crude Cell-Free Extract Obtained
from Disrupted Cells of *C. guilliermondii* FTI 20037

Temperature (°C)	Activity[a]		Stability[b]	
	XR	XD	XR	XD
25	300	84	244	87
30	353	202	271	121
35	—	290	337	152
40	740	352	365	197
45	—	—	378	178
50	975	306	340	151
55	—	252	312	81
60	1096	95	283	47
65	1118	—	—	—
70	1007	—	—	—
75	876	—	—	—

[a]Values represent the activities of XR and XD (expressed as U/mg of protein) at indicated temperature of the reaction medium.

[b]Values represent the residual activities of XR and XD (expressed as U/mg of protein) after an aliquot of the crude extract was left for 10 min at the desired temperature.

40°C, respectively (Table 4). Therefore, a compromise should be found between the temperatures of optimal activity and stability if XR was used in xylose-xylitol conversion in an enzyme reactor.

From Lineweaver-Burke plots, the apparent K_M for both enzymes against substrates and cofactors are as follows: for XR, $6.4 \times 10^{-2} M$ (xylose) and 9.5×10^{-3} mM (NADPH); for XD, $1.6 \times 10^{-1} M$ (xylitol) and 9.9×10^{-2} mM (NAD+). The straight-line fit was excellent for all parameters ($r > 0.99$). XR requires about 10-fold less xylose and cofactor than XD for the condition in which the reaction rate is half the V_{max}. This result could explain why xylitol can optimally be formed at an XR:XD ratio not so high, because in this study after 36 h of fermentation at 0.4 vvm the XR:XD ratio was 1.96 (Table 2) and xylitol production was 20 g/L (Fig. 1).

Acknowledgment

We wish to thank Fundação de Amparo à Pesquisa do Estado de São Paulo for financial support.

References

1. Taylor, K. B., Beck, M. J., Huang, D. H., and Sakai, T. T. (1990), *J. Ind. Microbiol.* **16,** 29–41.
2. Sene, L., Felipe, M. G. A., Vitolo, M., Silva, S. S., and Mancilha, I. M. (1998), *J. Basic Microbiol.* **38(1),** 61–69.
3. Hyvonen, L., Koivistoinen, P., and Voirol, F. (1982), *Adv. Food Res.* **28,** 373–403.

  4. Webb, S. R. and Lee, H. (1990), *Biotechnol. Adv.* **8,** 685–697.
  5. duPreez, J. C., Bosch, M., and Prior, B. A. (1986), *Appl. Microbiol. Biotechnol.* **23,** 228–233.
  6. Webb, S. R. and Lee, H. (1992), *J. Gen. Microbiol.* **138,** 1857–1863.
  7. Barbosa, M. F. S., Medeiros, M. B., Mancilha, I. M., Schneider, H., and Lee, H. (1988), *J. Ind. Microbiol.* **3,** 241–251.
  8. Alexander, N. J. (1985), *Biotechnol. Bioeng.* **27,** 1739–1744.
  9. Bradford, M. (1976), *Anal. Biochem.* **72,** 248–254.
 10. Felipe, M. G. A., Vitolo, M., and Mancilha, I. M. (1996), *Acta Biotechnologica* **16,** 73–79.
 11. Silva, S. S., Ribeiro, J. D., Felipe, M. G. A., and Vitolo, M. (1997), *Appl. Biochem. Biotechnol.* **63,** 557–564.
 12. Felipe, M. G. A., Vieira, D. C., Vitolo, M., Silva, S. S., Roberto, I. C., and Mancilha, I. M. (1995), *J. Basic Microbiol.* **35,** 171–177.
 13. van Zyl, C., Prior, B. A., and du Preez, J. C. (1991), *Enzyme Microb. Technol.* **13,** 82–86.
 14. Meyer, P. S., du Preez, J. C., and Kilian, S. (1992), *Appl. Microbiol.* **15,** 161–165.
 15. Almeida, S. J. B., Mancilha, I. M., Vanetti, M. C. D., and Teixeira, M. A. (1995), *Bioresour. Technol.* **52,** 197–200.
 16. Roberto, I. C., Mancilha, I. M., Souza, C. M. A., Felipe, M. G. A., Sato, S., and Castro, H. F. (1994), *Biotechnol. Lett.* **16,** 1211–1216.
 17. Vandeska, E., Kuzmanova, S., and Jeffries, T. W. (1995), *J. Ferment. Bioeng.* **80,** 513–516.
 18. Winkelhausen, E., Pittman, P., Kuzmanova, S., and Jeffries, T. W. (1996), *Biotechnol. Lett.* **18,** 753–758.
 19. Converti, A. and Del Borghi, M. (1996), *Acta Biotechnologica* **16,** 133–144.
 20. Girio, F. M., Peito, M. A., and Amaral-Collaço, M. T. (1989), *Appl. Microbiol. Biotechnol.* **32,** 199–204.
 21. Girio, F. M., Pelica, F., and Amaral-Collaço, M. T. (1996), *Appl. Biochem. Biotechnol.* **56,** 79–87.
 22. Yang, V. W. and Jeffries, T. W. (1990), *Appl. Biochem. Biotechnol.* **26,** 197–206.

Xylanase Production by *Aspergillus awamori* in Solid-State Fermentation and Influence of Different Nitrogen Sources

JUDITH L. S. LEMOS, MARIA C. DE A. FONTES,
AND NEI PEREIRA, JR.*

*Departamento de Engenharia Bioquímica, Escola de Química,
Universidade Federal do Rio de Janeiro, Ilha do Fundão,
CEP 21949-900 Rio de Janeiro, RJ, Brazil, E-mail: nei@eq.ufrj.br*

Abstract

The use of purified xylan as a substrate for bioconversion into xylanases increases the cost of enzyme production. Consequently, there have been attempts to develop a bioprocess to produce such enzymes using different lignocellulosic residues. Filamentous fungi have been widely used to produce hydrolytic enzymes for industrial applications, including xylanases, whose levels in fungi are generally much higher than those in yeast and bacteria. Considering the industrial importance of xylanases, the present study evaluated the use of milled sugarcane bagasse, without any pretreatment, as a carbon source. Also, the effect of different nitrogen sources and the C:N ratio on xylanase production by *Aspergillus awamori* were investigated, in experiments carried out in solid-state fermentation. High extracellular xylanolytic activity was observed on cultivation of *A. awamori* on milled sugarcane bagasse and organic nitrogen sources (45 IU/mL for endoxylanase and 3.5 IU/mL for β-xylosidase). Endoxylanase and β-xylosidase activities were higher when sodium nitrate was used as the nitrogen source, when compared with peptone, urea, and ammonium sulfate at the optimized C:N ratio of 10:1. The use of yeast extract as a supplement to the these nitrogen sources resulted in considerable improvement in the production of xylanases, showing the importance of this organic nitrogen source on *A. awamori* metabolism.

Index Entries: *Aspergillus awamori*; xylanases; nitrogen nutrition; solid-state fermentation; sugarcane bagasse.

Introduction

Filamentous fungi have been widely used to produce hydrolytic enzymes for industrial applications including xylanases. The genera

---

*Author to whom all correspondence and reprint requests should be addressed.

Aspergillus and *Trichoderma* have shown to be efficient producers of these enzymes on an industrial scale *(1)*. *Aspergillus awamori* has been used industrially for the production of several enzymes such as glucoamylase, α-amylase, and protease. Another important advantage of *A. awamori* is that it has a long history of safe use for the manufacture of food products destined for human consumption and is regarded as a nontoxigenic and nonpathogenic fungus *(2)*.

Xylanases have been considered for clarifying fruit juices and wines, food processing in combination with cellulases, and improving the nutritional properties of agricultural silage and grain feed. Its main potential application, though, relates to cellulosic pulp treatment in which the biocatalyst has already been incorporated into commercial bleach sequences *(3)*.

The bioconversion of lignocellulosic materials to fermentable sugars for fuel alcohol production has been hindered by economical and technical aspects, as well as the existence of more competitive sources of carbohydrates such as starch and sucrose. Nevertheless, there has been much effort to develop efficient bioprocesses using such raw materials, particularly those deriving from agricultural residues, which have no production cost attached to them, although costs for collection and transportation of these residues to a centralized processing location may be incurred.

Considering the industrial importance of xylanases, and the importance of nitrogen nutrition on fungus metabolism, in the present study we evaluated the effect of nitrogen sources and C:N ratio on xylanase production by *A. awamori* using milled sugarcane bagasse as the carbon source. Experiments were carried out in solid-state fermentation.

Materials and Methods

Maintenance and Propagation of Organism and Preparation of Inoculum

A. awamori NRRL 3112 was used as the xylanase-producing microorganism. Czapeck agar slants, incubated at 30°C for 5 to 6 d were used for conidia production to serve as inoculum in all fermentations. The conidia suspension was prepared by adding 4 mL of sterile water to a slope with a dense sporulation. Conidia enumeration was determined using a Neubauer camera (Assistent, West Germany). A standard inoculum of 5×10^6 conidia/g of bagasse was used in all cases.

Growth Media and Culture Conditions

Fermentations for enzyme production were carried out in 500-mL conical flasks, plugged with cotton corks, containing 4 g of roughly 1-mm-sized dry sugarcane bagasse, ground in a disc mill (Perten Laboratory mill 3600, Perten, SW). The sugarcane bagasse was moistened with 50 mL of an aqueous solution composed of 0.2 g of NaCl, 0.2 g of KH_2PO_4, 0.04 g of $MgSO_4 \cdot 7H_2O$, and different nitrogen sources at a C:N ratio of 14:1. Other

Table 1
Nitrogen Composition
of Different Media (C:N = 14:1)

Nitrogen source	Quantity (g/4 g bagasse)
NH_2CNH_2	0.24
$NH_2CNH_2{}^a$	0.14
$(NH_4)_2SO_4$	0.54
$(NH_4)_2SO_4{}^a$	0.30
$NaNO_3$	0.69
$NaNO_3{}^a$	0.39

[a]These media contained 0.5 g of yeast extract/ 4 g of bagasse.

C:N ratios were also investigated. The control experiment contained 0.5 g of peptone plus 0.5 g of yeast extract, providing 113.5 mg of total nitrogen, which is equivalent to a C:N ratio of 14:1. Fermentations were carried out in media containing individual nitrogen sources such as sodium nitrate, ammonium sulfate, urea, and combined nitrogen sources supplemented with yeast extract to the same total nitrogen content (Table 1). The flasks were incubated in a stove (FANEM, Brazil) at 30°C for 60 h, after which their whole contents were extracted by adding 100 mL of distilled water. After a period of 30 min under agitation at 150 rpm and at room temperature, the supernatant was separated by filtration to obtain the crude enzyme preparation.

Analysis

Extracellular protein was measured by the modified Lowry method (4) using bovine serum albumin as the standard. Elemental composition of sugarcane bagasse was determined using an Elementar Perkin-Elmer analyzer 2400 CHN.

Enzymatic Assays

Endoxylanase activity was measured by the formation of reducing sugars according to the following procedure. A 0.9-mL solution of 1% birchwood xylan (Sigma) in 50 mM citrate-phosphate buffer, pH 5.0, was preincubated for 2 min at 60°C followed by the addition of 0.1 mL of the diluted crude enzyme. The reaction was stopped after 2 min by the addition of 1 mL of the dinitrosalicylic acid reagent. One unit of activity was defined by the release of 1 µmol of reducing sugar (as xylose)/min. β-Xylosidase activity was determined by measuring the formation of p-nitrophenol according to the following procedure. A 0.5-mL solution of 1.25 mM p-nitrophenyl β-xylanopiranoside in 50 mM phosphate-citrate buffer, pH 5.0, was preincubated for 2 min at 55°C followed by the addition of 0.1 mL of the diluted crude enzyme. The reaction was stopped after

10 min by the addition of 4 mL of 0.25 M Na_2CO_3. One unit of enzyme activity was defined by the release of 1 µmol of p-nitrophenol/min.

pH Stability Assays

Endoxylanase and β-xylosidase stability experiments were carried out in phosphate-citrate buffer at different pH values. In each case, the supernatant of filtrated samples from *A. awamori* cultivation on peptone + yeast extract medium was adjusted to the selected condition by dilution with the corresponding buffer and incubation at 30 and 35°C, respectively. Enzymatic activities were measured after 1 h for endoxylanase and 0.5 h for β-xylosidase.

Assays were performed in duplicate, except for bagasse elemental composition, which was performed in triplicate. Each experiment was repeated twice.

Results and Discussion

Sugarcane bagasse, an abundant lignocellulosic residue produced in Brazil, was shown to be an adequate raw material for the production of xylanases. Its partial elemental composition was 40% carbon, 6% hydrogen, and 1.9% nitrogen, allowing the estimation of different C:N ratios for different nitrogen sources employed in the media composition.

According to the data presented in Fig. 1, the highest endoxylanase activity, approx 100 U/mL, was obtained when nitrate supplemented with yeast extract was used as the nitrogen source. However, the use of $NaNO_3$ as the sole nitrogen source provided similar results. The superiority of this inorganic source is also evidenced by the specific activity values as well as by the volumetric productivity (Table 2), pointing out that both media induce highest production of endoxylanases. Also, as judged by the specific activity figures, a lower amount of contaminant proteins is excreted. The majority of filamentous fungi are autotrophic with respect to nitrogen sources and are able to grow in synthetic media consisting basically of glucose, NH_4^+ or NO_3^-, and minerals. Therefore, *A. awamori* seems to have taken advantage of this widespread ability. The pathway for the reduction of nitrate to ammonium involves two enzymes: nitrate reductase and nitrite reductase *(5,6)*. In *Aspergillus nidulans*, synthesis of nitrate reductase and nitrite reductase is controlled by both nitrate induction and ammonium repression. Furthermore, nitrate reductase has been postulated to play an autogenous regulatory role, controlling its own synthesis and that of nitrite reductase *(7)*.

On the other hand, the medium with ammonium sulfate without yeast extract supplementation displayed the lowest activity value (13 U/mL). In this case, the crude enzyme preparation had a pH value of 2.9 (Table 2). This could have been in response to the low enzyme activity owing to pH inactivation. It is well known that ammonium is transported by the fungus as ammonia, leaving the hydrogen ion behind, resulting in acidification of

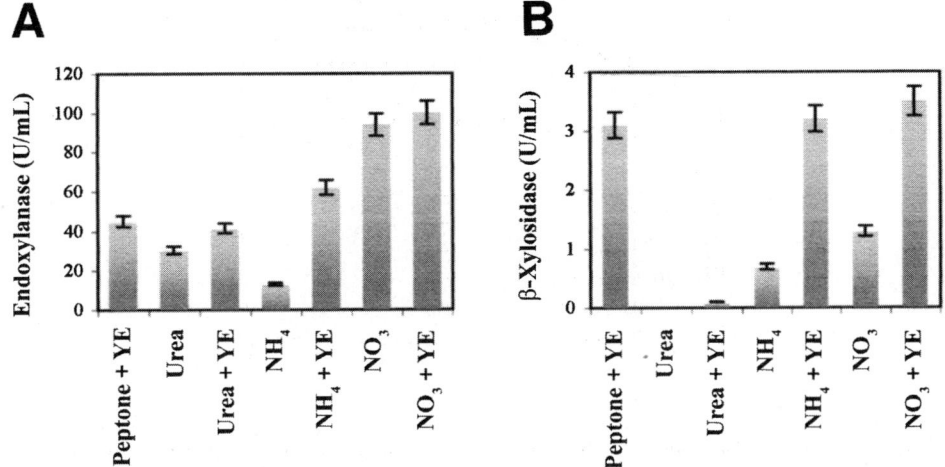

Fig. 1. Effect of organic and inorganic nitrogen sources on the production of **(A)** endoxylanase and **(B)** β-xylosidase (C:N = 14:1). Fermentations were carried out in media containing individual N sources such as sodium nitrate, ammonium sulfate, urea, and combined nitrogen sources supplemented with yeast extract (YE) to the same total nitrogen content.

Table 2
Effect of Nitrogen Sources on Production of Xylanases (C/N = 14/1)

Nitrogen source	Endoxylanase		β-Xylosidase			
	Volumetric productivity (U/[L·h])[a]	Specific activity (U/mg)	Volumetric productivity (U/[L·h])[a]	Specific activity (U/mg)	Initial pH	Final pH
Peptone + yeast extract	754	21.8	51.9	1.5	5.8	6.1
Urea	511	22.3	0.0	0.0	4.5	7.8
Urea + yeast extract	696	21.7	1.8	0.0	5.5	7.9
$(NH_4)_2SO_4$	216	20.9	11.5	1.1	4.4	2.9
$(NH_4)_2SO_4$ + yeast extract	1034	39.0	53.4	2.0	5.5	6.5
$NaNO_3$	1571	86.8	20.9	1.2	4.3	5.9
$NaNO_3$ + yeast extract	1666	60.5	57.7	2.1	5.5	6.1

[a]Calculated at 60 h of cultivation.

medium *(6)*. Nonetheless, the results obtained by Fernández-Espinar et al. *(8)* using different nitrogen sources on the production of α-arabino-furanosidase by *A. nidulans* showed that ammonium sulfate and ammonium chloride gave rise to notably higher enzyme activity than that reached in the presence of the other inorganic and organic compounds. Thus, it is clear that the ability of a cell to utilize a particular compound is a complex

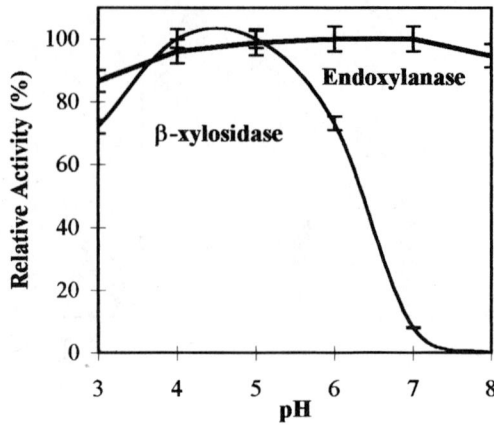

Fig. 2. pH stability of endoxylanase and β-xylosidase. Experiments were carried out with supernatants from *A. awamori* cultivation on peptone + yeast extract medium. The corresponding 100% of the relative activity was 42 U/mL for endoxylanase and 2.8 U/mL for β-xylosidase.

affair, depending on the genetic ability of the organism to synthesize the requisite enzyme in the first place, and then on its capacity to respond to induction.

In the present study, when ammonium sulfate was supplemented with yeast extract, the pH value of the enzyme preparation was about 6.5, indicating that the organic nitrogen source, of which approx 50% exists as free amino acid, was used preferentially. According to studies on *Saccharomyces cerevisiae* nitrogen consumption, ammonium is not absorbed from the wort until disappearance of the amino acids glutamate, glutamine, aspartate, asparagine, serine, threonine, arginine, and lysine *(9)*. The different effects of amino and ammonium nitrogen on product formation is the result of differences in the basic biochemical steps related to their use by the cell. As far as amino acids are concerned, they are assimilated and directly incorporated into proteins. This fact could be beneficial for all growth and enzyme production. Regarding ammonium, its use is limited by the rate at which it is incorporated into its organic counterparts *(10)*.

A rise in pH was observed when *A. awamori* was cultivated on urea and urea plus yeast extract. Urea is cleaved by urease and ammonia appears as result of the enzymatic hydrolysis, raising the pH of the medium *(11)*. It can be observed that the differences in pH were more disadvantageous for β-xylosidase activity than for endoxylanase when urea or urea plus yeast extract was used. This fact can be related to the lack of stability seen for β-xylosidase at basic pH (Fig. 2). To evaluate the possible effect of low and high pH values on the enzyme production, an experiment was conducted, and the results showed that β-xylosidase activity was greatly increased in media containing citrate-phosphate buffer, pH 5.0 (Fig. 3). However, the production of endoxylanase remained nearly constant in the same media, and its stability was higher and more constant in a broader pH

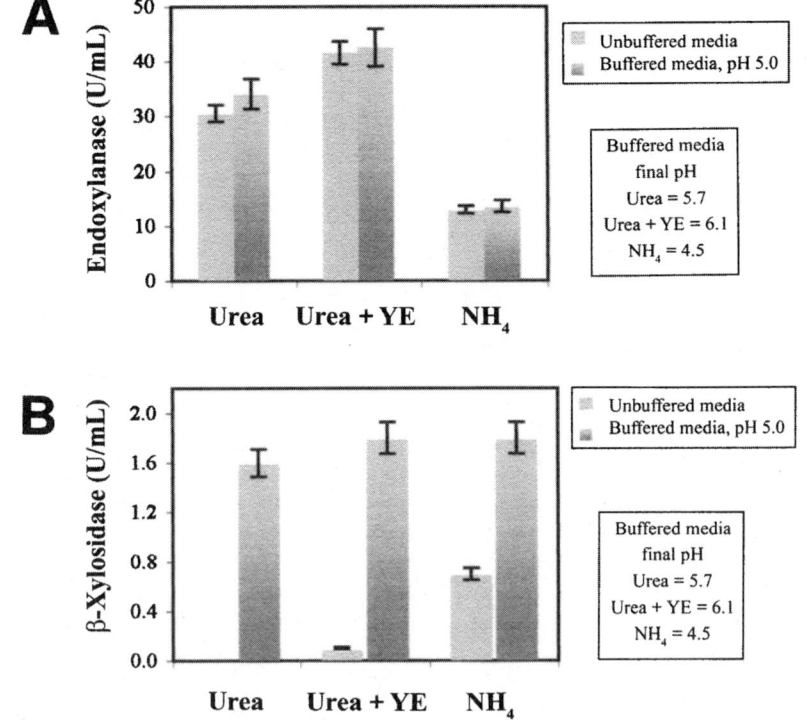

Fig. 3. **(A)** Endoxylanase and **(B)** β-xylosidase production in buffered (citrate-phosphate buffer, pH 5.0) and unbuffered media. YE, yeast extract.

range than β-xylosidase stability. Hence, the endoxylanase production did not increase as was expected in the buffered medium.

Peptone plus yeast extract has also been used to study the influence of organic nitrogen sources on endoxylanase and β-xylosidase activities. Our results show an endoxylanase activity comparable with that obtained in the presence of urea plus yeast extract. Nevertheless, both experimental data represent half of the activity obtained using $NaNO_3$ plus yeast extract medium. Generally, all fungi grow faster with complex organic nitrogen sources, but the effectiveness of these compounds varies considerably. In earlier studies, different organic nitrogen sources were found to improve xylanase production in several xylanase-producing fungi *(12–14)*. Their results are in disagreement with ours, which can be ascribed, particularly, to different species and even to variation in strain.

Concerning β-xylosidase production, $NaNO_3$, $(NH_4)_2SO_4$, and peptone, all supplemented with yeast extract, displayed the highest and similar results (3.5, 3.2, and 3.1 U/mL, respectively), with specific activity slightly better for the former.

Sodium nitrate plus yeast extract was used in further experiments, because this presented the most satisfactory results. These nitrogen sources were added on the basis of carbon percentage in bagasse (40%). Although this lignocellulosic residue contains nitrogen in its composition (2%),

A

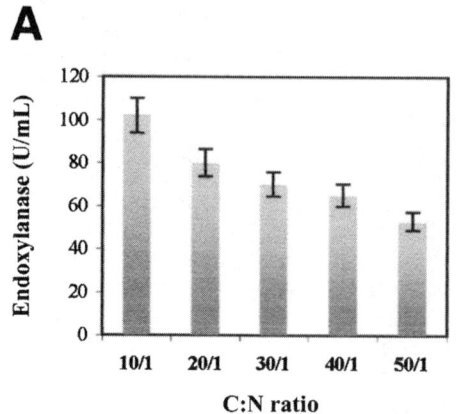

B

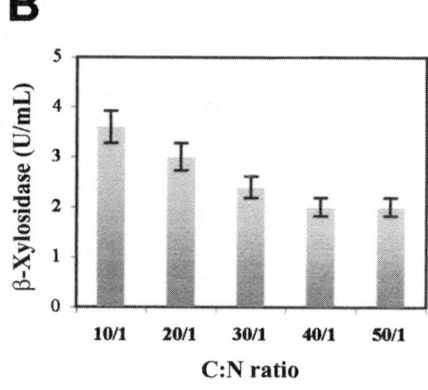

Fig. 4. Influence of C:N ratio on the synthesis of **(A)** endoxylanase and **(B)** β-xylosidase activities using NaNO$_3$ plus yeast extract. Of the total nitrogen added in each experiment, 57% proceeded from NaNO$_3$ and 43% from yeast extract.

A. awamori was unable to synthesize either endoxylanase or β-xylosidase from the basal medium, denoting the necessity of fortifying these media with nitrogen sources. The production of xylanolytic enzymes was maximized when the C:N ratio was adjusted to 10:1, as indicated in Fig. 4. The results show the importance of the C:N ratio on enzyme production. Considering the role of the nitrogen sources, one can observe that under nitrogen-limited conditions xylanase biosynthesis was less favored.

To determine the best proportions of sodium nitrate and yeast extract used as combined nitrogen sources, an experiment was performed using the best C:N ratio (10:1) obtained in the last experiment (Fig. 4). The highest values for xylanolytic enzyme activity (Fig. 5) were obtained for an initial sodium nitrate proportion of 40%, considering its nitrogen content. These results confirmed that yeast extract has an important role in enzyme synthesis by *A. awamori*, probably because a complex nitrogen source was essential for good growth and high enzyme activity, or because of some important elements contained in yeast extract that are necessary for the metabolism of fungus.

In conclusion, these results demonstrate that for the production of xylanases, sodium nitrate (to an initial proportion of 40%) supplemented with yeast extract appeared to be the best nitrogen source when added at a C:N ratio of 10:1, indicating that nitrogen source limitation is the cause for the decrease in xylanase yields. Considerations between organic and inorganic nitrogen sources are important, because their use is subject to metabolic versatility, as well as economical implications. The variation in pH media was shown to depend on nutrient composition, especially on nitrogen sources. Media containing proteins or related compounds showed a smaller variation in pH than those in which organic or inorganic nitrogen was used as the sole nitrogen source, except for NaNO$_3$. pH seems to have affected more β-xylosidase than endoxylanase activity.

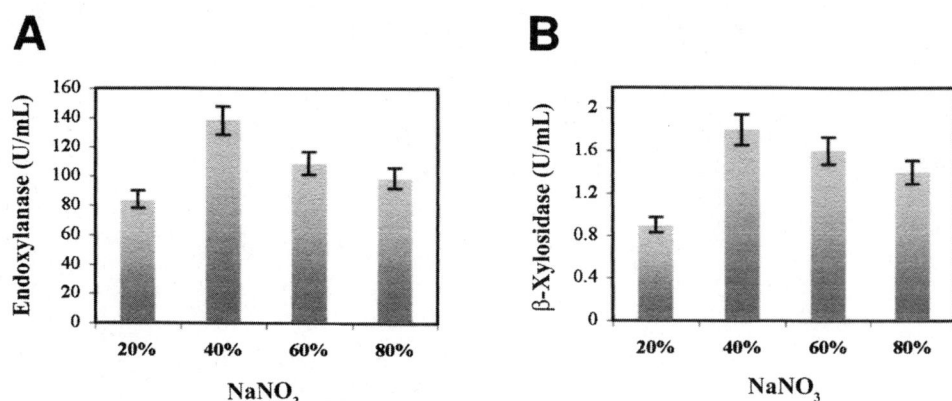

A

B

Fig. 5. Effect of proportions of sodium nitrate and yeast extract in **(A)** endoxylanase and **(B)** β-xylosidase production (C:N = 10:1).

References

1. Haltrich, D., Nidetzky, B., Kulbe, K. D., Steiner, W., and Zupancic, S. (1996), *Bioresour. Technol.* **58,** 137–161.
2. Cui, Y. Q., van der Lans, R. G. J. M., Giuseppin, M. L. F., and Luyben, K. C. A. M. (1998), *Enzyme Microb. Technol.* **23,** 157–167.
3. Carmona, E. C., Pizzirani-Kleiner, A. A., Monteiro, R. T. R., and Jorge, J. A. (1997), *J. Basic Microbiol.* **37,** 387–393.
4. Peterson, G. L. (1979), *Anal. Biochem.* **100,** 201–220.
5. Wiame, M. A. J., Grenson, M., and Arst, H. N. (1985), in *Advances in Microbial Physiology*, vol. 26, Rose, A. H. and Tempest, D. W., eds., Academic Press, London, pp. 1–87.
6. Griffin, D. H. (1994), *Fungal Physiology*, 2nd ed., J. Wiley and Sons, New York.
7. Marzluf, G. A. (1981), *Microbiol. Rev.* **45,** 437–461.
8. Fernández-Espinar, M. T., Peña, J. L., Piñaga, F., and Vallés, S. (1994), *FEMS Microbiol. Lett.* **115,** 107–112.
9. Slaughter, J. C. (1988), in *Physiology of Industrial Fungi*, Berry, D. R., ed., Blackwell Scientific Publications, Oxford, pp. 58–96.
10. Bon, E. P. S. and Webb, C. (1993), *Appl. Biochem. Biotechnol.* **39/40,** 349–369.
11. Schlegel, H. G. (1989), in *Basic Biotechnology—A Student's Guide*, Präve, P., Faust, U., Sittig, W., and Sukatsch, D. A., eds., VHC Publishers, pp. 67–101.
12. Smith, D. C. and Wood, T. M. (1991), *Biotechnol. Bioeng.* **38,** 883–890.
13. Haltrich, D., Laussamayer, B., and Steiner, W. (1994), *Appl. Microbiol. Biotechnol.* **42,** 522–530.
14. Gutierrez-Correa, M. and Tengerdy, R. P. (1998), *Biotechnol. Lett.* **20,** 45–47.

Kinetic and Mass Transfer Parameters of Maltotriose Hydrolysis Catalyzed by Glucoamylase Immobilized on Macroporous Silica and Wrapped in Pectin Gel

Luciana R. B. Gonçalves,*,[1] Glória S. Suzuki,[2] Roberto C. Giordano,[2] and Raquel L. C. Giordano[2]

[1]Departamento de Engenharia Química, Universidade Federal do Ceará, Campus do Pici, Bloco 710, sala 32, Fortaleza, CE, Brazil, CEP 60.455-760, E-mail: barios_goncalves@bol.com.br; and [2]Departamento de Engenharia Química, Universidade Federal de São Carlos, 13.565-905, C.P. 676, São Carlos, SP, Brazil, CEP 13.565-905

Abstract

Kinetic and mass transport parameters were estimated for maltotriose hydrolysis using glucoamylase immobilized on macroporous silica and wrapped in pectin gel at 30°C. Free enzyme assays were used to obtain the intrinsic kinetic parameters of a Michaelis-Menten equation, with product inhibition by glucose. The uptake method, based on transient experimental data, was employed in the estimation of mass transfer parameters. Effective diffusivities of maltotriose in pectin gel were estimated by fitting a classical diffusion model to experimental data of maltotriose diffusion into particles of pectin gel in the absence of silica. The effective diffusivities of maltotriose in silica were obtained after fitting a bidisperse model to experimental data of maltotriose hydrolysis using glucoamylase immobilized in silica and wrapped in pectin gel.

Index Entries: Maltotriose hydrolysis; kinetic parameters; effective diffusivities; macroporous silica; pectin gel.

Introduction

This work was part of a project that studied an unconventional continuous bioreactor configuration for the simultaneous hydrolysis and fermentation of liquefied starch designed to produce ethanol using

*Author to whom all correspondence and reprint requests should be addressed.

glucoamylase immobilized on silica and baker's yeast entrapped in pectin gel. In this process, the enzyme is covalently linked on controlled pore silica after silanization (with γ-aminopropylethoxysilane) and activation of the support with glutaraldehyde. The silica, containing enzyme, is later coimmobilized with baker's yeast in 4-mm-diameter spherical particles of pectin gel. This biocatalyst may be used in a continuous fixed-bed reactor *(1,2)*. To simulate this reactor, the reaction kinetic and mass transport parameters of the immobilized enzyme must be known. The process substrate, liquefied starch, is a mixture of glucose, maltose, maltotriose, maltotetraose, and maltopentaose. Ono et al. *(3)* report that maltose and maltotriose have the slowest rate of hydrolysis in this process. The present study focuses on the second substrate. The hydrolysis of maltose has been studied previously *(4,5)*.

Immobilization causes a deliberate restriction on the mobility of enzymes and cells, which can affect the mobility of solutes. These phenomena may reduce the apparent reaction rate and, consequently, may decrease the process efficiency, when compared to that for soluble enzymes *(6)*. A reduced apparent reaction rate can also result from external diffusion restrictions on the surface of supports. To prevent this effect, a batch system with vigorous agitation was used. Although silica particles are extremely fragile with respect to shear stress, the pectin gel that surrounds these particles protects them against the action of the stirrer. The uptake method, used here, is a classic methodology for the estimation of effective diffusivities in gel beads. It employs whole beads of the biocatalyst, preserving the shape they possess in the actual industrial process.

This approach, however, is transient in nature and requires a careful design of the experiments in order to minimize errors related to the time of sampling *(7)* and to the presence of adsorptive effects. Solute adsorption can lead to different concentrations inside and outside the support, therefore disguising the actual diffusion effects. This phenomenon, if present, must be taken into account for solutes that might interact with the gel by ionic or adsorptive forces. Gonçalves et al. *(5)*, however, observed that the adsorption of glucose, maltose, and lactose in pectin gel was not significant for the conditions used during the reaction experiments. Consequently, adsorption effects were not considered here.

The main purpose of the present study was to determine the kinetic parameters of maltotriose hydrolysis and effective diffusivities of maltotriose in macroporous silica containing immobilized enzyme wrapped in citric pectin gel. It was necessary to perform preliminary assays with free glucoamylase to estimate intrinsic kinetic parameters for maltotriose hydrolysis, under the operational conditions found in the bioreactor. It was assumed that only one amino group of the enzyme was bound to the activated support during the immobilization process. Therefore, the immobilization would not change significantly the conformation of the enzyme. So intrinsic and inherent kinetics would be essentially the same. This hypothesis was based on the reactivity of the amino and aldehyde groups present, respectively, in the enzyme and on the support. In this work, the immobilization reaction was proceeded at pH 4.2. Acidic pH favors the reactivity of aldehyde groups, but at low pH few amino groups will be uncharged and able to act as nucleophiles in the immobilization reaction. That low concentration prevents the formation of multipoint bonds between the same enzyme molecule and the activated support *(8)*.

Materials and Methods

Chemicals

Soluble starch (mol wt = 29,400) was purchased from Merck, and maltotriose was purchased from Omega. All other chemicals were of laboratory grade, from commercial suppliers.

Enzyme

Glucoamylase from *Aspergillus niger* was donated by Novo Nordisk Brazil (Araucária, Paraná, Brazil) (activity: 180 U/mL).

Supports

High-porosity silica, with an average diameter of 170 μm, porosity of 0.57, and mean pore diameter of 270 Å (measured by N_2 desorption, ASAP 2000; Micrometrics, Araraquara, São Paulo, Brazil) and low methoxylation citric pectin (type 8002) were donated by Braspectina do Brasil (Limeira, São Paulo, Brazil).

Enzyme Activity

Enzyme activity was evaluated by measuring the glucose released during hydrolysis of starch, 40 g/L in 100 m*M* acetate buffer, pH 4.2. The difference between enzymatic activities of the supernatant before and after immobilization was used to assess the silica enzyme loading. One unit of enzyme activity was defined as the amount of enzyme that yields 1 g of glucose/(L·h) from 40 g/L of soluble starch at 60°C and pH 4.2.

Enzyme Immobilization on Silica (5)

Silica was silanized with a solution of γ-aminopropyltriethoxysilane (0.5 [v/v]) at pH 3.3 and 75°C for 3 h with a liquid-solid ratio of 3.0 mL/g. Then, the catalyst was washed with water and acetone. It was later dried until no further change in weight occurred.

The support was activated with glutaraldehyde (in a 2.5% sodium hydrogenophosphate buffer, 0.1 *M* and pH 7.0) for 1 h at 20–25°C, with a liquid-solid ratio of 3.0 mL/g. It was washed again and reacted with the enzyme solution for 36 h at 20–25°C, pH 4.2, under stirring.

Preparation and Characterization of Biocatalyst

Six grams of citric pectin was added to 88 g of water and 6 mL of acetate buffer (1 *M* at pH 4.2). A known amount of silica containing immobilized enzyme was added to the resulting solution and trickled into 0.2 M CaCl$_2$ under mild stirring. The gel particles were then cured at 4°C for 18–24 h. Before each run, the mean particle diameter was calculated using a picnometer, with a sample of 500 spherical gel particles. The gel porosity was obtained by gravimetric analysis and its density using a picnometer.

Analytical Methods

Glucose was determined by the glucose-oxidase method (GOD-PAP; Merck) and maltotriose was measured by reducing sugar analysis (9).

Experimental Procedure

A batch reactor with magnetic stirring was used in all the experiments described herein. Enzymatic hydrolysis was carried out in 100 mM acetate buffer solution, pH 4.2, at 30°C. For all assays, 0.1-mL samples were taken periodically. The stirring was high enough to avoid the effects of external mass transport resistance. The stirring was increased until the concentration vs time curve of the product was no longer influenced by this variable.

Initial reaction rates of maltotriose hydrolysis were performed in the absence of glucose (product) in order to estimate the Michaelis-Menten parameters $V_m = k_3 E_t$ and K_m using linear regression (classic Lineweaver-Burk method [10]) and nonlinear regression (Marquardt algorithm). All other assays were performed in a batch reactor.

Mathematical Models

Three different models were used in this work: a homogeneous model to estimate intrinsic kinetic parameters (5), a diffusion model (11) for the effective diffusion coefficient of maltotriose in pectin gel, and a diffusive-reactive model (4) (bidisperse model) to estimate the effective diffusivity of maltotriose in silica wrapped in pectin gel in the presence of reaction. The diffusion and bidisperse models assume negligible external mass transfer resistance, invariant effective diffusivities, enzyme homogeneously immobilized on the silica, and constant temperature. Solute concentrations are based on the volume of the gel available for diffusion (volume of pores).

Intrinsic Kinetics of Maltotriose Hydrolysis

A Michaelis-Menten equation with competitive product inhibition by glucose was used to model the kinetics of maltotriose hydrolysis (Eq. 1):

$$-R_M = \nu R_G = \frac{k_3 E_t C_M}{K_m [1 + (C_G/k_i)] + C_M} \tag{1}$$

This is a simplified, lumped-parameter model of the hydrolysis.

Homogeneous Model

The mass balance for the batch reactor with free enzyme is as follows:

$$\frac{dC_{M1}}{dt} = -R_M \tag{2}$$

$$\frac{dC_{G1}}{dt} = \nu R_G \tag{3}$$

The ordinary differential Eqs. 2 and 3 were solved numerically, using the algorithm described earlier *(12)*. These equations were used to estimate the inhibition constant, k_i, in Eq. 1, using a maximum likelihood method *(13,14)*. This routine takes into account random errors, with normal distribution, in all measured variables (except time, which is assumed to be error free). The experimental variance is $0.25 \, (g/L)^2$ for glucose. Since the analytical method to measure maltotriose is less precise, only glucose data were used to estimate the kinetic parameters (k_3, K_m, k_i).

Diffusion Model

The fundamental equations are as follows:

Solute mass balance at the surface of the gel beads for component *i*:

$$\frac{dC_{il}}{dt} = \frac{\partial C_i}{\partial t}\bigg|_{\zeta=1} = -\frac{3}{R_p^2}\frac{(1-\varepsilon_r)}{\varepsilon_r}\varepsilon_{gp}D_{e,ig}\frac{\partial C_i}{\partial \zeta}\bigg|_{\zeta=1} \tag{4}$$

*i.c.: $t = 0$, $C_i = C_i(0)$

Solute mass balance inside the gel for component *i*:

$$\frac{\partial C_i}{\partial t} = \frac{1}{R_p^2}\frac{D_{e,ig}}{\zeta^2}\frac{\partial}{\partial \zeta}\left(\zeta^2\frac{\partial C_i}{\partial \zeta}\right) \tag{5}$$

i.c.: $t = 0$, $C_i = (0)$; **b.c.1: $\zeta = 0$, $\partial C_i/\partial\zeta = 0$; †b.c.2: $\zeta = 1$, $C_i(\zeta) = C_{il}$.

Equations 4 and 5 were solved numerically. The problem was discretized in the space coordinate using orthogonal collocation *(15)*, using Jacobi polynomials, $P^{(0,1/2)}$. Tests with different numbers of collocation points showed that five internal nodes gave accurate results.

When the solution is approximated by polynomials, the order of the problem is reduced. The resulting set of differential ordinary equations (which have time as independent variable) was solved numerically *(12)*. In all simulations, the volume retrieved when each sample is taken (0.1 mL) was accounted for. The fitting parameters in these equations were the effec-

*i.c., initial conditions; **b.c.1, boundary condition 1; †b.c.2, boundary condition 2.

tive diffusion coefficients. Again, a maximum likelihood approach was employed *(13,14)*.

Bidisperse Model

The fundamental equations are as follows:

Solute mass balance at the surface of the gel beads for component i:

$$\frac{dC_{il}}{dt} = \frac{\partial C_i}{\partial t}\bigg|_{\zeta=1} = -\frac{3}{R_p^2}\frac{(1-\varepsilon_r)}{\varepsilon_r}\varepsilon_p D_{e,ig}\frac{\partial C_i}{\partial \zeta}\bigg|_{\zeta=1} \tag{6}$$

i.c.: $t = 0$, $C_i = C_i(0)$

Solute mass balance inside the gel for component i:

$$\varepsilon_{gp}\frac{\partial C_{ig}}{\partial t} = \frac{\varepsilon_p}{R_p^2}\frac{De_{ig}}{\zeta^2}\frac{\partial}{\partial \zeta}\left(\zeta^2\frac{\partial C_{ig}}{\partial \zeta}\right) - \frac{3\varepsilon_S}{R_S^2}\left(\frac{\varepsilon_p - \varepsilon_{gp}}{\varepsilon_S - \varepsilon_{gp}}\right)\left(De_{iS}\frac{\partial C_{iS}}{\partial \zeta_S}\right)\bigg|_{\zeta=1} \tag{7}$$

i.c.: $t = 0$, $C_i = (0)$; b.c.1: $\zeta = 0$, $\partial C_{ig}/\partial \zeta = 0$; b.c.2: $\zeta = 1$, $C_{ig}(\zeta) = C_{il}$.

Solute mass balance inside the silica for component i:

$$\varepsilon_S\frac{\partial C_{iS}}{\partial t} = \frac{\varepsilon_S}{R_S^2}\frac{D_{e,iS}}{\zeta_S^2}\frac{\partial}{\partial \zeta_S}\left(\zeta_S^2\frac{\partial C_{iS}}{\partial \zeta_S}\right) - R_M \tag{8}$$

i.c.: $t = 0$, $C_{iS} = (0)$; b.c.1: $\zeta = 0$, $\partial C_{iS}/\partial \zeta = 0$; b.c.2: $\zeta = 1$, $C_{iS}(\zeta_S) = C_{ig}(\zeta)$.

Equations 6–8 were solved numerically, using the same approach described in the previous section. Effective diffusion coefficients were obtained using a direct search procedure.

Results and Discussion

Kinetic Parameters of Maltotriose Hydrolysis

To estimate the kinetic parameters of maltotriose hydrolysis, initial rates and transient assays were used. The initial rates of reaction were obtained after a linear regression of glucose concentration (product of this reaction) vs time for different initial concentrations of maltotriose ($S^0 = 1.78$, 4.52, 9.04, and 19.77 g/L). Values of V_m and K_m obtained from a Lineweaver-Burk plot (Fig. 1) were the initial guesses for a nonlinear, least-squares, Marquardt algorithm (Fig. 2). Finally, transient assays were performed to estimate k_i using the maximum likelihood method (Fig. 3). Table 1 shows the resulting parameters, together with some other values from the literature. The last ones correspond to different conditions and sub-

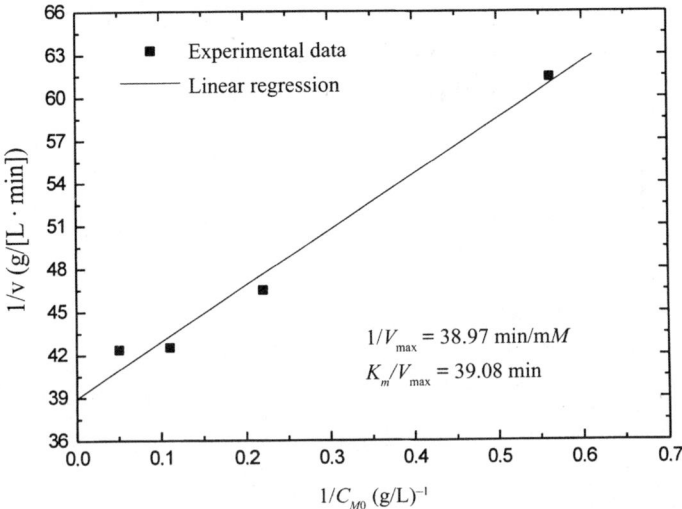

Fig. 1. Lineweaver-Burk diagram for maltotriose hydrolysis using free glucoamylase at pH 4.2 and 30°C.

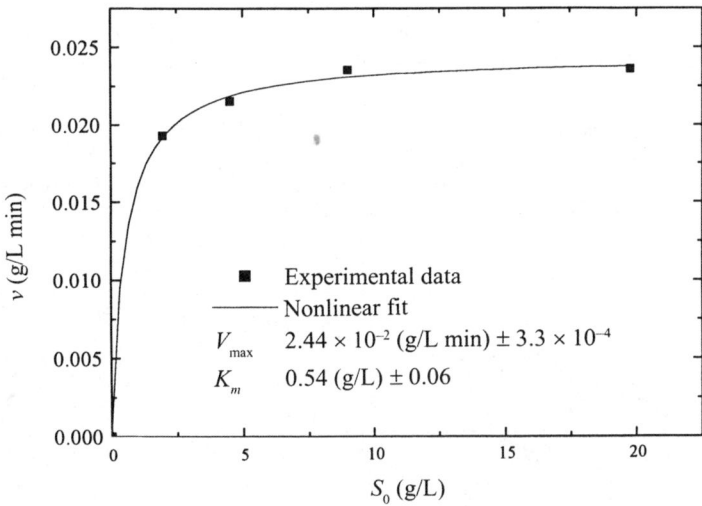

Fig. 2. Nonlinear fitting to maltotriose hydrolysis using free glucoamylase at pH 4.2 and 30°C.

strates, but their order of magnitude is in the same range. Although a lumped and simplified model was used to represent the hydrolysis reaction, it can be seen that K_m values, in molar basis, decrease sharply from maltose to starch, as already expected, and that k_3 values show the correct tendency (3). A sharp increase in this parameter from maltose to starch was expected. Since the fitted model represents the experimental data well, it can be used in the simulation of the global process, in which the biocatalyst contains enzyme and microorganism coimmobilized in pectin

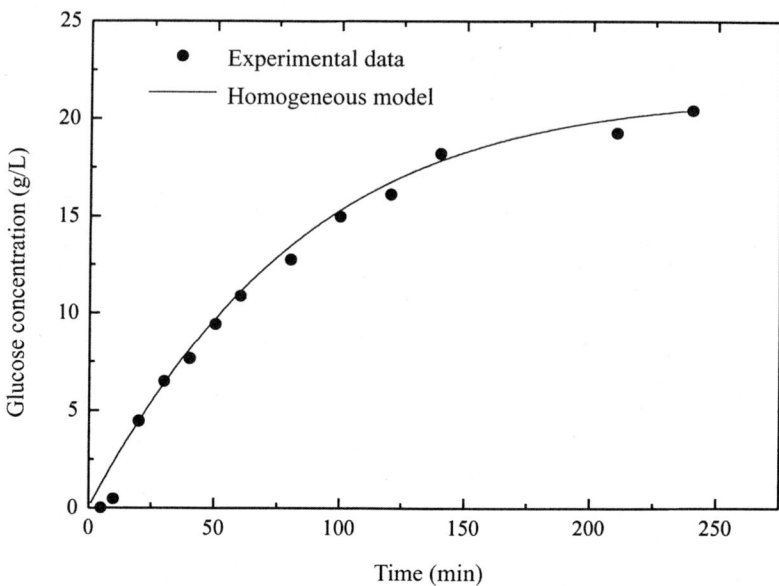

Fig. 3. Glucose concentration formed by maltotriose hydrolysis at pH 4.2 and 3°C using free enzyme (0.19 U/mL) and C_{M0} = 25.76 g/L.

Table 1
Kinetic Parameters of Maltotriose, Maltose, and Starch Hydrolysis
Using Glucoamylase at pH 4.2[a]

Parameter	Maltotriose, 30°C (this study)	Maltose, 30°C (5)	Starch, 30°C (16)	Maltotriose, 15°C (13)
K_m (g/mL)	$4.5 \times 10^{-4} \pm 6 \times 10^{-5}$	6.1×10^{-4}	5.0×10^{-4}	1.0×10^{-4}
K_m (M)	8.9×10^{-4}	17.8×10^{-4}	0.17×10^{-4}	2.0×10^{-4}
k_3 (g/[U·s])	$2.1 \times 10^{-5} \pm 3 \times 10^{-7}$	0.9×10^{-5}	3.7×10^{-5}	—
k_i (g/mL)	$5.2 \times 10^{-4} \pm 1 \times 10^{-6}$	1.2×10^{-1}	3.6×10^{-4}	—

[a]Confidence interval of 95%.

gel. Because the literature reports that glucoamylase has a multichain mechanism and the release of maltose was not considered here, more experiments measuring maltose and glucose concentrations are needed to confirm the obtained results.

The constants in Table 1 will be used to describe the kinetics of the immobilized enzyme; in other words, intrinsic and inherent rates are considered equal. This is a reasonable hypothesis, since immobilization was carried out at pH 4.2.

Under immobilization conditions, the low concentration of uncharged amine groups prevents the formation of multipoint bonds between the same enzyme molecule and the activated support (8).

Table 2
Effective Diffusion Coefficients in Pectin Gel Beads at 30°C

Substrate/gel	Effective diffusivity ($\times 10^6$ cm²/s)	Molecular diffusion coefficient in water ($\times 10^6$ cm²/s)[a]
Maltotriose/citric pectin (this work)	4.19 ± 0.1	4.63
Maltose/citric pectin *(4)*	4.49	5.98
Glucose/citric pectin *(4)*	5.29	
Glucose/κ-carrageenan *(17)*	3.83	6.29

[a]Calculated using Wilke-Chang method *(14)*.

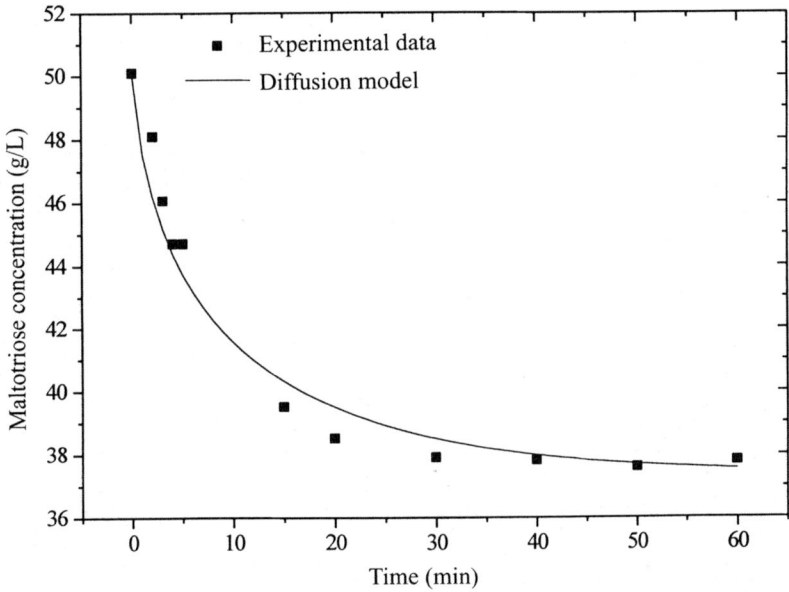

Fig. 4. Maltotriose diffusion into pure pectin particle when $D_p = 0.21$ cm and $\varepsilon_{gp} = 0.97$ at 30°C.

Consequently, there will be minimum conformational changes in the glucoamylase tertiary structure owing to the immobilization.

Effective Diffusion Coefficient of Maltotriose in Pectin Gel

The effective diffusion coefficient of maltotriose in pectin gel was estimated by fitting the diffusion model (Eqs. 4 and 5) to experimental data of maltotriose diffusion into particles of pectin gel in the absence of silica. Table 2 compares the effective diffusion coefficients of maltotriose, maltose, and glucose in two different gels with their respective molecular diffusion coefficients in water.

The fitting of the model for the maltotriose diffusion experiments is shown in Fig. 4. We can observe that effective diffusivities are slightly

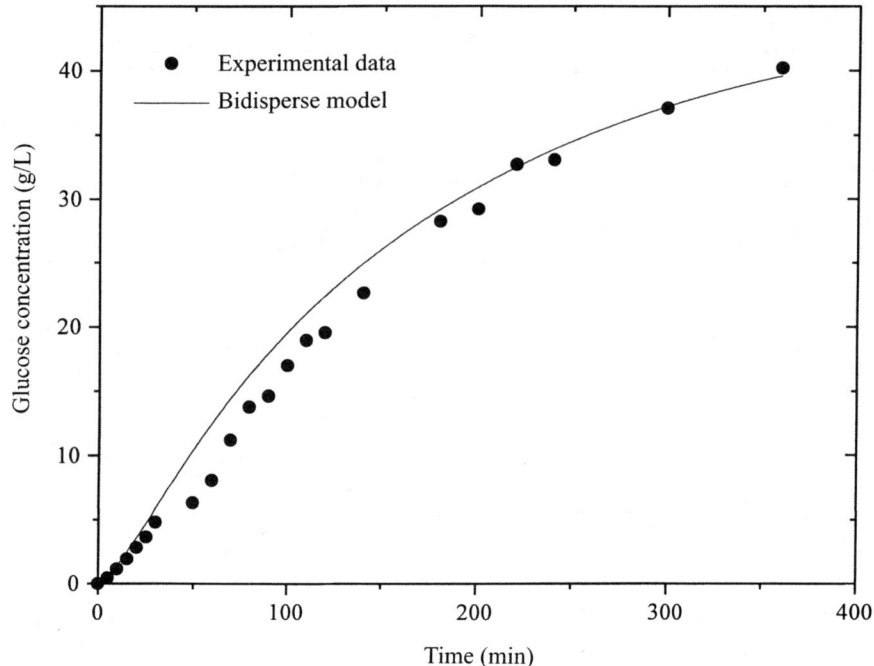

Fig. 5. Glucose production during maltotriose hydrolysis using glucoamylase immobilized on silica and wrapped in pectin gel at 30°C and pH 4.2.

Table 3
Effective Diffusivities
of Different Substrates in Silica and in Pectin Gel at 30°C

Substrate	Effective diffusivity in gel (cm²/s)	Effective diffusivity in silica (cm²/s)
Glucose	5.29×10^{-6} (8)	0.55×10^{-6} (5)
		1.08×10^{-6b}
		1.77×10^{-9c}
Maltose	4.49×10^{-6} (8)	0.50×10^{-6} (5)
Maltotriose	4.18×10^{-6a}	0.45×10^{-6a}

[a]Values estimated in this work.
[b]Silica porous diameter = 116 Å (18).
[c]Silica porous diameter = 74 Å (18).

lower than the molecular coefficient. This is an expected result, because the gel matrices have large-diameter pores.

Effective Diffusion Coefficient of Maltotriose in Macroporous Silica

Hydrolysis of maltotriose was performed using glucoamylase immobilized on silica and wrapped in pectin gel. The bidisperse model was used to estimate the effective diffusivity of maltotriose in the silica pores using the $D_{e,mg}$ obtained previously. The results are portrayed in Fig. 5 and the estimated coefficient is given in Table 3. The estimated coefficient was compatible with other researchers' values and one order of magnitude smaller than the molecular diffusion coefficient in water (Table 3). A good representation of the experimental data is observed. The small, systematic deviation of the model from the empirical data may be caused by the assumption that there is no release of maltose during the hydrolysis, as already discussed.

Conclusion

The estimated kinetic parameters for maltotriose hydrolysis catalyzed by gluoamylase and its effective diffusion coefficients in pectin gel and silica agreed with other researchers' data *(5,11,17,18)*. The bidisperse model represents the experimental data well, and it can be used in the simulation of the global process, in which the biocatalyst contains enzyme and micro-organism coimmobilized in pectin gel.

Acknowledgments

We thank CNPq, FAPESP, and CAPES for the sponsorship that made this work possible.

Nomenclature

C_G = glucose concentration (g/mL)
C_M = maltotriose concentration (g/mL)
$D_{e,mg}$ = effective diffusivity of maltose (or component i) in the gel (cm^2/s)
$D_{e,mS}$ = effective diffusivity of maltose (or component i) in the silica (cm^2/s)
E_t = enzyme reactor load (U/cm$^3_{reactor}$)
k_3 = kinetic constant (g/[U·s])
k_i = product inhibition constant (g/mL)
K_m = Michaelis-Menten constant (g/mL)
R_G = glucose rate of formation (g/[mL·s])
R_M = maltotriose rate of hydrolysis (g/[mL·s])
R_p = particle radius (cm)
$V_m = k_3 E_t$ = maximum rate (g/[mL·s])

Subscripts

i = component
l = liquid
g = pectin gel
S = silica
ε_{gp} = pure gel porosity
ε_p = biocatalyst porosity

ε_r = bed porosity
ε_s = silica porosity
ς = a dimensional radius
ν = stoichiometric coefficient (mass base)

References

1. Giordano, R. L. C., Hirano, P. C. N., Gonçalves, L. R. B., and Schmidell Netto, W. (2000), *Appl. Biochem. Biotechnol.* **84–86,** 643–654.
2. Gonçalves, L. R. B., Giordano, R. L. C., and Giordano, R. C. (1994), in *Proceedings of the 8th European Conference on Biomass for Energy, Environment, Agriculture and Industry*, vol. 2, Chartier, Ph., Beenackers, A. A. C. M., and Grassi, G., eds., Elsevier, Great Britain, pp. 1439–1446.
3. Ono, S., Hiromi, K., and Zimbo, M. J. (1954), *J. Biochem.* 55,315–55,320.
4. Gonçalves, L. R. B., Giordano, R. L. C., and Giordano, R. C. (1997), *Braz. J. Chem. Eng.* **14,** 333–339.
5. Gonçalves, L. R. B., Giordano, R. L. C., and Giordano, R. C. (1997), *Braz. J. Chem. Eng.* **14,** 341–346.
6. Tischer, W. and Kasche, V. (1999), *Tibtech.* **17,** 326–335.
7. Petersen, J. N., Davison, B. H., and Scott, C. D. (1991), *Biotechnol. Bioeng.* **37,** 386–388.
8. Guisan, J. M. (1987), *Enzyme Microb. Technol.* **10,** 375–382.
9. Somogyi, M. (1952), *J. Biol. Chem.* **195,** 19–23.
10. Segel, I. H. (1975), *Enzyme Kinetics: Behavior and Analysis of Rapid Equilibrium and Steady-State Enzyme Systems*, Wiley, New York.
11. Mattos, M. V. C., Giordano, R. C., and Giordano, R. L. C. (1996), *Braz. J. Chem. Eng.* **13,** 63–69.
12. Petzold, L. R. (1989), DDASSL code, version 1989, Computing and Mathematics Research Division, Lawrence Livermore National Laboratory, Livermore, CA.
13. Anderson, T. F., Abrams, D. S., and Grens, E. A. (1978), *AIChE J.* **24,** 20–29.
14. Himmelblau, D. M. (1970), *Process Analysis by Statistical Methods*, Wiley, New York.
15. Villadsen, J. V. and Michelsen, M. L. (1978), *Solution of Differential Equation Models by Polinomial Approximation*, Prentice Hall, Englewood Cliffs, NJ.
16. Giordano, R. L. C. and Schmidell Netto, W. (1989), *Revista Microbiologia* **20,** 376–381.
17. Nguyen, N.-L. and Luong, J. H. T. (1986), *Biotechnol. Bioeng.* **28,** 1261–1267.
18. Netrabukkana, R., Lourvanij, K., and Rorrer, G. L. (1996), *Ind. Eng. Chem. Res.* **35,** 458–464.

Selection of Stabilizing Additive
for Lipase Immobilization
on Controlled Pore Silica by Factorial Design

Cleide M. F. Soares,[1,2] Heizir F. De Castro,*[,1]
M. Helena A. Santana,[2] and Gisella M. Zanin[3]

[1]Departamento Engenharia Química,
Faculdade de Engenharia Química de Lorena, PO Box 116, 12600-000,
Lorena-SP, Brazil, E-mail: decastro@easygold.com.br;
[2]Departamento de Biotecnologia, Faculdade de Engenharia Química,
UNICAMP, PO Box 6066, 13081-970, Campinas-SP, Brazil;
and [3]Departamento de Engenharia Química,
Universidade Estadual de Maringá, 87020-900, Maringá-PR, Brazil

Abstract

Candida rugosa lipase was covalently immobilized on silanized controlled pore silica (CPS) previously activated with glutaraldehyde in the presence of several additives to improve the performance of the immobilized form in long-term operation. Proteins (albumin and lecithin) and organic molecules (β-cyclodextrin and polyethylene glycol [PEG]-1500) were added during the immobilization procedure, and their effects are reported and compared to the behavior of the immobilized biocatalyst in the absence (lacking) of additive. The selection of the most efficient additive at different lipase loadings (150–450 U/g of dry support) was performed by experimental design. Two 2^2 full factorial designs with two repetitions at the center point were employed to evaluate the immobilization yield. A better stabilizing effect was found when small amounts of albumin or PEG-1500 were added simultaneously to the lipase onto the support. The catalytic activity had a maximum (193 U/mg) for lipase loading of 150 U/g of dry support using PEG-1500 as the stabilizing additive. This immobilized system was used to perform esterification reactions under repeated batch cycles (for the synthesis of butyl butyrate as a model). The half-life of the lipase immobilized on CPS in the presence of PEG-1500 was found to increase fivefold compared with the control (immobilized lipase on CPS without additive).

Index Entries: Controlled pore silica; immobilization; lipase; additive; factorial design.

*Author to whom all correspondence and reprint requests should be addressed.

Introduction

Ester production by enzymatic reactions or biotransformations has stimulated the optimization of processes under nonaqueous media using several enzymes (1). In biotransformations, lipases stand out for carrying out esterification reactions with extreme process simplicity, superior quality of the final product, and high yields (2).

Lipase-catalyzed esterifications can be performed either with solid enzyme powder added directly to the solvent or with immobilized enzyme. The use of lipase immobilized on solid support is usually not necessary, because enzymes generally are insoluble in organic solvents. However, immobilization may protect the enzyme, to some extent, from solvent denaturation. It also helps in maintaining homogeneity of enzymes in the reaction media because it avoids aggregation of enzyme particles. In addition, this technique offers beneficial effect in the stability, as a function of the physicochemical interactions between support and enzyme (2,3).

Several studies have been conducted to establish methodologies for immobilizing lipases on different supports (3–5). Published results suggest that lipases are better immobilized on hydrophobic carriers owing to their peculiar physicochemical character (4). This property and the good characteristics demonstrated by silica-based carriers provided the basis of a choice for controlled pore silica (CPS) as immobilizing support for microbial lipase in previous studies developed by our group (6). In accordance with this work, Candida rugosa lipase was immobilized with high activity on silanized CPS activated with glutaraldehyde. Although this procedure has not considerably affected the catalytic properties of this immobilized system, decreased stability was noticed in the synthesis of butyl butyrate under repeated batches (half-life of 36 h) possibly owing to the lipase denaturation or desorption.

To overcome this limitation, several strategies have been reported. Among them, promising results have been achieved by the addition of stabilizing agents that protect the enzyme during the immobilization step (7–11). In the specific case of the lipases, which demand an interface for total catalytic activity, the use of macromolecular additives such as proteins, polyethylene glycol (PEG), and polyvinyl alcohol (7–9) have demonstrated stabilizing effects in the activity of the enzyme, avoiding changes in protein structure. On the other hand, the use of low molecular weight compounds such as monomeric carbohydrates (sorbitol, arabitol) and polysaccharides (dextran, starch) had no effect (8). Sometimes the role of these additives is masked by inert impurities included in commercial preparations (9).

The effect of additives on the activity of lipase preparation is not yet well understood. Probably they act through a combination of various effects including enzyme protection from inactivation during the immobilization step, retention of a water layer around the catalyst, and dispersing effects of the enzyme molecules that facilitate mass transport when additives are used together with immobilizing matrices. The kind of additive, its concentration, and the contact time are critical parameters that have to be optimized in each case *(7–11)*.

In the present study, the scope of this technique was explored and the influence of several additives on the activity of CPS-immobilized lipase was studied by employing statistical concepts *(12)*. In particular, emphasis was given to the selection of the most effective additive to improve the performance of the immobilized lipase-CPS in long-term operation. Based on data in the literature *(8–10)*, proteins (albumin and lecithin) and organic molecules (β-cyclodextrin and PEG) were tested as additives. Although polysaccharides are considered to have little effect on lipase activity, β-cyclodextrin was also tested as an additive to CPS derivatives because of its important properties of stabilization against chemicals, heat, light, and oxidation when used to encapsulate guest molecules *(13)*.

The immobilized systems on CPS in the presence of the additive were used in both hydrolysis of olive oil and synthesis of the butyl butyrate. Data were compared with those attained by CPS-immobilized lipase lacking additive under the same operational conditions *(6)*.

Materials and Methods

Materials

Commercial *C. rugosa* lipase (type VII) was purchased from Sigma (St. Louis, MO). The lipase was a crude preparation with a nominal specific activity of 1440 U/mg of protein based on the Bradford *(14)* method. CPS was supplied by Corning Glass Works, with the following characteristics: average particle porosity (ε) of 0.566, particle matrix density (ρs) of 2.178 g/cm^3, particle density (dry) (ρp) of 0.948 g/cm^3, particle size of 37.5 nm containing pores of 375 Å *(6)*. The silane γ-aminopropyltriethoxysilane (γ-APTS) and glutaraldehyde (25% solution) were from Sigma. Bovine serum albumin (BSA) (Sigma), soy lecithin (Sinthy), PEG-1500 (Sinthy), and β-cyclodextrin (Sumitomo, S.A.) were used as stabilizing agents. Olive oil (low acidity) was purchased at a local market. Substrates for esterification reactions (*n*-butanol and butyric acid) were from Merck and were dehydrated with 0.32-cm molecular sieves (aluminum sodium silicate, type 13 X, BHD Chemicals, Toronto, Canada). Solvents were standard laboratory grade and other reagents were purchased either from Aldrich (Milwaukee, WI) or Sigma.

Immobilization of Lipase on CPS

Lipase was immobilized by being covalently bound on CPS previously treated with γ-APTS, followed by the reaction of the pretreated beads with glutaraldehyde solution, according to the procedure previously described *(6)*. Suitable amounts of enzyme (0.1–0.3 g) were dissolved in 10 mL of distilled water and mixed with the support (1 g, dry wt) under low stirring for 2 h at room temperature. Proteins (albumin and lecithin) and organic molecules (β-cyclodextrin or PEG-1500) were added together with the enzyme solution at a fixed amount (5 mg/g of support, 200 µL of aqueous solution containing 50 mg of additive/mL). Next, 10 mL of hexane was added, and the mixture of enzyme, support, and additive was incubated overnight at 4°C. The immobilized lipase was filtered (Whatman filter paper 41) and thoroughly rinsed with hexane. Analyses of hydrolytic activities carried out on the lipase loading solution and immobilized preparations were used to determine the coupling yield (η%) according to Eq. 1:

$$\eta\ (\%) = \frac{Uads}{U_0} \times 100 \qquad (1)$$

in which $Uads$ is the total activity recovered on the support and U_0 is the units offered for immobilization.

Hydrolytic Activities

Hydrolytic activities of free and immobilized lipase were assayed by the olive oil emulsion method *(6)*. The substrate was prepared by mixing 50 mL of the olive oil with 50 mL of emulsification reagent. The reaction mixture consisting of 5 mL of the emulsion, 2 mL of 100 mM sodium phosphate buffer (pH 7.0), and either free (1 mL of lipase, 5 mg/mL) or immobilized (100–250 mg) lipase was incubated for 5 min at 37°C. The reaction was stopped by the addition of 10 mL of acetone-ethanol solution (1:1). The liberated fatty acid was titrated with 25 mM potassium hydroxide solution using phenolphthalein as an indicator. One unit of enzyme activity was defined as the amount of enzyme that produces 1 µmol of free fatty acid/min under the assay conditions.

Protein Assay

Protein was determined according to Bradford's *(14)* method using BSA as a standard. The amount of bound protein was determined indirectly by the difference between the amount of protein introduced into the coupling reaction mixture and the amounts of protein in the filtrate and washing solutions.

Esterification Reactions

Reaction systems consisted of heptane (10 mL), *n*-butanol (250 mM), butyric acid (250 mM), and immobilized lipase (0.75 g, dry wt). The mixture was incubated at 37°C for 24 h with continuous shaking at 150 rpm. The remaining butanol and the product formed were determined by gas chromatography using a 6-ft 5% DEGS on a Chromosorb WHP 80/10 mesh column (Hewlett Packard, Palo Alto, CA) and hexanol as the internal standard. Esterification activity was expressed as micromoles of butyl butyrate formed per minute per gram of dry support.

Operational Stability

The operational stability was assayed by the immobilized lipase in successive batches performed under the same conditions as described for esterification reactions. Twenty-four hours after starting each batch, the immobilized lipase was removed from the reaction medium and rinsed with heptane, to extract any substrate or product eventually retained in the matrix. One hour later (length of time required for the solvent to evaporate), the immobilized derivative was introduced into a fresh medium.

Experimental Design and Statistical Analyses

The selected variables were the additives (qualitative variable) and lipase loading (quantitative variable). Two 2^2 full factorial designs with two replicates at the center point were employed. For qualitative variables, in case of protein additives, (+) represents the presence of albumin and (–) represents the presence of lecithin, while in the case of the organic additives (–) represents the presence of PEG PW-1500 and (+) represents the presence of β-cyclodextrin. For the quantitative variable, (+) is the high level (450 U/g of support) and (–) is the low level (150 U/g of support). Four experiments were carried out at the center point level, coded as 0, for estimation of experimental error. The immobilization yield was used as response. Analysis of the results was based on data generated by the software Statistica (version 5.0).

Results and Discussion

Effect of Additive and Lipase Loading on Immobilization Yield

In the first factorial design, proteins (albumin and lecithin) were evaluated as qualitative factors. The design of this experiment is given in Table 1, together with the experimental results. Both additive (X_1) and lipase loading (X_2) affected the coupling yield. When lecithin was used as additive and lipase loading increased from the low level (150 U/g of support) to the high level (450 U/g of support) (runs 1 and 3), the coupling yield decreased from 18.0 to 10.8%. The replacement of lecithin for albumin (runs 2 and 4) promoted an increase in the yield and that effect was more pronounced at the high level than at the low level (32.2 vs 23.6%).

The statistical analysis for the response evaluated is summarized in Table 2. According to the student's t-test values, the additive (X_1) and the interaction between additives and lipase loading (X_1X_2) showed a significant effect at 95% confidence level whereas the lipase loading (X_2) did not present a significant influence at the same confidence level. In the experimental range studied, the best stabilizing effect was observed when albumin was used as the additive. Regardless of the lipase loading, runs using lecithin gave much lower coupling yields.

Based on the response evaluated, a mathematical model was developed. Table 3 gives the analysis of variance (ANOVA) for the model used

Table 1
Experimental Design and Coupling Yields
According to First 2^2 Full Factorial Design

Run no.	X_1	X_2	Additive	Lipase loading (U/g)	Coupling yield (%)
1	−1	−1	Lecithin	150	18.0
2	+1	−1	Albumin	150	23.6
3	−1	+1	Lecithin	450	10.8
4	+1	+1	Albumin	450	32.2
5	−1	0	Lecithin	300	12.5
6	+1	0	Albumin	300	26.9
7	−1	0	Lecithin	300	18.2
8	+1	0	Albumin	300	24.8

Table 2
Estimated Effects, Standard Errors, and Student's t-Test for Coupling Yield
According to 2^2 Full Factorial Design

Source	Effect	Standard error	t Value
Mean	20.88	±0.85	24.56[a]
X_1 (additive)	12.01	±1.71	7.02[a]
X_2 (lipase loading)	0.64	±2.42	0.26
X_1X_2	7.91	±2.42	3.27[a]

[a]$p < 0.05$.

Table 3
ANOVA for Model Regression Representing Coupling Yields
for Lipase on CPS in Presence of Proteins
According to 2^2 Full Factorial Design[a]

Source	Sum of square	Degrees of freedom	Mean square	F value	p Value
X_1	288.48	1	288.48	49.29	0.002
X_2	0.42	1	0.42	0.07	0.806
X_1X_2	62.49	1	62.49	10.68	0.031
Total	23.41	4	5.85	—	—

[a]$R^2 = 0.94$.

to estimate the coupling yield as a function of the experimental factors (X_1) and (X_2) and their interaction (X_1X_2). These data suggest that the coupling yield depends on the type of additive and the lipase loading. The existence of an interaction confirms published data in relationship to the stabilizing effects given by nonenzymatic protein, minimizing in this way the lipase denaturation during its fixation onto solid supports *(9)*.

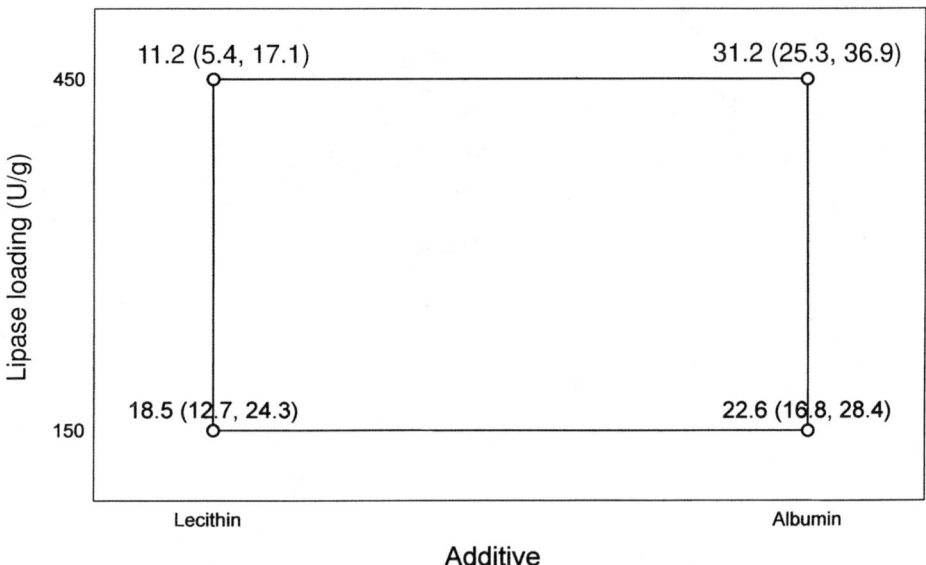

Fig. 1. Response predicted for lipase immobilization yields using proteins (albumin and lecithin) as additives.

The mathematical model representing the process in the experimental range studied (Eq. 2) was found to be appropriate, and the determination coefficient ($R^2 = 0.94$; Table 3) indicated that 94% of the variability in the coupling yield could be explained by this model.

$$\hat{y} = 20.87 + 6.0X_1 + 0.32X_2 + 3.95X_1X_2 \tag{2}$$

in which $\hat{y}$ is the value predicted for the coupling yield, and X_1 and X_2 are the coded values for additive and lipase loading, respectively.

Data obtained (Table 1) were compared with predicted values by the model (Fig. 1) and demonstrated that the model well represents the immobilization of lipase on CPS in the presence of protein as additives.

The response surface and the contour plot are shown in Fig. 2A and Fig. 2B, respectively. In agreement with the response surface, maximum coupling yield (31%) could be achieved working at high lipase loading (450 U/g of support) using albumin as an additive. The contour plot indicated the behavior for the variable response for future experiments, in which it can be statistically ensured that there is a need to increase the lipase loading if albumin is the additive of choice.

In the second experimental design, the influence of organic molecules (β-cyclodextrin or PEG-1500) on the coupling yield was studied. Table 4 indicates the experimental matrix together with the responses. The coupling yields varied from 7.8 to 59.5%, and the highest value was attained when a minimum level of lipase loading (95 U/g of support) was used in the presence of PEG-1500 (run 2). A decrease of 30% on the immobilization yield was attained at a lipase loading of 450 U/g of support (run 4), showing a negative effect of this additive for high lipase loadings. An inverse

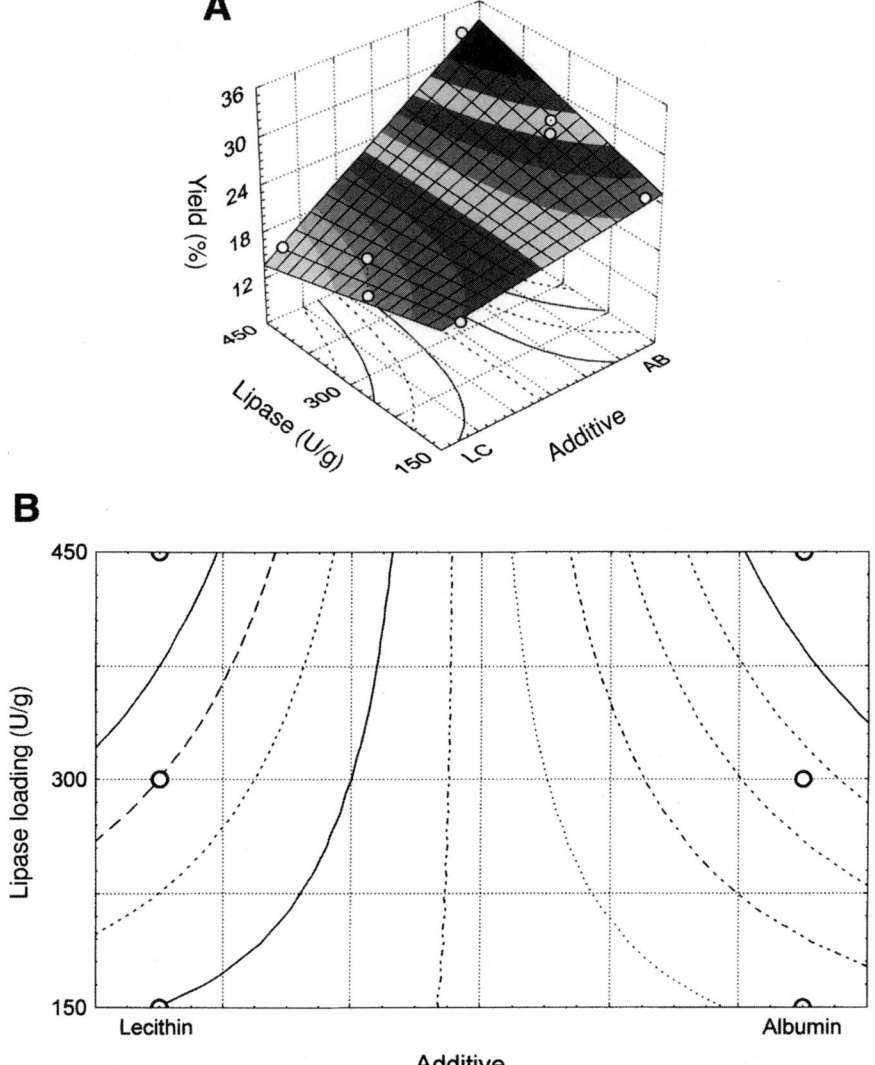

Fig. 2. **(A)** Surface response and **(B)** contour plot described by Eq. 2 using proteins (albumin [AB] and lecithin [LC]) as additives.

behavior was observed for β-cyclodextrin; the immobilization yield increased twice (7.8–14.3%) as the lipase loading increased from 95 to 285 U/g of dry support (runs 1 and 3).

Statistical analysis (Table 5) shows a significant effect for the variable additive (X_1) at a 95% confidence level, but not for lipase loading (X_2) and its interaction (X_1X_2).

The statistical model representing the immobilization coupling ($\hat{y}$) as a function of additive (X_1) and lipase loading (X_2) and their interaction (X_1X_2) can be expressed by Eq. 3:

$$\hat{y} = 26.36 + 14.95X_1 - 2.77X_2 - 6.02X_1X_2 \tag{3}$$

Table 4
Experimental Design and Immobilization Yields
According to Second 2^2 Full Factorial Design

Run no.	X_1	X_2	Additive	Lipase loading (U/g)	Coupling yield (%)
1	−1	−1	β-Cyclodextrin	150	7.8
2	+1	−1	PEG-1500	150	59.5
3	−1	+1	β-Cyclodextrin	450	14.3
4	+1	+1	PEG-1500	450	41.9
5	−1	0	β-Cyclodextrin	300	11.0
6	+1	0	PEG-1500	300	29.6
7	−1	0	β-Cyclodextrin	300	12.6
8	+1	0	PEG-1500	300	34.2

Table 5
Estimated Effects, Standard Errors, and Student's *t*-Test
for Immobilization Yield According to 2^2 Full Factorial Design

Source	Effect	Standard error	*t* Value
Mean	26.35	±3.38	7.79[a]
X_1 (additive)	29.90	±6.76	4.42[a]
X_2 (lipase loading)	−5.53	±9.56	−0.58
X_1X_2	−12.03	±9.56	−1.25

[a]Significant at 95% confidence level ($t = 4.30$).

Table 6
ANOVA for Model Regression Representing Coupling Yields
for Lipase on CPS in Presence of Organic Molecules
According to 2^2 Full Factorial Design[a]

Source	Sum of square	Degrees of freedom	Mean square	F value	*p* Value
X_1	1788.32	1	1788.32	19.55	0.011
X_2	30.63	1	30.63	0.33	0.59
X_1X_2	144.84	1	144.84	1.58	0.27
Total	365.89	4	91.47	—	—

[a]$R^2 = 0.84$.

The validity of this model was verified by the ANOVA (Table 6) where it can be observed that the regression was statistically significant ($p < 0.05$) with a determination coefficient of $R^2 = 0.84$. The parameters of the model indicated that the additive was the variable that showed the highest influence on the response variable. This was confirmed by the Fisher's F test, showing a value >2.0, which gave global validity of the resulting equation. According to this equation, the value predicted for immobilization cou-

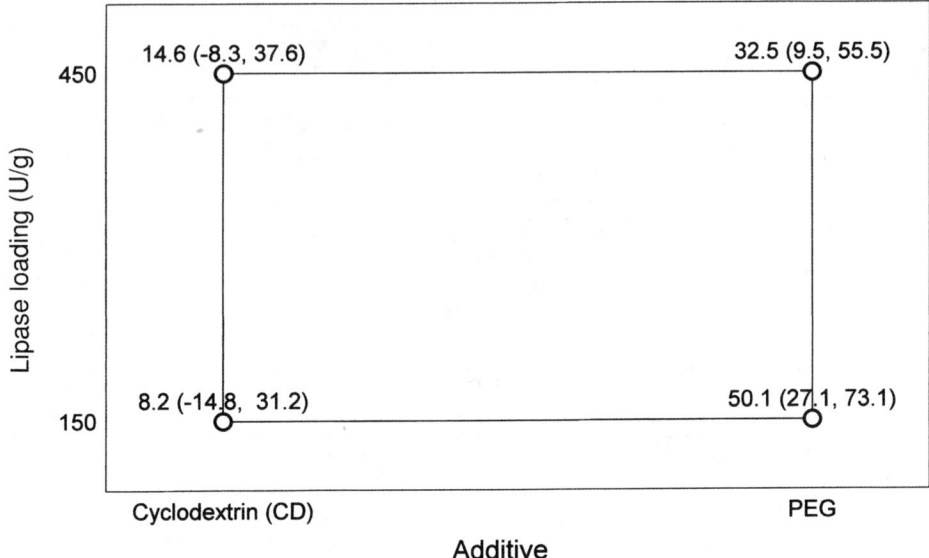

Fig. 3. Response predicted for lipase immobilization yields using organic molecules (β-cyclodextrin and PEG-1500) as additives.

pling was found to be 50% for lipase loading of 150 U/g of dry support in the presence of PEG-1500. The other predicted values for coupling yields according to the model proposed are shown in Fig. 3.

Figure 4 shows that it is interesting to work with β-cyclodextrin at a high level of lipase loading (450 U/g of support) while with PEG-1500 an inverse behavior was observed, a low level of lipase loading is sufficient to attain high immobilization yield.

Based on these results, a defined experimental behavior was observed, demonstrating in general terms that the use of PEG-1500 increases significantly the lipase immobilization yield on CPS. The contour plot justifies the need for performing an additional statistical design in order to optimize the conditions for lipase immobilization on CPS in the presence of PEG-1500. Further investigation of this optimization is under progress.

Comparison of Performance of Lipase Immobilized on Silica With and Without Additives in Hydrolysis Reactions

To allow a better evaluation of all tested additives, immobilization runs (controls) at lipase loading varying from 150 to 450 U/g of dry support without additives were also carried out. Figure 5 shows the hydrolytic activities for resulting derivatives (controls) together with the immobilization preparations with additives. Among the tested additives, lecithin was less effective than albumin, giving an immobilized lipase preparation with hydrolytic activity similar to that obtained without additive (controls).

β-Cyclodextrin showed positive and negative effects, depending on the lipase loading. The positive effect was observed for high lipase loading

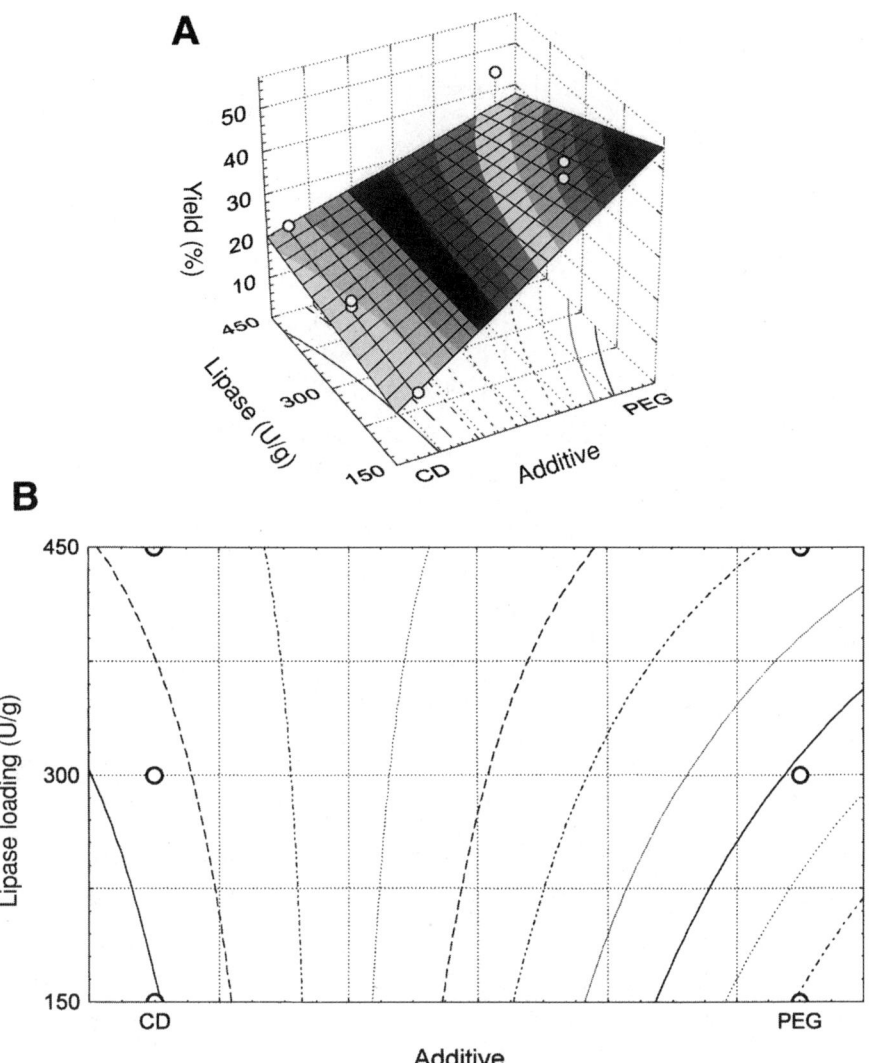

Fig. 4. **(A)** Surface response and **(B)** contour plot described by Eq. 3 using organic molecules (β-cyclodextrin [CD] and PEG-1500) as additives.

(450 U/g of support), and an activity of 70.8 U/mg of supports was obtained. For the other runs (low and medium levels), β-cyclodextrin provided immobilized derivatives with hydrolytic activities lower than those of the controls.

Albumin presented a positive effect for all lipase loadings, and a maximum value of 153.2 U/mg of dry support was obtained at a lipase loading of 450 U/g of support. The preparation of immobilized lipase that showed the highest hydrolytic activity (193 U/mg) was produced in the presence of PEG-1500, confirming the efficiency of this additive, as already reported by several researchers *(5,9)*. This result is quite effective when

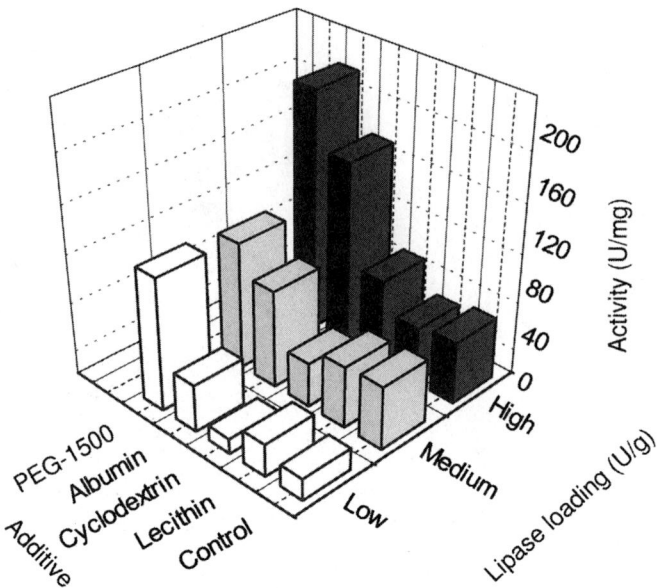

Fig. 5. Comparative hydrolytic activities for CPS immobilized lipase with and without additives for different lipase loadings.

compared to the control (50 U/mg), whose value was approximately four times lower than that obtained in the presence of PEG-1500.

Comparison of Performance of Lipase Immobilized on Silica With and Without Additives in Esterification Reactions

In another set of experiments, immobilized lipases were prepared from CPS such that the lipase loading was kept constant at 300 U/g. The esterification activity was then measured as described. A plot of the data is shown in Fig. 6A–E. Although some additives were not able to increase hydrolytic activities, additive treatment of lipase immobilized on CPS improved the reaction rate and the ester yields in all cases.

With the control (CPS immobilized lipase without additive) the esterification activity was 161 µmol/(g·min) (Fig. 6E). However, when the immobilized lipase was prepared in the presence of lecithin or albumin, the esterification activities increased up to twofold (Fig. 6A,B). CPS lipase derivatives with β-cyclodextrin or PEG-1500 also had superior behavior to that of the control, by promoting an increase of 1.6- and 2.7-fold, respectively, on esterification activity. Therefore, all additives exerted a positive influence on the esterification activities by increasing up to 2.7-fold the activity attained by the control. This beneficial effect is probably owing to the dispersing effects of the enzyme molecules that facilitate mass transfer when additives are used together with the immobilizing matrices. It is also probable that the additives improved the esterification activity by better preserving the native structure of the enzyme in organic media. When the

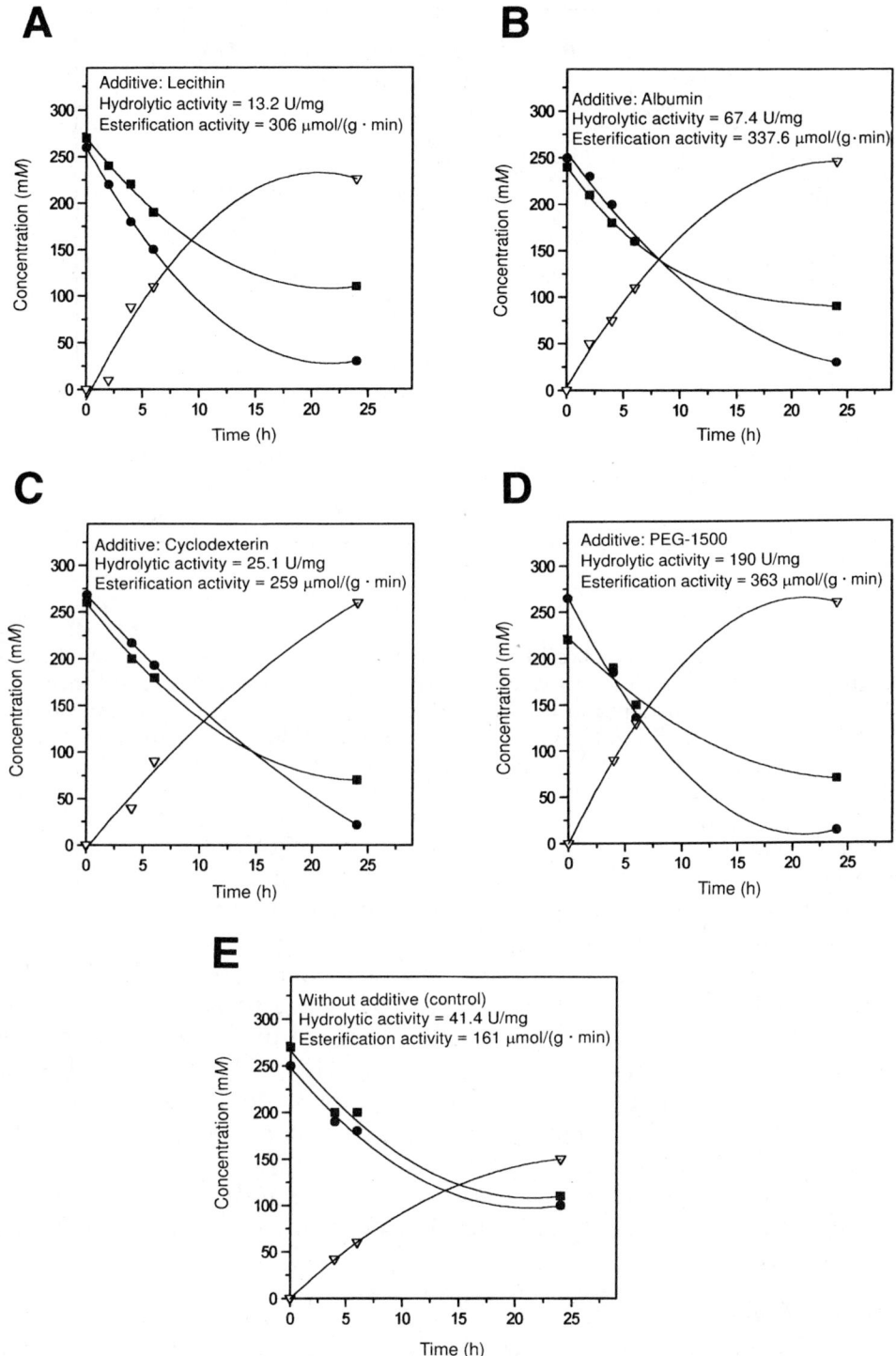

Fig. 6. Synthesis of butyl butyrate (▽) from butanol (●) and butyric acid (■) using lipase immobilized on CPS with additives **(A–D)** and without additive **(E)**. Procedure conditions are as described in Materials and Methods.

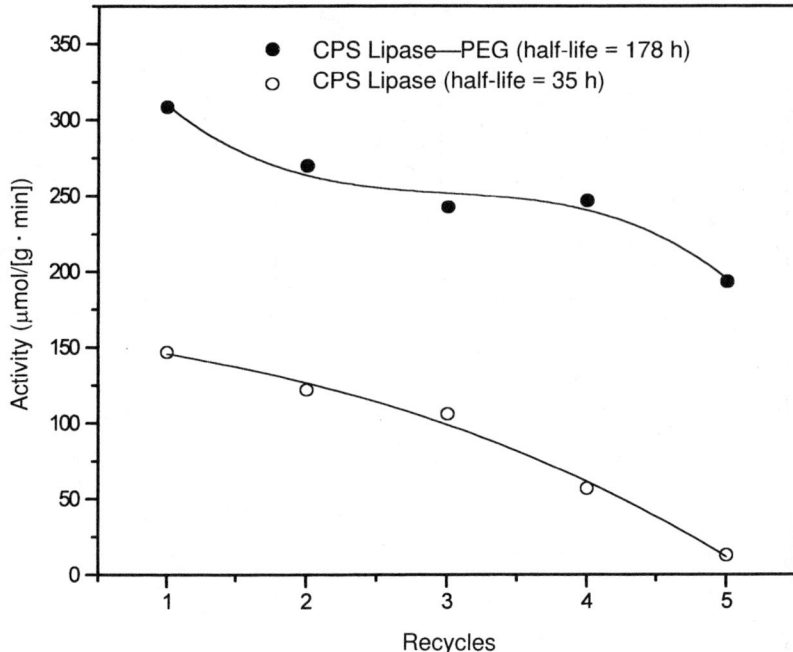

Fig. 7. Batch operational stability test of immobilized lipase derivatives. Esterification assay was carried out with substrate containing 250 m*M* butanol and 250 m*M* butyric acid in heptane. Initial esterification activities were 145 μmol/(g·min) for the lipase CPS (○) without PEG-1500 (control) and 310 μmol/(g·min) for the lipase-CPS with PEG-1500 (●).

water shell around the catalyst was maintained, a significant increase in the accessibility of active sites to the substrate was observed. Additionally, treatment of the immobilized derivatives also promoted a shift in the chemical equilibrium toward synthesis.

The best performance was obtained with PEG-1500 treatment (Fig. 6D). An ester yield of 56% could be obtained in 8 h using lipase immobilized on CPS with PEG-1500, whereas lipase-CPS lacking additive produced an ester yield of 50% in 15 h (Fig. 6E). These results might be explained by the fact that PEG is highly hygroscopic and water in the system might bind to the polyol. In this way, the nucleophilic attack by water was minimized, thus optimizing the butyl butyrate synthesis.

These results confirm data in the literature that indicate a considerable increase in catalytic activity of immobilized lipase by treating the support with additives *(5,9)*.

Recycle Potential

The operational stability of the most active immobilized derivative (CPS-lipase with PEG-1500) was determined during successive batch reactions at 37°C for 24 h using the esterification of *n*-butanol with butyric acid as a model, following methodology previously described *(6)*. Average values obtained in the operational stability runs are shown in Fig. 7 together with data obtained by the control.

While the lipase immobilized lacking PEG-1500 showed similar behavior to that previously attained (6) revealing a half-life of 35 h, the lipase immobilized on CPS with PEG-1500 gave a stable preparation and high esterification activity could be maintained for more than 120 h (five sequential batch reactions), which revealed a half-life of 178 h. As already emphasized, additives to the enzyme preparation are thought to affect stability primarily by affecting the distribution of water around the protein. The water content of the immobilized system has been found to be a decisive factor for the long-term stability of immobilized lipases, even for commercially available preparations, such as lipozyme (15).

Conclusion

The method for preparing the biocatalyst can influence the catalytic process. In agreement with previous studies (6), *C. rugosa* lipase can be immobilized with high activity on silanized CPS activated with glutaraldehyde. The present work aimed at improving the performance of the immobilized form in long-term operation. Four additives were tested in the immobilization step in order to select the most active derivative for both hydrolysis and esterification reactions. The methodology of experimental design was used to select the most efficient additive considering the coupling yield as a response variable.

The effect of immobilization with the additives on the esterification reaction was exceptionally large, compared with the effects exhibited by the additives on hydrolysis. This enhancement could be attributed to distinct effects of additives. It appears that a certain change in the protein conformation took place when additive bound to the biocatalyst or the retention of the water shell around the catalyst gave additional stability for the enzyme. In addition, by controlling the water system, a shift in the thermodynamic equilibrium toward the esterification reactions also could be obtained.

Among the tested additives, the most promising result was obtained with PEG-1500, which produced preparations with high hydrolysis (198 U/mg) and esterification activities (363 μmol/[g·min]). The performance of this derived in both hydrolysis and esterification reactions was twice superior to that attained with the lipase immobilized on CPS without additive. The half-life (178 h) of the lipase immobilized on CPS with PEG-1500 was found to increase fivefold when compared with previously published results.

Acknowledgments

We wish to acknowledge financial assistance from Fundação de Amparo a Pesquisa do Estado de São Paulo (FAPESP), Conselho Nacional de Desenvolvimento Científico e Tecnológico (CNPq), and Coordenação de Aperfeiçoamento de Pessoal de Nível Superior (CAPES) (Brazil).

References

1. Faber, K. (1997), *Biotransformation in Organic Chemistry: A Textbook*, 3rd ed., Springer-Verlag, Berlin.
2. Yahya, A. R. M., Anderson, W. A., and Moo-Young, M. (1998), *Enzyme Microb. Technol.* **23,** 438–450.
3. Reslow, M., Adlercreutz, P., and Mattiason, B. (1988), *Eur. J. Biochem.* **172,** 573–578.
4. Balcão, V. M., Paiva, A. L., and Malcata, F. X. (1996), *Enzyme Microb. Technol.* **18,** 392–416.
5. Bosley, J. A. and Peilow, A. D. (1997), *J. Am. Oil Chem. Soc.* **74,** 107–111.
6. Soares, C. M. F., De Castro, H. F., De Moraes, F. F., and Zanin, G. M. (1999), *Appl. Biochem. Biotechnol.* **77/79,** 745–758.
7. Reetz, M. T., Zonta, A., and Simpelkamp, J. (1996), *Biotechnol. Bioeng.* **49,** 527–534.
8. Wehtje, E., Adlercreutz, P., and Mattiasson, B. (1993), *Biotechnol. Bioeng.* **41,** 171–178.
9. Rocha, J. M. S., Gil, M. H., and Garcia, F. A. P. (1998), *J. Biotechnol.* **66,** 61–67.
10. Triantafyllou, A. O., Wehtje, E., Adlercreutz, P., and Mattiasson, B. (1995), *Biotechnol. Bioeng.* **45,** 406–414.
11. Triantafyllou, A. O., Wehtje, E., Adlercreutz, P., and Mattiasson, B. (1997), *Biotechnol. Bioeng.* **54,** 67–76.
12. Box, G. E. P., Hunter, W. G., and Hunter, J. S. (1978), *Statistics for Experimenters: An Introduction to Design, Data Analysis and Model Building*, Wiley & Sons, New York.
13. Szejtli, J. (1988), *Cyclodextrin Technology*, Kluwer Academic, Boston.
14. Bradford, M. M. A. (1976), *Anal. Biochem.* **72,** 248–254.
15. De Castro, H. F., Anderson, W. A., Legge, R. L., and Moo-Young, M. (1992), *Indian J. Chem.* **31,** 891–895.

Screening of Variables in β-Xylosidase Recovery Using Cetyl Trimethyl Ammonium Bromide Reversed Micelles

FRANCISLENE-ANDRÉIA HASMANN,[1] ADALBERTO PESSOA, JR.,[2] AND INÊS-CONCEIÇÃO ROBERTO*,[3]

[1]IPT/Butantan/USP, PO Box 66083, São Paulo/SP, Brazil;
[2]Biochemical and Pharmaceutical Department/FCF/USP, PO Box 66083, São Paulo/SP, Brazil; and [3]Department of Biotechnology, Faculty of Chemical Engineering of Lorena, PO Box 116, Lorena/SP, Brazil, E-mail: ines@debiq.faenquil.br

Abstract

β-Xylosidase recovery by micelles using cetyl trimethyl ammonium bromide (CTAB) cationic surfactant was verified under different experimental conditions. A 2^{5-1} fractional factorial design with center points was employed to verify the influence of the following factors on enzyme extraction: pH (x_1), CTAB concentration (x_2), electrical conductivity (x_3), hexanol concentration (x_4), and butanol concentration (x_5). Statistical analysis of the results shows that of the five variables studied only hexanol and electrical conductivity did not have significant effects on the recovery of β-xylosidase. The other factors had significant effects in increasing order: $(x_1) > (x_2) > (x_5)$. The model predicts a recovery value of about 45%, which is similar to that obtained experimentally (43.5%).

Index Entries: Reversed micelles; β-xylosidase; liquid-liquid extraction; statistical design.

Introduction

Xylans are major components of the hemicellulosic fraction of lignocellulosic biomass and their hydrolysis can be performed using xylanases *(1)*. The enzyme complex is composed of endoxylanases, which cleave internal xylosidic linkages on the xylan backbone, and β-xylosidase, which releases xylosyl residues by endwise attack of xylooligosaccharides *(1,2)*.

*Author to whom all correspondence and reprint requests should be addressed.

Plants are the source of renewable natural fibers used in the paper and textile industries. Normally, the cellulosic fibers are bleached by chemical processes, which are harmful to the environment. Environmental problems could be avoided by replacing the chemical bleaching processes with biologically oriented processes, such as the use of xylanase *(3–5)*. In fact, using xylanases facilitates pulp bleaching, lowers chlorine consumption, and reduces toxic discharges *(6,7)*. In addition, xylanases also can be employed in the clarification of beer and juice *(6–8)* and in baking processes, increasing the loaf volume *(8)*. At present, β-xylosidase is produced by *Penicillium janthinellum* from processed or refined substrates, such as sugars, cellulose, and xylan.

Some studies on xylanolytic complex recovery have been conducted employing techniques such as ethanol and salt precipitation, which have industrial applications, and liquid-liquid extraction by aqueous two-phase systems *(9–12)*. However, these purification techniques did not improve the purification factor of the enzyme satisfactorily.

Reverse micellar systems have been extensively studied as a technique for the extraction and purification of proteins *(13–15)*. This technique allows the recovery and concentration of proteins from a dilute aqueous solution containing other bioproducts *(13–16)*. A reverse micellar system consists of aggregates of surfactant molecules containing an inner water core dispersed in an organic solvent medium. The polar microenvironment inside the reverse micelle permits the solubilization of protein while maintaining its native structure. The overall liquid-liquid extraction process by reverse micelles is conducted in two fundamental steps: a forward extraction, by which a protein is transferred from an aqueous solution to a reverse micellar organic phase; and a back extraction, by which the protein is released from the reversed micelles and transferred to an aqueous phase, so that it can be recovered subsequently *(13–15)*. The extraction process is mainly governed by electrostatic interaction between the charged protein and the micellar wall, and protein transfer only takes place during the forward extraction, when the value of the pH of the aqueous phase is such that the net surface charge of the protein is electrically opposite of that of the surfactant head groups. Protein can also be extracted by hydrophobic interaction between the apolar regions of the molecule and the surfactant tail *(13)*. In the back-extraction, however, the pH value must allow the protein to have the same charge as the surfactant molecules and the ionic strength to be increased by the addition of salts. In this way, repulsion forces are created and the micellar diameter is diminished, causing the release of protein from the reverse micelles. Low ionic strength favors protein transfer to reverse micelles, and high values promote protein release *(17)*. This technique is therefore particularly interesting for the recovery of extracellular enzymes *(13–17)*.

The present study describes the transfer of extracellular β-xylosidase from *P. janthinellum* to a reversed micellar phase of the cetyl trimethyl ammonium bromide (CTAB) cationic surfactant and the influence of the following factors: pH, CTAB concentration, electrical conductivity, hexanol concentration, and butanol concentration. The extraction and recovery of β-xylosidase enzymic protein has been investigated with particular attention to the recovery of the enzymatic activity.

Materials and Methods

Chemicals

Birchwood 4-*O*-methyl-β-D-glucoroxylan (90% xylose) was obtained from Sigma (St. Louis, MO). The cationic surfactant CTAB was purchased from Merck (Darmstadt, Federal Republic of Germany) and used without further purification. All other chemicals were of analytical grade.

Preparation of Sugarcane Bagasse Hydrolysate

To prepare the hydrolysate for cultivation, 800 g of dry milled bagasse was mixed with 8 L of sulfuric acid solution (0.25%) and autoclaved for 45 min at 121°C. The liquid fraction was separated by filtration and the pH adjusted to 5.5 with 1.0 N NaOH.

Microorganism and Growth Conditions

The isolation of *P. janthinellum* (CRC 87M-115) from decaying wood was as described by Milagres *(18)*. The fungi, initially maintained in silica stocks and then transferred to agar slants, were cultivated at 30°C for 5 d in medium containing 2% glucose, 0.25% yeast extract, 2% concentrated salt solution (v/v) based on Vogel's *(19)*, medium and 2% agar-agar. The medium was autoclaved at 112°C for 5 min. To obtain the inoculum, the spores were suspended in water and the suspension was filtered through gauze placed in Erlenmeyer flasks. The final spore concentration was 10^5/mL.

Cultivation Medium and β-Xylosidase Production

The cultivation medium for enzyme production consisted of sugarcane bagasse hemicellulose hydrolysate supplemented with 2% concentrated salt solution (v/v) based on Vogel's medium and 0.1% yeast extract. The medium was autoclaved for 15 min at 121°C. The cultivation was carried out in Erlenmeyer flasks (125 mL) containing 25 mL of medium at an initial pH of 5.5 (uncontrolled). The flasks were agitated for 96 h (60 rpm) at 30°C.

β-Xylosidase Activity and Precipitation

β-Xylosidase activity was measured by the method described by Kumar and Ramón *(20)*. The enzyme was precipitated with ethanol at 20 and 60% (v/v). The pH value of the precipitation medium was adjusted to 4.5 by adding 1 M acetate buffer (pH 4.0) at a ratio of 9:1 (v/v). The ethanol was slowly mixed with the medium in a refrigerated bath at –4°C, and the mixture was centrifuged (2000g for 15 min) at 2°C. To prepare the aqueous phase, samples of the precipitate formed were separately solubilized in acetate buffer (0.03 M) to obtain pH 3.0 and 4.0, and in glycine/HCl buffer (0.03 M) to obtain pH 2.0. The electrical conductivity of the solubilized samples was adjusted to 4.0 mS/cm by adding NaCl.

Reversed Micellar Liquid-Liquid Extraction

Different values of CTAB concentration, electrical conductivity, hexanol and butanol concentrations, and pH were used in the enzyme extraction, according to the method of reversed micelles described by Pessoa and Vitolo *(15)* using CTAB prepared with a solution containing hexanol, isooctane, and butanol.

Experimental Design and Statistical Analyses

A 2^{5-1} fractional factorial design with center points was employed to evaluate the influence of the following factors on the enzyme extraction: pH, CTAB concentration, electrical conductivity, and hexanol and butanol concentrations (Table 1). For each of the factors, high (coded value: +1), center (coded value: 0) and low (coded value: –1) set points were selected. Extractions representing all 16 (2^{5-1}) set point combinations were made, as well as three extractions representing the center point (coded value: 0). Assays were conducted randomly. To analyze the results, STATGRAPHIC® software version 6.0 was used.

Results and Discussion

Table 1 shows the factor levels and the recovery results ($Y[\%]$) of β-xylosidase obtained from the CTAB reversed micelle extractions performed according to a 2^{5-1} fractional factorial design with center points. Experiments 4, 6, 10, 12, 14, and 16 (pH 8.0) provided recovery values >15%. On the other hand, all the experiments performed at pH 3.0 provided recovery values <8%, independently of the levels of the other factors. The highest enzyme recovery values (59.3, 32.3, and 35.3%) were obtained in experiments 4, 12, and 16, respectively, with a CTAB concentration of 0.2 M. Comparison of assays 12 and 4 shows that when butanol concentration was increased from 10 to 20% and hexanol concentration was decreased from 10 to 5%, the enzyme recovery increased by 84%.

The results show that different combinations among the variables studied (pH, CTAB concentration, hexanol and butanol concentrations, and electrical conductivity) can be better evaluated by statistical analysis.

Table 2 gives the estimated effects, standard errors, and student's *t*-test. According to the results, pH and butanol concentration had the most significant main effects (+20.26 and +8.86) on enzyme recovery ($p < 0.10$ and $R^2 = 0.95$). A significant interaction between the factors *AB*, with a high effect (+10.32), and *AE* (+10.92), can also be observed. The positive interaction means that the enzyme recovery was enhanced when high levels of pH (8.0), CTAB concentration (0.2 M), and butanol concentration (20%)

Table 1

Variables, Levels, and β-Xylosidase Recovery
Using a 2^{5-1} Statistical Factorial Design

Assay no.	pH	CTAB (M)	Electrical conductivity (mS/cm)	Hexanol (%)	Butanol (%)	Recovery (%)
1	3.0	0.1	4.0	5.0	20	0.0
2	8.0	0.1	4.0	5.0	10	0.6
3	3.0	0.2	4.0	5.0	10	0.0
4	8.0	0.2	4.0	5.0	20	59.3
5	3.0	0.1	10.0	5.0	10	7.0
6	8.0	0.1	10.0	5.0	20	18.7
7	3.0	0.2	10.0	5.0	20	1.7
8	8.0	0.2	10.0	5.0	10	0.0
9	3.0	0.1	4.0	10.0	10	7.3
10	8.0	0.1	4.0	10.0	20	17.5
11	3.0	0.2	4.0	10.0	20	2.9
12	8.0	0.2	4.0	10.0	10	32.3
13	3.0	0.1	10.0	10.0	20	1.5
14	8.0	0.1	10.0	10.0	10	18.8
15	3.0	0.2	10.0	10.0	10	0.0
16	8.0	0.2	10.0	10.0	20	35.3
17	5.5	0.15	7.0	7.5	15	5.1
18	5.5	0.15	7.0	7.5	15	4.5
19	5.5	0.15	7.0	7.5	15	9.5

Table 2

Estimated Effects, Standard Errors, and Student's *t*-Test
for 2^{5-1} Factorial Design

Variable	Estimated effect	Standard error	*t*-Value
A (pH)	20.26	±3.117	6.50[a]
B (CTAB)	7.50	±3.117	2.41[b]
C (electrical conductivity)	−4.61	±3.117	1.48
D (hexanol)	3.53	±3.117	1.13
E (butanol)	8.86	±3.117	2.84[b]
AB	10.32	±3.117	3.31[a]
AC	−4.61	±3.117	1.48
AD	2.79	±3.117	0.89
AE	10.92	±3.117	3.50[a]
BC	−9.77	±3.117	3.14[b]
BD	−1.17	±3.117	0.37
BE	7.88	±3.117	2.52[b]
CD	3.51	±3.117	1.12
CE	−1.02	±3.117	0.32
DE	−9.17	±3.117	2.95[b]
Average	11.68	±1.430	8.17

[a]Significant at the 5% level with 3 DF (*t* = 3.182).
[b]Significant at the 10% level with 3 DF (*t* = 2.353).

used. A possible explanation for this is the electrostatic interaction
ween the enzyme and the surfactant *(13,21)*. Concerning the interaction
ween CTAB and electrical conductivity (*BC*), the negative effect (–9.77)
ows that β-xylosidase recovery increased when a high CTAB concentra-
ion and a low electrical conductivity were used (0.2 *M* and 4 mS/cm,
respectively).

Figure 1A–E shows the response surfaces of the significant interactions
obtained after statistical analysis. Figure 1A, 1C, and 1D show response
surfaces that correlate the CTAB concentration with pH, electrical conductivity, and butanol concentration, respectively. The highest β-xylosidase
recovery value (59.3) was attained at the highest CTAB concentration
(0.2 *M*), probably because this enzyme has a high molecular weight
(~110 kDa), thus requiring reversed micelles with large diameters for its
encapsulation. This can be explained by increasing the surfactant concentration *(16)*. The influences of pH and electrical conductivity on extraction
suggest that the effect of the ionic strength primarily consists of mediating
the electrostatic interactions between the protein surface and the surfactant
head groups.

The interaction between butanol concentration (*E*) and pH (*A*)
(Fig. 1B) shows that β-xylosidase recovery increased when pH and butanol
concentration were maintained at the highest levels (8.0 and 20%, respectively). However, the interaction between butanol concentration (*E*) and
hexanol concentration (*D*) suggests that an antagonistic effect occurred,
owing to the addition of cosolvents to the cationic reversed micellar system
to increase the micellar size, which is also affected by the high ionic strength
(16,21). The Debye screening effect determines the electrical double-layer
properties adjacent to any charged surface and affects the range above
which electrostatic interactions can overcome thermal motion of the soluble
molecules. The characteristic distance for these electrostatic interactions is
the Debye length, which is inversely proportional to the square root of the
ionic strength. Thus, increased ionic strength will decrease this interaction
distance and, hence, inhibit the solubilization of the protein. In the present
study, the ionic strength also had influenced the micellar formation and
size. The extraction was conducted at a pH value higher than the enzyme
p*I* (isoelectric) value and its extraction depended only on electrostatic
interactions.

After analysis of variance (ANOVA) including the significant factors
(Table 3), a mathematical model was adjusted to the results (Eq. 1), and the
coefficient of correlation between the data and the model (R^2) was 0.90.
The validity of the model was verified by the ANOVA for total regression
(Table 4). According to the F test (F_{model} = 11.86 >> F_{table} = 3.35), the model
was highly significant. In addition, it had no lack of fit, since the differences
between the values experimentally obtained and the values predicted by
the model were owing to experimental errors:

$$Y = 11.69 + 10.13A + 3.75B + 4.43E + 5.16AB + 5.46AE - 4.89BC - 3.94BE - 4.58DE \quad (1)$$

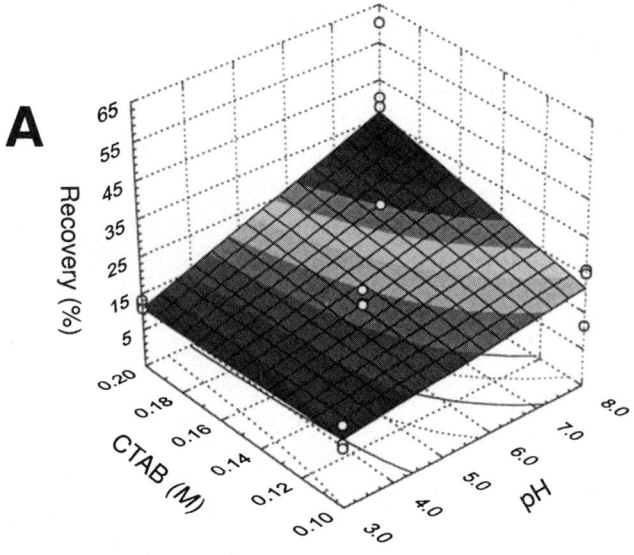

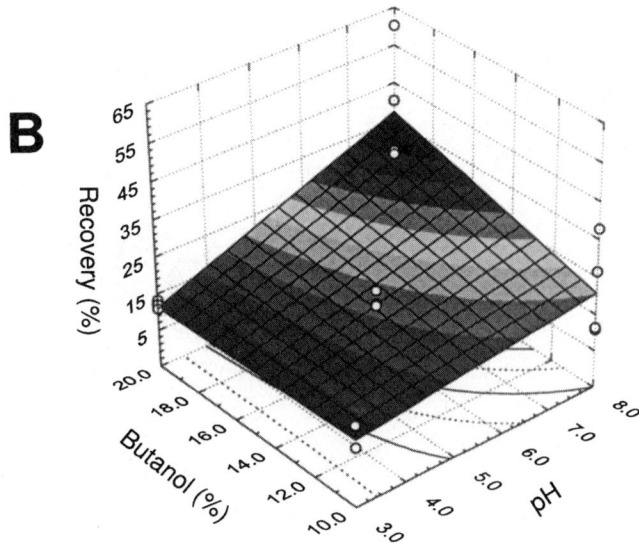

Fig. 1. **(A)** Effect of pH and CTAB concentration on β-xylosidase extraction by reversed micelles: electrical conductivity = 7.0 mS/cm, hexanol = 7.5%, and butanol = 15%; **(B)** effect of pH and butanol concentration on β-xylosidase extraction by reversed micelles: electrical conductivity = 7.0 mS/cm, hexanol = 7.5%, and CTAB = 0.15*M* (*continued on next page*).

in which *Y* is the β-xylosidase recovery (%), *A* is the pH, *B* is the CTAB concentration (*M*), *C* is the electrical conductivity (mS/cm) *D* is the hexanol concentration (%), and *E* is the butanol concentration (%).

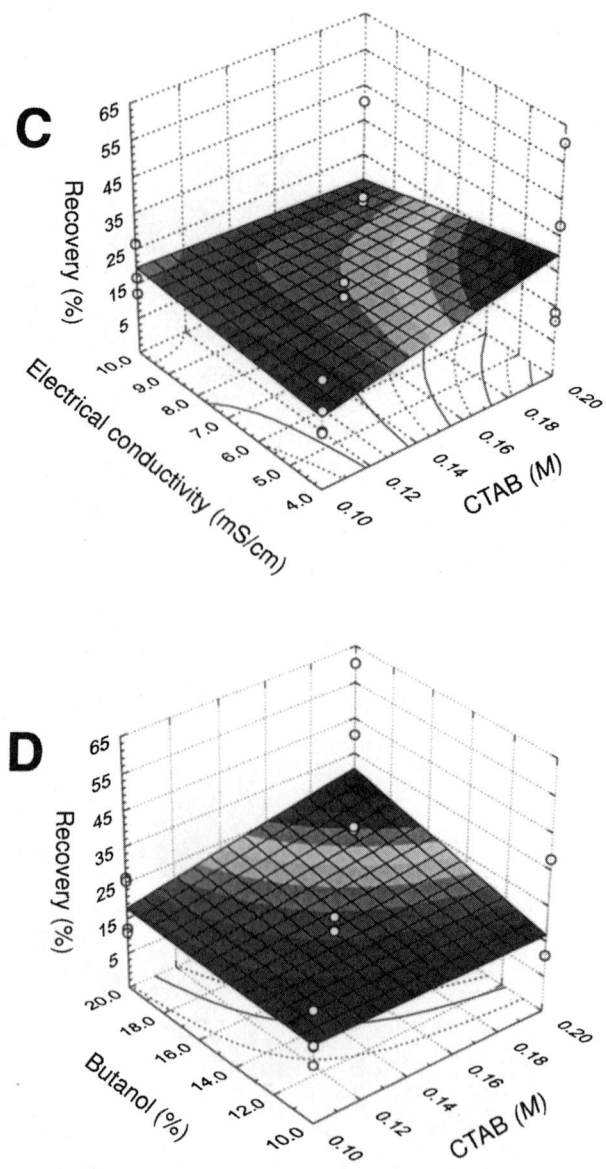

Fig. 1. *(continued)* **(C)** effect of electrical conductivity and CTAB concentration on β-xylosidase extraction by reversed micelles: pH = 5.5, hexanol = 7.5%, and butanol = 15%; **(D)** effect of CTAB concentration and butanol concentration on β-xylosidase extraction by reversed micelles: electrical conductivity = 7.0 mS/cm, pH = 5.5, and hexanol = 7.5% *(continued on next page).*

Equation 1 predicts a 44% β-xylosidase recovery (Y) under the following conditions: pH = 8.0, electrical conductivity = 4.0 mS/cm, CTAB concentration = 0.2 M, hexanol concentration = 5%, and butanol concentration = 20%. The result experimentally obtained (about 39%) strengthens the promise of validity of the model for interpolations.

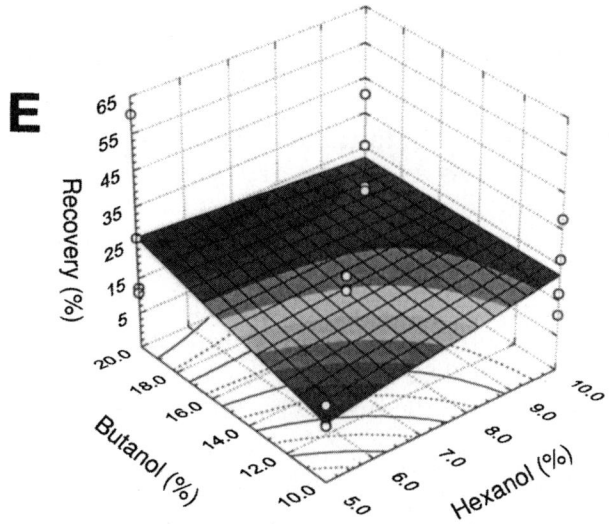

Fig. 1. *(continued)* **(E)** effect of butanol concentration and hexanol concentration on β-xylosidase extraction by reversed micelles: electrical conductivity = 7.0 mS/cm, pH = 5.5, and CTAB = 0.15M.

Table 3
ANOVA of Significant Factors

Source of variation	Sum of squares	Degrees of freedom	Mean square	F Values	p Values
A (pH)	1642.48	1	1642.48	218.49	0.004[c]
B (CTAB)	225.53	1	225.53	30.00	0.032[c]
E (butanol)	313.91	1	313.91	41.76	0.023[c]
AB	426.32	1	426.32	56.71	0.017[c]
AE	477.31	1	477.31	63.50	0.015[c]
BC^a	382.11	1	382.11	50.83	0.019[c]
BE	248.30	1	248.30	33.03	0.029[c]
DE^b	336.26	1	336.26	44.73	0.021[c]
Lack of fit	411.98	8	51.50	6.85	0.133
Pure error	15.04	2	7.52		
Total	4479.23	18			

[a]C, electrical conductivity.
[b]D, hexanol.
[c]Significant at the 5% level R^2 = 0.90.

Table 4
ANOVA for Total Regression[a]

Variation	Sum of squares	Degrees of freedom	Mean square	F Value
Model	4052.22	8	506.53	11.86
Error	427.02	10	42.70	—
Total	4479.24	18		

[a]R^2 = 0.90; F = 3.35.

Conclusion

A CTAB/isooctane/hexanol and butanol reversed micellar system was efficiently used as a recovery method of β-xylosidase from *P. janthinellum*. The recovery was controlled by electrostatic interactions. The response surface methodology helped (qualitatively) us to understand the mechanism of enzyme extraction. The recovery value obtained experimentally (39%) was very close to one predicted by the model (44%). This demonstrates that the use of reversed micellar solutions as extractants is an efficient way to recover the enzyme without reducing its activity, because the enzyme is not denatured by reagents.

Acknowledgments

We wish to thank Maria Eunice M. Coelho for revising the manuscript. This work was supported by a fellowship from FAPESP (Brazil). We also acknowledge the financial support of CAPES and CNPq.

References

1. Biely, P. (1985), *Trends Biotechnol.* **3**, 286–290.
2. Wong, K. K. Y., Tan, L. U. L., and Saddler, J. N. (1988), *Microbiol. Rev.* **52**, 305–317.
3. Durán, N., Milagres, A. M. F., Esposíto, E., Curotto, M. E. S., Carvalho, S. M. S., Femandes, E., Aguirre, C., and Teixeira, O. C. C. (1995), *Appl. Biochem. Biotechnol.* **53**, 55–162.
4. Milagres, A. M. F. and Prade, R. A. (1994), *Enzyme Microb. Technol.* **16**, 627–631.
5. Yang, J. L. and Eriksson, K. E. L. (1992), *Holzforschung* **46**, 481–488.
6. Costa, S. A., Pessoa, A. Jr., and Roberto, I. C. (1998), *Appl. Biochem. Biotechnol.* **70/72**, 629–639.
7. Brandani, V., Di Giacomo, G., and Spera, L. (1994), *Process Biochem.* **29**, 363–367.
8. Mutsaers J. H. G. M. (1991), in *Xylans and Xylanases International Symposium*, Elsevier, Amsterdam, p. 48.
9. Cortez, E. V. and Pessoa, A. Jr. (1999), *Process Biochem.* **35**, 271–283.
10. Pessoa, A. Jr. and Vitolo, M. (1998), *Process Biochem.* **33**, 291–297.
11. Costa, A. S., Pessoa, A. Jr., and Roberto, I. C. (2000), *J. Chromatogr. B* **743**, 339–348.
12. Hasmann, F. A., Pessoa, A. Jr., and Roberto, I. C., (2000), *Appl. Biochem. Biotechnol.* **84–86**, 1101–1111.
13. Rodrigues, E. M. G., Pessoa, A. Jr., and Milagres, A. M. F. (1999), *Appl. Biochem. Biotechnol.* **77–79**, 779–788.
14. Hasmann, F. A., Pessoa, A. Jr., and Roberto, I. C. (1999), *Biotechnol. Techn.* **13**, 239–242.
15. Pessoa, A. Jr. and Vitolo, M. (1991), *Biotechnol. Techn.* **11**, 421, 422.
16. Krei, G. A. and Hustedt, H. (1992), *Chem. Eng. Sci.* **47**, 99–111.
17. Pessoa, A. Jr. (1995), PhD thesis, University of São Paulo, São Paulo, Brazil.
18. Milagres, A. M. F. (1988), MS thesis, Federal University of Viçosa, Viçosa, Brazil.
19. Vogel, H. J. (1956), *Microbial. Genet. Bull.* **13**, 113–118.
20. Kumar, S. and Ramón, D. (1996), *FEMS Microbiol. Lett.* **135**, 287–293.
21. Pires, M. J. and Cabral, J. M. S. (1993), *Biotechnol. Prog.* **9**, 647–650.

Activity of Xylose Reductase from *Candida mogii* Grown in Media Containing Different Concentrations of Rice Straw Hydrolysate

ZEA D. V. L. MAYERHOFF,[1] INÊS C. ROBERTO,*,[2]
AND TELMA T. FRANCO[1]

[1]Faculty of Chemical Engineering, State University of Campinas,
PO Box 6066, Campinas, SP, Brazil; and [2]Department of Biotechnology,
Faculty of Chemical Engineering of Lorena, PO Box 116,
SP, Brazil, E-mail: ines@debiq.faenquil.br

Abstract

Xylose reductase (XR) activity was evaluated in extracts of *Candida mogii* grown in media containing different concentrations of rice straw hydrolysate. Results of XR activity were compared to xylitol production and a similar behavior was observed for these parameters. Highest values of specific production and productivity were found for xylose reductase (35 U/g of cell and 0.97 U/[g of cell·h], respectively) and for xylitol (5.63 g/g of cell and 0.13 g/[g of cell·h]) in fermentation conducted in medium containing 49.2 g of xylose/L. The maximum value of XR:XD ratio (1.82) was also calculated under this initial xylose concentration with 60 h of fermentation.

Index Entries: *Candida mogii*; hydrolysate concentration; rice straw; xylitol; xylose reductase.

Introduction

Xylose reductase (XR) is the enzyme responsible for the first step in xylose metabolism by yeasts *(1)*. In a reaction catalyzed by this enzyme, xylose is reduced to xylitol, which can be oxidized into xylulose or released into the environment, depending on the culture conditions of the microorganism. The oxidation of this polyalcohol is catalyzed by the enzyme xylitol dehydrogenase (XD). Xylitol is a sugar alcohol of economic interest because of its dietetic and anticariogenic properties *(2,3)*. Microbial and enzymatic processes have been studied as alternatives to the chemical

*Author to whom all correspondence and reprint requests should be addressed.

process currently employed for its production *(4–8)*. These biologic processes are expected to reduce the final cost of xylitol by allowing the utilization of xylose from hydrolysates of lignocellulosic materials, thus eliminating the need for the previous purification of this sugar *(9)*. Among several residues utilized as a xylose source for xylitol production by yeasts, rice straw has been deemed a potential substrate for this bioconversion *(6,9)*. Many yeasts have the capability of efficiently producing xylitol from hemicellulose hydrolysates. In a previous study, *Candida mogii* NRRL Y-17032 was selected among 31 yeast strains as a promising xylitol producer from rice straw hemicellulose hydrolysate *(10)*. Compared to microbiologic fermentation processes, the approach by enzyme technology employing isolated XR for xylitol synthesis should make a substantial increase in productivity because mass transfer limitations are avoided in homogeneous enzyme reactors *(7)*. Taking into account that XR is an inducible enzyme, enhancing its production depends on the optimization of microorganism culture conditions. Some factors have been studied in order to evaluate their influence on XR activity *(11,12)*. In the present study, the effect of initial substrate concentration was investigated in *C. mogii* fermentations employing rice straw hydrolysate as the source of carbon.

Materials and Methods

Preparation of Hemicellulose Hydrolysate

Rice straw hemicellulose hydrolysate was obtained by acid hydrolysis of the rice straw in an AISI 316 stainless steel 350-L stirred-tank reactor. The hydrolysis was run for 20 min at 120°C using 10 mL of 0.13 M H_2SO_4 solution/g of dry matter. The obtained hemicellulose hydrolysate was collected by centrifugation and homogenized. The following components were detected: 18.1 g/L of xylose, 3.5 g/L of glucose, 2.8 g/L of arabinose, 0.025 g/L of furfural, and 0.038 g/L of hydroxymethylfurfural (HMF). This hydrolysate was concentrated under vacuum at 70°C and the xylose concentration reached 220 g/L. Next, the pH was raised with NaOH pellets up to 8.0 and then lowered to 6.0 with 72% H_2SO_4 (w/w). Each time the pH level was changed, the precipitate was removed by centrifugation (2860g for 30 min).

Microorganism and Preparation of Inoculum

C. mogii NRRL Y-17032 obtained from Northern Regional Research Laboratory (Peoria, IL) was maintained at 4°C on nutrient agar slants. Inoculum was prepared by cultivating cells in 125-mL Erlenmeyer flasks containing 25 mL of medium. The medium was composed of rice straw hemicellulose hydrolysate diluted with distilled water to provide an initial xylose concentration of 30 g/L. Loopfulls of cells were suspended in a few milliliters of distilled water, and this suspension was pipetted into the flasks. The flasks were incubated in a rotary shaker at 200 rpm and 30°C. After 48 h, cells were harvested by centrifugation. The fermentation medium contained an initial cell concentration of 1 g/L.

Fermentation Conditions

Fermentation media were composed of rice straw hemicellulose hydrolysate containing 220 g/L of xylose diluted with distilled water to provide initial xylose concentrations of 30, 40, 50, 70, and 90 g/L. Twenty-five milliliters of these media was placed into 125-mL Erlenmeyer flasks that were inoculated and then incubated at 30°C at 200 rpm. Each sample was constituted by the total volume of one Erlenmeyer flask.

Preparation of Cell-Free Extracts

Cells were harvested by centrifuging at 2860*g* for 30 min at 4°C, washed with 0.1 *M* potassium phosphate buffer (pH 7.2), and resuspended to a concentration of 15 g of dry cell weight/L with the same buffer. Cell disruption was conducted by sonication for 35 min in 1-s pulses with 1-s intervals in Sonics & Materials Disrupter equipment. Samples were centrifuged for 10 min at 6000*g* in a Jouan MR1812 centrifuge, and the cell-free extract was used for enzymatic tests.

Xylose Reductase Assay

Enzyme activities were determined spectrophotometrically by following the oxidation or reduction of the coenzyme at 340 nm. The assays were performed as described by Bolen and Detroy *(13)* except for cofactor concentration and buffer pH. The XR assay reaction mixture contained 50 m*M* potassium phosphate buffer, pH 7.2; 10 m*M* mercaptoethanol; 0.12 m*M* NADPH; 50 m*M* D-xylose; cell extract; and distilled water in a total volume of 1 mL. The reaction was started by the addition of D-xylose. XD activity was determined in a similar manner with phosphate buffer, D-xylose, and NADPH substituted by 75 m*M* Tris buffer, pH 8.6; 50 m*M* xylitol; and 0.12 m*M* NAD$^+$, respectively. One unit was defined as the amount of enzyme catalyzing the oxidation or reduction of 1 μmol of cofactor/min. Specific activities were expressed as units per gram of cell.

Analytical Methods

Cell mass was determined by using a calibration curve that correlates optical density at 600 nm and dry cell weight. Concentrations of sugars were determined by high-performance liquid chromatography (HPLC) using a Shimadzu C-R7A chromatograph equipped with a refractive index detector and a Bio-Rad Aminex HPX-87H column. Concentrations of furfural and HMF were determined by HPLC using a Waters chromatograph equipped with an ultraviolet detector and an RP18 column.

Results and Discussion

Table 1 presents the partial composition of fermentation media. Xylose concentrations were close to those expected from the dilutions, varying from 29.2 to 88.3 g/L. Concentrations of glucose and arabinose were both

Table 1
Partial Composition of Fermentation Media
with Hydrolysate Concentrations Equivalent to Different Xylose Contents

Component (g/L)	Xylose content (g/L)				
	29.2	38.6	49.3	69.8	88.3
Glucose	5.5	7.3	9.2	13.0	15.6
Arabinose	5.1	6.7	8.6	12.1	14.6
Furfural	0.017	0.021	ND[a]	ND[a]	0.028
HMF	0.026	0.032	0.042	0.062	0.076

[a]ND, not determined.

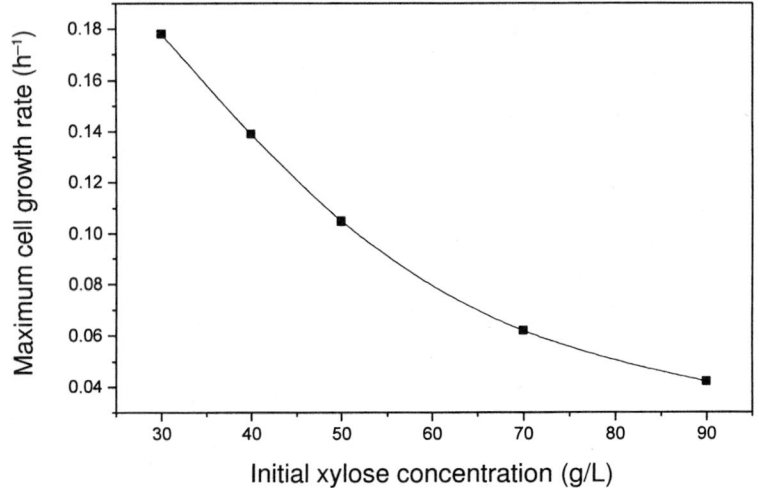

Fig. 1. Effect of hydrolysate concentration expressed as initial xylose content on maximum cell growth rates of the yeast *C. mogii*.

about 18% in relation to xylose levels. In addition to sugars, other substances could also be detected. Furfural and HMF are frequently found in hemicellulose hydrolysates *(14,15)*. Acetic acid and lignin degradation products were not analyzed and may have volatilized during vacuum concentration. Initially, we intended to study a range of hydrolysate concentrations larger than that presented. However, in another study for xylitol production employing the yeast *Candida guilliermondii* and rice straw hydrolysate *(16)*, no significant substrate consumption was detected above 90 g/L of xylose. Because the yeast *C. mogii* has presented a behavior similar to *C. guilliermondii* in all the studies conducted previously *(10)*, the range was restricted up to 90 g/L.

Figure 1 shows maximum cell growth rates. This parameter was strongly influenced by rice straw hydrolysate concentration. A gradual decrease in maximum cell growth rate resulted by increasing hydrolysate

Table 2
Fermentative and Enzymatic Parameters Determined
Along Fermentation Time with Initial Rice Straw Hydrolysate Contents Equivalent
to Different Xylose Concentrations

S_0 (g/L)	Run time (h)	Xylose used (%)	$Y_{P/S}$ (g/g)	$Y_{X/S}$ (g/g)	q_{P1} (g/[g·h])	q_{P2} (U/[g·h])	P_{xtol} (g/g)	P_{XR} (U/g)	XR:XDH ratio
29.2	24	84	0.43	0.19	0.09	0.79	2.23	19	0.69
	36	97	0.42	0.24	0.05	0.36	1.78	13	0.89
	48	97	0.39	0.32	0.03	0.15	1.24	7	1.02
38.6	24	69	0.47	0.17	0.11	0.46	2.74	11	0.88
	36	93	0.46	0.16	0.08	0.56	2.95	20	0.81
	48	97	0.44	0.22	0.04	0.33	2.03	16	0.80
49.3	36	82	0.59	0.13	0.13	0.97	4.64	35	1.70
	48	95	0.58	0.10	0.12	0.63	5.55	30	1.29
	60	99	0.54	0.10	0.09	0.50	5.63	30	1.82
69.8	48	37	0.65	0.15	0.09	0.21	4.32	10	1.38
	72	66	0.53	0.11	0.07	0.18	4.82	13	0.58
	96	99	0.45	0.09	0.05	0.22	5.01	21	0.75
88.3	96	43	0.32	0.09	0.04	0.16	3.79	15	1.75
	120	49	0.30	0.12	0.02	0.08	2.60	9	1.68
	144	65	0.24	0.08	0.02	0.08	3.25	11	1.51

[a]$Y_{P/S}$, xylitol yield (g/g of xylose used); $Y_{X/S}$, cell yield (g/g of xylose used); q_{P1}, specific xylitol productivity (g/[g of cell·h]); q_{P2}, specific XR productivity (U/[g of cell·h]); P_{xtol}, specific xylitol production (g/g of cell); P_{XR}, specific XR production (g/g of cell). Values are average of duplicates.

concentrations. The highest value (0.18 h^{-1}) was achieved with an initial xylose concentration of 29.2 g/L, and the lowest value (0.04 h^{-1}) was found in fermentation with 88.3 g of xylose/L. Several researchers have reported growth inhibition by rising substrate concentration on yeast physiology. A decline of yield and specific rate of cell production when the amount of xylose initially present in the culture increased was found in a study for xylitol production from *C. guilliermondii* in synthetic medium *(17)*. The yeasts *C. guilliermondii* and *Candida parapsilosis* showed the same relationship between cell growth and xylose concentration under different aeration conditions *(11)*. Growth inhibition in fermentations using hemicellulose hydrolysates as substrate has been reported at lower levels of xylose concentration than in synthetic media *(16,18)*. du Preez et al. *(18)* observed an increase in maximum specific growth rate from 0.21 to 0.35 h^{-1} by diluting to half the hydrolysate concentration in fermentation medium. These results suggest that in the present study inhibition has occurred as a function of the increase in concentration of toxic compounds.

Table 2 shows the results of the fermentative and enzymatic parameters monitored for each initial xylose concentration along fermentation time. The highest values of specific production of xylitol (5.63 g/g of cell) and XR (35 U/g of cell) were found with an initial xylose concentration of

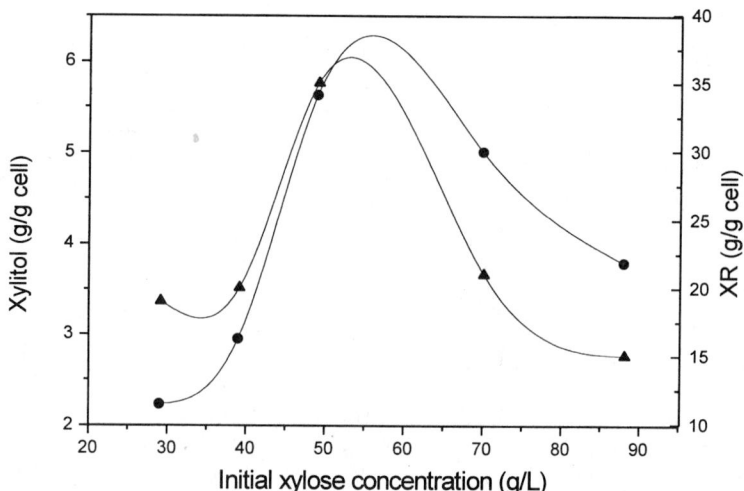

Fig. 2. Highest values of specific production for xylitol (—●—) and XR (—▲—) in fermentations with different hemicellulosic hydrolysate concentrations expressed as initial xylose content.

49.3 g/L. The lowest values for both xylitol (1.24 g/g of cell) and XR (7 U/g of cell) were determined at a fermentation condition of 29.2 g of xylose/L and 48 h. The effect of initial substrate concentration condition on specific production is shown in Fig. 2, where points of maximum can be observed at about 50 g of xylose/L. Time course profiles suggested a relationship between xylitol and XR for this parameter (Fig. 3). In most cases, values increased or decreased with xylitol-specific formation. Nolleau et al. *(11)* reported similar behavior when evaluating the influence of xylose concentration on xylitol production from strains of *C. guilliermondii* and *C. parapsilosis* in synthetic medium. According to them, maximum values of specific enzyme activity were found for *C. guilliermondii* (0.63 U/mg of total protein) and *C. parapsilosis* (0.42 U/mg of total protein) at their optimal initial xylose concentrations (100–150 and 200–300 g/L, respectively) for xylitol accumulation.

According to Table 2, values of specific productivity varied from 0.02 to 0.13 g/(g of cell·h) for xylitol and from 0.08 to 0.97 U/(g of cell·h) for XR. For this parameter, a relationship between xylitol and XR along the fermentation time was found only in fermentations with 29.2, 49.3, and 88.3 g of xylose/L. Figure 4 shows the effect of initial substrate concentration condition on specific productivity. The highest values for both products were found in fermentation with 49.3 g of xylose/L. Specific production rates of xylitol and XR have already been associated in *C. mogii* fermentation using synthetic media with an initial xylose concentration varying from 5.3 to 53 g/L, although values have not been presented *(12)*. In the present study, the maximum values for total formation of xylitol and enzyme were found with an initial xylose concentration of 49 g/L in a 60-h

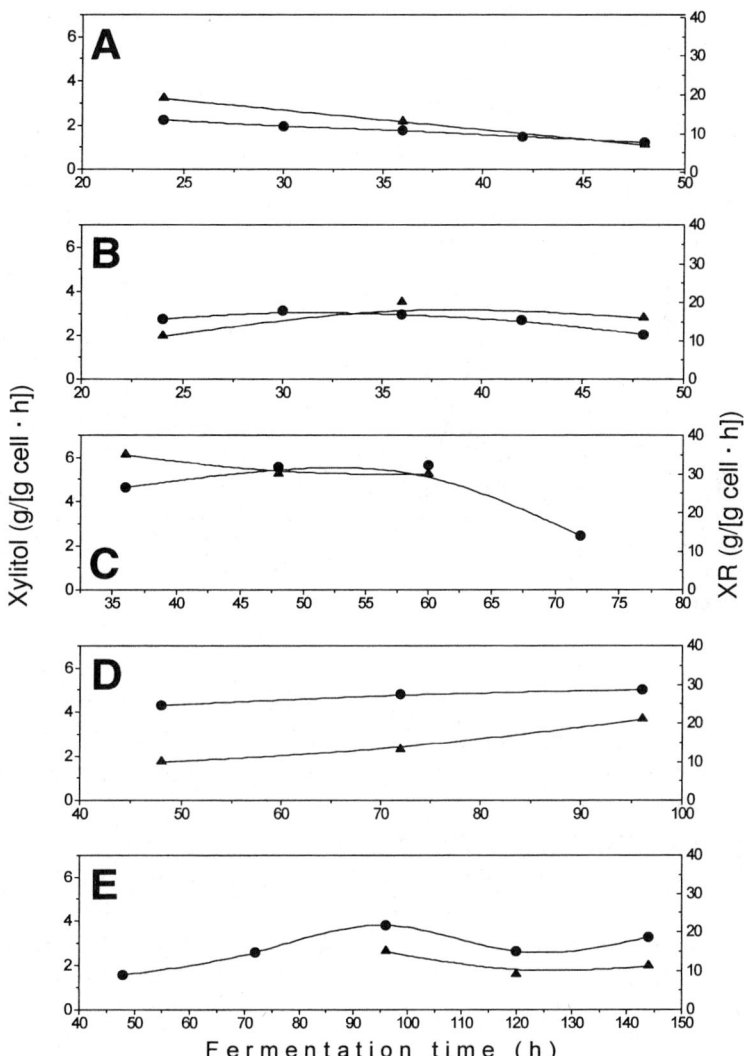

Fig. 3. Specific production of xylitol (—●—) and XR (—▲—) along fermentation time with hydrolysate concentrations equivalent to **(A)** 29.2, **(B)** 38.6, **(C)** 49.3, **(D)** 69.8, and **(E)** 88.3 g/L.

fermentation (data not shown). Under this condition, the XR:XD ratio was the highest (1.82). Similar values were calculated for the yeasts *Candida boidinii (19)* and *Debaryomyces hansenii (20)*. For these microorganisms, XR:XD ratio varied from 1.10 to 2.10 and 1.14 to 2.26, respectively.

Xylitol and XR production was positively influenced by initial hydrolysate concentration up to a certain level, whereas cell growth was strongly limited. Nevertheless, under higher initial hydrolysate concentrations, inhibition of xylitol and XR production was also observed. The inhibitory effect of hydrolysate concentrations in yeast fermentations has been attributed to toxic chemicals often formed during acid hydrolysis

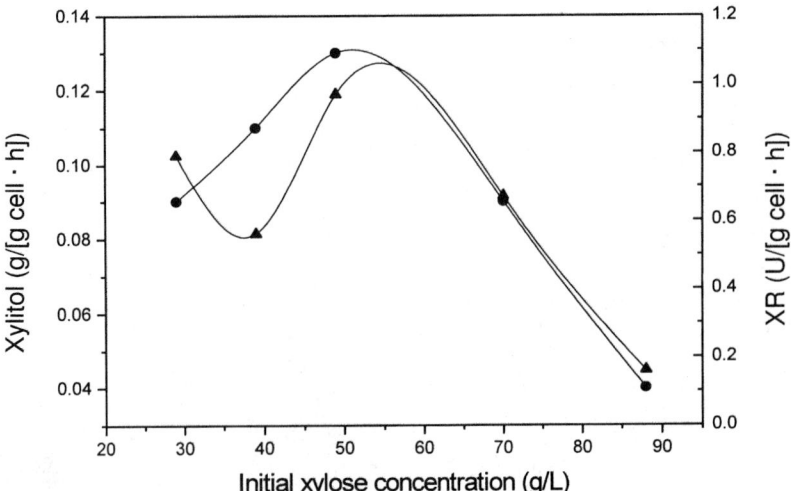

Fig. 4. Highest values of specific productivity for xylitol (—●—) and XR (—▲—) in fermentations with different hemicellulosic hydrolysate concentrations expressed as initial xylose content.

(16,21,22). In a study of the effects of lignocellulosic degradation products on xylose fermentation, Delgenes et al. *(23)* found that the intensity of growth inhibition was closely related to the initial concentration of tested inhibiting molecules.

Results drawn from the experiments in the present study showed an optimal initial hydrolysate concentration condition for xylitol and XR production with about 50 g of xylose/L. However, because the effect of substrate concentration seems to be dependent on aeration conditions *(11),* it is possible that the formation of these products might be improved by using different aeration levels.

Acknowledgments

We wish to thank Maria Eunice M. Coelho for revising the manuscript. We also acknowledge the financial support from Fundação de Amparo à Pesquisa do Estado de São Paulo and Conselho Nacional de Desenvolvimento Científico e Tecnológico, Brazil.

References

1. Chiang, C. and Knight, S. G. (1960), *Nature* **188,** 79–81.
2. Mäkinen, K. K. (1992), *J. Appl. Nutr.* **44,** 16–28.
3. Yilikari, R. (1979), *Adv. Food Res.* **25,** 159–180.
4. Heikkilä, H., Nurmi, J., Rahkila, L., Töyrylä, M., and Kirkkonummi, M. T. (1992), US patent 5,081,026.
5. Strehaino, P. and Dupuy, M.-L. (1990), FR patent 2,641,545.
6. Roberto, I. C., Felipe, M. G. A., Mancilha, I. M., Vitolo, M., Sato, S., and Silva, S. S. (1995), *Biores. Technol.* **51,** 255–257.

7. Nidetzky, B., Neuhauser, W., Haltrich, D., and Kulbe, K. D. (1996), *Biotechnol. Bioeng.* **52,** 387–396.

8. Kitpreechavanich, V., Hayashi, M., Nishio, N., and Nagai, S. (1984), *Biotechnol. Lett.* **6,** 651–656.

9. Roberto, I. C., Mancilha, I. M., de Souza, C. A., Felipe, M. G. A., Sato, S., and Castro, H. F. (1994), *Biotechnol. Lett.* **16,** 1211–1216.

10. Mayerhoff, Z. D. V. L., Roberto, I. C., and Silva, S. S. (1997), *Biotechnol. Lett.* **19,** 407–409.

11. Nolleau, V., Preziosi-Belloy, L., Delgenes, J. P., and Navarro, J. M. (1993), *Curr. Microbiol.* **27,** 191–197.

12. Sirisansaneeyakul, S., Staniszewski, M., and Rizzi, M. (1995), *J. Ferment. Bioeng.* **80,** 565–570.

13. Bolen, P. L. and Detroy, R. W. (1985), *Biotechnol. Bioeng.* **27,** 302–307.

14. Rodrigues, D. C. G. A., Silva, S. S., and Felipe, M. G. A. (1998), *J. Biotechnol.* **62,** 73–77.

15. Alves, L. A., Felipe, M. G. A., Almeida e Silva, J. B., Silva, S. S., and Prata, A. M. R. (1998), *Appl. Biochem. Biotechnol.* **70/72,** 89–98.

16. Silva, C. J. S. M. and Roberto, I. C. (1999), *Biotechnol. Technique* **13,** 743–747.

17. Meyrial, V., Delgenes, J. P., Moletta, R., and Navarro, J. M. (1991), *Biotechnol. Lett.* **13,** 281–286.

18. du Preez, J. C. (1994), *Enzyme Microb. Technol.* **16,** 944–956.

19. Vandeska, E., Kuzmanova, S., and Jeffries, T. W. (1995), *J. Ferment. Bioeng.* **80,** 513–516.

20. Girio, F. M., Roseiro, J. C., Sá-Machado, P., Duarte-Reis, A. R., and Amaral-Collaço, M. T. (1994), *Enzyme Microb. Technol.* **16,** 1074–1078.

21. Dominguez, J. M., Gong, C. S., and Tsao, G. T. (1996), *Appl. Biochem. Biotechnol.* **57/58,** 49–56.

22. Preziosi-Belloy, L., Nolleau, V., and Navarro, J. M. (1997), *Enzyme Microb. Technol.* **21,** 124–129.

23. Delgenes, J. P., Moletta, R., and Navarro, J. M. (1996), *Enzyme Microb. Technol.* **19,** 220–225.

Kinetic Studies of Lipase from *Candida rugosa*

A Comparative Study Between Free and Immobilized Enzyme onto Porous Chitosan Beads

Ernandes B. Pereira,[1] Heizir F. De Castro,[2] Flávio F. De Moraes,[1] and Gisella M. Zanin*,[1]

[1]Department of Chemical Engineering, Maringá State University, 87020-900, Maringá-PR, Brazil, E-mail: gisellazanin@ maringa.com.br; and [2]Department of Chemical Engineering, Faculty of Chemical Engineering of Lorena, PO Box 116, 12600-000, Lorena-SP, Brazil

Abstract

The search for an inexpensive support has motivated our group to undertake this work dealing with the use of chitosan as matrix for immobilizing lipase. In addition to its low cost, chitosan has several advantages for use as a support, including its lack of toxicity and chemical reactivity, allowing easy fixation of enzymes. In this article, we describe the immobilization of *Candida rugosa* lipase onto porous chitosan beads for the enzymatic hydrolysis of olive oil. The binding of the lipase onto the support was performed by physical adsorption using hexane as the dispersion medium. A comparative study between free and immobilized lipase was conducted in terms of pH, temperature, and thermal stability. A slightly lower value for optimum pH (6.0) was found for the immobilized form in comparison with that attained for the soluble lipase (7.0). The optimum reaction temperature shifted from 37°C for the free lipase to 50°C for the chitosan lipase. The patterns of heat stability indicated that the immobilization process tends to stabilize the enzyme. The half-life of the soluble free lipase at 55°C was equal to 0.71 h ($K_d = 0.98$ h^{-1}), whereas for the immobilized lipase it was 1.10 h ($K_d = 0.63$ h^{-1}). Kinetics was tested at 37°C following the hydrolysis of olive oil and obeys the Michaelis-Menten type of rate equation. The K_m was 0.15 mM and the V_{max} was 51 µmol/ (min·mg), which were lower than for free lipase, suggesting that the apparent affinity toward the substrate changes and that the activity of the immobilized lipase decreases during the course of immobilization.

Index Entries: Lipase; immobilization; chitosan; physical adsorption; characterization; hydrolysis.

*Author to whom all correspondence and reprint requests should be addressed.

Introduction

One of the main limitations to obtaining precursors and products of commercial interest can be associated with the use of chemical catalysts, which are not very versatile and require high temperatures to attain satisfactory reaction rates. In addition to possessing low specificity, generally these catalysts provide mixed chemical compounds or byproducts that require further purification steps *(1,2)*. The cost of energy is increasing, and this will increase the cost of existing energy chemical processes, making the enzyme-catalyzed processes an alternative route to compete with the current practice of chemical synthesis *(3)*. Enzymes are being examined intensively for the preparation of new classes of reagents, especially sugars, chiral synthons, metabolites, and food components *(1,3)*. Emerging technology for the production of such compounds employs lipases that are abundant, stereospecific, stable, and versatile enzymes. In addition to the lipolytic reactions, lipases catalyze a variety of synthetic transformations such as esterifications and interesterifications *(4–7)*.

The industrial use of lipases as catalysts depends on their efficient immobilization and the employment of appropriate supports in such a way that the initial investment in raw material (enzyme and support) is compensated by the high activity and stability of the derivative *(7)*. The high cost of popular supports (silica-based carriers and synthetic polymers) leads many researchers to search for cheaper substitutes such as $CaCO_3$ *(8)*, rice husk *(9)*, chitin, and chitosan *(10,11)*. Of these alternatives, the derivative of chitin, chitosan, appears to be more attractive since chitin is the second most abundant biopolymer in nature after cellulose *(10)*. In addition, this support presents several advantages as enzyme immobilization carrier. Among the most prevalent are versatility in the physical forms that are available (flakes, porous beads, gel, fiber, and membrane); scarce biodegradability, low cost; ease of handling; high affinity toward the proteins, and above all nontoxicity *(11)*. Moreover, good results were obtained in a number of previous studies in which chitosan was used to immobilize lipase *(12,13)* and other hydrolases such as amyloglucosidase, papain, β-glucosidase, and α-L arabinofuranosidase *(10,11)*.

In pursuing our interest in the immobilization and subsequent use of lipases *(14–16)*, we have investigated the feasibility of using chitosan as the matrix for immobilizing microbial lipases. The immobilization criteria were based on the use of a low-cost method of loading enzyme into the support. The enzyme used was nonspecific lipase from *Candida rugosa*. The chosen method of immobilization was simple adsorption, whereby the enzyme adheres to the surface of the support particles by van der Waals forces of attraction.

Two physical forms of chitosan (flakes and porous beads) were tested as supports following previously described methodology *(14)*, and the immobilization efficiency was assessed with respect to the recovery of both protein and hydrolytic activity. The best chitosan form with the highest immobilization efficiency was selected for further studies, including full characterization of the immobilized derivative under aqueous medium (hydrolysis of olive oil as a model). Comparative studies of free and immobilized enzyme were conducted in terms of pH, temperature, and thermal stability. The enzymatic hydrolysis with the immobilized enzyme in the framework of the Michaelis-Menten mechanism was also analyzed.

Materials and Methods

Materials

Commercial *C. rugosa* lipase (type VII) and bovine serum albumin (BSA) were purchased from Sigma (St. Louis, MO). The lipase was a crude preparation with a nominal activity of 1440 U/mg and 16.2 mg of protein/g of powder based on the Bradford *(17)* protein assay method. Chitosan in two different forms was tested: flakes (analytical grade chitosan C-3646 obtained starting from crab shell containing 85% deacetylation; Sigma) and porous beads (pharmaceutical grade supplied by SP Chemical Farma, SP, Brazil) having a purity of 93%, moisture of 6%, and 40-mesh granulometry according to the manufacturer's information. Olive oil (low acidity) was purchased at a local market. Solvents were standard laboratory grade and other reagents were purchased from either Aldrich (Milwaukee, WI) or Sigma.

Immobilization of Lipase onto Chitosan

Lipase was immobilized by physical adsorption on chitosan following previous methodology *(14)* with slight modifications. Chitosan (3 g) was initially soaked in hexane, under agitation (100 rpm) for 1 h. Then, the excess of hexane was drained and 0.5 g of lipase previously dissolved in 10 mL of distilled water was added. The fixation of lipase onto support proceeded under agitation for 3 h at room temperature followed by 18 h at 4°C without agitation. The derivative was filtered (Whatman filter paper 41) and thoroughly rinsed with hexane. The enzyme activity before and after immobilization was determined by measuring the hydrolytic activities of the supernatant liquid solutions, which allowed calculation of the activity yield of the immobilized enzyme preparation.

The effect of enzyme loading on immobilized enzyme activity was studied by varying the amount of lipase offered (0.1–1.0 g of lipase) to a fixed amount of support (3.0 g of chitosan).

Hydrolysis Assay

Hydrolytic activities of free and immobilized lipase were assayed by the olive oil emulsion method according to the modification proposed by Soares et al. *(16)*. The substrate was prepared by mixing 50 mL of olive oil with 50 mL of gum arabic solution (7% [w/v]). The reaction mixture containing 5 mL of the emulsion, 2 mL of 100 m*M* sodium phosphate buffer (pH 7.0), and either free (1 mL, 5 mg/mL) or immobilized (250 mg) lipase was incubated for 5 min at 37°C. The reaction was stopped by adding 10 mL of acetone-ethanol solution (1:1). The liberated fatty acids were titrated with 25 mM potassium hydroxide solution in the presence of phenolphthalein as an indicator. One unit of enzyme activity was defined as the amount of enzyme that liberated 1 μmol of free fatty acid/min under the assay conditions.

Catalytic Properties of Lipase Preparations

The estimation of free and immobilized hydrolytic activities at different pH values were carried out with reaction mixtures containing 100 mM of the sodium phosphate buffer at a pH range from 3.0 to 9.0 at 37°C. The effect of temperature in both lipase activities was determined from 30 to 60°C under the assay conditions. For the determination of thermal stability, either free or chitosan lipase preparations were incubated in 2 mL of sodium phosphate buffer (pH 7.0) at different temperatures (40–60°C) for 1 h. Samples were withdrawn and assayed for residual activity as previously described, taking an unheated control to be 100% active.

Protein Assay

Protein was determined according to Bradford's *(17)* method using BSA as the standard. The amount of bound protein was determined indirectly from the difference between the amount of protein introduced into the coupling reaction mixture and the amount of protein in the filtrate and in the washing solutions.

Results and Discussion

Selection of Chitosan Form
and Determination of Lipase Loading on Support

Two chitosan forms (flakes and porous beads) were tested as supports for immobilization of *C. rugosa* lipase by physical adsorption. The structure of chitosan form significantly interfered with both the protein fixation and activity yield, as shown in Table 1. The highest percentage of protein recovery (78.4%) and catalytic activity (14.7%) was obtained by using porous chitosan beads (PCB). It appears that the structure in the form of small granules, characteristic of the pharmaceutical chitosan grade, provided a better distribution of the lipase on the support surface, improving the contact between the interface water/oil, which is necessary for the expression of the activity of immobilized lipases.

The PCB were used to obtain PCB-lipase preparations of different lipase loadings by changing the concentration of lipase in aqueous solution in which chitosan was immersed. The influence of lipase loading in the range of 0.1–1.0 g of lipase/g of chitosan was studied; Figure 1 presents the results.

Table 1
Immobilization of *C. rugosa* Lipase onto Different Physical Forms of Chitosan[a]

Chitosan form	Bound protein (%)	Activity yield (%)	Hydrolytic activity (μmol/[mg·min])
Flakes	73.10	7.10	22.88
Porous beads	78.40	14.70	42.67

[a]Total amount of protein offered to the immobilization: 16 mg; lipase loading: 360 U/g of support.

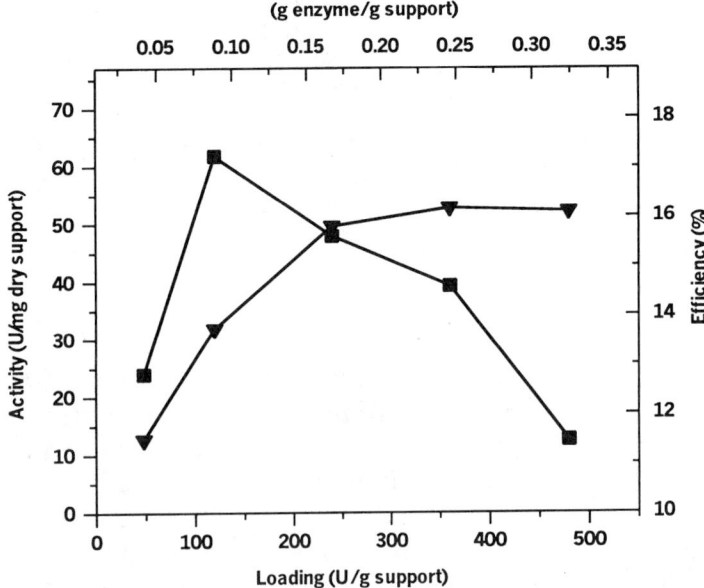

Fig. 1. Effect of lipase loading on hydrolytic activity (■) and activity yield (▼) for *C. rugosa* immobilized onto PCB. Efficiency was calculated by dividing the catalytic activity by lipase loading (see Materials and Methods).

The hydrolytic activity of the PCB-lipase increased from 12.7 to 52.2 U/mg of dry support as more lipase was loaded onto the support (48–480 U/g of support). However, when converted to an efficiency plot (activity/loading), higher efficiencies were obtained at lower lipase loading (120 U/g of support). This may suggest that at loadings over 240 U/g of support, instead of obtaining high lipase fixation on the support surface, multilayer adsorption might occur that could block or inhibit the substrate access (emulsion of olive oil) to the lipase-active sites at the lower layers. It is likely that only the enzyme molecules fixed onto the external layers of chitosan beads are responsible for the detected activity. Therefore, most of the other experiments were carried out using immobilized preparation at the lower lipase loadings of 120 U/g of dry chitosan.

These results are favorable compared with those reported in the literature *(6)* in terms of activity yield. Such yields are probably owing to the methodology used; i.e., the use of an organic medium (hexane) on the coupling step of the enzyme on the support. Similar results have been described by several researchers, indicating a new trend in the use of organic nonpolar solvents as dispersion media for lipase immobilization on different supports *(13,17,18)*. The mechanism responsible for such an improvement has been not fully understood. According to Oliveira et al. *(15)*, two hypotheses can be put forward. First, an expansion of the supports may occur in nonpolar solvents that can promote better distribution of the enzyme on the support surface. Second, the low polarity of solvents such as hexane helps maintain the enzyme's protective layer of water.

Physicochemical Properties and Kinetics for Free and Immobilized Lipase

Methods for immobilizing enzymes should preserve, as much as possible, their original activities and specificities. However, immobilization seems to either inhibit or enhance the action of lipase. To verify changes occurring in the original kinetics and physicochemical properties of the free lipase on immobilization on chitosan, a comparative study between free and immobilized lipase was carried out in terms of pH, temperature, and thermal stability.

Figure 2 shows the variation in relative activity as a function of the pH for free lipase and PCB-lipase. On immobilization, the optimum pH (7.0) for free lipase was shifted for more acidic values (pH 6.0), indicative of the matrix behaving as a polycation *(11)*. It may be assumed that hydrogen and hydroxyl ions are distributed differently between the area close to the surface and the remainder of the solution, with negative charges clustering close to the immobilized enzyme. A microenvironment is thus formed close to the immobilized enzyme with a higher pH than that of the external solution so that the optimal pH becomes lower than that of the free enzyme. It can also be seen that the immobilization procedure conferred higher stability to pH for the immobilized enzyme, since at pH 3.0 the immobilized enzyme still presented 21% of its activity. This fact corroborates with that observed in the literature, as in many cases the immobilization procedure increases pH stability *(20)*.

Figure 3 illustrates the dependence of temperature on both free and immobilized lipase. Maximum activity of the free lipase occurred at 37°C (3400 U/mg of solid), and PCB-lipase showed a maximum activity at 45°C (70.7 U/mg of dry support). It was observed that the immobilization procedure increased the optimum temperature of the enzyme, and this is highly desirable, because higher operational temperatures would lead to lower risks of microbial contamination.

The temperature data were replotted in the form of Arrhenius plots (Figs. 4A,B), from which the energy of activation was calculated by Eq. 1, according to the fitted Eqs. 2 and 3 giving 9.90 and 7.26 kcal/mol for the free and immobilized enzyme, respectively. Therefore, in the immobilized

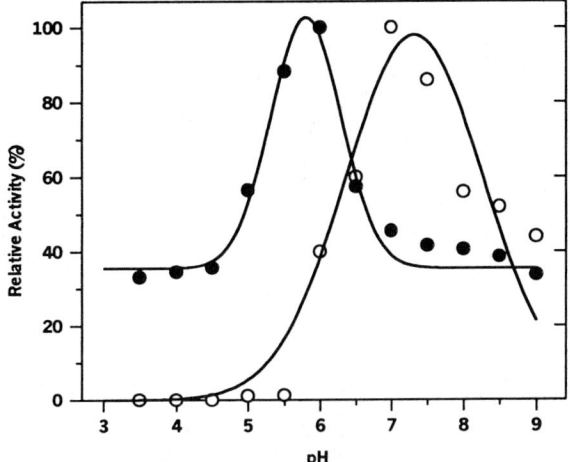

Fig. 2. Effect of reaction pH on hydrolytic activities of lipase preparations. Enzymes were assayed with olive oil emulsion as substrate at 37°C: (○) free lipase; (●) PCB-lipase. Starting activities (free lipase: 3400 U/mg; PCB-lipase: 51.4 U/mg) were taken as 100%.

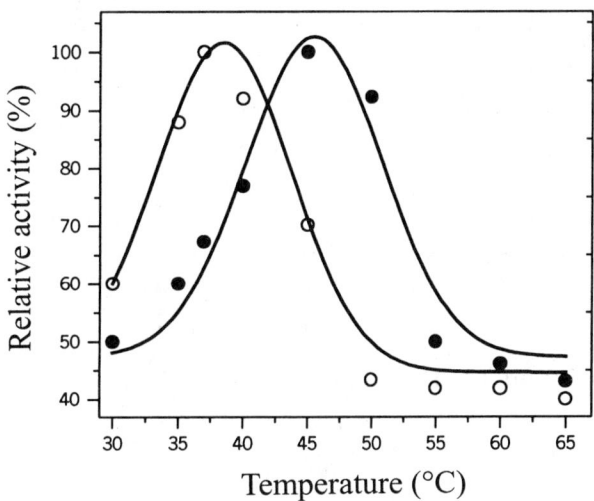

Fig. 3. Effect of reaction temperature on hydrolytic activities of lipase preparations. Enzymes were assayed with olive oil emulsion as substrate at pH 7.0: (○) free lipase; (●) PCB-lipase. Starting activities (free lipase: 3400 U/mg; PCB-lipase: 51.4 U/mg) were taken as 100%.

enzyme form the energy of activation of lipase is 27% lower. Lower activation energy in relation to the free enzyme can be considered an indicative of diffusion resistance of product and substrate in the case of the immobilized enzyme *(21)*. The values of activation energy found for free and immobilized lipases are in the range found for most of the enzymes.

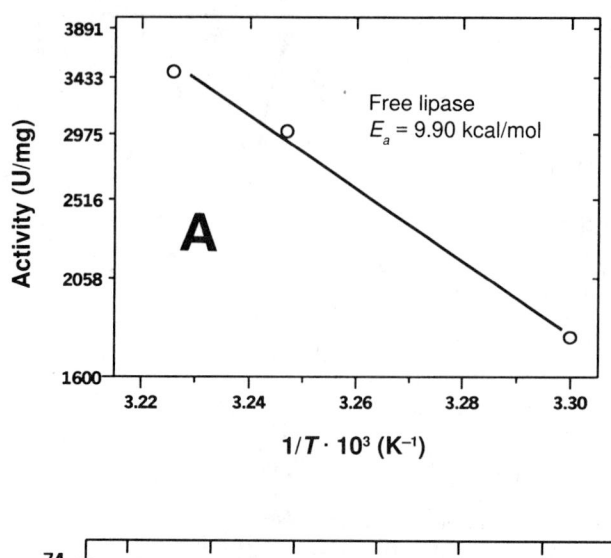

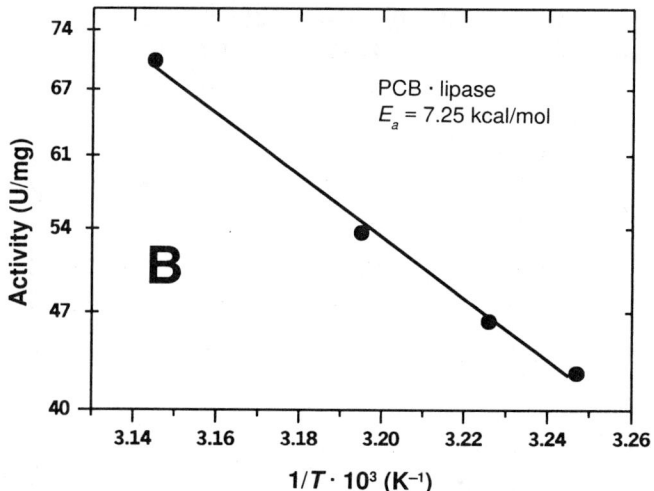

Fig. 4. Arrhenius plots for **(A)** free lipase and **(B)** PCB-lipase using olive oil emulsion 50% (v/v), pH 7.0.

$$A = A_0 \exp(-E_a/RT) \tag{1}$$

$$\text{Free lipase: } A = 1.33 \times 10^6 \exp(-9900/RT) \quad r^2 = 0.999 \tag{2}$$

$$\text{PCB-lipase: } A_i = 6.47 \times 10^6 \exp(-7260/RT) \quad r^2 = 0.954 \tag{3}$$

in which A_0 is the constant of Arrhenius, A is the activity of the enzyme, E_a is the energy of activation of the hydrolysis reaction (cal/g-mol), R is the universal gas constant (1.987 cal/g-mol K), and T is the absolute temperature (K).

Figure 5 presents a comparison of thermal stability for both lipase forms. The immobilized lipase in PCB is more stable than the free one at temperatures ≥50°C. For temperatures ≤45°C, both preparations exhibited

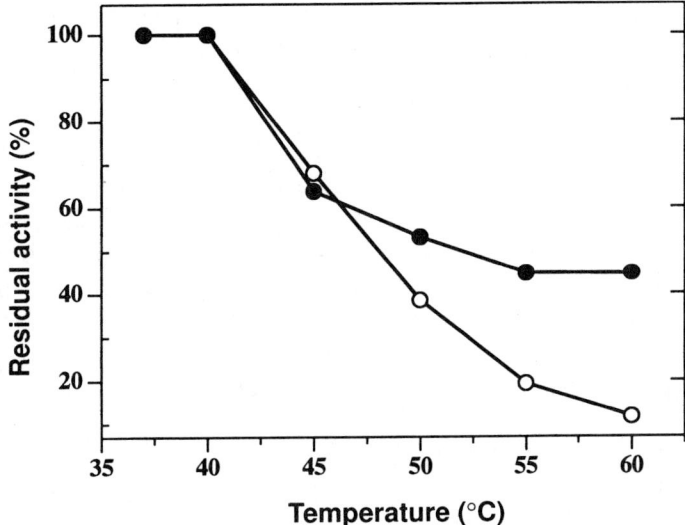

Fig. 5. Temperature deactivation for ($\bigcirc$) free lipase and ($\bullet$) PCB-lipase at different temperatures. Experiments were carried out in 100 m*M* phosphate buffer (pH 7.0). Starting activities (free lipase: 3400 U/mg; PCB immobilized lipase: 51.4 U/mg) were taken as 100%.

Table 2
Thermal Deactivation of *C. rugosa* Lipase Thermally Denatures
in Olive Oil Emulsion (50% [v/v]), pH 7.0, at Different Temperatures for 1 h

Temperature (°C)	Free lipase		PCB-lipase	
	K_d (h^{-1})	$t_{1/2}$ (h)	K_d (h^{-1})	$t_{1/2}$ (h)
45	0.52	1.33	0.45	1.54
50	0.98	0.70	0.63	1.10
55	1.69	0.41	0.81	0.85
60	2.45	0.28	0.83	0.83
65	2.50	0.28	1.00	0.69

a similar trend. The PCB-lipase treated at 60°C for 1 h still held significant activity (on the order of 36% in relation to its activity at 37°C), whereas the free lipase lost 94% of its original activity.

Based on these results, thermal inactivation constants were calculated (K_d) for free and immobilized lipase by using Eq. 4, and the results are given in Table 2:

$$A_{in} = A_{in_0} \exp(-K_d \cdot t) \tag{4}$$

As can be seen, the loss of the catalytic activity for the free enzyme is higher than for the immobilized enzyme. This fact demonstrates that the immobilization procedure on chitosan gave to the lipase higher thermal stability (i.e., lower values of $-K_d$). Note, however, that the higher thermal stability of the immobilized lipase could be a consequence of mass transfer resistance. Since the activity yield was 14.7% for lipase immobilized onto chitosan beads, it is likely that mass transfer resistances are contributing to lowering the activity yield and increasing the apparent thermal stability of the immobilized enzyme.

Data of thermal inactivation constants (K_d) as a function of the temperature allow the calculation of the energy of deactivation by applying a similar approach to the one used for the energy of activation. In most of the available literature (21), it is normally assumed that the enzyme thermal denaturation is a reaction with the rate of enzyme deactivation (r_d) being first order in relation to the concentration of the active enzyme (E):

$$r_d = -K_d E \tag{5}$$

and the deactivation constant (K_d) being a function of temperature as given by the Arrhenius equation:

$$K_d = K_d^0 \exp(-E_d/RT) \tag{6}$$

in which E_d is the energy of deactivation, R is the universal gas constant (1.987 cal/mol K), and T is the absolute reaction temperature.

Values obtained for K_d for various test temperatures are plotted in the form of an Arrhenius plot, i.e., log of K_d against the inverse of absolute temperature, yielding the energy of deactivation (E_d), as the angular coefficient of the adjusted straight lines, multiplied by R, the universal constant. Curves were plotted for the free and immobilized lipase (Figs. 6A,B). Analysis of Fig. 6A shows that lipase thermal deactivation as either free or immobilized enzyme is more accentuated at higher temperatures and a single straight line cannot be fitted to the data, showing that the simple assumptions leading to Eqs. 4–6 do not hold in these enzyme preparations. The Arrhenius type of equation fitted to the experimental data shown in Fig. 6A for the temperature range from 40 to 50°C is as follows:

$$K_d = 9.96 \times 10^{18} \exp(-29,221/RT) \quad r^2 = 0.9999 \tag{7}$$

It was observed that the PCB-lipase shows thermal inactivation at temperatures higher than 55°C, and for this case the Arrhenius type of equation adjusted to the results shown in Fig. 6B for the temperature range from 50 to 60°C is as follows:

$$K_d = 5.15 \times 10^{18} \exp(-29,615/RT) \quad r^2 = 0.9032 \tag{8}$$

The value of the energy of deactivation for free lipase $(E_d = 29.22$ kcal/mol) was practically equal to the PCB-lipase $(E_d = 29.62$ kcal/mol), indicating that the lipase is quite sensitive to denaturation temperatures higher than 50°C.

All enzymatic reactions yield data that can be analyzed in the framework of the Michaelis-Menten mechanism. To examine whether or not the rate of hydrolysis obeys this type of kinetics, a study was carried out by varying the proportion of olive oil in the substrate emulsion from 10 to

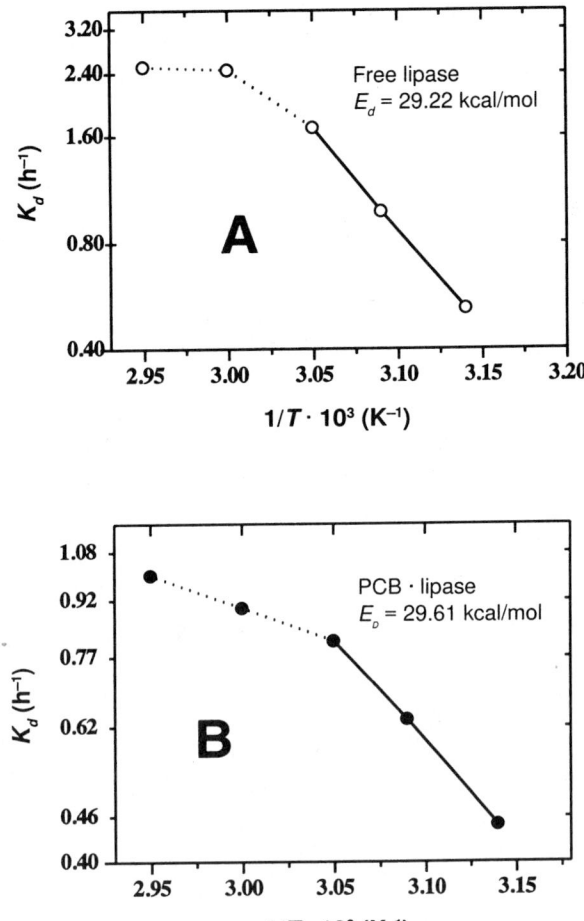

Fig. 6. Energy of deactivation for **(A)** free lipase and **(B)** PCB-lipase.

70% (v/v). This gave a concentration in fatty acids from 280 to 2000 mM. K_m and V_{max} values were determined from the Lineweaver-Burke plots derived from Fig. 7A and Fig. 7B, in which hydrolytic activities for free and immobilized lipase, respectively, are presented as a function of the substrate concentration. Values of V_{max} were 38.46 and 51 µmol/(mg·min) for free and immobilized lipase, respectively, while the K_m value for free enzyme was 0.42 mM and for the immobilized lipase was 0.15 mM. The results suggest that the activity of the free lipase as a function of the concentration of fatty acid follows kinetics of the Michaelis-Menten type, indicating that in the range studied, a possible inhibition by reaction products was not detected. For the case of the PCB-lipase, different behavior was observed and a slight reduction in the activity was detected for substrate concentration >50% (v/v). This can be indicative of substrate inhibition or diffusional resistance. We chose to study the immobilized lipase with 120 U/g of support since it showed the highest catalytic efficiency (Fig. 1);

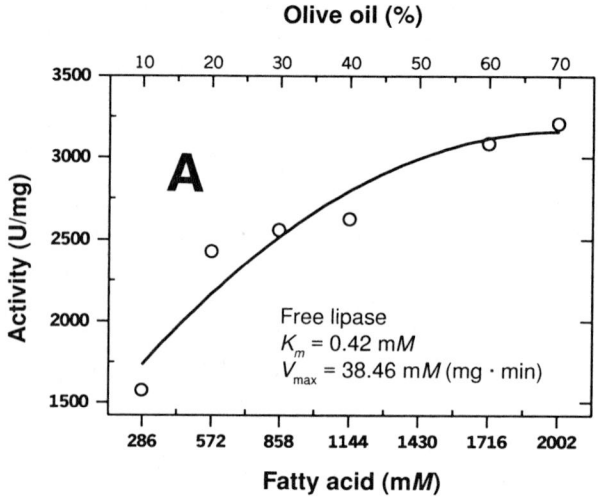

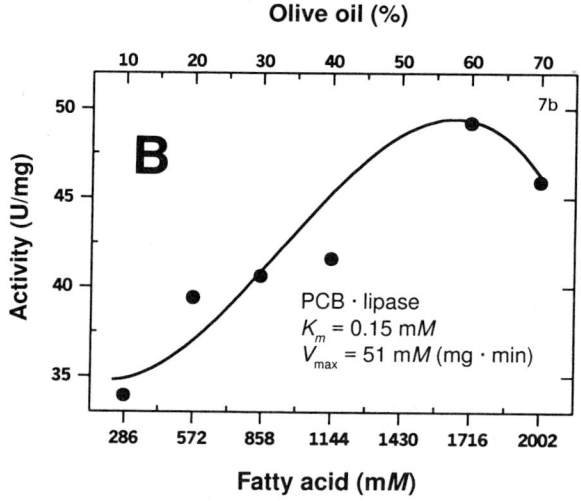

Fig. 7. Hydrolytic activities for **(A)** free lipase and **(B)** PCB-lipase as a function of substrate concentration at pH 7.0 and 37°C.

however, this preparation also has demonstrated the possible presence of mass transfer resistance, and, therefore, it should be emphasized that the kinetic parameters obtained with this preparation can be influenced by enzyme load, particle diameter, porous diameter, and so on. If we were interested in obtaining true intrinsic kinetic parameters of the immobilized lipase and not the ones at the highest catalytic efficiency, a lower enzyme load would have been used. Data of this nature in the literature are lacking; however, such data would help researchers access the relative importance of the phenomena that lower activity yield, such as external and internal mass transfer limitations, sterical hindrances, and conformational changes.

Table 3
Comparison of Physicochemical Properties
and Kinetics Parameters for Free and Immobilized Lipase

Parameter	Free lipase	PCB-lipase
Activity (U/mg)	3700	51.4
Water content (%)	5	20
Optimum pH	7.0	6.0
Optimum temperature (°C)	37	45
Energy of activation (kcal/mol)	9.90	7.26
Thermal inactivation constant (K_d, h^{-1}) at 50°C	0.98	0.63
Half-life (h) at 50°C	0.63	1.1
Energy of deactivation (kcal/mol)	29.22	29.62
K_m (mM)	0.42	0.15
V_{max} (μM/[mg·min])	38.46	51

Table 3 summarizes the general characteristics of the free lipase and PCB-lipase.

Conclusion

Enzyme immobilization by physical adsorption traditionally refers to binding of the enzymes via weak attractive forces to an inert carrier that has not been chemically derivatized. Because the carrier is directly involved in binding to the enzyme, both morphologic and chemical characteristics play important roles. Of the two chitosan forms (flakes and porous beads), PCB showed the most favorable morphologic properties for the immobilization of *C. rugosa* lipase. It is likely that internal mass transfer occurred with both chitosan forms, but the flakes displayed the lowest catalytic activity yield. The methodology for immobilizing lipase on PCB by adsorption using hexane as the dispersion medium gave high protein retention (80%), and compared with other lipase immobilization methods, relatively high activity yield of 17% at lipase loading of 120 U/g of dry support. This is still low, however, in comparison with other enzymes and may reflect the presence of mass transfer limitations owing to multilayer enzyme load. The thermal stability of the immobilized lipase was higher than that for the free enzyme, although this effect may be a consequence of multilayer enzyme immobilization and mass transfer resistance. Kinetic parameters obtained based on the curve of the activity as a function of the substrate concentration (fatty acid) indicated that the free enzyme presents a kinetic of the Michaelis-Menten type, while the apparent kinetic behavior of PCB-lipase resulted in enzyme rates at higher substrate concentration, which may be a consequence of diffusional resistance limitations or the presence of inhibition by substrate. These facts need further study, particularly the relative contribution of different phenomena that reduce the yield of immobilized lipase activity.

Acknowledgments

This work was supported by research grants from Brazilian research agencies CNPq, CAPES, and FAPESP.

References

1. Leuenberger, H. G. W. (1990), *Pure Appl. Chem.* **62,** 753–768.
2. Faber, K. (1997), *Biotransformation in Organic Chemistry: A Textbook*, Faber, K., ed., Springer-Verlag, Berlin, pp. 3–25.
3. Yahya, A. R. M., Anderson, W. A., and Moo-Young, M. (1998), *Enzyme Microb. Technol.* **23,** 438–450.
4. Roberts, S. M. and Turner, M. K. (1998), *Enantiomer* **3,** 9–18
5. Jarger, K. E. and Reetz, M. T. (1998), *TIBTECH* **16,** 396–403.
6. Gandhi, N. N. (1997), *J. Am. Oil Chem. Soc.* **74,** 621–634.
7. Balcão, V. M., Paiva, A. L., and Malcata, F. X. (1996), *Enzyme Microb. Technol.* **18,** 392–416.
8. Rosu, R., Iwasaki, Y., Shimizu, N., Doisaki, N., and Yamane, T. (1998), *J. Biotechnol.* **66,** 51–59.
9. Tantrakulsiri, J., Jeyashoke, N., and Krisanangkura, K. (1997), *J. Am. Oil Chem. Soc.* **74,** 173–175.
10. Krajewska, B. (1991), *Acta Biotechnol.* **11,** 269–277.
11. Felse, P. A. and Panda, T. (1999), *Bioprocess Eng.* **20,** 505–512.
12. Itoyama, K., Tokura, S., and Hayashi, T. (1994), *Biotechnol. Prog.* **10,** 225–229.
13. Carneiro da Cunha, M. G., Rocha, J. M. S., Garcia, F. A. P., and Gil, M. H. (1999), *Biotechnol. Technol.* **13,** 403–409.
14. De Castro, H. F., Oliveira, P. C., Soares, C. M. F., and Zanin, G. M. (1999), *J. Am. Oil Chem. Soc.* **76,** 147–152.
15. Oliveira, P. C., Alves, G. M., and De Castro, H. F. (2000), *Biochem. Eng. J.* **5,** 63–71.
16. Soares, C. M. F., De Castro, H. F., De Moraes, F. F., and Zanin, G. M. (1999), *Appl. Biochem. Biotechnol.* **77–79,** 745–757.
17. Bradford, M. M. A. (1976), *Anal. Biochem.* **72,** 248–254.
18. Mustranta, A., Forssell, P., and Poutanen, K. (1993), *Enzyme Microb. Technol.* **15,** 133–139.
19. Fukunaga, K., Minamijima, N., Sugimura, Y., Zhang, Z., and Nakao, K. (1996), *J. Biotechnol.* **52,** 81–88.
20. Zanin, G. M., Calsavara, L. P. V., Kambara, L. M., and De Moraes, F. F. (1995), *Appl. Biochem. Biotechnol.* **51/52,** 253–262.
21. Zanin, G. M. and De Moraes, F. F. (1998), *Appl. Biochem. Biotechnol.* **70/72,** 383–394.

Extraction by Reversed Micelles
of the Intracellular Enzyme Xylose Reductase

Ely V. Cortez,[1] Maria das Graças de Almeida Felipe,[2]
Inês C. Roberto,[2] Adalberto Pessoa, Jr,[1]
and Michele Vitolo*[,1]

[1]Faculdade de Ciências Farmacêuticas/USP, PO Box 66083,
São Paulo, SP, Brazil, E-mail: ines@debiq.faenquil.br;
and [2]Faculdade de Engenharia Química de Lorena,
CEP 12.600-970, PO Box 116, Lorena, SP, Brazil

Abstract

Xylose reductase enzyme (EC 1.1.1.21) produced by *Candida guilliermondii* in sugarcane bagasse was extracted by reversed micelles of *N*-benzyl-*N*-dodecyl-*N*-*bis* (2-hydroxyethyl) ammonium chloride cationic surfactant. An experimental design was employed to evaluate the influences of the following factors on the enzyme extraction: temperature, cosolvent, and surfactant concentration. A model was used to represent the enzyme recovery and fit of the experimental data. The extraction yielded a total recovery of 130%, and the purity increased 4.8-fold. This study demonstrates that liquid-liquid extraction by reversed micelles is a process able to recover and increase the enzymatic activity and purity of XR produced by *C. guilliermondii*.

Index Entries: Reversed micelles; xylose reductase; liquid-liquid extraction.

Introduction

Liquid-liquid extraction by reversed micelles is a useful and very versatile tool for separating biomolecules. This process shows a close similarity to the liquid-liquid extraction process, because both are biphasic and consist in partitioning a targeted solute between an aqueous feed phase and an organic phase, with a subsequent back transfer to a second aqueous stripping phase *(1)*.

Reversed micellar systems have great potential for industrial application, because they provide a favorable environment for protein solubilization in the organic phase with preservation of biologic activity *(2)*. A number of recent studies on reversed micellar methodology clearly demonstrate

*Author to whom all correspondence and reprint requests should be addressed.

the interest in reversed micelles for the separation of biotechnologic products. Both intra- and extracellular biomolecules can be extracted from various sources and at the same time purified and concentrated to the same extent by relatively simple means, using processes that are easy to scale up *(1)*.

Xylitol is a sweetener with important properties such as anticariogenicity, low caloric value, and negative dissolution heat. Because it can be used successfully in food formulations and in the pharmaceutical industry, its production is in great demand *(3)*. Xylitol can be obtained by microbiologic processes, because many yeasts and filamentous fungi synthesize the xylose reductase (XR) enzyme, which catalyzes the xylose reduction into xylitol as the first step in xylose metabolism. Xylitol production by biotechnologic means has several economic advantages in comparison to the conventional process based on the chemical reduction of xylose. The efficiency and productivity of this fermentation chiefly depend on the microorganism and the process conditions employed *(3)*. The present study evaluates the effectiveness of reversed micelles in extracting XR enzyme produced by *Candida guilliermondii* grown in sugarcane bagasse hydrolysate.

Materials and Methods

Microorganism

The cells were obtained from fermentations conducted with *C. guilliermondii* FTI 20037 as described by Barbosa et al. *(4)*. The yeast was maintained on malt-extract agar slants at 4°C.

Preparation of Hemicellulosic Hydrolysate

Sugarcane bagasse was hydrolyzed in a 250-L reactor at 121°C for 20 min with H_2SO_4 (solid:liquid ratio of 1:10). A portion of the hydrolysate was further concentrated under vacuum at 70°C to increase xylose concentration fourfold. The vacuum procedure was used to avoid sugar degradation. The hydrolysate was then treated as described by Alves et al. *(5)*, to reduce the concentrations of toxic substances.

Inoculum Preparation, Medium, and Fermentation Conditions

A medium containing 3.0 g/L of xylose supplemented with 20.0 g/L of rice bran extract, 2.0 g/L of $(NH_4)_2SO_4$, and 0.1 g/L of $CaCl_2 \cdot 2H_2O$ was used for growing the inoculum. Erlenmeyer flasks (125 mL), each containing 50 mL of medium with inoculum (initial pH 5.5), were incubated on a rotary shaker (200 rpm) at 30°C for 24 h.

For the fermentation, concentrated bagasse hemicellulosic hydrolysate containing 42 g/L of xylose was employed. The hydrolysate was supplemented with the same nutrients as used for the inoculum preparation. Cultivation was done by a batch process in a 1.25-L fermentor (BIOFLO III; New Brunswick, Edison, NJ) under agitation of 300 rpm and aeration rate of 0.6 vvm ($K_L a = 22.5$ h^{-1}) at 30°C and an initial pH of 5.5.

Preparation of Cell-Free Extracts

Cells were harvested by centrifuging at 800g and washed in phosphate buffer (50 mM, pH 7.2). Cell pellets were stored in a freezer. For enzymatic analysis, cell extracts were thawed and disrupted by a sonic disruption technique using a Sonics & Materials disrupter. The cell homogenate was then centrifuged at 10,000g (Jouan MR 1812) at 4°C for 10 min, and the supernatant solution was used for enzymatic assays.

Enzyme Assays

XR activity was determined spectrophotometrically at 340 nm at room temperature *(6)*. One enzyme unit was defined as 1 μmol of NADPH oxidized per minute using an extinction coefficient of 6.22×10^{-3} M^{-1}/cm^{-1}. Specific activity was expressed as units/milligram of protein based on protein determination, according to the method of Lowry et al. *(7)*, using bovine serum albumin as the standard.

Liquid-Liquid Extraction

Liquid-liquid extraction was performed using an experimental design. The enzyme, from the crude extracts, was extracted by N-benzyl-N-dodecyl-N-*bis* (2-hydroxyethyl) ammonium chloride (BDBAC) reversed micelles in isooctane, by a two-step procedure. In the first step (forward extraction), 3.0 mL of the crude extract (containing XR) was mixed with an equal volume of micellar microemulsion (BDBAC in isooctane/hexanol/water). This mixture was agitated on a vortex for 1 min, to obtain the equilibrium phase, and again separated into two phases by centrifuging at 657g for 10 min (Jouan Centrifuge Model 1812; Saint-Herblain, France). Next, 2 mL of XR-BDBAC-micellar phase was mixed with 2.0 mL of fresh aqueous phase (1.0 M acetate buffer at pH 5.5 with 1.0 M NaCl), to transfer the enzyme from the micelles to this fresh aqueous, called the second aqueous phase (backward extraction), which was finally collected by centrifugation (657g for 10 min). Both aqueous phases (first and second) and the crude extract were assayed to determine enzyme activity and protein concentration. The extraction results are reported in terms of total activity recovered (percent) in the second aqueous phase using the XR content of the crude extract as a reference.

Experimental Design and Statistical Analysis

To verify the influence of temperature, cosolvent, and surfactant concentrations on the enzyme recovery (Y), a 2^3 full factorial design with centered face and three repetitions at the center point was employed (Table 1). For each of the three factors, high (coded value: +1), center (coded value: 0) and low (coded value: –1) set points were selected. Extractions representing all 14 set point combinations (2^3 + centered face) were performed, as well as three extractions representing the center point (coded value: 0). Assays were conducted randomly.

Table 1
Matrix for a 2^3 Full Factorial Design with Centered Face

Run no.	Coded values			Actual values		
	Temp. (°C)	Cosolvent (%)	Surfactant (%)	Temp. (°C)	Cosolvent (%)	Surfactant (%)
1	−1	−1	−1	5	6	0.1
2	+1	−1	−1	30	6	0.1
3	−1	+1	−1	5	9	0.1
4	+1	+1	−1	30	9	0.1
5	−1	−1	+1	5	6	0.2
6	+1	−1	+1	30	6	0.2
7	−1	+1	+1	5	9	0.2
8	+1	+1	+1	30	9	0.2
9	0	0	0	17.5	7.5	0.15
10	0	0	0	17.5	7.5	0.15
11	0	0	0	17.5	7.5	0.15
12	−1	0	0	5	7.5	0.15
13	+1	0	0	30	7.5	0.15
14	0	−1	0	17.5	6	0.15
15	0	+1	0	17.5	9	0.15
16	0	0	−1	17.5	7.5	0.1
17	0	0	+1	17.5	7.5	0.2

Results and Discussion

Table 2 gives the results of experiments based on a 2^3 full factorial matrix. A high recovery of XR activity could be observed in many experiments.

The isoelectric point (p*I*) of XR produced by *C. guilliermondii* in sugarcane bagasse hydrolysate is unknown. However, XR produced by *Pachysolen tannophilus* NRRL Y-2460 has a p*I* equal to 4.9 *(8)* and XR produced by *Candida tropicalis* has a p*I* between 4.1 and 4.15 *(9)*. Therefore, the XR described in the present study could have the same p*I* range and a negative global charge at pH 7.0. This would improve the electrostatic interaction between enzyme and cationic surfactant BDBAC *(1)*. The electrostatic interaction is one of the most predominant factors in the reversed micelle extraction, and this explains the high recovery of XR in our experiments (Table 2). This interaction can cause the enzyme migration to the micellar core, when it is opposite to the electrical charge between enzyme and surfactant. The XR recovery was above 100% since all enzyme activity present in the crude extract was transferred to the fresh aqueous phase after backward-extraction, and the process reduced the concentration of several enzyme inhibitors (mainly hydrophobic compounds such as furfural, hydroxymethyl furfural, phenols) present in the crude extract. Using BDBAC reversed micelles, Hasmann et al. *(10)* recovered about 50% of β-xylosidase from fermented medium and Pessoa and Vitolo *(2)* recovered 90% of inulinase.

Table 2
Results of a 2^3 Full Factorial Design with Centered Face

Experiment no.	Initial activity (U/mL)	Backward extraction activity (U/mL)	Recovery (%)
1	0.23	0.36	140
2	0.52	0.42	81
3	0.35	0.56	145
4	0.41	0.49	118
5	0.45	0.21	47
6	0.52	0.0	0
7	0.54	0.63	116
8	0.38	0.38	94
9	0.55	0.66	122
10	0.55	0.68	123
11	0.55	0.72	131
12	0.31	0.37	122
13	0.39	0.43	110
14	0.28	0.40	142
15	0.35	0.41	120
16	0.35	0.44	126
17	0.39	0.41	104

Table 3
Variance Analysis of Significant Factors and Interactions
for XR Extraction Process by Reversed Micelles of BDBAC[a]

Source of variation	Sum of squares	Degrees of freedom	Mean square	F Value	p Value
X_1: temperature	2798.9	1	2798.9	6.28	0.0407
X_2: cosolvent	3319.7	1	3319.7	7.45	0.0294
X_3: surfactant	6170.2	1	6170.2	13.84	0.0075
X_2X_3	1830.1	1	1830.1	4.10	0.0824
X_1X_1	2924.4	1	2924.4	8.20	0.0169
X_3X_3	1020.7	1	1020.7	2.86	0.1216
Residue	3567.2	10	356.7		
Total (corr.)	21,631.37	16			

[a]$R^2 = 0.84$.

Considering that there is no literature on XR extraction with reversed micelles, we can say that our results were quite good.

In our work the pH was maintained at 7.0 and electrical conductivity at 14 mS/cm. Our objective was to avoid loss of activity, and with these two values we attained good results, with no need to test other values.

Table 3 gives the variance analysis of the factors and interactions that were important for the XR extraction process by reversed micelles. All the factors (X_1, X_2, and X_3) and the interaction X_1X_1 were significant

Table 4
Regression Coefficients

Average	129.24
Temperature—X_1	−16.73
Cosolvent—X_2	18.22
Surfactant—X_3	−24.84
X_1X_1	−17.21
X_2X_3	15.12
X_3X_3	−18.36

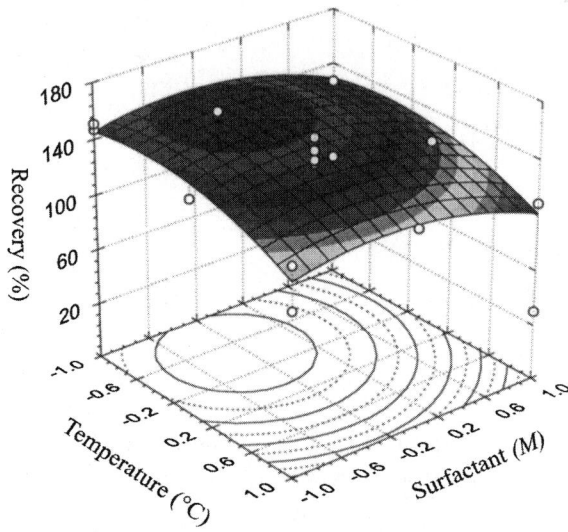

Fig. 1. Surface response described by Eq. 1 for temperature (X_1) and surfactant (X_3). Cosolvent concentration was maintained at superior level ($X_2 = 1$).

at a 95% confidence level. The interactions X_2X_3 and X_3X_3, although less significant (Table 4), were also taken into account for establishing the model.

Table 4 gives the regression coefficients of the second-order model (quadratic), which was used to estimate the percentage of enzyme recovery as a function of temperature, cosolvent, and surfactant concentration.

The statistical significance of the quadratic model was evaluated by the F test, which revealed that this regression is statistically significant at the 5% probability level. The model did not show lack of fit, and the determination coefficient ($R^2 = 0.84$) indicates that 84% of the variability in the recovery can be explained by the model. Thus, the mathematical model representing the XR extraction process in the experimental region studied can be expressed by Eq. 1 and by Fig. 1:

$$Y = 129.2 − 16.7X_1 + 18.2X_2 − 24.8X_3 + 15.1X_2X_3 − 17.2X_1^2 − 18.4X_3^2 \quad (1)$$

in which Y is the enzyme recovery (%); and X_1, X_2, and X_3 are coded values for temperature, cosolvent, and surfactant concentrations, respectively.

Table 5
Purification of XR Produced by *C. guilliermondii* Using Reversed Micelles

Purification step	Total protein (mg/mL)	Total activity (U/mL)	Specific activity (U/mg)	Purification factor
Crude extract	1.38	0.41	0.30	1.0
Aqueous phase after back-extraction	0.39	0.55	1.42	4.8

The maximum recovery value of XR (152%) corresponded to the point defined by the temperature of 11.4°C ($X_1 = -0.49$), cosolvent concentration of 9% ($X_2 = 1$), and surfactant concentration of $0.14\,M$ ($X_3 = -0.26$). Triplicate XR extraction runs were performed at these optimum conditions predicted by the model, and the average recovery value was about $135 \pm 3\%$, which is close to that predicted by the model ($152 \pm 23\%$).

Table 5 gives the results of total protein analysis regarding a point of maximal extraction. The purification factor increased 4.8-fold.

Conclusion

This study demonstrates that liquid-liquid extraction by reversed micelles is a process able to recover and increase the enzymatic activity and purity of XR produced by *C. guilliermondii* cultivated in sugarcane bagasse hydrolysate. It was possible to obtain a representative model for the extraction of XR.

Acknowledgments

We are grateful to Maria Eunice M. Coelho for revising the manuscript. This work was supported financially by FAPESP and CNPq, Brazil.

References

1. Kilikian, B. V., Bastazin, M. R., Minami, N. M., Gonçalves, E. M. R., and Pessoa A. Jr. (2000), *Braz. J. Chem. Eng.* **17,** 29–38.
2. Pessoa A. Jr. and Vitolo, M. (1998), *Process Biochem.* **33,** 291–297.
3. Silva, S. S., Felipe, M. G. A., and Mancilha, I. M. (1998), *Appl. Biochem. Biotechnol.* **70–72,** 331–339.
4. Barbosa, M. F. S., Medeiros, M. B., Mancilha, I. M., Schneider, H., and Lee, H. (1988), *J. Ind. Microbiol.* **3,** 241–251.
5. Alves, L. A., Felipe, M. G. A., Almeida e Silva, J. B., Silva, S. S., and Prata, A. M. R. (1998), *Appl. Biochem. Biotechnol.* **70/2,** 89–98.
6. Alexander, N. J. (1985), *Biotechnol. Bioeng.* **37,** 1739–1744.
7. Lowry, O. H., Rosebrough, N. J., Farr, A. L., and Randall, R. J. (1951), *J. Biol. Chem.* **193,** 265–275.
8. Ditzelmüeller, G., Kubicek, C. P., Woehrer, W., and Roehr, M. (1984), *Can. J. Microbiol.* **30,** 1330–1336.
9. Yokoyama, S. I., Suzuki, T., Kawai, K., Horitsu, H., and Takamizawa, K. (1995), *J. Ferment. Bioeng.* **79,** 217–223.
10. Hasmann, F. A., Pessoa A. Jr., and Roberto, I. C. (1999), *Biotechnol. Technique* **13,** 239–242.

Controlled Hydrolysis of Cheese Whey Proteins Using Trypsin and α-Chymotrypsin

Célia Maria A. Galvão,[1] Astréa F. Souza Silva,[1] Marcos Franqui Custódio,[2] Rubens Monti,[2] and Raquel de Lima C. Giordano*,[1]

[1]Departamento de Engenharia Química,
Universidade Federal de São Carlos, C.P. 676, CEP 13565-905,
São Carlos, SP, Brazil, E-mail: raquel@deq.ufscar.br;
and [2]Departamento de Alimentos e Nutrição, UNESP,
CEP 14801-902, Araraquara, SP, Brazil

Abstract

This study examined the production of protein hydrolysates with controlled composition from cheese whey proteins. Cheese whey was characterized and several hydrolysis experiments were made using whey proteins and purified β-lactoglobulin, as substrates, and trypsin and α-chymotrypsin, as catalysts, at two temperatures and several enzyme concentrations. Maximum degrees of hydrolysis obtained experimentally were compared to the theoretical values and peptide compositions were calculated. For trypsin, 100% of yield was achieved; for α-chymotrypsin, hydrolysis seemed to be dependent on the oligopeptide size. The results showed that the two proteases could hydrolyze β-lactoglobulin. Trypsin and α-chymotrypsin were stable at 40°C, but a sharp decrease in the protease activity was observed at 55°C.

Index Entries: Cheese whey; β-lactoglobulin; protein hydrolysates; trypsin; α-chymotrypsin.

Introduction

Cheese whey is the most abundant protein subproduct of the dairy industry, representing about 85–95% of the volume of the milk. It contains about 55 g/L of lactose and 7 g/L of proteins. The main proteins present in the whey are α-lactalbumin (16%, weight basis), β-lactoglobulin (49%), bovine serum albumin (BSA) (5%), and immunoglobulins (10%). Worldwide production of whey is about 145 million tons, with 60% recovered by

*Author to whom all correspondence and reprint requests should be addressed.

several methods and 40% discarded directly into rivers without previous treatment. The development of products with high commercial value from this residue seems to be an attractive solution for the environmental problem caused by disposal of cheese whey *(1)*.

Controlled enzymatic hydrolysis of proteins (under mild conditions) leads to hydrolysates that possess improved functional properties or a pleasant taste, as well the elimination of allergenic characteristics. There are numerous published works in which these hydrolysates have been investigated *(2–5)*.

Cheese whey was initially characterized to determine appropriate pretreatment before reaction. Then, several hydrolysis experiments were made using whey proteins and purified β-lactoglobulin, as substrates, and trypsin and α-chymotrypsin, as catalysts. Temperatures were 40 and 55°C and pH was 8.0. The influence of enzyme concentration on the degree of hydrolysis, stability of the enzymes under operational conditions, and, finally, specificity of both proteases were studied. Maximum experimental hydrolysis degrees were compared to theoretical values and the resulting peptide compositions calculated.

Materials and Methods

Materials

Sweet cheese whey was donated by different cheese manufacturers of São Carlos, SP, Brazil. Trypsin (EC 3.4.21.4), from bovine pancreas, was donated by Novo Nordisk of Brazil or purchased from Sigma (treated to eliminate α-chymotrypsin activity). α-Chymotrypsin (EC 3.4.21.1), from bovine pancreas and treated to eliminate trypsin activity; purified β-lacto-globulin; and the synthetic substrates benzoyl-L-arginine ethyl ester (BAEE) and benzoyl-L-tyrosine ethyl ester (BTEE) were purchased from Sigma. All other reagents were of analytical grade from several commercial brands.

Methods

Lyophilization and Filtration of Whey

After cooling at –50°C in a Bio-Freezer (Forma Scientific), 200 mL of whey was placed in an Edwards lyophilizer (model L4RL) for 8 h, resulting in 0.03 g/mL of powder whey per volume of *"in natura"* whey. For small volumes, whey was filtered through filter paper and under vacuum in Millipore membranes with a diameter of 47 mm and several porosities (1.2, 0.45, and 0.22 μm). For volumes above 1 L, an ultrafiltration hollow-fiber unit was used 0.45 μ, CFP-4-E-6A; A/G Technology coupled to a microfiltration unit (model Pellicon; Millipore S/A).

Proteins

The protein contents in the liquid or lyophilized (after dissolution) cheese whey were analyzed by the Kjeldahl method using a Büchi unit (model 323).

Enzymatic Activity

First of all, two specifics solutions were made (for trypsin, 2 mL of 0.5 mM BAEE solution, dissolved in 50 mM phosphate buffer, pH 7.6; for α-chymotrypsin, 100 μL of 8 mM BTEE in ethanol and 2 mL of 100 mM phosphate buffer, pH 7.0). These solutions were transferred to two differents quartz cuvets (at 25°C, in a spectrophotometer, model Ultrospec 2000; Pharmacia Biotech). To these solutions, 100 and 200 μL of enzymatic solution (diluted, if necessary) of α-chymotrypsin and trypsin, respectively, were added. The increase in absorbance that occurs during the hydrolysis of the synthetic substrates BAEE at 253 nm, for trypsin, and BTEE at 254 nm, for α-chymotrypsin, was followed *(6,7)*. The activity was calculated after determining the initial reaction rate.

Enzymatic Hydrolysis

Hydrolysis experiments were developed in a Metrohn pHstat (model Titrino). The protein hydrolysis reaction was followed through base consumption (NaOH), which occurs to keep the reaction pH constant. When a peptide bond is broken, proton liberation occurs. Ionic equilibrium was taken into account in the calculations. The whey was transferred to a jacketed glass reactor at 40 or 55°C and the pH adjusted to 8.0. Then, the enzyme mass dissolved in 1 mL of 0.1 mol/L of $CaCl_2$ was added to the reactor, which contained a magnetic stirrer, and was coupled to a Neslab thermostatic bath, with recirculation (model RTE 111). In the sequential hydrolysis, trypsin was inactivated (80°C for 15 min). After cooling to 40°C, α-chymotrypsin was added.

Stability of Proteases

The stability of the enzymes was followed along the whole period of the reaction. At different times samples were withdrawn and assayed at 25°C, as described in the standard procedure.

Experimental Hydrolysis Degree

The experimental hydrolysis degree was calculated according to Adler-Nissen *(8)* using the following equation:

$$\%GH = C_b \times V_b \times \frac{1}{\alpha} \times \frac{1}{h_{total}} \times \frac{1}{M_p} \times 100\%$$

in which C_b = concentration of the base (mol/L); V_b = volume of the base (mL); $1/\alpha = 1.2$; $1/h_{total} = \Sigma$ mmol of individual amino acids by protein gram = 8.8; and M_p = protein mass.

Theoretical Hydrolysis Degree

The theoretical hydrolysis degree was calculated from knowledge of the primary sequence *(9,10)* and the specificity of the used proteases *(11)*. Trypsin is assumed to hydrolyze peptide bonds where lysine and arginine residues are in the carboxyl side of the bond; α-chymotrypsin, tryptophan, phenylalanine, tyrosine, and leucine residues were taken into account. The theoretical hydrolysis degree for cheese whey was calculated as follows:

$$\%HD_{th} = \frac{n_\alpha \times C_\alpha + n_\beta \times C_\beta + n_{BSA} \times C_{BSA}}{n_{T\alpha} \times C_\alpha + n_{T\beta} \times C_\beta + n_{TBSA} \times C_{BSA}} \times 100$$

in which $n_{\alpha, \beta, BSA}$ is the number of specific residues in the α-lactalbumin, β-lactoglobulin, and BSA proteins, respectively, for trypsin or α-chymotrypsin; $n_{T\alpha, T\beta, TBSA}$ is the total number of residues in the α-lactalbumin, β-lactoglobulin, and BSA proteins, respectively; and $C_{\alpha, \beta, BSA}$ is the concentration of α-lactalbumin, β-lactoglobulin, and BSA in the cheese whey, respectively.

The theoretical hydrolysis degree for cheese whey considered that the protein composition was 3.0 g/L of α-lactalbumin, 1.2 g/L of β-lactoglobulin, and 0.4 g/L of BSA. It is difficult to ascertain a more accurate value for this variable. The whey protein concentration depends on the source of the product; in addition, whey contains other proteins such as immunoglobulins, membrane proteins, and peptones in small concentrations. Table 1 shows the calculated values for whey proteins and purified β-lactoglobulin.

Theoretical Peptide Composition ($\%PC_{th}$)

The theoretical peptide distribution percentile was calculated for the maximum degree of hydrolysis as follows:

$$\%PC_{th} = \frac{p_{T\alpha} \times C_\alpha + p_{T\beta} \times C_\beta + p_{TBSA} \times C_{BSA}}{p_{th\alpha} \times C_\alpha + p_{th\beta} \times C_\beta + p_{thBSA} \times C_{BSA}} \times 100$$

in which $p_{th\alpha, th\beta, thBSA}$ is the theoretical number of peptides obtained from α-lactalbumin, β-lactoglobulin, and BSA hydrolysis, respectively, using trypsin and α-chymotrypsin alone or for sequential action (α-chymotrypsin after trypsin action), for several molecular mass ranges; $p_{T\alpha, T\beta, TBSA}$ is the total number of peptides obtained from the maximum hydrolysis degree of each situation; and $C_{\alpha, \beta, BSA}$ is the concentration of α-lactalbumin, β-lactoglobulin, and BSA in cheese whey, respectively.

Sodium Dodecyl Sulfate Polyacrylamide Gel Electrophoresis

Sodium dodecyl sulfate polyacrylamide gel electrophoresis (SDS-PAGE) of β-lactoglobulin after hydrolysis with α-chymotrypsin at different reaction times was performed on a Hoefer SE 200 vertical slab gel according to the manufacturer's instructions. The gel was stained with Coomassie blue and fixed with methanol. Molecular weights were estimated using a Pharmacia PMW calibration kit (14,000–94,000 Daltons).

Table 1
Theoretical Hydrolysis Degree (HDth) Calculated for Main Proteins in Cheese Whey[a]

Protein	Trypsin			α-Chymotrypsin							
				woLeu[b]		wLeu[c]		woLeu[d]		wLeu[d]	
	Nt	Ns	HDth (%)	Ns	HDth (%)	Ns	HDth (%)	Ns	HDth (%)	Ns	HDth (%)
α-Lactalbumin	123	14	11.38	12	9.76	24	19.51	11	8.94	21	17.07
β-Lactoglobulin	162	18	11.11	10	6.17	32	19.75	7	4.32	27	16.67
BSA	582	82	14.09	49	8.42	110	18.30	41	7.04	90	15.46
Whey[e]	—	—	11.96	—	7.39	—	19.48	—	5.84	—	16.41

[a]Nt = total number of residues; Ns = number of specific residues for each protease; $HDth = (Ns/Nt) \times 100$; wLeu and woLeu = Ns and $HDth$ calculated with and without considering peptide bonds where leucine is present at the bond C-side.
[b]Not considering leucine residues.
[c]Considering leucine residues.
[d]After total hydrolysis with trypsin.
[e]α-Lactalbumin (1.2 g/L), β-lactoglobulin (3.0 g/L), BSA (0.4 g/L).

Table 2
Proteic Content in *In Natura*
and Filtrate Whey and in Retentate

Material analyzed	Concentration of protein (g/L)
In natura whey	8.6 ± 0.12
Filter paper	0.63
Membrane (1.2 mm)	0
Membrane (0.45 µm)	0
Membrane (0.22 µm)	0.24
Filtrate whey	8.1 ± 0.10

Results and Discussion

Characterization of Cheese Whey

Cheese whey contains many insoluble particles represented mainly by insoluble globules of fat and casein. To verify whether the particulate protein could be hydrolyzed, *in natura* whey was submitted to successive microfiltration steps using membranes of different porosities. Observation under microscope of all the obtained permeates (×400 magnification) showed that most of the particulate material, as well visible fat, was retained on filter paper. Table 2 presents the results of protein analysis of the different whey fractions obtained during the filtration steps. It can be observed that only the fractions retained in membranes of 1.2 (insoluble globules of casein) and 0.22 µm (bacteria, probably) had protein content. Two kinds of experiments were then made. In the first, the particulate material was submitted to hydrolysis with trypsin and no base consumption was observed. In the second, *in natura* and filtrated whey were hydrolyzed using trypsin. The hydrolysis degree obtained at the same operational conditions for the two substrates and the protein content of the retained fractions on the filter paper, after hydrolysis, were compared. The results (not shown here) indicated that the particulate material (probably casein precipitated that is kept in the whey after the separation of the solid fraction in the process of cheese production) is inaccessible to the catalytic attack of trypsin. Therefore, there is no advantage to keeping those particles in the whey and two steps of microfiltration are recommended: the first one using a membrane with high porosity to remove casein and fat, and the second one, to remove bacteria. Although the low-porosity membrane could retain both solids, that would imply a great increase in the filtration time, with consequent risk of whey contamination.

Hydrolysis of Cheese Whey Proteins
with Trypsin and α-Chymotrypsin

The hydrolysis of whey proteins using increasing enzyme:substrate ratios (E:Ss) was an important point to determine which E:S ratio would

Table 3
Experimental Hydrolysis Degree (*HDexp*) for Filtrate Whey Using Trypsin[a]

Time (s)	(E:S)	Cenz (U/mL)[b]	%HDexp	HDth (%)	η[b] (%)[c]
0	1:100	12.96	—	—	—
1800	—	11.55	5.57 ± 0.61	11.96	46.57
1860	1:50	—	—	—	—
3600	—	20.98	8.8 ± 0.95	11.96	73.58
0	1:20	64.78	—	—	—
1800	—	54.98	8.13 ± 0.48	11.96	67.98
1860	1:16.7	—	—	—	—
3600	—	60.38	11.28 ± 0.6	11.96	94.31
0	1:4	323.89	—	—	—
1800	—	256.41	11.93 ± 1.19	11.96	99.75

[a]E:S = enzyme:substrate ratio; η = (*HDexp*/*HDth*) × 100; NaOH = 0.027 *M*.
[b]The yield was not calculated at the time when the enzyme was added.
[c]Calculated yield.

Table 4
Experimental Hydrolysis Degree (*HDexp*) for Filtrate Whey Using α-Chymotrypsin

Time (s)	(E:S)	Cenz (U/mL)[b]	%HDexp	HDth woLeu (%)[c]	HDth WLeu (%)[d]	η (%)[e]
0	1:100	11.98	—	—	—	—
1800	—	10.28	7.53 ± 0.56	7.39	19.48	38.66
1860	1:50	—	—	7.39	—	—
3600	—	18.62	11.20 ± 0.79	7.39	19.48	57.34
0	1:20	56.92	—	7.39	—	—
1800	—	49.17	7.60	7.39	19.48	39.01
1860	1:16.7	—	—	7.39	—	—
3600	—	54.15	10.71	7.39	19.48	54.97
0	1:4	284.62	—	7.39	—	—
1800	—	241.83	8.56	7.39	19.48	43.89
1860	1:3.8	—	—	7.39	—	—
3600	—	234.63	10.62	7.39	19.48	54.52

[a]E:S = enzyme:substrate ratio; η = (*HDexp*/*HDth*) × 100; NaOH = 0.027 *M*.
[b](BTEE-U)/mL of solution.
[c]Not considering leucine residues.
[d]Considering leucine residues.
[e]Calculated yield.

lead to the maximum experimental hydrolysis degree, for a predetermined time, i.e., one that would allow us to control the reaction degree. Experiments with three E:S ratios were accomplished, using trypsin and α-chymotrypsin, separately; the results are given in Tables 3 and 4.

For trypsin, the maximum experimental hydrolysis degree was reached within 0.5 h of reaction with an initial E:S of 0.25 (initial concentration of enzyme about 324 BAEE-U/mL of solution). The agreement between experimental and theoretical values of the maximum hydrolysis degree indicates a high specificity of trypsin for lysine and argentine residues, as reported in the literature *(11)*. However, because the theoretical value calculated for whey is not exact, this conclusion has to be confirmed using purified β-lactoglobulin as the substrate. In the case of α-chymotrypsin, the maximum experimental hydrolysis degree seems to be, in fact, about 11%, reached after 60 min of reaction. This value did not change significantly, within the experimental error, when enzyme concentrations were four times higher.

Concerning α-chymotrypsin specificity, Abeles et al. *(11)* stated that this enzyme attacks preferentially peptide bonds with bulky nonpolar aromatic groups, but it also attacks, although more slowly, nonpolar groups such as leucine. The experimental value obtained for the maximum hydrolysis degree is smaller than the theoretical calculated value, when the possibility of hydrolysis of all peptide bonds with leucine residues on the carboxyl side of the bonds is considered. A possible explanation would be that a fraction of the enzyme molecules binds to the proteins at the points where phenylalanine, tyrosine, and tryptophan residues are in the required positions and breaks the peptide bonds quickly. Meanwhile, other enzyme molecules bind to protein molecules at leucine residues, but the reaction rate is small in this case. That behavior could explain the observed experimental hydrolysis degree. However, if the reaction rate were the only restriction, when the enzyme concentration is increased, the experimental degree should also be higher, which was not observed. For all the investigated conditions, the hydrolysis degree reached approx 56% of the maximum theoretical degree. Therefore, this "kinetic" explanation for the uncompleted hydrolysis is not reasonable. In addition, thermal inactivation of the enzyme cannot be responsible for this behavior. The activities of the proteases were followed during the reaction and were equal to 100% of the initial activity for the whole reaction time, at 40°C.

Another possible explanation for this phenomenon could be the incomplete hydrolysis of β-lactoglobulin (about 50% of the protein content of cheese whey). Reddy et al. *(12)* affirm that α-chymotrypsin does not attack this protein, owing to its high conformational stability. To verify this hypothesis, α-chymotrypsin was used to hydrolyze purified β-lactoglobulin.

Hydrolysis of β-Lactoglobulin: Specificity of Proteases Trypsin and α-Chymotrypsin

Table 5 presents the results obtained in independent experiments using trypsin and α-chymotrypsin. It is observed that, for trypsin, the maximum experimental hydrolysis degree is approached. An experiment with trypsin, treated to eliminate α-chymotrypsin activity, was also

Table 5
Experimental Hydrolysis Degree (*HDexp*)
of β-Lactoglobulin Using Trypsin and α-Chymotrypsin[a]

Trypsin[b]			
Time (s)	%*HDexp*	*HDth* (%)	η (%)[c]
0	—	11.11	—
3600	7.29 ± 0.06	11.11	65.62
7200	8.77 ± 0.06	11.11	78.94
10,800	9.51 ± 0.44	11.11	85.60
14,400	10.05 ± 0.33	11.11	90.46

α-Chymotrypsin				
Time (s)	%*HDexp*	*HDth* woLeu (%)[d]	*HDth* wLeu (%)[e]	η (%)[c]
0	—	—	—	—
3600	12.62	6.17	19.75	63.90
7200	14.96	6.17	19.75	75.75
10,800	14.96	6.17	19.75	75.75
14,400	14.96	6.17	19.75	75.75

[a]Mass of protein = 0.175 g; mass of enzyme = 0.002 g; C_{enz} = 12.5 (BAEE- or BTEE-U/mL of solution in the reactor); NaOH = 1 M.
[b]Trypsin from Novo Nordisk of Brazil.
[c]Calculated yield.
[d]Not considering leucine residues.
[e]Considering leucine residues.

performed to verify the influence of protease purity on hydrolysis degree. The results showed that the experimental hydrolysis degree is similar to the calculated, within the experimental error, confirming the previous results and validating, therefore, the methodology used in this work to study protease specificity.

The results using α-chymotrypsin allowed us to conclude that this enzyme is able to attack peptide bonds of β-lactoglobulin protein, within its specificity. A yield of 75.75% was observed, and this value is higher than that obtained for cheese whey, in which other proteins are also present. Figure 1 shows the disappearance of the β-lactoglobulin band in SDS-PAGE after 4 h of reaction, confirming that this protein is hydrolyzed by α-chymotrypsin. Thus, the inaccessibility of the protein to this protease is not the reason for the low yields that were observed. Proceeding with the experimental investigation, sequential hydrolysis using the two proteases was performed. First of all, cheese whey and purified β-lactoglobulin were submitted to trypsin action. After inactivation of this protease, the substrate was submitted to α-chymotrypsin. The experimental hydrolysis degrees obtained were compared with the theoretical values calculated for

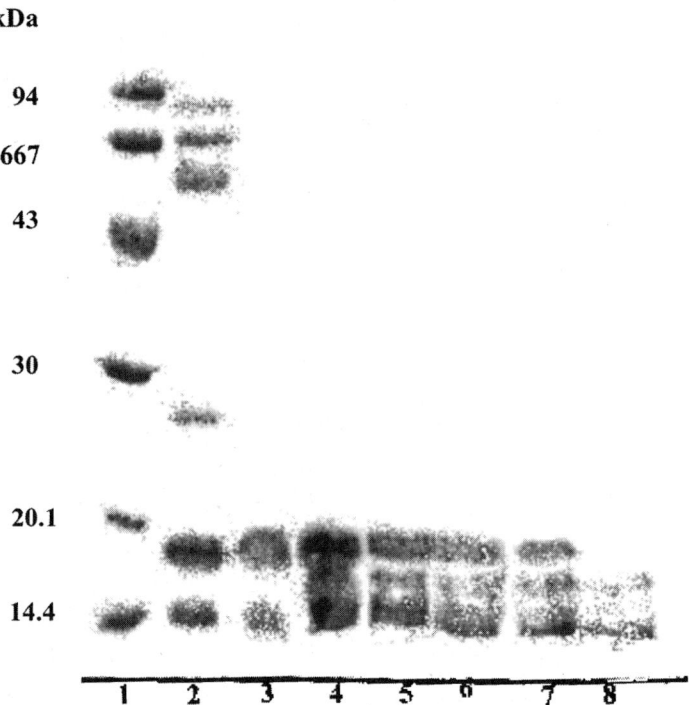

Fig. 1. Incubation of β-lactoglobulin with α-chymotrypsin. Lane 1: Molecular weight marker (14,000–94,000 Daltons); lane 2: unhydrolyzed cheese whey; lane 3: unhydrolyzed β-lactoglobulin (4 g/L); lanes 4, 5, 6, 7, and 8: hydrolysates after 30 min, 2 h, 4 h, 6 h, and 24 h, respectively.

the sequential action. Table 6 presents the obtained results, for filtrated whey and β-lactoglobulin.

The yields obtained with the sequential action of α-chymotrypsin are even smaller than those obtained using this protease alone. For whey, HDexp of 11 and 6.4% (yields of about 56 and 39%) were obtained for individual and sequential action of α-chymotrypsin, respectively. For β-lactoglobulin, HDexp of 15 and 5.14% (yields of 75.75 and 30.83%) were obtained, respectively. Since treated proteases were used and trypsin has a high specificity, an increase in the yield was expected after the sequential action experiments. However, a reduction was observed for whey and β-lactoglobulin (note that for the purified protein the theoretical degree can be calculated with accuracy).

We finally conclude that the decrease in length of the polypeptide chain prevents the hydrolysis of some bonds, probably the ones that have leucine residues, since the affinity of the enzyme for this residue may be smaller than for the aromatic ones. Positive confirmation of this conclusion will require size identification of the liberated oligopeptides after the hydrolysis reaction. This analysis should also include the identification of peptides' terminal residues.

Table 6
Sequential Hydrolysis of Cheese Whey Proteins
and Purified β-Lactoglobulin Using Trypsin and α-Chymotrypsin[a]

		Filtrated whey				
Enzyme	Time (h)	Cenz	%HDexp	HDth (%)		η (%)[b]
Trypsin	4	23.57	11.09	11.96		92.73
				HDth woLeu (%)[c]	HDth wLeu (%)[d]	
Chymotrypsin (after action of the trypsin)	4	10.92	6.43	5.84	16.41	39.18

		β-Lactoglobulin				
Enzyme	Time (h)	Cenz	%HDexp	HDth (%)		η (%)[b]
Trypsin	4	12.25	9.82	11.11		88.39
				HDth woLeu (%)[c]	Hdth wLeu (%)[d]	
Chymotrypsin (after action of the trypsin)	4	10.98	5.14	4.32	16.67	30.83

[a]NaOH = 1 M.
[b]Calculated yield.
[c]Not considering leucine residues.
[d]Considering leucine residues.

Theoretical Peptide Compositions (%PC$_{th}$)

Probable distributions of the molecular mass of the peptides for maximum degree of hydrolysis with trypsin, chymotrypsin, and chymotrypsin after previous action of trypsin (sequential hydrolysis) were calculated. This assessment was based on the primary sequence of proteins, on the specificity of the proteases, and on the assumed protein composition for cheese whey. Figure 2 shows the action of chymotrypsin after total hydrolysis of purified β-lactoglobulin. Table 7 presents the simulated percentile compositions calculated for different cases. These simulated results were compared with those reported in the literature (13–15), in which the authors analyzed, by high-performance liquid chromatography and mass spectrometry, the molecular masses of the peptides obtained under similar operational conditions.

Chobert et al. (16) studied the hydrolysis of β-lactoglobulin with trypsin. They found that 20% of the resulting peptides were about 2700 Daltons.

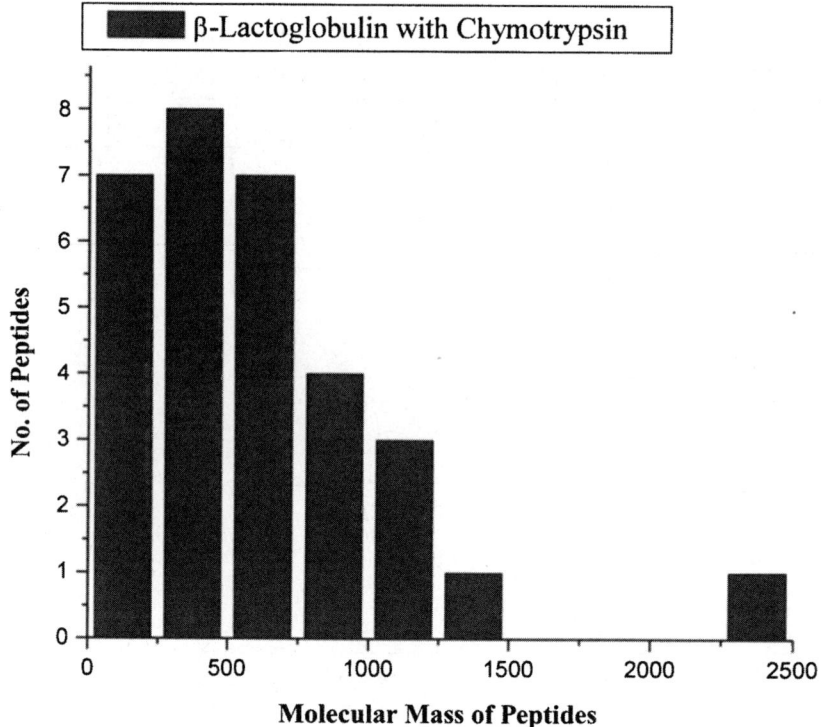

Fig. 2. Simulated distribution of peptide size after the hydrolysis of β-lactoglobulin with chymotrypsin.

Our theoretical distribution shows 10.5% of peptides in the range 2500–2750 Daltons. However, probably the value of 20% obtained by Chobert et al. *(16)* also took into account the peptides obtained in the 1250- to 2750-Daltons range, because the molecular weight standards usually have a larger range of values. It is also possible that they did not achieve the maximum degree of hydrolysis in their experiments.

After 4 h of hydrolyzing β-lactoglobulin with trypsin, a peptide with molecular mass of 1658 Daltons was identified by Otte et al. *(13)* using mass spectrometry. This fragment was also theoretically identified when simulating the peptide composition in this work.

Iung et al. *(17)*, after hydrolysis of β-lactoglobulin with trypsin, obtained two fragments with molecular mass between 4000 to 5000 Daltons and 14,000 Daltons (10% yield); molecular mass was estimated by SDS-PAGE. These components were not identified in our maximum hydrolysis profile. Iung et al.'s *(17)* experiments probably did not achieve maximum reaction conversion, either.

The analytical characterization of the peptide size distribution obtained after controlled proteolytic hydrolysis will be made in continuing studies of this work.

Table 7
Probable Distributions of Molecular Mass of Peptides for Maximum Degree of Hydrolysis[a]

Molecular mass range (Daltons)	α-Lactalbumin			β-Lactoglobulin			BSA			Cheese whey		
	Tryp	Chymo	Chymo after trypsin action	Tryp	Chymo	Chymo after trypsin action	Tryp	Chymo	Chymo after trypsin action	Tryp	Chymo	Chymo after trypsin
0–250	13.3	24.0	36.1	15.8	22.6	41.3	8.4	23.4	30.6	13.1	23.1	37.5
250–500	20.0	36.0	41.7	10.5	25.8	19.6	26.5	31.5	43.4	17.0	29.1	30
500–750	20.0	20.0	11.1	15.8	22.6	26.1	20.5	15.3	17.9	17.9	20.2	21.2
750–1000	20.0	8.0	5.6	21.1	12.9	8.7	12.1	13.5	5.7	18.1	12.2	7.3
1000–1250	13.3	4.0	2.8	15.8	9.7	2.2	7.2	2.7	1.2	12.8	6.8	2.0
1250–1500	—	—	—	5.3	3.2	2.2	13.3	8.1	1.2	6.8	3.9	1.5
1500–1750	6.7	4.0	2.8	5.3	—		7.2	1.8	—	6.1	1.2	0.5
1750–2000	—	—	—	—	—		3.6	—	—	1.1	—	—
2000–2250	—	—	—	—	—		—	3.6	—	0	1	—
2250–2500	—	4.0	—	—	3.2		1.2	—	—	0.4	2.5	—
2500–2750	—	—	—	10.5	—		—	—	—	5.5	—	—
4750–5000	6.7	—	—	—	—		—	—	—	1.1	—	—

[a]Tryp, trypsin; Chymo, chymotrypsin.

Table 8
Experimental Hydrolysis Degrees
of Filtered Whey at 40 and 55°C Using Trypsin and α-Chymotrypsin

			Trypsin		
Time (s)	%HDexp		HDth (%)	η (%)[a]	
	$T = 55°C$	$T = 40°C$		$T = 55°C$	$T = 40°C$
0	—	—	—	—	—
3600	7.51	4.23	11.96	62.79	35.37
7200	7.91	5.51	11.96	66.14	46.07
10,800	8.18	6.71	11.96	68.40	56.10
14,400	8.43	8.05	11.96	70.48	67.31

			α-Chymotrypsin			
Time (s)	%GHexp		HDth (%)		η (%)[a]	
	$T = 55°C$	$T = 40°C$	woLeu (%)[b]	wLeu (%)[c]	$T = 55°C$	$T = 40°C$
0	—	—	—	—	—	—
3600	8.70	7.07	7.39	19.48	44.66	36.29
7200	9.29	8.76	7.39	19.48	47.69	44.97
10,800	9.58	9.82	7.39	19.48	49.18	50.41
14,400	9.71	10.60	7.39	19.48	49.85	54.41

[a]Calculated yield.
[b]Not considering leucine residues.
[c]Considering leucine residues.

Hydrolysis of Cheese Whey Proteins with Trypsin and α-Chymotrypsin at 55°C

The use of temperatures higher than 40°C is important to avoid contamination as well as to increase reaction rates. Studies comparing evaporation and ultrafiltration to concentrate raw whey concluded that the latter displayed higher economic viability (11). The permeation rate, however, is a critical variable. Viotto (18) studied several whey pretreatments in order to sustain high permeation flows during the concentration process. Keeping whey at 55°C, for 44 min, was found to be the most effective treatment. For this reason, hydrolysis experiments using the two proteases were conducted at 55°C; the results are given in Table 8. Values obtained at 40°C are also presented to facilitate discussion.

We can note that the values of hydrolysis degree obtained in the first hour of reaction are higher at 55°C, as expected, for the two proteases. Since the same trypsin concentration was used, it was expected to reach a higher hydrolysis degree for the highest temperature, provided that enzymatic deactivation was negligible. Nevertheless, the conversion achieved for both temperatures, 40 and 55°C, after 4 h of reaction, is very close—and still far from the theoretical maximum degree. The explanation for these results is

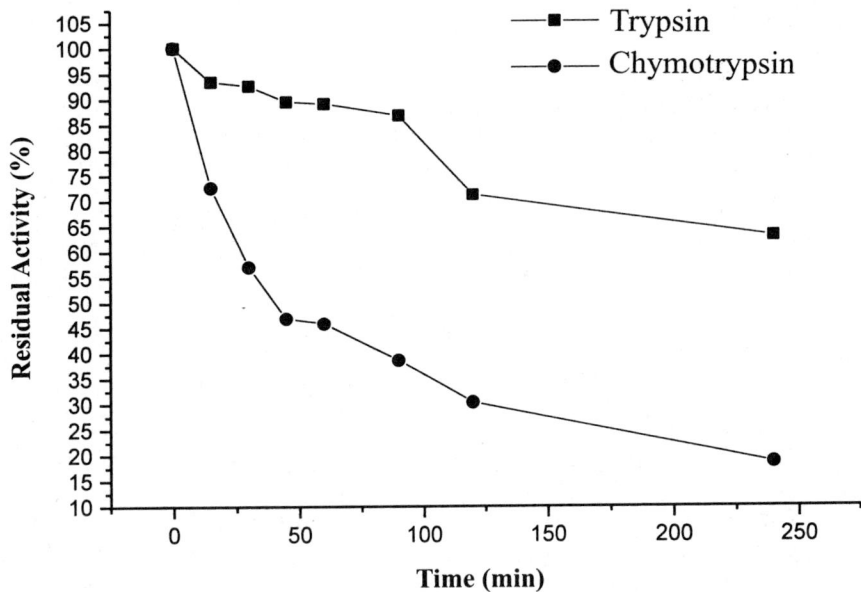

Fig. 3. Residual activity of proteases during hydrolysis of cheese whey proteins at 55°C.

the sharp decrease in the protease activity at 55°C, as can be seen in Fig. 3. The use of high temperatures for this kind of process is strongly recommended to avoid contamination, but the low operational stability of the enzymes in such conditions has to be overcome. Multipoint immobilization of the proteases on insoluble supports as agarose might significantly improve the thermal stability of these enzymes, and this will be investigated in the continuation of this work.

Conclusion

The maximum hydrolysis degree for sweet cheese whey (11.9%) using trypsin was reached after 1800 s of reaction, with an initial concentration of enzyme equal to 324 (BAEE-U)/mL of solution; one hundred percent yield with respect to the theoretical value was obtained, confirming the high specificity of this protease. Using α-chymotrypsin the obtained yield was 56% for cheese whey and 75.75% for β-lactoglobulin, showing that the enzyme does hydrolyze this protein. SDS-PAGE analysis confirms this result. The sequential use of α-chymotrypsin after trypsin indicated that the action of α-chymotrypsin seems to depend on the size of the oligopeptide: the proteases were stable when operating at 40°C but had severe thermal inactivation at 55°C.

Acknowledgments

This work was supported by PADCT/CNPq, CAPES, CNPq, FAPESP, and PRONEX.

References

1. Morr, C. V. and Ha, E. Y. W. (1993), *Crit. Rev. Food Sci. Nutr.* **33**, 431–476.
2. Moulin, G. and Galzy, P. (1984), *Biotechnol. Genet. Eng. Rev.* **1**, 347–374.
3. Margot, A., Flaschel, E., and Renken, A. (1994), *Proc. Biochem.* **29**, 257–262.
4. González-Tello, F. C., Jurado, E., Páez, M. P., and Guadix, E. M. (1994), *Biotechnol. Bioeng.* **44**, 523–528.
5. González-Tello, F. C., Jurado E., Páez, M. P., and Guadix, E. M. (1994), *Biotechnol. Bioeng.* **44**, 529–532.
6. Blanco, R. M., Calvete, J. J., and Guisán, J. M. (1989), *Enzyme Microb. Technol.* **11**, 353–359.
7. Guisán, J. M., Bastida, A., Cuesta, C., Lafuente, R. F., and Rosell, C. M. (1991), *Biotechnol. Bioeng.* **38**, 1144–1152.
8. Adler-Nissen, J. (1986), *Enzymic Hydrolysis of Food Proteins*, Elsevier Applied Science.
9. Hirayama, K., Akashi, S., Furuya, M., and Fukuhara, K. (1990), *Biochem. Biophys. Res. Commun.* **173(2)**, 639–646.
10. <http://class. fst. ohio-state. edu/FST822/lectures/milk2. htm>, last verification in May 2000.
11. Abeles, R. H., Frey, P. A., and Jencks, W. P. (1996), *Biochemistry*, Jonis and Bartlett.
12. Reddy, I. M., Kella, N. K. D., and Kinsella, J. E. (1988), *J. Agric. Food Chem.* **36**, 737–741.
13. Otte, J., Zacora, M., Qvist, K. B., Olsen, C. E., and Barkholt, V. (1997), *Int. J. Dairy* **7**, 835–848.
14. Dalgalarrondo, M., Chobert, J.-M., Dufour, E., Bertrand-Harb, C., Dumont, J.-P., and Haertlé, T. (1990), *Milchwissen-schaft* **45**, 212–216.
15. Turgeon, S. L., Gauthier, S. F., Mollé, D., and Léonil, J. (1992), *J. Agric. Food Chem.* **40**, 669–675.
16. Chobert, J.-M., Dalgalarrondo, M., Dufour, E., Bertrand-Harb, C., Dumont, J.-P., and Haertlé, T. (1991), *Biochimica Biophysica Acta* **1077**, 31–34.
17. Iung, C., Paquet, D., and Linden, G. (1991), *Le Lait* **71**, 385–394.
18. Viotto, W. H. (1993), PhD thesis, UNICAMP, Campinas, Brazil (in Portuguese).

Solid-State Fermentation of Phytase from Cassava Dregs

KUI HONG,*,1 YAN MA,2 AND MEIQIU LI2

*1Key Lab of Plant Protection Research Institute,
National Key Biotechnology Laboratory for Tropical Crops,
Chinese Academy of Tropical Agricultural Sciences, Chenxi,
Haikou, Hainan, 571101, China, E-mail: hongk@public.dzptt.hi.cn;
and 2School of Engineering, South China University
of Tropical Agriculture, Danzhou, Hainan, 571737, China*

Abstract

Phytases produced by numerous microorganisms and plants degrade phytic acid that has chelated with metal ions in food and feed. It is important to study phytase for the role of metal ions in nutrition of animals and humans as well as in the reduction of organic phosphate content of aqueous environment. This article reports on solid-state fermentation of phytase from a new substrate of cassava dregs. Large quantities of cassava dregs are produced in tropical areas as a byproduct of cassava starch processing. Protein and inorganic salts were found to be low in cassava dregs. Cassava dregs could be employed for phytase synthesis after the addition of a nitrogen source and mineral salts. Ammonium nitrate was the best nitrogen source among the nitrogen sources investigated, including beef extract, yeast extract, urea, ammonium nitrate, sodium nitrate, and ammonium sulfate. Sodium dodecyl sulfate promoted phytase production from cassava dregs. A maximum phytase yield of 6.73 U/g of dry mass was obtained. The obtained phytase was stable at feed-processing temperature, since 70% of initial enzyme activity was maintained after 30 min of treatment at 75°C.

Index Entries: Phytase; cassava dregs; nitrogen source; thermostability.

Introduction

Phytase, *myo*-inositol hexakisphosphate phosphohydrolase (EC 3.1.3.8), is an acid phosphatase capable of hydrolyzing phytic acid (*myo*-inositol hexakisphosphate) as well as a number of other organophosphate substrates *(1)*. Microbial and plant phytases are utilized to enhance the absorption of metal nutrients such as iron, calcium, and zinc from food and feed

---

*Author to whom all correspondence and reprint requests should be addressed.

for humans and monogastric animals (2,3), thereby decreasing the potential organic phosphate pollution of surface water. Since the first discovery of phytase from rice bran in 1907 and later from microorganisms, *Aspergillus spp.*'s phytases have been widely investigated and applied as feed additives. The most commonly used preparation of extracellular phytase and phosphatase was derived from *Aspergillus niger (ficuum)* NRRL 3135 and utilized for dephosphorylation of cottonseed meal, soya bean meal, and rapeseed meal, which are used as major protein supplements in poultry feeds (4). Two phytases, PhyA (with a pH optimum of 5.0) and PhyB (with a pH optimum of 2.5) were isolated from the culture filtrate of *A. niger* NRRL 3135. These two enzymes were biochemically characterized, and partially sequenced, and the genes for these two enzymes were cloned (1). A phytase with a high affinity for phytic acid was detected and purified to homogeneity in *Aspergillus niger* SK-57. A significant difference between a low-K_m phytase from *A. niger* SK-57 and a high-K_m phytase from *A. ficuum* was recognized (5).

Phytase thermostability is of great interest because the enzyme should withstand the high temperature of feed processing when used as feed additive. Phytases produced from *Aspergillus fumigatus* and *A. niger* and a pH 2.5 acid phosphatase from *A. niger* were compared for their efficiency at high temperatures and their ability to withstand the heat generated during industrial processing. The phytases of *A. fumigatus* and *A. niger* were both denatured at temperatures between 50 and 70°C. In contrast to these two phytases, *A. niger* pH 2.5 acid phosphatase displayed considerably higher thermostability (6). Pasamontes et al. (7) reported a heat-stable phytase able to withstand temperatures up to 100°C over a period of 20 min with a loss of only 10% of the initial enzymatic activity. The *phyA* gene encoding this heat-stable enzyme has been cloned from *A. fumigatus* and overexpressed in *A. niger*. The enzyme showed high activity with 4-nitrophenyl phosphate at a pH range of 3.0–5.0 and with phytic acid at a pH range of 2.5–7.5 (7). A phytase gene was cloned from the thermophilic fungus *Thermomyces lanuginosus* and heterologously expressed in a *Fusarium oxysporum* strain. The *Thermomyces* phytase retained activity at a temperature up to 75°C (8). A thermal-stable phytase was rapidly designed from DNA sequence using protein sequence comparisons to improve the phytase's stability (9).

Nitrogen and phosphorus in media are two important factors in phytase production. Generally, inorganic nitrogen sources are more easily assimilated than organic nitrogen sources by fungi. Ammonium nitrate has found extensive application in a fungal nutrition source for a large number of imperfect fungi. Gibson (10) found that cornstarch was a good source of carbon and phosphorus for phytase production by *A. niger* NRRL 3135. Because the phosphorus that bound with the cornstarch was connected with C_3 of glucose in the starch and was released with difficulty, it had little effect on phytase production. On the other hand, phytase accumulation of *A. niger* NRRL 3135 was inhibited when the concentration of phosphorus was higher than 10 mg/100 g of feedstock using potato starch (11).

Commercially, phytases are produced by aqueous fermentation. However, molds are naturally grown on solid-state substrates. Wheat bran and rice bran have been employed for phytase production *(12)*. Cassava is a starch crop planted in tropical and subtropical areas. Large quantities of dregs are produced annually in the cassava starch-processing factory. In China alone, >950,000 t of cassava dregs are produced each year. Apart from their partial use for ethanol production and feed, most of the dregs are discarded as landfill (13). In the present study, we report on phytase production from cassava dregs using *Aspergillus niger* PD by solid-state fermentation. We discuss nitrogen and surface tension reducing agents regarding their effect on phytase production from cassava dregs. We also note that the heat stability of the phytase obtained is suitable for feed processing.

Materials and Methods

A. niger PD is a strain stored in our laboratory. It was isolated in a rice field in Hainan, China, formerly for amylase production and now selected for phytase production. It was maintained at 4°C on potato dextrose agar (PDA) slants. Cassava dregs were obtained from a cassava starch–producing company in Qiongzhong County, Hainan, China. They were sieved to 40 meshes before using.

Inocula were prepared in 500-mL conical flasks containing 20 g of wheat bran and mineral salts solution containing 0.5 g/L of $MgSO_4 \cdot 7H_2O$, 0.5 g/L of KCl, and 0.1 g/L of $FeSO_4 \cdot 7H_2O$ at 60–65% moisture content. Growth media were sterilized at 121°C for 30 min and then inoculated with spores grown on PDA slants. The cultures were grown for 3 d at 30°C.

Cassava dregs mixed with various nitrogen sources and optimized mineral salts solution were prepared with (g/kg of cassava dregs) K_2HPO_4 (0.1), $MgSO_4$ (0.5), KCl (0.5), and $FeSO_4$ (0.1), at 65% moisture content *(14)*. The media were used to produce phytase after being sterilized at 121°C for 30 min. Fermentation was carried out at 28–30°C.

Phytase activity was assayed in 1-mL samples extracted from the fermented cassava dregs with 2% $CaCl_2$ solution added, 1 mL of 0.2 M citric acid and sodium citrate buffer, and 5 mmol/L of phytate at pH 5.0. The mixture was incubated at 37°C for 15 min, and the reaction was terminated with 2 mL of trichloroacetic acid. The released phosphoric acid was estimated by adding 4 mL of freshly prepared solution containing 1 M H_2SO_4, 2.5% ammonium molybdate, and 10% ascorbic acid at a ratio of 3:1:1. The mixture was incubated at 50°C for 20 min and measured at 660 nm. A unit of phytase activity is defined as 1 µmol of phosphate/min under assay condition.

The fermented cassava dregs were dried at 42 ± 2°C for 8–10 h, comminuted, and stored at 4°C for the stability tests. We termed the enzyme obtained with the solid substances *koji*. Thermal stability at different temperatures was carried out as follows: The koji was treated at various temperatures (25, 40, 50, 60, 70, 80, and 90°C for 30 min), and the phytase activities were analyzed after the koji had cooled. Thermal stability at 75°C was detected every 20 min after the koji was treated at 75°C for 30 min. The koji stored at 25°C was used as control during the analysis.

Results and Discussion

The composition of cassava dregs differed considerably from the conventionally used solid-state fermentation substrates. Compared to rice chaff and wheat bran, cassava dregs contain higher starch and cellulose but lower protein and minerals. Nitrogen sources and other necessary minerals should be added if these dregs are employed as phytase production substrates. The composition of three batches of cassava dregs varied in this study. No growth was observed using cassava dregs as a sole substrate, even though phytase was produced using rice chaff or wheat bran as sole substrates. On the other hand, phosphate content in cassava dregs was found to be very low compared to rice chaff and wheat bran (Table 1). This is a useful property for phytase production, since it was reported that high phosphate inhibited phytase accumulation *(10,11)*.

The nitrogen source is an important substrate that affects enzyme production. Various nitrogen sources were investigated for their effects on phytase accumulation. We found that higher phytase activities were produced when adding inorganic rather than organic nitrogen sources to cassava dregs. Urea gave the highest phytase accumulation among organic nitrogen sources, while NH_4NO_3 produced the highest accumulation among the inorganic nitrogen sources. The highest phytase activity was obtained at 1 to 2% NH_4NO_3 among the nitrogen sources tested (Fig. 1). It was suggested that when grown on cassava dregs, *A. niger* PD preferred ammonium (both from inorganic sources and organic sources) over other nitrogen sources, and ammonium nitrate was the most preferred salt among the ammonium salts used. Sodium nitrate was reported to be the most positive nitrogen for *A. niger* HZ-94 in phytase production on starch liquid culture, and no phytase activity was detected when using urea as the nitrogen source *(15)*. That finding clearly contradicts our result. Urea was found to promote α-galactosidase production by *A. niger* MRSS 234 in solid-state fermentation and its effect increased with the fermentation time, in contrast to the negative effect of all the ammonium salts used *(16)*. The effect of nitrogen sources on enzyme production by fungi is complicated. It is generally accepted that inorganic nitrogen sources are more easily assimilated than organic nitrogen sources by fungi; ammonium nitrate has been applied in a fungal nutrition source for many imperfect fungi. This may be owing to the fact that the principal metabolite of nitrogen metabolism is glutamic acid, and a nitrogen source assimilation that can easily lead to the formation of glutamic acid will be the most positive source. In our study, urea and ammonium nitrate may afford ammonium a more easily formed glutamic acid than other sources. When other mineral salts, including K_2HPO_4,

Table 1
Components of Cassava Dregs, Rice Chaff, and Wheat Bran and Phytase Activity Produced When Used as Sole Substrate

	Starch (%)	Protein (%)	Cellulose (%)	Minerals (%)	Fatty acids (%)	Total phosphorus (%)	Phytase activity (U/dry mass)[a]
Cassava dregs							
I	45.8	2.12	42.9	2.63	2.47	0.006	No growth
II	55.9	2.97	34.7	2.32	2.71	0.008	No growth
III	70.1	4.37	16.5	3.05	2.95	0.012	No growth
Rice chaff	39.1	12.1	15.6	13.6	14.7	1.21	1.63
Wheat bran	47.3	17.3	17.8	7.57	3.6	0.33	2.16

[a]Fermentations were carried out at 30°C for 7 d.

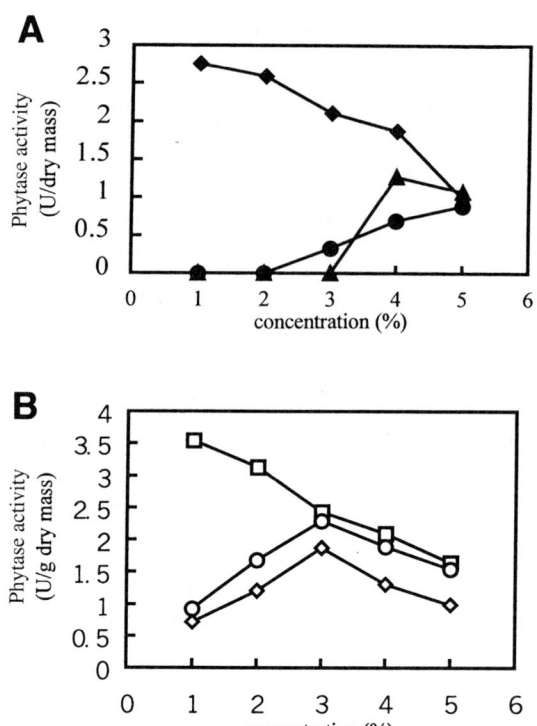

Fig. 1. Effect of **(A)** organic nitrogen sources of urea (♦), beef extract (●), and yeast extract (▲) and **(B)** inorganic nitrogen sources of NH_4NO_3 (□), $(NH_4)_2SO_4$ (○), and $NaNO_3$ (◇) on phytase synthesis from cassava dregs, with mineral salts of (g/kg of cassava dregs) K_2HPO_4 (0.1), $MgSO_4$ (0.5), KCl (0.5), and $FeSO_4$ (0.1) added at 65% moisture content and 30°C for 7 d.

$MgSO_4$, KCl, and $FeSO_4$, were investigated, together with ammonium nitrate for the purpose of optimizing their quantity mixed to cassava dregs, by $L_9(3)^4$ orthogonal experiments, the optimized concentration of ammonium nitrate was 2% *(14)*.

Surface tension–reducing agents at concentrations of 0–1.5% were investigated for their effect on phytase accumulation. Sodium dodecyl sulfate (SDS) promoted phytase accumulation by more than double, but Tween-80 and oleic acid had no effect (Fig. 2). Since SDS has a strong anionic surface tension–reducing activity, this activity may be owing to the effect of SDS on permeation of cell membranes. To prove this, further experiments will be necessary.

The maximum phytase activity was produced with the following added minerals (g/kg of cassava dregs): K_2HPO_4 (0.1), $MgSO_4$ (0.5), KCl (0.5), and $FeSO_4$ (0.1), at 65% moisture content. Under these optimized conditions, 6.73 U/g of dry mass phytase was obtained on d 8, and the maximum protein content presented on d 7. High total sugar content still remained at 31.9% on d 11 (Fig. 3). The high content of total sugar may owing to the unused starch in the dregs. Further investigation is in progress

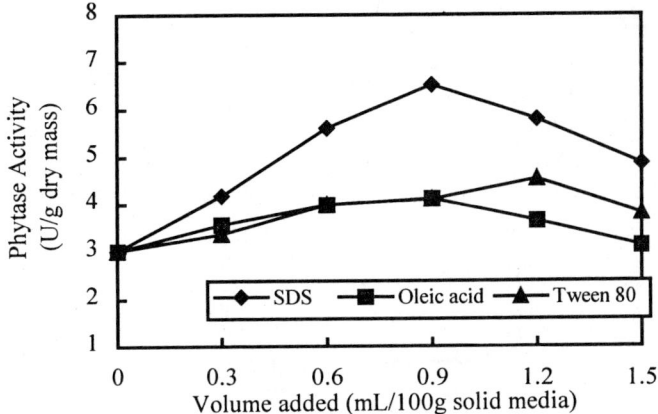

Fig. 2. Effect of surface tension reducers: SDS (◆), Tween-80 (▲), and oleic acid (■) on solid-state fermentation of phytase synthesis from cassava dregs, with mineral salts of (g/kg of cassava dregs) K_2HPO_4 (0.1), $MgSO_4$ (0.5), KCl (0.5), and $FeSO_4$ (0.1) added at 65% moisture content and cultured at 30°C for 7 d.

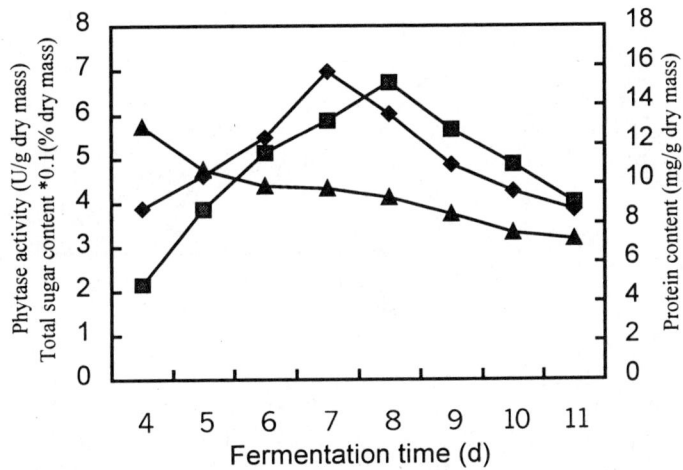

Fig. 3. Solid-state fermentation time course of phytase from cassava dregs, with mineral salts of (g/kg of cassava dregs) K_2HPO_4 (0.1), $MgSO_4$ (0.5), KCl (0.5), and $FeSO_4$ (0.1) added at 65% moisture content and 30°C for 7 d. Phytase activities (■), total sugar contents × 0.1 (▲), and protein contents (◆) vary with fermentation time.

using a mixed culture of *A. niger* PD with another amylase-producing strain that should improve phytase production by consuming more starch.

Phytase thermal stability was investigated to determine its utility as a feed additive. Enzymes used as animal feed supplements should be able to withstand temperatures as high as 60–90°C, which may be reached during the feed-pelleting process *(7–9)*. We treated the obtained solid phytase at temperatures of 25, 40, 50, 60, 70, 75, 80, and 90°C for 30 min and found that *A. niger* PD phytase maintained 50–80% activity between 60 and 90°C over

A

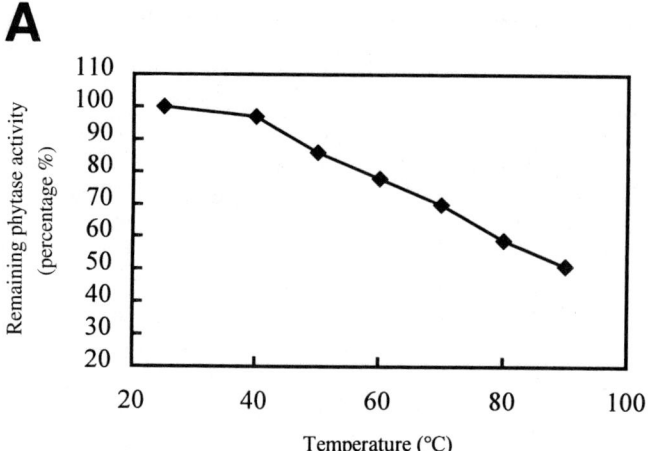

B

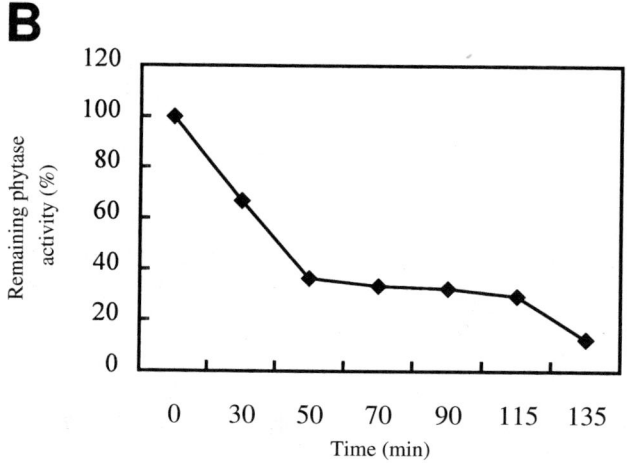

Fig. 4. **(A)** Thermal stability of phytase produced from cassava dregs. The obtained solid phytase was treated at various temperatures (25, 40, 50, 60, 70, 80, and 90°C for 30 min), and the phytase activities were analyzed after cooling. **(B)** Thermal stability at 75°C was detected every 20 min after the solid phytase was treated at 75°C for 30 min.

30 min, while 70% phytase activity was maintained at 75°C (the feed-pelleting temperature) over 30 min (Fig. 4). These findings suggest that this enzyme could be used in feed additives.

Conclusion

We have attempted to find a new application for cassava dregs by producing phytase from the dregs. The content of phosphorus in cassava dregs was lower compared to other conventionally used solid-state fermentation substrates, a feature useful for phytase production. Mineral salts, especially ammonium salt, should be added as a nitrogen source because nitrogen content is low in cassava dregs. Ammonium nitrate was found to be the most effective nitrogen source among the nitrogen sources tested. SDS promoted phytase production. The thermal stability of the obtained phytase adapted to the temperature of feed processing. Further investigation will explore additional improvement in starch utilization of the dregs and the efficiency of application.

Acknowledgment

This work was supported in part by a Research and Education Grant for Young and Mid-Career Scholars, Province of Hainan, China.

References

1. Ehrlich, K. C., Montalbano, B. G., Mullaney, E. J., Dischinger, H. C., and Ullah, A. H. J. (1993), *Biochem. Biophys. Res. Commun.* **195**, 53–57.
2. Sandberg, A. S. and Anderson, H. (1996), *J. Nutr.* **126**, 476–480.
3. Han, Y. M., Yang, F., Zhou, A. G., Miller, E. R., Ku, P. K., Hogberg, M. G., and Lei, X. G. (1997), *J. Anim. Sci.* **75**, 1017–1025.
4. Zyla, K. and Koreleski, J. (1993), *J. Sci. Food Agric.* **61**, 1–6.
5. Nagashima, T., Tange, T., and Anazawa, H. (1999), *Appl. Environ. Microbiol.* **65**, 4682–4684.
6. Wyss, M., Pasamontes, L., Rémy, R., et al. (1998), *Appl. Environ. Microbiol.* **64**, 4446–4451.
7. Pasamontes, L., Haiker, M., Wyss, M., Tessier, M., and van Loon, A. P. (1997), *Appl. Environ. Microbiol.* **63**, 1696–1700.
8. Berka, R. M., Rey, M. W., Brown, K. M., Byun, T., and Kloz, A. V. (1998), *Appl. Environ. Microbiol.* **64**, 4423–4427.
9. Lehmann, M., Kostrewa, D., Wyss, M., et al. (2000), *Protein Eng.* **13**, 49–57.
10. Gibson, D. M. (1987), *Biotech. Lett.* **9**, 305–310.
11. Howson, S. J. and Davis, R. P. (1983), *Enzyme Microb. Technol.* **5**, 377.
12. Han, Y. W. and Gallagher, D. J. (1987), *J. Ind. Microbiol.* **2**, 195–200.
13. Chen, G. G., Pang, C. W., and Liang, J. J. (1997), *Food Feed Ind.* **6**, 23, 24.
14. Ma, Y., Hong, K., Li, M. Q., and Zhao, J. T. (2000), *Chin. J. Trop. Crops* **21(2)**, 58–63.
15. Zhen, W. D. (1996), MS thesis, Najing Agricultural University, China.
16. Srinivas, M. R. S., Chand, N., and Lonsane, B. K. (1994), *Bioprocess Eng.* **10(3)**, 139–144.

The Effect of Additional Autopolyploidization in a Slow Growing Cellulase Hyperproducer of *Trichoderma*

Hideo Toyama* and Nobuo Toyama

*Department of Food Science and Technology,
Faculty of Horticulture, Minamikyushu University, Takanabe,
Miyazaki 884-0003, Japan, E-mail: gaf00771@nifty.com*

Abstract

M14-2 is a cellulase hyperproducer derived from *Trichderma reesei* QM 6a, but with a growth rate lower than that of the original strain. When M14-2 was autopolyploidized followed by haploidization and selection, the strain with both a higher cellulase productivity per mycelia and a higher growth rate could be obtained as M14-2B. This strain seemed to be constructed using gene sources amplified by additional autopolyploidization.

Index Entries: *Trichoderma*; cellulase; colchicine; cellulose; benomyl.

Introduction

Trichoderma reesei is a cellulolytic fungus and produces stable cellulase in high yield *(1)*. This fungus is widely utilized for industrial production of cellulase *(2)*. However, to expand the use of cellulase, the productivity of this fungus should be increased and the cost of cellulase should be reduced. Therefore, we attempted to develop a method for construction of cellulase hyperproducers for use in food processing using a model strain of *T. reesei* QM 6a without gene cloning techniques.

The M14-2 strain, a cellulase hyperproducer, was obtained from the conidia of M14 strain by chemical mutation *(3)*. The M14 strain is an auto-polyploid and derived from the mycelial mat of *T. reesei* QM 6a treated with 0.1% (w/v) colchicine (Wako, Osaka, Japan) solution for 14 d at 26°C. The cellulase productivity per mycelia of M14-2 increased considerably, but the growth rate was lower than that of the original strain. Therefore, we attempted to improve the growth rate of M14-2 using haploidization and autopolyploidization techniques.

*Author to whom all correspondence and reprint requests should be addressed.

Methods, Results, and Discussion

M14-2 was incubated on a potato dextrose agar (PDA) medium (BBL, Cockeysville, USA) at 26°C and preserved at 4°C. A mycelial mat (2 mm × 2 mm) of M14-2 was directly incubated on the haploidizing medium for 3 wk at 26°C. PDA medium containing 0.6 µg/mL benomyl [1-(butyl-carbamoyl)-2-benzimidazolecarbamate] (Sigma, St. Louis, MO) was used for haploidization *(4)*. During the incubation, fan-shaped sectors were produced from the colony. The selection of cellulase hyperproducers was carried out using those conidia generated on the colony. A Mandels's medium containing 1.0% (w/v) glucose (Wako, Osaka, Japan) and 0.5% (w/v) peptone (Difco, Detroit, MI) (pH 6.0) was used as the basic liquid medium *(5)*. For the selection of cellulase hyperproducers, conidia were added to 30 mL of Mandels's medium (bottom layer medium) containing 1.0% (w/v) Avicel (Funakoshi, Tokyo, Japan), 1.5% (w/v) agar (Difco, Detroit, MI), 0.5% peptone, and 0.1% (v/v) polyoxyethyleneoctylphenylether (Triton X-100) (Wako, Osaka, Japan) in a deep plate [6 cm (=depth) × 9 cm] and left for 1 h at 4°C to harden the agar. After hardening the agar, 200 mL of the same medium (upper layer medium) was poured on the bottom layer medium and left for 1 h at 4°C to harden the agar again. After 5 d of incubation, 16 colonies appeared on the surface of the selection medium and the estimation of cellulase productivity was carried out. Mandels's medium containing 1.0% (w/v) carboxymethylcellulose sodium salt (CMC-Na), degree of substitution 0.7 (Wako, Osaka, Japan), 0.5% peptone, 1.5% agar, and 0.1% Triton X-100 was used for the estimation of cellulase productivity (pH 6.0). The mycelial mat (2 mm × 2 mm) was put on the medium for estimation of cellulase productivity and incubated for 6 d at 26°C. After incubation, 0.1% (w/v) Congo red (Merck, Darmstadt, Germany) solution was poured on plates and left them for 1 h followed by washing with 1 M NaCl (Wako, Osaka, Japan) solution. After washing, a clear zone appeared around a colony by cellulase formation. Diameters of the clear zone and colony were measured by a digital caliper (Mitsutoyo, Kawasaki, Japan) and cellulase productivity was estimated using the ratio between the two diameters. After estimation of cellulase productivity, one colony was isolated as M14-2A. The growth rate of the strain, M14-2A, was almost the same with that of *T. reesei* QM6a but M14-2A lost hypercellulase productivity per mycelia.

Next, the second autopolyploidization of M14-2A was attempted. A mycelial mat (10 mm × 10 mm) was incubated stationarily in Mandels's medium containing 0.1% (w/v) colchicine (Wako, Osaka, Japan), 1.0% glucose, and 0.5% peptone (pH 6.0) for 35 d at 26°C. After incubation, haploidization was carried out on those colchicine-treated mycelial mats for 3 wk at 26°C, and fan-shaped sectors were also produced. The conidia were incubated in the selection medium mentioned above. After 4 d of incubation, 11 colonies appeared on the surface of the medium. After estimation of cellulase productivity, one colony was isolated as M14-2B. More-

Table 1
Cellulase Production on CMC-Na Plates

Strains	Diameter of clear zone (mm)[a]	Diameter of colony (mm)[a]	Ratio
T. reesei QM 6a	26.80 ± 0.62	16.47 ± 0.31	1.61
T. reesei Rut C-30	31.80 ± 0.65	14.30 ± 0.35	2.22
M14	29.90 ± 0.67	27.09 ± 0.36	1.10
M14-2	33.94 ± 0.65	9.71 ± 0.36	3.50
M14-2A	33.13 ± 0.63	20.98 ± 0.34	1.58
M14-2B	46.67 ± 0.63	15.05 ± 0.38	3.10

[a]The diameter of clear zone and the diameter of colony were measured by a digital caliper after 6 d-incubation at 26°C. The ratio between the diameter of clear zone and the diameter of colony was used for estimation of cellulase productivity.

over, cellulase productivities of *T. reesei* QM 6a, *T. reesei* Rut C-30 (ATCC 56765), M14, M14-2, M14-2A, and M14-2B were estimated using the medium for estimation of cellulase productivity. From these results, it appeared that the growth rate of M14-2B increased extremely compared with that of M14-2 and cellulase productivity was also maintained as shown in Table 1. Cellulose hydrolyzing activities of *T. reesei* QM 6a, M14, M14-2, M14-2A, and M14-2B were measured using the wheat bran culture. For the measurement of cellulose hydrolyzing activities, 7.5 g of wheat bran was used containing 7.5 mL of Mandels's medium and 4.5 g of peptone in a 100 mL Erlenmeyer flask. One loopful of conidia was added to a flask of the solid medium for the measurement of cellulose hydrolyzing activity and incubated at 26°C for 6 d. Those flasks were shaken once a day. After incubation, 15 mL of 0.1 *M* acetate buffer (pH 5.0) was added, stirred using a glass rod, and left to stand for 1 h. The enzyme solution was then extracted from the wheat bran culture using a nylon cloth. The extracts were centrifuged at 5510*g*, and the top clear portion was used as the enzyme solution. As the substrates of enzyme reaction, 1.0% (w/v) of Avicel, CM-cellulose (Wako, Osaka, Japan), or Salicin (Wako, Osaka, Japan) was added to 100 mL of 0.1 *M* acetate buffer (pH 5.0). Three tenths of a milliliter of enzyme solution and 3.0 mL of substrate were mixed and incubated for 60 min at 40°C using a reciprocal shaker (Thomastat T-22S, Tokyo, Japan). The agitation speed was 125 strokes/min. After enzyme reaction, two drops of 0.1 *N* HCl (Wako, Osaka, Japan) solution were then added to the mixture in order to stop the reaction. The reaction mixture was filtrated with filter paper (No. 2, Whatman, Maidstone, UK). The amount of reducing sugar in the reaction mixture was measured using 3,5-dinitrosalicylic acid (DNS) (Wako, Osaka, Japan) *(6)*. IU was based on the amount of enzyme producing reducing sugar equivalent to 1 µmol of glucose per minute. As shown in Table 2, it appeared that the hydrolyzing activities of Avicel, CMC, and Salicin in M14-2B were about 2, 3, and 3 times higher, respectively, than those of *T. reesei* QM 6a, and 1.4, 2, and 2 times higher than those of M14-2. The proliferation of M14-2B on the wheat bran medium was superior to that

Table 2
Cellulose Hydrolyzing Activities in Wheat Bran Culture

Strains	Hydrolyzing activities (IU/mL) of[a]			Growth
	Avicel	CMC	Salicin	
T. reesei QM 6a	53	45	26	++
M14	68	55	36	++
M14-2	88	74	44	+
M14-2A	93	88	51	++
M14-2B	120	136	70	++

[a]For the measurement of cellulose hydrolyzing activities, 7.5 g of wheat bran was used containing 7.5 mL of Mandels's medium and 4.5 g of peptone in a 100 mL Erlenmeyer flask. One loopful of conidia was added to a flask and incubated at 26°C for 6 d. The amount of reducing sugar in the reaction mixture was measured using 3,5-dinitrosalicylic acid (DNS) (Wako, Osaka, Japan) (6). IU was based on the amount of enzyme producing reducing sugar equivalent to 1 μmol of glucose per minute.

of M14-2. The distribution of nuclear diameters of *T. reesei* QM 6a, M14, M14-2, and M14-2B were compared. A mycelial mat was directly stained by Giemsa solution (Merck, Darmstadt, Germany) and photomicrographs were taken using a microscope (BH-2, Olympus, Tokyo, Japan) with an automatic exposure meter (PM-CBAD, Olympus, Tokyo, Japan) and a camera (C35AD, Olympus, Tokyo, Japan). Those photomicrographs were enlarged and the nuclear diameter of 100 nuclei per sample was measured by a digital caliper. The distribution of nuclear diameter per sample was investigated from those measured values. From the results, M14-2B was found to be still polyploid in addition to M14 and M14-2. We suspected that M14-2B seemed to be constructed through chromosomal recombination from the colchicine-treated M14-2 (autopolyploid). From the above results, it was concluded that a cellulase hyperproducer which has higher growth rate and higher cellulase productivity per mycelia, M14-2B, might be constructed using gene sources amplified by additional autopolyploidization from a low growing cellulase hyperproducer, M14-2.

Acknowledgments

We would like to thank Yakult Pharmaceutical Industry Co. Ltd., and Tsukishima Kikai Co. Ltd., for useful discussions and financial support.

References

1. Evans, E. T., Wales, D. S., Sagar, B. F., and Bratt, R. P. (1992), *J. Gen. Microbial.* **138,** 1639–1646.
2. Nieves, R. A., Ehrman, C. I., Adney, W. S., Elander, R. T., and Himmel, M. E. (1998), *World J. Microbiol. Biotechnol.* **14,** 301–304.
3. Toyama, H. and Toyama, N. (1999), *Microbios* **100,** 7–18.
4. Peterbauer, C. K., Heidenreich, E., Kubicek, C. P., and Baker, R. T. (1992), *Can. J. Microbiol.* **38,** 1292–1297.
5. Mandels, M. and Sternberg, D. (1976), *J. Ferment. Technol.* **54,** 267–286.
6. Miller, G. L. (1959), *Anal. Chem.* **31,** 426–428.

Author Index

A
Aden, A., 253
Adney, W. S., 99
Agblevor, F. A., 423
Aguiar, W. B., 655
Akpa, E., 551
Almeida, R. M. R. G., 537
Alves, T. L. M., 459
Andrade, C. M. M. C., 655
Antranikian, G., 655
Araujo, M. L. G. C., 537
Atala, D. I. P., 353

B
Baker, J. O., 99
Bala, G. A., 487
Balasubramanian, N., 367
Ballesteros, I., 237
Ballesteros, M., 237
Baron, M., 469
Budzikiewicz, H., 551

C
Calsavara, L. P. V., 615
Cherry, J., 405
Cheryan, M., 283
Chiang, V., 3
Chua, H., 171, 447, 515
Cockrem, M. C. M., 485
Conte, R. A., 413
Cortez, E. V., 753
Costa, A. C., 353
Cruz, A. J. G., 341, 537
Custódio, M. F., 761

D
Dale, B. E., 269, 481
Davis, R. H., 297
Davison, B. H., 195, 205
De Castro, H. F., 413, 703, 739
De Moraes, F. F., 615, 643, 739
Decker, S. R., 99
Dias, M. A., 413
Ding, Y., 387
Dinus, R. J., 23
Doran-Peterson, J. B., 269
Du, L., 387

E
Eddy, F. P., 51
Escobar, J. M., 283
Esteghlalian, A. R., 575
Evans, B. R., 97

F
Farmer, J., 627
Felipe, M. G. A., 671, 753
Ferraz, A., 563
Fontana, J. D., 469
Fontes, M. C. de A., 681
Foster, B. L., 269
Fox, J. W., 99
Fox, S. L., 487
Franco, T. T., 729
Freire, D. M. G., 459
Fuchs, R., 551

G
Gaffney, A. M., 185
Galvão, C. M. A., 761
Gedvilas, L. M., 51
Gilkes, N. R., 575
Giordano, R. C., 341, 537, 691
Giordano, R. L. C., 341, 691, 761
Gonçalves, A. R., 63, 563
Gonçalves, L. R. B., 691
Guedes, T. A., 437
Gunasekaran, M., 185

H
Hamilton, J., 627
Hamilton, M. A., 503
Hanley, T. R., 235
Harris, R. A., 481
Hasegawa, A., 155
Hasmann, F.-A., 719
Hensirisak, P., 423
Himmel, M. E., 99
Hokka, C. O., 341, 537
Hong, K., 777

J
Jacobsen, S. E., 377
Jacques, P., 219, 551
Jönnson, L. J., 35

Subject Index